The art of control engineering

The art of
control engineering

Ken Dutton

Steve Thompson

Bill Barraclough

 **Addison-Wesley**

Harlow, England • Reading, Massachusetts • Menlo Park, California
New York • Don Mills, Ontario • Amsterdam • Bonn • Sydney
Singapore • Tokyo • Madrid • San Juan • Milan • Mexico City
Seoul • Taipei

Addison Wesley Longman Limited
Edinburgh Gate
Harlow
Essex
CM20 2JE
England

and Associated Companies throughout the world.

Text design by Sally Grover Castle
Typeset by 54
Printed in Great Britain by Henry Ling Ltd., at the Dorset Press,
Dorchester, Dorset.

First printed 1997

ISBN 0-201-17545-2

British Library Cataloguing-in-Publication Data
A catalogue record for this book is available from the British Library

Library of Congress Cataloging-in-Publication Data is available

The companion software available with this book may be obtained from The MathWorks anonymous FTP site at ftp.mathworks.com in the directory /pub/books/dutton/.

The MathWorks, Inc.
24 Prime Park Way
Natick, MA 01760-1500
Phone: (508) 647-7000
Fax: (508) 647-7001
E-mail: information@mathworks.com
WWW: http://www.mathworks.com

Contents

Preface

The material in this book is based upon undergraduate, postgraduate and industrial courses taught at Sheffield Hallam University (England), and at The Queen's University of Belfast (Northern Ireland). It may be used to take a student from zero knowledge of control systems, by easy stages, up to an appreciation of some advanced control schemes. A feature of the text is that it considers the *implementation* of the various control schemes; although it would still be necessary to be guided by a good practitioner of the 'Art of control engineering', if applications are to be implemented safely and efficiently. Other features of the text are its layered approach to the teaching of control and its treatment of the mathematics essential to an understanding of the subject. Both of these aspects are expanded later in the preface. In brief, the pragmatic approach and the practical comments given throughout the book should prove useful to the student studying control, the educator planning a course in control studies and to practising control engineers. This preface contains the following sections, giving useful information about this book, and how it can best be used:

- A broad overview of the book
- Mathematics (notes on the book's approach to mathematics, and how it is included)
- Computer-aided control system design (CACSD) and MATLAB®
- How to use the book
- Acknowledgements (with an email address for comments)

A broad overview of the book

This book is structured to suit the way control engineering is taught in European institutions (but it is sufficiently flexible to be easily used elsewhere). It assumes little previous knowledge other than familiarity with basic physics, algebraic manipulation and elementary calculus. The major additional topics in mathematics necessary to study control engineering (such as the Laplace transform, the z-transform and matrix algebra) are covered as they are needed.

In the past, students who have been motivated to study all the control engineering options (electives) their programme of study offers, have typically needed several textbooks (including, perhaps, texts on classical control, multi-variable control, digital control and nonlinear control), some of which might be used for only a very small percentage of their course. An ambitious aim in writing this book has been to provide, in a single volume, a complete undergraduate coverage of control engineering (and also to cover some postgraduate courses).

Typically, an undergraduate student undertaking a three academic year course of study with a high control engineering content might well encounter control lectures each year. Accordingly, the chapters in this book are aimed at three different *levels*, corresponding roughly to these three years of study, as indicated in Figure P1. This means that some of the major topics might occur in each of the three levels. Of course, the higher the level, the greater the depth of treatment. A further benefit of this 'layered' approach is that some of the fundamental divisions of control engineering (such as frequency-domain methods versus time-domain methods) are largely overcome, since all these methods are studied side by side at each level, thus providing a well-integrated approach to the subject as a whole. 'Maps' of how some major topics can be followed through the text in a less integrated fashion, if desired, are given later in this preface.

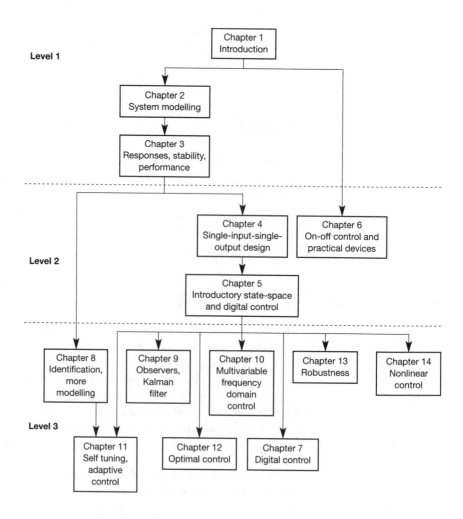

Figure P1 The layered approach of the text.

Mathematics

Control engineering can easily become a highly mathematical discipline (and there are several very mathematical control texts on the market). This text is designed to be different, although it is inevitable that some parts of it will look rather mathematical on a cursory inspection. However, even in the most intensively mathematical parts, if the text is read in the order suggested, then sufficient steps of the mathematics are given to allow it to be followed.

The mathematics in this text is presented as it arises naturally, and is usually limited to that which is *necessary* for an understanding of the particular topic under consideration. For example, the Laplace transform, which is the cornerstone of the frequency-domain approaches, is introduced in a 'user-friendly' manner within the text, at the point where it becomes necessary to use it.

Sometimes, if a substantial amount of mathematical background is considered helpful, an appendix is used. For example, Appendix 1 contains all the matrix algebra necessary to follow the time-domain and multivariable control methods described in the text. In addition, it treats matrices and vectors in a helpful, 'control-orientated' manner.

Our aim, therefore, has been to integrate the mathematics with the text, since we are, after all, interested in promoting control engineering and, in our experience, seemingly abstract mathematics can 'turn off' many a student. Rigorous mathematical proofs are avoided, except where they aid an understanding of how to make the best use of the method in question. In general, the treatment is more pragmatic and practical than will be found in many books. Even so, there is always sufficient theoretical background to allow an understanding of each technique.

As one example of a full derivation which *has* been included, Appendix 6 (combined with part of Chapter 9) gives a mathematical derivation of the Kalman filter. To some, this will seem the most complicated mathematics in the book, but it has been included because the derivation is more complete than that provided by any other textbook known to the authors, and is therefore useful background material for an undergraduate text. However, 'background' is the appropriate word, and illustrates one aspect of the authors' pragmatic approach since, for an engineering course (unlike a mathematics course), it is unlikely that a full derivation of the Kalman filter (or any other very complex algorithm) would be taught line by line. Rather, it resides in the appendix where interested readers may study it, but only the outline steps would be mentioned in class, as an aid to understanding where the algorithm comes from, why it works and how it might best be used in practice.

In other places where a proof might be of interest, but the proof is given adequately elsewhere, appropriate references are quoted. However, the text never subsequently relies on knowledge of such a proof.

Each chapter in the book has an introductory section listing any new mathematical ideas which will be encountered in that chapter. In this way, the student can see where the need for some greater mathematical knowledge is imminent. However, since the new mathematics is always integrated with the text, the motivation for the study of the new topic will be clear.

Computer-aided control system design (CACSD) and MATLAB®

No new control engineering text can ignore the extent to which the availability of cheap computing power, and powerful software packages, has revolutionized the subject. This text acknowledges this by including some relatively modern analysis and design techniques which cannot be applied 'by hand' because of the volume of calculation involved (see, for example, Chapter 10). The problem is that there are many excellent software packages which can be used for control system design studies, each having its own particular strengths and weaknesses. In producing this text, it was necessary to opt for just one package, for consistency, and we chose to use MATLAB – see (The Mathworks Inc., 1993a, 1993b) in the references. For those readers not familiar with the power of modern CACSD software packages, it will be helpful to know that MATLAB was used to produce almost all of the system response plots, of every kind, shown in the book.

On the other hand, it is not assumed that the reader *must* have access to MATLAB in order to be able to make use of the book. MATLAB has been used simply as one representative CACSD package. For this reason, full details of MATLAB code are not given in the text (although full details are available in Appendix 3, and in the various files on the accompanying disk). What we have done is to include some sample MATLAB code in simple cases, to indicate how each design method might be specified for computer assistance. In this way, the reader who does not have access to MATLAB (or, indeed, to any other CACSD environment) can still use the vast majority of the text, while simply ignoring all references to the files on the disk.

For those readers who do have access to MATLAB, several MATLAB m-files are provided on the accompanying disk, which support the written text. Their file names generally link up with the numbers of the figures in the text, so it is easy to find the m-file that will generate a particular figure. By modifying copies of these files, the reader is able to experiment with controller parameter values, or to try other modifications to the various control system designs and simulations presented in the book. Several of the m-files thus constitute extremely useful templates for many of the techniques of control system design and simulation. Appendix 3 describes MATLAB, and the MATLAB software configuration needed to run these m-files. Appendix 4 describes the related package SIMULINK® (The Mathworks Inc., 1994a), which is used for digital computer simulation (although most of the simulations on the disk just use MATLAB). The ASCII text file *readme.txt* on the disk contains further information.

Details of one suitable philosophy for including MATLAB- and SIMULINK-based material into the teaching of this text can be found in Dutton and Barraclough (1996).

How to use the book

Figure P1 gives an overview of how the material in the book is structured. It also shows broadly which chapters contain some material necessary for the study of later chapters. At the introductory level, it will be necessary to study the first three chapters (or parts of them, at least) for most purposes. Conversely, at the highest level, Chapters 7 to 14 are almost independent of each other, making it very easy to select topics as required. Also, parts of some of these chapters can easily be omitted if desired, thus widening the choice even more. The divisions shown between the levels can be moved to some extent, to suit any particular programme of study.

As a guide to the selection of material for various purposes, we offer the following suggestions (which can be extended or contracted as required, to suit the course of study). Whatever aspects of control engineering are to be studied, our basic belief is that an underlying knowledge of the frequency-domain techniques (often called 'classical control') is the best foundation from which to begin. Even if a course concentrates on the time-domain (state-space) methods, the importance of the 'feel' which comes from the frequency-domain study of the input–output behaviour of systems, and of the effects of adding simple controllers into feedback loops, cannot be overestimated. This means that, in our opinion, most of Chapter 1, plus a selection of topics from Chapters 2, 3 and 4, will almost always be desirable.

If a course heavily biased towards the frequency-domain methods is required, then the selection shown in Figure P2 is suggested. The various optional sections indicated can be chosen to suit the time available. The course could terminate at any stage after the 'Level 2' material.

For a course biased towards the time-domain methods, taking into account the previous comments, Figure P3 suggests a suitable route through the text. Again, the course could terminate at any stage after the 'Level 2' material.

For a purely digital control course, it will still be necessary to study the background of frequency-domain and/or time-domain methods, so as to be able to design that which is to be digitally implemented (apart from the purely digital

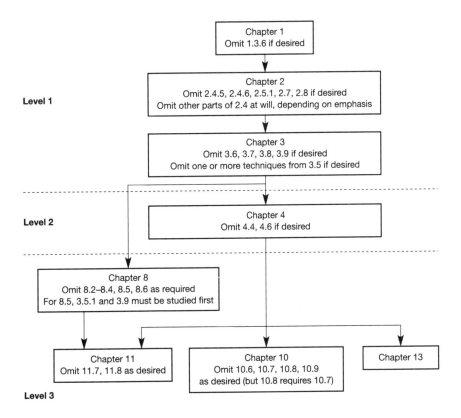

Figure P2 A mainly frequency-domain control course.

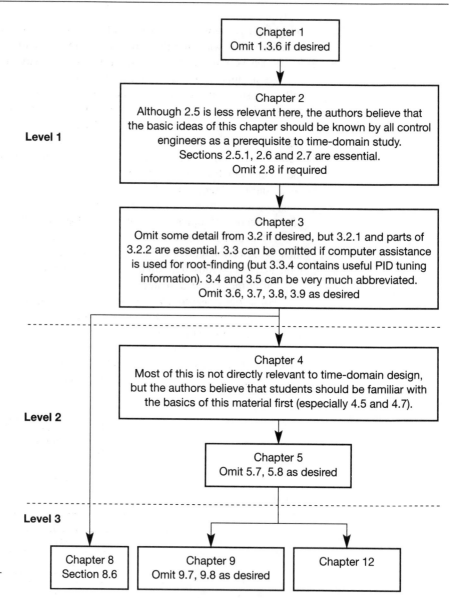

Figure P3 A mainly time-domain control course.

controllers in Chapter 7). With that proviso, Figure P4 suggests suitable routes. Note that Section 9.7 contains some useful comments on general matters of implementation, and these could usefully be extracted, even if the subject matter on which that section is based is not understood.

Finally, for those requiring a nonlinear control course, Figure P5 illustrates the possibilities. Again, it is necessary to study various basic aspects of frequency-domain and/or time-domain control (as indicated in Figure P5) depending upon the nonlinear techniques chosen for study.

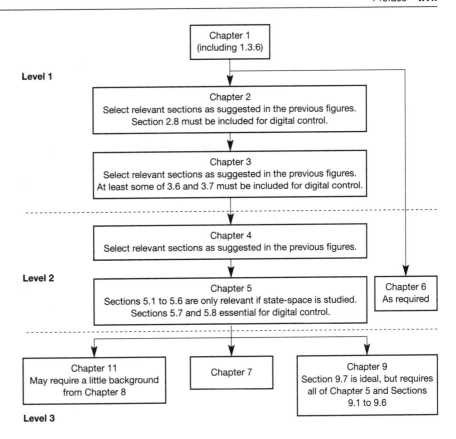

Figure P4 A digitally biased control course.

Acknowledgements

The production of a tome such as this is impossible without the contributions of many people. Those to be thanked for helping to shape our views and for their helpful suggestions include our own teachers, all our former industrial colleagues and our present industrial contacts and academic colleagues (in the case of Ken Dutton, this includes thanking Bill Barraclough, who taught some parts of Ken's control engineering degree in the early 1970s!). Not to be forgotten are several generations of students (undergraduate and postgraduate), from whose questionings we never cease to learn, and whose comments have helped in organizing the presentation of much of the material. A special thanks goes to those academics at other institutions who have tested some of our material on their own students and provided invaluable feedback.

We are also grateful to the School of Engineering of Sheffield Hallam University for providing the necessary computing and printing facilities to allow production of the draft. The staff of Addison Wesley Longman in the UK have also been extremely helpful whenever required, and the helpful comments of the referees of the proposal and the draft were also very gratefully received (and many were acted upon!).

Lastly, but certainly not least, we heartily thank all our families and friends who endured several hundred hours of our time being diverted away from them and from our normal activities, towards the writing of this book. Without their

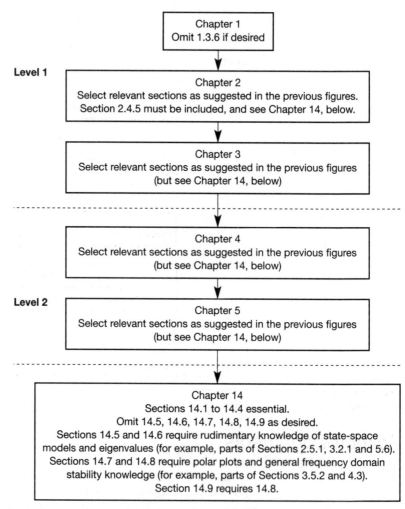

Figure P5 A nonlinear control course.

continued patience and understanding, the writing of the book really would have been impossible.

The authors would be very happy to receive any comments or suggestions. These can be sent to Ken Dutton via the email address: k.dutton@shu.ac.uk

Ken Dutton
School of Engineering
Sheffield Hallam University

Steve Thompson
Department of Mechanical and Manufacturing Engineering
The Queen's University of Belfast

Bill Barraclough
School of Engineering
Sheffield Hallam University

September 1996

1 Introduction

1.1   Preview

Each chapter in this book begins with a section like this. Usually (especially in the later chapters) it tells the reader what he or she needs to know before reading the chapter, in order to be able to understand it fully. Generally, once the reader has read Chapter 1, further chapters will only need knowledge gained earlier in the text. These 'preview' sections also indicate what the reader should expect to learn by studying the chapter. In addition, they list any mathematical techniques which will be introduced for the first time during the chapter.

Sometimes, the preface to a book is there because people expect there to be a preface. In this book, the preface also contains much useful information about the structure and use of the book. It shows that major topics have been divided between several chapters. When this occurs, reference is made to the next chapter in which the topic is developed.

In this chapter the reader will learn:
- the fundamental concepts and terminology of control engineering – some of which are necessary to understand fully the remaining items in this list!
- that *models* in the form of mathematical equations and block diagrams can often be found, which adequately describe how a real-world 'system' (yet to be defined) behaves
- that these models might be useful in designing a control system – which will alter the behaviour of the real-world system, in some desired manner
- how the performance of control systems can be specified
- that there are very many different aspects of control engineering that can be studied.

NEW MATHEMATICS FOR THIS CHAPTER

Many control engineering texts have early chapters which introduce all the mathematical techniques needed in the book. This text is not like that. Instead, we do not introduce any mathematical tool until it is needed in a control engineering context. In this way, the reader only covers the mathematics actually required in order to understand his or her chosen set of topics; and covers it at the time it is used. Sometimes, the new mathematics is introduced in the text, where it is needed. At other times, for more complicated or involved aspects of mathematics, an appendix is used so as not to upset the flow of the text too much.

There is very little mathematics in this introductory chapter, but we do mention one or two mathematical ideas, and use one or two equations. To understand this chapter, the reader need only be familiar with the basic ideas of algebra (forming and manipulating equations) and calculus (simple integration and differentiation). Matrix algebra is mentioned in passing, but not actually used. If required, Appendix 1 gives a unique slant on matrix algebra from a control engineering viewpoint. However, it might be better to delay study of that until it is really needed.

1.2 Control engineering – terminology

If a dozen control engineers are asked to define 'control engineering', they will produce a dozen different answers. Most will probably contain some comment about altering the natural behaviour of some system or other, so that it behaves as required. However, due to the vagueness of such attempts at formal definitions, it is perhaps better to let the subject evolve naturally from the consideration of a control problem. As an illustrative system, we choose to consider the control of an inverted pendulum. One version of this problem will be familiar, in that it requires someone to balance a broom handle vertically on the palm of his or her hand (Figure 1.1). The

Figure 1.1 Solving the broom-balancing problem.

task is not particularly difficult. After placing the point of the handle on the palm in approximately the required vertical position, appropriate eye–hand coordination, which can be achieved by most people after a little practice, produces the desired result. Despite the relative simplicity of the task, the mechanisms involved in producing the solution are worthy of further examination. Indeed the study of such mechanisms is the theme of all control texts.

Before looking at the mechanisms, let us look at the broom handle. Its shape is fixed. If its shape changed it would cease to be a broom handle. In control engineering terms the handle is the *plant* (or *process*) to be controlled. Like many engineering plants and processes, it evolved. This evolutionary path determined the best material and the length and diameter of the handle in order to minimize cost but, not surprisingly, gave little thought to the possibility of it being balanced in a vertical position. Fortunately the control mechanisms within humans are very adaptable and can compensate for this oversight in the evolutionary development of the handle.

In general, most industrial control problems (simple examples include controlling the speed of a motor, or the level of fluid in a tank, or the temperature of a furnace) involve the fitting of controls to plants and processes that started their evolution before control concepts were well understood. This invariably complicates the control problem and reduces the potential benefits. However, the ingenuity of the control engineer can often overcome these difficulties and, in most cases, produce a well-behaved piece of equipment. When control becomes an integral part of the design process, the ultimate control scheme is invariably less complex, simplifying maintenance and reducing operator training. In addition, a good control scheme will maximize the plant's efficiency.

Control engineers like to use *block diagrams* to simplify the understanding of complex systems by breaking them down into smaller, interconnected subsystems: see Figure 1.2. If the broom handle is the plant, then the *controls* are all the

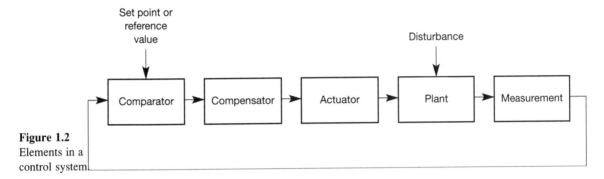

Figure 1.2
Elements in a
control system

elements, or *blocks* in Figure 1.2, which are external to the plant and which are there to ensure properly organized behaviour. Given that the objective is to keep the handle vertical, the first block to be considered is the detection, or *measurement* element, which indicates the position of the handle. In Figure 1.1 the measurements are made by the eyes which send an appropriate message to the brain. There is a communications system, the neurological system which sends messages, or *signals*, from the eye to the brain. In Figure 1.2 the signals are indicated by the connecting lines entering or leaving the various blocks. The brain performs two main functions. It compares the measured position of the handle with the desired position (the *setpoint* or *reference value*), and it calculates the appropriate action to be taken. In the language of control, the brain performs the function of both a *comparator* and a *compensator* (the two together forming a *controller* – see below). Finally, the signal emanating from the compensator portion of the brain activates the muscles in the arm which position the hand carrying the lower end of the handle. The arm is the *actuator*, providing both power amplification and motion.

Comment: Note that strictly, the *compensator* in a control loop is the block shown as such in Figure 1.2. The *controller* is taken to include the compensator, the comparator and any other blocks and connections added to the plant in order to control it in the required manner. However, while the term *compensator* is almost always used with its proper meaning given above, the term *controller* is very often used (in this book, too!) to refer only to the compensator! This is of only semantic importance, as the context should always make the meaning clear, but it is worth noting, in order to save confusion over the common usage of two different words for the same thing.

When looking at Figure 1.2, notice how information circulates around the *control system*. In this context a *control system* refers to the plant and all the associated blocks (that is, the plant plus the controller). Measurement information taken from the plant is, in turn, acted on by the comparator, compensator and actuator, before being fed back to the plant. The resulting control system is called a *closed-loop* system and is characterized by the *feedback* signal which carries information from the measurement device to the comparator. In contrast, an *open-loop* control system does not have such a feedback path.

To show the importance of feedback in control it is possible to break the feedback path in the broom handle balancing problem, Figure 1.1, by blindfolding the person balancing the handle. All the control system elements are present, but the circulation of information is lost and consequently the handle falls. It may quite reasonably be argued that if the room is still (equivalent to no external *disturbance* – see below), and if the experimenter has a steady hand (equivalent to there being no *noise* signals within the loop – see below), then provided the handle was originally vertical it will remain vertical. This is true, and gives the justification for many open-loop systems. However, for systems sensitive to noise or disturbances (and the broom handle balancing problem is very sensitive to both), feedback control is essential.

Note that 'noise' means unwanted signals corrupting the normal signals within the loop. Noise is generally random in nature and will often have frequency components much higher than those associated with the signal it corrupts. The most common exception to this statement is regular 'hum' caused by a.c. mains power supplies.

Disturbances are generated externally to the loop and are signals which act on the plant to cause definite changes in operation. They may be caused by such things as changes in atmospheric conditions, variability in some product entering the plant, variations in the power supply, or the ageing of components in the system.

One of the signals, the *setpoint*, or *reference value*, indicates the desired position of the handle. Although this signal is external to the information circulating around the closed-loop system, it nevertheless plays an important role. If the setpoint is fixed, that is, the control objective is to maintain the top end of the handle at some fixed point in space, then the control problem is said to be one of *regulation*, and the control system is a *regulator*. When the control objective is to move the handle along a predefined trajectory, the control problem is said to be one of *tracking*, and the system might be called a *tracker*. These distinctions are discussed at greater length in Section 5.4, and in the examples below.

With these terms defined, there is perhaps sufficient information to start looking at familiar pieces of equipment and identifying the various control elements.

Example 1.1 *Domestic oven temperature control*

The temperature control of the oven in an electric cooker is an example of a closed-loop regulatory system. Here, the control objective is to maintain the oven temperature at, or near, the manually adjusted setpoint temperature. A sensor within the oven measures the oven temperature. The resulting temperature signal is compared with the setpoint and a very simple compensator provides a signal which is used to switch (actuate) power to the oven's heating elements. The heating elements are switched on if the oven temperature is somewhat less than the setpoint temperature, and off if the temperature is somewhat above the setpoint. For a domestic cooker this strategy appears adequate and has, for many years, been the main method of oven temperature control.

Example 1.2 *Domestic washing machine*

A basic domestic automatic washing machine typifies an open-loop system. Here the pre-set washing cycle activates various washing operations for a set period of time. It is assumed that if the set sequence of operations is followed correctly, then the clothes inserted into the machine will emerge clean and undamaged. However, because the system is essentially open-loop the manufacturers provide copious instructions to try to deal with all the variables that can occur in a typical wash.

Example 1.3 *Domestic central heating*

For those living in a colder climate, examination of a domestic central heating system reveals a number of control loops. Water heated by the boiler is continuously pumped round the house. There is invariably regulatory control on the exit temperature of the water from the boiler. If the boiler is oil, or gas fired, this temperature control system is similar to that found in the oven of an electric cooker (Example 1.1).

Solid fuel boilers use a different closed-loop control strategy to regulate circulating water temperatures. A bimetallic strip, whose initial position can be set manually, senses temperature changes. Deflections of this strip are used to open, or close, a damper (basically a flap acting as a throttle valve) altering the flow of air into the combustion chamber. The control signals are mechanical and the control action is continuous (like the handle balancing problem, but unlike the on–off control in Examples 1.1 and 1.2).

Heat carried by the circulating water is dissipated by means of heat radiators. When a radiator is fitted with a thermostatically controlled valve, the heat dissipation process from that radiator is closed-loop. If the radiator has no means of detecting the room temperature, the process is open-loop.

Example 1.4 *The McDonnell Douglas DC-X*

Aircraft have a large number of control systems. Cabin pressurization and the automatic pilot are examples of regulatory control systems, whereas an automatic landing system is an example of a tracking system. Even our old friend the broom handle has a serious counterpart. During 1993 the unmanned McDonnell Douglas Delta Clipper Experimental (DC-X), a prototype, single-stage-to-orbit launch vehicle, was seen on international television news rising smartly from its launch pad to an altitude of 50 metres, where it then stopped in mid-air, moved 100 metres down range and then descended for a perfect landing. This rocket performed all the functions demonstrated by the handle balancing problem. However, in the rocket the sensors are more likely to be gyros and accelerometers, the brain is replaced by a computer with flight and navigational control algorithms comprising over 30 000 lines of Ada language code, and the actuators are the rocket motors (unfortunately, the rocket suffered a serious accident at the end of July 1996, when one of its four landing 'legs' failed to deploy, and it toppled over on landing and caught fire).

1.3 A stroll through the field of control

This section will introduce many of the concepts and topics developed in the text, without the complication of a formal mathematical base. Again the inverted pendulum will, when appropriate, be used to illustrate the various subjects.

1.3.1 Mainly on inputs and outputs

Looking more closely at the blocks in Figure 1.2, it can be seen that information enters and leaves each element. Information entry points are called *inputs* and exit points are called *outputs*. Each block could be said to 'transform' its input signal(s) into an output signal. Because of this action of transferring a signal from input to output, one of the many possible mathematical descriptions of a block is referred to as a *transfer function*, but more will be said on this topic in Chapter 2. At this point no explicit assumptions regarding the amount or type of information circulating in the control system have been made (we shall return to this point shortly).

Once the numbers of inputs and outputs associated with the plant have been defined, a structure is required for the control scheme we intend to design. For closed-loop systems, this structure is often based on that shown in Figure 1.2, although other structures are possible. Regardless of which control structure is used, the inputs and outputs associated with the controls are defined once the plant's inputs and outputs are known.

Finding the best, or most appropriate, inputs and outputs for a given plant is part of the art of control system modelling. In general, outputs are the measurable quantities to be controlled. In the examples of simple systems mentioned earlier, these might include:

- the position of the top of the inverted broom handle of Section 1.2 (but see Example 1.5, below, for a fuller consideration of this particular system)
- the temperature in an oven
- the water temperature in a washing machine
- the time for which the washing machine performs various operations (washing, rinsing, spinning and drying, for example)
- the temperature of the water leaving a central heating boiler
- the temperature of a room containing a central heating radiator
- the cabin pressure in an airliner
- the heading maintained by an aircraft
- the position (flight path) of an aircraft during landing
- the vertical attitude and height of hover of the DC-X rocket.

Sometimes the quantity to be controlled cannot be measured directly and must be inferred from available measurements (of other signals) using an *estimator* or *observer* (developed in Chapter 9).

The inputs of a plant are the variables which, if adjusted, would alter the measured outputs. When the inputs can be manipulated directly they are called *manipulable inputs* or simply *inputs*. In the above examples, these might include:

- the nerves that cause the arm muscles to operate, for balancing the broom handle
- on/off signals to the heating elements of the oven and the washing machine; and also to the various washing machine motors and water valves
- on/off signals to the central heating boiler's fuel source and ignition system
- the rod determining valve position in a thermostatically operated radiator valve
- control signals to an airliner's cabin pressurization pump
- control signals to the control surfaces of an aircraft
- control signals to the valves determining the amount of fuel entering the rocket motors; and position signals to direct the rocket motors' thrust.

Inputs that cannot be manipulated directly are called *disturbance inputs*, or simply *disturbances*. There are many examples of these for the systems mentioned above. Some are:

- involuntary muscle spasms (for example, sneezing) while balancing the broom handle
- opening of the oven door
- variation of the amount of clothes, and the material from which they are made, in the washing machine
- variations in the calorific value of the fuel in the central heating boiler
- doors or windows being opened in the centrally heated room, or the external temperature changing
- changes of altitude of the pressurized aircraft cabin
- effects of wind gusts on all the aerospace examples.

Sometimes there is a choice of output measurements, because measuring instruments may be available to measure one signal just as well as another, with little difference in terms of cost or installation effort. Inputs, however, are normally associated with actuators providing power amplification (electric drives, hydraulic rams, and so on), and therefore tend to be fixed plant items. They also add to the running costs. For most processes and plants this means that the number and location of the manipulable inputs are fixed, and well defined.

The outline block diagram of Figure 1.2 is a convenient way of picturing a control loop, but before continuing it is necessary to consider briefly how (for analysis and design purposes) the words in the blocks might be replaced with some information relating to the behaviour of the real-world elements. Once the various blocks and their connecting signals have been identified, it is the transforming action of each block on its input signals that becomes important. For consistency, the transforming action of each block will often be represented using a common basis; normally a mathematical one.

The equation, or set of equations, providing the relationship between the block's input(s) and output(s) is called the *mathematical model*, or simply the *model* (the topic of modelling is considered at many points during the text, but it is next covered in more detail in Sections 1.3.4 and 1.3.8).

Linearity and superposition

If the describing equations of a block (system element) are linear, that is, they obey the *superposition principle* (explained below), then the block is said to be *mathematically linear*, or simply *linear*.

The superposition principle is based upon what happens to the output of a block (or a whole system), when two different signals are applied to its input – firstly one at a time, and then summed together. Consider a simple electronic amplifier with an input voltage u, an output voltage y and a voltage gain (amplification factor) K. For the present, it can be assumed that the dynamic response of the amplifier is so fast that it can be ignored when compared with the dynamics of other components connected to it. Even so, the modelling of such a component may not be as simple as it seems.

The simplest possible model of this amplifier assumes that the output is given by $y = Ku$. The input–output characteristic of such a model is shown in Figure 1.3(a). From that figure it is apparent that if two separate inputs u_1 and u_2 are

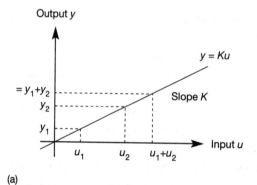

(a)

(b)

Figure 1.3 Illustrations of superposition and linearity. (a) Input–output characteristic of a linear system model; (b) input–output characteristic of a nonlinear system model; (c) input–output characteristic of a more obviously nonlinear model.

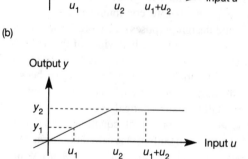

(c)

applied, giving rise to outputs y_1 and y_2 respectively, then the output which arises when these inputs are *superimposed* (that is, the sum of the two inputs is applied as $u_1 + u_2$) is equal to the sum of the two individual outputs ($y_1 + y_2$). Such a system model is said to obey the principle of superposition, and is linear.

If the amplifier is non-ideal, and adds a constant offset c to the output, so that $y = Ku + c$ as shown in Figure 1.3(b), then this property is lost. Even though the characteristic is a straight line, and the output equation is the normal mathematical equation of a straight line, this system is *nonlinear* in a control engineering sense, as the principle of superposition does not apply (the output corresponding to the superimposed input $u_1 + u_2$ is clearly *not* equal to $y_1 + y_2$ in this case).

Finally, it is inevitably the case that, if the amplifier is driven hard enough, a maximum output will be reached which cannot be exceeded (the output will be limited by the amplifier's power supply voltage). If this behaviour is introduced into the model, the characteristic becomes very obviously nonlinear, as shown in Figure 1.3(c). Two fairly general requirements for obtaining linear models (ones which obey the principle of superposition) are suggested by this simple investigation. Namely, offsets must be removed and signal excursions must be small.

These requirements are more stringent for some systems than for others, and Chapter 14 provides a reasonably thorough introduction to nonlinear systems, whereas the rest of the text concentrates mainly on linear system models. Strictly, we should always use the term 'mathematically' linear, to denote that we may well use a linear mathematical model of a plant, even though the real plant will, to some degree, inevitably be nonlinear (as noted above, and discussed more thoroughly in Chapter 14). Thus, linear models are always approximations to reality but, fortunately, often of sufficient accuracy to allow successful use.

Linear models can be much more complicated than the simple gain considered above. They can include the dynamic behaviour of the system, and many other aspects, as we shall see. Traditionally, a study of control engineering starts with the study of linear models of plants, or processes, having one input and one output – the so-called *single-input-single-output*, or *SISO*, system. The majority of the techniques studied for such systems are based on a linear *Laplace transform* model of the plant (the Laplace transform is a very useful mathematical technique and is introduced in Chapter 2). On the other hand, a mathematically linear plant having more than one input and/or output (called a *multi-input-multi-output* or *MIMO* or *multivariable* system) may be modelled using a set of first-order, ordinary, linear differential equations presented in matrix form. This is known as a *state-space* model and is also introduced in Chapter 2 (such models can equally well be used for SISO systems). Another representation of a MIMO plant is by a matrix of Laplace transforms, the *transfer function matrix* form (see Chapter 2), but this is used only in more advanced studies (for example, in Chapter 10).

For engineering plant, the physical laws governing the system's behaviour are constant, so the (linear) transfer function and state-space models mentioned above are interchangeable for many purposes. Furthermore, a SISO system is just one special case of the more general MIMO system. However, the authors believe that a firm grasp of the concepts involved in the control of SISO systems is a prerequisite for the study of MIMO systems. That is the approach adopted in this text, and the reason why MIMO systems are not treated in any great detail until Chapter 10.

Example 1.5 *Inputs and outputs of the inverted pendulum*

It is instructive to re-examine the inverted pendulum problem in an attempt to identify the various inputs and outputs. The handle (see Figure 1.4) has six 'degrees of freedom'. It can move up and down, from side to side, backwards and forwards (that is, it has straight line, or *translational*, motion) along the z-, y- and x-axes respectively. Further, it could rotate about any of the three axes. The eye detects all these translations and rotations, so there are at least six measurable outputs. Since the eye also detects the angular and linear velocities, accelerations, and so on, many other measurements could be included.

To analyse the system, it would be desirable to minimize (for ease of analysis) the number of signals coming out of the measurement block in Figure 1.2 and being fed back to the comparator input. The number of outputs could be reduced by noting that significant rotation about the z-axis (vertical) can be ignored, since this would imply that the person balancing the handle was rotating – a movement which has no effect on the vertical balance. Also, translation along the z-axis requires that the hand be raised or lowered, and this could similarly be ignored from a balancing viewpoint, as it has no effect on the balance. Finally, it might be assumed that motion of the handle in the z–x plane is independent of motion in the z–y plane. For analysis purposes, only one of these planes need then be considered (the model for the other plane being the same), so that the system now has just two measurable outputs (one displacement and one rotation in the same plane) together with their various derivatives.

Turning our attention to the inputs, since whole-body rotations are being neglected, and other rotations of the hand are likely to cause the rod to slip and fall, only the three translational inputs would be applied to the base of the handle. As discussed above, the input in the z-axis direction is not required, reducing the number to two. When, for analysis purposes, it is assumed that the handle rotates or translates in a single plane, the system then has just one input. In control terms, if we wish to control both the rotation and the translation of the handle, the analysis system (model) is SIMO (single-input-multi-output), whereas the physical system is MIMO. If, in the analysis system, we only wish to keep the handle vertical (that is, to control the rotation without worrying about lateral movements) then the system could be considered to be SISO. If the handle was being balanced outside, any breeze would alter the position of the handle and hence become a disturbance input.

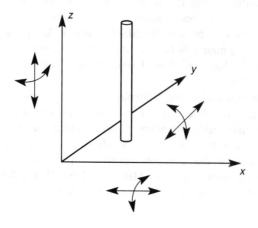

Figure 1.4 The degrees of freedom of an inverted pendulum.

1.3.2   On how the plant dictates the controller

Figure 1.5 shows a standard arrangement known as a two-degree-of-freedom feedback configuration (the reason for the name is that the designer has freedom to choose the contents of two blocks, P and K). The plant is a physical piece of hardware, and will be assumed to have fixed dynamics. By this, we mean that its response to an input today will be the same as its response to the same input applied

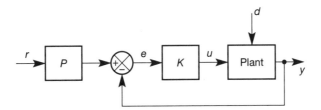

Figure 1.5 Standard two-degree-of-freedom feedback configuration.

last week – this is called a *stationary* system. Note that not all plants are stationary systems in this sense. For example, vehicles whose initial fuel load is a significant proportion of their mass (such as rockets and racing cars) have dynamic behaviour which changes as the fuel is used up. Returning to our assumed stationary system, its output y is a function of the manipulable input u and any external disturbance d. Note that conventionally the disturbance d is shown modifying the plant's output signal y (as shown in Figure 1.6, for example). A perfect transducer (one having an instantaneous response) measures the output variable at y, this measurement being

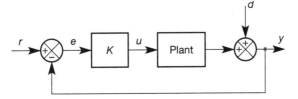

Figure 1.6 Standard single-degree-of-freedom feedback configuration.

conventionally shown by a dot on the plant output signal (but this is often omitted when dealing with control theory, rather than practical control engineering drawings). Note also, that a transducer is strictly any device that converts one form of energy to another. However, in control systems terminology, it has become commonplace to refer to measuring transducers (pressure gauges, flowmeters, loadcells, tachometers, accelerometers, voltmeters, current transformers, gyros, and so on) simply as 'transducers'; and actuating transducers (motors, hydraulic cylinders, solenoids, valves, and so on) as 'actuators'.

The resulting measurement, which at this stage is assumed to be noise free, is fed back to the controller. Ignoring transducer dynamics is permissible so long as the speed of response of the transducer is significantly faster than that of the plant, or if the transducer dynamics are assumed to be included with those of the plant (that is, they are inside the 'Plant' block). At this point it will suffice to say that this is often the case. The controller consists of a pre-filter P which modifies the setpoint r, a comparator (again shown in its conventional form) and a forward path compensator K which generates the plant actuating signal u.

In Chapter 2, the modelling of systems is considered. In the early parts of the book, the pre-filter P is seldom required, leading to the single-degree-of-freedom configuration shown in Figure 1.6. Assume that a plant is actually represented by a block G, but that the methods to be studied in Chapter 2 lead to a nominal model of the plant called G_m. What is not shown in Figure 1.5 is how the nominal plant model G_m might conceivably be used to design the controller elements P and K. However, Section 13.3 analyses Figure 1.5 further, in terms of the differences that will inevitably exist between the real plant G and its model G_m. Although we are not yet in a position to understand the analysis of Chapter 13, it is useful to note that some fundamental points arise from it.

Firstly, the next few chapters show how to design the controller elements (typically, the contents of the block K, but also P if needed), using information contained in the plant model G_m. The contents of K (and P) are therefore determined by the plant model. If G_m is identical (or close) to the true representation G, then the plant dictates the controller.

The achievable performance of the closed-loop scheme will also be limited by the plant, which will have certain maximum signal levels, and maximum allowable rates of change of signals. For example, even the world's most accomplished control engineer could not design a physically realizable control system to cause a supertanker to adopt a course at $90°$ to its present one within 10 seconds. For example, neither the colossal amounts of energy required from the engines nor the effective transmission of that energy to the water would be remotely achievable. Note that such a scheme *could* be designed theoretically on paper, and could even appear to work in the simplest computer simulation. However, if the simulation was made to display the demands being placed on the supertanker's engines and its propulsion and manoeuvring systems, and these were compared with a practical study of the limitations of the real-world plant items, the impossibility of the operation would immediately become clear. These aspects of computer simulation are encountered in several chapters.

Chapter 13 shows that, if the plant model is a perfect representation of the plant, the function of the feedback loop is solely to compensate for disturbances, while the pre-filter is responsible for the accuracy with which the plant output y follows changes in the reference input r (that is, the tracking accuracy). Of course, in reality, various approximations are always necessary in obtaining plant models (as Chapter 2 explains). Therefore, the plant and nominal plant model are never identical ($G_m \neq G$). The result of this is that both plant disturbance and modelling errors are fed back, and the compensator must be modified to accommodate the additional modelling errors. This indicates the need for 'good' models. To design a control system which is tolerant, or *robust* to modelling errors (or other design approximations), is the subject of Chapter 13.

1.3.3 A simple control philosophy – if it works

In general, the more stringent the control requirements for a plant, the more complex the control system will need to be, and the larger the required number of inputs and outputs. An often-employed control strategy for MIMO systems is therefore to assume that the system is linear, and then reduce the complexity by assuming that the MIMO system can be treated as a number of linear SISO systems.

If it works, this simple control philosophy has a number of advantages. Techniques for designing linear SISO systems are well understood. A few basic compensators may be used to control the majority of plants and processes successfully. Plant operators are familiar with SISO controllers and therefore operator training time for new plant is minimized. Similarly, plant maintenance is simplified and the range of spares minimized. The problem is that it is sometimes not possible to control MIMO systems in this way.

In the domestic central heating system, Example 1.3, this strategy is adequate and the boiler temperature can be controlled as though it were independent of room temperature. If the occupants demand a higher room temperature and turn up the radiator's setpoint, the water flow rate will increase, and so more heat will be extracted from the circulating water. Cooler water returning to the boiler reduces the boiler's exit temperature and activates the boiler's temperature control system. When a control action in one loop causes a change in another loop like this, the loops are said to be *coupled*, or to *interact*. MIMO systems invariably interact but the question is, 'when do these interactions become unacceptable?'

Consider again the two-dimensional inverted pendulum of Example 1.5. Attempts to build a control scheme for this system consisting of two separate loops, one controlling rotation and the other controlling translation in the plane, would fail. Assume the handle was vertical, but forward of its desired position. The correcting action in the position control loop would be to move the base of the handle *backwards*. However, this would induce a rotation in the handle (making the top fall forwards) for which the correcting action in the rotation control loop would be to move the base of the handle *forwards*. With each action, the correcting action in the other loop would become larger and the handle would soon fall. A system in which either a disturbance, or a setpoint change, causes ever-increasing control actions within the loop is said to be *unstable*. Instability can occur for reasons other than loop cross-coupling, and is common to SISO, SIMO, MIMO and MISO systems. A prime objective of control system design is to produce closed-loop systems that are *stable*.

The reason why the domestic central heating system can be treated as though it were a number of single loops and the inverted pendulum system cannot, is due to the strength of the coupling between the loops. *Decoupled*, or weakly coupled systems, like the domestic central heating system, are amenable to SISO control strategies, whereas strongly coupled systems, like the inverted pendulum, require an additional decoupling controller (see Chapter 10), or a truly SIMO, or MIMO, control scheme.

1.3.4 The design problem

It should now be evident that the first design requirement is a model that describes the relationship between a plant's input and output signals. If we do not have such a model, how can we decide on the required control signals to apply to the inputs, so as to obtain some desired outputs? In general, the better the model, the more likely it is that the actual response of the system will match the designed response. So what is a 'good' model?

For control system design, a good model must satisfy two basic properties:

(1) It must be a representation of the plant that will enhance the ability to understand, explain, predict, change and control the behaviour of the system. Normally, the model is a mathematical representation, although graphical

representations which show the plant's output(s) to a well-defined input(s), and even physical models, can also be used (for example, scale models of cars, aircraft and bridges, for aerodynamic testing in wind tunnels; or transparent plastic scale models of furnaces, with injected smoke for examining gas flows).

(2) It must contain only the essential aspects of an existing plant (or of a plant that is to be built). If the model is an over-simplified representation of the plant it will lose the properties indicated in (1). Alternatively, if the model contains an excess of information, it becomes unwieldy and complicates the design process. Further, constructing highly complex models is uneconomical (exceedingly time-consuming) and often impractical. Producing and ensuring that a model is a good representation of the plant can account for over 90 per cent of the effort expended on most control system design studies. The model, however, only provides the means by which it becomes possible to design the controls. More is said on this topic in Section 1.3.8.

Aspects of design – stability of linear systems

After a 'good' model is obtained, the most fundamental design requirement is that the control system is stable. That is, the outputs of the system and all of its components should exhibit bounded responses to bounded input signals. A sinewave having amplitude $= A \sin \omega t$ (see Figure 1.7(a)) is bounded, since it is possible to fix limits at $\pm A$ which will not be exceeded (so $A = 2$ in this case). A signal which can be described by a negative exponential, amplitude $= Ae^{-xt}$ (Figure 1.7(b)), is also bounded; but a positive exponential, amplitude $= Ae^{xt}$ (Figure 1.7(c)), is unbounded. In engineering systems, any unbounded response invariably means plant misbehaviour, breakdown or component failure, with inevitable economic and safety implications.

Ideally the plant should be open-loop stable (that is, stable before the addition of any control loops). A stable plant is desirable since loop failure within the closed-loop system then leaves the stable open-loop plant. Further, in an emergency, any malfunction of the closed-loop system can be corrected by deliberately breaking the

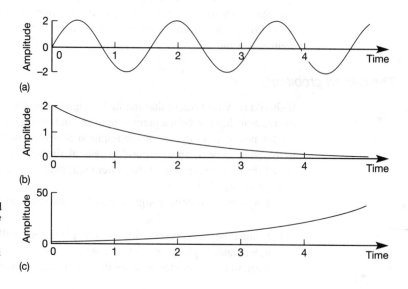

Figure 1.7 Bounded and unbounded signals. (a) The signal $2 \sin(4t)$ – a bounded signal. (b) The signal $2e^{-0.6t}$ – a bounded signal. (c) The signal $2e^{0.6t}$ – an unbounded signal.

loop. It is important to realize that it is possible for a plant to be open-loop stable, but for the closed-loop system in which it is incorporated to be unstable (several such examples occur later in the text). Invariably for the engineer, all closed-loop control systems must be designed for stability, irrespective of the plant's open-loop stability.

Unstable open-loop plants do occur. Sometimes the plant instability is unavoidable, but sometimes the designer deliberately introduces open-loop instability into the plant. With some chemical reactions, for example, the reaction process is inherently unstable and product manufacture depends on the stabilizing influence provided by the closed-loop control system. Note that the unbounded nature of the output, often associated with rapidly increasing temperatures, indicates that the energy of the reaction is increasing.

Modern fighter aircraft are a further example of open-loop unstable plant. Here the instability is deliberately introduced so that the plane will respond quickly to pilot commands. Since control action is needed continuously to stabilize the aircraft, it follows that a slackening of that action (induced by the pilot altering the control inputs) causes a rapid and, if unchecked, unstable response.

There are a number of ways of examining system stability. Perhaps the most intuitive is to apply a bounded input to the system (or, more likely, apply a simulated bounded input to the system model) and observe the nature of the response. Bounded-input-bounded-output (*BIBO*) stability investigation normally uses a mathematically well-defined input such as an *impulse*. As the name suggests, this effectively gives the system a sudden sharp jolt which acts over as short a period of time as is physically possible (infinitesimally short, in theory). This is used to perturb a system which is initially at rest; that is, the values of the signals flowing round the closed-loop are not changing with time or, to use the language of control, the system is initially under *steady-state* conditions. Note that the term 'steady state' can also be used in a frequency-domain analysis (Chapter 3) to describe a system having a constant (or 'steady') oscillatory response. In general no distinction is made between this form of 'steady' oscillatory motion and a 'steady' rest state (zero motion), since this will be clear from the context.

Following the application of the impulse, if, after some time, the system settles at its original steady-state level, then the system is said to be *asymptotically* stable. If, however, the system produces a bounded response other than the original steady-state level, it is *marginally* stable. For linear systems, any other response to an impulse input is unstable. BIBO stability is illustrated in Figure 1.8.

Stability can also be tested by examining the linear equations describing the system. Linear dynamic systems are described by ordinary, linear, differential equations. A dynamic (or 'dynamical') system is any system in which signal values can change with time. This means that we can consider not only engineering systems, but populations, weather systems, economic systems and so on. The fundamental mathematical analysis of any linear, dynamic system will always result in differential equations, containing terms such as d/dt (signal value), and the reader may recall that an entity known as the 'complementary function' crops up in the solution of such equations (don't be too alarmed if this doesn't come to mind at the moment). In control engineering, this complementary function becomes the *characteristic equation* of the system, as we shall see in Chapter 3.

For stability, the complementary function must consist of mathematically bounded terms, so we shall have to choose the contents of our controller to make

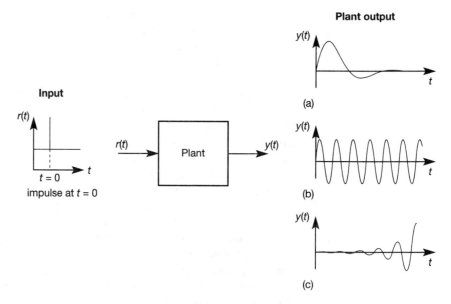

Figure 1.8 Plant stability.
(a) An asymptotically stable
response; (b) a marginally
stable response; (c) an
unstable response.

this happen, irrespective of whatever else we hope to achieve. Although the complementary function (or the system's characteristic equation) of a closed-loop control scheme is usually easy enough to find, finding the range of possible solutions for which the system will be stable is more problematic.

Fortunately, there are techniques for testing a system's characteristic equation for variations in control parameters – and hence for establishing ranges of stable solutions. The most commonly used techniques for linear systems are *Routh's stability criterion*, the *Nyquist* stability criterion and the *inverse Nyquist* criterion (the latter is used mainly for SISO systems having compensators in the feedback paths, and for MIMO systems). These are all developed in Chapter 3. There is also the *Jury* stability test, which is specific to digital systems.

Rules established from these various stability criteria are often incorporated within control system design techniques – for example, *Bode plots* and the *polar* and *inverse polar plots* (introduced in Chapter 3), Evans' *root locus* method (Chapter 4) and, for MIMO systems, the *characteristic locus* and *inverse Nyquist array* methods (Chapter 10).

Aspects of design – stability of nonlinear systems

Nonlinear systems (systems that fail to obey the principle of superposition) can behave very differently from linear systems, to the extent that most of the techniques mentioned so far no longer apply. The classification of stability for nonlinear systems is more involved, and is often achieved using techniques originally developed by Lyapunov. *Lyapunov stability* analysis (discussed in Chapter 14) can be used to analyse the stability of several kinds of nonlinear systems, but is actually based on the pioneering work of Lagrange on conservative mechanical systems. These are systems in which energy is conserved – such as a pendulum, constructed using a light rigid rod and swinging in a vacuum with negligible pivot friction, where the energy is repeatedly converted from potential to kinetic and back again, as the pendulum swings. Lagrange was able to show that an

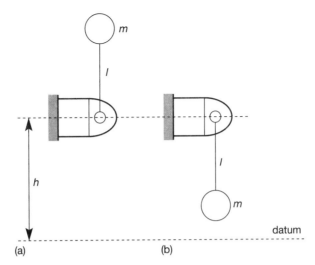

Figure 1.9 Pendulum equilibrium positions. (a) Inverted equilibrium position; (b) pendant equilibrium position.

equilibrium point of such a system (that is, one of the points where it could possibly remain at rest – see Figure 1.9) will be stable if it corresponds to a local minimum of the potential energy function for the system (as in Figure 1.9(b)) and unstable if the local potential energy function is a maximum (as in Figure 1.9(a)).

These ideas about nonlinear systems are not really used again until Chapter 14, although there is one section in Chapter 2 which expands them a little. This may seem strange, as we hinted earlier that all real systems are inevitably nonlinear to some extent. However, it turns out that it is very often possible to work with linear mathematical models, and the techniques that make this possible are introduced in Section 2.4.5. Nevertheless, the reader should be aware that all linear analysis methods are approximations that may break down (for example, if the system is to be operated away from the conditions under which the linear mathematical model was obtained).

Where linear approximations *can* be used, it is wise to do so, because there are standard solution methods to most linear control problems. For nonlinear systems, it is more often the case that every system has to be treated on its own merits, so methods that will work for one system will not work for another. The study of nonlinear control is therefore conceptually harder than that of linear control.

Aspects of design – performance

Although the dynamics of a plant may be fixed, its apparent behaviour can often be changed by incorporating it within a control system. For example, the plant may have unstable dynamics, which the addition of a controller might stabilize. Alternatively, the response of the plant to input signals may be too slow, and might be improved by adding a controller. Another common problem is that the steady-state value of a plant's output may be incorrect following an input change, and such a *steady-state error* might be corrected by adding a suitable controller.

For the closed-loop system of Figure 1.2, the measurement and actuation devices will, to a large extent, be defined by the plant. Indeed, for design and analysis purposes, these hardware elements are often assumed to form a non-dynamic part of the plant itself. This is the case in Figure 1.5, where these devices do not appear in their own right, but are assumed to be inside the block labelled 'Plant'.

It must be stressed that it is physically impossible for a hardware device, such as an instrument measuring the output of a plant, or an actuator manipulating an input, to respond instantaneously. All measurement and actuation devices are independent dynamical systems in their own right. Neglecting their dynamics simplifies the design process, but occasionally at the cost of producing unexpected behaviour when the controls are eventually introduced to the plant. To avoid this, if the sensor or actuator dynamics are significant, they are either included explicitly as blocks in their own right (as in Figure 1.2), or their dynamics can be included in the plant model (as in Figure 1.5). However, it is important to realize that the only elements that can normally be *adjusted* by the designer are the compensator and, if present, the pre-filter.

It is also evident that the basis of any closed-loop performance requirement will be stability. That is, the chosen compensator must eventually return the plant to an equilibrium position following a disturbance. In general, there will be many compensators, each with a wide range of parameter settings, that would satisfy this requirement. However, stability alone is too vague a requirement since 'eventually' could mean years or microseconds and, for most closed-loop systems, both of these extremes would be inappropriate. To be of practical value, the desired performance must be specified. That is, for a known disturbance, the time taken to reach equilibrium and the manner in which the closed-loop system reaches equilibrium must closely approximate that defined by the designer.

Consider the closed-loop system shown in Figure 1.10. The important quantities for measuring closed-loop system performance are $e(t)$, $u(t)$ and $y(t)$. The error signal $e(t)$ is the difference between the desired input $r(t)$ and measured

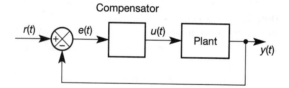

Figure 1.10 A closed-loop control system.

output $y(t)$. Since the error signal changes with time, and can be positive or negative, one commonly used performance measure is the area enclosed when the square of the error signal is plotted against time, see Figure 1.11. If the same

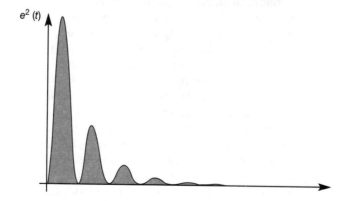

Figure 1.11 A plot of the square of the error signal with time. The shaded area is $\int_0^\infty e^2(t)\, dt$.

excitation is used, the integral of the error squared (IES – shaded in Figure 1.11) gives a number which quantifies the selected compensator's performance. In theory, the smaller the IES, the smaller the error and the better the performance. In practice, controllers that minimize this type of performance criterion can produce very oscillatory responses, which makes them of only limited value. Variations exist and these will be described in subsequent chapters – particularly in Chapter 12, on optimal control.

The changes in the actuation signal $u(t)$ are a measure of the energy expended in compensating for a given disturbance. They can therefore indicate the monetary costs associated with a particular control action. Monetary costs of control action alone are a poor indicator of performance, since zero change in the control action produces the least cost. For this reason, if a function of the actuation signal is to be included, it is invariably added to the previously defined error-based performance function. For example, an oscillatory response gives a low IES, but increases the required actuation (and monetary) cost, whereas a sluggish response gives a high IES value, but a low actuation cost. A controller that minimizes a combined actuation and error-based performance measure should, in theory, produce an acceptable system response. In practice, minimization of these combined performance measures can be difficult (see Chapter 12).

From experience, it is known that the way a system responds can be quantified, and hence specified, by examining the output response $y(t)$ produced when the input $r(t)$ suddenly changes from one steady level to another at some point in time. Such a change in the input is referred to as a step change, and if the input changes from zero to one it is called a *unit step*. The change this produces in $y(t)$ is known as a *unit step response*, see Figure 1.12. Also shown in Figure 1.12 are some of the performance criteria associated with a unit step response, namely steady-state error, settling time t_s, peak overshoot $(y(t_p) - y_{ss})$, time to peak overshoot t_p, frequency of oscillation and rise time t_r (which is usually measured between the times corresponding to 10 and 90 per cent of the steady-state value). These parameters are usually specified in terms of maximum or minimum values, and can be used to define preferred

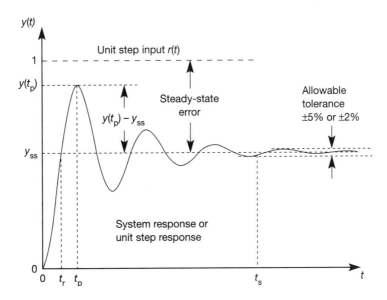

Figure 1.12 Commonly used step response criteria.

response characteristics. It should be noted that they are not independent. For example, a fast rise time is usually obtained by making the system more oscillatory. However, an oscillatory system will tend to have a large peak overshoot.

Choosing performance criteria is an important aspect of control system design. Too tight a set of specifications may be impossible to attain, or lead to control systems which are too costly to realize. Conversely, an incorrectly specified set of performance criteria results in a system which is inadequate. Remember that the plant dynamics are normally fixed and the object of control system design is to find a compensator that will manipulate the plant's input so as to produce acceptable output behaviour. That is, the plant, and to a lesser extent the compensator, dictate the possible range of closed-loop behaviour.

1.3.5 Compensators

Most modern compensators are realized using digital equipment which either mimics analog control components, or can provide unique discrete control actions having no analog counterpart. The compensator shown in Figure 1.10 operates on the error signal, $e(t)$, and generates the actuation signal, $u(t)$, which drives the plant. It may be noted that although this is by no means the only possible control system structure, it is the most common. Also, in order to understand the function of the compensator, it is simpler to deal with a SISO rather than MIMO system.

Although an infinite number of compensator types can be imagined, a typical analog compensator will have three basic modes of operation (some, or all, of which may be present): *proportional action* (often referred to as *P-action*) in which the actuation signal is proportional to the error signal, *integral action* (or *I-action*) when the actuation signal is proportional to the time integral of the error signal and *derivative action* (*D-action*) where the actuation signal is proportional to the time derivative (rate of change) of the error signal. The combination of these actions gives the ubiquitous PID (or 'three term') controller, which is discussed more fully in Section 4.5.2.

When a proportional compensator is used, the closed-loop system's response can often be well behaved and reasonably fast relative to that of the open-loop plant; that is to say, it can have improved steady-state error and rise times. Typically, low values of proportional gain (*low gain*) will give rise to a heavily damped response and a large steady-state error (for example, if a unit step is applied, the plant output may settle a long way away from unity). Increasing the gain will speed up the response by reducing the damping (hence making the system more oscillatory), and reduce the steady-state error. Further increases in gain will further reduce the steady-state error but can make the system too oscillatory. Too high a gain can produce continuous oscillations in the closed-loop system, or may result in instability.

When too large a steady-state error becomes a problem, this can be eliminated with the addition of integral action. However, introducing I-action slows the closed-loop response, that is, it increases the system's settling and rise times. Because of the detrimental effect of integral action on response times it is seldom used on its own for compensation purposes. In industry the vast majority of controllers will employ a combination of proportional and integral action, the so-called PI controller.

The final compensator action, D-action, has no effect on steady-state error (because then the derivative, that is, the rate of change of error, is zero) but during a transient it will tend to speed up the response times. Derivative action is never used on its own. It is most often used to complement proportional and integral action to give the full PID control action and occasionally with proportional control to produce PD-action. Derivative action must be used with care, since any noise on the error signal will be amplified (because signal noise typically has high rates of change) and produce erratic behaviour in the measured system response.

With analog components, proportional and integral action can be realized, but pure derivative action is impossible and therefore approximations are used. This inability to implement pure derivative control is explained properly, with supporting mathematics, in Chapter 4. For the present, all we can say easily is that it effectively takes two values of a signal to calculate its rate of change by either analog or digital means, so that the exact value of the derivative at any point in time cannot be known. It may be thought that this could be circumvented by the use of a sensor to measure the derivative directly where possible (a tachometer measuring angular velocity, which is the rate of change of angular position, for example). Although this is done, and does help, such a sensor is also a real device, having its own dynamics of response, so again the output would not be exactly equal to the derivative at any given time point.

These derivative approximations, often called *lead compensators*, do have the properties associated with derivative action, that is, they tend to speed up the system's response times and amplify noise. However, the increase in speed and the susceptibility to noise will be less than that of pure derivative action. Similarly, a *lag compensator* approximates integral action by reducing (but not eliminating) steady-state error and increasing response times. Since the designer can select the amount of speed-up in a lead compensator, or the reduction in steady-state error provided by lag compensation, these elements are extremely useful. They are also discussed in Chapter 4.

Example 1.6 *Time responses of PID compensator terms*

Assume that a compensator can have either proportional action $u(t) = Ke(t)$, or integral action $u(t) = \int e(t).dt$, or derivative action $u(t) = de(t)/dt$. What is the compensator output when the input is (a) a unit impulse, (b) a unit step and (c) a sinewave of the form $e(t) = A \sin \omega t$?

(a) An impulse may be thought of as a transient perturbing signal. A unit impulse, see Figure 1.13, has an amplitude A, which tends to infinity, and a duration T, which tends

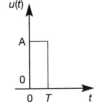

Figure 1.13 A rectangular pulse of amplitude A and duration T.

to zero, such that the enclosed area AT is unity. If a unit impulse enters a proportional compensator, the output is an impulse with enclosed area equal to the constant of proportionality K. With I-action the output is a unit step (because the integral of the impulse is the area under it). With D-action the output signal is a positive-going impulse coinciding with the rising edge of the applied impulse, followed after the time T by a similar negative-going impulse on the falling edge of the impulse. In theory, all these impulses are of infinite amplitude and zero width, so the actual output cannot be specified.

(b) When a unit step enters a compensator having P-action, the output is a step of amplitude equal to the constant of proportionality K. The integral action compensator produces a ramp of K units s^{-1}, see Figure 1.14, and the D-action compensator a unit impulse, coinciding with the rising edge of the step.

(c) With proportional action and a sinewave input, the output from the compensator is $u(t) = KA \sin \omega t$. The I-action compensator produces the output $u(t) = -A \cos \omega t = A \sin(\omega t - \pi/2)$. That is, the output lags the input by 90° (that is, by $\pi/2$ radians) and consequently an I-action compensator is a lag compensator. D-action gives the output $u(t) = A \cos \omega t = A \sin(\omega t + \pi/2)$ and the output leads the input by 90°.

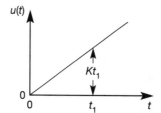

Figure 1.14 A ramp input of gain K units s^{-1}.

1.3.6 Digital control

Most newly implemented controllers will now be digital in nature; that is, they will be implemented within a digital computer which controls the plant. In a digital compensator, the error signal $e(t)$ of Figure 1.10 is formed inside a computer program. It therefore consists of a stream of sequential (or *discrete-time*) data, usually in binary form. This enters the compensator, which uses a numeric algorithm (within the computer program) to process the data. The limitations on the algorithm's complexity are dictated only by the memory size of the processor and its speed of computation. This means that discrete compensators are, in theory, much more versatile than their analog counterparts. In practice there are a few useful digital SISO compensators that do not have a continuous analog counterpart, such as the Dahlin and Kalman controllers (see Chapter 7). However, many digital controllers are 'simply' computer implementations of analog controller designs, as discussed towards the end of Chapter 5. The main advantage of discrete computer-based systems is then their ability to provide additional information which can be used for plant monitoring, fault detection and alarms, and also their flexibility in terms of ease of program modification. This becomes particularly important with MIMO plant, or systems having a large number of control loops.

To control the inverted pendulum, the person balancing the handle is performing the measurement, comparison, compensation and actuation tasks simultaneously. That is, he or she is involved in the *parallel processing* of analog information. It is a function that comes naturally and consequently is taken for granted. However, most digital computers work *sequentially*, using digital signals sampled at discrete time instants. In addition, there is a set of instructions (that is, the program) which the computer systematically obeys, one at a time, starting with the first and ending with the last. This means that a single computing element cannot, for example, observe a measurement and provide an actuation signal at the same time. It is the examination of this discrete, or sequential, mode of operation exhibited by digital computers that gives the impetus for studying digital control.

Assume that the person balancing the handle is in a dark room and that a stroboscopic light flashes briefly once every half-second. In control terms, the position of the handle is *sampled* every half-second. This process of sampling an analog signal in order to obtain a digital signal is known as *A–D* (pronounced A-to-D) conversion. Using the sampled measurement signal it is, with practice, still possible to balance the handle. However, if the period between flashes is gradually increased, a point is reached when it becomes impossible to maintain balance. Stability in digital systems is therefore not only a function of the closed-loop system parameters, but also of the sampling rate. See Chapter 5 for further details.

When a signal is sampled there is a loss of information; but is all the information in a signal required for closed-loop control? Clearly not, as the experiment with the stroboscope demonstrates. What is required is sufficient information to reconstruct the signal. Figure 1.15 shows the same sampled sinewave using different sampling rates. Remember that once sampling has taken place, all information between the sampling intervals is lost – all we know is the value of

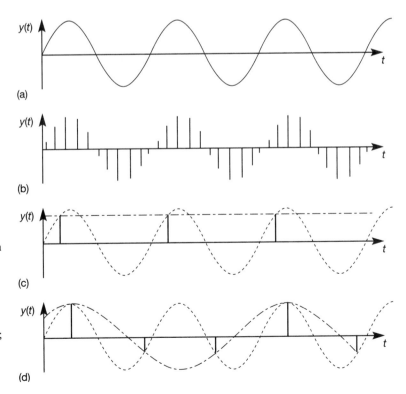

Figure 1.15 The effects of varying sampling rate when generating a discrete-time signal from a continuous-time one. (a) Original sinewave given by $y(t) = \sin \omega t$; (b) sampled sinewave – fast sample rate; (c) sampled sinewave – slow sample rate; (d) aliasing.

each sample. With a relatively fast sampling rate (Figure 1.15(b)) the original sinewave (Figure 1.15(a)) is apparent. As the sampling rate is decreased (so that the time between samples is increased) more information is lost. When the sampling rate equals the frequency of the oscillation (Figure 1.15(c)), the signal appears to be of constant amplitude. With longer sampling periods, the apparent frequency of oscillation appears lower than that of the original signal (Figure 1.15(d)), a phenomenon known as *aliasing*. The problem of signal reconstruction from samples is dealt with in Chapter 5.

Digital control systems have the same closed-loop elements as analog control systems. In the context of real-world control engineering, the plant is always likely to be of a continuous-time nature. The terms 'digital' and 'analog' therefore distinguish between control systems that have *some* discrete-time (or digital) signals and those in which *every* signal is a continuous-time (or analog) one. Typically, a measurement sampled in a computer-based digital control system passes sequentially through a set of instructions which perform the comparison and compensation functions. The discrete output signal from the compensator is then converted to analog form (*D–A* conversion) for the actuator. Since the computer is operating sequentially, the control action resulting from a given measurement is delayed by a period of time approximately equal to the measurement sampling interval. Note that digital sensors and actuators are gradually becoming available, which remove the need for A–D and D–A conversion.

Further, the process of analog signal reconstruction will also introduce delays. Assume that a sinewave is sampled and then fed directly to a D–A converter. The converter only has past information and, under normal circumstances, could not predict into the future. In this situation, one solution is for the D–A converter to hold the current sample until the next sample appears, see Figure 1.16. Comparing the output with the input will show that moving from a digital to an analog signal therefore introduces a delay. Also, the analog output is not smooth. If a smooth signal is required, the necessary filtering could result in further delays, compared with the 'instantaneously reconstructed' output in the figure.

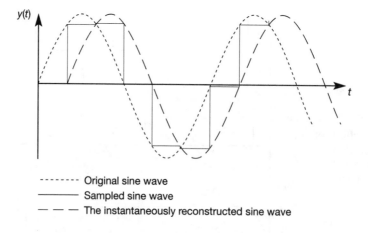

- - - - - - - Original sine wave
———————— Sampled sine wave
— — — The instantaneously reconstructed sine wave

Figure 1.16 Sampling delays.

Despite the accumulation of delays and loss of information associated with sampling, which adds a level of complexity to the design process, most digital design techniques are based on continuous linear system strategies. What changes when using the various techniques is the interpretation of the results. Also, because there is the possibility of using a computer to implement the controls, digital control is more flexible and often superior (in terms of achievable performance characteristics) to analog control.

1.3.7 Beautiful theories and ugly facts

In this stroll in the field of control it was generally assumed that the model, usually a linear mathematical model, was a true representation of the plant and its associated measurement and actuation equipment. Any mismatch between the model and plant could be ignored since the modelling errors would somehow be compensated for by the controls. Stability and performance of the controlled model were both necessary and sufficient to guarantee the closed-loop behaviour of the physical (real-world) system. This concept of control system design, which was prevalent for many years, also assumed that any small discrepancies between the predicted and actual responses could be eliminated by minor online adjustment of the controls.

The various techniques mentioned in Section 1.3.4 of this chapter, for the analysis and design of linear systems, were developed broadly between 1920 and 1950. They are loosely called *frequency-domain* methods, because they work by investigating the behaviour of the system as frequency changes. This will probably seem an unusual concept at present, but it will be explained in Chapter 3. As you will discover later in the text, they are largely based on graphical design techniques, which are therefore expected, from the outset, to be used with inherent approximation. They work very well for well-behaved plants which are passably linear, and they are still routinely applied to such plants all over the world. The PID controller, for example, is to be found in the vast majority of factories, laboratories and similar installations where there are industrial control loops. That is why it was mentioned in more detail than might be expected in this introductory chapter.

The methods used in designing controllers by these approaches are called true *design* (or *analysis*) methods, as the process is an iterative one involving design decisions by the engineer at every stage. The designer typically makes an initial design decision, based upon viewing various graphical plots arising from an analysis of the system. The effects of the design on these plots are then noted, and an improved design is developed. This process is repeated until satisfactory results are achieved.

The iterative process is greatly aided by computer-aided control system design (CACSD) software packages. Indeed, one such package, MATLAB (The Mathworks Inc., 1993a, 1993b – see reference list, and Appendix 3), together with its Control Systems Toolbox, is used extensively throughout the text. It was used to produce most of the response plots in the book, and is necessary to run all the files on the accompanying computer disk – which allow the reader to repeat and modify the examples. Although the uses of MATLAB (or another CACSD package) suggested throughout the text can definitely help the reader's understanding of several aspects of control engineering, we do not actually rely on the reader being able to use CACSD at any point (except in a few end-of-chapter problems). The reader can therefore use the text whether or not he or she has access to MATLAB.

In the 1960s and 1970s a rapid new development of control theory began, based upon more analytical methods which directly use the differential equation models of the plant. These are *time-domain* methods, because they are concerned with the behaviour of the system as time passes. They are broadly called *state-space* methods, and are described in several chapters of the text, beginning with Chapter 2. Many books call the frequency-domain techniques 'classical control' and the state-space-based methods 'modern control'. However, the original 'modern' control is not very 'modern' now and, besides, these terms introduce a false distinction between the approaches. The authors' view is that all the approaches have their own advantages and disadvantages. A knowledge of as many approaches as possible, together with some knowledge of digital and nonlinear control, contributes to a well-rounded control engineer, who is then able to make an informed decision as to the best methods to use in any given situation.

Part of the impetus for the development of the state-space methods was to try to solve more difficult control problems, for which the previous methods had often proved inadequate (for example, multivariable control). These methods, in contrast to the earlier *analysis* methods, might be called *synthesis* methods, because they directly synthesize the required controller from the plant model and information about the required performance. In other words, one plugs the details of the plant and the required performance into a suitable CACSD program, and out pops the required controller. This is obviously of great appeal – if it works. However, the growth in these strategies and their use in academic institutions was not matched by a similar growth in their use by industry, with a few exceptions such as the aerospace industry. Perhaps the main reason for this technology gap was that the beautiful theories had ignored some ugly facts. Only in the past 10–15 years has much effort been made to correct this.

The first ugly fact is that the implicit assumption that model–plant mismatch could be easily catered for by online tuning, is false. Furthermore, plants are not linear, instrumentation is not perfect, actuators invariably have a limited operating range and, for maintenance and development purposes, control strategies should be easily understood.

To study the effects of model–plant mismatch opens up the subject of *robust control*, see Chapter 13. For now, suffice it to say that it is often possible to allow for the effects of inaccuracies in a plant model when designing a controller using the nominal model. It may be possible to design a controller that maintains stability and performance for a range of models (obtained for operating conditions which deviate from those for which the nominal model was found). The controller then has in-built robustness and tolerance of modelling errors.

To avoid clouding the development of the 'normal' design process, however, this matter is not often mentioned explicitly except in Chapter 13, and in frequent reminders that models are only approximate, and should not therefore be trusted too far. For the interested reader, the basic ideas of modelling errors (Sections 13.1 and 13.2) can be understood without reading more of the text first (although one or two details might remain vague).

1.3.8 Mathematical modelling

Amongst other things, Section 1.3.7 considered (briefly) the problem of mismatch between the plant and its nominal model. However, this begs the question of how to obtain the nominal model in the first place, and how to ensure its suitability.

As previously stated (Section 1.3.4), ensuring that a control model is a 'good' representation of the plant can account for over 90 per cent of the effort expended on a control system design study. Without a 'good' model, most control system designs are doomed to fail, or at best to produce a performance significantly inferior to that predicted using simulation. Simulation is effectively the art of programming a computer to solve the equations which make up the mathematical model of the plant and the designed control elements. The predicted response to various inputs (including disturbances and modelling errors) can then be viewed on the computer screen, before expensive errors are made on the real-world plant. With the ready availability of CACSD software, this is now standard practice. Most of the time-response plots in this text were produced by digital computer simulations. Most of the MATLAB files that did this are on the disk accompanying the book, and more details are given throughout the text.

To complicate matters further, mathematical modelling is not an exact science, in that it depends on the skills and experience of the modeller. The purpose of this section is therefore to provide the reader with some insight into the subject area before dealing with it in more detail in Chapter 2.

Mathematical modelling for control system design studies could be thought of as consisting of a number of distinct stages:

(1) Understanding the plant.

(2) Identifying the various inputs, outputs and disturbances.

(3) Producing an idealized representation of the plant in terms of elements which can be described mathematically.

(4) Developing the equations (that is, the model) which describe the idealized representation. Initially, linear approximations will be made here in the interests of simplicity and ease of future analysis and design.

(5) Obtaining data, often by performing experiments, in order to establish unknown parameter values.

(6) Checking and adjusting the model (stages 3–5), until it produces responses which satisfactorily represent the plant behaviour. With this stage of the modelling process there is no substitute for experience. However, as a rule of thumb, the more demanding the design specifications, the greater the amount of modelling detail required. If linear models finally fail to give satisfactory results, nonlinear behaviour will have to be introduced.

(7) Finally, simplifying the model for the proposed control study. This can include linearizing the equations of a nonlinear model about a specified operating point and/or removing redundant information.

Although the above list suggests that the modelling process is progressive, in practice it is iterative with several stages often being performed in parallel. The understanding of a plant, or process, will be improved by performing the other modelling stages. For complex plant having many inputs, those that should be used for control purposes and those that should be considered disturbances are often not obvious. In a complex plant it may sometimes take many years and some external factor, such as the need for improved efficiency, before the best inputs and outputs for achieving a particular control objective are established (the need to make efficiency savings being necessary to justify the high cost of such an investigation).

Selecting the number of idealized elements to include in a model, and the interconnections between them in order to represent the plant, is also a matter of choice.

Also, the level of complexity of the elements themselves (lumped-parameter models versus distributed-parameter models, for example) is again a matter of choice. A *lumped-parameter* model is one in which certain aspects of the system being modelled are imagined to be lumped (concentrated) at a single location. For example, in Figure 1.9, we imagine the pendulum rod to be relatively massless, and we say that the entire mass of the system is lumped at a point at the end of the rod. Such simplifications make modelling a much easier proposition. To make an extremely accurate model of the pendulum, we would need to consider that the rod also has mass, and that this mass is distributed along the rod, not concentrated at a single point. Representation of this kind of information requires a *distributed-parameter* model, and usually leads to relatively complex models involving partial differential equations and being notoriously difficult to solve.

An examination of the remaining modelling stages also demonstrates the need for experience and judgement. In this respect, modelling is an art rather than a science. Chapter 2 will concentrate on simple lumped-parameter models which can be used to represent various kinds of plant.

1.4   Conclusions

In this chapter, the basic ideas of control engineering have been introduced, and will be much extended in subsequent chapters. Block diagram representations have been mentioned together with mathematical models of systems. Real engineering systems are nonlinear and have distributed parameters, but by using linear approximations and lumped-parameter assumptions in generating models, the control system design problem can be simplified. Basic ideas about the stability of systems, and other aspects of system performance influenced by adding closed-loop control, have been introduced. There are different approaches to control system design, each of which has its own advantages and disadvantages, and some of these approaches have been mentioned in readiness for further study.

It was noted that computer-aided control system design (CACSD) software can assist in the design process, and that MATLAB (The Mathworks Inc., 1993a, 1993b), as a representative CACSD environment, is used throughout the text. There are many files on the computer disk accompanying the text, which can be run via MATLAB and its Control Systems Toolbox.

The subsequent chapters build on these introductory ideas by first establishing simple lumped-parameter models for the basic building blocks of standard system components. Techniques are then provided which show how the equations describing these lumped-parameter models may be found. By manipulating these equations into one of a number of standard forms (continuous-time models, Laplace transformed models, discrete-time models or state-space models, for example), it becomes possible to standardize the control system design procedures. For this reason, the chapters dealing with control system design will often take as their starting position one of these standard forms. This does not limit the proposed design techniques, since most of the various standard model forms are interchangeable.

1.5 Problems

Problems 1 and 2 are intended for discussion. In both cases the problem descriptions are deliberately vague, so that the reader may explore different scenarios.

1.1 Assume that an archer fires an arrow at a target.
   (a) Identify the plant.
   (b) Identify the reference value(s), input(s), disturbance(s) and output(s). Would there be any system noise, if so what?
   (c) Is the system closed- or open-loop? Would it make any difference if the archer fired two arrows at the target? Would it make any difference if the archer repeatedly fired at the target?
   (d) Is the system tracking or regulating?
   (e) Is the system continuous or discrete?

1.2 A conductor directs the orchestra at a public venue to play Ludwig van Beethoven's Symphony No 3, 'Eroica'.
   (a) Is the system consisting of the conductor and orchestra open- or closed-loop?
   (b) Identify the reference value(s), input(s), disturbance(s) and output(s). Would there (in the control sense) be any system noise, if so what?
   (c) Is this system continuous or discrete?
   (d) Is there a larger system which includes the audience? Is this system open- or closed-loop? Do the two systems interact?
   (e) Is this larger system continuous or discrete?

1.3 For the level control system shown in Figure P1.3, the operator is attempting to maintain a constant head h of liquid in the tank. For this system:
   (a) define the elements which make up the plant, actuator, measuring device, comparator and controller.

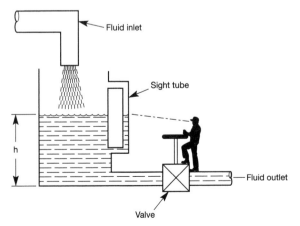

Figure P1.3 Level control system for Problem 1.3.

(b) Is the system SISO or multivariable?
(c) Define the input(s), output(s) and reference value(s).
(d) Why is the fluid inlet a disturbance input?

1.4 Figure P1.4 shows a counterflow heat exchanger which uses cooling water to reduce the oil temperature in a piece of machinery. There is ample cooling water, although the mean inlet temperature will fluctuate. Also, the flow rate and temperature of the hot oil flowing into the heat exchanger can vary significantly.
   (a) What are the disturbance and manipulable inputs?
   (b) What are the measurable and controllable outputs?
   (c) Describe a control scheme which might be used to maintain a constant outlet oil temperature.

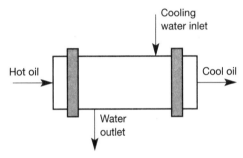

Figure P1.4 Counterflow heat exchanger for Problem 1.4.

1.5 A typical industrial flow control scheme is shown in Figure P1.5. Draw a block diagram of the system, using a schematic notation similar to that used in Figure 1.2.

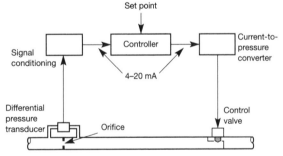

Figure P1.5 Industrial flow control scheme for Problem 1.5.

1.6 (a) Test the equations for the three terms in a PID controller (given in Example 1.6), to show that they are all linear. *Hint*: Reread the note about the superposition principle in Section 1.3.1.
   (b) Does this mean that the entire PID controller, whose output is always the *sum* of these three terms, is therefore a linear system element?
   (c) Would you change your opinion in the case of a controller whose output was the *product* of two of the terms (it is not suggested that such a controller would be a useful one)?

2 An introduction to control system modelling

2.1 Preview

Chapter 1 stressed that a *model* of a system is required for use in controller design. Mathematical models were mentioned (an equation, or set of equations, which adequately describes the behaviour of the system). In particular, 'lumped-parameter' models were briefly discussed, in which the equations making up the model are derived after some simplifying assumptions which 'lump' certain parameters of the system together at some point.

In the vast majority of control system design, lumped-parameter models are used for simplicity. This chapter introduces the formation of this type of model. However, it must not be forgotten that other kinds of model are possible. Chapter 1 mentioned distributed-parameter models. Also, physical modelling (by the construction of scale models) was briefly discussed.

There are yet more possibilities, but these are not discussed in this text. For example, *fuzzy logic control*, which has appeared in items as diverse as cars, cameras and washing machines, allows the blurring of sharp

values – so a system might be required to maintain a temperature of 'about 60 degrees' rather than the precise requirement, '60 degrees', which is usual when using mathematical models.

Some material in this chapter is marked as optional. The rest follows a logical sequence and, if studied in order, should present few difficulties.

In this chapter, the topics covered are:

- an overview of mathematical modelling
- simple, linear, lumped-parameter modelling
- modelling of basic electrical, mechanical, hydraulic and thermal systems
- linearization of some nonlinear equations
- state-space and Laplace transform model representations
- block diagram representation of linear models
- block diagram algebra and reduction
- model formulation for continuous-time and discrete-time models
- the conversion of models from one form to another.

NEW MATHEMATICS FOR THIS CHAPTER

In addition to the basic algebra and calculus already assumed in Chapter 1, some elementary matrix algebra is used. The *Laplace transform* is introduced, as an alternative means of handling the differential equations which constitute a mathematical model. These topics are fully covered by the text and by Appendices 1 and 2. Some knowledge of complex numbers will also be needed for the later parts. For the optional section on *Taylor series* expansion of a general function, some elementary knowledge of *partial derivatives* is necessary.

2.2 An introduction to control system modelling

Control system modelling is a subject in its own right. Essentially, there are two approaches to finding the model. In the first, the system is broken down into smaller elements. For each element a mathematical description is then established by working from the physical laws which describe the system's behaviour. The simplest such technique is *lumped-parameter modelling*, which is considered in this chapter.

The second approach is known as *system identification* (introduced in Section 3.9), in which it is assumed that an experiment can be carried out on the system, and that a mathematical model of the system can be found from the results. This approach can clearly only be applied to existing plant, whereas lumped-parameter modelling can additionally be applied to a plant yet to be built, working purely from the physics of the proposed plant components.

Control system modelling is a specialization of the more general area of mathematical modelling. Like all mathematical models, a control system model provides insight into the operation of the system and defines the cause and effect relationships between variables. In a control system model, the important relationship is that between the manipulated inputs and the measurable outputs. Ideally this relationship should be linear (see Section 1.3.1), and capable of being described by an expression of low order (that is, an equation, or a set of equations, containing as few terms as possible). Most commonly, a low-order, linear differential equation model is used.

The basis for the development of any mathematical model is provided by the fundamental physical laws describing the behaviour of the system. For control system modelling, it is usual to analyse an idealized equivalent of the physical system, in order to simplify the task. In this ideal equivalent system, each element has a single property or function. For example, an actual mass becomes a mass concentrated at a point, with no compressibility (stiffness) or damping effects associated with it. An electrical inductor is assumed to have pure inductance with no resistance or capacitance. If the resistance is significant, it must be represented by a separate model of a pure resistor. The advantage of considering such an idealized (or lumped-parameter) system is that each element has only one independent variable (time), so that the system can then be described using an ordinary differential equation model. If more than one independent variable is considered, partial differential equations arise, making the modelling procedure much harder.

Probably the most important task in lumped-parameter modelling is determining which assumptions can validly be made. Obviously an extremely rigorous model that includes every phenomenon in microscopic detail would take a long time to develop and might be impossible to solve. It might also prove very difficult to achieve. For example, a vast amount of effort has been expended on atmospheric modelling for weather forecasting. The resulting models are extremely complex, requiring the ultimate in computing power to run. Nevertheless, the results are still imperfect.

On the other hand, an over-simplified model might bear no dynamic resemblance to the original system. In practice, the amount of detail incorporated tends to be a function of the available resources in time, funding, solution techniques and hardware. Nevertheless, any assumption that is made should be carefully considered and listed, since it will impose limitations on the model, which must be borne in mind when evaluating the system's predicted behaviour.

When fundamental physical laws are applied to the lumped-parameter model, the resulting equations may be nonlinear (see Sections 1.3.1 and 1.3.4), in which case further assumptions may have to be made in order to produce an ordinary linear differential equation model which is soluble. In such cases, it is not unusual to assume that system operation will be restricted to small perturbations about a given operating condition. If the assumed operating region is small enough, most nonlinear plants may be adequately described by a set of linear equations. If these assumptions cannot be made, nonlinear control techniques may be appropriate (Chapter 14).

Once all the equations of the mathematical model have been written, it is a good idea, particularly with complex systems of equations, to make sure that the number of variables equals the number of equations. If it does not, the system is either under-specified or over-specified and something is wrong with the formulation of the problem. This kind of consistency check may seem trivial, but it can save many hours of frustration and confusion. Checking that the units (or dimensions) of all terms and all equations are consistent is perhaps another trivial and obvious step, but one that is often overlooked.

Finally, one of the more important parts of model development is that of model validation. In this context, validation is the art of testing that the mathematical model does indeed describe the real-world situation. Sometimes this cannot be done at the design stage, because the system has not yet been built. However, even in this situation, there is usually either a similar existing system, or perhaps a pilot plant, from which some experimental data can be obtained for comparison purposes.

2.3 Lumped-parameter models

This section expands on the ideas developed in Section 1.3.8. In particular, it will define some simple elements which can be used to produce an idealized representation of the plant or process. The entities considered are the basic components of many physical plants and include masses, springs, dampers, resistors, capacitors, conductors, inductors, cross-sectional areas of fluid tanks and thermal capacities. The approach taken is one that can be applied to the modelling of many simple systems, provided that the system is:

Linear. It must obey the principle of superposition as outlined in Section 1.3.1. That is, if an input $I_1(t)$ causes an output $O_1(t)$, and an input $I_2(t)$ causes an output $O_2(t)$, then an input $I_1(t) + I_2(t)$ causes an output $O_1(t) + O_2(t)$ if the system is linear. Consider the element shown in Figure 2.1 which has an input and output. If the element described a translational damper (such as an automobile shock absorber) then the input would be chosen as the force applied to the damper (in newtons, N) and the output would be the change in velocity across the damper (in

Figure 2.1 A general
input–output element.

metres per second); that is, the equation for the damper is given by $v(t) = (1/B)f(t)$, where B is the damping coefficient in units of N / (m s^{-1}), or N s m^{-1} if preferred. If, however, the element in Figure 2.1 described an electrical resistor, the input could be defined to be the current $i(t)$ (in amperes, A) and the output the potential difference across the resistor $v(t)$ (in volts, V). The describing equation would then be $v(t) = Ri(t)$, where R is the resistance in units of ohms (Ω).

In either case, the principle of superposition holds. In the case of the resistor, for example, if an input of 1 A produced an output of 3 V, and an input of 2 A produced an output of 6 V, then an input of 3 A would produce an output of 9 V.

Stationary (or time invariant). The parameters inside the element, for example the resistance or damping, must not vary with time. In other words, an input applied today must give the same result as the same input applied yesterday or tomorrow. A vehicle that burns large masses of fuel, such as a racing car or a space vehicle, is an example of a system which is *not* stationary in this sense (or in any other sense!). Its dynamic behaviour will alter significantly as its mass decreases.

Deterministic. The outputs of the system at any time can be determined from a knowledge of the system's inputs up to that time. In other words, there is no random (or *stochastic*) behaviour in the system, since its outputs are always a specific function of the inputs. The term *causal* is also used for such systems. The precise arrival time of a bus (to within 30 seconds, say) is an example of a stochastic (that is, non-deterministic) event, because it is unpredictable from past behaviour. Industrial systems in which measured input and output signals are corrupted with large amounts of noise, thus making the measured values uncertain, are also stochastic systems.

Consider an element representing an idealized component such as the resistor shown in Figure 2.2. If the resistor is linear then

$$i(t) = \frac{1}{R} v(t) \tag{2.1}$$

where

$$i(t) = \text{the current (A)}$$

$$v(t) = \text{the potential difference (V)}$$

$$R = \text{the resistance } (\Omega)$$

Note that the currents at both a and b are equal to $i(t)$. It is possible to think of the current as flowing *through* the resistor and hence to define the current as a *through variable*. The voltage (or, to be pedantic, the potential difference) is measured *across* the resistor and can therefore be defined as an *across variable*. Since the voltage at point b will be less than the voltage at a, this is illustrated by an arrow from b to a as shown in Figure 2.2.

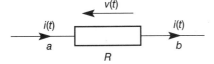

Figure 2.2 An ideal resistor.

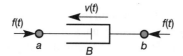

Figure 2.3 An ideal damper.

Now consider an element representing an ideal damper as shown in Figure 2.3. For this damper:

$$f(t) = Bv(t) \tag{2.2}$$

where

$$f(t) = \text{the force (N)}$$

$$v(t) = \text{the velocity difference across the damper (m s}^{-1})$$

$$B = \text{the damping coefficient (N/(m s}^{-1}))$$

The force $f(t)$ on either side of the damper must be the same and hence force could be thought of as a *through variable*. Due to the rate of change of compression of the damper, there will be a change in velocity *across* the damper, which may be represented by the arrow from b to a as shown in Figure 2.3. Equation (2.1) describing the resistor is similar to Equation (2.2) describing the damper, in that both have the form:

$$(\text{through variable}) = (\text{constant}) \times (\text{across variable}) \tag{2.3}$$

Since Equation (2.3) could be used to describe either the resistor or the damper, these two elements may be thought of as being analogous to each other. Further, Equation (2.3) is clearly a linear equation in that it obeys the superposition principle. Any element that can be described by Equation (2.3) will itself be linear and produce linear equations.

What has been developed here is one type of element in the so-called 'force–current analogy'. Other analogies are possible, in particular the force–voltage analogy. However, in this text only the force–current analogy will be considered.

Now consider elements which can be described by an equation of the form:

$$(\text{through variable}) = \text{constant} \times \int (\text{across variable}) \, dt \tag{2.4}$$

Equation (2.4) is again linear and can be used to describe many different elements. For example the spring shown in Figure 2.4 is described by the equation:

$$f(t) = K \int v(t) \, dt$$

where

$$K = \text{the spring stiffness (N m}^{-1})$$

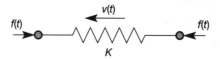

Figure 2.4 An ideal spring.

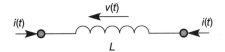

Figure 2.5 An ideal inductor.

The inductor shown in Figure 2.5 is described by the equation:

$$i(t) = \frac{1}{L} \int v(t) \, dt$$

where

$$L = \text{the inductance in henry (H)}$$

So both fall into this category.

Some elements can be described by a linear equation of the form:

$$(\text{through variable}) = \text{constant} \times \frac{d}{dt} (\text{across variable}) \tag{2.5}$$

Both a perfect mass (Figure 2.6) and a perfect capacitor (Figure 2.7) fall into this category. The equation describing the mass is given by:

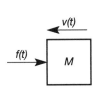

Figure 2.6 An ideal mass.

$$f(t) = M \frac{d}{dt} v(t) \tag{2.6}$$

where

$$M = \text{the mass (kg)}$$

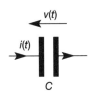

Figure 2.7 An ideal capacitor.

and the capacitor is described by:

$$i(t) = C \frac{d}{dt} v(t) \tag{2.7}$$

where

$$C = \text{the capacitance in farads (F)}$$

At this stage, three idealized electrical elements (the resistor, inductor and capacitor) have been modelled, and shown to be analogous to three idealized mechanical elements (the viscous damper, spring and mass respectively). By grouping the electrical and mechanical elements as shown in Tables 2.1 and 2.2 the force–current analogy becomes evident.

The analogies developed in Tables 2.1 and 2.2 can be extended to rotational mechanical systems. In rotational systems the through variable is torque and the across variable is angular velocity. Using Newton's second law for rotational systems gives, for a constant inertia, the equation:

$$T(t) = I \frac{d}{dt} \omega(t) \tag{2.8}$$

Table 2.1 Ideal electrical system elements.

Across variable: potential difference v (V)
Through variable: current i (A)

Component	Circuit symbol	Defining equation
Resistor	$v(t)$ $i(t)$ R (Ω)	$i(t) = \dfrac{v(t)}{R}$
Inductor	$v(t)$ $i(t)$ L (H)	$i(t) = \dfrac{1}{L}\displaystyle\int v(t).dt$
Capacitor	$v(t)$ $i(t)$ C (F)	$i(t) = C\,\dfrac{dv(t)}{dt}$

Table 2.2 Ideal rectilinear (translational) mechanical system elements.

Across variable: linear velocity v $(\mathrm{m\,s^{-1}})$
Through variable: force f (N)

Component	Circuit symbol	Defining equation
Linear damper (viscous friction too)	$v(t)$ $f(t)$ B $(\mathrm{N/m\,s^{-1}})$	$f(t) = Bv(t)$
Linear spring (stiffness also)	$v(t)$ $f(t)$ K $(\mathrm{N\,m^{-1}})$	$f(t) = K\displaystyle\int v(t).dt$
Mass (rectilinear motion)	$v(t)$ $f(t)$ M (kg)	$f(t) = M\,\dfrac{dv(t)}{dt}$

where

$$T(t) = I \frac{d}{dt} \omega(t)$$

$$T(t) = \text{the torque (N m)}$$

$$\omega(t) = \text{the angular velocity (rad s}^{-1})$$

$$I = \text{the moment of inertia (kg m}^2) \text{ (or N m}/(\text{rad s}^{-2}))$$

Again it is evident that Equation (2.8) is a particular realization of Equation (2.5) and therefore inertia could be considered analogous to both mass and capacitance (cf. Equations (2.6) and (2.7)).

The stiffness of an elastic shaft is given by the equation:

$$T(t) = \frac{GI_p}{l} \theta(t) \tag{2.9}$$

where

$$G = \text{the modulus of rigidity (N m}^{-2} \text{ rad}^{-1})$$

$$I_p = \text{the polar second moment of area. For a circular shaft of}$$

$$\text{diameter } d \text{ (m)}, I_p \text{ is given by } \frac{\pi d^4}{32} \text{ (m}^4)$$

$$l = \text{the length of the shaft (m)}$$

$$\theta(t) = \text{the angle of twist (rad)}$$

In terms of the angular velocity, Equation (2.9) becomes:

$$T(t) = \frac{GI_p}{l} \int \omega(t) \, dt$$

and therefore becomes a special case of Equation (2.4). This means that the torsional spring is analogous to the linear spring and the inductor.

Similarly, an ideal linear torsional damper (that is, one offering only viscous friction) can be described by an equation of the form:

$$T(t) = B\omega(t)$$

where

$$B = \text{the torsional damping (N m}/(\text{rad s}^{-1}))$$

and, as such, is a special case of Equation (2.3). Again all these elements and their defining equations are summarized in Table 2.3.

Further extensions to the tables of analogies are possible which will encompass simple fluid and thermal elements as shown in Tables 2.4 and 2.5. The tank shown in Figure 2.8 is an example of fluid capacity. For this system, assuming the fluid density is constant, the conservation of volume equation states that 'the time rate of change of fluid volume within the tank is equal to the volumetric flow rate into the tank less the volumetric flow rate from the tank'. Therefore, for Figure 2.8:

$$\frac{dV(t)}{dt} = q_i(t) - q_o(t) \tag{2.10}$$

Table 2.3 Ideal rotational mechanical system elements.

Across variable: angular velocity ω (rad s^{-1})
Through variable: torque T (N m)

Component	Circuit symbol	Defining equation
Damper (viscous coupling, bearing friction)	$\omega(t)$ $T(t)$ ◄ B (Nm/rad s^{-1})	$T(t) = B\omega(t)$
Torsion spring (shaft stiffness also)	$\omega(t)$ $T(t)$ ◄ K (Nm rad^{-1})	$T(t) = K\int\omega(t).dt$
Inertia (any rotating mass)	$\omega(t)$ $T(t)$ ◄ I (Nm/rad s^{-2})	$T(t) = J\dfrac{d\omega(t)}{dt}$

Table 2.4 Ideal fluid system elements.

Across variable: pressure head h (m)
Through variable: volume flow rate q (m^3 s^{-1})

Component	Circuit symbol	Defining equation
Fluid resistance (in pipes and valves)	$h(t)$ $q(t)$ ◄ R (s m^{-2})	$q(t) = \dfrac{h(t)}{R}$
Fluid capacity (e.g. stored fluid in accumulator or tank)	$h(t)$ $q(t)$ ◄ A (m^2) (cross-sectional area)	$q(t) = A\dfrac{dh(t)}{dt}$

$$(m)\,h = \tfrac{1}{?}R = m^3 \tfrac{s}{?} \times R$$
$$R = \frac{s}{m^2}$$
$$m^2 \times \frac{m}{s} = \frac{m^3}{s}$$

where

$$V(t) = \text{the tank fluid volume (m}^3)$$

$$q_i(t) = \text{the volumetric flow rate into the tank (m}^3\,\text{s}^{-1})$$

$$q_o(t) = \text{the volumetric flow rate out of the tank (m}^3\,\text{s}^{-1})$$

but

$$V(t) = Ah(t)$$

Table 2.5 Ideal thermal system elements.

Across variable:	temperature difference θ (K)	
Through variable:	heat flow rate q (W)	
Component	Circuit symbol	Defining equation
Thermal resistance (of interfaces too)	$\theta(t)$ $q(t) \leftarrow$ $\rightarrow$ R (K W⁻¹) (or °C W⁻¹)	$q(t) = \dfrac{\theta(t)}{R}$
Thermal capacity (ability to store heat energy)	$\theta(t)$ $q(t) \leftarrow$ $\rightarrow$ C (J K⁻¹)	$q(t) = C\,\dfrac{d\theta(t)}{dt}$

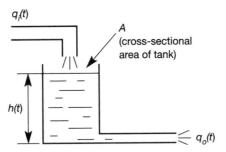

Figure 2.8 A simple tank system.

where

$$A = \text{the tank cross-section area } (\text{m}^2)$$

$$h(t) = \text{the head of fluid in the tank (m)}$$

and hence Equation (2.10) may be written as

$$A\,\frac{dh(t)}{dt} = q(t)$$

where

$$q(t) = q_i(t) - q_o(t)$$

$$= \text{the net volumetric flow rate into the tank } (\text{m}^3\,\text{s}^{-1})$$

The flow from the tank $q_o(t)$ will be dependent on $h(t)$, the head of fluid in the tank. For small changes in head the flow from the tank can be expressed as:

$$q_o(t) = \frac{1}{R}\,h(t)$$

where

R is the resistance to fluid flow, due to the pipework $(s\,m^{-2})$

R is therefore analogous to mechanical damping or electrical resistance.

Finally, for thermal systems, the first law of thermodynamics may be used to determine the temperature changes in a body; that is, 'the rate of change of a body's internal energy is equal to the flow of heat into the body less the flow of heat out of the body'. This may be expressed as

$$C\,\frac{dT(t)}{dt} = q_i(t) - q_o(t)$$

where

C = thermal capacity of body (joules per kelvin) $(J\,K^{-1})$

$T(t)$ = temperature (K)

$q(t)$ = heat flow rate (watts) (W)

The thermal capacity of a body may be found from the equation:

$$C = MS$$

where

M = mass of body (kg)

S = specific heat capacity of the material $(J\,(kg\,K)^{-1})$

The heat flow rate through a body is a function of the thermal resistance of the body. This is normally assumed to be linear, and therefore:

$$q(t) = \frac{T_1(t) - T_2(t)}{R}$$

where

R = thermal resistance $(K\,W^{-1})$

$T_1(t) - T_2(t)$ = temperature difference across the body (K)

thus giving the elements shown in Table 2.5.

In engineering systems there are sources of energy, which are either across variable sources (voltage, velocity, pressure head or temperature gradient), or through variable sources (current, force, torque, fluid flow rate or heat flow rate). The symbols for both the across and through variable generators are shown in Table 2.6.

Note that other analogies are possible. For example, an electrical transformer, a lever, a gearbox and a hydraulic jack could all be considered analogous. However, the use of analogies in modelling is rather limited, and is only included to demonstrate the use of linear differential equations across a spectrum of engineering disciplines.

Table 2.6 Ideal through and across variable generators.

	Electrical systems	Translational systems	Rotational systems	Hydraulic systems	Thermal systems
Across variable generator	v	v	ω	h	θ
Through variable generator	i	f	T	q	q

Example 2.1 *An automobile suspension system*

Consider the development of a lumped-parameter model for the suspension system of the automobile shown in Figure 2.9.

The first stage, according to the procedure listed in Section 1.3.8, is to understand the plant. In this case the suspension system defines the movement of the car body relative to the movement of the wheels. If the motion of the body is deemed unsatisfactory (perhaps it oscillates for too long after hitting a bump), then the suspension system (that is, the springs and dampers used to support the car body on the wheels) must be adjusted until the system gives the appropriate ride or change of position.

Stage 2 requires the identification of the various inputs, outputs and disturbances. Clearly the output to be controlled is the motion of the car body relative to the wheels. The input, in this case a disturbance input, is caused by undulations in the road which will alter the wheel positions on the car. There will be no forcing input, unless an actuator is used which will apply an additional force, or forces, to the car body (for example, an active suspension system).

Stage 3 requires the production of an idealized representation of the plant which can be described mathematically. This is the lumped-parameter model. It will be assumed that the total mass of the car body can be grouped into one lump, as shown in Figure 2.10. All the stiffness and damping effects provided by the various shock absorbers are lumped into the ideal spring and dashpot (or damper) respectively, shown in Figure 2.10. Also shown in this figure is a forcing function which represents the force which could be supplied by an actuator. Clearly the car of Figure 2.9 and the lumped-parameter model shown in Figure 2.10 bear little resemblance to one another. However, it is anticipated that the lumped-parameter model will contain the essential dynamic information relating to the car, and therefore by analysing this simple system the motions, or some of the motions, exhibited by the car may be inferred.

The next stage of the analysis, Stage 4, is to produce the equations of motion which describe this idealized representation (or lumped-parameter model) and this is the topic covered in Section 2.4.

Figure 2.9 An automobile.

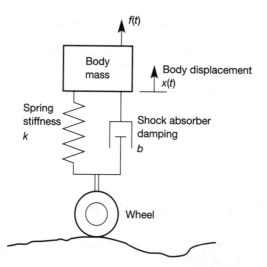

Figure 2.10 A lumped-parameter model for an automobile suspension system.

Example 2.2 *An internal combustion engine's inlet manifold – emissions reduction*

Reduction of vehicle emissions is an area of increasing importance, particularly for the automotive control engineer. It can be shown that emissions are related to the in-cylinder air to fuel ratio which, in turn, is determined by the amount of fuel and air leaving the engine's inlet manifold. This example will develop a lumped-parameter model for the inlet manifold of a spark ignited internal combustion engine.

According to Section 1.3.8, the first requirement in lumped-parameter modelling is to understand the plant. In this case this leads to the simple manifold schematic shown in Figure 2.11. In this system, the pumping action of the engine produces a vacuum within the manifold, thus drawing air through the manifold and into the cylinder. The flow of air through the manifold and into the cylinder is regulated by the throttle valve. Fuel is injected into the air stream. It is assumed that some of the fuel entering the manifold will immediately vaporize due to the reduced pressure levels, and then flow with the air stream directly into the cylinder. The fraction of injected fuel not vaporizing, impacts with the manifold wall to form a puddle which, due to the elevated temperature of the manifold wall, will eventually evaporate and then join with the air–fuel mixture flowing into the cylinder. It is implicitly assumed that the cylinder valves, not shown, do not affect the flow of air and fuel into the cylinder, and that the fuel is injected in a constant stream.

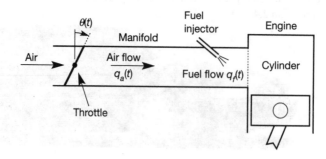

Figure 2.11 An engine inlet manifold schematic diagram.

Reading Section 1.3.8 further, the various inputs and outputs need to be identified next. In the manifold, the output is the air to fuel ratio entering the cylinder since, for emissions reduction, this is the quantity that must be controlled. The input to the system is the fuel injection rate, and the quantity of injected fuel can be freely manipulated. Throttle position, which is normally manipulated by the vehicle driver to adjust the engine speed, regulates the air flow. Therefore, for this application, air flow (or throttle movement) is a disturbance input.

To develop the lumped-parameter model, use will be made of the fluid elements depicted in Table 2.4. Note that for this application, the air to fuel ratio is normally calculated in terms of *mass* flow rates. Therefore both sides of the defining equations in Table 2.4 will eventually need to be multiplied by the fluid density to convert from volumetric flow to mass flow.

Air flow past the throttle valve might be represented by a modified fluid resistance, having a defining equation of the form:

$$q_a(t) = \frac{1}{R}\, h(t)\, \frac{k}{h(t)}\, \theta(t)$$

or

$$q_a(t) = \frac{k}{R}\, \theta(t)$$

where

k is a constant relating pressure head to throttle angle $(\mathrm{m\,rad}^{-1})$

$h(t)$ is the pressure head (m) forcing the air into the cylinder

In fact, to assume that the air flow past the throttle valve is a linear function of the throttle angle is a gross over-simplification, and this assumption needs to be noted carefully. Its validity can only be assessed by a more thorough attempt at modelling, or by comparing the model and plant responses (Stage 4 of Section 1.3.8).

Now consider the fuel dynamics. The injected fuel splits into two streams, one vaporizing and entering the cylinder directly, and the other impinging with the manifold wall to form a puddle. The puddle could conceivably be represented by a tank similar to that shown in Figure 2.8. The outflow from the 'tank' would be the evaporation rate from the puddle, and could be assumed to be related only to the size of the puddle. That is, in the lumped-parameter model, it will be related to the head of fluid in the 'tank'. Consequently, evaporation can be represented by a linear valve restricting the flow of fluid from the 'tank'.

The lumped-parameter model is completed by defining three more elements: one which splits the injected fuel into the two streams, another which will add the vaporized and evaporated fuel entering the cylinder, and the third element which will divide the air mass flow rate by the total amount of fuel in the cylinder. These additional elements complete the lumped-parameter model of the manifold shown in Figure 2.12. When examining this figure, note that the amount of vaporized fuel $q_{fv}(t)$ is assumed to be a function of the throttle angle $\theta(t)$. This is perhaps reasonable, since the amount of vaporized fuel will largely depend on the manifold pressure, which in turn is related to throttle angle. Also note that since there are two nonlinear elements, the split element and the division element, the resulting model will be nonlinear.

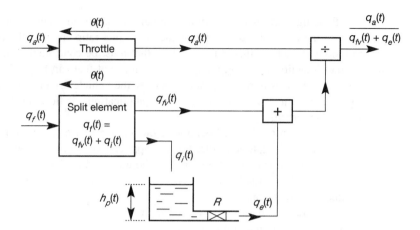

Figure 2.12 A lumped-parameter engine inlet manifold model.

2.4   Equations of motion from lumped-parameter models

Since each engineering discipline has its own techniques for finding equations of motion, it is appropriate first to show how a lumped-parameter model, based on one set of elements, can be converted into an equivalent model using another set of elements. There are some restrictions, particularly with thermal and fluid elements for which the equivalents of the spring or inductance have not been defined here (for example, there is a concept of 'inertance' in fluid dynamics which models the inertial tendency of a moving plug of fluid to continue moving, and is roughly analogous to inductance).

2.4.1   Drawing equivalent circuit diagrams

This approach may be appreciated most by readers with an electrical or electronic bias (Sections 2.4.3 and 2.4.6 give approaches which may find more favour with mechanically biased readers). Nevertheless, it is worth all readers gaining an appreciation of the behaviour of the 'analogs' discussed here. The idea of this approach is that systems represented by collections of the simple lumped-parameter models discussed above can be represented as something very like electrical circuit diagrams – even if they are not electrical systems. If this can be achieved, the models can be analysed by the usual techniques of analysing electrical circuits (even if they are mechanical or thermal models, for example). One suitable approach, which follows a systematic procedure, is given below and then illustrated by an example.

(1) Draw a reference node representing the reference value of the across variable. A 'node' in this context is simply a point to which the terminals of two or more system elements will be connected in the drawing, and therefore has a single value of across variable associated with it. The reference node is therefore likely to be the $0\,\text{V}$ line in an electrical circuit, or a similar 'datum line' representing zero velocity in an equivalent circuit for a mechanical system, for example. Everything else is modelled with respect to this reference node.

(2) Provide a generator to supply the through or across variable of interest, that is, to generate the voltage, current, force, velocity etc. that drives the system, and which is the cause of the system response. This generator will usually (but not always) have one end connected to the reference node.

(3) Identify groups of element terminals in the system which experience the same value of across variable (voltage, velocity and so on). Connect each group to one node in the equivalent circuit. There will therefore be as many nodes (connection points) as there are different values of across variable in the system.

(4) All element terminals not yet connected to anything after step 3 should be connected to the reference node.

Point 4 needs some explanation. Firstly, note that as each 'floating' connection is tied to the reference node, it must be verified that this is indeed correct and that something has not been overlooked in step 3.

Secondly, there is a conceptual problem with masses and inertias in mechanical systems. When analysing electrical systems, fluid systems or thermal systems, the analogous elements to mass and inertia (that is, the various capacitances) clearly have an identifiable value of across variable at each side of the component; that is, there is, as expected, a value of across variable across them. This seems obvious, but it is not the case for masses and inertias since these are rigid, and each side of them therefore moves with the same velocity.

As a result, when constructing the equivalent circuit (step 3), one end of the symbol for each mass or inertia is connected to a node having the correct value of velocity associated with it, but the other end will always be left 'floating'. Step 4 then indicates that all these 'floating' connections be tied to the reference node, which implies that one end of the symbol for *every* mass or inertia in a system will always be connected to the reference node. This is, in fact, correct. The way to visualize why, is to regard the entire mass, or inertia, as moving with a certain velocity with respect to the reference node. One end of the symbol for each mass and inertia will therefore be connected to the reference node, and the other to a node having the correct velocity for the element in question. This is clarified in the following example.

Example 2.3 **An equivalent electrical circuit for a mechanical slipping clutch**

Figure 2.13 shows a simple mechanical coupling with input velocity $\omega_i(t)$ and output velocity $\omega_o(t)$. It might represent a torque converter, or a mechanical automobile speedometer without the restraining spring. Find an analogous electrical circuit.

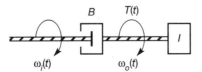

Figure 2.13 A torsional coupling.

For this mechanical rotational system, Table 2.3 gives the through variable as torque and the across variable as angular velocity. Applying the steps given above, and using the symbols from the table, the system circuit diagram of Figure 2.14(a) is obtained. The steps of the procedure are:

Step 1 Draw a reference node to represent zero angular velocity.

Step 2 The input variable is $\omega_i(t)$, so draw an across variable generator to apply this with respect to the reference node (see Table 2.6).

Step 3 There is only one other value of across variable in the system, namely $\omega_o(t)$, so there will be a node in the equivalent circuit having this value of across variable. The components to be connected to this node are those that move with velocity $\omega_o(t)$ in the lumped-parameter model, that is, one end of the damper and one end of the inertia. The other end of the damper moves with velocity $\omega_i(t)$, so it will be connected to the node having that value of across variable, as shown.

Step 4 There are no more nodes (different values of across variable) in the lumped-parameter model, so now tie any 'floating' connections to the reference node (with careful consideration in case anything has been overlooked). In this case, there is only one such connection – the spare end of the inertia symbol. As discussed before, it is correct to tie this to the reference node, and the system circuit diagram (Figure 2.14(a)) is complete.

To apply electrical circuit analysis techniques to this system, the defining equations in column three of Tables 2.1 and 2.3 show that the damper may be regarded as a resistor of value $1/B$ and the inertia as a capacitor of value I. This gives Figure 2.14(b).

Equally, the inertia–damper system of Figure 2.13 could be regarded as being a mechanical analog of the electrical R–C circuit of Figure 2.14(b).

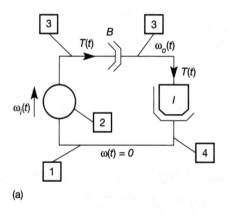

(a)

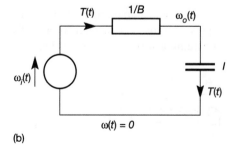

Figure 2.14 (a) System circuit diagram for the torsional coupling. (b) Equivalent electrical circuit diagram for the torsional coupling.

(b)

It has been shown that physical systems can be represented by lumped-parameter models and that, for simple systems, analogies exist. These analogies permit conversion of lumped-parameter elements from one system type to another. However, once the lumped-parameter model is fixed, the equations of motion describing that lumped-parameter model (not the physical system) need to be extracted. In the following subsections, a variety of techniques for finding the equations of motion from lumped-parameter models is presented.

2.4.2 Mathematical models for electrical systems

In electrical systems, Kirchhoff's current law (stating that the currents flowing into a node sum to zero) provides the continuity equations, and applies to through variables in general. Kirchhoff's voltage law (which, stated loosely, says that the voltages around a closed path sum to zero), provides the compatibility equations which similarly apply to across variables. The continuity and compatibility equations which define the laws governing the system, together with the physical laws of the elements, are combined to produce the mathematical model.

Example 2.4 *An electrical lead compensator*

Find the mathematical model for the lumped-parameter lead compensator network shown in Figure 2.15. The lead compensator (in mechanical, electrical, electronic or pneumatic form) is one of the basic building blocks of controller design, and is discussed in Section 4.5.4. For the present, it is simply used as an example system for analysis.

The input to this network is shown as a voltage source $v_i(t)$ and the output is the voltage $v_o(t)$ across resistor R_2. From Table 2.1, the element equations are:

$$i_{R_1}(t) = \frac{1}{R_1} v_{R_1}(t) \tag{2.11}$$

$$i_C(t) = C \frac{d}{dt} v_C(t) \tag{2.12}$$

$$i_{R_2}(t) = \frac{1}{R_2} v_o(t) \tag{2.13}$$

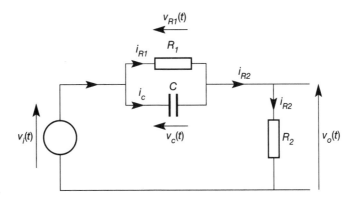

Figure 2.15 A passive electrical lead compensator.

The compatibility equations are:

$$v_i(t) = v_{R_1}(t) + v_o(t)$$

$$v_{R_1}(t) = v_C(t) \tag{2.14}$$

The continuity equation is:

$$i_{R_2} = i_C + i_{R_1} \tag{2.15}$$

Substituting the element equations into the continuity equation and then using the compatibility equations to eliminate $v_C(t)$ and $v_{R_1}(t)$ gives the ordinary linear differential equation:

$$R_1 R_2 C \frac{d}{dt} v_o(t) + (R_1 + R_2) v_o(t) = R_1 R_2 C \frac{d}{dt} v_i(t) + R_2 v_i(t) \tag{2.16}$$

which is a mathematical model relating $v_o(t)$ to $v_i(t)$.

2.4.3 Mathematical models for mechanical systems

Most mechanical systems consist of masses and/or inertias, which are connected together by springs and dampers. For simple systems, there is no reason why the methods of Example 2.3 should not be followed directly by those of Example 2.4, to find the required model. However, for more complex systems this approach is insufficient.

Perhaps the simplest structured analytical method for finding the equations of motion is that based on free-body diagrams. In this method, each mass or inertia is imagined to be displaced from its equilibrium position, and then isolated from the surrounding system. Each individual mass or inertia is then drawn, and the forces or torques driving it back to its equilibrium position are indicated. Newton's second law of motion is then applied to each body to yield the required equations of motion. Since the method assumes that the masses and inertias are displaced from an equilibrium position, it is normal to work with displacements rather than velocities. That is, in Table 2.2, since velocity is the time rate of change of the displacement $x(t)$ (m), $v(t)$ is replaced by $dx(t)/dt$. Similarly, in Table 2.3 $\omega(t)$ is replaced by $d\theta(t)/dt$, where $\theta(t)$ is the angular displacement (rad).

For a rectilinear (that is, translational) system having constant mass, Newton's second law indicates that, for a consistent system of units: *the sum of forces equals the mass times the acceleration.*

In the SI system of units, force is measured in newtons (N), mass in kilograms (kg) and acceleration in metres per second squared (m s^{-2}).

For a rotational system Newton's second law becomes: *the sum of the moments equals the moment of inertia times the angular acceleration.*

The moment, or torque, has units of newton-metres (N m), the inertia units of kilogram metres squared (kg m^2) and the angular acceleration units of radians per second squared (rad s^{-2}).

When there are no masses or inertias in the system, the mechanical impedance method is often used. In this method, the object is to replace a system of mechanical elements by an idealized element Z such that

$$f = Zx$$

where

$$f = \text{the force acting on the system}$$

$$Z = \text{the impedance of the system}$$

$$x = \text{the displacement across the system}$$

There are two rules, one for mechanical elements in series, and one for mechanical elements in parallel.

Mechanical series laws

From Newton's third law (*to every action there is an equal and opposite reaction*) the force acting on each element equals the total force applied to the series system.

From the geometric constraints, the total displacement of the series system is equal to the sum of the individual displacements across each element.

Mechanical parallel laws

Again from Newton's third law, the sum of the forces acting on the various elements must equal the total force applied to the parallel system.

From the geometric constraints, the displacement of the parallel system is equal to the displacement of each element.

Example 2.5 *Mathematical model of an accelerometer*

The lumped-parameter model of a simple mechanical accelerometer is shown in Figure 2.16. The displacement $x(t)$ of the mass M with respect to the accelerometer casing is related to the acceleration of the case. k is the spring stiffness and b the damping coefficient. Determine the relationship between the displacement $x(t)$ and acceleration $d^2y(t)/dt^2$. Note that the time dependence of x and y will now be omitted to aid clarity. It may be assumed that all motions take place in the directions shown.

The input to the system is the acceleration d^2y/dt^2 of the accelerometer casing, and the output is the deflection x of the mass. Since the accelerometer mass M is constrained to linear motion, the system has one degree of freedom (that is, one measurement is

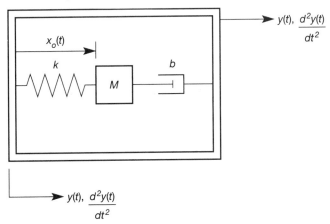

Figure 2.16 A mechanical transitional accelerometer.

sufficient to specify the position of all elements in the system) and so one free-body diagram is required. Giving the mass M a displacement x from its equilibrium position produces the free-body diagram shown in Figure 2.17. Note that the distance between the mass and the casing (x_o) is equal to x plus some fixed initial displacement.

Applying Newton's second law of motion gives

$$Ma = -kx - b\frac{dx}{dt}$$

where a is the acceleration of the mass relative to the earth and is given by:

$$a = \frac{d^2x}{dt^2} - \frac{d^2y}{dt^2}$$

Substituting for a gives the required equation of motion (mathematical model) as:

$$M\frac{d^2x}{dt^2} + b\frac{dx}{dt} + kx = M\frac{d^2y}{dt^2}$$

Note that each term in the above equation has units of newtons (N).

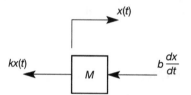

Figure 2.17 Free-body diagram for the accelerometer mass.

Example 2.6 *Modelling a rotational system with two inertias*

A rotational system with two degrees of freedom is described by the lumped-parameter model shown in Figure 2.18. The shafts have torsional stiffnesses k_1 and k_2 (in units of N m rad^{-1}), while the linear dashpot has an operating radius r (metres) and damping b (in N/(m s^{-1})). The system is forced by the input torque $T(t)$ and has two outputs $\theta_1(t)$ and $\theta_2(t)$, which are the angular displacements of inertias I_1 and I_2 respectively. Find the equations of motion.

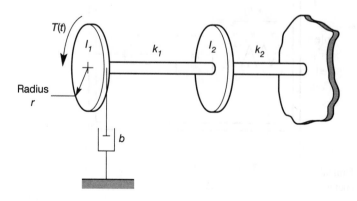

Figure 2.18 A rotational system.

Two free-body diagrams are required as shown in Figure 2.19: one for inertia I_1, which is given an angular displacement $\theta_1(t)$, and the other for inertia I_2, with its angular displacement $\theta_2(t)$.

The dashpot acting on inertia I_1 produces a linear retarding force of $b \times$ (linear velocity). The linear velocity is equal to the tangential velocity of the inertia I_1. Combining these, the linear retarding *force* is $br \, d\theta_1/dt$. For inclusion in the torque equations, this linear force must be multiplied again by r, to give the resulting retarding (damping) *torque*.

Applying Newton's second law of motion for rotational systems to both free-body diagrams yields the two equations of motion as:

$$I_1 \frac{d^2\theta_1}{dt^2} + (br)r \frac{d\theta_1}{dt} + k_1(\theta_1 - \theta_2) = T(t)$$

and

$$I_2 \frac{d^2\theta_2}{dt^2} + k_2\theta_2 - k_1(\theta_1 - \theta_2) = 0$$

In this example there are two output variables, $\theta_1(t)$ and $\theta_2(t)$, and two equations of motion. Note also that each term in each equation has units of N m.

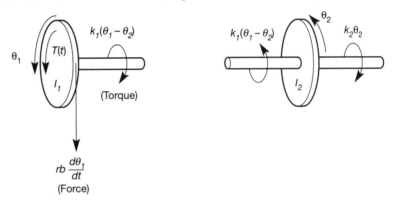

Figure 2.19 Free-body diagrams for the rotational system of Figure 2.18.

Example 2.7 *Modelling a mechanical lag-lead compensator*

Find the equations of motion for the mechanical lag-lead compensator shown in Figure 2.20. (Lag-lead compensators are control system elements whose design and purpose are discussed in Section 4.5.4. For the present, it simply provides a useful mechanical system example for analysis – think of it as part of a suspension system.) The input to the compensator is the displacement $u(t)$ and the output is the displacement $y(t)$, so a model is required which gives the relationship between these quantities.

Consider the mechanical impedance Z_1 of the parallel dashpot b_1 and spring k_1. From the first parallel mechanical law, the force equation is:

$$f_{z_1} = f_{b_1} + f_{k_1} \tag{2.17}$$

and from the second parallel law

$$(u - y)_{system \; z_1} = (u - y)_{dashpot \; b_1} = (u - y)_{spring \; k_1} \tag{2.18}$$

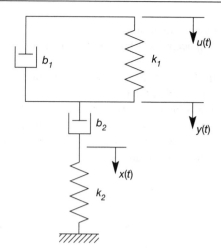

Figure 2.20 A mechanical lag-lead compensator.

The equations for the parallel elements are:

$$f_{b_1} = b_1 \frac{d}{dt}(u - y) \tag{2.19a}$$

$$f_{k_1} = k_1(u - y) \tag{2.19b}$$

Substituting the element Equations (2.19) into the force Equation (2.17) and noting the geometric constraint Equation (2.18), the impedance Z_1 of the parallel elements may be found such that:

$$f_{z_1} = Z_1(u - y) \tag{2.20}$$

Now consider the series impedance Z_2 of dashpot b_2 and spring k_2. Let $x(t)$ be the displacement at the point between the dashpot b_2 and spring k_2. The first series law for forces indicates that:

$$f = f_{z_1} = f_{b_2} = f_{k_2}$$

and the second series law that

$$u = (u - y) + (y - x) + x \tag{2.21}$$

The element equations are Equation (2.20) together with:

$$f_{b_2} = b_2\left(\frac{dy}{dt} - \frac{dx}{dt}\right) \tag{2.22a}$$

and

$$f_{k_2} = k_2 x \tag{2.22b}$$

Solving Equations (2.20) and (2.22) for $u - y$, $y - x$ and x and then substituting the results into the second series law, Equation (2.21) will, after some manipulation, produce an equation of the form

$$f = Zu \tag{2.23}$$

The required relationship between $y(t)$ and $u(t)$ is obtained by eliminating f from Equation (2.23) by substitution of Equation (2.20).

At this point, the reader may wonder how the dashpot equation is solved to give the relationship between f and $y - x$, and how Equation (2.20) is obtained to find the relationship between f and $u - y$. Fortunately, such problems are easily dealt with using Laplace transforms, which are described in Section 2.5.

2.4.4 Mathematical models for fluid and thermal systems

Fluid systems

Many fluid systems consist of tanks connected together by pipes and valves. The equations describing such systems are found by applying the fundamental laws of fluid mechanics. By making various simplifying assumptions, the idealized lumped-parameter components found in Table 2.4 can be generated. The system equations are then obtained from the system geometry and the lumped-parameter component equations.

Example 2.8 *Modelling a two-tank hydraulic system*

Determine the equations describing the fluid system shown in Figure 2.21, in terms of the relationship between the inflow q_i and outflow q_o.

In Figure 2.21 R_1 and R_2 are the resistances of the valves and pipework indicated, C_1 and C_2 are the capacitances (cross-sectional areas) of the two tanks with fluid heads h_1 and h_2 respectively, and q_1 is the flow between the tanks. For tank 1, the rate of change of fluid volume in the tank is equal to the flow in minus the flow out (the fluid capacity equation in Table 2.4):

$$\frac{dV_1}{dt} = q_i - q_1$$

But, since $V_1 = $ volume of tank $1 = C_1 h_1$, then

$$C_1 \frac{dh_1}{dt} = q_i - q_1 \tag{2.24}$$

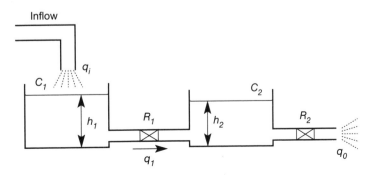

Figure 2.21 A representative fluid system.

The flow between the tanks is given by

$$q_1 = \frac{h_1 - h_2}{R_1}$$

which, on substitution into Equation (2.24) yields

$$C_1 R_1 \frac{dh_1}{dt} + h_1 = R_1 q_i + h_2 \qquad (2.25)$$

For tank 2 the continuity equation is

$$C_2 \frac{dh_2}{dt} = q_1 - q_o \qquad (2.26)$$

Also

$$q_o = \frac{h_2}{R_2} \qquad (2.27)$$

Substituting into Equation (2.26) for q_1 and q_o produces

$$R_1 C_2 R_2 \frac{dh_2}{dt} + R_1 h_2 + R_2 h_2 = R_2 h_1 \qquad (2.28)$$

Equations (2.25), (2.27) and (2.28) are the required system equations.

Thermal systems

Thermal systems include such things as the mixing of hot and cold streams, heat transfer, combustion and chemical reactions. Most systems have some thermal component, and the various modelling techniques are well covered in the literature.

For simple, lumped-parameter models, the first law of thermodynamics may be used to determine the temperature changes in a body. One statement of this law is that the net heat supplied to a system from its surroundings is equal to the net work done by the system on its surroundings. Application of this law and the law governing thermal resistance produces the components and equations given in Table 2.5.

Example 2.9 *Modelling a simple thermal system*

The insulated tank of water shown in Figure 2.22 is heated by an electrical element. Develop an equation for the rise in water temperature.

Assume that the water has a uniform temperature and that there is no heat storage in the insulation. Let:

$$T_w = \text{temperature of the water (K)}$$

$$T_a = \text{temperature of the air (K)}$$

$$q = \text{rate of heat supply (W)}$$

$$q_o = \text{rate of heat loss (W)}$$

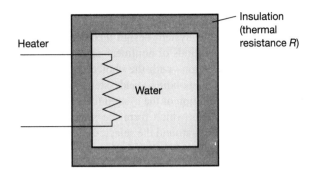

Figure 2.22 A simple thermal system.

From the thermal capacity equation (Table 2.5),

$$C \frac{dT_w}{dt} = q - q_o$$

where

$$C = \text{thermal capacity of the water } (\mathrm{J\,K^{-1}})$$

but, from the thermal resistance equation (Table 2.5)

$$q_o = \frac{T_w - T_a}{R}$$

where

$$R = \text{thermal resistance of the insulation } (\mathrm{K\,W^{-1}})$$

Hence

$$\frac{dT_w}{dt} + \frac{T_w}{RC} = \frac{q}{C} + \frac{T_a}{RC}$$

which is the required equation.

2.4.5 Linearization

Advanced section

This section may be omitted, but it contains necessary background material for a study of Chapter 14.

All the system components found in Tables 2.1 to 2.5 have been linear, and it has been assumed, with the exception of Example 2.2, that these components interconnect to form lumped-parameter models using linear connections. The mathematical models resulting from these lumped-parameter models are also linear, and must therefore satisfy the superposition principle (see Section 2.3). Any equation that is not linear is said to be nonlinear and any system that has a nonlinear equation in its mathematical description is also said to be nonlinear.

In general, most lumped-parameter modelling of physical systems will involve the use of some nonlinear components and nonlinear connections. However, as indicated in Chapter 1, there is considerable advantage if a physical system can be modelled using linear differential equations, since this permits use of the many powerful linear control system design techniques to be studied later. Fortunately, linearization of many kinds of nonlinear equations is fairly straightforward.

The procedure begins with the choice of a set of steady-state input value(s) together with the corresponding steady-state output value(s), representative of the normal operating condition of the system (often simply called the *operating point*). The nonlinear functions which form the system model are then expanded using *Taylor series* expansion around the selected operating point. Implicit in the analysis is the fundamental assumption that the linearized model is only going to be used in the vicinity of the operating point. This means that all deviations from the operating point should be small (although the acceptable definition of 'small' changes from system to system).

A Taylor series expansion is written in terms of deviations from the operating point, and consists of an infinite series of terms in which these deviations are linear, squared, cubed and so on as the series progresses. By neglecting all terms higher than first order (because the small deviations, raised to any power higher than unity, are assumed to be relatively insignificant) an approximate linear equation, representing the full nonlinear series, is obtained.

Assume, for example, that a model comprising some nonlinear function $f(x, y)$ of the process variables x and y has been obtained, and that the original steady-state values of these variables are x_o and y_o respectively.

Since, in this example, the function $f = f(x, y)$ is a function of two variables, the Taylor series expansion will include terms which are the partial derivatives of f with respect to x and y. Performing the expansion around the steady-state operating point, and truncating the Taylor series after the first partial derivatives, gives the required linear approximation thus:

$$f(x, y) = f(x_o, y_o) + \frac{\partial f}{\partial x}\bigg|_{x_o, y_o} (x - x_o) + \frac{\partial^2 f}{\partial x^2}\bigg|_{x_o, y_o} \frac{(x - x_o)^2}{2!}$$

$$+ \frac{\partial^3 f}{\partial x^3}\bigg|_{x_o, y_o} \frac{(x - x_o)^3}{3!} + \cdots + \frac{\partial f}{\partial y}\bigg|_{x_o, y_o} (y - y_o)$$

$$+ \frac{\partial^2 f}{\partial y^2}\bigg|_{x_o, y_o} \frac{(y - y_o)^2}{2!} + \frac{\partial^3 f}{\partial y^3}\bigg|_{x_o, y_o} \frac{(y - y_o)^3}{3!} + \cdots$$

Or, on neglecting terms higher than first order:

$$f(x, y) \approx f(x_o, y_o) + \frac{\partial f}{\partial x}\bigg|_{x_o, y_o} (x - x_o) + \frac{\partial f}{\partial y}\bigg|_{x_o, y_o} (y - y_o) \qquad (2.29)$$

where $f(x, y)$ is the approximate linearized model at some new operating point represented by x and y.

The first term on the right-hand side (RHS) of Equation (2.29) is the value of f at the original operating point (x_o, y_o) and will be a constant. The second term on the RHS is read as, 'the partial derivative of f with respect to x, evaluated at the original operating point, and multiplied by the deviation of x from its original operating point value'. The effect of this may be easier to visualize in another way. The partial derivative is actually the slope of the graph of f vs. x (with y held constant at y_o) at the operating point x_o. It thus gives us a tangent to f at x_o. Moving along this tangent by an amount $x - x_o$ (which is implied by multiplying the slope by $(x - x_o)$) therefore predicts the change in the value of f from its value at the operating point. When the first term in Equation (2.29) is taken across to the LHS, as below, this is precisely what is required. See also Example 2.10. The third RHS term is the obvious rewording of this for y.

In practice, the system performance relative to the new operating point (x, y) is required, rather than that relative to the original point (x_o, y_o). Therefore, Equation (2.29) is rewritten as:

$$[f(x, y) - f(x_o, y_o)] \approx \frac{\partial f}{\partial x}\bigg|_{x_o, y_o} (x - x_o) + \frac{\partial f}{\partial y}\bigg|_{x_o, y_o} (y - y_o)$$

Example 2.11 illustrates the importance of this modification.

Example 2.10 *A linearized model of a simple pendulum*

A lumped-parameter model of such a pendulum comprises a massless, rigid rod of length l (m), hinged at one end to a frictionless bearing which is fixed in space, as shown in Figure 2.23(a). The other end of the rod is attached to a point mass m (kg).

To obtain the mathematical model, it is assumed that the pendulum can oscillate only in the plane of the paper. With this constraint, only one coordinate – the angular displacement of the rod, θ (rad) – is required for a complete specification of the pendulum's geometric location. Such a mechanical system is said to have one degree of freedom. A further assumption is that the only force acting on the system is gravitational.

There is no input that can be manipulated, but there are two possible disturbance inputs: the initial angular displacement of the rod and the initial angular velocity.

The free-body diagram in Figure 2.23(b) shows the rotational mechanical system displaced by an angle θ from its equilibrium position. There is a moment of $-mgl \sin \theta$ due to the displacement of the mass and the gravitational force mg (where g is the

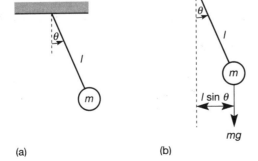

Figure 2.23 (a) A simple pendulum – lumped-parameter model. (b) A simple pendulum – free-body diagram.

(a)

(b)

acceleration due to gravity, $g \approx 9.81 \text{ m s}^{-2}$). The negative sign indicates that the direction of the moment is in the opposite direction to θ. From Newton's second law for rotational systems:

$$I \frac{d^2\theta}{dt^2} = -mgl \sin \theta$$

Noting that, for this system, the moment of inertia, I, is given by $I = ml^2$, then:

$$m \frac{d^2\theta}{dt^2} + \frac{mg}{l} \sin \theta = 0 \tag{2.30}$$

Equation (2.30) is nonlinear because of the sinusoidal term, but it may be linearized by assuming that only small perturbations (swings) will occur about the operating point. Selecting the equilibrium position $\theta = 0$ rad, and making a Taylor series expansion of $\sin \theta$, gives:

$$\sin \theta = \theta - \frac{\theta^3}{3!} - \frac{\theta^5}{5!} - \cdots$$

Hence, for small angles measured in radians, $\sin \theta \approx \theta$. Using a calculator to test this approximation shows that, for small angles in radians, it is indeed the case that $\sin \theta \approx \theta$. It also shows that the resulting model will be at least 10 per cent in error by the time θ reaches $\pm \pi/4$ radian (45°).

To visualize what is happening here, sketch a plot of $\sin \theta$ vs. θ (for θ in radians) and draw a straight line of unity slope through the origin ($\theta = 0$). The sinewave represents the full nonlinear equation, while the straight line is the approximation $\sin \theta \approx \theta$. It becomes obvious that the further from the operating point (the origin) one moves, the less accurate the linear model becomes.

Making the proposed substitution $\sin \theta \approx \theta$ in Equation (2.30) gives the linear model:

$$\frac{d^2\theta}{dt^2} + \frac{g}{l} \theta \approx 0$$

Checking the units for consistency indicates that each term has units of rad s^{-2}.

Since the operating point was at zero, there was no constant offset to complicate things. It is interesting to note that the standard equation of a *straight line*, $y = mx + c$, fails the superposition test, and is therefore *nonlinear* according to the given definition. To linearize $y = mx + c$, the constant offset c must be removed. This is easily achieved by redefining y in terms of its deviations from c, which is effectively shifting the origin to the point $(0, c)$, that is, $(y - c) = mx$. In the general case, this is what happens when the first RHS term of Equation (2.29) is moved to the LHS. Example 2.11 offers further clarification.

Example 2.11 *A linearized model of fluid flow through a valve*

From Bernoulli's law the flow through a valve q (m^3 s^{-1}) is related to the pressure head across the valve h (m) by the following equation, in which g is the acceleration due to gravity, and C_d is the coefficient of discharge (m^2):

$$q = C_d \sqrt{(2gh)}$$

Find a linear relationship between q and h (at present, the square root makes it nonlinear).

For this equation, the truncated Taylor series expansion of q around the steady-state value of h, namely h_o, is found using Equation (2.29) with $x = h$ and $y = 0$ (since q is a function of one variable only in this case):

$$q|_h \approx q|_{h_o} + \frac{\partial q}{\partial h}\bigg|_{h_o} (h - h_o)$$

$$= q|_{h_o} + \left(\frac{1}{2} C_d h^{-1/2} \sqrt{2g}\right)\bigg|_{h_o} (h - h_o)$$

$$= q|_{h_o} + \left(C_d \sqrt{\frac{g}{2h_o}}\right)(h - h_o)$$

Notice that the above equation defines a straight line, since the term multiplying $h - h_o$ is a constant and $q|_{h_o}$ is a constant, equal to the value of q evaluated at h_o, that is

$$q|_{h_o} = C_d \sqrt{(2gh_o)}$$

To produce the required linear flow equation, the constant $q|_{h_o}$ is moved to the LHS to give:

$$[q|_h - q|_{h_o}] = \left(C_d \sqrt{\frac{g}{2h_o}}\right)(h - h_o)$$

When $h = h_o$, there is no change in the head (that is, no deviation from the original operating point) and hence no change in flow from the value $q|_{h_o}$. This may also be stated (using Δh to represent a change in h and so on) as, when $\Delta h = 0$ then $\Delta q = 0$. For clarity, the Δ is usually omitted and the linearized flow model about the operating point h_o is simply written as:

$$q = \left(C_d \sqrt{\frac{g}{2h_o}}\right)h$$

where it is to be understood that q and h now represent *deviations* (changes) from their operating point values.

Example 2.12 *Linearizing a product of two variables*

Linearize the model consisting of the product of two dependent variables x and y, such that

$$f(x, y) = xy$$

Directly from the Taylor series of Equation (2.29), we obtain:

$$xy \approx x_o y_o + y_o(x - x_o) + x_o(y - y_o)$$

which has converted the nonlinear product into a 'straight line' function in x and y. Again, if the first term on the RHS is moved over to the LHS, everything is expressed in terms of changes from the original operating point, and the result becomes linear.

2.4.6 Lagrange's equations

Advanced section

This section is typical of the approaches normally used to model mechanical translational and rotational systems. It is, however, optional if the emphasis is on control system design, rather than control system modelling.

As the number of components in a lumped-parameter model increases, the techniques described so far in Section 2.4 become more unwieldy. Alternative techniques do exist. One such technique is based on Lagrange's equations, which provide a very powerful means of determining the equations of motion of a dynamic system. They are easily shown to be based on Hamilton's principle, which can be loosely interpreted as:

For a dynamic system in which the work of all forces is accounted for in the Lagrangian (see below), an admissible motion between specific configurations of the system at times t_1 and t_2 is a natural motion if, and only if, the energy of the system remains constant.

Although this is phrased in terms of 'forces' and 'motions', the underlying ideas are concerned with *energy*, and are therefore equally applicable to electrical systems.

For conservative systems (see Section 1.3.4), Lagrange's equation may be written as

$$\frac{d}{dt}\left(\frac{\partial L}{\partial \dot{q}_i}\right) - \frac{\partial L}{\partial q_i} = 0$$

where

> $L = T - V$ is the Lagrangian (T and V are the kinetic and potential energies in the system, respectively)
>
> q_i = generalized coordinates (see below)

For more general systems (that is, ones including power dissipation):

$$\frac{d}{dt}\left(\frac{\partial L}{\partial \dot{q}_i}\right) - \frac{\partial L}{\partial q_i} + \frac{\partial P}{\partial \dot{q}_i} = Q_i \tag{2.31}$$

where

> P = power function, describing the dissipation of energy by the system
>
> Q_i = generalized external forces acting on the system

As seen in the previous examples, the number of degrees of freedom of a body is the number of independent quantities that must be specified if the position of the body is to be uniquely defined. Any unique set of such quantities is referred to as a set of generalized coordinates for the system. In Example 2.5, the position of the mass of the accelerometer is defined by the displacement x and therefore x is a generalized coordinate for this system. The system in Example 2.6 has two degrees of freedom, and so two generalized coordinates are required; these could be θ_1 and

Table 2.7 Energy expressions for electrical and mechanical elements.

Energy type	Mechanical	Electrical
Kinetic energy T	Mass $T = \frac{1}{2}\, m\dot{x}^2$	Inductor L $T = \frac{1}{2}\, L\dot{q}^2$
Potential energy V	Spring k $V = \frac{1}{2}\, kx^2$	Capacitor C $V = \frac{1}{2}\, Cv^2 = \dfrac{1}{2C}\, q^2$
	Gravitational m $V = mgh$	–
Dissipative energy P	Damper b $P = \frac{1}{2}\, b\dot{x}^2$	Resistor R $P = \frac{1}{2}\, R\dot{q}^2$

θ_2, the angular displacements of the two inertias. Clearly, from Section 2.4.1, there must be an electrical analogy. Loop currents (or, more usually, the charges in the various loops) could form a set of generalized coordinates.

The various energy forms for linear mechanical and electrical elements are summarized in Table 2.7. Note that the current i is expressed in terms of charge q as

$$i = \frac{dq}{dt} = \dot{q} \tag{2.32}$$

and that linear velocity v is expressed in terms of displacement x as

$$v = \frac{dx}{dt} = \dot{x} \tag{2.33}$$

Lagrange's equations have many modelling applications, although the intention in this section is to demonstrate their effectiveness in developing mathematical models for electromechanical systems.

Example 2.13 *A model of a capacitor microphone*

The lumped-parameter model of a capacitor microphone is shown in Figure 2.24. In the equilibrium position and with no external force applied to the moving plate, there is a charge q_o (coulombs) on the capacitor. This charge results in a force of attraction between the plates which pre-tensions the spring. A sound wave applies a force to the moving plate, resulting in displacement x (m) from the equilibrium position. This motion alters the capacitance C of the capacitor and results in a change in charge. This example uses Lagrange's equations to produce a mathematical model for this lumped-parameter system.

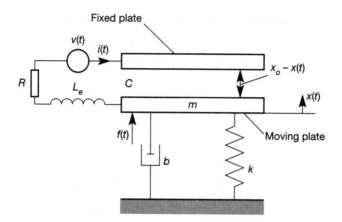

Figure 2.24 A capacitor microphone.

The system has two degrees of freedom, one mechanical and the other electrical. Two coordinates are therefore required, and the charge q and displacement x from equilibrium are selected.

The kinetic energy function for the system must contain a term for the electrical equivalent of 'kinetic energy' as well as the more usual mechanical component, and is:

$$T = \frac{1}{2} L_e i^2 + \frac{1}{2} mv^2 = \frac{1}{2} L_e \dot{q}^2 + \frac{1}{2} m\dot{x}^2 \tag{2.34}$$

where

$$L_e = \text{the inductance of the arrangement (H)}$$

$$i = \text{the current flowing in } L_e \text{ (A)}$$

$$m = \text{the mass of the moving plate (kg)}$$

$$v = \text{the velocity of the moving plate (m s}^{-1})$$

the dot notation is used to represent the time derivative, $\dot{q} = \dfrac{dq(t)}{dt}$ etc.

The potential energy function is also required. The mechanical part of this is found by integrating the force stretching or compressing the spring, over the deflection of the spring. This yields units of N m, which are easily shown to have the dimensions of energy. Thus:

$$\int kx \, dx = \frac{1}{2} kx^2$$

where $k = $ the spring stiffness (N m^{-1})

The electrical part of the 'potential energy' function is found as follows. If C (F) $=$ the capacitance between the plates, then the voltage appearing across the plates is given by $v_c = q/C$. At the same time, the current in the circuit is given by $i = dq/dt$. The instantaneous power is therefore $v_c i = (q/C) \, dq/dt$. Power is rate of energy expenditure, so the energy is found by integrating this expression with respect to time as follows:

$$\int v_c i \, dt = \int \frac{q}{C} \frac{dq}{dt} \, dt = \int \frac{q}{C} \, dq = \frac{1}{2} \frac{q^2}{C}$$

The potential energy function is therefore:

$$V = \frac{1}{2}\frac{q^2}{C} + \frac{1}{2}kx^2$$

However, the capacitance C is a function of plate separation $x_o - x$ given by:

$$C = \frac{\varepsilon A}{x_o - x}$$

where

$$A = \text{area of plates } (\text{m}^2)$$

$$\varepsilon = \text{dielectric constant of air } (\text{F m}^{-1})$$

so:

$$V = \frac{1}{2\varepsilon A}(x_o - x)q^2 + \frac{1}{2}kx^2 \tag{2.35}$$

Since this system is not conservative (the resistance R (Ω) and the damping b $(\text{N/(m s}^{-1}))$) dissipate energy) a power dissipation function for the system is required. This comprises the integral of the voltage across the resistor with respect to current (thus giving units of watts for the electrical part of the system), and the integral of the force exerted by the damper with respect to velocity (thus giving units of watts for the mechanical part of the system):

$$P = \int Ri\,di + \int bv\,dv = \frac{1}{2}Ri^2 + \frac{1}{2}bv^2$$

Changing from i and v to the chosen coordinates using Equations (2.32) and (2.33):

$$P = \frac{1}{2}R\dot{q}^2 + \frac{1}{2}b\dot{x}^2 \tag{2.36}$$

Applying Lagrange's equation (2.31) to the coordinate x gives:

$$\frac{d}{dt}\left(\frac{\partial L}{\partial \dot{x}}\right) - \frac{\partial L}{\partial x} + \frac{\partial P}{\partial \dot{x}} = f(t) \tag{2.37}$$

where $L = T - V$ and $f(t)$ is the external force at x. Hence from Equations (2.34) and (2.35):

$$L = \frac{1}{2}L_e\dot{q}^2 + \frac{1}{2}m\dot{x}^2 - \frac{1}{2\varepsilon A}(x_o - x)q^2 - \frac{1}{2}kx^2 \tag{2.38}$$

From Equations (2.36) and (2.38), the terms needed for Equation (2.37) are:

$$\frac{\partial L}{\partial \dot{x}} = m\dot{x}, \qquad \frac{\partial L}{\partial x} = \frac{q^2}{2\varepsilon A} - kx \qquad \text{and} \qquad \frac{\partial P}{\partial \dot{x}} = b\dot{x}$$

Lagrange's equation (2.37) therefore yields:

$$m\ddot{x} + b\dot{x} + kx - \frac{1}{2\varepsilon A}q^2 = f(t) \tag{2.39}$$

Similarly, applying Lagrange's equation to coordinate q yields

$$L\ddot{q} + R\dot{q} + \frac{1}{\varepsilon A}(x_o - x)q = f(t) \tag{2.40}$$

Equations (2.39) and (2.40) form the required mathematical model.

2.4.7 Models of interconnected systems

With complex systems, a sensible modelling technique is to reduce the system to a number of manageable lumped-parameter elements, which are interconnected. The mathematical model for each element is then determined, and an overall system model developed by combining all the equations for all the elements in a way which represents the interconnections. However, when adopting this approach, care should be taken to ensure that one element does not interact with, or load, another in the real plant. In Example 2.8 (Figure 2.21), the two tanks do interact with each other, since the head of fluid in tank 1 is partially dependent on the head of fluid in tank 2. The mathematical model describing the changing head in tank 1 thus requires a feedback of information from the head of fluid in tank 2 (see Example 2.22). If the tanks were non-interacting (that is, if the outlet flow from tank 1 simply poured into tank 2, as in Figure 2.25), then the two tanks could be modelled independently.

In general then, if elements are non-interacting, the output from one stage becomes the input to the next. When interactions occur they must be accounted for in the model, which usually requires the feedback of information from a later stage to an earlier stage.

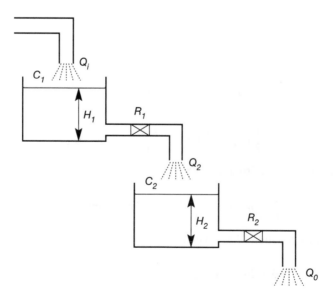

Figure 2.25 Two non-interacting tanks.

Example 2.14 *Modelling an interconnected network – incorrectly*

Demonstrate that the electrical network shown in Figure 2.26(a) cannot, in general, be modelled by the independent networks shown in Figures 2.26(b) and (c). Under what conditions *could* independent modelling be carried out? Assume that the line L carries no current.

 The equation relating the input voltage v_i to the output voltage v_o in Figure 2.26(a), for no current flow along L, is given by (proof of which is left as an exercise for the reader – see Problem 2.1):

$$\frac{R_1}{R_3}(R_2+R_3)C\frac{dv_o}{dt}+\frac{R_1+R_2+R_3}{R_3}v_o=v_i$$

For Figure 2.26(b) the equation relating v_i and v is:

$$R_1C\frac{dv}{dt}+v=v_i \tag{2.41}$$

For Figure 2.26(c) the equation relating v_o and v, again for no current flow in L, is:

$$\frac{R_2+R_3}{R_3}v_o=v \tag{2.42}$$

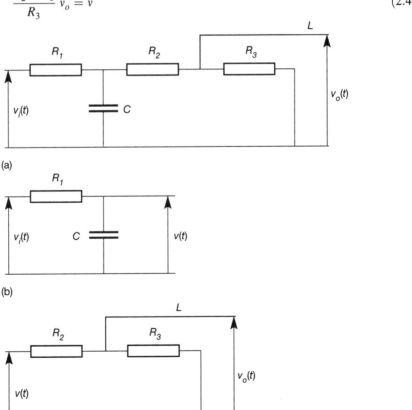

(a)

(b)

Figure 2.26 (a) A complete electrical circuit. (b) First subsystem of the circuit of Figure 2.26(a). (c) Second subsystem of the circuit of Figure 2.26(a).

(c)

Eliminating v from Equations (2.41) and (2.42) yields

$$\frac{R_1}{R_3}(R_2 + R_3)C\frac{dv_o}{dt} + \frac{R_2 + R_3}{R_3}v_o = v_i$$

which demonstrates the inadequacy of independent modelling for interacting systems, as it does not agree with the original (correct) model.

The two modelling procedures would give identical results if the resistance R_3 were made very large. Under these conditions there would be a negligible flow of current through R_3 and, since by definition there is no current flow through L, the circuit in Figure 2.26(c) would then not load the circuit in Figure 2.26(b) – that is, the two circuits would not interact.

2.5 Differential equations and Laplace transforms

From the preceding sections, it is evident that a mathematical model, describing small perturbations about a system's operating point, usually consists of ordinary linear differential equations. The *solution* of these differential equations yields the system's dynamic and steady-state characteristics. Unfortunately it is not easy to deal with differential equations directly. Although it may be useful to solve the equations and hence find the system's response (or responses) to a given input (or inputs), such a solution provides little information about the changes required to the system if the responses are unsatisfactory.

For this reason, alternative systematic methods of solution have been developed, which are also capable of providing the basis for various analysis and design techniques. This section will concentrate on two such methods, namely, the *state-space* and *Laplace transform* methods of representing system models, and the relationship between these methods.

2.5.1 An introduction to state-space models

The differential equation model found in Example 2.14 is said to be of *first order*, because it involves only a first derivative. On the other hand, the total model of Example 2.13 is of *fourth order*, because it involves two separate equations, each of which contains a second derivative. In general, the *order* of a model will be equal to the number of separate energy storage mechanisms in the system being modelled. In Example 2.14, there was just one – the capacitor. In Example 2.13, there were four – the capacitance of the microphone plates, the inductance of the electrical circuit, the spring acting on the moving plate and the mass of the moving plate (which 'stores' potential energy). Note that all the elements mentioned store energy, but that electrical resistance and mechanical damping do not – they *dissipate* it. Theoretically, if an electrical circuit could be built containing only parallel capacitance and inductance, it would be a conservative system and would oscillate continuously. Since there is no resistance in the circuit, all the energy would be continuously exchanged between the inductor and the capacitor. Similarly, a theoretical mass–spring system (one with zero damping, friction and air resistance) would also continuously oscillate, since the energy would repeatedly be exchanged between the mass and the spring without loss.

It is possible to solve ordinary, linear, first- and second-order differential equations directly. However, complex systems such as aircraft and chemical plants (for example) can easily reach orders of several tens (or hundreds) and direct solutions become impractical. Even computers are not very good at reliably solving a 100th-order equation directly – it is a difficult problem from a numerical analysis viewpoint. It follows that, in the analysis and design of control schemes, a 100th-order differential equation could not easily be dealt with directly. For these reasons, a systematic approach, such as the state-space methodology, is required.

The state-space approach to modelling works by replacing high-order differential equations with a set of simultaneous first-order differential equations. For example, a 100th-order differential equation is replaced by a set of 100 simultaneous first-order differential equations. This might not seem much of a step forward, but it actually provides a sound framework both for obtaining solutions and for control system analysis and design. Computers are very good at solving any number of first-order simultaneous equations. Furthermore, simultaneous first-order differential equations can be written in a standard *vector—matrix* form, providing a compact notation with which to work. This means that matrix algebra is the primary analytical tool for this approach. Appendix 1 gives a unique viewpoint of matrix algebra for control engineering and should contain all the information required. In particular, Section A1.1 is sufficient for the present chapter.

The remainder of this section gives a definition of the standard state-space model of all linear systems (Equations (2.43) and (2.44) below), followed by some methods of generating such models, and a couple of examples. Section 2.5.2 then begins the study of the alternative Laplace transform approach.

The *state* of a system

The *state* of a system may be thought of as a vector quantity. In the context of control engineering, the state is specified by a set of variables arranged in vector form (Appendix 1). Specifically, the state vector of a system is defined as a set of variables (the *state variables*), the choice of which is free so long as they obey the following general rules:

- They must be linearly independent. That is, one cannot simply be a multiple of another, or a weighted sum of two others, for example.
- They must be sufficient in number to specify completely the dynamic behaviour of the system.
- They may not be system inputs, or linear combinations of system inputs (since inputs are impressed from outside the system, and are therefore not states of the system itself).

It transpires that, for an nth-order system, at least n state variables are needed. In practice, it is usual to choose precisely n state variables, so as to avoid possible computational difficulties.

There are many ways of choosing the state variables. Some lead to states which are physically meaningful and measurable in the real world. Others lead to states which are abstract (that is, they do not correspond to physically measurable signals), but give the resulting model neat mathematical forms which are easy to work with. In this chapter, both approaches are introduced. In any case, the states for an nth-order system are conventionally called $x_1, x_2, x_3, \ldots, x_n$, and are written as the column vector $x = [x_1 \ x_2 \ x_3 \ \ldots \ x_n]^{\mathrm{T}}$ (note the transpose operator).

Similarly, if the system has m inputs and p outputs, then the input and output vectors u and y are the column vectors $u = [u_1 \ u_2 \ u_3 \ \ldots \ u_m]^{\mathrm{T}}$ and $y = [y_1 \ y_2 \ y_3 \ \ldots \ y_p]^{\mathrm{T}}$.

A state-space model of *any* delay-free, linear system, of *any* order, is then made up of the following pair of time-domain, differential, vector–matrix equations (a pure time delay affecting a signal can only be included in a state-space model approximately – see Chapter 8 – but is easily handled by Laplace transform methods):

$$\dot{x} = Ax + Bu \tag{2.43}$$

$$y = Cx + Du \tag{2.44}$$

Equation (2.43) is the *state equation* and Equation (2.44) the *output equation*. The vector $\dot{x}$ is the time derivative of x, comprising the derivative of each individual element. The quantities A, B, C and D are all purely numerical matrices, as listed below.

$x =$ the $n \times 1$ state vector for an nth-order system

$u =$ the $m \times 1$ input vector for a system with m inputs

$y =$ the $p \times 1$ output vector for a system with p outputs

$A =$ the $n \times n$ system (or plant) matrix for an nth-order system

$B =$ the $n \times m$ input matrix for an nth-order system with m inputs

$C =$ the $p \times n$ output matrix for an nth-order system with p outputs

$D =$ the $p \times m$ feed forward matrix for a system with p outputs and m inputs

Equation (2.43) (the state equation) describes the dynamic behaviour of the system's states in response to applied inputs (and initial conditions – see Sections 3.2.1 and 3.6.2). Equation (2.44) (the output equation) is non-dynamic, and simply maps the state variables and inputs onto the outputs.

Equations (2.43) and (2.44) are two of the most important in control engineering. The whole of the time-domain expansion of control theory since the 1960s is based upon them. One of the attractive aspects of the state-space approach is that Equations (2.43) and (2.44), with no alteration whatsoever, describe all linear systems. The only things that change from system to system are the sizes of the matrices and vectors (according to the list above) and the numerical contents of A, B, C and D, none of which affects the appearance of Equations (2.43) and (2.44).

Conversion of a system's mathematical model, described by ordinary linear

Figure 2.27 The relationship between $u(t), z(t)$ and $y(t)$.

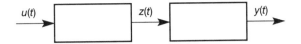

differential equations, into a state-space model can be illustrated by considering the equation:

$$\frac{d^3y}{dt^3} + a_1 \frac{d^2y}{dt^2} + a_2 \frac{dy}{dt} + a_3y = b_0 \frac{d^2u}{dt^2} + b_1 \frac{du}{dt} + b_2u \qquad (2.45)$$

which represents a third-order system (the highest derivative of the output is third order) with one output $y(t)$ and one input $u(t)$ – the dependency on t is omitted from the analysis for clarity. As part of a systematic conversion approach, consider an intermediate variable $z(t)$ such that the system is conceptually split into two blocks as shown in Figure 2.27. Note, however, that the approach to be described is never actually followed by hand. Once it has been used to prove the method, the result can be found by *inspection* of the differential equation model (such as that of Equation (2.45)), as described below. Returning to Figure 2.27, the first block produces an output $z(t)$ from the input $u(t)$. This intermediate output $z(t)$ then becomes the input to the second block which produces the required output $y(t)$. Therefore

$$\frac{d^3z}{dt^3} + a_1 \frac{d^2z}{dt^2} + a_2 \frac{dz}{dt} + a_3z = u \qquad (2.46)$$

and

$$y = b_0 \frac{d^2z}{dt^2} + b_1 \frac{dz}{dt} + b_2z \qquad (2.47)$$

Equations (2.46) and (2.47) can be derived by various means. Use of the Laplace transform, studied next in this chapter, is perhaps the easiest, so the reader might like to return to the derivation of these two equations after studying it (see Problem 2.2). For the moment, the important thing to notice is the form of the resulting equations, and their relationship to the original model (Equation (2.45)). Since the system is third order, three state variables are required, and these may be selected according to the following systematic pattern (for example):

$$x_1 = z$$

$$\dot{x}_1 = x_2 = \dot{z} \qquad (2.48)$$

$$\dot{x}_2 = x_3 = \ddot{z} \qquad (2.49)$$

Differentiating x_3 and using Equation (2.46) yields:

$$\dot{x}_3 = \dddot{z} = -a_3x_1 - a_2x_2 - a_1x_3 + u \qquad (2.50)$$

Writing Equations (2.48) to (2.50) as the three rows of the state equation gives:

$$\begin{bmatrix} \dot{x}_1 \\ \dot{x}_2 \\ \dot{x}_3 \end{bmatrix} = \begin{bmatrix} 0 & 1 & 0 \\ 0 & 0 & 1 \\ -a_3 & -a_2 & -a_1 \end{bmatrix} \begin{bmatrix} x_1 \\ x_2 \\ x_3 \end{bmatrix} + \begin{bmatrix} 0 \\ 0 \\ 1 \end{bmatrix} u \qquad (2.51)$$

The output equation is obtained from Equations (2.47) to (2.49) and $x_1 = z$, and is:

$$y = [b_2 \ b_1 \ b_0] \begin{bmatrix} x_1 \\ x_2 \\ x_3 \end{bmatrix} + [0]u \qquad (2.52)$$

This particular state-space realization has various names. The A matrix is in a *canonical* form, sometimes known as the *companion form* or the *controllable canonical form* for reasons which will emerge in Chapter 5. In this form, the state variables are the intermediate variable z and its derivatives.

Note that there is no unique state-space model of any given system. The canonical form of Equation (2.51) has some mathematical advantages. However, there is actually an infinite number of other possible state-space models *of the same system*. These are obtained by choosing different sets of states, of which some are even more mathematically convenient (for example, the A matrix can be made diagonal by suitable choice of the states, thus making certain types of analysis relatively easy).

The main advantage of Equations (2.51) and (2.52) over other forms is that by comparing the ordinary linear differential equation, Equation (2.45), with the state Equation (2.51) and the output Equation (2.52), it can be seen that the differential equation can be converted into this canonical form by simple substitution of the differential equation coefficients into the matrices A and C. The conversion can be performed by inspection, with none of the intervening mathematics being necessary, provided that in Equation (2.45) the order of the differential expression containing the output is greater than the order of the differential expression containing the input (this should always be the case for real-world systems, for reasons which are explained in Section 4.5.2). If, in an approximate model, the orders of the two differential expressions are the same, then a further substitution is required as shown in Example 2.15. It will never be the case, for a believable model of a real system, that an equation of the form of Equation (2.45) has an input polynomial of *higher* order than the output polynomial.

The rules for obtaining Equations (2.51) and (2.52) from Equation (2.45) by inspection, for a SISO system, with a higher order output than input are as follows:

- Ensure that the differential equation of the system is of the form of Equation (2.45), *and that the coefficient of the highest order term in the output (on the LHS) is unity* (divide the entire equation by it, if this is not initially the case).

- The required sizes of A, b and c (d will always be a scalar zero under these circumstances) are found from the list following Equation (2.44). b will always be a single column vector, because there is a single input, c will always be a single row vector for a single output, A is always square. The unspecified dimension of all these is always equal to the order of the system, n.

- A always takes the form of a square matrix of zeros, except for a diagonal of unity elements above the leading diagonal (the *leading diagonal* runs from the top left-hand corner to the bottom right), and a non-zero last row. The last row comprises the coefficients of the LHS (output terms) of Equation (2.45) (excluding the first coefficient, which is why it must be unity), written down in reverse order, and with their signs changed. If any term is absent from Equation

(2.45), a zero coefficient must be inserted. For first-order systems, *a* becomes scalar, and is regarded as a single-element 'last row'.

- *b* is always a column vector of zeros, except for the last element, which is always unity.
- *c* is a row vector comprising the coefficients of the RHS (input terms) of Equation (2.45), written down in reverse order, but with the signs preserved. Again, if any term is absent from Equation (2.45), a zero coefficient must be inserted.

The only major disadvantage of this form of state-space model is that the states are not physically meaningful. There is no reason why the variable $z(t)$ and its derivatives should either correspond to any meaningful signal in the real world, or be physically measurable. Nevertheless, if no mistakes have been made, the resulting state-space model should replicate the input–output behaviour of the modelled system. From that viewpoint, the states are not particularly important, as they are purely a matter of how we choose to represent the *internal* behaviour of the system (but they become important when measurements of them are needed for control purposes, which will be the case in Chapter 5).

Example 2.15 *A state-space model of an electrical lead compensator*

Find a set of state-space equations for the lead compensator of Example 2.4 and Figure 2.15 which is described by the ordinary differential equation (from Equation (2.16)):

$$\frac{dv_o(t)}{dt} + \frac{R_1 + R_2}{CR_1R_2} v_o(t) = \frac{dv_i(t)}{dt} + \frac{1}{CR_1} v_i(t) \tag{2.53}$$

The controllable canonical state-space form is obtained analytically (if desired) by introducing an intermediate variable $z(t)$ thus:

$$\frac{dz(t)}{dt} + \frac{R_1 + R_2}{CR_1R_2} z(t) = v_i(t) \tag{2.54}$$

$$v_o(t) = \frac{dz(t)}{dt} + \frac{1}{CR_1} z(t) \tag{2.55}$$

By letting $x_1(t) = z(t)$ and $u(t) = v_i(t)$ then from Equation (2.54), the state equation is given by

$$\dot{x}_1 = \left[-\frac{R_1 + R_2}{CR_1R_2}\right] x_1 + [1]u \tag{2.56}$$

Substituting $y(t)$ for $v_o(t)$ and $x_1(t)$ for $z(t)$ in Equation (2.55), and replacing dz/dt by the expression for $\dot{x}_1(t)$ from Equation (2.56), yields the output equation:

$$y = \left[-\frac{1}{CR_2}\right] x_1 + [1]u \tag{2.57}$$

Note that in this particular realization the system state $(x_1(t) = z(t))$ has no direct physical value, and cannot be identified in Figure 2.15. In fact, analysis of Equation (2.57), together with Figure 2.15, shows that $x_1(t) = CR_2v_c(t)$. Also notice that the 'by inspection' rules, given above, need modifying for this system, since the output and input portions of Equation (2.53) are of the same order.

When there is a term in the input of the original differential equation model which is of the same order as the highest order term in the output, proceed as follows. Let b_{-1} be the coefficient of the new highest-order term in the input of the differential equation model (for example, in Equation (2.53), $b_{-1} = 1$). The rules for A and b are unchanged. c is obtained as before, but then b_{-1} times the last row of A is added to it. d will be a scalar, equal in value to b_{-1}. The reader should try to obtain Equations (2.56) and (2.57) using this method.

Example 2.16 *An alternative state-space model for Example 2.15*

Demonstrate that other state-space realizations are possible, by letting the state variable in the lead compensator of Example 2.4 be the voltage across the capacitor (see Figure 2.15). In many ways, this is a more important example than the last one, because it begins to show how the engineer can often choose system states to be physically meaningful and measurable quantities. In practical control system design, as opposed to theoretical analysis, it will usually be desirable to have access to the state variables for use in control schemes (see Chapter 5), so this will almost always be the approach used in selecting the states.

Hence, let:

$$x_1 = v_c(t)$$

Now using this in Equations (2.12) and (2.15),

$$C\dot{x}_1 = i_{R_2}(t) - i_{R_1}(t)$$

The currents can be eliminated by using the element equations (2.14), (2.11) and (2.13) thus:

$$C\dot{x}_1 = \frac{1}{R_2} v_o(t) - \frac{1}{R_1} x_1 \qquad (2.58)$$

but the compatibility equation states that

$$v_o(t) = v_i(t) - v_c(t) = v_i(t) - x_1 \qquad (2.59)$$

Setting $y(t) = v_o(t)$, $u = v_i(t)$ and then substituting Equation (2.59) into Equation (2.58) gives the state equation

$$\dot{x}_1 = \left[-\frac{R_1 + R_2}{CR_1R_2} \right] x_1 + \left[\frac{1}{CR_2} \right] u \qquad (2.60)$$

The output equation is obtained directly from Equation (2.59).

$$y = -[1]x_1 + [1]u \qquad (2.61)$$

Equations (2.60) and (2.61) are the required state and output equations. They are clearly different from Equations (2.56) and (2.57), but note that the 'A matrices' (cf. Equation (2.43) – but they are scalars in this case) are identical. This is to be expected in a first-order system because it is this quantity which fixes the system dynamics, which ought to be unchanged between the two representations. (In a higher order system, it turns out to be the *eigenvalues* of A which remain constant between different state-space models of the same system; but that is a matter for later.) In this new realization, the state variable is the voltage across the capacitor and therefore should be measurable. This may well be a useful attribute for use in a control system.

It may occur to the reader that, since it was shown in Example 2.15 that $x_1(t) = CR_2v_c(t)$, the present example could equally well have been solved by making that substitution into Equations (2.56) and (2.57), and then relabelling $v_c(t)$ as the new $x_1(t)$. While this is true for this simple example, in general such a line of attack would not be worth pursuing.

Various aspects of state-space models are pursued further in Section 2.7 and in the problems. Problems 2.9 and 2.11 address the combination of state-space models in series and parallel.

2.5.2 Laplace transforms

Whereas the state-space model is the most widely used *time-domain* model for linear systems, the Laplace transform provides the mathematical foundation for most of the *frequency-domain* control techniques, that is, most control analysis and design techniques developed in the West prior to the 1950s, plus more modern frequency-domain methods such as the inverse Nyquist array and characteristic locus. (In the former Eastern bloc countries, time-domain methods had generally been preferred, due to different historical areas of work. In fact, it was largely the success of the Soviet Sputnik space programme in 1957 that caused the renewal of interest in time-domain analysis in the West, and thus led to the development of the state-space methods.)

The Laplace transform method is a substitutional one, in which the linear differential equation model of the system is transformed into the complex frequency, or Laplace domain. This greatly simplifies the mathematics, since the operation of integration associated with the time-domain solution is replaced by algebraic manipulation of the transformed equations. After this manipulation, the required time-domain solution is obtained by making the inverse Laplace transformation. This process is shown schematically in Figure 2.28.

When using Laplace transforms, the requirement is a good set of Laplace transform tables. It is useful to know how these tables are formulated, but such knowledge is not essential. Each entry in the table could be derived afresh by the engineer, but the important thing is the application of the tables to specific problems.

This section, together with Appendix 2, defines the Laplace transform and examines its operations in order to demonstrate how tables of Laplace transforms, such as Tables 2.8 and 2.9, are produced. Also, and possibly more importantly, the derivations should aid in the understanding of how these tables are to be applied.

Time domain

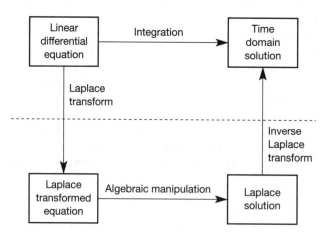

Figure 2.28 Laplace transformation.

Laplace domain or
complex frequency domain

Definition of the Laplace transform

The basic Laplace transform of a time signal $f(t)$ is defined as:

$$F(s) = \int_0^\infty f(t)\, e^{-st}\, dt \tag{2.62}$$

and written symbolically as

$$F(s) = \mathscr{L}[f(t)]$$

Table 2.8 Laplace transform operations.

Operation	$f(t)$	$F(s)$
1. Transform integral	$f(t)$	$\int_0^\infty f(t)e^{-st}\, dt$ or $\mathscr{L}\, f(t)$
2. Linearity	$f_1(t) \pm f_2(t)$	$F_1(s) \pm F_2(s)$
3. Constant multiplication	$af(t)$	$aF(s)$
4. Complex shift theorem	$e^{\pm at} f(t)$	$F(s \mp a)$
5. Real shift theorem	$f(t - T)$	$e^{-Ts} F(s)\ (T \geq 0)$
6. Scaling theorem	$f(t/a)$	$aF(as)$
7. First derivative	$\dfrac{d}{dt} f(t)$	$sF(s) - f(0)$
8. nth derivative	$\dfrac{d^n}{dt^n} f(t) \equiv f^n(t)$	$s^n F(s) - \displaystyle\sum_{r=1}^{n} \dfrac{d^{r-1}}{dt^{r-1}} f(0) s^{n-r}$
9. First integral	$\displaystyle\int_0^t f(t)\, dt$	$\dfrac{1}{s} F(s)$
10. Convolution integral	$\displaystyle\int_0^t f_1(\tau) f_2(t - \tau)\, d\tau$	$F_1(s) F_2(s)$

Table 2.9 Laplace transforms of common functions.

Time function $f(t)$	Laplace transform $\mathscr{L}[f(t)] = F(s)$
1. $\delta(t)$: unit impulse	1
2. $u(t)$: unit step	$1/s$
3. t	$1/s^2$
4. t^n	$n!/s^{n+1}$
5. e^{at}	$1/(s-a)$
6. $\cos \omega t$	$s/(s^2 + \omega^2)$
7. $\sin \omega t$	$\omega/(s^2 + \omega^2)$
8. $e^{-at} \cos \omega t$	$\dfrac{s+a}{(s+a)^2 + \omega^2}$
9. $e^{-at} \sin \omega t$	$\dfrac{\omega}{(s+a)^2 + \omega^2}$
10. $\dfrac{e^{-at} - e^{-bt}}{b-a}$	$\dfrac{1}{(s+a)(s+b)}$
11. $\dfrac{\omega}{\sqrt{(1-\zeta^2)}} e^{-\zeta\omega t} \sin[\omega\sqrt{(1-\zeta^2)}t]$	$\dfrac{\omega^2}{s^2 + 2\zeta\omega s + \omega^2}$
12. $\dfrac{1}{T^n(n-1)!} t^{n-1} e^{-t/T}$	$\dfrac{1}{(1+sT)^n}$
13. $1 - \cos \omega t$	$\dfrac{\omega^2}{s(s^2 + \omega^2)}$
14. $1 - e^{-t/T}$	$\dfrac{1}{s(1 + Ts)}$
15. $1 - \dfrac{t+T}{T} e^{-t/T}$	$\dfrac{1}{s(1 + Ts)^2}$
16. $\dfrac{\omega^2}{\sqrt{(1-\zeta^2)}} e^{-\zeta\omega t} \sin[\omega\sqrt{(1-\zeta^2)}t + \phi]$ where $\phi = \tan^{-1} \dfrac{\sqrt{(1-\zeta^2)}}{-\zeta}$	$\dfrac{s\omega^2}{s^2 + 2\zeta\omega s + \omega^2}$

It is common practice to use a capital letter F which is a function of the new variable s (see below) for the transform of the time signal $f(t)$. Also, it is assumed that $f(t)$ is zero for all times before $t = 0$.

In Equation (2.62) the exponent st must be dimensionless, otherwise the expression e^{-st} is meaningless. Thus the variable s has dimensions of 1/time, which is the dimension of frequency. Since s is also a complex quantity, it is often referred to as the *complex frequency*.

Two useful properties follow directly from the definition of Equation (2.62). These are the properties of linearity and constant multiplication, respectively:

$$\mathscr{L}[f_1(t) + f_2(t)] = \mathscr{L}[f_1(t)] + \mathscr{L}[f_2(t)]$$

and

$$\mathscr{L}[af(t)] = a\mathscr{L}[f(t)]$$

Such properties are referred to as Laplace transform operations. A list of such operations is given in Table 2.8.

Laplace transforms of some common functions

The Laplace transforms of a number of common functions are given in Table 2.9. For example, Equation (2.62) may be used to find the Laplace transform of

$$f(t) = e^{at}$$

as

$$\mathcal{L}[e^{at}] = \int_0^\infty e^{at} e^{-st} \, dt$$

or

$$F(s) = \int_0^\infty e^{-(s-a)t} \, dt$$

which becomes

$$F(s) = -\frac{1}{s-a} [e^{-(s-a)t}]_0^\infty$$

Since s is complex, the term within the square brackets will approach zero for increasing t provided the real part of $s - a$ is positive, leaving

$$F(s) = \frac{1}{s-a} \tag{2.63}$$

which, from entry 5 of Table 2.9, is the expected solution.

Equation (2.63) may be used to find other standard transforms. If the constant a equals zero, then

$$\mathcal{L}[e^{0t}] = \mathcal{L}[1] = \frac{1}{s} \tag{2.64}$$

Since it is assumed that $f(t)$ is zero for all time before $t = 0$, Equation (2.64) defines a unit step (see entry 2 of Table 2.9).

Setting $a = j\omega$ in Equation (2.63) gives

$$\mathcal{L}[e^{j\omega t}] = \frac{1}{s - j\omega}$$

Since

$$e^{j\omega t} = \cos \omega t + j \sin \omega t \tag{2.65}$$

then

$$\mathcal{L}[\cos \omega t + j \sin \omega t) = \frac{1}{s - j\omega} \tag{2.66}$$

Applying the property of linearity to the left-hand side of Equation (2.66) and rationalizing the right-hand side produces

$$\mathscr{L}[\cos \omega t] + j\mathscr{L}[\sin \omega t] = \frac{s}{s^2 + \omega^2} + j\,\frac{\omega}{s^2 + \omega^2} \tag{2.67}$$

Equating the real and imaginary parts of Equation (2.67) gives entries 6 and 7 of Table 2.9.

Further useful results are obtained if a is set equal to unity in Equation (2.63). Then

$$\mathscr{L}[e^t] = \frac{1}{s - 1} \tag{2.68}$$

Now, both sides of Equation (2.68) have power series expansions, namely

$$e^t = 1 + t + \frac{1}{2!}\,t^2 + \frac{1}{3!}\,t^3 + \cdots$$

and

$$\frac{1}{s - 1} = \frac{1}{s} + \frac{1}{s^2} + \frac{1}{s^3} + \cdots$$

Using the property of linearity and equating corresponding terms in the two series produces entries 3 and 4 in Table 2.9.

To generate further entries for Table 2.9, the best procedure is to build on previous results in preference to applying the defining integral. To this end, further Laplace transform operations are required.

Laplace transform operations

Apart from linearity and constant multiplication, there are a number of other useful Laplace transform operations (see Table 2.8) which can be used to establish other entries in Table 2.9. Each of the operations listed in Table 2.8 follows from the Laplace transform definition. Rather than proving each entry, this section concentrates on a few of their applications.

By the complex shift theorem, for example, the Laplace transform of $e^{-3t}\sin 4t$ becomes

$$\mathscr{L}[e^{-3t}\sin 4t] = \frac{4}{(s + 3)^2 + 16}$$

which agrees with entry 9 of Table 2.9.

The real shift theorem is useful for systems containing a *time delay*. Time delays, or *dead times*, are frequently encountered in chemical engineering systems in which a process stream is flowing through a pipe in essentially plug flow. If an individual element of fluid takes time T to flow from the entrance to the exit of a pipe, the pipe acts as a time delay. More is said on time delays in subsequent chapters.

For control purposes, entries 7 and 8 of Table 2.8 are probably the most frequently used operations. They are best illustrated by means of an example.

Assume that the mathematical model of an unforced (that is, there is no driving input) lumped-parameter system is given by

$$\frac{d^2x}{dt^2} + 5\frac{dx}{dt} + 6x = 0 \tag{2.69}$$

In general, the response of a system consists of two parts: the *forced* response due to the applied input (there is none in this example), and the *free* response due to the initial conditions. For the sake of this illustration let the initial conditions be:

$$x(0) = 0$$

and

$$\frac{dx}{dt}(0) = 7$$

Using linearity to find the Laplace transform of each term in Equation (2.69) together with the constant multiplication and derivative operations, yields

$$\left(s^2X(s) - sx(0) - \frac{dx}{dt}(0)\right) + 5(sX(s) - x(0)) + 6X(s) = 0$$

which, with some algebraic manipulation, becomes

$$X(s) = \frac{s+5}{s^2+5s+6}x(0) + \frac{1}{s^2+5s+6}\frac{dx}{dt}(0)$$

Inserting the initial conditions gives

$$X(s) = \frac{7}{(s+2)(s+3)} \tag{2.70}$$

The corresponding time response can, in this case, be obtained directly from Table 2.9 (entry 10) as

$$x(t) = 7\frac{e^{-2t} - e^{-3t}}{3 - 2}$$

or

$$x(t) = 7e^{-2t} - 7e^{-3t}$$

Clearly the Laplace transform technique is very powerful, as this solution was found much more easily than by directly solving Equation (2.69). The ability to convert a constant-parameter, ordinary linear differential equation in x into an algebraic equation in s simplifies the solution procedure considerably. Broadly, derivatives are replaced with multiplications by s, while integrals are replaced with divisions by s.

For control system design using frequency-domain techniques, however, most of the design procedure is actually performed *in the complex frequency domain* (or, more loosely, the *Laplace domain*, or simply, as above, *the frequency domain*) using the Laplace transform model itself. Only when a design is deemed acceptable in the Laplace domain should the solution of the equations be sought in the time domain. However, rather than follow a chronological design sequence, subsequent sections consider the various techniques that are used to find time-domain solutions.

Poles and zeros

The *poles* of a Laplace function, such as that in Equation (2.70), are the particular values of s that make the function evaluate to infinity. They are therefore equal to the roots of the *denominator* polynomial of the function. For example, the function of Equation (2.70) has two poles, one at $s = -2$ and one at $s = -3$.

The *zeros* of a Laplace function, such as that in Equation (2.70), are the particular values of s that make the value of the function go to zero. They are therefore equal to the roots of the *numerator* polynomial of the function. Equation (2.70) apparently has no zeros of this type, because there are no terms in s in the numerator. However, if the order of the numerator is lower than that of the denominator, then such a function always has one or more zeros at $s = \infty$. This is because if $s = \infty$, the *denominator* becomes infinite and so the entire function tends to zero. Equation (2.70) could therefore be said to have two zeros at $s = \infty$, one corresponding to each denominator s. In general, the number of such zeros at infinity is equal to the number of poles minus the number of numerator zeros. Further, in such a Laplace function, the total number of zeros (including those at infinity) will always equal the number of poles.

2.5.3 Inverse Laplace transforms

The formal definition of the inverse Laplace transform, which converts $F(s)$ to the corresponding time function $f(t)$, is

$$f(t) = \frac{1}{2\pi j} \int_{\sigma-j\omega}^{\sigma+j\omega} F(s)e^{st} \, ds$$

and is written symbolically as

$$f(t) = \mathscr{L}^{-1}[F(s)]$$

This integral is difficult to evaluate directly, and the normal procedure is to manipulate $F(s)$ into one of the standard forms for which the inverse Laplace transform is known. Table 2.9, for example, provides a number of Laplace transforms for common functions. This table can be used in either direction: just as, given $f(t)$, it is possible to find $F(s)$ so, given $F(s)$, it is possible to find $f(t)$. The amount and type of manipulation applied to $F(s)$ depends on the proposed method of solution – graphical, hand calculation or by computer. Heaviside's partial-fraction expansion method (see Appendix 2) provides the basis for the development of all three approaches.

Example 2.17 *Time solution of a Laplace function*

Use Tables 2.8 and 2.9 to find the time functions corresponding to the transforms

$$\frac{10}{s^2 + 9} \quad \text{and} \quad \frac{2e^{-2s}}{s + 3}$$

With reference to Table 2.9, the first function can be manipulated into the transform of $\sin 3t$ with the multiplying factor $10/3$. The corresponding time function is therefore:

$$\mathscr{L}^{-1}\left[\frac{10}{3}\frac{3}{s^2+3^2}\right] = \frac{10}{3}\sin 3t$$

The second transform can be seen to be that for $2e^{-3t}$ multiplied by e^{-2s}, and the inverse Laplace transform then follows from the real shift theorem as:

$$f(t) = 2e^{-3(t+2)}$$

Example 2.18 *Another inverse Laplace transform*

Find the inverse Laplace transform of

$$F(s) = \frac{s+5}{s^2+2s+5}$$

This example demonstrates that manipulation is often required to change the transform into a standard form. As it stands, it is not to be found in the tables (but could perhaps be found in more comprehensive tables). In this transform the denominator can be rewritten to give

$$F(s) = \frac{s+5}{(s+1)^2+4}$$

or

$$F(s) = \frac{s+1}{(s+1)^2+4} + \frac{4}{(s+1)^2+4}$$

The complex shift theorem and the standard forms for sine and cosine functions in Table 2.9 then allow the inverse transform to be found by inspection:

$$\mathscr{L}^{-1}[F(s)] = e^{-t}\cos 2t + 2e^{-t}\sin 2t$$

Partial fractions

The Laplace transform solution of an ordinary linear differential equation often takes the form of a rational polynomial:

$$F(s) = \frac{N(s)}{D(s)} \tag{2.71}$$

where $N(s)$ and $D(s)$ are polynomials. Hence Equation (2.71) could be rewritten as

$$F(s) = \frac{b_m s^m + b_{m-1}s^{m-1} + \cdots + b_0}{s^n + a_{n-1}s^{n-1} + \cdots + a_0}$$

For many purposes, this rational polynomial must be *strictly proper* – that is to say, the degree m of the numerator must be less than the degree n of the denominator. If m is equal to n, then the system is said to be *proper*. If m is greater than or equal to

n, the numerator can always be made of lower degree by dividing the denominator into the numerator. Such a division process will produce a proper rational polynomial, plus a remainder polynomial. However, for engineering systems, the Laplace transform solution is almost always a strictly proper rational polynomial, and the problem is one of finding the inverse transform. To do so, a partial fraction decomposition of $F(s)$ into a sum of simpler terms is required. This can be illustrated by considering the terms

$$F(s) = -\frac{1}{s+1} - \frac{3}{s-2} + \frac{4}{s-3} \tag{2.72}$$

The inverse transform of Equation (2.72) may be obtained directly from Tables 2.8 and 2.9. However, if Equation (2.72) is written with a common denominator – that is, in rational polynomial form – then

$$F(s) = \frac{7s - 5}{s^3 - 4s^2 + s + 6}$$

and the inverse transform is not immediately obvious.

The normal procedure for dealing with a rational polynomial is first to factorize the denominator polynomial $D(s)$ and then to split it into partial fractions. The particular type of partial-fraction expansion to be employed is determined by the types of root involved and is discussed (with worked examples) in Appendix 2.

2.5.4 The convolution integral

The convolution integral (entry 10 of Table 2.8) is defined by

$$\mathcal{L}^{-1}[F_1(s)F_2(s)] = \int_0^t f_1(\tau)\, f_2(t - \tau)\, d\tau \tag{2.73}$$

where $F_1(s)$ and $F_2(s)$ are the transforms of $f_1(t)$ and $f_2(t)$ respectively. Proof of this equation may be obtained from the definition of the Laplace transform (Equation (2.62)), and will not be given here. It is, however, instructive to examine a graphical interpretation of the convolution process.

Suppose, for the sake of illustration, that $f_1(\tau)$ and $f_2(\tau)$ are the time signals shown in Figure 2.29. The term $f_2(t - \tau)$ in Equation (2.73) is obtained by delaying $f_2(\tau)$ by a time t to give $f_2(\tau - t)$, followed by reflection about a vertical axis at $\tau = t$, as shown in Figure 2.30. The convolution integral involves the product of $f_1(\tau)$ and $f_2(t - \tau)$. Figure 2.31 shows this product for various values of t. In this figure, the shaded areas represent the values of the convolution integral for the two signals at the specified times. Combining all the values for each time instant allows the convolution integral to be generated, as shown in Figure 2.32.

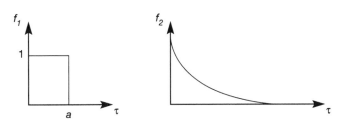

Figure 2.29 Time signals $f_1(\tau)$ and $f_2(\tau)$.

Figure 2.30 The signal $f_2(t - \tau)$.

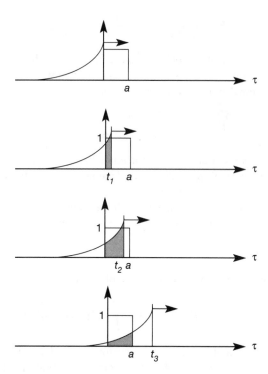

Figure 2.31 The product $f_1(\tau) f_2(t - \tau)$ for various values of t.

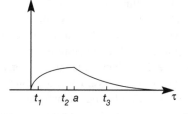

Figure 2.32 The convolution integral $\int f_1(\tau) f_2(t - \tau) \, d\tau$.

Convolution and the related technique of deconvolution are particularly useful in signal processing and in some forms of identification (Chapter 8). However, it has been introduced here for two reasons: to illustrate that the inverse transform of a Laplace transform product is not the product of the individual time functions, and to demonstrate the use of the integral in finding inverse transforms.

Example 2.19 *An inverse Laplace transform via the convolution integral*

Find the inverse Laplace transform of

$$F(s) = \frac{2}{(s+3)(s^2+4)}$$

using the convolution integral.

Let $F(s) = F_1(s)F_2(s)$, where

$$F_1(s) = \frac{1}{s+3} \quad \text{with inverse} \quad f_1(t) = e^{-3t}$$

and

$$F_2(s) = \frac{2}{s^2+4} \quad \text{with inverse} \quad f_2(t) = \sin 2t$$

The convolution integral indicates that

$$\mathscr{L}^{-1}[F_1(s)F_2(s)] = \int_0^t \sin 2\tau \, e^{-3(t-\tau)} \, d\tau$$

$$= e^{-3t} \int_0^t \sin 2\tau \, e^{3\tau} \, d\tau$$

Integrating by parts twice gives

$$f(t) = \frac{1}{13} \left(2e^{-3t} + 3 \sin 2t - 2 \cos 2t \right)$$

This answer may be checked by expanding $F(s)$ by partial fractions (see Appendix 2).

2.5.5 Initial and final value theorems

These two further properties of Laplace transforms are useful for checking the solution of a differential equation, and also for extracting information about, for example, the steady-state response of a system without having to perform the inverse Laplace transform operation.

Initial value theorem

Suppose $f(t)$ and $F(s)$ are a Laplace transform pair. Then, provided the limit of $sF(s)$ as $s \to \infty$ exists, the initial value of the time function is given by

$$f(0) = \lim_{t \to 0} [f(t)] = \lim_{s \to \infty} [sF(s)] \tag{2.74}$$

Final value theorem

Suppose $f(t)$ and $F(s)$ are a Laplace transform pair and that $F(s)$ is stable. (Stability is dealt with in Chapter 3, but essentially the restriction implies that $f(t)$ is bounded – that is, it does not become infinite as t tends to infinity.) Under these conditions

$$f(\infty) = \lim_{t \to \infty} [f(t)] = \lim_{s \to 0} [sF(s)] \tag{2.75}$$

Example 2.20 *Initial and final value theorems*

Find the initial and final values of the time function whose Laplace transform is

$$Y(s) = \frac{2s^2 + 7s + 4}{s(s+2)(s+1)}$$

From the initial value theorem, Equation (2.74),

$$y(0) = \lim_{s \to \infty} [sY(s)]$$

Hence

$$y(0) = \lim_{s \to \infty} \left[\frac{2s^2 + 7s + 4}{(s+2)(s+1)} \right]$$

which, on dividing the numerator and denominator by s^2 and letting $s \to \infty$, gives

$$y(0) = 2$$

The final value theorem applies, since the poles of $Y(s)$ are non-positive (one is zero), and will therefore not give rise to unbounded terms in $y(t)$. Hence, from Equation (2.75):

$$y(\infty) = \lim_{s \to 0} [sY(s)] = 2$$

If a partial-fraction expansion (Appendix 2) is performed on $Y(s)$, the resulting inverse transform is found to be

$$y(t) = 2 - e^{-2t} + e^{-t}$$

Letting $t = 0$, and then $t = \infty$, produces results that agree with those already obtained.

2.5.6 Transfer function models

A transfer function model is usually obtained from the linear differential equations representing the lumped-parameter model. For a system having a single input $u(t)$, and a single output $y(t)$, the transfer function model is defined as the Laplace transformed output $Y(s)$ divided by the input $U(s)$, such that

$$\frac{Y(s)}{U(s)} = F(s)$$

or, alternatively, $Y(s) = F(s)U(s)$. In the general block diagram element of Figure 2.1, the block would contain the transfer function $F(s)$, the output would be $Y(s)$ and the input $U(s)$. The block diagram now contains all the information given in the transfer function model, that is, '*output = contents × input*'. This is why $F(s)$ is called a *transfer function* – it shows how the input of the block (and, hopefully, the real system element represented by the block) is transferred to the output. Section 2.6 takes this idea further.

For real physical systems the function $F(s)$ is a rational polynomial such that

$$F(s) = \frac{N(s)}{D(s)} \qquad (2.76)$$

with the order of the denominator polynomial $D(s)$ being greater than or equal to the order of the numerator polynomial $N(s)$. If the denominator polynomial is extracted and set equal to zero, that is

$$D(s) = 0$$

the resulting equation is called the system's *characteristic equation*, since it can be shown to characterize the system's dynamics (see Chapter 3). Its roots (or zeros) are the poles of $F(s)$ (the values of s that make the rational polynomial infinite). The roots of the numerator polynomial $N(s)$ are the zeros of $F(s)$ (the values of s that make the rational polynomial zero).

Both the poles and zeros of $F(s)$ can be complex values of s and, as such, will have real and imaginary parts. When only physical engineering systems are being considered, any complex pole or zero must have a complex conjugate. In general any root will be of the form

$$s = \sigma + j\omega$$

and may therefore be plotted on an Argand diagram. Such a plot, with σ on its real axis and ω on its imaginary axis, is referred to as an *s-plane plot*. It is common practice to identify poles of $F(s)$ by a superimposed cross and zeros by an enclosing circle. All the pole–zero plots in this book follow this convention.

For many design applications, it is desirable to present the transfer function, Equation (2.76), in factored form. The standard (or 'root locus') form (see Chapter 4) is

$$F(s) = \frac{K(s + z_1)(s + z_2) \cdots (s + z_m)}{s^r(s + p_1)(s + p_2) \cdots (s + p_{n-r})}$$

or

$$F(s) = \frac{K \displaystyle\prod_{i=1}^{m} (s + z_i)}{s^r \displaystyle\prod_{i=1}^{n-r} (s + p_i)} \tag{2.77a}$$

where $m \leq n$, and $-z_i$ and $-p_i$ are the non-zero finite zeros and poles respectively of $F(s)$. This transfer function represents a type r system. The *type number* of a system is defined as the number of poles at the origin of the s-plane ($s = 0$). Since a division by s is equivalent to the time-domain operation of integration, it is also equal to the number of pure integrators (that is, open-loop integrators) in the forward path of the system.

The system is also said to be of *rank R*, where $R = m - n$, the difference between the total number of poles and zeros. This rank number is equal to the number of zeros at infinity (see Section 2.5.2). However, it must not be confused with the *rank of a matrix*, used later in the state-space sections.

The importance of the type number is that it indicates a stable closed-loop system's steady-state response to various forcing inputs (see Chapter 3). Rank is important in producing Bode and Nyquist plots (see Chapter 3) and root locus diagrams (Chapter 4).

The Bode form of the transfer function $F(s)$ is obtained by factoring out all the z_i and p_i in Equation (2.77a) to give

$$F(s) = \frac{K_B \prod_{i=1}^{m} \left(1 + \frac{s}{z_i}\right)}{s^r \prod_{i=1}^{n-r} \left(1 + \frac{s}{p_i}\right)} \tag{2.77b}$$

In this form the coefficient K_B is called the Bode gain. Bode plots are covered in Chapter 3.

Finally, when a system has several inputs and/or outputs the *transfer function matrix* form can be used. For example, for a system having p outputs and m inputs:

$$\begin{bmatrix} Y_1(s) \\ Y_2(s) \\ \vdots \\ Y_p(s) \end{bmatrix} = \begin{bmatrix} F_{11}(s) & F_{12}(s) & \cdots & F_{1m}(s) \\ F_{21}(s) & \ddots & & \vdots \\ \vdots & & F_{ij}(s) & \vdots \\ F_{p1}(s) & \cdots & \cdots & F_{pm}(s) \end{bmatrix} \begin{bmatrix} U_1(s) \\ U_2(s) \\ \vdots \\ U_m(s) \end{bmatrix} \tag{2.78}$$

where the element $F_{ij}(s)$ is a rational polynomial giving the relationship between output $Y_i(s)$ and input $U_j(s)$, such that:

$$\frac{Y_i(s)}{U_j(s)} = F_{ij}(s) \tag{2.79}$$

Example 2.21 *A transfer function model of the lead compensator of Example 2.4*

Find the transfer function model for the lead compensator described in Example 2.4. Taking the Laplace transforms of Equation (2.16) and assuming zero initial conditions gives

$$sR_1R_2CV_o(s) + (R_1 + R_2)V_o(s) = sR_1R_2CV_i(s) + R_2V_i(s)$$

and after some algebraic manipulation

$$\frac{V_o(s)}{V_i(s)} = \alpha \frac{Ts + 1}{\alpha Ts + 1} \tag{2.80}$$

where

$$\alpha = \frac{R_2}{R_1 + R_2} \quad \text{and} \quad T = R_1C$$

Since α must be less than one, Equation (2.80) defines a lead compensator and has the pole–zero plot shown in Figure 2.33.

Again, Equation (2.80) can be represented by the block diagram of Figure 2.1, with $V_o(s)$ as output, $V_i(s)$ as input, and the RHS of Equation (2.80) as the block contents. This is a great step forward in simplicity, compared with Equation (2.16).

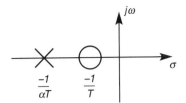

Figure 2.33 Pole–zero plot of a lead compensator.

2.6 Block diagrams

Previous sections have covered modelling and the mathematical foundation of the Laplace transform method. In this section the concepts of a transfer function and its related block diagram are developed.

Block diagrams provide a pictorial representation of a system and its associated control structure and compensators. The application of block reduction techniques, or '*block diagram algebra*', condenses the Laplace transform system equations, together with any controller equations, into a form suitable for either design studies, or inverse transformation to investigate responses in the time domain. Block diagram algebra may also be used to demonstrate many of the advantages of feedback control systems.

2.6.1 Transfer functions and block diagrams

The transfer function concept, introduced in Section 2.5.6, can be extended by considering the electrical circuit shown in Figure 2.34. Assume the current response $i(t)$ due to a change in applied voltage $v(t)$ is required.

Application of Kirchhoff's voltage law around the circuit gives

$$v(t) = Ri(t) + \frac{1}{C} \int_0^t i(t) \, dt + L \frac{di(t)}{dt}$$

Ignoring initial conditions, this equation transforms to

$$V(s) = RI(s) + \frac{I(s)}{sC} + LsI(s) \tag{2.81}$$

and Equation (2.81) may be rearranged to give

$$\left(sL + R + \frac{1}{sC} \right) I(s) = V(s) \tag{2.82}$$

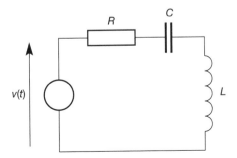

Figure 2.34 An RLC electrical circuit.

In general, any system model ought to predict the behaviour of the system in response to the applied forcing input. However, a system will also exhibit an additional transient response due to any non-zero initial conditions existing at the time the input is applied.

Control system studies are usually concerned more with the forced response than with the transient due to the initial conditions. If a linear system is stable, the influence of the initial conditions on the output becomes negligible as time progresses. It is therefore common practice to assume all the initial conditions to be zero, in which case any system model could be written as the ratio of output over input. Hence Equation (2.82) may be written as

$$\frac{I(s)}{V(s)} = \frac{Cs}{CLs^2 + CRs + 1} \tag{2.83}$$

and represented pictorially as in Figure 2.35.

Figure 2.35 A block diagram of the RLC circuit of Figure 2.34.

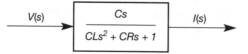

$$V(s) \longrightarrow \boxed{\dfrac{Cs}{CLs^2 + CRs + 1}} \longrightarrow I(s)$$

Note that in a *nonlinear* system, it does not follow that the response as time increases is independent of the initial conditions (see Chapter 14). Also, linear systems with non-zero initial conditions cannot be arranged in the form of Equation (2.83), because the inverse Laplace transform then gives rise to non-vanishing terms in $i(0)$ and $di(0)/dt$ in Equation (2.81), which would prevent this.

Recall from Section 2.5.6 that the transfer function of a constant linear system is defined as

$$F(s) = \frac{\mathscr{L}[\text{output}]}{\mathscr{L}[\text{input}]}$$

with all initial conditions set to zero. The input–output relationship defined by Equation (2.83) and the block diagram in Figure 2.35 provide the same transfer function information. Note that it is only possible to draw the single-block arrangement of Figure 2.35 *because* the initial conditions were assumed to be zero. If this is not the case, then Equation (2.83) will have output contributions both from the input (as shown) *and* from the initial conditions. This gives rise to two separate LTF blocks, whose outputs must be added together. The two inputs would be $V(s)$ and the initial value $I(0)$. In state-space methods, the initial conditions are included automatically (see Sections 3.2.1 and 3.6.2).

Most complex modelling exercises result in a number of subsystems, each of which has a transfer function representation. The use of block diagrams is a convenient method of pictorially grouping these subsystems such that the system is represented in a mathematically meaningful way, and this is best demonstrated by an example (see Example 2.27, too).

Example 2.22 **A block diagram model for Example 2.8**

Figure 2.36 shows a system of coupled tanks (the same system that was analysed in Example 2.8). Derive an equation for each of the three subsystems – the two tanks and the outflow – and put them into block diagram form. Combine the three diagrams to form a block diagram of the overall system.

For tank 1, the governing equation is

$$C_1 R_1 \frac{dH_1(t)}{dt} + H_1(t) = R_1 Q_i(t) + H_2(t)$$

and with zero initial conditions its Laplace transform is

$$(1 + C_1 R_1 s)H_1(s) = R_1 Q_i(s) + H_2(s)$$

or

$$H_1(s) = \frac{R_1 Q_i(s) + H_2(s)}{1 + C_1 R_1 s} \tag{2.84}$$

Equation (2.84) has one output, $H_1(s)$, and two inputs, $Q_i(s)$ and $H_2(s)$. The block diagram representation of this element is shown in Figure 2.37. The circle with a cross in it represents a comparator, which takes the sum of the incoming signals. The two plus signs indicate the sign of the signals. Sometimes they might be drawn beside the connections to the comparator, rather than inside the symbol.

For tank 2, the governing equation is

$$R_1 C_2 R_2 \frac{dH_2(t)}{dt} + R_1 H_2(t) + R_2 H_2(t) = R_2 H_1(t)$$

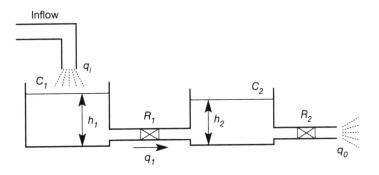

Figure 2.36 A system of coupled tanks.

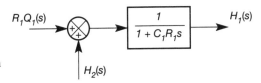

Figure 2.37 Block diagram for tank 1 of Figure 2.36.

Its Laplace transform form, with zero initial conditions, is

$$(R_1 + R_2 + C_2 R_1 R_2 s)H_2(s) = R_2 H_1(s) \qquad (2.85)$$

The block diagram representation is shown in Figure 2.38. The equation describing the outflow from tank 2 is

$$\frac{H_2(s)}{R_2} = Q_o(s) \qquad (2.86)$$

and may be represented by the block diagram in Figure 2.39.

All three block diagrams may now be combined into a single diagram for the system, as shown in Figure 2.40, from which the transfer function for tank 1, for example, is obtained as:

$$\frac{H_1(s)}{E_1(s)} = \frac{1}{1 + C_1 R_1 s}$$

where

$$E_1(s) = R_1 Q_1(s) + H_2(s)$$

This could, of course, have been obtained directly from Equation (2.84). However, if the transfer function between $Q_o(s)$ and $Q_i(s)$ were required, Equations (2.84), (2.85) and (2.86) would have to be manipulated algebraically. The advantage of the block diagram representation of transfer functions is that it makes possible a structured approach to the manipulation of such system equations.

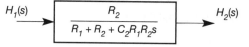

Figure 2.38 Block diagram for tank 2 of Figure 2.36.

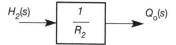

Figure 2.39 Block diagram for the outflow equation of Figure 2.36.

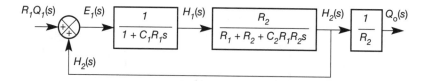

Figure 2.40 Complete block diagram for Figure 2.36.

2.6.2 Block diagrams and block diagram algebra

In the examples given so far, block diagrams have been used to represent the Laplace transform equations for the plant. However, the technique is easily extended to include the control equipment and its associated feedback paths, as shown in Figure 2.41, which contains all the basic elements associated with a single-input-single-output control system. This figure uses a standard notation for

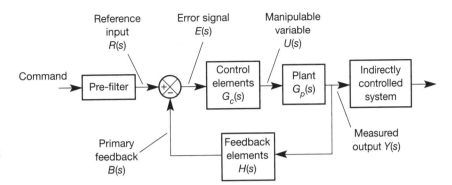

Figure 2.41 Block diagram of a general feedback control scheme.

paths and elements: $R(s)$ denotes the Laplace transform of the actual reference or demand signal applied to the system; $E(s)$ is an error signal, or the output of a comparator (in this case $E(s)$ is the difference between $R(s)$ and the feedback signal, $B(s)$); $U(s)$ and $Y(s)$ are, respectively, the Laplace transformed versions of the plant's manipulable input and its measured output. All forward path elements are denoted by G. The subscripts used to distinguish between the various blocks, for example the control element and the plant, are denoted by $G_c(s)$ and $G_p(s)$ respectively in Figure 2.41. Similarly, all feedback blocks are denoted by H and distinguished by means of subscripts. Typically, all the Gs and Hs are rational polynomials. Note that all of the equations are in the Laplace domain – there is no time-domain signal in the figure.

For control purposes, a representation with only one block between some input and output is often all that is required. So, for example, if the object were to design the compensating elements $G_c(s)$ in Figure 2.41, an open-loop transfer function between $U(s)$ and $B(s)$ would be needed. If, however, a closed-loop time response were required, a single block would be required which would give the relationship

Table 2.10 Block diagram reduction rules.

	Rule	Original diagram	Equivalent diagram	Equation
1	Blocks in series	$u \to \boxed{G_1} \to \boxed{G_2} \to y$	$u \to \boxed{G_1 G_2} \to y$	$y = G_1 G_2 u$
2	Blocks in parallel	$u \to \boxed{G_1}$, $\boxed{G_2} \to \otimes \to y$	$u \to \boxed{G_1 \pm G_2} \to y$	$y = (G_1 \pm G_2)u$
3	Blocks in a feedback loop	$u \to \otimes \to \boxed{G_1} \to y$, $\boxed{G_2}$	$u \to \boxed{\dfrac{G_1}{1 \mp G_1 G_2}} \to y$	$y = G_1(u \pm G_2 y)$

Table 2.11 Block manipulation rules.

	Manipulation	Original diagram	Equivalent diagram	Equation
1	Moving a summing point ahead of an element			$y = Gu_1 - u_2$
2	Moving a summing point beyond an element			$y = G(u_1 - u_2)$
3	Moving a take-off point ahead of an element			$y = Gu$
4	Moving a take-off point beyond an element			$y = Gu$ $u = y/G$
5	Removing an element from a forward path			$y = (G_1 - G_2)u$
6	Inserting an element in a forward path			$y = G_1 u - u$
7	Removing an element from a feedback path			$y = \dfrac{Gu}{1 + GH}$
8	Inserting an element in a feedback path			$y = \dfrac{G_1 u}{1 + G_1}$
9	Rearrangement of summing points			$y = u_1 - u_2 + u_3$
10	Interchange of summing points			$y = u_1 - u_2 + u_3$
11	Moving a take-off point ahead of a summing point			$y = u_1 - u_2$
12	Moving a take-off point beyond a summing point			$y = u_1 - u_2$ $u_1 = y + u_2$

between the command input and the output of the indirectly controlled variable. The technique of condensing a number of blocks into a single block is called block reduction.

Block reduction techniques are based on three simple rules. The first deals with the combination of two blocks on the same path, the second with the combination of two blocks, both of which are on forward paths and the third with the combination of two blocks, one of which is on a feedback path. These rules are summarized in Table 2.10 where, for convenience, the variable s has been omitted and, for clarity, signals are given lower-case letters. Note that, for each rule, the governing equation must hold for both the original and the equivalent diagram.

Block reduction occasionally requires some block manipulation. Table 2.11 shows most of the manipulations that are likely to be encountered in practice. Again, the variable s has been omitted and, contrary to normal convention, the Laplace domain signals are given lower-case letters.

An alternative to block reduction is block diagram algebra. Using this technique, the procedure involves the following steps:

(1) Label the output from each comparator: $E_1(s), E_2(s), \ldots$

(2) Establish a transfer function equation for each system output and comparator output.

(3) Eliminate those variables which are not required.

In step 2, the transfer function equations are obtained by tracing the signal, or signals, back to either a comparator or an output position. This, and the block reduction method are best illustrated by examples.

Example 2.23 *A block diagram reduction exercise*

The block diagram of a multiple-loop feedback control system is shown in Figure 2.42. Use block diagram reduction to simplify this to a single block relating $Y(s)$ to $R(s)$. Note that, for clarity, the dependency upon s has been omitted from the transfer functions within the blocks.

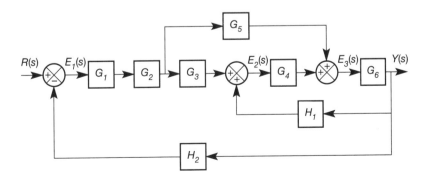

Figure 2.42 A block diagram for reduction.

This system contains a feedforward loop G_5, as well as a positive feedback loop $G_4 G_6 H_1$. The first step is to combine G_1 and G_2 using entry 1 in Table 2.10, and to move the summing junction of the feedforward loop in front of the G_4 block, as shown in Figure 2.43 (the latter manipulation may not be immediately obvious: it is the result of combining entries 1 and 10 from Table 2.11).

Block diagram manipulation rules can be combined in any convenient way. Indeed, any manipulation or combination of manipulations is valid, so long as the overall effects along the various paths between input and output, and feedback paths back to comparators, remain unchanged.

At this stage blocks G_4 and G_6 may be combined and the feedback loop $G_4 G_6 H_1$ reduced using entry 3 in Table 2.10. Also, from entry 2 of Table 2.10 the parallel blocks G_5/G_4 and G_3 may be combined, as shown in Figure 2.44. The forward path blocks may now be combined and the feedback loop H_2 eliminated to produce the required solution, which is shown in Figure 2.45.

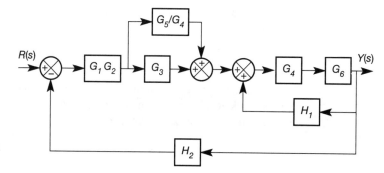

Figure 2.43 First step in reducing the block diagram of Figure 2.42.

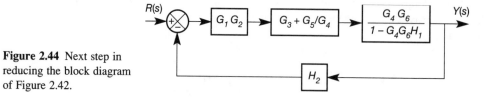

Figure 2.44 Next step in reducing the block diagram of Figure 2.42.

Figure 2.45 Input–output transfer function for the block diagram of Figure 2.42.

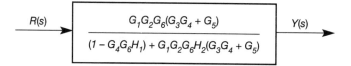

Example 2.24 *Block diagram algebra for Example 2.23*

Use block diagram algebra to solve Example 2.23.

In Figure 2.42, there are three comparators, with outputs $E_1(s)$, $E_2(s)$ and $E_3(s)$, and one system output, $Y(s)$. The four transfer function equations are therefore

$$Y(s) = G_6 E_3(s) \tag{2.87}$$

$$E_1(s) = R(s) - H_2 Y(s) \tag{2.88}$$

$$E_2(s) = G_3 G_2 G_1 E_1(s) + H_1 Y(s) \tag{2.89}$$

$$E_3(s) = G_4 E_2(s) + G_5 G_2 G_1 E_1(s)$$

The comparator outputs $E_1(s)$, $E_2(s)$ and $E_3(s)$ may now be eliminated, as follows. First, $E_3(s)$ is eliminated from Equation (2.87):

$$Y(s) = G_6 G_4 E_2(s) + G_6 G_5 G_2 G_1 E_1(s) \tag{2.90}$$

Next, Equations (2.88) and (2.89) are substituted for $E_1(s)$ and $E_2(s)$ in Equation (2.90):

$$Y(s) = G_6 G_4 \{G_3 G_2 G_1 [R(s) - H_2 Y(s)] + H_1 Y(s)\} + G_6 G_5 G_2 G_1 [R(s) - H_2 Y(s)]$$

Grouping terms gives

$$(G_6 G_4 G_3 G_2 G_1 H_2 - G_6 G_4 H_1 + G_6 G_5 G_2 G_1 H_2 + 1)Y(s)$$
$$= (G_6 G_4 G_3 G_2 G_1 + G_6 G_5 G_2 G_1)R(s)$$

from which the required solution is obtained as

$$\frac{Y(s)}{R(s)} = \frac{G_1 G_2 G_6 (G_3 G_4 + G_5)}{(1 - G_6 G_4 H_1) + G_1 G_2 G_6 H_2 (G_3 G_4 + G_5)}$$

Example 2.25 *Block diagram manipulation using superposition for multiple inputs*

This example shows how superposition may be used to handle systems with more than one input: determine the output $Y(s)$ in the system shown in Figure 2.46.

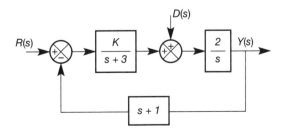

Figure 2.46 Block diagram of a two-input system in the Laplace domain.

Setting $D(s) = 0$ gives the transfer function between $Y(s)$ and $R(s)$ as

$$\frac{Y(s)}{R(s)} = \frac{2K}{s(s+3) + 2K(s+1)}$$

Setting $R(s) = 0$ gives the transfer function between $Y(s)$ and $D(s)$ as

$$\frac{Y(s)}{D(s)} = \frac{2(s+3)}{s(s+3) + 2K(s+1)}$$

Since a Laplace transfer function is a linear operator, the principle of superposition is used to generate the overall output as the sum of the two input contributions:

$$Y(s) = \frac{2KR(s)}{s(s+3) + 2K(s+1)} + \frac{2(s+3)D(s)}{s(s+3) + 2K(s+1)}$$

or

$$Y(s) = \frac{2KR(s) + 2(s+3)D(s)}{s(s+3) + 2K(s+1)}$$

2.7 Some relationships between transfer function and state-space models

The technique used in Section 2.5.1 to transform an ordinary linear differential equation representing a SISO system into an equivalent state-space equation is easily modified to transform a transfer function equation into a state-space equation. Hence, the transfer function

$$F(s) = \frac{b_0 s^m + b_1 s^{m-1} + \cdots + b_m}{s^n + a_1 s^{n-1} + \cdots + a_n} \qquad m < n \qquad (2.91)$$

(which would be the transfer function for Equation (2.45) if $m = 2$ and $n = 3$), may be transformed by inspection into the state-space equations:

$$\dot{x} = \begin{bmatrix} 0 & 1 & 0 & \cdots & 0 & 0 \\ 0 & 0 & 1 & \cdots & 0 & 0 \\ \vdots & \vdots & \vdots & & \vdots & \vdots \\ 0 & 0 & 0 & \cdots & 0 & 1 \\ -a_n & -a_{n-1} & -a_{n-2} & \cdots & -a_2 & -a_1 \end{bmatrix} x + \begin{bmatrix} 0 \\ 0 \\ \vdots \\ 0 \\ 1 \end{bmatrix} u$$

and

$$y = [b_m \quad b_{m-1} \quad \cdots \quad b_1 \quad b_0] x$$

Note that this only works for transfer functions of the precise form of Equation (2.91), with the highest order denominator coefficient equal to unity. Zero entries must be made in place of any absent a or b terms. The rules are those listed in Section 2.5.1, with the 'output' and 'input' of Equation (2.45) replaced by the denominator and numerator of Equation (2.91) respectively.

To transform the general state space equation

$$\dot{x}(t) = Ax(t) + Bu(t) \tag{2.92}$$

and

$$y(t) = Cx(t) + Du(t) \tag{2.93}$$

into transfer function form is achieved by taking Laplace transforms of Equations (2.92) and (2.93). In the ensuing text, when Laplace transforms are taken, the upper-case notation $X(s)$, $U(s)$ and $Y(s)$ could be used. However, the lower-case notation for vectors (as opposed to upper-case for matrices) takes precedence here, so the dependency on (s) is included for clarity in this case. Taking Laplace transforms of Equation (2.92) (zero initial conditions) gives:

$$sx(s) = Ax(s) + Bu(s) \qquad \text{so} \qquad [sI - A]x(s) = Bu(s)$$

or

$$x(s) = [sI - A]^{-1}Bu(s) \tag{2.94}$$

Substitution of Equation (2.94) into the Laplace transform of Equation (2.93) yields:

$$y(s) = \{C[sI - A]^{-1}B + D\}u(s) \tag{2.95}$$

For MIMO systems, this may be written in the shortened form:

$$y(s) = G(s)u(s) \tag{2.96}$$

where $G(s)$ is the *transfer function matrix* of the system, having one row per output, and one column per input. For SISO systems, a division by $U(s)$ can be carried out, leading to the usual rational SISO transfer function:

$$\frac{Y(s)}{U(s)} = c[sI - A]^{-1}b + d \tag{2.97}$$

Note that, for SISO systems, c and b are row and column vectors respectively, and d is a scalar, so the overall result is scalar, as expected.

Example 2.26 *Converting a state-space model to a transfer function*

Find the transfer function for the SISO system having the state-space model:

$$A = \begin{bmatrix} -6 & -11 & -6 \\ 1 & 0 & 0 \\ 0 & 1 & 0 \end{bmatrix}, \qquad b = \begin{bmatrix} 1 \\ 0 \\ 0 \end{bmatrix}, \qquad c = [0 \quad 1 \quad 6], \qquad d = 0$$

First, for use in Equation (2.97), note that:

$$[sI - A]^{-1} = \begin{bmatrix} s+6 & 11 & 6 \\ -1 & s & 0 \\ 0 & -1 & s \end{bmatrix}^{-1}$$

$$= \frac{1}{s^3 + 6s^2 + 11s + 6} \begin{bmatrix} s^2 & -11s-6 & -6s \\ s & (s+6)s & 6 \\ 1 & s+6 & s^2+6s+11 \end{bmatrix}$$

(see Section A1.1.3 for definition of the inverse).

Now, from equation (2.97)

$$\frac{Y(s)}{U(s)} = c[sI - A]^{-1}b + d$$

Hence

$$\frac{Y(s)}{U(s)} = \frac{1}{s^3 + 6s^2 + 11s + 6} \begin{bmatrix} 0 & 1 & 6 \end{bmatrix} \begin{bmatrix} s^2 & -11s-6 & -6s \\ s & (s+6)s & 6 \\ 1 & s+6 & s^2+6s+11 \end{bmatrix} \begin{bmatrix} 1 \\ 0 \\ 0 \end{bmatrix} + 0$$

or

$$\frac{Y(s)}{U(s)} = \frac{s+6}{s^3 + 6s^2 + 11s + 6}$$

which is the required transfer function.

A system introducing some aspects of control

This introductory material on state-space and transfer function models ends with a model of a system to be used at several points during the text. In the interests of tractability, some aspects of the model are approximate. Nevertheless, it is useful to have one model of appropriate complexity, on which various techniques can be tried. The lack of modelling rigour in no way affects the application of the various control methods used – but it *would* affect the accuracy of control achieved on the real system.

Figure 2.47 shows an open-loop arrangement for controlling the azimuth (rotational position) of a satellite receiving aerial. The basic scheme is typical of many remote-positioning systems, including steerable satellite dishes used by domestic television receivers, systems for manoeuvring ships and aircraft, screwdown (thickness control) systems on metal rolling mills, head positioning in computer disk drives, automatic camera focus, or cable-free throttle control in automobiles. The power involved in moving such systems varies from a fraction of a watt to several kilowatts, depending upon the application, but the basic block diagram is the same.

In the particular system of Figure 2.47, the motor drive electronics converts the mains electricity supply into a suitable supply for the motor, and sets the level of

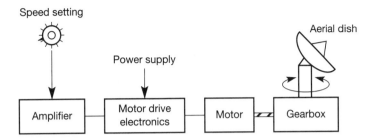

Figure 2.47 A satellite antenna azimuth control scheme – open loop.

this supply according to the 'speed setting' knob. The motor drives a gearbox, which turns the aerial.

In order to obtain a desired aerial position, a human operator will have to take several factors into account. Not the least of these is the fact that nothing in the real world happens instantaneously. In particular, if the aerial is large, and thus possesses significant inertia, it will take an appreciable time both to accelerate from rest, and to decelerate to rest. Depending on the type of gearbox construction (pinion drive vs. worm drive, for example), the operator may therefore have to advance the speed control setting from zero, wait a while for the aerial to accelerate, and then reduce the speed setting to a smaller value *in anticipation* of the deceleration time of the aerial, finally reducing the setting to zero such that the aerial comes to rest at the correct position. Clearly, it will not be easy to achieve rapid and accurate position control in this way. In particular, if a *tracking* system is required, in order to follow a non-geostationary satellite (or other vehicle) accurately, some form of automatic closed-loop control must be used.

In order to design a controller using the methods described in this book, a mathematical model of the system is required. Converting Figure 2.47 into a suitable block diagram model, as shown in Figure 2.48(a), is the first stage in modelling the system.

In this figure the 'positioning signal' indicates the desired position of the aerial and, as such, replaces the operator's speed setting knob. A voltage signal which is proportional to the desired angular position of the aerial has been assumed, and this could very easily be generated using a simple potentiometer.

The 'drive system' block includes the amplifier, the motor drive electronics

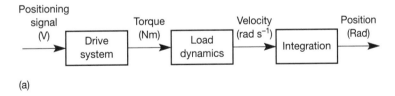

(a)

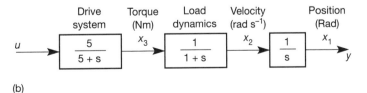

Figure 2.48 (a) Simplified block diagram model for the system of Figure 2.47. (b) Numerical values for a frequency domain (LTF) model of Figure 2.47.

(b)

and the motor. To assume that all these elements can be represented by a single block having a simple transfer function is a gross over-simplification. However, the speed of response of the amplifier is likely to be as fast as (or faster than) that of the drive electronics which, in turn, will be much faster than the speed of response of the motor. For this reason, the dynamics of the amplifier and drive electronics have been ignored. Furthermore, if the aerial is large, the response of the motor will, in turn, be much faster than the load dynamics of the mechanical elements in the gearbox and the aerial inertia, and this should be reflected in the model of Figure 2.48(b). The response of both the drive and load blocks has been assumed to be of an overdamped appearance, leading to approximate first-order models for both.

The final integrator block converts the aerial's speed of rotation into angular position. Note that the measured angular position (not explicitly shown in the figure) would be represented by a proportional voltage signal (from a potentiometric transducer, perhaps) compatible with that used as the 'positioning signal'.

Although the various simplifications will have produced a model of little practical value, it is nevertheless of value for demonstrating various control system design techniques at a level of complexity which permits hand calculation of the results. With computer assistance, more realistic models of higher order could simply be substituted for the model used here, but the design and analysis methods would remain the same.

Figure 2.48(b) shows the same block diagram as Figure 2.48(a), but with the appropriate Laplace transfer functions (LTFs) inserted (suitably approximated to nearest integers, as the primary purpose of this particular model is in demonstrating control principles later on). These Laplace transfer functions could be obtained either by modelling the components of lumped-parameter models of the various parts of the system, using the methods of this chapter, or from the results of tests on the real plant, as outlined in Chapter 3. Note that, conventionally, the input has become the signal u (but it remains a controlling voltage signal), and the output position becomes the signal y.

Since this is a third-order system (comprising three cascaded first-order LTF blocks which, of course, multiply together to give the overall LTF – Section 2.6), three independent state variables are needed for a state-space description. Indicated in Figure 2.48(b) are three meaningful and measurable signals labelled x_1, x_2 and x_3. Any other suitable signals may be selected, provided that they are independent (that is, one is not a combination of some others) and are not equal to the input signal (because that is not an internal state of the system).

The overall transfer function of the system of Example 2.27 (from Figure 2.48(b)) is:

$$\frac{Y(s)}{U(s)} = \frac{5}{s(s+1)(s+5)} = \frac{5}{s^3 + 6s^2 + 5s} \tag{2.98}$$

and will be used to apply the frequency-domain analysis and control methods of later chapters. It is worth noting that all the information which the block diagram contained about the internal structure of the system has been lost. For example, from Equation (2.98), it is not possible to say how the motor torque or the load velocity will behave, because they do not appear in the equation. Only the output (load position) and input (control voltage) appear. This is a general feature of frequency-domain methods – they use input–output models. However, as indicated in the subsequent text, a state-space description can maintain the internal information too.

In addition, without extra work, Equation (2.98) does not reveal how the system will respond if the initial conditions are not zero (for example, if the system is already in motion when the control voltage is changed). This is also a general feature of frequency-domain methods, and arises because initial conditions are often set to zero when deriving LTF models. Non-zero initial conditions may be included in the model of Figure 2.48(b), only by including them as disturbance signals acting on the plant (see Example 2.25). For use in control system design, it is not easy to decide how to handle LTF models with several inputs and/or outputs (see Chapter 10). The state-space model automatically includes initial conditions (Sections 3.2.1 and 3.6.2), and caters for multiple inputs and outputs (Section 2.5.1).

From such comments, it seems that the state-space model 'wins' every time. This is what the early workers in the 1960s thought, and it led to a lot of controversy about the 'best' methods. There are, however, some balancing disadvantages of using state-space models, which are listed in Section 2.9, and it is really a matter of knowing sufficient to be able to choose the best approach for the system in hand. It transpires that, in general, state-space methods work well (for example) in aerospace and in some other areas where system models are relatively accurate, but frequency-domain methods work better in many other industrial plants.

Example 2.27 *A state-space model of the antenna-positioning system – by inspection*

A state-space equivalent (in the form of Equations (2.43) and (2.44)) to the frequency-domain model of Figure 2.48(b) may be obtained by inspection, as outlined at the beginning of Section 2.7. This method is called *direct programming*, because it is used to program analog computer simulations.

Equation (2.98) gives the overall LTF of the system, and it is in an appropriate form, in that the numerator is of lower order than the denominator (that is, it is strictly proper) and the coefficient of the highest-order denominator term is unity. Applying the rules given in Section 2.5.1 yields the following model (note that the system for this example (Figure 2.48(b)) was first used in an excellent text by Blackman (1977) which is now out of print. However, he did not attach to it any physical function, but presented it as a purely numerical example. An equally excellent current text specialising in state-space methods is Friedland (1987)).

$$A = \begin{bmatrix} 0 & 1 & 0 \\ 0 & 0 & 1 \\ 0 & -5 & -6 \end{bmatrix}, \qquad b = \begin{bmatrix} 0 \\ 0 \\ 1 \end{bmatrix}, \qquad c = [5 \ \ 0 \ \ 0], \qquad d = 0 \quad (2.99)$$

This is a perfectly valid and easily obtained state-space model of the system. However, it shares the disadvantage of the LTF model, in that any information about the originally chosen internal states has been lost. The states in the model of Equation (2.99) are different from those in Figure 2.48(b).

To illustrate these points consider Figure 2.49(a), which depicts a general block diagram of any state-space model of any SISO system defined by Equations (2.43) and (2.44). In such figures, the thick lines represent multivariable signal paths, and the thin lines single variables. Blocks such as integrators and summers in multivariable paths are taken to act individually on each signal in the path, while matrices and vectors perform multiplications on the signals in the path as a set (vector), as described in Section A1.1.1. Thus, for example, the forward path integral represents a set of independent integrators, one per signal in the state vector, generating the state vector from its derivative.

This particular form of block diagram, containing only integrators, summers and gain elements, is known as a *simulation diagram*, because it is used in programming dynamic system simulations on analog computers (which contain these very elements).

Figure 2.49(b) is the specific simulation diagram corresponding to Equation (2.99). In drawing it, it is only necessary to realize that an integrator is required to convert the derivative of a variable into the variable itself. Otherwise, the structure follows naturally from the equations.

Note that x_3 in this model does not represent any measurable signal on the plant, while in this particular case, x_1 and x_2 are 1/5 of the position and velocity signals (x_1 and x_2 would normally also turn out to be unmeasurable in a general case). Problem 2.3 shows that the input–output relationship remains unchanged.

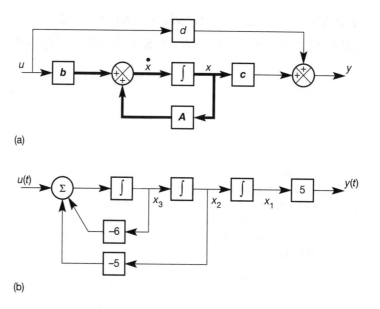

(a)

(b)

Figure 2.49 (a) General simulation diagram for a SISO state-space model. (b) A simulation diagram for direct programming of Figure 2.48(b).

Example 2.28 *A state-space model of the antenna-positioning system – preserving the states*

In Figure 2.48(b), a physically meaningful state vector is indicated as:

$$x = \begin{bmatrix} x_1 \\ x_2 \\ x_3 \end{bmatrix} = \begin{bmatrix} \text{angular position (rad)} \\ \text{velocity (rad s}^{-1}) \\ \text{torque (N m)} \end{bmatrix}$$

To generate a state-space model which preserves these states will be more complicated than the direct programming method (Example 2.27). The method to be presented is generally applicable, relying on the fact that a LTF block can be written as output = contents × input. Initially, the dependence of the variables on time (t) and the Laplace operator (s) is indicated, but it is dropped later for reasons of clarity, once the context is clear.

The integrator which converts load angular velocity to position can be written as:

$$X_1(s) = \frac{1}{s} X_2(s)$$

therefore $sX_1(s) = X_2(s)$. From Table 2.8, multiplication by s in a LTF is equivalent to the time-domain operation of differentiation (with zero initial conditions), and therefore, transforming to the time domain:

$$\frac{dx_1(t)}{dt} = x_2(t) \qquad \text{or} \qquad \dot{x}_1 = x_2 \tag{2.100}$$

For the block representing the load dynamics:

$$X_2(s) = \frac{1}{1+s} X_3(s)$$

therefore $X_2 + sX_2 = X_3$, and so inverse Laplace transforming gives:

$$x_2(t) + \dot{x}_2(t) = x_3(t) \qquad \text{or} \qquad \dot{x}_2 = -x_2 + x_3 \tag{2.101}$$

Similarly, for the drive system:

$$X_3 = \frac{5U}{5+s} \qquad \text{so, as above,} \qquad \dot{x}_3 = -5x_3 + 5u \tag{2.102}$$

Equations (2.100) to (2.102) are the *state equations* for the system. Arranging in the usual form (Equation (2.43)):

$$\begin{bmatrix} \dot{x}_1 \\ \dot{x}_2 \\ \dot{x}_3 \end{bmatrix} = \begin{bmatrix} 0 & 1 & 0 \\ 0 & -1 & 1 \\ 0 & 0 & -5 \end{bmatrix} \begin{bmatrix} x_1 \\ x_2 \\ x_3 \end{bmatrix} + \begin{bmatrix} 0 \\ 0 \\ 5 \end{bmatrix} u$$

This can be written in the more compact form, as required: $\dot{x} = Ax + bu$, with

$$A = \begin{bmatrix} 0 & 1 & 0 \\ 0 & -1 & 1 \\ 0 & 0 & -5 \end{bmatrix} \quad \text{and} \quad b = \begin{bmatrix} 0 \\ 0 \\ 5 \end{bmatrix} \tag{2.103}$$

In this case, the *output equation* is simply $y = x_1$. However, to conform with Equation (2.44):

$$y = \begin{bmatrix} 1 & 0 & 0 \end{bmatrix} \begin{bmatrix} x_1 \\ x_2 \\ x_3 \end{bmatrix} \quad \text{or} \quad y = cx \quad \text{where} \quad c = \begin{bmatrix} 1 & 0 & 0 \end{bmatrix} \tag{2.104}$$

Equations (2.103) and (2.104) now constitute a different state-space model of the system. Note that the d quantity does not exist in either model, as there is no direct coupling from input to output. A simulation diagram corresponding to this model appears in Figure 2.50. Replacing the integrators by $1/s$, and performing a block diagram reduction exercise, shows that the transfer function relating the input and output is identical to that of Equation (2.99) (see Problem 2.4).

Figure 2.50 A simulation diagram for a state-space model of Figure 2.48(b), maintaining the specified states.

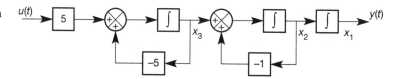

Different state-space models

Two state-space models for the system of Figure 2.48(b) have been generated. The freedom of choice of state variables suggests that by swapping around the positions of x_1, x_2 and x_3, there are immediately five other possible state-space descriptions of the system. Each of these would have a set of measurable and meaningful state variables, which is an advantage. However, they all have the disadvantages that their derivation is tedious, and the resulting form of the quantities A, b and c arising in such a model is random. For example, there is no particular pattern in the contents of the A matrix in Equation (2.103), which might assist in future analysis. The form of A, b and c is dependent entirely upon the layout of the original system.

Often, for the sake of easier mathematical manipulation, and a more intuitive form of model, it is convenient if these quantities (especially the A matrix) have some particular predefined structure. For example, a diagonal A matrix (Appendix A1.1.3) would represent a system whose state variables are non-interacting. It is very easy to transpose, invert, or find the determinant of such a matrix; and it also turns out to have other useful properties (discussed later).

Apart from direct programming, there are two other analog computer programming techniques (parallel programming and iterative programming) which give other definite forms to the state-space model. These are not described here, but their existence confirms that there are many possible choices of state-space model (parallel programming is outlined in Problem 2.5). In fact, using a linear algebra device known as the '*similarity transform*' (see Section A1.6) an infinite number of different models can be generated. However, apart from the six possible models obtained by rearranging the order of the states in Figure 2.48(b), none would have a complete set of physically meaningful states.

Finally, note that Problem 2.6 introduces the use of MATLAB for performing such model conversions.

2.8 *Simple discrete-time models for digital control and simulation*

All the time-domain models discussed so far have contained *continuous* functions of time. That is, their time responses to (say) a step input will be smooth curves, similar to those in Figure 1.12. Many (probably most) modern control systems are implemented using digital computers and, as indicated in Section 1.3.6, the digital control algorithms use *samples* of the continuous data taken at *discrete instants in time* (because the time between the samples is required for digital processing and to send the calculated control signals to the plant). Section 1.3.6 also noted that care is needed in selecting the rate at which signals are sampled. To design such control systems directly (see, for example, Chapters 7, 9 and 12), or to carry out digital computer simulations of continuous-time designs, *discrete-time models* of the systems are required.

Often, a discrete-time model is obtained by converting the *differential* equations of the continuous-time model into *difference* equations. Essentially, time is divided into discrete intervals, or *steps*, with a sample of the signals taken only at the start and end of each time step. Discrete-time models then allow the latest values of the system outputs (or states) (that is, at the present discrete-time step) to be calculated from the known values of the outputs and inputs at previous time steps. They are called difference equations because (in the simplest conversion technique) the derivative in a continuous-time model is replaced by the *difference* between successive samples, divided by the sampling period (which is assumed constant). This approximates the slope of the continuous response at the sampling instant, as shown in Figure 2.51:

$$\left.\frac{dy}{dt}\right|_k \approx \frac{\Delta y}{\Delta t} = \frac{y_k - y_{k-1}}{h}$$

This approach has the drawback of only giving a series of 'snapshots' of the behaviour of the system rather than a continuous representation. A film or video recording, however, suffers from the same deficiency; but the lack of intermediate information is not a problem, provided that the time between samples is short enough. Figure 1.15 in Chapter 1 demonstrated one consequence of getting it wrong.

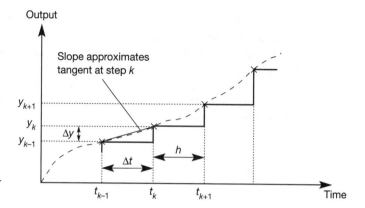

Figure 2.51 The simplest continuous-time to discrete-time conversion method.

2.8.1 Difference equation models

A difference equation model is normally expressed in the form:

$$y_n = A_1 y_{n-1} + A_2 y_{n-2} + A_3 y_{n-3} + \cdots + B_0 u_n + B_1 u_{n-1}$$
$$+ B_2 u_{n-2} + \cdots$$

where the A and B values are constants, y is the output and u is the input. The nth output or input is the latest one, the $(n-1)$th is the previous one, and so on; the meanings of the symbols are further explained by Figure 2.52. This type of model is known as an ARMA model (Auto-Regressive Moving Average) because the coefficients A give a dependence on previous outputs (the *auto-regressive* part) and the coefficients B give a dependence on present and past inputs (the *moving-average* part; it is, in fact, a weighted average). In Figure 2.52, it is assumed that the latest sample of the input u (namely u_n) has been taken at a time t equal to 3 seconds, resulting in an output value at that time of y_n. The sampling interval has been assumed to be 0.25 s for demonstration purposes, so the previous samples of y and u relate to the previous sample times at 2.75 s $(n-1)$, 2.5 s $(n-2)$, 2.25 s $(n-3)$, and so on.

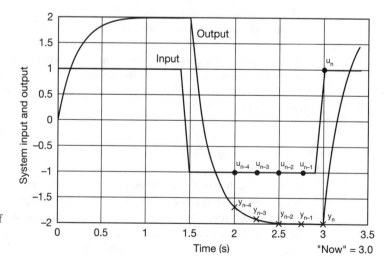

Figure 2.52 Illustration of sampled-signal terminology.

How such models work will be demonstrated by calculating their time responses in Chapter 3 (this is the basis of digital computer simulation). Methods of converting from a continuous-time representation to a discrete-time one are discussed in Section 5.8.2 and the alternative numerical-integration approach is discussed in Sections 3.6.1 and 5.7.

2.8.2 Discrete-time state-space models

All the state-space models considered so far have been continuous-time models, involving time derivatives of the state variables. However, a discrete-time model of a system, which fulfils the same functions as a state-space model, would be of advantage for the design of digital computer-based controllers. Such models exist, and are used in the state-space control system analysis and design methods developed in later chapters. In fact, such models are readily derived from continuous-time state-space models and, for some problems, have considerable advantage (see for example Section 9.8 and Chapter 12).

For the present, note that the discrete-time state-space model is a recursive model, in which the state vector at any given sampling instant (instant $k + 1$, say) depends only on the input vector and state vector at the previous instant (instant k). The general form of the model is:

$$x_{k+1} = \boldsymbol{\Phi} x_k + \boldsymbol{\Delta} u_k \tag{2.105}$$

$$y_{k+1} = C x_{k+1} + D u_{k+1} \tag{2.106}$$

The output equation (Equation (2.106)) is identical to the continuous-time version (Equation (2.44)). The new matrices $\boldsymbol{\Phi}$ and $\boldsymbol{\Delta}$ in the state equation (Equation (2.105)) are functions of the continuous-time A and B matrices (Equation (2.43)), and the sampling interval h (where $h = t_{k+1} - t_k$, as in Figure 2.51). The derivation of $\boldsymbol{\Phi}$ and $\boldsymbol{\Delta}$ is given in Section 3.6.2. Once such a model is obtained, it can be initialized using x_0 and u_0 (that is, the values of $x(t)$ and $u(t)$ at the start of the simulation), and then repeatedly applied to generate the state and output responses as time passes, one sampling instant after another.

2.9 Concluding remarks

This chapter has introduced simple lumped-parameter modelling and generated linear differential equation models to represent lumped-parameter models mathematically. Further, it has been shown how these models may be converted into Laplace transfer function models (for use in the frequency domain) or state-space models (for use in the time domain). These two forms of model cover the requirements for most of the continuous-time analysis and design of control systems. For discrete-time designs, the basic ideas of difference equation models have been introduced.

Linearization of nonlinear models and Lagrange's technique for finding mathematical models from lumped-parameter models have also been covered. However, these topics might be considered optional in a 'first course' in control.

Given a LTF or state-space model, it has been shown how the system can be represented as a block diagram. These diagrams may be manipulated to obtain different forms of model, or relationships between given signals in the model.

With regard to state-space models, there is an infinite number of possible models for a given system, of which some will have physically meaningful states, and some will not. There are certain constant mathematical attributes which link all state-space models of the same system, and these will be studied further in Section 3.2.1. Also, whatever state-space representation of a system is used, the input–output relationship between u and y (for example, the LTF) is the same. Therefore, if only the system's input–output behaviour is of interest, the most easily obtained, or the most mathematically tractable state-space model should be used. How to select this model will be revealed in later chapters.

The following list summarizes the major differences between LTF and state-space models:

- A LTF model relates only the system output and input. Any internal structure and behaviour of the system is lost. State-space representations can preserve such internal information, if the state variables are appropriately chosen.

- When working with LTFs, one may have to solve high-order equations. In the state-space approach, only first-order equations appear in the model. However, for a tenth-order system, 10 first-order equations must be solved simultaneously. The solution of simultaneous first-order differential equations is, however, well understood, and can be performed using standard computer packages. It could be argued that a tenth-order LTF equation could also be solved easily by computer packages, but in terms of the numerical methods used, the balance swings in favour of the simultaneous first-order equations as the order increases.

- The LTF approach as it existed in the early 1960s (that is, using LTF models with the intention of using frequency-domain methods) was only easily applicable to SISO systems. For a *multivariable* system having many inputs and/or outputs (MIMO), there would be a transfer function relating each input to each output. Thus for a system with four inputs and five outputs there would be an arrangement of 20 transfer functions. This is not easily used in designing controllers, although Chapter 10 addresses the problem using more recent techniques. In the state-space approach, the system model is arranged in a vector–matrix form. It therefore matters little how many inputs or outputs a system has, the matrices just change in size.

- A Laplace transfer function model is only defined for zero initial conditions. To include initial condition effects in frequency-domain analysis, one must go back a step to the original differential equations, and include the initial condition terms rigorously when taking the Laplace transforms. The resulting block diagram model then has disturbance inputs appearing which represent the effects of the initial conditions. Again this is then difficult to handle from a frequency-domain standpoint. The state-space approach uses the differential equations directly, so all initial conditions are automatically included (Sections 3.2.1 and 3.6.2) and are simply set to zero if not required.

- Allowing the contents of the A, b, C and/or D quantities to be time-varying enables the state-space model to handle time-varying systems, whereas a LTF model is only for stationary systems.

But there are drawbacks to state-space models, some of which can be serious:

- Many processes involve pure time delays which, although readily accommodated using LTFs (because the Laplace transform of a time delay is linear), are difficult to handle using continuous-time state-space models (see Chapter 8).

- The methods of control system design based on frequency-domain models are true *design* methods. A controller is designed, and its effects may be investigated using one of the standard frequency-domain methods to be studied later. The controller is then repeatedly refined until the results appear acceptable. Control system design using state-space models, on the other hand, tends to use *synthesis* methods. That is to say, an algorithm will exist for designing a certain kind of controller (these will also be studied later). The designer simply plugs the system model (A, B, and so on) into the algorithm, and out comes the design of the controller – it is synthesized from the model parameters. This sounds an excellent approach, but it relies on having very good plant models – if the model is wrong, the controller will be wrong. In practice, it is hard to get good models of many industrial plants.

- Since the LTF methods tend to be graphical, and work on the plant model's inputs and outputs, the approximations involved are apparent. Further, the results tend to be robust in the face of poorly modelled plants and unexpected disturbances. The state-space synthesis methods depend not only upon the correct input–output model, but also on correctly modelled internal behaviour. State-space solutions to real control problems therefore tend to be more 'fragile'. A lot of the more recent research is aimed at overcoming this deficiency of the state-space approach.

- The mathematical techniques of Laplace transforms are widely known, and the use of frequency response plots (Chapter 3) can be easily tied in with system behaviour which is readily visualized. Frequency-domain controllers can often be tuned using 'rules of thumb', and a little engineering intuition. Frequency-domain control is therefore a relatively accessible discipline, with many practitioners.

- The matrix algebra required for state-space methods is, perhaps, less well known (but not intrinsically difficult); and the relationship of the numbers in a matrix to physically measurable effects is less clear than using frequency-response plots. If a controller is designed which is effectively just a matrix of coefficients, tuning it requires an in-depth understanding of how a variation of any given parameter will affect the plant.

These factors, together with the pragmatic observation that frequency-domain control seems able to cope with the majority of real control problems, have led to a generally slow rate of introduction of the state-space techniques. Nevertheless, in areas where plant models are good (such as aerospace), state-space-based control has found ready acceptance and wide application. There are also many problems which are unapproachable by frequency-domain methods, such as large MIMO systems, or systems whose dynamics vary significantly with time. Again, state-space methods have found wide application in such areas.

As a gross generalization, frequency-domain control is likely to be the best solution to those problems to which it is readily applicable (that is, most SISO loops, some multivariable systems using the methods of Chapter 10, some systems with significant time delays, some systems where the models are uncertain), whereas state-space methods can be tried for all other systems (any system where the models are trustworthy, MIMO systems in general, systems corrupted by significant noise (see Chapter 9), and so on).

2.10 Problems

2.1 Prove that the model of the system of Figure 2.26(a) is given by the equation preceding Equation (2.41).

2.2 Use Laplace transforms to derive Equations (2.46) and (2.47) from Equation (2.45), with z defined according to Figure 2.27.

2.3 (a) In Figure 2.49(b), replace the integrators with the Laplace domain equivalent ($1/s$) and use block diagram reduction techniques to prove that the overall Laplace transfer function relating y to u is correct (compare your result with Equation (2.98)).

(b) Use Equation (2.97) to obtain the same result.

2.4 Repeat Problem 2.3 for the system of Figure 2.50. Confirm that the results are identical not only with each other, but with the results of Problem 2.3.

2.5 (a) Use the parallel programming approach (given below) to obtain a state-space representation of the LTF model of Equation (2.98). What do you notice about the form of the resulting model?

(b) Repeat Problem 2.3(a) using the simulation diagram from part (a), and 2.3(b) using the model parameters (A, b, c, d) from part (a).

The parallel programming approach is carried out as follows:

● Split the LTF (Equation (2.98)) into partial fractions.

● Carry out a direct programming operation on each of the individual (first-order) partial-fraction terms.

● Draw a separate simulation diagram for each of the direct programming results.

● Connect these first-order simulation diagrams (three of them for this problem) in parallel – all the inputs will be connected to u, and the outputs must be algebraically summed to give y according to the partial-fraction expansion.

● Label the integrator outputs as the state variables, and hence construct the state and output equations from the diagram.

2.6 Throughout the text, MATLAB is used as a representative CACSD (computer-assisted control system design) environment. Appendix 3 introduces MATLAB, and this problem suggests some use of it relevant to Chapter 2. If you do not have access to MATLAB and its Control Systems Toolbox, skip this problem.

Write an m-file, to check that Examples 2.15 and 2.16 give the same LTF model, and that it agrees with that of Example 2.21. Read Appendix 3 for details of how to write the m-file. Assume component values as follows: $C = 1$ μF, $R_1 = 330$ kΩ and $R_2 = 510$ kΩ.

Mathematical statements *, /, +, – and so on are written more or less as you would on paper, with parentheses ( ) to force the order of execution where it differs from that just given, for example, alpha $= r2/(r1 + r2)$.

The MATLAB *ss2tf* command will give the LTFs from the state-space models (use *help ss2tf*).

To get the numerator and denominator of the LTF in Example 2.21, write the vector explicitly, for example, num21 = alpha*[tau 1].

Some of the following problems also make use of MATLAB.

2.7 Using the direct programming rules, obtain state-space models of the following systems. (If MATLAB and the Control Systems Toolbox are available, check the results using your state-space models in the *ss2tf* command – but note that trying to use the complementary *tf2ss* command to find the state-space models in preference to doing the direct programming will give unexpected results – see Problem 2.8.)

(a) $\dfrac{Y(s)}{U(s)} = \dfrac{3}{s^3 + s^2 + 6s + 2}$

(b) $\dfrac{Y(s)}{U(s)} = \dfrac{1}{s^3 + 4s^2 + 8}$

(c) $\dfrac{Y(s)}{U(s)} = \dfrac{4}{2s^2 + 5s + 4}$

(d) $\dfrac{Y(s)}{U(s)} = \dfrac{3s^2 + 1}{6s^2 + s + 4}$

(e) $\dfrac{Y(s)}{U(s)} = \dfrac{s^2 + 1}{s^2 - 1}$

2.8 Despite appearances, MATLAB is not required to do most of this question.

(a) MATLAB also uses a direct programming ('by inspection') method to obtain its state-space models. However, it uses a different approach to that given in the text. MATLAB starts with the direct programming simulation diagram of (for example) Figure 2.49(b), but numbers the states in reverse order (that is, left to right, rather than right to left).

Using this information, obtain the state-space model MATLAB would obtain for the antenna-positioner (Figures 2.48(b) and 2.49(b)). Deduce the rules that MATLAB uses to obtain SISO state-space models from LTF models by inspection. Check your result (if you have MATLAB) as in part (b), below.

(b) Repeat Problems 2.7(a) to 2.7(e) using the MATLAB direct programming rules derived above. If MATLAB and the Control Systems Toolbox are available, check the results using the *tf2ss* command. The method is illustrated here for (d):

$\gg num = [3\ 0\ 1];$ % note the zero coefficient of s

$\gg den = [6\ 1\ 4];$
$\gg [a,b,c,d] = tf2ss(num,den)$ % no semicolon, so results are displayed

2.9 Obtain a state-space model of the plant in Figure P2.9(a), using the signal between the blocks as one of the state variables. Note that for the third-order block, three more states must be introduced. Use approaches (a) and then (b), given below:

(a) Choose the output as one of the states, and use the rule 'output = contents × input' on each block. For each block, cross-multiply and then take inverse Laplace transforms with zero initial conditions. This will give the state equation of the first-order block directly.

For the third-order block, get rid of all derivatives higher than the first, by defining two extra states x_2 and x_3 (assuming x_1 has been chosen at the output). For example, let $\dot{x}_1 = x_2$ and $\dot{x}_2 = x_3$ (so $\ddot{x}_1 = x_3$).

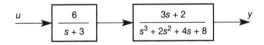

(a)

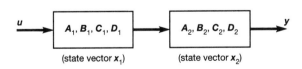

(b)

Figure P2.9 (a) System for Problem 2.9, part (a). (b) Series connection of two state-space models.

Combine the simulation diagrams of the first- and third-order blocks and hence obtain the overall (fourth-order) state-space model.

(b) Use the same method as (a) for the first-order block, but use direct programming on the third-order block. Then show that the series connection of two state-space models, as depicted in Figure P2.9(b), is given by (see Section A1.3 on partitioned matrices):

$$\dot{x} = \begin{bmatrix} \dot{x}_1 \\ \dot{x}_2 \end{bmatrix} = \begin{bmatrix} A_1 & 0 \\ B_2C_1 & A_2 \end{bmatrix} \begin{bmatrix} x_1 \\ x_2 \end{bmatrix} + \begin{bmatrix} B_1 \\ B_2D_1 \end{bmatrix} u$$

and

$$y = [D_2C_1 \mid C_2] \begin{bmatrix} x_1 \\ x_2 \end{bmatrix} + D_2D_1u$$

Substitute the separate state-space models of the two blocks into this to obtain the overall model.

2.10 (a) Draw the simulation diagram corresponding to the system model:

$$A = \begin{bmatrix} 0 & 1 & 0 \\ 0 & -1 & 1 \\ 0 & 0 & -5 \end{bmatrix}, \quad B = \begin{bmatrix} 0 & 2 \\ 0 & 0 \\ 5 & 1 \end{bmatrix},$$

$$C = \begin{bmatrix} 1 & 0 & 0 \\ 0 & 0 & 1 \end{bmatrix}, \quad D = \begin{bmatrix} 0 & 1 \\ 0 & 0 \end{bmatrix}$$

Note that this is an extension of the model of Equations (2.103) and (2.104), so the result can be compared with Figure 2.50. In this way, confirm that the extra column in B represents the addition of an extra input, and the extra row in C represents the addition of an extra output. Also, show that the non-zero element in D represents direct input–output coupling (as in Figure 2.49(a)).

(b) Use Equation (2.95) to obtain a transfer function matrix model of the system of part (a). Compare the result with that of Problem 2.4(b). *Hint*: It is possible to reuse a lot of the working of Problem 2.4(b).

(c) By comparing the TFM model of part (b) with the block diagram of part (a), confirm that $g_{11}(s)$ (which should be the same as Equation (2.98)) is the correct LTF relating output 1 to input 1. Similarly confirm that $g_{22}(s)$ is the LTF relating output 2 to input 2.

(d) Use block diagram reduction techniques on the result of part (a) to confirm the results for $g_{12}(s)$ and $g_{21}(s)$ in part (b).

2.11 (a) Obtain separate state-space models of the systems in Figure P2.11(a), using the indicated state variables (plus one extra state for the second-order block).

(b) Show that the parallel connection of two state-space models, as shown in Figure P2.11(b), is given by (see Appendix A1.3 on partitioned matrices):

$$\dot{x} = \begin{bmatrix} \dot{x}_1 \\ \dot{x}_2 \end{bmatrix} = \begin{bmatrix} A_1 & 0 \\ 0 & A_2 \end{bmatrix} \begin{bmatrix} x_1 \\ x_2 \end{bmatrix}$$

$$+ \begin{bmatrix} B_1 & 0 \\ 0 & B_2 \end{bmatrix} \begin{bmatrix} u_1 \\ u_2 \end{bmatrix} \quad \text{and}$$

$$y = [C_1 | C_2] \begin{bmatrix} x_1 \\ x_2 \end{bmatrix} + [D_1 | D_2] \begin{bmatrix} u_1 \\ u_2 \end{bmatrix}$$

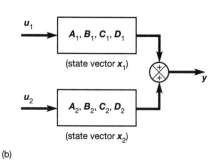

(a)

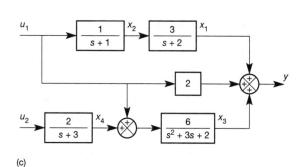

(b)

(c) Using a slightly modified form of the result of part (b), and the state-space models of part (a), obtain an overall state-space model of the system shown in Figure P2.11(c).

2.12 Obtain a state-space model for the system shown in Figure P2.12, using the indicated state variables wherever possible. Notice that the sizes of the matrices in the model should confirm the descriptions in Section 2.5.1.

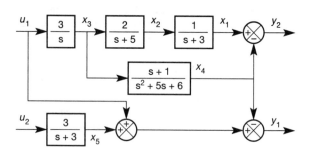

Figure P2.12 Multivariable system for Problem 2.12.

2.13 Determine the equations of motion for the mechanical systems shown in Figure P2.13. In each case the input displacement is $u(t)$, and the output displacement is $y(t)$.

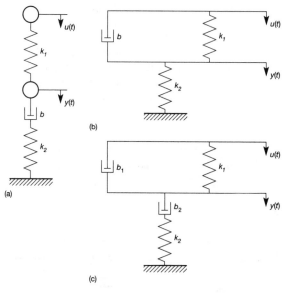

Figure P2.13 (a) Mechanical lag compensator; (b) mechanical lead compensator; (c) mechanical lag-lead compensator.

Figure P2.11 (a) Two systems for Problem 2.11, part (a). (b) Parallel connection of two state-space models at their outputs. (c) System for Problem 2.11, part (c).

2.14 Determine the equations relating the input voltage $v_i(t)$ to the output voltage $v_o(t)$ of the electrical networks shown in Figure P2.14. Loading may be ignored.

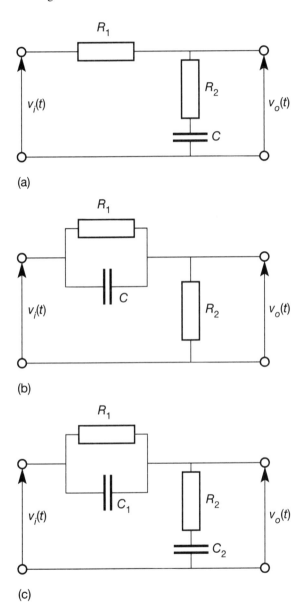

(a)

(b)

(c)

Figure P2.14 (a) Electrical lag compensator; (b) electrical lead compensator; (c) electrical lag-lead compensator.

2.15 Show that the corresponding pairs of mechanical and electrical compensators shown in Figures P2.13 and P2.14 are analogous.

2.16 The tank shown in Figure P2.16 consists of three compartments. Compartment 1 is heated electrically to a constant temperature T_1, and compartments 2 and 3 are at temperatures T_2 and T_3 respectively. Heat can flow only from compartment 1 to compartments 2 and 3, as shown.

There will be an exchange of heat between compartments 2 and 3, both of which can also lose heat to the environment, which is at temperature T_a.

Develop equations for the change in temperature of fluid in tanks 2 and 3, given that their respective thermal capacities are C_2 and C_3.

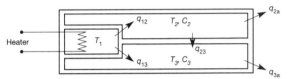

Figure P2.16 A three-compartment heating tank.

2.17 Using the current–force analogy, find the mechanical system equivalent to the electrical circuit shown in Figure P2.17.

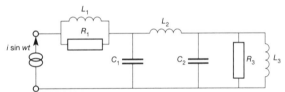

Figure P2.17 Electrical circuit for Problem 2.17.

2.18 The loudspeaker model shown in Figure P2.18 is based on an ideal moving coil transducer, a device which converts electrical energy into mechanical energy without loss. A change in current on the electrical side produces a change in force on the mechanical side. Given that the magnetic energy for the transducer is

$$T = \frac{1}{2} L(x) \dot{q}^2$$

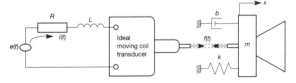

Figure P2.18 A loudspeaker model.

where q = charge = current × time, produce a mathematical model for this system relating the current i to the displacement x.

2.19 Find the inverse Laplace transforms of the following functions:

$$Y_1(s) = \frac{1}{s(s+2)}$$

$$Y_2(s) = \frac{10}{(s+4)(s+10)^3}$$

$$Y_3(s) = \frac{1}{s^2 + 2s + 3}$$

2.20 Calculate the initial and final values of the functions having the following Laplace transforms:

$$F_1(s) = \frac{2(s+1)}{s(s+3)(s+5)^2}$$

$$F_2(s) = \frac{4}{s^3 + 5s^2 + 12s + 8}$$

Check that the values are correct by finding the inverse Laplace transforms and evaluating them at $t = 0$ and $t = \infty$.

2.21 Use Laplace transforms to find the responses of the systems governed by the following equations:

(a) $\dfrac{d^2x}{dt^2} + 3\dfrac{dx}{dt} + 2x = 1$

given that $x(0) = 1$ and $\dot{x}(0) = 1$, and

(b) $\dfrac{d^2x}{dt^2} + 4\dfrac{dx}{dt} + 4x = 3\dfrac{dy}{dt} + 2y$

given that $x(0) = \dot{x}(0) = 0$ and $y = e^{-3t}$ for $t \geq 0$.

2.22 The following equations represent various systems, each having an input $r(t)$ and output $y(t)$. Put these equations in block diagram form.

(a) $\dfrac{d^2y}{dt^2} + 6\dfrac{dy}{dt} + 3y = r$

(b) $\dfrac{d^2y}{dt^2} + 6\dfrac{dy}{dt} + 3y = 5\dfrac{dr}{dt} + 4r$

(c) $y = \dfrac{d^2r}{dt^2} + \dfrac{dr}{dt} + r$

(d) $\dfrac{dy}{dt} + Ay = v$ and $\dfrac{d^2v}{dt^2} + B\dfrac{dv}{dt} = C\dfrac{dr}{dt} + Dr$

2.23 A simplified model of an aircraft's pitch control system is shown in Figure P2.23. In this model the pitch angle is $\theta_o(s)$, $\theta_i(s)$ is the pilot's input signal

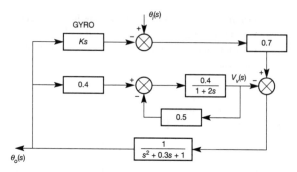

Figure P2.23 Aircraft pitch control system – simple model.

and $V_v(s)$ is the vertical velocity. Determine the differential equation relating $\theta_o(t)$ and $\theta_i(t)$.

2.24 A two-input, two-output multivariable plant together with its controller and feedback loops is shown in Figure P2.24. The controller consists of two dynamic compensators, $C_1(s)$ and $C_2(s)$, and two forward path gains, K_1 and K_2. In this arrangement the compensators are used to decouple the open-loop plant – that is, for the decoupled plant a disturbance at $E_i(s)$ produces a response only at output $Y_i(s)$ ($i = 1$ or 2), and the other output is unaffected. In this way the multivariable plant behaves as though it were two independent single-input, single-output plants. The forward path gains act as simple proportional controllers which, together with their corresponding feedback loops, are used to control the decoupled plant.

For this system, find the transfer function equations $C_1(s)$ and $C_2(s)$, which decouple the plant. Also determine the closed-loop transfer functions for both decoupled loops, and reduce each transfer function to its simplest form. Although Chapter 10 discusses the design of such controllers, this problem can be answered at this stage by writing the open-loop transfer function to each output from each error signal, and choosing the contents of $C_1(s)$ and $C_2(s)$ to remove the interaction.

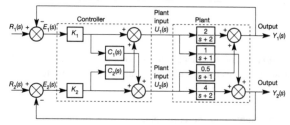

Figure P2.24 A multivariable system with decoupling compensator.

2.25 Determine the overall transfer function of the multiloop control system shown in Figure P2.25.

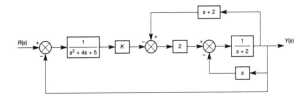

Figure P2.25 Control system for Problem 2.25.

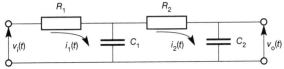

Figure P2.26 Electrical network for Problem 2.26.

2.26 Show that for zero initial capacitor charges, the Laplace transformed loop equations of the network shown in Figure P2.26 are

$$V_i(s) = R_1 I_1(s) + \frac{1}{C_1 s} (I_1(s) - I_2(s))$$

$$0 = \frac{1}{C_1 s} (I_2(s) - I_1(s)) + \left(R_2 + \frac{1}{C_2 s}\right) I_2(s)$$

$$V_0(s) = \frac{I_2}{C_2 s}$$

Put these equations into block diagram form, and hence determine the overall transfer function $V_0(s)/V_i(s)$.

2.27 Determine the transfer function models for the mechanical and electrical compensators in Problems 2.13 and 2.14.

3 System responses, stability and performance

3.1   Preview

In Chapter 2 the steps involved in finding the mathematical model of a plant have been examined. In particular, in Section 2.5, it was shown that if the system is passably linear, the plant can be represented by a linear transfer function model, or by a linear state-space model. Historically these two model forms were produced independently and used to develop various control system design strategies. However, for modelling the input–output behaviour of a system, the linear transfer function and linear state-space representations are interchangeable (Section 2.7), so many of these design strategies have much in common. Indeed, the insight gained from (say) a study of transfer function techniques helps in the understanding of state-space techniques and vice versa. This chapter lays the foundation for these design studies by examining single-input-single-output (SISO) systems represented using transfer function models (although state-space models are mentioned, too). Further, rather than considering specific systems, the approach adopted is an analysis of transfer function models in general. The intent is to

identify the various general characteristics of linear SISO systems, from which those pertinent to a particular system can be extracted.

In this chapter, the topics covered are:

- an introduction to the time and frequency response analysis of linear systems
- the relationship between the poles of a Laplace transfer function (LTF) model or the eigenvalues of the plant matrix in a state-space model, and its performance
- relationships between the frequency- and time-domain responses
- simple stability analysis
- the generation of the time response, from simple discrete-time models
- how to find models of systems from simple plant tests.

> **NEW MATHEMATICS FOR THIS CHAPTER**

The additional mathematics used in this chapter is complex algebra (that is, involving $j = \sqrt{(-1)}$). Topics covered include rectangular to polar conversion and vice versa and the identity $e^{j\theta} = \cos\theta + j\sin\theta$. Some trigonometric identities are also used, and are stated as required.

3.2 Some basic design requirements

In Section 1.3.2, it was stated that the plant dictates the controller. That is, the control engineer initially assumes that the plant dynamics $G(s)$ are fixed and then uses the open-loop plant's transfer function, Figure 3.1, to design a controller. This controller will be designed to give the closed-loop system, Figure 3.2, the required response characteristics. As an aside, such design studies may suggest plant modifications, that is, changes in the plant's dynamics $G(s)$. Often this is impossible. Even in cases where it is possible to modify the plant, any modifications will be specific to that particular plant and therefore cannot be discussed here.

Figure 3.1 A general Laplace transfer function block.

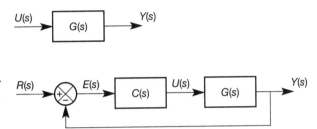

Figure 3.2 Laplace transfer function representation of a common feedback control arrangement.

Once the controller $C(s)$ has been designed, the closed-loop system needs to be tested to ensure that the desired response characteristics have been achieved. To this end, the block diagram reduction techniques of Section 2.6.2 are employed to provide the relationship between the input $R(s)$ and output $Y(s)$. The result of this reduction procedure is a Laplace transfer function (LTF) which may be represented in one of the standard forms, Equation (2.77).

Since both the open- and closed-loop transfer functions have the same standard forms, their analysis is identical. This can cause confusion during the design process. It is therefore important to remember that in general, open-loop transfer functions are used to design closed-loop controllers and closed-loop transfer functions are used to test response characteristics.

Perhaps the most basic design requirement is the ability to specify a system's performance. The requirement is to be able to quantify the output resulting from some input stimulus applied to the system. Since there is an infinite number of possible different inputs, and each will produce a different response, only a few standard inputs are normally considered. Mathematically, these inputs are those which are easily manipulated, and for which the resulting system responses can be shown to be readily quantified. The inputs considered are the following:

- *Step input.* Figure 3.3 depicts the time response of a step function having magnitude A acting on the system's input $u(t)$. Mathematically, the step input is defined as

$$u(t) = \begin{cases} 0 & \text{for } t < 0 \\ A & \text{for } t > 0 \end{cases}$$

At $t = 0$ the function is undefined, but it is normal practice to assume that $u(t) = A$ when $t = 0$. The Laplace transform of a step input of height A is A/s. In

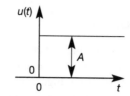

Figure 3.3 A step input
signal of height A units.

calculating the system output, it therefore adds a pole to the system's LTF (since output = LTF × input), which in turn maps into a point at the origin of the *s-plane* (see Section 2.5.6).

- *Ramp input.* The time response of a ramp function is shown in Figure 3.4. Mathematically, this function is defined as

$$u(t) = \begin{cases} 0 & \text{for } t \leq 0 \\ kt & \text{for } t > 0 \end{cases}$$

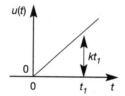

Figure 3.4 A ramp input
signal of slope k unit s^{-1}.

Its Laplace transform is given by k/s^2, and results in a double pole at the origin of the *s*-plane. Since the input to the system is unbounded (that is, this input signal continuously increases with time), the output of the linear model will also be unbounded (but not that of the real plant, which will eventually encounter a saturation limit – see Chapter 14) and the system's response is said to be unstable. Although ramp inputs are useful for determining the performance of certain systems (for example, missiles and machine tools), their application to physical plant must be carried out with care, precisely because of this capacity to cause saturation.

- *Pulse inputs.* There are many kinds and shapes of pulses. Two are particularly useful, one for experimental testing and the other for mathematical convenience. These are the rectangular pulse, which is formed by two successive steps of equal magnitude but opposite sign, shown in Figure 3.5 and the unit impulse, which is a pulse of unit integral area but zero duration and is shown in Figure 3.6. The Laplace transform of the unit impulse is unity, and consequently it does not affect the system's poles and zeros. This response is often implicitly assumed, by

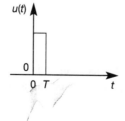

Figure 3.5 A rectangular
pulse signal formed from a
rising step signal followed
by a falling step signal.

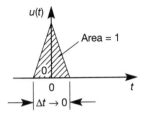

Figure 3.6 A unit impulse.

evaluating the system's time response directly from its open-loop LTF. If this is done, then a unit impulse input must be being assumed, because the output is the open-loop LTF × unity, and the only input with a Laplace transformed value of unity is a unit impulse.

Pulse inputs are particularly useful when dealing with an operational system, because there is no prolonged disturbance of the output variable, and also because pulses are often readily superimposed on the input variable. (Pulse testing is considered in Chapter 8.)

- *Steady-state sinusoidal input.* From the point of view of frequency-domain control system design techniques, this is the most useful of the forcing inputs. The relevant response for analysis is the sinusoidal steady-state response, which is observed only after all transient effects have disappeared, so the timescale becomes arbitrary. However, it is normal to assume that under steady-state conditions a sinewave

$$u(t) = A \sin \omega t$$

is injected into the system (see Figure 3.7), where A is the zero-to-peak amplitude and ω is the angular frequency in rad s^{-1}.

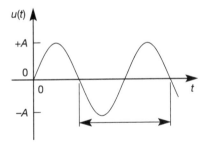

Figure 3.7 A sinusoidal input signal.

The Laplace transform of this function is

$$U(s) = \frac{A\omega}{s^2 + \omega^2}$$

which adds two imaginary poles to the s-plane, one at $s = j\omega$ and the other at $s = -j\omega$. The steady-state sinusoidal (or harmonic) response of a system and its relationship with the s-plane is considered in Section 3.2.3.

- *Random input.* In operational systems, all variables are continually changing. Although for well-controlled systems these changes are small and random, they are the actual forcing inputs. The responses of the operating plant and these random inputs may be correlated statistically and used to analyse the system dynamics. To reduce and interpret this data, extensive machine computation is normally required. Normal operating signals will often not contain sufficient information to allow proper determination of the dynamics; in which case small-amplitude, artificially generated, pseudo-random signals can be used instead (see Chapter 8). However, in principle, for operating plants the use of random signals offers a means of dynamic analysis which eliminates the need for significant process disturbance.

Figure 3.8 A general time-domain block diagram element representing a system.

In order to develop other basic design requirements, consider the system shown schematically in Figure 3.8, with input $u(t)$ and output $y(t)$. This system is assumed to be governed by an ordinary linear differential equation which is Laplace transformable, so that

$$Y(s) = \frac{N(s)}{D(s)} U(s) + \frac{I(s)}{D(s)} \tag{3.1}$$

where $I(s)$ is a function of the initial conditions $y(0)$, and

$$G(s) = \frac{N(s)}{D(s)}$$

is the system's transfer function.

The system's output $y(t)$ is the result of two terms: a transient, produced by the initial conditions on the system, and a forced response due to the input $u(t)$. In mathematical terms this is equivalent to saying that the solution of an ordinary linear differential equation is given by the complementary function (the transient response) and the particular integral (the forced response). When a system is in a steady-state condition, all the transients due to the initial conditions have subsided and $I(s)$ becomes zero.

3.2.1 Stability

Stability and Laplace transfer function models
A unit impulse is often used to determine a system's stability (see Section 1.3.4). For a system initially in steady-state conditions $(I(s) = 0)$ the output $Y(s)$ due to a unit impulse $(U(s) = 1)$ is obtained from Equation (3.1) as

$$Y(s) = \frac{N(s)}{D(s)}$$

Factorizing the characteristic function $D(s)$ and using a partial-fraction expansion (Appendix 2) gives the output

$$Y(s) = \frac{A_1}{s+p_1} + \frac{A_2}{s+p_2} + \cdots \frac{A_n}{s+p_n} \qquad (3.2)$$

Since it was implicitly assumed that the characteristic function of $D(s)$ was of order n, this factorization has produced n poles $-p_1, -p_2, \ldots, -p_n$, which could be real or occur in complex conjugate pairs. When a pole p_j is real then the coefficient A_j with which it is associated will be real. However, when there is a pair of complex conjugate poles p_j and p_{j+1} then the coefficients A_j and A_{j+1} with which they are associated will form a complex conjugate pair. The inverse Laplace transform of Equation (3.2) is found by taking the inverse Laplace transform of each term on the right-hand side of Equation (3.2) and then summing all the time responses to find the required response $y(t)$.

For a real pole at $s = -p_i$, the corresponding time response is (see Table 2.9):

$$\mathscr{L}^{-1}\left[\frac{A_i}{s+p_i}\right] = A_i e^{-p_i t} \qquad (3.3)$$

If the root of $(s+p_i)$ is negative (that is, p_i is positive) then the exponent in the time response is also negative and this element decays exponentially. If the root of $(s+p_i)$ is zero (that is, $p_i = 0$) the corresponding time response reduces to the constant A_i. When the root of $(s+p_i)$ is positive (p_i is negative), the exponent is positive and the response increases exponentially. The various responses and their relationship to the pole position in the s-plane can be seen from Figure 3.9.

With a pair of complex conjugate poles the inverse Laplace transform becomes

$$\mathscr{L}^{-1}\left[\frac{g+jh}{s+a-jb} + \frac{g-jh}{s+a+jb}\right]$$

$$= (g+jh)e^{-(a-jb)t} + (g-jh)e^{-(a+jb)t}$$

$$= (g+jh)e^{-at}e^{jbt} + (g-jh)e^{-at}e^{-jbt}$$

$$= e^{-at}[(g+jh)(\cos(bt) + j\sin(bt))$$

$$+ (g-jh)(\cos(bt) - j\sin(bt))] \qquad (3.4)$$

After multiplying out the brackets, and collecting terms, Equation (3.4) becomes

$$\mathscr{L}^{-1}\left[\frac{g+jh}{s+a-jb} + \frac{g-jh}{s+a+jb}\right] = 2e^{-at}[g\cos(bt) - h\sin(bt)] \qquad (3.5)$$

Recalling that

$$M\cos(bt + \theta) = M(\cos bt \cos\theta - \sin bt \sin\theta)$$

letting $g = M\cos\theta$ and $h = M\sin\theta$ gives the required rectangular to polar coordinate transformation enabling Equation (3.5) to be written as

$$\mathscr{L}^{-1}\left[\frac{g+jh}{s+a-jb} + \frac{g-jh}{s+a+jb}\right] = 2Me^{-at}\cos(bt + \theta) \qquad (3.6)$$

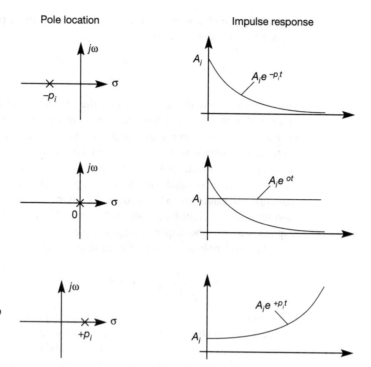

Figure 3.9 The relationship between pole location and impulse response for first-order systems.

Equations (3.5) and (3.6) will be considered in greater detail in the following section.

At the moment, it is important to note that each of the exponential terms in Equations (3.5) and (3.6) is the product of the real part of the complex poles and time. From this it can be seen that it is the location of the real part of the complex conjugate poles (Equation (3.4)) that determines whether the response decreases with time or not.

For completeness poles having the same s-plane location need to be considered. Assume that m of the n system poles ($m < n$) all have the same value p_i. These poles may be real or complex, but if the poles are complex then (for physical systems) there will be m corresponding complex conjugate poles. Using partial-fraction expansion (Appendix 2) produces terms of the form

$$\frac{A_{im}}{(s+p_i)^m} + \frac{A_{i(m-1)}}{(s+p_i)^{m-1}} + \cdots + \frac{A_{i1}}{s+p_i}$$

for which the corresponding inverse Laplace transform yields the time response:

$$A_{im}\frac{t^{(m-1)}}{(m-1)!}e^{-p_it} + A_{i(m-1)}\frac{t^{(m-2)}}{(m-2)!}e^{-p_it} + \cdots A_{i1}e^{-p_it} \tag{3.7}$$

and again it is seen to be the pole location that determines the dynamics of the response.

The time response of Equation (3.2) may now be obtained by summing the individual impulse response contributions from all the poles. That is

$$y(t) = A_1 e^{-p_1 t} + A_2 e^{-p_2 t} + \cdots + A_n e^{-p_n t} \tag{3.8}$$

Examining the system's impulse response (Equation (3.8)) gives the relationship between a system's pole positions and its stability. From Section 1.3.4 a system whose impulse response $y(t)$ decays to its original steady-state value with time is said to be asymptotically stable. This will occur if the real parts of *all* the system poles have negative real parts. If *any* of the poles has a positive real part then the system is unstable, and the response $y(t)$ will increase with time. When there are unrepeated system poles having a real part of zero and no poles with positive real parts, then the system is marginally stable. Systems which have repeated poles with zero real parts are always unstable. This can be seen from Equation (3.7), since the t terms are unbounded and the exponential terms will become constants. The s-plane stability regions are indicated in Figure 3.10.

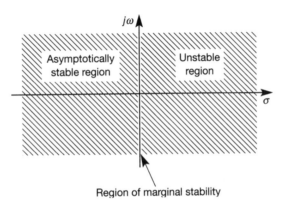

Figure 3.10 Stability regions in the s-plane.

Region of marginal stability

Stability and state-space models

In Section 2.5.1, the standard linear state-space model (Equations (2.43) and (2.44)) was introduced, and its solution was found in Section 2.7 by taking Laplace transforms. This solution will now be extended by including the initial conditions. Note that only the solution to the (dynamic) state equation is required, since the output equation is non-dynamic and simply maps the state responses to the outputs.

Taking Laplace transforms of the state equation $\dot{x}(t) = Ax(t) + Bu(t)$ gives:

$$sx(s) - x(0) = Ax(s) + Bu(s), \quad \text{or}$$

$$[sI - A]x(s) = x(0) + Bu(s), \quad \text{or}$$

$$x(s) = [sI - A]^{-1}x(0) + [sI - A]^{-1}Bu(s) \tag{3.9}$$

Equation (3.9) can be solved by analogy with the equivalent scalar equation:

$$X(s) = \frac{1}{s-a} x(0) + \frac{1}{s-a} bU(s) \tag{3.10}$$

Taking inverse Laplace transforms of Equation (3.10) gives the time-domain solution:

$$x(t) = e^{at}x(0) + \int_0^t e^{a(t-\tau)} b.u(\tau).d\tau$$

Now, by analogy, the solution of the multivariable state equation, Equation (3.9), must be:

$$x(t) = e^{At}x(0) + \int_0^t e^{A(t-\tau)} B.u(\tau).d\tau \tag{3.11}$$

which is sometimes written as:

$$x(t) = \boldsymbol{\Phi}(t)x(0) + \int_0^t \boldsymbol{\Phi}(t-\tau).B.u(\tau).d\tau \tag{3.12}$$

where $\boldsymbol{\Phi}(t)$ is called the *transition matrix* and is equal to e^{At}.

Note that the transition matrix can also be specified in the Laplace domain by rewriting Equation (3.9) as:

$$x(s) = \boldsymbol{\Phi}(s)x(0) + \boldsymbol{\Phi}(s)Bu(s) \tag{3.13}$$

where $\boldsymbol{\Phi}(s) = [sI - A]^{-1}$.

Additionally, in Section 5.6.2, a third specification of the transition matrix is given, in terms more suited to matrix algebra applications.

Equation (3.12) (and Equation (3.13)) can be compared with Equation (3.10). Note that both solutions have a free-response term (due to the initial conditions on the states $x(0)$) and a forced response term (due to the applied inputs $u(t)$). Earlier in this section, it was shown that it is the denominator terms $(s - a)$ in Equation (3.10) that fix the dynamic response of the system and hence determine its stability. By analogy, in Equations (3.12) or (3.13), it is the transition matrix that carries this information and, since this is a function of the plant matrix A, it must be this A matrix that determines the stability of the state-space model (this may have been deduced from Figure 2.49(a), since A is in the only dynamic loop).

To discover what it is about the A matrix that carries the same information as the poles of the LTF model, consider the rôle played by the transition matrix $\boldsymbol{\Phi}(s)$ in Equation (3.13). From the standard state-space model output equation (Equation (2.44)):

$$y(s) = Cx(s) + Du(s)$$

which, on combining with Equation (3.13), gives:

$$y(s) = C\boldsymbol{\Phi}(s)Bu(s) + Du(s) + C\boldsymbol{\Phi}(s)x(0)$$

This was the method used in Section 2.7 (Equation (2.95) and Example 2.26) to convert from a state-space model to a transfer function model. In particular, note that to evaluate $\boldsymbol{\Phi}(s)$ (the inverse of $[sI - A]$) required the determinant $|sI - A|$, which was seen to contain all the system poles. That is, for the equivalent LTF model, $|sI - A|$ is the system's characteristic function and therefore defines the system's dynamics.

Now, the *eigenvalues* of the A matrix (see Section A1.2 and Equation (A1.7), repeated below), are the values of the scalar quantity λ that satisfy the characteristic equation (CE):

$$|\lambda I - A| = |A - \lambda I| = 0 \tag{3.14}$$

The eigenvalues of a plant matrix A are therefore identical to the roots of the determinant $|sI - A|$, and are therefore the characteristic values for that plant. Further, irrespective of the state-space representation selected for any particular system, the *eigenvalues are the same.* Changing the selected states does not change the system dynamics. The following example demonstrates that the eigenvalues of a state-space model, and the poles of the corresponding LTF model, are identical.

Example 3.1 *Poles and eigenvalues of the antenna-positioning system*

Figure 2.48(b) (see Section 2.7) introduced a system whose Laplace transfer function is:

$$\frac{Y(s)}{U(s)} = \frac{5}{s(s+1)(s+5)} = \frac{5}{s^3 + 6s^2 + 5s} \tag{3.15}$$

By inspection, the poles of this model are $s = 0$, $s = -1$ and $s = -5$. Further, in Example 2.28, a state-space model of the system, using the set of state variables indicated in Figure 2.48(b), is given by (see Equation (2.103)):

$$A = \begin{bmatrix} 0 & 1 & 0 \\ 0 & -1 & 1 \\ 0 & 0 & -5 \end{bmatrix}, \qquad b = \begin{bmatrix} 0 \\ 0 \\ 5 \end{bmatrix} \qquad \text{and} \qquad c = \begin{bmatrix} 1 & 0 & 0 \end{bmatrix}$$

Using Equation (3.14), the characteristic equation (CE) is therefore:

$$|\lambda I - A| = \left| \lambda \begin{bmatrix} 1 & 0 & 0 \\ 0 & 1 & 0 \\ 0 & 0 & 1 \end{bmatrix} - \begin{bmatrix} 0 & 1 & 0 \\ 0 & -1 & 1 \\ 0 & 0 & -5 \end{bmatrix} \right| = \begin{vmatrix} \lambda & -1 & 0 \\ 0 & \lambda+1 & -1 \\ 0 & 0 & \lambda+5 \end{vmatrix} = 0$$

Expanding the determinant (see Section A1.1.2) gives:

$$\lambda(\lambda + 1)(\lambda + 5) = 0 \tag{3.16}$$

The eigenvalues are the roots of this equation, namely,

$$\lambda_1 = 0, \quad \lambda_2 = -1 \quad \text{and} \quad \lambda_3 = -5$$

It can be seen that the CE in Equation (3.16) is the same as the CE of the original transfer function; it does not matter that one is written in terms of λ and one in terms of s. Both provide the same values, and give the same information.

Therefore, for stability of a state-space model, no eigenvalue of the A matrix must have a positive real part. The same restrictions about eigenvalues on the imaginary axis apply, as discussed for poles earlier.

Looking back at the examples in Section 2.7, note that each state-space representation of the *same system* gives the *same eigenvalues*. A discussion of the part played by the corresponding *eigenvectors* in forming the time response is postponed until Section 5.6.

3.2.2 Performance and pole positions

The system shown in Figure 3.8 and described by Equation (3.1) will now be examined with regard to its performance. Ultimately, the performance of a system (see Section 1.3.4) will be judged by its time-domain response. The time response considered in this section is the step response. However, to simplify the analysis the concept of dominant poles will be introduced.

Dominant poles

Most stable linear systems could be represented by a transfer function having distinct poles and zeros which are all contained in the left half of the *s*-plane. The time-domain response of such a system may be found by performing a partial-fraction expansion, and then taking and summing the inverse Laplace transform of each term. However, each term has a dynamic part consisting of a negative exponential in time, the magnitude of which is determined by the real part of the system's poles, and an amplitude which is a function of the relative position of the poles and zeros (see Equations (3.2) and (3.8)). If the relative amplitude associated with a given term is small, or if the magnitude of the negative real part of any pole is large (thus producing a rapid exponential decay), the effect of removing that term from the system's time response is likely to be negligible. For this reason, it is often possible to produce a close approximation to the time response by considering only those terms in the partial-fraction expansion whose poles are most positive. This type of analysis is called a dominant pole analysis.

Dominant pole analysis requires the transient contribution from the non-dominant system poles to be small. This in turn implies either of the following:

- The non-dominant poles are well to the left of the dominant pole(s), so that the corresponding transients die away relatively rapidly.
- Any pole near the dominant pole(s) is close to a zero, so that the magnitude of its transient response will be very small.

In general these conditions may be verified by visual inspection of the pole–zero map. Figure 3.11 shows two typical pole–zero maps, one having a dominant single pole and the other a dominant pair of complex conjugate poles. The corresponding time-domain responses of systems having pole–zero maps similar to these, appear to be predominantly first or second order.

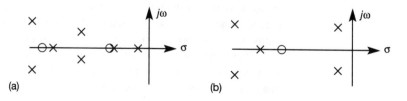

Figure 3.11 Illustrating the idea of dominant poles.

(a)

(b)

*Example 3.2 **A dominant pole analysis of a sixth-order system***

Find the step response $y(t)$ of a system having the transfer function

$$G(s) = \frac{2810.1(s+4)}{(s+3.8)(s+6)(s^2+2s+17)(s^2+10s+29)}$$

and compare it with the response obtained from a dominant pole analysis.

The pole–zero map for this system (see Figure 3.12) indicates that the dominant poles are a complex pair at $s = -1 + 4j$ and $s = -1 - 4j$. Performing a partial-fraction expansion and then grouping terms, the inverse Laplace transform of the forced system is (Problem 3.1):

$$y(t) = 1 + e^{-t}(0.6225 \cos 4t - 0.5714 \sin 4t)$$

$$+ e^{-5t}(0.9728 \cos 2t - 3.3296 \sin 2t) - 0.5184e^{-3.8t} - 2.0796e^{-6t}$$

and its step response is shown in Figure 3.13 (produced by the MATLAB m-file *fig3_13.m* on the accompanying disk).

In a dominant pole analysis, the step response could be obtained from the first two terms in the above equation, namely

$$y(t) = 1 + e^{-t}(0.6225 \cos 4t - 0.5714 \sin 4t)$$

This response is also shown in Figure 3.13. Note that for small values of t there is a large discrepancy between the two responses, but as t increases they become indistinguishable. Theoretically they would be identical only when t becomes infinite, but for all practical purposes they would, in this case, be considered to have become identical after about 1 second.

An alternative and more frequently used approach is to remove all but the dominant poles (and zeros) from the transfer function. For the transfer function $G(s)$, above, this gives

$$G_2(s) \approx \frac{17}{s^2+2s+17}$$

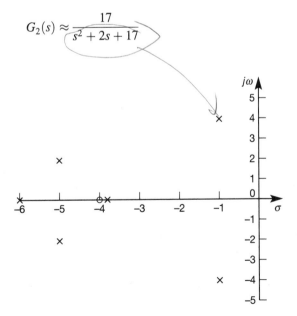

Figure 3.12 *s*-plane pole and zero locations for Example 3.2.

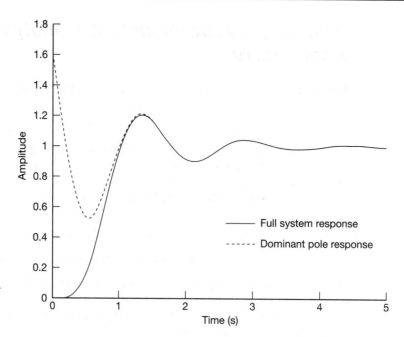

Figure 3.13 Comparison of full and dominant pole analyses for Example 3.2.

Note that the numerator term has been adjusted so that the steady-state value of this dominant pole analysis and the original step response are the same. That is, the steady-state gains $G(0)$ and $G_2(0)$ are both unity. For a type 0 system (Section 2.5.6) the steady-state gain is the factor by which a steady input must be multiplied to give the system's final steady-state output. When $G(s)$ and $G_2(s)$ are excited by a unit step input the final value theorem (Section 2.5.5) indicates that

$$y(\infty) = y_2(\infty) = 1$$

For a unit step input, the inverse Laplace transform of $G_2(s)U(s)$ is

$$y_2(t) = 1 - e^{-t}(\cos 4t - 0.25 \sin 4t)$$

and produces the dashed response shown in Figure 3.14. This response may be compared with the full response $y(t)$ which is also shown in Figure 3.14. Superficially the two responses look very different. However, a closer examination of the responses will show that the main difference is a phase shift.

Normally, a dominant pole analysis is carried out by inspection. Given a full pole–zero map, the designer uses the dominant pole (or poles) to estimate the system's response (the subsequent text deals with dominant pole analysis). If it appears that the estimated response matches the desired response, then this would be checked using the full model.

The approximation of this sixth-order model by a second-order model might suggest the use of dominant poles for model reduction. However, as demonstrated by Figures 3.13 and 3.14, model reduction is always accompanied by a loss of system information. In some situations the lost information might be relatively unimportant, but in others it is critical, and can lead to erroneous conclusions.

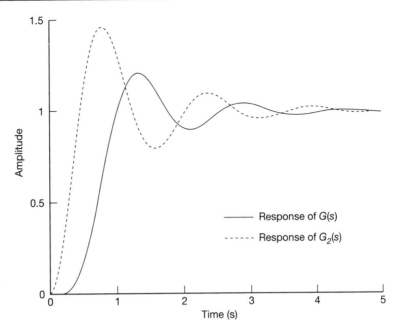

Figure 3.14 Comparison of full and truncated LTF analyses for Example 3.2.

A first-order dominant system

A system with a negative dominant first-order pole can be represented by the transfer function

$$G(s) = \frac{A}{s + a} \tag{3.17}$$

Since it is a type 0 system, the steady-state gain is

$$G(0) = A/a$$

and, for a unit step input $(U(s) = 1/s)$, the output response is

$$y(t) = \frac{A}{a} - \frac{A}{a} e^{-at}$$

Setting the system's steady-state gain to unity $(A = a)$ produces the time response

$$y(t) = 1 - e^{-at}$$

and it is evident that the pole position $(s = -a)$ determines the shape of the response.

Figure 3.15(a) shows two first-order systems' step responses. One has a pole at -0.5 which could be considered close to the unstable right-hand half of the s-plane, and one has a pole at -5.0, which could be considered well into the stable left-hand half of the s-plane. The further the pole is moved into the left-hand half of the s-plane, the closer the output matches the input. The file *fig3_15.m* on the accompanying disk produces the plot. To use a design strategy which simply moves a system's closed-loop poles well into the left-hand half of the s-plane may, at this point, seem desirable. It must be remembered, however (Section 1.2), that the inputs to real systems will be contaminated with noise. If both systems are now

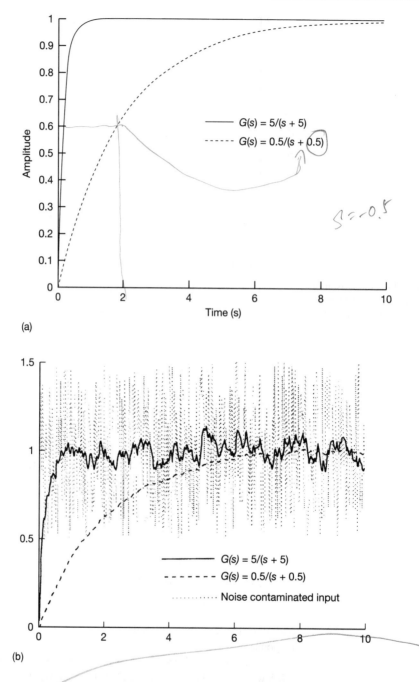

Figure 3.15 (a) Fast and slow first-order step responses. (b) The systems of Figure 3.15(a) with superimposed input noise.

excited with unit step inputs contaminated with noise (see Figure 3.15(b)), the responses shown in Figure 3.15(b) are produced. The system with its pole at -5.0 transmits a large fraction of the noise with the signal, whereas the system with its pole at -0.5 removes (or filters out) much of the noise. This design conflict between a system's speed of response and its noise filtering properties will be considered in Section 3.4.

Before leaving this section, note that there is another very common standard form of the first-order LTF. This is the so-called 'gain–time constant' (or Bode) form, which is obtained by dividing Equation (3.17) by the magnitude of the pole (that is, by a):

$$G(s) = \frac{K}{1 + s\tau} \tag{3.18a}$$

where

K is the gain (sometimes called the d.c. gain, or low-frequency gain), $K = A/a$

τ is the *time constant* in seconds – defined below, $\tau = 1/a$

If a unit step input is applied as before $(U(s) = 1/s)$, the time response is now given by:

$$y(t) = K(1 - e^{-t/\tau}) \tag{3.18b}$$

which gives the response in Figure 3.16, for $K = 1$. Any first-order system ('first-order lag', or 'simple lag') represented in the form of Equation (3.18a) will have the unit step response of Figure 3.16. The vertical scale only needs multiplying by the gain, K, and the horizontal scale is easily converted into seconds by knowledge of the time constant, τ. The time constant is the time taken by the output to achieve 63.2 per cent of the remaining distance to the final value, from any given time (for proof, set $t = \tau$ in Equation (3.18b)). From this normalized form, it can be seen that any stable first-order step response reaches 95 per cent of its final value after three time constants (put $t = 3\tau$ in Equation (3.18b)), and 99 per cent after five time constants.

Figure 3.16 A standard (normalized) first-order step response.

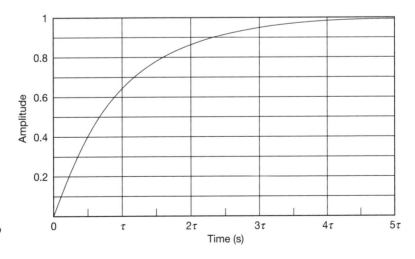

A second-order dominant system

A system having a pair of dominant complex conjugate poles may be approximated by the standard second-order equation:

$$\frac{Y(s)}{U(s)} = \frac{\omega_n^2}{s^2 + 2\zeta\omega_n s + \omega_n^2} \tag{3.19}$$

Note that ω_n is called the *undamped natural frequency* (rad s^{-1}), and ζ is called the *damping ratio* (dimensionless). Often, a gain factor K will be included in the numerator, as in Equation (3.18). For the present, this is assumed to be unity in Equation (3.19).

When excited by a unit step input Equation (3.19) becomes

$$Y(s) = \frac{\omega_n^2}{s(s^2 + 2\zeta\omega_n s + \omega_n^2)} \tag{3.20}$$

This system has three poles at $s = 0$, $s = -\zeta\omega_n + j\omega_d$ and $s = -\zeta\omega_n - j\omega_d$, where ω_d is the *damped frequency* of the system response, given by

$$\omega_d = \omega_n\sqrt{1 - \zeta^2} \tag{3.21}$$

The pole–zero map for this system is shown in Figure 3.17. To find the time response a partial-fraction expansion of Equation (3.20) is required. This takes the form

$$Y(s) = \frac{A_1}{s} + \frac{A_2}{s + \zeta\omega_n - j\omega_d} + \frac{A_3}{s + \zeta\omega_n + j\omega_d} \tag{3.22}$$

Using the cover-up rule (Appendix 2)

$$A_1 = \left.\frac{\omega_n^2}{s^2 + 2\zeta\omega_n s + \omega_n^2}\right|_{s=0} = \frac{\omega_n^2}{\omega_n^2} = 1$$

$$A_2 = \left.\frac{\omega_n^2}{s(s + \zeta\omega_n + j\omega_d)}\right|_{s=-\zeta\omega_n + j\omega_d} = \frac{-\zeta}{2\sqrt{1 - \zeta^2}} - \frac{j}{2} = -\frac{1}{2}\left(\zeta\frac{\omega_n}{\omega_d} + j\right)$$

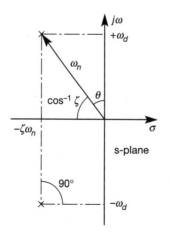

Figure 3.17 Pole locations in the *s*-plane for underdamped second-order systems.

and therefore A_3 will be:

$$A_3 = -\frac{1}{2}\left(\zeta\frac{\omega_n}{\omega_d} - j\right)$$

Substituting the values of A_1, A_2 and A_3 into Equation (3.22) and then by comparison with Equations (3.3), (3.4) and (3.5) the inverse Laplace transform is given by

$$y(t) = 1 - e^{-\zeta\omega_n t}\left[\zeta\frac{\omega_n}{\omega_d}\sin\omega_d t + \cos\omega_d t\right] \tag{3.23}$$

and a typical time response for this system is shown in Figure 3.18. It may be noted that the response is contained within two exponential envelopes at $1 + \exp(-\zeta\omega_n t)$ and $1 - \exp(-\zeta\omega_n t)$. From the pole–zero map of Figure 3.17, the complex poles have their real parts at $-\zeta\omega_n$. Also, the periodic time of the response is given by $2\pi/\omega_d$ (that is, the frequency of oscillation is ω_d rad s^{-1}). Again from the pole–zero map, the imaginary parts of the complex poles are at $-\omega_d$ and $+\omega_d$. The damping ratio ζ which, in Figure 3.17, is found from the angle between the ω_n vector and the negative real axis, can take on any value between 0 and 1, and determines the general shape of the response.

Figure 3.19(a) gives a set of standard second-order response curves. The vertical axis assumes that the d.c. gain (the factor K which might have appeared in the numerator of Equation (3.20)) is unity. If not, then simply multiply the scale by K. The horizontal axis has also been normalized to apply to any system, by plotting against the dimensionless quantity $\omega_n t$. The axis could therefore be read as if it were in seconds, for a system in which $\omega_n = 1$ rad s^{-1}. ζ takes the values indicated in the

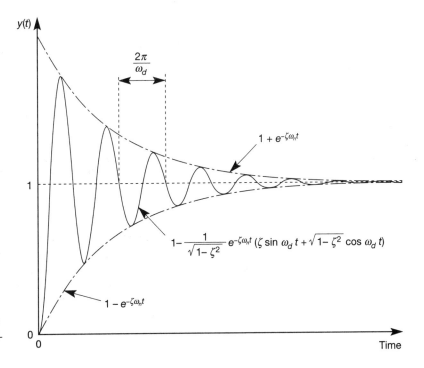

Figure 3.18 Step response of an underdamped second-order system.

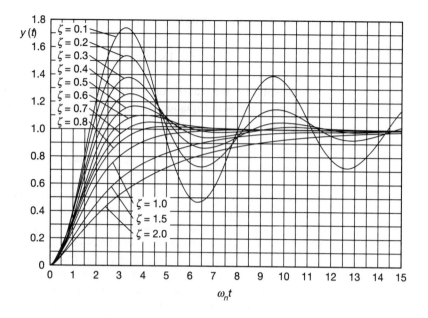

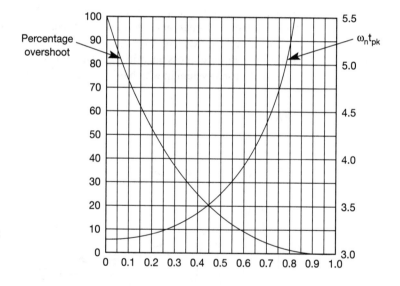

Figure 3.19 Standard (normalized) second-order system step responses. (a) Step responses. (b) Relationship between percentage overshoot, peak parameters and damping ratio.

figure. The smaller the damping ratio, the more oscillatory the response until, when $\zeta = 0$, there will be sustained oscillations. When the damping ratio is unity, the response is said to be *critically damped*. In this condition, there is a double pole at $-\zeta\omega_n$ and the response is the fastest possible without overshooting (that is, without the response exceeding its final steady-state value). For values of ζ greater than unity, there would be two distinct real poles centred on $-\zeta\omega_n$ and the response is then said to be over-damped (and cannot overshoot, as it consists of two cascaded first-order systems).

Step response performance criteria

The object of defining performance criteria is to establish a workable set of standards which can be applied to achieve an optimal response. However, there are problems in determining what aspect of the response makes it optimal. For example, in the automatic positioning of an astronomical telescope, a steady-state error would be unacceptable, and any setpoint change must be performed gradually in order to minimize shock loadings on the structure. In missile tracking systems, speed of response (that is, a fast rise time) is essential, and *small* steady-state errors would be permissible. These two examples illustrate some of the diversity of control problems. Each problem has its own peculiar requirements, and consequently the designer must determine those aspects which are critical and ensure that appropriate criteria are selected. For this reason it is usual to identify certain parameters which describe the main attributes of the response, and to use them as the performance criteria. A design which best meets these criteria may then be thought of as optimal, although in reality the design will only be as good as the selected criteria.

With step response performance criteria, two approaches are possible. One specifies limits on certain features of the response and the other, which uses integral performance indices, tries to quantify in a single positive measure the whole of the response. Both have advantages and disadvantages, and both have been widely used. Integral performance indices tend to be more appropriate in simulation work and state-space analysis.

The step response shown in Figure 3.20 indicates a number of commonly used performance criteria:

(1) *Steady-state error.* The difference between the demand input $r(t)$ and the steady-state output y_{ss}. Given a stable model of the system, this would normally be found using the final value theorem (see Example 3.3, below).

(2) *Rise time*, t_r. The shortest time required for the response to achieve some specified percentage of its final value, y_{ss}. Sometimes, the 100 per cent rise time is used, as shown in Figure 3.20, although the rise time is often taken as the time from 10 per cent of the final value of the response to when it first achieves 90 per cent of its final value. This latter method allows the rise time of non-overshooting responses to be quoted.

(3) *Peak overshoot*, $y(t_{p1}) - y_{ss}$. Sometimes referred to as the initial or maximum overshoot, peak overshoot is the amplitude of the first peak. This is normally expressed as a percentage of the final (steady-state) value. Figure 3.19(b) shows the variation of percentage overshoot with damping ratio, for damping ratios between zero and unity.

(4) *Peak time*, t_{p1}. The time from the initiation of the response to peak overshoot. Figure 3.19(b) gives a graph showing the variation of the normalized time to the first peak, against damping ratios between zero and unity.

(5) *Subsidence ratio.* In a decaying oscillation this is the ratio of the amplitudes of successive cycles. A subsidence ratio of 4 : 1 or 3 : 1 and a peak overshoot of 30 per cent would provide a practical optimal response for many process control systems.

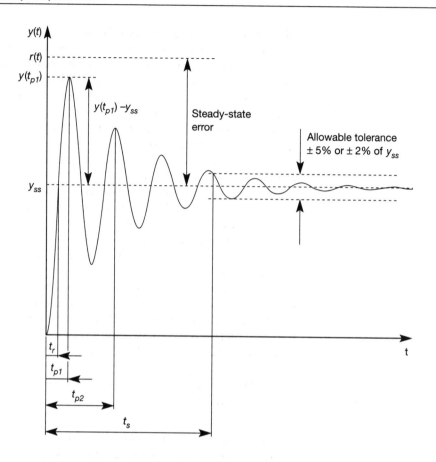

Figure 3.20 Performance measures on second-order system step responses.

(6) *Settling time*, t_s. The time taken for the response to reach and remain within some specified range of its final value. An allowable tolerance of between ± 2 and ± 5 per cent is usual.

(7) *Number of oscillations to settling time*. This is self-explanatory.

The formulation of these specifications in the s-plane is considered later.

Example 3.3(a) *Steady-state error calculation by the final value theorem*

Section 3.4.1 below investigates the effect of system *type* number (recall from Chapter 2 that this is the number of poles at the origin of the s-plane) on steady-state performance. It is worth pre-empting that discussion with an example of how the final value theorem is used to evaluate steady-state error.

Consider a pressure control system. The output is the pressure in a cylinder, and the input is a voltage proportional to the desired pressure, applied to a servovalve. It is therefore required that the output follows the input with a gain of unity at steady state. The open loop system is modelled by a simple lag with a d.c. gain of 0.6 and time constant

35 ms, cascaded with a second-order element of damping ratio 0.71 and undamped natural frequency 38.9 rad s^{-1}, as follows:

$$G(s) = \frac{0.6}{(0.035s + 1)} \cdot \frac{1513}{(s^2 + 55s + 1513)}$$

The final value theorem (see Section 2.5.5) states that the final value of this system's output ($y(\infty)$, say) in response to a unit step input (which has a Laplace transform $1/s$ from entry 2 of Table 2.9) is given by:

$$y(\infty) = \lim_{t \to \infty} [y(t)] = \lim_{s \to 0} [sY(s)] = \lim_{s \to 0} \left[s \frac{1}{s} G(s) \right]$$

$$= \lim_{s \to 0} [G(s)] = \frac{0.6 \times 1513}{1513} = 0.6$$

This might have been expected since the d.c. gain (which is effectively the same thing as the steady-state gain) was specified as 0.6 above. The final value of 0.6, compared with the input step of 1.0, means that there is a steady-state error of 40 per cent.

What might not be so easy to visualize without calculation, is what will happen if the pressure is measured using a transducer of negligible dynamics, and the measurement is used in a closed-loop control scheme to try to reduce this steady-state error. The closed-loop arrangement is exactly that shown in Figure 3.2, where $G(s)$ is the open-loop transfer function from above, and $C(s)$ is going to be a simple gain, K, for the purposes of this example (that is, a simple proportional-only controller is being used).

The closed-loop transfer function of the arrangement of Figure 3.2, with $C(s) = K$, is given by:

$$\frac{Y(s)}{R(s)} = \frac{KG(s)}{1 + KG(s)} = \frac{K \times 0.6 \times 1513}{(0.035s + 1)(s^2 + 55s + 1513) + K \times 0.6 \times 1513}$$

Applying the final value theorem to this, for a unit step input on R, gives the result:

$$y(\infty) = \lim_{t \to \infty} [y(t)] = \lim_{s \to 0} [sY(s)] = \frac{K \times 0.6 \times 1513}{1513 + K \times 0.6 \times 1513} = \frac{0.6K}{1 + 0.6K}$$

It can be seen that with a gain of $K = 2$, for example, the steady-state error will actually be *worse* than that of the open-loop scheme! As K is increased the steady-state error improves, until at $K = 2.5$ the closed-loop system has the same steady-state error as the open-loop system. As K is increased further, the steady-state error decreases, but the result above shows that it can never become zero, even if K could be made very large.

What is worse is that the dynamic behaviour of the system (that is, the amount of overshoot, the decay ratio and the settling time, all of which were acceptable in open-loop) is actually very poor for most values of K. This is examined in Example 3.3(b), where it is found that for values of K greater than about 8.3, the closed-loop system becomes unstable.

The conclusion is that a proportional-only controller is not good enough for this system. At the very least, proportional plus integral control is needed to get rid of the steady-state error (this idea was introduced in Section 1.3.5, and is discussed more fully in Section 4.5.2).

Returning to the discussion of performance, there are also many integral performance indices which, if minimized, will optimize the response in some sense.

Although not necessarily related to a system's step response, it is convenient to introduce performance indices at this point. Typically, integral performance indices measure a system's closed-loop performance by measuring 'the area' between the time axis and the system's error response on an error response plot. Some of the more common ones are:

- *The integral of the squared error (ISE)*. The ISE is one of the more popular measures, as it lends itself most readily to mathematical manipulation. Its major disadvantage is that it can produce an unacceptably oscillatory response. To overcome this problem, the measure may be modified to include additional terms such as the integral of the squared time rate of change of error (or the squared error velocity) or the integral of the error squared acceleration. As well as including these extra terms, each term may have associated with it a scalar weighting, w say, as shown below. However, the problem is one of interpreting the modified measure in terms of the expected response.

$$ISE = \int_0^\infty e^2(t)\, dt$$

$$ISE_{mod} = \int_0^\infty \left(w_1 e^2(t) + w_2 \dot{e}^2(t) + w_3 \ddot{e}^2(t)\right) dt$$

- *The integral of the absolute value of error (IAE)*. This measure places equal weighting on all deviations from the final steady-state value. Its main disadvantage is the mathematical determination of a minimum value for all except the simplest of systems.

$$IAE = \int_0^\infty |e(t)|\, dt$$

- *The integral of time by absolute error (ITAE)*. For the engineer this is probably the most acceptable of the indices since it most nearly matches intuitive expectations. Following a step change in demand, it is inevitable that there will be a large error in the response, and to penalize this would place an artificial bias on the measure. However, at some later time a smaller error should be heavily penalized. In general, for a system of any order, the transfer function minimizing the ITAE will have an acceptable form of transient response. Again, the main disadvantage is in the analytical determination of a minimum value of the index.

$$ITAE = \int_0^\infty t|e(t)|\, dt$$

Chapter 12 looks in more detail at time domain optimal control.

Performance specifications in the *s*-plane

In a development of some simple rules for *s*-plane performance specifications, only the dominant poles of a system need be considered. Desirable responses tend to be ones which are under-damped (but not to excess), and this occurs if the system has a dominant pair of complex conjugate poles. The corresponding time-domain step response may therefore be approximated by a second-order system, with the response (from Equation (3.23)):

$$y(t) = 1 - e^{-\zeta\omega_n t}\left[\zeta\frac{\omega_n}{\omega_d}\sin\omega_d t + \cos\omega_d t\right] \tag{3.24}$$

Note that if a non-unity gain factor K multiplies ω_n^2 in the numerator of Equation (3.20), then the values of $y(t)$ given by Equation (3.24) (and (3.25), below) must also be multiplied by K.

By noting that $\omega_d = \omega_n\sqrt{1-\zeta^2}$ (Equation (3.21), for $\zeta < 1$), the expression for $y(t)$ becomes

$$y(t) = 1 - \frac{e^{-\zeta\omega_n t}}{\sqrt{1-\zeta^2}}[\zeta\sin\omega_d t + \sqrt{1-\zeta^2}\cos\omega_d t] \tag{3.25}$$

Now from Figure 3.17

$$\cos\theta = \frac{\omega_d}{\omega_n} = \sqrt{1-\zeta^2} \quad \text{and} \quad \sin\theta = \zeta\omega_n/\omega_n = \zeta$$

which on substitution into Equation (3.25) and application of the cosine formula

$$\cos(\omega_d t - \theta) = \cos\omega_d t\cos\theta + \sin\omega_d t\sin\theta$$

gives

$$y(t) = 1 - \frac{e^{-\zeta\omega_n t}}{\sqrt{1-\zeta^2}}\cos(\omega_d t - \theta) \tag{3.26}$$

(see Equation (3.6)). From this equation it is possible to estimate the positions in which the dominant poles must be in order to meet a particular performance specification. Again note that if a non-unity gain factor K multiplies ω_n^2 in the numerator of Equation (3.20), then the values of $y(t)$ given by Equation (3.26) must also be multiplied by K, and that this will additionally apply to some of the results below (such as that of Equation (3.27)). Consider each of the following specifications described above:

(1) *Steady-state error*. This would normally be found using the final value theorem, see Example 3.3(a), above. For the system described by Equation (3.24) or (3.26), setting t to infinity gives the value $y(\infty) = 1$ and, since the input was a unit step, the steady-state error $r(\infty) - y(\infty) = 0$.

(2) *100 per cent rise time*. This will occur the first time $\cos(\omega_d t - \theta)$ becomes equal to zero, that is when

$$\omega_d t - \theta = \pi/2$$

Hence the rise time is

$$t_r = (\pi/2 + \theta)/\omega_d$$

The rise time is a function of the damped frequency ω_d.

(3) *Peak overshoot.* This is obtained by substituting the expression for the peak time into the response $y(t)$ (see the note following Equation (3.26), above):

$$y(t_p) = 1 + e^{-\zeta \omega_n t_p} \qquad (3.27)$$

and is seen to be a function of the real part of the complex poles. Since

$$t_p = \frac{\pi}{\omega_d} = \frac{\pi}{\omega_n \sqrt{(1 - \zeta^2)}}$$

then the peak overshoot is a function of ζ only. Equation (3.27) gives the maximum or peak value of the response. The percentage maximum overshoot (Figure 3.19(b)) is given by

$$\left(\frac{y(t_p)}{y(\infty)} - 1 \right) \times 100 \text{ per cent}$$

(4) *Peak time.* The peak time may be found by differentiating the response $y(t)$ and setting the derivative dy/dt equal to zero. This will be found to occur when $\sin(\omega_d t)$ first becomes zero, hence the peak time is

$$t_p = \pi / \omega_d \qquad (3.28)$$

which is again a function of the damped frequency.

(5) *Subsidence ratio.* Assume that the subsidence ratio is $R_s : 1$, such that the first peak $y(t_{p1}) - y(\infty)$ is R_s times greater than the second peak, $y(t_{p2}) - y(\infty)$. It may be shown that for any two adjacent peaks the subsidence ratio of a second-order system will be the same. One cycle later than Equation (3.28) the second peak occurs when $t_{p2} = 3\pi / \omega_d$. Hence

$$y(t_{p1}) - 1 = R_s \left(y(t_{p2}) - 1 \right)$$

Substituting from Equation (3.27) and taking natural logarithms yields

$$\frac{\zeta}{\sqrt{(1 - \zeta^2)}} = \frac{\log_e (R_s)}{2\pi}$$

which indicates that the subsidence ratio is a function of the damping ratio ζ alone.

(6) *Settling time.* This is obtained by considering the decay of the response envelope. For a 2 per cent settling time,

$$0.02 = \exp(-\zeta \omega_n t_s)$$

or, taking natural logarithms,

$$t_s = 4/(\zeta \omega_n)$$

For a 5 per cent settling time the expression becomes (see Example 3.3(b), below):

$$t_s = 3/(\zeta \omega_n) \qquad (3.29)$$

Settling time is a function of the real part of the dominant poles. The reciprocal of $\zeta\omega_n$ has units of time, and is referred to as the equivalent time constant of the second-order system. Equation (3.29) may also be used to define dominance. Non-dominant poles decay within their own response envelope and consequently, if there were a real pole at $-\sigma$, the transient due to this pole would have the form $Ke^{-\sigma t}$. This term would decay to within 5 per cent of its initial value when $t = 3/|\sigma|$. If the system rise time t_r is greater than or equal to $3/|\sigma|$, then the exponential term will have only a small effect on the performance measures. A similar argument holds for complex poles. Therefore, in general, a dominant pole analysis requires the non-dominant poles to be close to a zero, or to have a negative real part of magnitude greater than or equal to $3/t_r$.

(7) *Number of oscillations to settling time.* Given the periodic time of the damped oscillations and the settling time of the system, then

$$\text{Number of oscillations} = \frac{\text{Settling time}}{\text{Periodic time}}$$

and is a function of the damping ratio.

Example 3.3(b) *Settling time and overshoot calculation for the system of Example 3.3(a)*

An examination of the open-loop transfer function of the system of Example 3.3(a) suggests that a dominant pole analysis of the type outlined above is impossible, because the real pole is at about $s = -28.6$ and the second-order part has complex poles whose real parts are $s = -27.5$. No part can therefore be classed as 'dominant' over the other.

However, in closed-loop, the situation changes dramatically. Example 3.3(a) gives the closed-loop transfer function as:

$$\frac{Y(s)}{R(s)} = \frac{KG(s)}{1 + KG(s)} = \frac{K \times 0.6 \times 1513}{(0.035s + 1)(s^2 + 55s + 1513) + K \times 0.6 \times 1513}$$

A general expression could be written for the poles of this LTF, and evaluated at various values of K. However, it is far easier to use a package such as MATLAB (Appendix 3) to do it for us numerically. The MATLAB commands to achieve this are as follows:

```
≫ num = 0.6*1513;              % open-loop numerator

≫ den = conv([0.035 1], [1 55 1513]);   % open-loop denominator

≫ k = 2;                       % controller gain

≫ [numc, denc] = cloop(k*num, den, −1);   % closed-loop LTF

≫ damp(denc)                   % see below
```

The command *damp(denc)* is issued with no semicolon, so the results are displayed. It shows every closed-loop pole (it calls them eigenvalues), together with its associated damping ratio and undamped natural frequency.

(Aside: Readers who do not have access to such a computer package must insert a chosen value of K into the closed loop LTF given above, factorize the denominator to find the first-order root and the second-order complex pair of roots, and fit values of damping ratio and undamped natural frequency to the resulting second-order term, by comparison with the standard form (see, for example, Equation (3.19)).)

Using $K = 2$, as in the MATLAB example above, reveals that the first-order closed-loop pole is at about $s = -58.6$, while the real parts of the two complex poles are at about $s = -12.5$. The second-order poles can therefore be regarded as dominant, and the associated values of damping ratio and undamped natural frequency can be used to evaluate the performance measures outlined above.

In this case ($K = 2$), these are approximately $\zeta = 0.31$ and $\omega_n = 40.3 \text{ rad s}^{-1}$. The corresponding values of overshoot and 5 per cent settling time, from above, are therefore about 36 per cent and 241 ms respectively. Remember to include the closed-loop d.c. gain (equivalent to the value of y_{ss} for a unit step input) as a multiplying factor on the value of y_{tp} given by Equation (3.27). Of course, if MATLAB is available, then

$$\gg \text{step(numc, denc)}$$

will display the closed-loop step response for confirmation (the values will be found to be approximate, due to the dominant pole approximation).

As K increases, the values of these performance indicators get progressively worse. At $K = 7$, the dominant second-order approximation is almost perfect, because the real parts of the complex roots are very small compared with that of the real pole. This time, the values are $\zeta = 0.03$ and $\omega_n = 53 \text{ rad s}^{-1}$. The overshoot and 5 per cent settling time are therefore about 90 per cent and 1728 ms respectively (there are also about 15 oscillations during this settling time, and the steady-state error is still about 19 per cent). By the time K is increased to 8.3, the closed-loop system has become unstable (the *damp* command reports positive real parts for the complex roots). Example 3.7(b), confirms this value analytically.

Using dominant pole analysis, the above performance specifications have all been shown to be related to either the damped frequency, the damping ratio or the equivalent time constant of the dominant poles. In the s-plane, lines of constant damped frequency ω_d are lines parallel to the real axis of the s-plane. A line of constant damping ratio is a radial line emanating from the origin of the s-plane and is at an angle ϕ to the negative real axis; that is,

$$\zeta = \cos \phi \qquad \text{or} \qquad \phi = \cos^{-1} \zeta$$

Lines of constant time constant, $\zeta\omega_n$ lines, are parallel to the imaginary axis. For completeness, note that lines of constant ω_n describe circles centred on the origin of the s-plane. All these lines are shown in Figure 3.21.

In order to obtain an optimal step response it would be normal to impose limits on the value of ω_d, which defines bands on the rise time, the peak time and the damped frequency. A minimum and a maximum value of the damping ratio ζ specify the subsidence ratio, the peak overshoot and the number of oscillations to the settling time. A minimum time constant defines the peak overshoot and, probably more importantly, the settling time. The maximum time constant specifies the system's noise-rejection properties.

The design of a system using s-plane performance criteria typically reduces to the problem of ensuring that the system's dominant poles are not within the shaded

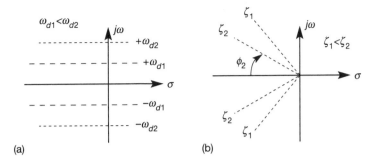

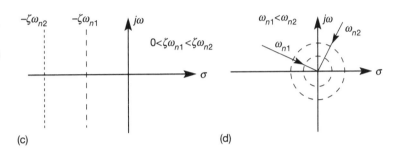

Figure 3.21 Constant parameter contours in the *s*-plane. (a) Lines of constant damped frequency; (b) lines of constant damping ratio; (c) lines of constant 'equivalent time constant'; (d) lines of constant undamped natural frequency.

area in Figure 3.22. Again, in terms of the design problem, it is desirable to keep the specifications to a minimum. Over-specification may create unnecessary difficulties, and seldom leads to any significant improvement in the transient response. Removal of the constraints on the maximum value of ζ, and/or on the minimum value of ω_d, would ease the design problem considerably (see Figure 3.22).

Before leaving this section, it will be useful to consider the effect on the time response of shifting the dominant poles along the *s*-plane performance lines. Figure 3.23 illustrates this shift for two second-order systems. When the systems have the same $\zeta\omega_n$ value, the responses are contained within the same exponential

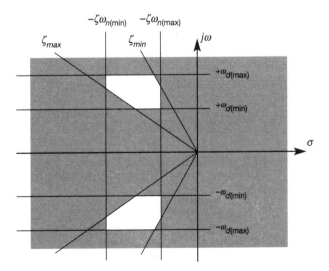

Figure 3.22 The effects of performance specifications in the *s*-plane.

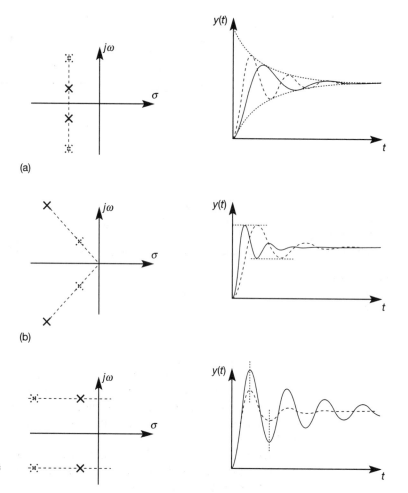

Figure 3.23 The relationship between pole locations and step response for second-order systems. (a) Systems with the same value of $\zeta\omega_n$; (b) systems with the same value of damping ratio (ζ); (c) systems with the same value of damped frequency (ω_d).

envelopes, but the frequency of oscillation is greater for the system having the larger ω_d value (see Figure 3.23(a)). The same ζ value indicates that the peak overshoot in the two systems is the same (see Figure 3.23(b)), but the speed of response will be faster for the system whose poles have the more negative real part. Figure 3.23(c) shows two systems with the same value of ω_d. Although the frequency of the response is the same, the more negative poles produce the faster and more heavily damped response.

3.2.3 Frequency response methods

Design methods based on a system's steady-state frequency response characteristics have many attractions. High-order systems, complicated controller dynamics and time delays are easily handled. The main disadvantage is that for systems higher than second order, there is no direct relationship between the transient response in the time domain and the various frequency response plots. However, experience shows that, provided certain frequency conditions are met, good transient response characteristics may be expected. These conditions are usually known as frequency performance criteria.

In the subsequent sections the relationship between the Laplace and frequency domains is explored, conditions for closed-loop instability are established and used to gain insight into various frequency performance criteria. The Nyquist, Bode and Nichols plots are also examined.

The steady-state frequency response

Consider again the open-loop system shown in Figure 3.8 which is governed by Equation (3.1). To determine the system's steady-state response to a harmonic forcing input let

$$u(t) = a \sin \omega t \tag{3.30}$$

which, on Laplace transformation, becomes

$$U(s) = \frac{\omega a}{s^2 + \omega^2}$$

Substituting for $U(s)$ in Equation (3.1) gives

$$Y(s) = \frac{N(s)}{D(s)} \frac{\omega a}{s^2 + \omega^2} + \frac{I(s)}{D(s)}$$

Since the system is assumed to be stable, the effect of the initial conditions diminishes with time and

$$\mathcal{L}^{-1} \left[\frac{I(s)}{D(s)} \right] \to 0 \quad \text{as } t \to \infty$$

Therefore

$$y(t) = \mathcal{L}^{-1} \left[\frac{N(s)}{D(s)} \frac{\omega a}{s^2 + \omega^2} \right] \quad \text{as } t \to \infty \tag{3.31}$$

In order to solve for the steady-state response of Equation (3.31) it is first necessary to make a partial-fraction expansion of the terms requiring inverse transformation, namely the terms within the square brackets:

$$\frac{N(s)}{D(s)} \frac{\omega a}{s^2 + \omega^2} = \frac{A_1}{s - j\omega} + \frac{A_2}{s + j\omega} + \text{(all terms arising from } D(s)) \tag{3.32}$$

As the system is stable, all the terms arising from the system's characteristic function $D(s)$ must be functions which disappear as $t \to \infty$. Hence, the steady-state response may be found by solving for A_1 and A_2. Using the Heaviside formula, Appendix 2:

$$A_1 = \left(\frac{N(s)}{D(s)} \frac{\omega a (s - j\omega)}{(s - j\omega)(s + j\omega)} \right) \Bigg|_{s=j\omega} \tag{3.33a}$$

which reduces to

$$A_1 = \frac{a}{2j} \frac{N(j\omega)}{D(j\omega)} = \frac{a}{2j} G(j\omega) \tag{3.33b}$$

Similarly,

$$A_2 = -\frac{a}{2j} G(-j\omega) \tag{3.34}$$

The terms $G(j\omega)$ and $G(-j\omega)$ are complex, and their real and imaginary parts are functions of ω. Hence, by the normal rules of complex algebra, these terms may be written in polar form as

$$G(j\omega) = M(\omega)e^{j\phi(\omega)} \quad \text{or} \quad Me^{j\phi} \tag{3.35a}$$

and

$$G(-j\omega) = M(\omega)e^{-j\phi(\omega)} \quad \text{or} \quad Me^{-j\phi} \tag{3.35b}$$

Thus, from Equations (3.33) and (3.34)

$$A_1 = \frac{a}{2j} Me^{j\phi} \tag{3.36a}$$

$$A_2 = \frac{a}{2j} Me^{-j\phi} \tag{3.36b}$$

Substituting the expansion of Equation (3.32) back into Equation (3.31) yields the steady-state response as

$$y(t)_{ss} = \mathscr{L}^{-1}\left[\frac{A_1}{s - j\omega} + \frac{A_2}{s + j\omega}\right]$$

which, on inverse Laplace transformation, becomes

$$y(t)_{ss} = A_1 e^{+j\omega t} + A_2 e^{-j\omega t} \tag{3.37}$$

Substituting Equations (3.36) for A_1 and A_2 into Equation (3.37) gives

$$y(t)_{ss} = \frac{a}{2j} Me^{+j\phi}e^{j\omega t} - \frac{a}{2j} Me^{-j\phi}e^{-j\omega t}$$

or

$$y(t)_{ss} = \frac{a}{2j} M(e^{j(\phi+\omega t)} - e^{-j(\phi+\omega t)})$$

Since $\sin x = (1/2j)(e^{jx} - e^{-jx})$, then

$$y(t)_{ss} = aM \sin(\omega t + \phi) \tag{3.38}$$

On comparing the steady-state output (Equation (3.38)) with the input (Equation (3.30)), it is evident that the input signal has been multiplied by the factor M, known as the magnitude ratio or magnification factor. As M (or, more correctly, $M(\omega)$) is a function of frequency, for any given frequency it may be determined directly from the open-loop transfer function $G(s)$. Also, the steady-state input and output signals have the same frequency of oscillation, but the output is phase-shifted by an amount ϕ. This phase shift $\phi(\omega)$, like the magnification factor, is frequency-dependent and may be determined directly from the open-loop transfer function $G(s)$.

The above analysis is simply a calculation of the residues associated with a pair of complex conjugate poles on the imaginary s-plane axis. Provided the system is linear and stable, the steady-state response is characterized by $M(\omega)$ and $\phi(\omega)$ which, in turn, depend on the transfer function $G(s)$. From Equations (3.33) and (3.35),

$$G(s)|_{s=j\omega} = G(j\omega)$$

and

$$G(j\omega) = \frac{N(j\omega)}{D(j\omega)}$$

One way to find $G(j\omega)$ is to replace every value of s in $G(s)$ by $j\omega$, and then to reduce $G(j\omega)$ to its simplest form using the normal rules of complex algebra. For example, if

$$G(s) = \frac{s+4}{s^2 + 5s + 6}$$

then

$$G(j\omega) = \frac{j\omega + 4}{(j\omega)^2 + 5j\omega + 6}$$

which may be 'simplified' to give

$$G(j\omega) = \frac{\omega^2 + 24}{\omega^4 + 13\omega^2 + 36} + j\,\frac{\omega(\omega^2 - 14)}{\omega^4 + 13\omega^2 + 36}$$

This equation uses rectangular coordinates – that is, it has the form

$$G(j\omega) = U(\omega) + jV(\omega)$$

To convert this equation to polar form, let

$$G(j\omega) = M(\omega)e^{j\phi(\omega)}$$

where

$$M(\omega) = \sqrt{(U(\omega)^2 + V(\omega)^2)}$$

and

$$\phi(\omega) = \tan^{-1}[V(\omega)/U(\omega)]$$

This procedure can become quite involved, particularly if the transfer function is of high order. Fortunately, there is a simpler way to find $M(\omega)$ and $\phi(\omega)$.

In general a transfer function in which s is set equal to $j\omega$ may be represented by an expression of the form

$$G(j\omega) = K\,\frac{\displaystyle\prod_{i=1}^{m}(a_i + jc_i)}{\displaystyle\prod_{i=1}^{n}(g_i + jh_i)}$$

where m and n are the numbers of terms in the numerator and denominator, respectively. For this system

$$M(\omega) = K \frac{\prod\limits_{i=1}^{m} \sqrt{(a_i^2 + c_i^2)}}{\prod\limits_{i=1}^{n} \sqrt{(g_i^2 + h_i^2)}} \tag{3.39}$$

and

$$\phi(\omega) = \sum_{i=1}^{m} \tan^{-1}(c_i/a_i) - \sum_{i=1}^{n} \tan^{-1}(h_i/g_i) \tag{3.40}$$

Before leaving this section note that, in general, a forcing input excites many harmonic signals, producing a response which, depending on the dynamics of the plant, is either attenuated or amplified. Furthermore, these signals will be phase-shifted relative to each other, with the effect that the recombined signals will tend to weaken or strengthen each other.

Example 3.4(a) *Finding the magnitude and phase responses of a LTF*

For the transfer function

$$G(s) = \frac{s+4}{s^2 + 5s + 6}$$

find the magnitude $M(\omega)$ and phase $\phi(\omega)$.

First find $G(j\omega)$ by setting $s = j\omega$, hence

$$G(j\omega) = \frac{j\omega + 4}{6 - \omega^2 + 5j\omega}$$

From Equation (3.39)

$$M(\omega) = \frac{\sqrt{(\omega^2 + 4^2)}}{\sqrt{[(6 - \omega^2)^2 + 25\omega^2]}} \tag{3.41}$$

and from Equation (3.40)

$$\phi(\omega) = \tan^{-1}\left(\frac{\omega}{4}\right) - \tan^{-1}\left(\frac{5\omega}{6 - \omega^2}\right) \tag{3.42}$$

Note that when $\omega^2 > 6$ in the last term, the *arctan* of a negative number is required. The normal procedure is to rewrite Equation (3.42) as:

$$\phi(\omega) = \tan^{-1}\left(\frac{\omega}{4}\right) - \left[180° - \tan^{-1}\left(\frac{5\omega}{\omega^2 - 6}\right)\right] \quad \text{for} \quad \omega^2 > 6 \tag{3.43}$$

Equations (3.41), (3.42) and (3.43) provide the required solution. Alternatively, the denominator of $G(s)$ could be factorized to give

$$G(s) = \frac{(s+4)}{(s+2)(s+3)}$$

for which

$$G(j\omega) = \frac{(j\omega+4)}{(j\omega+2)(j\omega+3)}$$

Then, from Equation (3.39):

$$M(\omega) = \frac{\sqrt{(\omega^2+4^2)}}{\sqrt{(\omega^2+2^2)}\sqrt{(\omega^2+3^2)}} \tag{3.44}$$

and from Equation (3.40)

$$\phi(\omega) = \tan^{-1}\left(\frac{\omega}{4}\right) - \tan^{-1}\left(\frac{\omega}{2}\right) - \tan^{-1}\left(\frac{\omega}{3}\right) \tag{3.45}$$

By substituting various values of ω into the expressions for $M(\omega)$ and $\phi(\omega)$ it is readily checked that Equations (3.41) and (3.44) are equivalent; and that Equation (3.45) is equivalent to Equation (3.42) when $\omega^2 < 6$, and to Equation (3.43) when $\omega^2 > 6$.

Example 3.4(b) *Steady-state frequency analysis of a recording system*

A temperature recording system with transfer functions as shown in Figure 3.24 is used to measure a time-varying fluid temperature in an experiment. For steady-state operation a Fourier analysis of the recorder output trace yields the expression

$$Q_o = 1.5 + \sin 10t + 0.2 \sin 60t$$

Determine the corresponding expression for the true input temperature.
    The overall transfer function for the temperature recording system is

$$G(s) = \frac{50 \times 10^{-3}}{(0.05s+1)\left(\dfrac{s^2}{1600} + \dfrac{s}{40} + 1\right)} = \frac{Q_o(s)}{Q_i(s)}$$

Since the steady-state output was given by

$$Q_{oss} = 1.5 + \sin 10t + 0.2 \sin 60t$$

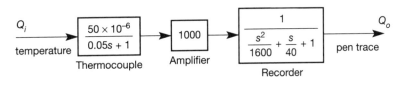

Figure 3.24 Block diagram of the temperature recording system for Example 3.4.

the corresponding input must have been of the form

$$Q_i = A + B \sin(10t - \phi(10)) + C \sin(60t - \phi(60))$$

In general, for an input of the form

$$q_i(t) = a \sin \omega t$$

the corresponding steady-state output will be (see Equation (3.38))

$$q_o(t)_{ss} = aM(\omega) \sin(\omega t + \phi(\omega))$$

To find $M(\omega)$ and $\phi(\omega)$ from $G(s)$, set $s = j\omega$. Then

$$G(j\omega) = \frac{50 \times 10^{-3}}{(0.05j\omega + 1)\left(-\dfrac{\omega^2}{1600} + \dfrac{j\omega}{40} + 1\right)}$$

$$= \frac{50 \times 10^{-3}}{(1 + 0.05j\omega)\left(1 - \dfrac{\omega^2}{1600} + \dfrac{j\omega}{40}\right)}$$

Hence, from Equation (3.39)

$$M(\omega) = \frac{50 \times 10^{-3}}{\sqrt{[1 + (0.05\omega)^2]}\sqrt{\left[\left(1 - \dfrac{\omega^2}{1600}\right)^2 + \dfrac{\omega^2}{1600}\right]}}$$

and from Equation (3.40)

$$\phi(\omega) = 0 - \tan^{-1}(0.05\omega) - \tan^{-1}\left[\frac{\dfrac{\omega}{40}}{1 - \dfrac{\omega^2}{1600}}\right]$$

For $\omega = 0$

$$M(\omega) = 50 \times 10^{-3}$$

which is the system's steady-state gain. Hence, to obtain a constant output signal of 1.5 the input signal must be

$$1.5/(50 \times 10^{-3}) = 30$$

For $\omega = 10$ rad s^{-1}

$$M(10) = \frac{50 \times 10^{-3}}{\sqrt{(1.25)}\sqrt{(0.9414)}} = 46.1 \times 10^{-3}$$

and

$$\phi(10) = -0.465 - 0.261 = -0.725 \text{ rad}$$

For $\omega = 60$ rad s^{-1}

$$M(60) = \frac{50 \times 10^{-3}}{\sqrt{(10)}\,\sqrt{(3.8125)}} = 8.10 \times 10^{-3}$$

and

$$\phi(60) = -1.249 + 0.876 = -0.373 \text{ rad}$$

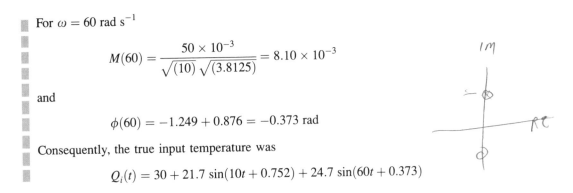

Consequently, the true input temperature was

$$Q_i(t) = 30 + 21.7 \sin(10t + 0.752) + 24.7 \sin(60t + 0.373)$$

Conditions for closed-loop marginal stability

Marginal stability in a closed-loop system occurs when the controller and other control elements in the loop are adjusted so as to produce a self-sustained steady-state cycling in the response. Such a condition requires the closed-loop system to have two complex conjugate poles on the imaginary axis of the s-plane, and all its other poles to have negative real parts. By quantifying this condition in terms of a system's frequency response, it is shown to provide a basis for the definition of performance criteria.

As an aside, note that in practice it would be hard to build a reliable oscillator from a linear system with two complex conjugate poles on the imaginary axis of the s-plane. This is because the slightest disturbance (for example, the effects of temperature on component values) would cause the poles to move slightly to the left (in which case the oscillations would eventually die away) or slightly to the right (in which case, they would gradually grow). To maintain reliable oscillations (for use in a laboratory signal generator, for example) actually requires a deliberately *nonlinear* system, undergoing a phenomenon known as a *limit cycle* – see Chapter 14.

Consider the marginally stable closed-loop system shown in Figure 3.25 which is performing self-sustained oscillations. A time profile of the system's response is shown in Figure 3.26 in which the first half-cycle of a sinewave is shown as a solid line in the $e(t)$ plot. This signal passes through the various elements in the loop and emerges at $b(t)$ with the same amplitude but with a phase lag of 180°, as shown. The signal then enters the comparator, which in this case inverts the signal $b(t)$. Wave inversion is equivalent to a phase shift of $-180°$ (or $-\pi$ radians), since

$$-b = -A \sin \omega t = A \sin(\omega t - \pi)$$

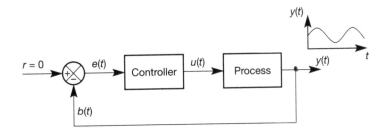

Figure 3.25 A marginally stable control loop, exhibiting self-sustained oscillations.

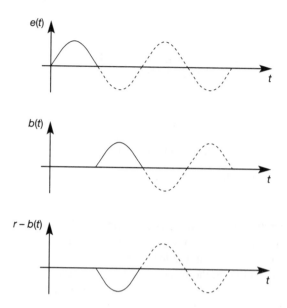

Figure 3.26 Analysis of the phasing of signals in Figure 3.25.

The output of the comparator, $r - b(t)$, is in phase with the error signal $e(t)$ and provides the second half of the sinewave. This process repeats itself continuously, and the system performs self-sustained oscillations.

Two conditions must be met if sustained oscillations are to occur: at the signal frequency ω, $M(\omega)$ must be unity and $\phi(\omega)$ must be $-180°$ (that is, the phase lag must be 180°). If $M(\omega)$ were less than unity, 0.5 say, then the feedback signal $b(t)$ would be half the amplitude of the error signal $e(t)$, phase-shifted by $-180°$. This signal would be inverted to form the new error signal at $e(t)$. As the process continued, each successive peak in the error signal would be a quarter the height of its predecessor, and the system would have quarter-amplitude damping – that is, the subsidence ratio would be $4:1$ (see the subsection on 'step response performance criteria' in Section 3.2.2). If $M(\omega)$ were greater than unity, the amplitude of successive peaks would increase and the system would become unstable. By similar arguments, when $M(\omega) = 1$, $\phi(\omega)$ must be $-180°$ for sustained oscillations to occur. Under these conditions the total phase shift, including that of the comparator, is $-360°$, and signals going around the closed-loop are in phase.

Since the above conditions will make a closed-loop system marginally stable, the design problem is reduced to the relatively simple task of adjusting the loop characteristics so as to produce a stable condition which, from experience, is known to provide satisfactory responses. Typically, the process elements are known but are not easily accessible and cannot be altered to any great extent. Consequently, the problem is principally one of specifying the controller and associated hardware required for given specifications of control system performance.

Experience shows that a good, stable control system will usually be obtained if the overall open-loop magnitude ratio $M(\omega)$ is set equal to some value between 0.4 and 0.5 when the phase angle $\phi(\omega)$ is $-180°$. A companion rule is that the phase angle should be between $-115°$ and $-135°$ when the magnitude ratio $M(\omega)$ is unity. In general, for an asymptotically stable, open-loop system, the application of these rules produces a slightly under-damped closed-loop control system that responds to a forcing step input with a 20–30 per cent overshoot, followed by decaying oscillations having a subsidence ratio of about 3 : 1.

It is useful at this point to introduce some terminology:

- *Phase crossover* is the frequency at which the overall open-loop phase angle $\phi(\omega)$ first reaches the critical value of $-180°$.
- *Gain crossover* is the frequency at which the overall open-loop magnitude ratio $M(\omega)$ first reaches the value of unity.
- *Phase margin* is the number of degrees by which the phase angle is numerically smaller than the critical angle of $(-)180°$ at gain crossover.
- *Gain margin* is the factor by which the magnitude ratio must be multiplied at phase crossover to make it unity.

The empirically derived design rules may now be restated in terms of the more commonly used measures of gain and phase margin, as:

Gain margin between 2 and 2.5, phase margin between 45° and 65°.

In using these rules the following points should be noted:

- The above design rules do not give accuracy of control. It is possible to produce a satisfactory transient response, but still have an unacceptable steady-state error.
- The rules say nothing about desirable speeds of response.
- The rules do not guarantee closed-loop stability for all systems. However, using Nyquist's stability criterion (Section 4.3), it can be shown that for open-loop stable systems of type 0, 1 or 2, the rules will produce closed-loop stability.

It is also worth noting that consideration of system stability can broadly be divided into two aspects. Firstly, there is the question of whether the system is, or is not, stable. Such analysis is considered further in Section 3.3. Given that a system is stable, it is then necessary to consider *how* stable it is. Such analysis is considered in the frequency domain by gain and phase margin measures.

Example 3.5 *Evaluation of gain and phase margins from frequency response data*

The open-loop frequency response characteristics of a simple single-loop control system, including the controller but excluding the comparator, are found to be as follows:

Frequency, cycles/min	0.01	0.06	0.10	0.6	1.0
Magnitude ratio	4.82	1.00	0.47	0.08	0.02
Phase angle	$-10°$	$-122°$	$-180°$	$-272°$	$-316°$

What are the gain margin and phase margin for this system?

From the data in the table, the gain margin (which is the safety factor in the magnitude ratio when the phase angle is $-180°$) is 1.00/0.47 or 2.13, and the phase margin (which is the margin of safety in the phase angle at gain crossover, overall magnitude equal to unity) is $180° - 122°$, or 58°.

3.3 Routh stability

In Section 3.2 it was shown that the response of a system to an impulse could be approximated by a dominant pole analysis. Dominant poles are those roots of the system's LTF denominator, Equation (3.1), which have the most positive real parts. System zeros, particularly if they are close to, or to the right of, the dominant poles (that is, have more positive real parts than the dominant poles) will alter the shape of the response. However, the role of zeros in shaping a system's response will not be covered until Chapter 4.

Some simple stability definitions for linear systems were provided in Section 3.2.1. It was seen that for asymptotic stability all the system's poles must be contained within the open left-hand half of the s-plane. Any pole in the right-hand half of the s-plane makes the system unstable, see Figure 3.27. Unrepeated dominant poles on the imaginary axis will make the system marginally stable, while multiple (repeated) dominant poles on the imaginary axis make the system unstable. The stability of a system in terms of its dominant poles and their impulse response is summarized in Table 3.1, in which multiple poles (two or more poles occupying the same location) are indicated by a number above the pole. For example, system number 3 has two poles occupying the location (0, 0) in the s-plane.

3.3.1 Direct stability tests

If the root locations of a system's characteristic equation are known, then it is a simple matter to establish stability by visually checking that all the roots have negative real parts. Also, if the zero locations between the forcing input and the measured output are known, the response may be determined. Indeed, it is normal practice, when finding a system's time response using inverse Laplace transform techniques, to start by factoring the system's denominator polynomial and checking for asymptotic stability. Such a check is called a direct stability test. When suitable computing equipment and software is available (for example, MATLAB – Appendix 3, or any other suitable package), direct stability checks are the preferred solution (as was done using MATLAB in Example 3.3(b)). However, care should be exercised if poles appear very close to the imaginary s-plane axis. Due to numerical rounding errors, these poles may appear stable, but actually be unstable, and vice versa.

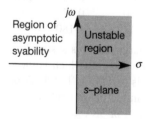

Figure 3.27 Unstable region for poles in the s-plane.

Table 3.1 Impulse responses and s-plane pole plots.

	Pole locations in s-plane	Impulse response	Comment
1			Asymptotically stable
2			Marginally stable
3			Unstable
4			Unstable
5			Asymptotically stable
6			Marginally stable
7			Unstable

3.3.2 A necessary condition for asymptotic stability

In general a characteristic equation may be written in the form

$$D(s) = \sum_{i=0}^{n} a_i s^{n-1} = 0$$

or

$$D(s) = a_0 s^n + a_1 s^{n-1} + \cdots + a_{n-1}s + a_n = 0$$

If $D(s)$ has real coefficients, then its complex roots will occur in conjugate pairs. Let $D(s)$ have p real roots, $-\alpha_1, -\alpha_2, \ldots, -\alpha_p$ and q pairs of complex roots, $-\beta_1 \pm j\phi_1, -\beta_2 \pm j\phi_2, \ldots, -\beta_q \pm j\phi_q$, so that $n = 2q + p$. In factored form $D(s)$ may be written as

$$D(s) = a_0(s + \alpha_1) \cdots (s + \alpha_p)(s^2 + 2\beta_1 s + \beta_1^2 + \phi_1^2)$$
$$\cdots (s^2 + 2\beta_q s + \beta_q^2 + \phi_q^2)$$

All the α and β values will be positive if the system is stable. By expansion, it is found that each of the a values must exist, and all must be of the same sign. This is a necessary condition for stability, but it may not be sufficient. However, if one of the a values is negative or zero then the system is unstable (if $a_n = 0$ the system is at best marginally stable).

Example 3.6 **Stability by inspection of the characteristic equation**

Comment on the stability of the following systems' characteristic equations:

(a) $D(s) = s^5 + 5s^4 + 3s^3 - 2s^2 + s + 8 = 0$

(b) $D(s) = s^5 + 5s^4 + 3s^3 + 2s^2 + 8 = 0$

(c) $D(s) = s^5 + 5s^4 + 3s^3 + 2s^2 + s = 0$

(d) $D(s) = -s^5 - 5s^4 - 3s^3 - 2s^2 - s - 8 = 0$

System (a) is unstable, since the coefficient associated with s^2 is negative and therefore it does not satisfy the necessary condition for stability.

The s term is missing in system (b), so the system is unstable.

In system (c) there is a root at $s = 0$, which means that the system is at best marginally stable. The remainder polynomial does satisfy the necessary condition for stability and therefore Routh's criterion should be used to test for sufficiency (see Section 3.3.3). Routh's criterion would indicate that all the remaining roots are in the left half of the s-plane.

System (d) satisfies the necessary condition for stability. The system may or may not be stable, and a further test is required. Routh's criterion will show that the system is unstable.

Note that MATLAB (Appendix 3) can very easily find the roots of such equations, which can then be inspected to see whether or not all the real parts are negative. For example, to solve (b) using MATLAB, use the single command (note the zero coefficient of s):

$$\gg \text{roots}([1 \ 5 \ 3 \ 2 \ 0 \ 8])$$

This immediately confirms the instability, as there is a complex pair of roots with real parts of $+0.62$.

3.3.3 The Routh stability criterion

The Routh stability criterion provides a quick and easy method of establishing a system's stability. It involves calculating the number (not the location) of characteristic roots within the unstable right half of the s-plane. The number of roots in the stable left half of the s-plane and the number of roots on the imaginary axis may also be found. Apart from the stability checks, usually carried out before determining root locations, the method may also be used to establish the limiting values for a variable parameter beyond which a system would become unstable. This section does not attempt to prove Routh's stability criterion, but just gives the basic results.

The characteristic equation of the system whose stability is to be tested must be expressed in the form

$$D(s) = \sum_{i=0}^{n} a_i \, s^{n-i} = 0$$

before Routh's criterion may be applied. For this polynomial, the *Routh array* is shown below. The first two rows are made up from the characteristic equation coefficients in the obvious manner. The elements in the third row onwards are obtained from elements in the previous two rows in the following way:

$$b_1 = \frac{a_1 a_2 - a_0 a_3}{a_1}, \qquad b_2 = \frac{a_1 a_4 - a_0 a_5}{a_1}, \cdots$$

$$c_1 = \frac{b_1 a_3 - a_1 b_2}{b_2}, \qquad c_2 = \frac{b_1 a_5 - a_1 b_3}{b_1}, \cdots$$

Row						
0	s^n	a_0	a_2	a_4	a_6	$\cdots$
1	s^{n-1}	a_1	a_3	a_5	a_7	$\cdots$
2	s^{n-2}	b_1	b_2	b_3	b_4	$\cdots$
3	s^{n-3}	c_1	c_2	c_3	c_4	$\cdots$
$\vdots$	$\vdots$	$\vdots$	$\vdots$	$\vdots$	$\vdots$	
$n-1$	s^1	y_1	y_2			
n	s^0	z_1				

(3.46)

The number of sign changes in the first column $(a_0, a_1, b_1, c_1, \ldots, y_1, z_1)$ of the Routh array is equal to the number of roots of $D(s)$ with positive real parts. The criterion and the array are named after Edward John Routh (1831–1907), in whose 1877 paper the array first appeared.

Example 3.7(a) **A simple Routh stability test**

Use the Routh array to test the stability of a system having the characteristic equation:

$$D(s) = s^4 + s^3 + s^2 + 2s + 3 = 0$$

The system satisfies the necessary conditions for stability, so the Routh array (Equation (3.46)) may be used to test for sufficiency. The array appears below and the first column of the array has two sign changes, so the system is unstable and has two poles with positive real parts.

Solving for the roots of $D(s)$ produces $-1.074 \pm j0.706$ and $+0.574 \pm j1.219$.

Row				
0	s^4	1	1	3
1	s^3	1	2	
2	s^2	-1	3	
3	s	5	0	
4	s^0	3		

Example 3.7(b) **A Routh stability test for the system of Example 3.3**

The pressure control system investigated in Example 3.3 was said to be unstable by the time the controller gain K reached 8.3. From that example (either 3.3(a), or 3.3(b)), the denominator of the closed-loop LTF (which gives the characteristic equation when equated to zero) leads to:

$$D(s) = (0.035s + 1)(s^2 + 55s + 1513) + K \times 0.6 \times 1513$$

$$= 0.035s^3 + 2.925s^2 + 107.955s + 1513 + 907.8K = 0$$

Forming the Routh array:

Row			
0	s^3	0.035	107.955
1	s^2	2.925	$1513 + 907.8K$
2	s	$89.85 - 10.86K$	0
3	s^0	$1513 + 907.8K$	

The last element in the first column is positive for any positive value of K. However, the third element is only positive for $K < 89.85/10.86$. If K exceeds this value, there will be two sign changes in the first column, implying two roots in the right-hand half of the s-plane, and an unstable system. This limiting value of K is about 8.27, confirming the findings of Example 3.3(b), that the pair of complex roots became unstable for K values greater than about 8.3.

The Routh array: special cases

A zero appearing in the first element of any of the n rows indicates instability or, at best, marginal stability. The array could not be continued by the normal method of construction, since this would involve division by zero. If the root distribution is required an alternative technique must be adopted, and this gives rise to the following special cases.

The first column term of any row vanishes, but some of the remaining terms in the row are not zero. There are several methods available for dealing with this particular case; some are better suited to hand calculation, and some to machine computation. The method given here is best used with hand calculation and, if applied with care, will always yield the correct result.

In this method, the first-column zero is replaced with an arbitrarily small number, δ say, and the array is continued in the normal way. The limit as $\delta \to 0$ is then found, and the first column of the array is checked for sign changes. Again, the number of sign changes equals the number of roots with positive real parts.

Example 3.8 *A Routh array with a first-column zero*

Test the stability of a system having the characteristic equation

$$D(s) = s^5 + 2s^4 + 2s^3 + 4s^2 + s + 1 = 0$$

For this polynomial the Routh array is

Row				
0	s^5	1	2	1
1	s^4	2	4	1
2	s^3	0	0.5	

The zero is replaced with δ and the array continued. After taking the limit as $\delta \to 0$, the final array, from row 2 onwards, becomes:

2	s^3	δ	0.5
3	s^2	$-1/\delta$	1
4	s	0.5	0
5	s^0	1	

Since there are two sign changes in the first column, the system is unstable with two positive roots. A root-solving routine indicates that the roots of $D(s)$ are $-0.090 \pm j0.533$, $+0.069 \pm j1.274$ and -1.957.

All the coefficients of a row become zero. This indicates the presence of a divisor polynomial $F(s)$ whose roots are all symmetrically located about the origin of the s-plane.

Assume that, in the general Routh array of Section 3.3.3, row 4 is found to be an all-zero row. The coefficients of the required divisor polynomial are obtained from the previous row, in this case row 3, to give:

$$F(s) = \sum_{i=0}^{(n-3)/2} c_{i+1} s^{(n-3)-2i} = 0$$

Since the roots of $F(s)$ are symmetrically located around the real and imaginary axes, they will be of the form

$$s = \pm\sigma \quad \text{or} \quad s = \pm j\beta \quad \text{or} \quad s = -\sigma \pm j\beta \quad \text{and} \quad s = +\sigma \pm j\beta$$

Clearly $F(s)$ will always be of even order, and consequently the all-zero row will always be associated with an odd power of s.

In order to complete the array, $F(s)$ is differentiated with respect to s and the coefficients of $dF(s)/ds$ substituted into what was the all-zero row. Using these new coefficients, the array is completed in the normal way.

The array may now be interpreted as follows. As far as the all-zero row, the number of sign changes in the first column indicates the number of roots of the remainder polynomial with positive real parts. From what was the row before the all-zero row, each change in sign in the first column of the array indicates the number of roots of the divisor polynomial with positive real parts. However, as the roots are symmetrical, any sign changes associated with the divisor polynomial $F(s)$ will indicate the number of roots in the right half of the s-plane, and also the number in the left half of the s-plane. Roots of the divisor polynomial which are not accounted for in this way must lie on the imaginary axis.

Example 3.9 **A Routh array with a zero row**

A closed-loop control system with unity negative feedback has an open-loop transfer function as follows, in which K is the controller gain, G is everything else in the forward path and H is the unity feedback gain:

$$KGH(s) = \frac{K}{s(s+1)(s^2+s+1)}$$

For what value of K will the system be marginally stable, and what is the corresponding frequency of oscillation?

For this system with the feedback loop *closed*, block diagram reduction (Section 2.6.2) produces the closed-loop characteristic equation:

$$D(s) = s^4 + 2s^3 + 2s^2 + s + K = 0$$

and the corresponding Routh array is

Row				
0	s^4	1	2	K
1	s^3	2	1	
2	s^2	1.5	K	
3	s	$1 - 4K/3$		

For $K = 3/4$, row 3 becomes an all-zero row and the divisor polynomial of row 2 is

$$\tfrac{3}{2}s^2 + \tfrac{3}{4} = 0$$

or

$$F(s) = 2s^2 + 1 = 0$$

By dividing $F(s)$ into $D(s)$ it is found that $F(s)$ is the required divisor polynomial, and that

$$D(s) = (2s^2 + 1)(\tfrac{1}{2}s^2 + s + \tfrac{3}{4})$$

In order to complete the array when $K = 3/4$, $F(s)$ is differentiated with respect to s and the coefficients of $dF(s)/ds$ are used to replace the zero coefficients of row 3. Now $dF(s)/ds = 4$, and the completed array is

Row				
0	s^4	1	2	0.75
1	s^3	2	1	
2	s^2	1.5	0.75	
3	s	4		
4	s^0	0.75		

Since there are no sign changes up to and including row 2, there are no roots of the remainder polynomial in the right half of the s-plane. Since there are no sign changes from row 2, all roots of the divisor polynomial must lie on the imaginary axis.

Hence when $K = 3/4$ the system will have a pair of complex conjugate roots on the imaginary axis, and will therefore be marginally stable. To find the location of these roots, and hence the frequency of oscillation, s is set equal to $j\omega$ and the value of ω found from $F(j\omega)$:

$$F(j\omega) = -2\omega^2 + 1 = 0$$

hence

$$\omega = \frac{1}{\sqrt{2}} \ \text{rad s}^{-1}$$

which is the required frequency of oscillation.

3.3.4 The Routh array and PID controller design

Many industrial controllers for single input/single output systems consist of three elements: proportional (P), integral (I) and derivative (D) action (Section 1.3.5). One idealized transfer function of a controller which includes all three items, the so-called three-term controller, is given below (the physical significance of the terms is discussed in Chapter 4).

$$G_c(s) = K[1 + T_d s + 1/(T_i s)]$$

where

K = gain of the proportional term

T_d = derivative action time (or rate time) (seconds)

T_i = integral action (or reset time) (seconds)

There are a number of controllers based on the three-term (PID) controller. The most common are the P or PI type, which together account for the majority of industrial control elements. These controllers are derived from the three-term controller $G_c(s)$ by making adjustments to T_d and T_i:

$T_d = 0$ and $T_i = \infty$ gives a P controller

$T_d = 0$ and T_i finite gives a PI controller

Commercially available PID controllers are normally electronic or, for hazardous environments, pneumatic. It is important to note that derivative action (represented by the term $T_d s$) cannot be implemented exactly by these technologies in practice, which is why this PID representation is an *idealized* LTF. PID controllers are considered further in Chapter 4.

The next section examines the empirical Ziegler and Nichols methods for establishing settings for PID controllers, as it makes a good example of the use of the Routh criterion (even if a full understanding of the PID controller itself must wait until Chapter 4).

Ziegler–Nichols rules for controller tuning

These results were designed to be used when a process or model is available and amenable to a few simple experiments (J. G. Ziegler and Nathaniel Burgess Nichols, 1942).

(i) First approach

The process to be controlled is shown in Figure 3.28. Assume that it has the property that, under purely proportional control, it is asymptotically stable in the range $0 \leq K < K_c$ and becomes unstable in an oscillatory manner for $K > K_c$. For this type of system the following experimental procedure is specified:

(1) Turn up the gain K until the onset of continuous oscillations. At this critical gain K_c the closed-loop system is marginally stable – on the boundary between stable and unstable behaviour – so any gain adjustments must be carried out with extreme care.

(2) Note the value K_c and the period of the oscillations T.

(3) The recommended settings are given by:

P control	$K = 0.5K_c$	(3.47)
PI control	$K = 0.45K_c$	(3.48a)
	$T_i = 0.833T$	(3.48b)
PID control	$K = 0.6K_c$	(3.49a)
	$T_i = 0.5T$	(3.49b)
	$T_d = 0.125T$	(3.49c)

If a Laplace transform model of the plant is available, Routh's array may be used to establish the critical gain K_c and the corresponding period of oscillation T. The procedure is:

(1) Find the system's closed-loop characteristic equation under purely proportional control.

(2) Form the Routh array and establish the gain K_c that produces an all-zero row (see earlier). If the system becomes unstable in an oscillatory manner, the all-zero row will often be the row associated with s^1. In this case the divisor polynomial will be second order and there will be no roots of the remainder polynomial with positive real parts. Note that the system should remain stable for all positive values of K below the critical value.

(3) Use the divisor polynomial to find the period of oscillation T, and apply the recommended initial settings given above.

Figure 3.28 A plant with a three-term controller in a feedback arrangement (proportional action is shown, but integral and derivative action are added in the text).

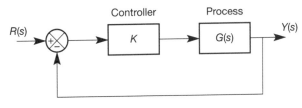

Controller   Process

$R(s)$ K $G(s)$ $Y(s)$

Example 3.10 *Initial three-term controller settings by Routh's array*

Use the Ziegler–Nichols rules to find P, PI and PID controller settings for a plant having the open-loop transfer function

$$G(s) = \frac{6}{(s+1)(s+2)(s+3)}$$

Under proportional control (the approach is the same as in Examples 3.3 and 3.7(b)), the closed-loop characteristic equation is:

$$D(s) = s^3 + 6s^2 + 11s + 6(1+K) = 0$$

and the Routh array for this system is

Row			
0	s^3	1	11
1	s^2	6	$6(1+K)$
2	s	$10-K$	

For row 2 to be an all-zero row, K must equal 10 (that is, $K_c = 10$). To understand how K_c and ω are found, refer to Example 3.9 and the associated description. The divisor polynomial is obtained from row 1 and is

$$s^2 + 11 = 0$$

Letting $s = j\omega$ and solving for ω gives the frequency of oscillation as

$$\omega = \sqrt{11} \text{ rad s}^{-1}$$

and

$$T = 2\pi/\omega = 1.895 \text{ s}$$

The recommended settings are therefore as follows:
For proportional control $K = 10/2 = 5$; this gives the closed-loop transfer function as

$$\frac{Y(s)}{R(s)} = \frac{30}{s^3 + 6s^2 + 11s + 36}$$

For P+I control $K = 4.5$ and $T_i = 1.57$, to give the closed-loop transfer function as

$$\frac{Y(s)}{R(s)} = \frac{42.5s + 27}{1.57s^4 + 9.434s^3 + 17.3s^2 + 51.89s + 27}$$

For PID control $K = 6, T_i = 0.947$ and $T_d = 0.237$, which gives the closed-loop transfer function as

$$\frac{Y(s)}{R(s)} = \frac{8.073s^2 + 34.1s + 36}{0.947s^4 + 5.683s^3 + 18.5s^2 + 39.78s + 36}$$

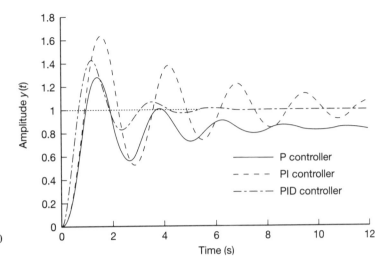

Figure 3.29 Responses of the system of Example 3.10 with various controllers.

Figure 3.29 shows the output response $y(t)$ produced by a forcing step input $r(t)$ for this system under P, PI and PID control. The MATLAB m-file *fig3_29.m* on the accompanying disk allows the entry of any required K, T_i and T_d settings, so any other values can be tried, as well as those suggested above (as can a different open-loop system, of course). Note the relatively fast response of this system under proportional control, and that the final steady-state output $y(t)$ is offset from the demanded input of 1. In order to eliminate this steady-state error, the PI controller may be used. However, as shown, the introduction of integral action reduces the stability of the system (the real part of the dominant poles is closer to the imaginary axis). The inclusion of derivative action has a stabilizing effect on the plant, but derivative action could not be used if the controller were being fed with measurement noise.

Using MATLAB (as in Appendix 3), the unit step response of the system with the PID controller (as an example) can be simply produced by the following commands (although the m-file *fig3_29.m* on the accompanying disk is obviously a more flexible way to do it):

$\gg$ num = [8.073   34.1   36];

$\gg$ den = [0.947   5.683   18.5   39.78   36];

$\gg$ step(num,den);grid

Note that a better response is obtainable by 'tuning' the initial parameters suggested by the Ziegler–Nichols methods – see Chapter 4.

(ii) Second approach

This tuning method is included for the sake of completeness; it is not based on any stability test. The method is particularly suitable for open-loop systems having a measured step response containing appreciable time lag. This response is often referred to as the plant's signature (or process reaction curve). For a process plant, which will usually have large inertias, this is normally non-oscillatory, with a fairly well-defined point of inflection.

The test procedure is normally carried out as follows:

(1) Generate a step response from the open-loop plant. Typically, the response obtained will be similar to that shown in Figure 3.30.

(2) Measure the slope of the response R and the lag time L (see Figure 3.30). Note that R should pass through the response's point of inflection, and should therefore have the maximum possible slope.

(3) The recommended initial settings are:

$$\text{P control} \qquad K = 1/(RL) \qquad\qquad (3.50)$$
$$\text{PI control} \qquad K = 0.9/(RL) \qquad\qquad (3.51a)$$
$$T_i = 3.3L \qquad\qquad (3.51b)$$
$$\text{PID control} \qquad K = 1.2/(RL) \qquad\qquad (3.52a)$$
$$T_i = 2L \qquad\qquad (3.52b)$$
$$T_d = 0.5L \qquad\qquad (3.52c)$$

The types of response obtained from the various PID controller combinations will be similar in form to the responses shown in Figure 3.29. This PID choice is aimed at giving a subsidence ratio of 4:1 (Section 3.2.2) (sometimes called a 'quarter decay ratio'), whereas Figure 3.29 exhibits a subsidence ratio of about 6:1 in the PID trace.

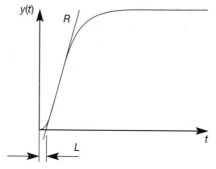

Figure 3.30 Illustrating the Ziegler–Nichols parameters for controller tuning.

3.3.5 Concluding comments

The various stability methods, the s-plane performance criteria, or the Ziegler–Nichols rules for producing PID controller settings, are of limited use. They will indicate whether a particular design meets some given performance specification, or provide initial settings for a PID controller. However, if the design does not meet the specifications, or if the initial controller settings are inadequate, they provide no indication of what modifications are required. In the following sections, design techniques are introduced which not only indicate stability, but also provide information on the controller structure and adjustments needed to meet given performance specifications.

3.4 Performance specifications in the frequency domain

So far, all the performance and stability techniques which have been introduced are equally applicable to open- or closed-loop systems. Only when the open-loop system fails to meet the desired stability or performance criteria is there a design problem, and then it is the closed-loop performance that is important. In the frequency domain, the gain and phase margin specifications (see Section 3.2.3) relate specifically to the open-loop transfer function, while the other common specifications normally relate to the closed-loop system. To understand why there is a need for these various performance specifications, consider the open-loop plant of Figure 3.31. This plant has an actuator input $u(s)$ and an output $y(s)$ which consists of the plant output plus some external disturbance $d(s)$. For clarity, lower-case letters which are functions of s represent signals, and upper-case letters represent transfer functions. Assuming that the open-loop plant $G(s)$ does not meet the required performance specifications then there is a control problem.

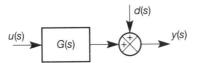

Figure 3.31 A plant with an output disturbance.

Typically, a closed-loop solution is sought by adopting the closed-loop structure shown in Figure 3.32. Comparing Figure 3.32 with Figure 1.5 shows that the pre-filter (P in Figure 1.5, but $P(s)$ generally) has been set at unity. This is justifiable, since a pre-filter is invariably designed after the design of the standard single degree of freedom system is complete. A further difference between the two figures is that Figure 3.32 includes the signal $n(s)$ which represents measurement errors or noise on the feedback signal.

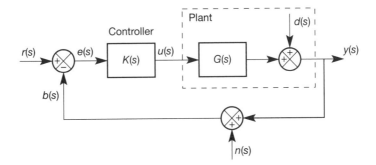

Figure 3.32 The plant of Figure 3.31 in a feedback loop with measurement noise.

Block diagram reduction on Figure 3.32 indicates that the closed-loop output for the system is given by:

$$y(s) = \frac{G(s)K(s)}{1 + G(s)K(s)} \left[r(s) - n(s) \right] + \frac{1}{1 + G(s)K(s)} d(s) \qquad (3.53)$$

and the controller output by

$$u(s) = \frac{K(s)}{1 + G(s)K(s)} [r(s) - n(s) - d(s)] \qquad (3.54)$$

Rather than simply writing down the LTF for $e(s)$ from Figure 3.32, consider the LTF for the actual *tracking error* $e_T(s) = r(s) - y(s)$. Note that $e_T(s)$ differs slightly from $e(s)$ (Figure 3.32) in that the noise term $n(s)$ in the feedback loop shown in the figure is not included. This modification is made so that the actual error in the tracking of $r(s)$ by $y(s)$ is evaluated. The noise term $n(s)$ is included indirectly, by virtue of its appearance in Equation (3.53), which can be substituted into the definition of $e_T(s)$ above. Doing this, and rearranging terms, gives:

$$e_T(s) = \frac{1}{1 + G(s)K(s)} [r(s) - d(s)] + \frac{G(s)K(s)}{1 + G(s)K(s)} n(s) \qquad (3.55)$$

Although Equations (3.53) to (3.55) refer to the SISO case, a similar analysis could be produced for the multivariable case. However, in this section only the SISO system, shown in Figure 3.32, is considered. The right-hand sides of Equations (3.53) to (3.55) contain a common denominator term, namely:

$$F(s) = 1 + G(s)K(s) \qquad (3.56)$$

For SISO systems this is often called the system's 'characteristic polynomial' since it contains the dynamics (exponential exponents) that characterize the system's response. If the feedback loop in the closed-loop system of Figure 3.32 is broken at $b(s)$ then the system's open-loop transfer function, the transfer function between $r(s)$ and $b(s)$, is given by

$$b(s) = G(s)K(s)r(s) \qquad (3.57)$$

By comparison with Equation (3.56), it may be seen that the closed-loop characteristic polynomial is

$$F(s) = 1 + \textit{the system's open-loop transfer function}$$

When dealing with systems in general (SISO and multivariable) the quantity $G(s)K(s)$, Equation (3.57), gives the difference between the input $r(s)$ and feedback $b(s)$. For this reason $F(s)$, Equation (3.56), is called the *return difference*.

The system's closed-loop transfer function (the relationship between $y(s)$ and $r(s)$) is given by

$$T(s) = \frac{G(s)K(s)}{1 + G(s)K(s)}$$

There is an interesting relationship between the closed-loop transfer function and its sensitivity to specific parameter changes. Let the controller of Figure 3.32 consist of a pure gain K, so that the closed-loop transfer function becomes

$$T(s) = \frac{KG(s)}{1 + KG(s)} \qquad (3.58)$$

The differential sensitivity or Bode sensitivity of $T(s)$ with respect to K is defined as

$$\frac{\partial T(s)/T(s)}{\partial K/K} = \frac{K}{T(s)} \frac{\partial T(s)}{\partial K}$$

Now

$$\frac{\partial T(s)}{\partial K} = \frac{\partial}{\partial K} \{KG(s)[1 + KG(s)]^{-1}\}$$

$$= G(s)[1 + KG(s)]^{-1} - KG^2(s)[1 + KG(s)]^{-2}$$

$$= \frac{G(s)}{[1 + KG(s)]^2}$$

and so

$$\frac{K}{T(s)} \frac{\partial T(s)}{\partial K} = \frac{1}{1 + KG(s)}$$

which is the reciprocal of the return difference, see Equation (3.56). Hence the quantity

$$S(s) = \frac{1}{1 + G(s)K(s)} \tag{3.59}$$

is called the *sensitivity function*. Also, since from Equations (3.58) and (3.59),

$$T(s) + S(s) = 1 \tag{3.60}$$

the closed-loop transfer function $T(s)$ is also called the *complementary sensitivity function*.

Using the sensitivity and complementary sensitivity functions, Equations (3.53) to (3.55) may be written as

$$y(s) = T(s)[r(s) - n(s)] + S(s) \, d(s) \tag{3.61}$$

$$u(s) = K(s)S(s)[r(s) - n(s) - d(s)] \tag{3.62}$$

$$e_T(s) = S(s)[r(s) - d(s)] + T(s)n(s) \tag{3.63}$$

For disturbance rejection Equation (3.61) indicates that $S(s)$ should be small. This can be achieved by making the system's open-loop transfer function (the return transfer function) large. A small sensitivity function will also give good tracking, since this would reduce the error signal $e_T(s)$ in Equation (3.63). However, for noise rejection $T(s)$ must be small, Equation (3.63), and since it is impossible from Equation (3.60) for both $T(s)$ and $S(s)$ to be small, there is an unavoidable trade-off between tracking (or attenuating disturbances) and filtering out measurement noise. This trade-off was alluded to in Section 3.2.2 ('A first-order dominant system'), and is further discussed in Chapter 13.

Given these desirable performance characteristics, it is now possible to consider how these conflicting design requirements could be satisfied. Within the frequency domain, the solution is to shape the frequency response so that $S(j\omega)$ is small at low frequencies and $T(j\omega)$ is small at high frequencies. Which frequencies are defined as being 'high', and which frequencies are defined as 'low', depends on the particular problem.

For $S(j\omega)$ to be small, Equation (3.59) indicates that $G(j\omega)K(j\omega)$ must be large. In particular, at zero frequency (that is, steady-state d.c.), if the error is to be minimized then $G(0)K(0)$ must tend to infinity. The relationship between the open-loop system's zero frequency response and the closed-loop system's final response error will be considered further in Section 3.4.1.

When the open-loop system's frequency response produces a phase shift of $-180°$ there is a potential closed-loop stability problem if the gain is greater than 1. The open-loop system's gain and phase margins (see towards the end of Section 3.2.3) are therefore good indicators of closed-loop stability.

One way to make $G(j\omega)K(j\omega)$ large, and hence make $S(j\omega)$ large, is to make the controller $K(j\omega)$ large. However, this can cause actuator problems. Equation (3.62) gives the control signal fed to the plant's actuator. Since all physical systems will have limits on their inputs and outputs it is desirable to keep $u(s)$ within specified limits in order to prevent saturation, that is, $K(j\omega)S(j\omega)$ must be limited. However, for large $K(j\omega)$

$$K(j\omega)S(j\omega) = \frac{K(j\omega)}{1 + G(j\omega)K(j\omega)} \approx \frac{1}{G(j\omega)}$$

Now, with physical systems, $G(j\omega)$ will tend to become small with increasing frequency (because no real system can follow an infinite frequency). Under these circumstances the actuator signals can become large. This is a further complication during the design stage.

To minimize noise $T(j\omega)$ must be small (Equation (3.63)). Since noise tends to be a high-frequency phenomenon, the requirement is that $T(j\omega)$ and hence $G(j\omega)K(j\omega)$ is small at high frequencies, see Equation (3.58). A performance measure which indicates when the closed-loop frequency response starts to become small would therefore be particularly useful. Just such a measure exists, and is referred to as the system's bandwidth.

To summarize, the system's open-loop transfer function should have a large gain at low frequencies, to satisfy tracking requirements and rejection of low-frequency disturbance. At high frequencies, the gain is kept low to filter out high-frequency noise. Over the mid-range of frequencies, the frequency response should satisfy the gain and phase margin requirements. Consequently, the control system design problem is one of shaping the open and closed-loop frequency responses by correctly selecting the controller transfer function $K(s)$, a topic dealt with in Chapter 4. From experience, satisfactory frequency response shaping is achieved by satisfying the frequency-domain specifications listed below.

Open-loop frequency response specifications. For open-loop stable systems of types 0, 1 and 2, a gain margin which is greater than unity, combined with a phase margin greater than $0°$, will guarantee closed-loop stability. With unstable open-loop systems, or systems having any other type number, alternative methods for checking closed-loop stability must be employed.

(1) *Gain margin.* For a stable closed-loop system the gain margin gives the factor by which the loop gain may be increased before the system becomes unstable, see Section 3.2.3.

(2) *Phase margin.* For second-order systems, phase margin and damping ratio can be shown to be related. A simple rule of thumb states that the numerical value of the phase margin in degrees divided by 100 gives the closed-loop damping ratio. With systems higher than second order this rule only indicates the damping ratio, and should be used in conjunction with the other frequency-domain specifications.

Closed-loop frequency response specifications

(3) *Bandwidth.* Bandwidth is normally defined as the frequency at which the magnitude ratio drops to 0.707 of its zero-frequency level (given that the gain characteristic is 'flat' at zero frequency). The factor 0.707 is $1/\sqrt{2}$, and arises because the fundamental definition of bandwidth is the 'half power point', that is, the frequency at which the power in the signal has halved. Since, in electrical systems for example, signal levels are normally measured in terms of voltages or currents, the square law relationship between either of these quantities and signal power gives rise to the $\sqrt{2}$.

On the decibel scale, a factor of 0.707 is equivalent to a fall of 3 dB from the zero frequency gain. A magnitude ratio M is usually converted to decibels by the formula $20 \log_{10}(M)$ dB, see Section 3.5.1. (Note that this decibel formula is derived from the formula $10 \log_{10} P$, where P is a *power* gain, and strictly applies only in rather tightly defined circumstances. However, it is normally used in a fairly cavalier manner throughout control systems work.)

The bandwidth gives a measure of the transient response properties. A large bandwidth corresponds to a faster rise time, since higher-frequency signals are passed to the output. If the bandwidth is small, only signals of relatively low frequencies are passed, and the time response will generally be slow and sluggish. The bandwidth of a system is also an indicator of its noise filtering characteristics: an unnecessarily wide bandwidth would produce a system with poor noise rejection characteristics.

(4) *Maximum magnitude ratio.* This is usually referred to as 'M peak' (M_p), and gives an indication of a system's damping. For a second-order system, it may be shown that

$$\zeta^2 = \frac{1}{2} - \frac{1}{2}\sqrt{\left(1 - \frac{1}{M_p^2}\right)} \text{ for } M_p \geq 1 \text{ so } M_p = \frac{1}{2\zeta\sqrt{(1 - \zeta^2)}}$$

$$(3.64)$$

Normally, large M_p corresponds to a large peak overshoot in the step response. For most design problems an optimum value of M_p would be somewhere between 1.1 and 1.5 which, for a second-order system, would give a damping ratio between 0.54 and 0.36, respectively.

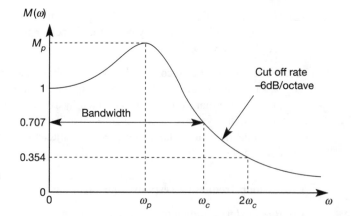

Figure 3.33 Performance criteria on a frequency response plot.

(5) *Frequency at M_p.* For a second-order system, the resonant frequency ω_p is given by

$$\omega_p = \omega_n \sqrt{(1 - 2\zeta^2)} \tag{3.65}$$

and, like bandwidth, indicates the system's speed of response.

(6) *Cut-off rate (or roll-off rate).* This is the rate of decrease in the magnitude ratio outside the system's bandwidth. A large high-frequency cut-off rate would indicate a system with good signal-to-noise ratio. However, high cut-off characteristics may be accompanied by a large M_p which, in turn, corresponds to a system with low damping. A typical cut-off rate specification is -6 dB/ octave of frequency. That is, doubling the frequency (increasing it by one octave) should halve the magnitude ratio. It turns out to be the same thing as -20 dB/decade of frequency (that is, multiplying the frequency by a factor of 10 should cause the magnitude ratio to decrease by a factor of 10).

The closed-loop performance criteria (3)–(6) are illustrated on the frequency response plot shown in Figure 3.33.

3.4.1 Tracking errors and system type

If the system's open-loop transfer function has n poles at the origin, then it is said to be of type n. With no noise and no disturbance Equation (3.55) indicates that the closed-loop error signal is given by

$$e(s) = \frac{1}{1 + G(s)K(s)} \, r(s)$$

Assuming that the closed-loop system is stable, then the final value theorem can be used to find the steady-state error. For example, if the open-loop transfer function is type 0 and $r(s)$ is a unit step $(r(s) = 1/s)$ then the final value theorem indicates that

$$e(\infty) = \lim_{s \to 0} (se(s)) = \frac{1}{1 + G(0)K(0)}$$

which in this case is the system's sensitivity function evaluated at $s = 0$ or $\omega = 0$. The steady-state error of stable closed-loop systems to various forcing inputs is shown in Table 3.2.

Table 3.2 Steady-state error for system types 0, 1 and 2.

System type	Unit step $(1/S)$	Unit ramp $(1/S^2)$	Unit parabola $(1/S^3)$
Type 0	$\dfrac{1}{1 + G(0)K(0)}$	∞	∞
Type 1	0	$\dfrac{1}{G(0)K(0)}$	∞
Type 2	0	0	$\dfrac{1}{G(0)K(0)}$

3.5 Frequency response plots

For the standard single degree of freedom control system configuration, see Figure 3.32, the system's open-loop transfer function is $G(s)K(s)$. To obtain the system's open-loop frequency response, s is set equal to $j\omega$, see Section 3.2.3. This results in an expression having the general form

$$G(j\omega)K(j\omega) = \frac{K \displaystyle\prod_{i=1}^{m}(j\omega + z_i)}{(j\omega)^r \displaystyle\prod_{i=1}^{n-r}(j\omega + p_i)}$$

which may be reduced to

$$G(j\omega)K(j\omega) = M(\omega)e^{j\phi(\omega)}$$

The frequency response consists of three variables – frequency ω, magnitude $M(\omega)$ and phase $\phi(\omega)$. The various graphical representations of these variables for a range of frequencies ω produce the required frequency response plots.

The intention of this section is to introduce the various frequency plots and then show how a system may be analysed using the frequency-domain performance specifications of Section 3.4. To aid in this analysis, consider a system having the open-loop transfer function

$$G(s)K(s) = \frac{K}{s(s + 1)(s + 2)}$$

where K is the controller gain which, for the sake of argument, could take the values 4, 1.5 or 0.4. Note that the open-loop system is of type 1 (1 pole at the origin of the s-plane) and of rank 3 (3 more poles than finite zeros). Since the open-loop system is of type 1 and stable, Section 3.4.1 shows that any stable closed-loop realization would track a step with zero steady-state error. Furthermore, the gain and phase margin measures can be used to test for closed-loop stability.

The open-loop frequency response for this system is given by

$$G(j\omega)K(j\omega) = \frac{K}{j\omega(j\omega + 1)(j\omega + 2)}$$ (3.66)

which has magnitude

$$M(\omega) = \frac{K}{\omega\sqrt{(\omega^2 + 1)}\sqrt{(\omega^2 + 4)}}$$ (3.67)

and phase

$$\phi(\omega) = -90° - \tan^{-1}(\omega) - \tan^{-1}\left(\frac{\omega}{2}\right)$$ (3.68)

The closed-loop transfer function for this system is given by

$$T(s) = \frac{K}{s^3 + 3s^2 + 2s + K}$$

and from Routh's stability criterion will be stable for values of K in the range of $0 < K < 6$. Finding the closed-loop system poles when $K = 4$ gives $s = -2.80$ and $s = -0.10 \pm j1.20$. This system is second-order dominant with an approximate damping ratio ζ of 0.085 and a $\zeta\omega_n$ value of 0.10 rad s^{-1}. Therefore, the response to a unit step would be oscillatory, with a decay rate of $e^{-0.10t}$, see Figure 3.34 (the m-file *fig3_34.m* on the accompanying disk will duplicate the plots of Figure 3.34 and confirm all the given values).

When $K = 1.5$ the poles are at $s = -2.43$ and $s = -0.28 \pm j0.73$. Again the system is second-order dominant, but now the approximate damping ratio ζ is 0.36 and the $\zeta\omega_n$ value is 0.28 rad s^{-1}. For many applications, the damping ratio would be considered acceptable, and the decay rate would be faster than in the previous response. This new response is also shown in Figure 3.34.

Finally, when $K = 0.4$ (or 0.385 to be more precise) there is one pole at $s = -2.15$ and a double pole at $s = -0.422$. The dominant double pole makes the system critically damped and produces the step response shown dashed in Figure 3.34.

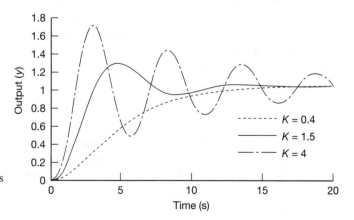

Figure 3.34 Step responses of a system for various controller gain settings.

The *closed-loop* frequency response can be used to evaluate parameters such as bandwidth, and is given by

$$T(j\omega) = \frac{K}{(K - 3\omega^2) + j\omega(2 - \omega^2)}$$

with magnitude

$$M_{CL}(\omega) = \frac{K}{\sqrt{\{(K - 3\omega^2)^2 + \omega^2(2 - \omega^2)^2\}}}$$

and phase

$$\phi_{CL}(\omega) = -\tan^{-1}\left(\frac{\omega(2 - \omega^2)}{K - 3\omega^2}\right)$$

3.5.1 *Logarithmic plots or Bode diagrams*

These are two plots in rectangular coordinates, in which the magnitude is expressed in decibels (dB), and the phase angle in degrees, both plotted as functions of the logarithm of frequency in rad/unit time (normally, rad s^{-1}).

Bode diagrams are normally plotted on semi-logarithmic graph paper, so that the dB values plotted on the linear vertical axis have the effect of producing a logarithmic scale, while the frequency values can be plotted directly on the horizontal axis, allowing the logarithmic axis to do the work of conversion.

The frequency response magnitude ratio is expressed in decibels, as noted earlier, using

$$M(\omega)_{\text{dB}} = 20 \log_{10} M(\omega)$$

where $M(\omega)_{\text{dB}}$ is the log modulus in decibels and $M(\omega)$ is the magnitude ratio.

Typical Bode plots are shown in Figure 3.35. For convenience, the two types of diagram are shown together. As usual, the magnitude ratio falls off and the phase angle becomes increasingly lagging with increasing frequency.

One of the main reasons for using a logarithmic scale for the magnitude ratio is the ease with which the dynamic elements in a control loop can be manipulated. At a given frequency, the magnitude ratio is obtained by multiplying together the individual magnitude ratios of the elements (which becomes *adding* in dB – see

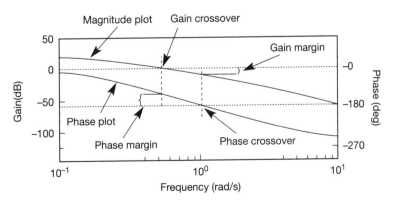

Figure 3.35 Stability margins on a Bode diagram.

below), and the phase angle is obtained by summing individual phase angle elements.

Consider the Bode form of a transfer function $F(s)$ (see Equation (2.77b)) for which the frequency response is

$$F(j\omega) = K_B \frac{\displaystyle\prod_{i=1}^{m} [1 + j(\omega/z_i)]}{(j\omega)^r \displaystyle\prod_{i=1}^{n-r} [1 + j(\omega/p_i)]} \tag{3.69}$$

In this form, the coefficient K_B is called the Bode gain.

The magnitude of $F(j\omega)$ for any frequency ω is the product of the Bode gain, the magnitude of each factor in the numerator and the reciprocal of the magnitude of each factor in the denominator. However, if the magnitude of $F(j\omega)$ is expressed in decibels, then

$$20 \log_{10} |F(j\omega)| = 20 \log_{10} |K_B| + \sum_{i=1}^{m} 20 \log_{10} |(1 + j(\omega/z_i))|$$

$$- 20 \log_{10} |(j\omega)^r| - \sum_{i=1}^{n-r} 20 \log_{10} |(1 + j(\omega/p_i))| \tag{3.70}$$

(Note that in logarithmic form all products become summations.) The phase angle of $F(j\omega)$ at any frequency is found in the normal way as

$$\arg F(j\omega) = \sum_{i=1}^{m} \arg(1 + j(\omega/z_i)) - \arg(j\omega)^r$$

$$- \sum_{i=1}^{n-r} \arg(1 + j(\omega/p_i)) \tag{3.71a}$$

or

$$\arg F(j\omega) = \sum_{i=1}^{m} \tan^{-1}(\omega/z_i) - r(90°) - \sum_{i=1}^{n-r} \tan^{-1}(\omega/p_i) \tag{3.71b}$$

Note that if K_B is negative then a further $180°$ phase lag (that is, $-180°$) must be included.

Both the magnitude and phase plots are obtained by summing the contributions provided by each term in the open-loop transfer function. If all the coefficients of the unfactorized transfer function are real, then each complex pole or zero in the factored transfer function will have a complex conjugate. This means that there are only four possible terms that need to be considered: the gain K_B, poles (or zeros) at the origin, real poles (or zeros) not at the origin and complex conjugate poles (or zeros). It also means that for a compensator in series with a plant, the overall frequency response plot in Bode form is simply the sum of those of the compensator and plant at each frequency value.

Bode plots for a gain term

A constant K_B provides a magnitude contribution of $20\log_{10}|K_B|$ and a phase angle of $0°$ if K_B is positive, or $-180°$ if K_B is negative. If $K_B = 1$, then $20\log_{10}|K_B|$ is zero dB; if $K_B = 2$, then the dB magnitude is 6 and if $K_B = 0.5$, the dB magnitude is -6. In all three cases the phase angle contribution is $0°$. For $K_B = -2$ or -0.5, the dB magnitude would be 6 or -6, respectively, but the phase contribution would be $-180°$. A Bode plot for a positive gain term is shown in Figure 3.36.

Bode plots for poles or zeros at the origin of the *s*-plane

A system with poles or zeros at the origin of the *s*-plane will have an expression in its transfer function of the following form, in which r is a positive integer equal to the number of poles at the origin of the *s*-plane (that is, the system's type number is r). Also, r is a negative integer for zeros at the origin.

$$1/s^r$$

The magnitude and phase contributions will be, respectively,

$$20\log_{10}\frac{1}{(j\omega)^r} = -20r\log_{10}\omega \text{ dB} \tag{3.72}$$

and

$$\arg\frac{1}{(j\omega)^r} = -r90° \tag{3.73}$$

The equation for the Bode magnitude plot, Equation (3.72) describes a straight line of slope $-20r$ dB/decade of frequency, passing through the 0 dB point when $\omega = 1$ (a 'decade' indicates a tenfold increase in frequency). The Bode phase plot, obtained from Equation (3.73), indicates that the phase angle is independent of frequency, and has a value which is dependent on r. Bode plots for one, two and three poles at the origin are shown in Figure 3.37.

Equations (3.72) and (3.73) are also valid for zeros at the origin of the *s*-plane. The Bode plots for a system with one, two or three zeros at the origin of the *s*-plane would be the reflections of the Bode plots in Figure 3.37 about the 0 dB and $0°$ lines.

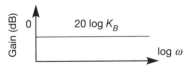

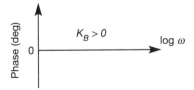

Figure 3.36 A Bode plot for a pure gain term.

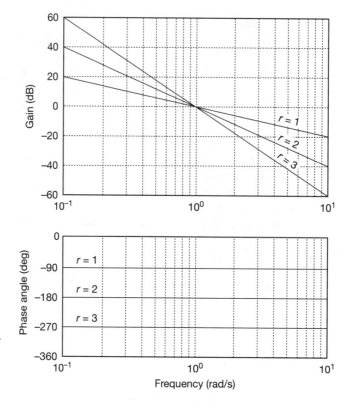

Figure 3.37 Bode plots for one, two and three pure integrators (poles at the origin of the s-plane).

Bode plots for real poles and zeros

A real pole or zero not at the origin of the s-plane contributes to the system's transfer function a term of the form

$$\frac{1}{(1 + s/\omega_c)^r} \tag{3.74}$$

The form of Equation (3.74) may be compared with Equation (3.18). In Equation (3.74), r is any positive or negative integer. Its introduction serves two purposes. If r is positive, the contribution is from system poles, and if it is negative the contribution is from system zeros. Also, if the magnitude of r is greater than unity, there are r poles or zeros occupying the same s-plane location, and the pole or zero is said to be of order r. The quantity ω_c is known as the 'corner frequency', for reasons which will soon become clear. For ω_c in rad s^{-1}, it is the reciprocal of the time constant (τ in Equation (3.18)) in seconds.

The dB magnitude and phase angle given by Equations (3.70) and (3.71) are now

$$-10r \log_{10}(1 + (\omega/\omega_c)^2) \tag{3.75}$$

and

$$-r \tan^{-1}(\omega/\omega_c) \tag{3.76}$$

Asymptotic approximations for the magnitude plot are obtained by considering the normalized frequency ω/ω_c. When ω/ω_c is very small, the dB magnitude (Equation (3.75)) approximates to

$$-10r \log_{10} 1 = 0 \text{ dB} \tag{3.77a}$$

and for large values of ω/ω_c to

$$-20r \log(\omega/\omega_c) \text{ dB} \tag{3.77b}$$

The corresponding phase angle for small ω/ω_c from Equation (3.76), is approximately

$$-r \tan^{-1}(0) = 0° \tag{3.78a}$$

and when ω/ω_c is very large, the corresponding phase angle tends to

$$-r\, 90° \tag{3.78b}$$

From Equations (3.77), it is apparent that the Bode magnitude plot asymptotically approaches a horizontal straight line at 0 dB as $\omega/\omega_c \to 0$ and $-20r \log_{10}(\omega/\omega_c)$ dB as $\omega/\omega_c \to \infty$. When plotted on a logarithmic frequency scale, the high-frequency asymptote is a straight line with a slope of $-20r$ dB/decade, or $-6r$ dB/octave. The low- and high-frequency asymptotes intersect at the corner frequency $\omega = \omega_c$ rad s^{-1}. Figure 3.38 shows the asymptotic Bode gain plots for real poles of order one, two and three.

Asymptotic Bode phase angle plots may be obtained from Equation (3.78). An asymptote for the mid-range of frequencies may be obtained by drawing a tangent to the exact phase curve at the corner frequency of $\omega = \omega_c$ rad s^{-1}. This asymptote would intersect the low-frequency asymptote at $\omega = \omega_c/5$, and the high-frequency

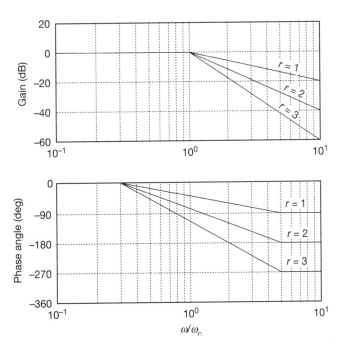

Figure 3.38 Asymptotic approximations to the Bode plots for one, two and three first-order poles at the same s-plane location.

asymptote at $\omega = 5\omega_c$. Figure 3.38 shows the asymptotic phase plots for real poles of order one, two and three.

The errors produced by using the given asymptotic approximations for the real poles and zeros, rather than an accurate curve, are given in Table 3.3, in which a pole of order r gives the factor $+r$, and a zero the factor $-r$. Again, the Bode plots for zeros of order r will be the reflections of the plots in Figure 3.38 about the 0 dB and $0°$ lines.

Table 3.3 Asymptotic errors for a real pole (or zero) of order r and corner frequency ω_c, with the mid-range phase asymptote plotted between $\omega_c/5$ and $5\omega_c$.

	$\omega_c/10$	$\omega_c/5$	$\omega_c/2$	ω_c	$2\omega_c$	$5\omega_c$	$10\omega_c$
Magnitude error (dB)	$-0.043r$	$-0.17r$	$-0.96r$	$-3r$	$-0.96r$	$-0.17r$	$-0.043r$
Phase angle error	$-5.7r°$	$-11.3r°$	$-0.8r°$	$0°$	$+0.8r°$	$-11.3r°$	$+5.7r°$

It should be noted that other mid-range phase asymptotes have been proposed. A common approximation is to use an asymptote which intersects the low-frequency asymptote at $\omega = \omega_c/10$, and the high-frequency asymptote at $\omega = 10\omega_c$. The phase angle errors produced using this approximation are given in Table 3.4.

Table 3.4 Asymptotic phase angle error for a real pole (or zero) of order r and corner frequency ω_c, with the mid-range phase asymptote plotted between $\omega_c/10$ and $10\omega_c$.

	$\omega_c/10$	$\omega_c/5$	$\omega_c/2$	ω_c	$2\omega_c$	$5\omega_c$	$10\omega_c$
Phase angle error	$-5.7r°$	$-2.3r°$	$-4.9r°$	$0°$	$+4.9r°$	$+2.3r°$	$+5.7r°$

Bode plots for complex conjugate poles and zeros

Complex conjugate poles or zeros are normally multiplied together and expressed in the standard form

$$\frac{1}{\left(1 + \dfrac{2\zeta s}{\omega_n} + \dfrac{s^2}{\omega_n^2}\right)^r} \qquad 0 \le \zeta \le 1 \tag{3.79}$$

Equation (3.79) can be compared with Equation (3.19), where ω_n is the undamped natural frequency, ζ the damping ratio and r an integer indicating the number of complex conjugate roots at a given location. Again, if r is positive the roots are poles, and if r is negative the roots are zeros.

The frequency response function corresponding to Equation (3.79) is given by

$$\frac{1}{\left[1 + j2\zeta\left(\dfrac{\omega}{\omega_n}\right) - \left(\dfrac{\omega}{\omega_n}\right)^2\right]^r}$$

which has a dB magnitude ratio of

$$-10r \log_{10}\left[\left(1 - \frac{\omega^2}{\omega_n^2}\right)^2 + 4\zeta^2 \frac{\omega^2}{\omega_n^2}\right] \tag{3.80}$$

and a phase angle of

$$-r \tan^{-1} \frac{2\zeta(\omega/\omega_n)}{1 - (\omega/\omega_n)^2} \tag{3.81}$$

An asymptotic magnitude plot is obtained by considering the frequency ratio ω/ω_n. When ω/ω_n is very small, Equation (3.80) gives the low-frequency asymptote as 0 dB. The high-frequency asymptote, obtained by letting ω/ω_n become very large, is

$$-40r \log_{10}(\omega/\omega_n)$$

which is line of slope $-40r$ dB/decade emanating from the corner frequency ω_n.

Use of the asymptotic magnitude approximation will result in an error in this plot around the corner frequency, the size of which depends on the damping ratio, ζ. These errors are given in Table 3.5. For low values of ζ the magnitude ratio tends to a peak value near the corner frequency. It can be shown that the exact curve peaks when (Equation (3.65)):

$$\omega = \omega_n\sqrt{(1 - 2\zeta^2)} \tag{3.82}$$

and that at this frequency the peak value of the dB magnitude ratio (from Equation (3.64)) is

$$20 \log_{10} \frac{1}{2\zeta\sqrt{(1 - \zeta^2)}}$$

Since frequency must be real, Equation (3.82) indicates that a specific maximum value occurs only if $\zeta < 1/\sqrt{2}$ – that is, less than 0.707. If the damping ratio is greater than 0.707, the maximum value of the magnitude curve is along the 0 dB line. Figures 3.39(a) and (b) show the Bode plots and asymptotic Bode plots for several second-order systems (the MATLAB m-file *fig3_39.m* on the accompanying disk draws these curves).

The asymptotic phase plot is obtained from Equation (3.81). At low frequencies (when the frequency ratio ω/ω_n is small), the phase angle is approximately 0°; at high frequencies the phase angle tends to $-r180°$. For the frequencies around the corner frequency ω_n it is normal to draw an asymptote

Table 3.5 Asymptotic magnitude error for complex conjugate poles (or zeros) of order r and corner frequency ω_n.

ζ	$\omega_n/10$	$\omega_n/5$	$\omega_n/2$	ω_n	$2\omega_n$	$5\omega_n$	$10\omega_n$
1	$-0.086r$	$-0.34r$	$-1.94r$	$-6r$	$-1.94r$	$-0.34r$	$-0.086r$
0.707	0	$-0.007r$	$-0.263r$	$-3r$	$-0.263r$	$-0.007r$	0
0.5	$+0.043r$	$+0.17r$	$+0.902r$	0	$+0.902r$	$+0.17r$	$+0.043r$
0.3	$+0.071r$	$+0.287r$	$+1.85r$	$+4.44r$	$+1.85r$	$+0.287r$	$+0.071r$
0.2	$+0.08r$	$+0.325r$	$+2.2r$	$+7.96r$	$+2.2r$	$+0.325r$	$+0.08r$

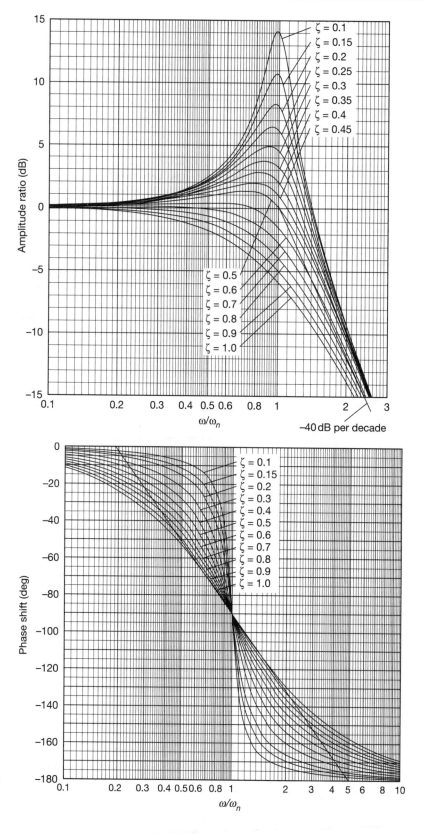

Figure 3.39 Standard (normalized) second-order system frequency responses. (a) Magnitude responses; (b) phase responses.

consistent with those drawn for real poles and zeros. Consequently, the mid-range frequency asymptote crosses the low-frequency asymptote when $\omega = \omega_n/5$, and crosses the high-frequency asymptote when $\omega = 5\omega_n$ (see Figure 3.39(b)). Again, the discrepancy between the exact and asymptotic phase plot is a function of the damping ratio, as indicated in Table 3.6 and Figure 3.39. For completeness, Table 3.7 gives the errors in the asymptotic phase plot when the mid-range asymptote crosses the low-frequency asymptote at $\omega_n/10$, and crosses the high-frequency asymptote at $10\omega_n$.

Table 3.6 Asymptotic phase angle error for complex conjugate poles (or zeros) of order r and corner frequency ω_n, with the mid-range phase asymptote plotted between $\omega_n/5$ and $5\omega_n$.

ζ	$\omega_n/10$	$\omega_n/5$	$\omega_n/2$	ω_n	$2\omega_n$	$5\omega_n$	$10\omega_n$
1	$-11.4r°$	$-22.6r°$	$-1.6r°$	$0°$	$+1.6r°$	$+22.6r°$	$+11.4r°$
0.707	$-8.1r°$	$-16.4r°$	$-7.9r°$	$0°$	$+7.9r°$	$+16.4r°$	$+8.1r°$
0.5	$-5.8r°$	$-11.8r°$	$-17.5r°$	$0°$	$+17.5r°$	$+11.8r°$	$+5.8r°$
0.3	$-3.5r°$	$-7.1r°$	$-29.4r°$	$0°$	$+29.4r°$	$+7.1r°$	$+3.5r°$
0.2	$-2.3r°$	$-4.8r°$	$-36.3r°$	$0°$	$+36.3r°$	$+4.8r°$	$+2.3r°$
0.1	$-1.2r°$	$-2.4r°$	$-43.6r°$	$0°$	$+43.6r°$	$+2.4r°$	$+1.2r°$

Table 3.7 Asymptotic phase angle error for complex conjugate poles (or zeros) of order r and corner frequency ω_n, with the mid-range phase asymptote plotted between $\omega_n/10$ and $10\omega_n$.

ζ	$\omega_n/10$	$\omega_n/5$	$\omega_n/2$	ω_n	$2\omega_n$	$5\omega_n$	$10\omega_n$
1	$-11.4r°$	$-4.6r°$	$-9.8r°$	$0°$	$+9.8r°$	$+4.6r°$	$+11.4r°$
0.707	$-8.1r°$	$-10.7r°$	$-19.6r°$	$0°$	$+19.6r°$	$+10.7r°$	$+8.1r°$
0.5	$-5.8r°$	$-15.3r°$	$-29.2r°$	$0°$	$+29.2r°$	$+15.3r°$	$+5.8r°$
0.3	$-3.5r°$	$-20.0r°$	$-41.1r°$	$0°$	$+41.1r°$	$+20.0r°$	$+3.5r°$
0.2	$-2.3r°$	$-22.3r°$	$-48.0r°$	$0°$	$+48.0r°$	$+22.3r°$	$+2.3r°$
0.1	$-1.2r°$	$-25.9r°$	$-55.3r°$	$0°$	$+55.3r°$	$+25.9r°$	$+1.2r°$

Asymptotic Bode plots are important in that they help in understanding how various compensators can be used to shape the frequency response of a system (see Chapter 4) and in system identification (see Section 3.9). Nowadays, Bode and the other frequency response plots are normally found using appropriate software as the following examples show. Nevertheless, the authors believe that to plot a few examples by hand provides considerable insight.

Example 3.11 *Bode plots for a fourth-order system with one finite zero*

Draw the asymptotic Bode plots for a system having the open-loop transfer function

$$G(s)K(s) = \frac{320(1+s)}{s(s+2)(s^2+4s+16)}$$

The open-loop frequency response of $G(s)K(s)$ is obtained by setting $s = j\omega$, hence

$$G(j\omega)K(j\omega) = \frac{320(1 + j\omega)}{j\omega(j\omega + 2)[(j\omega)^2 + 4j\omega + 16]}$$

In Bode form, this is

$$G(j\omega)K(j\omega) = \frac{10(1 + j\omega)}{j\omega\left(1 + \dfrac{j\omega}{2}\right)\left[1 + \dfrac{j\omega}{4} - \left(\dfrac{\omega}{4}\right)^2\right]}$$

From Equation (3.70) the dB magnitude is obtained as

$$20 \log_{10}|G(j\omega)K(j\omega)| = 20 \log_{10} 10 + 20 \log_{10}|1 + j\omega| - 20 \log_{10}|j\omega|$$

$$- 20 \log_{10}\left|1 + \frac{j\omega}{2}\right| - 20 \log_{10}\left|1 + \frac{j\omega}{4} - \left(\frac{\omega}{4}\right)^2\right|$$

and from Equation (3.71a) the phase angle is

$$\arg G(j\omega)K(j\omega) = \arg(1 + j\omega) - 90° - \arg\left(1 + \frac{j\omega}{2}\right) - \arg\left[1 + \frac{j\omega}{4} - \left(\frac{\omega}{4}\right)^2\right]$$

As $\omega \to \infty$, the phase angle tends to $-270°$. This could be deduced without resorting to the phase angle equation, by noting that the difference between the number of poles and zeros is 3; in other words, the system's rank is 3. For minimum phase systems (systems having no poles or zeros in the right half of the s-plane) with a positive loop gain, the ultimate phase shift is given by

$$\arg G(j\infty)K(j\infty) = -R90°$$

where R is the system's rank.

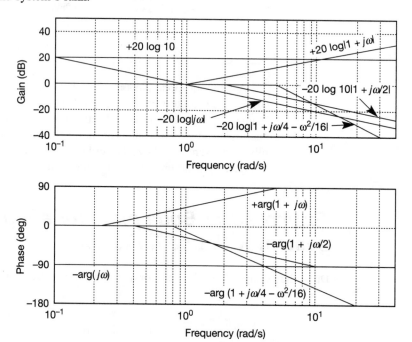

Figure 3.40 Asymptotic approximation Bode plot components for Example 3.11.

An asymptotic plot for each of the terms in the magnitude and phase expressions may now be drawn (see Figure 3.40). The asymptotic plots of each element may be summed graphically to produce the overall asymptotic Bode plot shown in Figure 3.41. For reference, the actual Bode plot is also shown in this figure.

Using MATLAB, the actual Bode plot can be very easily drawn as follows (see Appendix 3 for more detail):

$\gg$ num = 320*[1   1];

$\gg$ den = conv([1   0], conv([1   2],[1   4   16])); % the *conv*olution
% command multiplies
$\gg$ bode(num,den)   % out two polynomials

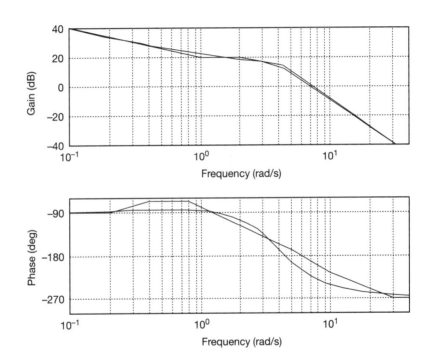

Figure 3.41 Full Bode plot for Example 3.11.

Example 3.12 *Bode plots for a third-order system with variable d.c. gain*

Use Bode plots to analyse the system having the open-loop transfer function

$$G(s)K(s) = \frac{K}{s(s+1)(s+2)}$$

when K can take the values 4, 1.5 and 0.4. Note that this is the system introduced in Section 3.5, and for which the closed-loop step responses for the three gains are shown in Figure 3.34.

Equations (3.67) and (3.68) may be used to find the open-loop system's frequency data over the range $0.1 \, \text{rad s}^{-1} \leq \omega \leq 10 \, \text{rad s}^{-1}$. With the magnitude data in dB and the phase angle in degrees, the open-loop Bode plots of Figure 3.42 are produced (the m-file *fig3_42.m* on the accompanying disk will do this).

In Figure 3.42 changes in the gain K alter the level (but not the shape) of the magnitude plot. For example, when $K = 0.4$ it is 0.267 times its value when $K = 1.5$. Converting to dB, $20 \log_{10} 0.267 = -11.5$ dB, so the magnitude plot for $K = 0.4$ is 11.5 dB lower than that for $K = 1.5$ at every frequency value.

This is in contrast to the phase plot which, as expected, is identical for all three gains (the phase in Equation (3.68) is independent of the gain K). Further, with the open-loop system being of type 1, the trend in the dB magnitude is always to reduce with increasing frequency. Also, the phase plot has a phase shift which becomes closer to $-90°$ as the frequency reduces. With increasing frequency the phase plot tends to $-270°$ (that is, $-90°$ times 3, the rank number of this minimum phase system; see Section 3.8 for some notes on non-minimum-phase systems).

The open-loop Bode plots are used to find the gain and phase margins as illustrated in Figure 3.43, for the case where $K = 1.5$ (the m-file *fig3_42.m* also lists these for each gain). Projecting a line from the phase plot at the point where the phase first becomes $-180°$ onto the gain plot, as shown in Figure 3.43, enables the gain margin to be measured. The gain margin is the length of the projected line between the magnitude plot and the 0 dB line, and in this case is $+12.04$ dB. Converting this to a gain gives the value 4 which indicates that K could be increased by a factor of 4 before the closed-loop system became unstable. This is in agreement with the value previously found using Routh's stability criterion (Section 3.3.3 and early in Section 3.5).

The phase margin is found by projecting down from the point where the magnitude plot first crosses the 0 dB line. Since the distance of the phase plot from the $-180°$ line at this frequency is $42°$, the phase margin is $+42°$. From the rule of thumb introduced in Section 3.4, an estimate of the closed-loop damping ratio is 0.42 and this may be compared with the damping ratio for the dominant poles which was found to be 0.36 (see Section 3.5 following Equation (3.68)).

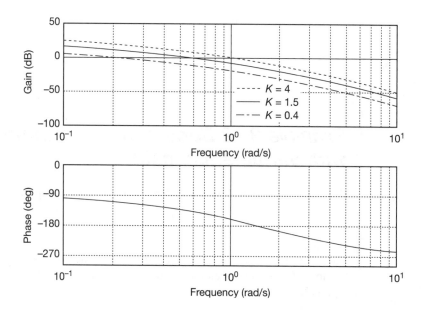

Figure 3.42 Bode plots for the variable-gain system of Example 3.12.

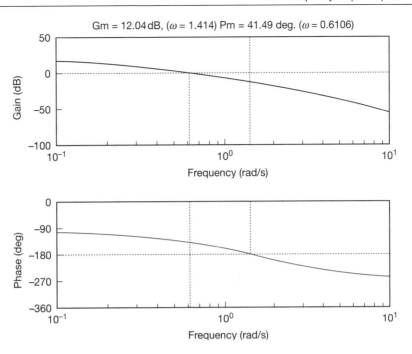

Figure 3.43 Stability margins on the Bode plot for Example 3.12 ($K = 1.5$).

Although a gain margin of 4 is higher than that recommended (2 to 2.5) this is not necessarily a problem (but a *lower* gain margin might well be). In this case a high gain margin makes the system robust in that it would be tolerant of gain errors (see Chapter 13). The phase margin is just outside the generally accepted range ($45°$–$65°$). It may be noted, however, that the acceptability, or otherwise, of a closed-loop response depends on the application. For the process industries, a slightly lower phase margin (indicating a lower damping ratio) may be desirable, whereas for a profile cutting machine a higher phase margin would be required, since no overshoot in the time response would be allowable.

When the system gain K is 4, the gain margin is 3.52 dB (or 1.5) and the phase margin is $11°$. Both the gain and phase margins are positive and the closed-loop system is stable. However, the low gain and phase margins suggest an oscillatory response close to the stability boundary. Reducing the system gain K to 0.4 gives a gain margin of 23.9 dB (or 15) and a phase margin of $73°$. Although the system is stable, these values are well outside the accepted range with the high phase margin suggesting a very overdamped response.

Figure 3.44 gives the *closed-loop* Bode plots for gains $K = 4$, $K = 1.5$ and $K = 0.4$ (and the m-file *fig3_44.m* on the disk will duplicate these). The plots are typical of second-order dominant, type 0 systems. For low frequencies the magnitude plot has a constant value equal to the system's steady-state gain, and the phase plot starts from $0°$.

All the closed-loop frequency-domain specifications can be found from the closed-loop Bode magnitude plot. When $K = 4$ the peak magnification, M_p, is 14.7 dB or 5.43 and from Equation (3.64) this indicates a dominant ζ value of 0.092. M_p occurs at the frequency ω_p of 1.25 rad s^{-1} and using Equation (3.65) to find ω_n, suggests a $\zeta\omega_n$ value of 0.11 rad s^{-1}. The bandwidth is 1.8 rad s^{-1} and suggests that when the gain K is 4, the system will have the worst noise filtering characteristics. However, the cut-off rate is -18 dB per octave (or -60 dB per decade) which would normally be satisfactory.

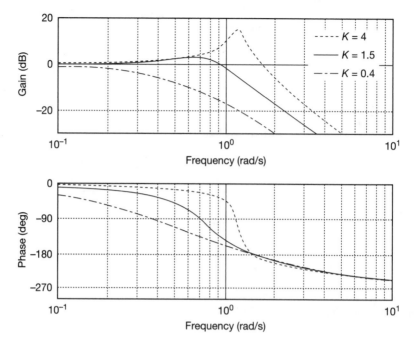

Figure 3.44 Closed-loop Bode plots for the variable-gain system of Example 3.12.

When $K = 1.5$, M_p is 3.1 dB (1.428) and ω_p is 0.66 rad s^{-1} which, from Equations (3.64) and (3.65), gives a ζ value of 0.38 and a $\zeta\omega_n$ value of 0.29 rad s^{-1}. The bandwidth is 1.07 rad s^{-1} and the cut-off rate is again -18 dB per octave.

After adjusting the system gain K to 0.4, the gain plot does not exhibit a peak magnification, suggesting that the response is overdamped. The bandwidth is 0.28 rad s^{-1}, indicating good noise rejection properties but a relatively slow response.

Comparison should be made between all the predicted response characteristics and those obtained from a dominant pole analysis (earlier in Section 3.5).

3.5.2 Polar and inverse polar plots (Nyquist and inverse Nyquist diagrams)

Polar plots in their various guises provide a powerful set of stability and performance design techniques. The open-loop polar plot is a plot of the gain (magnitude) and phase locus of the open-loop LTF $G(s)K(s)$ for all frequencies. The open-loop inverse polar plot is similar and shows the reciprocal function, the locus of the inverse of the gain and negative phase for all frequencies.

Both diagrams may be plotted on polar or linear graph paper and provide the basis for the Nyquist stability criterion (Chapter 4). This method is more powerful than the Routh criterion (Section 3.3.3) in that it indicates the degree of stability, or the adjustments required to produce stability, and also provides exact stability information for systems containing time delays (Section 3.7). For now, note that Nyquist diagrams, named after Harry Nyquist (1889–1976), are mappings of the 'Nyquist contour' (see Section 4.3.1 for more detail) from the s-plane onto the $G(s)K(s)$-plane (the $G(s)K(s)$-plane is effectively an Argand diagram for the real and imaginary parts of $G(s)K(s)$). The Nyquist contour in the s-plane is 'D'-shaped, and the vertical of the 'D' is the imaginary s-plane axis, extending to infinity in each direction. Mappings along this axis produce frequency response data from which the open-loop polar plots are obtained. These polar plots therefore form part of the Nyquist diagram. Furthermore, after testing for stability, only the polar plot portion of the Nyquist diagram is used for design work. Similarly, the inverse Nyquist diagram contains the inverse open-loop polar plot. Again, after checking for stability, only that portion of the diagram containing the inverse polar plot is used.

Nyquist diagrams are particularly useful as they can be used to determine the closed-loop stability of a system directly from an open-loop plot. It should be noted that, except in specific cases, satisfactory gain and phase margins do not *guarantee* closed-loop stability. This means that when using Bode diagrams (see Section 3.5.1), for example, it may still be necessary to test the stability of a particular design using the Nyquist or Routh stability criteria.

Inverse Nyquist diagrams are particularly useful when the system has a minor feedback loop, or if feedback compensation is required (feedback compensation is the deliberate introduction of dynamic elements into the feedback path in order to meet a particular design specification). Familiarity with inverse Nyquist diagrams is a prerequisite for certain frequency domain studies of multivariable systems (see Section 10.7).

A typical polar plot is shown in Figure 3.45 and a typical inverse polar plot in Figure 3.46, for a system having an open-loop transfer function $G(s)$. In both the Nyquist and inverse Nyquist diagrams (the more general forms of the polar and inverse polar plots) a unit circle is the locus of all points at which the respective magnitudes will be unity, and the negative real axis is the locus of all points at which the phase lag is 180°. Arrows on the $G(j\omega)$ and $G^{-1}(j\omega)$ loci indicate the directions of increasing frequency.

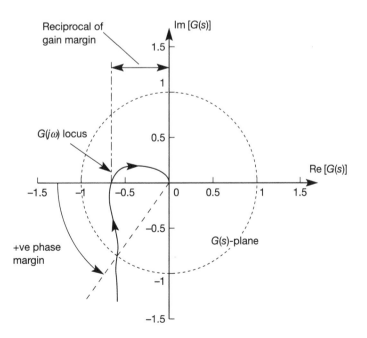

Figure 3.45 Stability margins on the direct polar (Nyquist) plot.

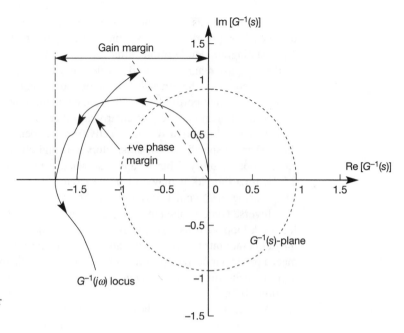

Figure 3.46 Stability margins on the inverse polar (inverse Nyquist) plot.

Polar plots

A system's type number and the related steady-state error information (see Section 3.4.1) may be determined directly from the open-loop polar plot. Figure 3.47 shows a selection of polar plots for systems of different type number. Again it is assumed that the system's open-loop transfer function $G(s)K(s)$ may be simply represented by the transfer function $G(s)$. The $G(j\omega)$ locus for a type 0 system starts from a point on the $0°$ line, at $\omega = 0$. A type 1 system has a locus of infinite magnitude at

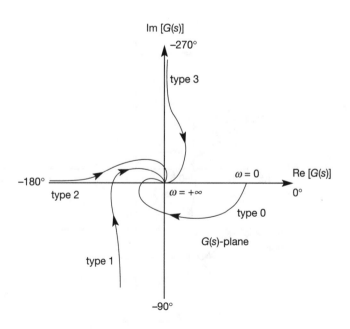

Figure 3.47 The general effect of system type number on direct polar (Nyquist) plots.

$\omega = 0$ and starts from a line asymptotic to the $-90°$ axis. Type 2 and 3 systems start from lines asymptotic to the $-180°$ and $-270°$ axes, respectively. Incidentally, as $\omega \to \infty$, the magnitude ratio tends to zero and the phase angle to $-R90°$, for a minimum phase system (one with all poles and zeros in the left-half s-plane). Here R is the rank of the system (excess of poles over zeros). If the closed-loop system is stable, the error in a type 0 system for a unit step input is given by (Section 2.5.5):

$$e(t)_{ss} = \frac{1}{1 + KG(0)}$$

where K is any additional loop gain not included in $G(s)$, and $G(0)$ is the magnitude of the system's open-loop frequency response evaluated at zero frequency. The magnitude of $G(0)$ may be measured directly from the polar plot, and indicates that the further the starting point of the locus is from the origin of the $G(s)$ plane, the smaller is the steady-state error.

For systems with higher type numbers the $G(j\omega)$ locus becomes infinite at zero frequency. Therefore, if a particular forcing input produces a steady-state error, the magnitude of the error cannot be calculated directly from the polar plot. There are techniques for quantifying the error, using polar data, in which the location of the zero-frequency asymptote relative to the appropriate axis is found. The greater the distance between the axis and the asymptote, the greater the reduction in any steady-state error.

For a system with open-loop LTF $G(s)$, the closed-loop frequency response may be expressed as

$$T(j\omega) = \frac{G(j\omega)}{1 + G(j\omega)}$$

Now, $T(j\omega)$ has magnitude

$$|T(j\omega)| = \frac{|G(j\omega)|}{|1 + G(j\omega)|} \tag{3.83}$$

and argument

$$\arg T(j\omega) = \arg G(j\omega) - \arg(1 + G(j\omega)) \tag{3.84}$$

In the polar plot of the open-loop system transfer function (see Figure 3.48) the modulus of $G(j\omega)$ is the length of the vector from the origin of the plot to the point on the $G(j\omega)$ locus corresponding to the frequency value ω. The argument of $G(j\omega)$ is the angle this vector makes with the positive real axis. By simple vector addition, the modulus of $1 + G(j\omega)$ is the length of the vector from the $-1 + j0$ point to the point on the $G(j\omega)$ locus corresponding to the frequency value ω. Again, its argument is the angle this vector makes with the positive real axis.

By selecting a number of points on the $G(j\omega)$ locus and applying Equations (3.83) and (3.84), the closed-loop frequency response is obtained. With the use of a computer even this simple procedure becomes unnecessary. However, it is worth exploring the method further, since it illuminates the development of the Nichols chart (see Section 3.5.3).

The magnitude of the closed-loop frequency response at a given frequency is expressed as the ratio of the length of two vectors, see Equation (3.83). Clearly, an infinite number of vectors could be found which would give the same magnitude

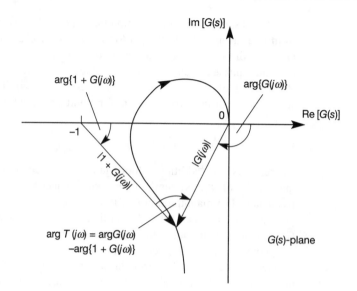

Figure 3.48 Closed-loop information from an open-loop direct polar (Nyquist) plot.

value. Let us find the contour in the $G(j\omega)$ plane which corresponds to these values. Let the constant closed-loop magnitude in question have the value of M, and let the open-loop transfer function $G(j\omega)$ have the coordinate points $x + jy$. From Equation (3.83),

$$M = \frac{|x + jy|}{|1 + x + jy|}$$

Squaring both sides and rearranging yields

$$\left(x + \frac{M^2}{M^2 - 1}\right)^2 + y^2 = \frac{M^2}{(M^2 - 1)^2}$$

which is the equation of a circle having its centre at $(-M^2/(M^2 - 1), 0)$ and a radius of $|M/(M^2 - 1)|$.

For each value of M, such an 'M-circle' may be drawn in the $G(j\omega)$-plane. Every time the open-loop $G(j\omega)$ locus intersects such a circle, the M value of the circle gives the gain of the corresponding closed-loop frequency response at frequency ω.

A similar analysis may be carried out on the argument of the closed-loop frequency response. Let N be some constant value of angle, and let the coordinates of the $G(j\omega)$ locus be $x + jy$ as before. Then, from Equation (3.84):

$$\tan^{-1} N = \tan^{-1}(y/x) - \tan^{-1}[y/(1 + x)]$$

$$\tan^{-1} N = \tan^{-1} \frac{(y/x) - y/(1 + x)}{1 + (y/x)[y/(1 + x)]}$$

from which it may be deduced that

$$N = \frac{y}{x^2 + x + y^2}$$

After further manipulation, the above equation can be written in the form

$$\left(x+\frac{1}{2}\right)^2+\left(y-\frac{1}{2N}\right)^2=\frac{1}{4}\left(\frac{N^2+1}{N^2}\right)$$

which is also the equation of a circle, but this time with its centre at $(-1/2, 1/(2N))$ and radius

$$\frac{1}{2N}\sqrt{(N^2+1)}$$

For each value of N a circle could be drawn in the $G(j\omega)$-plane. This time, an intersection of the open-loop $G(j\omega)$ locus with an 'N-circle' gives the corresponding closed-loop phase angle.

Before dealing with the various closed-loop stability criteria, it is interesting to note that the Nichols chart (described in Section 3.5.3) is obtained by mapping these M and N circles in polar coordinates, onto M and N loci in rectangular magnitude (dB) and phase coordinates.

The open-loop polar plot of the $G(j\omega)$ locus with superimposed M circles is used to determine the closed-loop peak magnification, bandwidth and cut-off rate. In Figure 3.49, the $G(j\omega)$ locus touches the $M=2$ circle at frequency ω_p. This indicates that $M_p=2$ and the peak frequency is ω_p rad s^{-1}. Equation (3.64) may be used to approximate the effective damping ratio from M_p, which will be found to be 0.26. A further test would be to measure the phase margin, and use the rule of thumb given in Section 3.4 to find an approximate value for the damping ratio, namely (phase margin)/100. The two methods should give similar results, although the estimated damping ratio found using the M_p value will be more representative of the response.

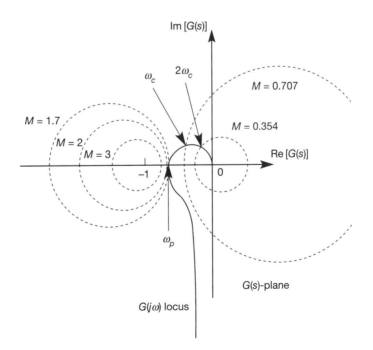

Figure 3.49 Circles of closed-loop peak magnification on the open-loop direct polar (Nyquist) plot.

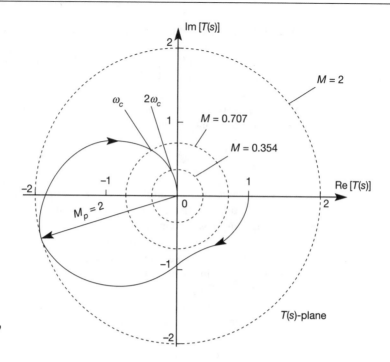

Figure 3.50 The information of Figure 3.49 transferred to a *closed-loop* direct polar (Nyquist) plot.

For a zero-frequency magnitude ratio of unity, the bandwidth may be determined from the frequency at which the $G(j\omega)$ locus intersects the $M = 0.707$ circle (see the comments on bandwidth following Equation (3.63)). In Figure 3.49 the bandwidth is the value of the frequency at ω_c. The cut-off rate is the rate of decrease in the magnitude ratio. In Figure 3.49, the frequency doubles if the magnitude ratio is halved, which indicates a cut-off rate of 6 dB/octave (or 20 dB/decade).

In the $T(s)$-plane (the plane for the polar plot of the *closed-loop* LTF), the M circles become circles of radius M with their centre at the origin of the plane, and the N circles become radial lines emanating from the origin of the plane. The closed-loop frequency measures of bandwidth, peak magnification and cut-off rate may all be made directly. Figure 3.50 shows the closed-loop $T(j\omega)$ locus corresponding to the open-loop $G(j\omega)$ locus in Figure 3.49.

Example 3.13 *Example 3.12 repeated using the direct polar plot*

Use polar plots to analyse the response of a system having the open-loop transfer function

$$G(s)K(s) = \frac{K}{s(s+1)(s+2)}$$

K can take the values 4, 1.5 and 0.4.

This system was analysed in Example 3.12 and therefore the only requirement is to show how the various frequency-domain specifications may be obtained from the open-loop and closed-loop polar plots, Figures 3.51 and 3.52 respectively.

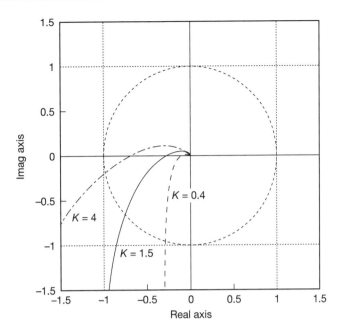

Figure 3.51 Direct polar (Nyquist) plots for the variable-gain system of Example 3.13.

Figure 3.45 shows how the gain and phase margins are determined. In Figure 3.51 these are found to be as follows for the various values of K:

$K = 4$ gain margin 1.5, phase margin $11°$

$K = 1.5$ gain margin 4, phase margin $42°$

$K = 0.4$ gain margin 15, phase margin $73°$

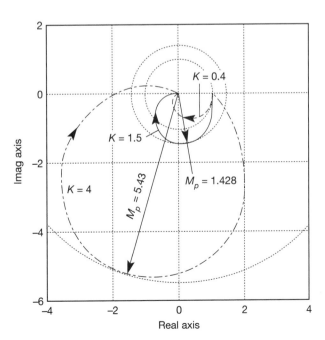

Figure 3.52 The information of Figure 3.51 transferred to a *closed-loop* direct polar (Nyquist) plot.

For this system, only one open-loop polar plot need be drawn, since only $M(\omega)$ changes with frequency (see Equations (3.66) to (3.68)). Different gains, K, are accommodated by simply rescaling the real and imaginary axes (alternatively, this can be imagined as moving all points on the plot radially away from the origin to increase the gain, or towards the origin to decrease it). The frequencies at which gain crossover and phase crossover occur are not directly measurable from the open-loop polar plot and if required, must be obtained from the frequency response data (MATLAB can do this – see below).

Since it is normally easier to produce a closed-loop polar plot than draw M and N circles on the open-loop polar plot, the closed-loop polar plot is shown in Figure 3.52. By drawing circles centred at the origin of the closed-loop polar plot as shown in Figure 3.49, M_p, ω_p, bandwidth and cut-off rate may all be determined. Again some reference to the frequency response data will be required. As expected, the performance measures are found to be the same as those found in Example 3.12.

Using MATLAB (Appendix 3), Figure 3.51 (for example) can be produced by using the m-file *fig3_44.m*, and substituting the command *nyquist* for the command *bode*. This is done in the file *fig3_51.m* on the accompanying disk (also, three times as many points are plotted to give smooth curves). With the Multivariable Frequency Domain Toolbox (MVFDTB–Ford *et al.* (1990)) for MATLAB, M-circles can be added, and this is also done in *fig3_52.m*. Note that both these files additionally generate mirror images of the curves in the text, corresponding to 'negative frequencies' – these are explained later. In the case of $K = 0.4$, the closed-loop system is overdamped (see Example 3.12), so the appropriate M-circle should be the one corresponding to the low-frequency gain. From the final value theorem applied to the closed-loop LTF for this case, this ought to be unity (the MATLAB command *dcgain(nc,dc)* inserted into the m-file will confirm this).

Without the rather specialized MVFDTB, an alternative is to use MATLAB to scan the frequency response data and find the maximum closed-loop gain. For this to be successful, there must be enough frequency points to be sure of 'catching' a value more or less at the peak. It is therefore wise to set up a frequency vector, containing at least a few hundred points (a Bode plot can be produced and checked for smoothness around the peak). For example, for $K = 4$, proceed as follows (the *bode* command is used because it provides magnitude and phase data; the *nyquist* command used in this way produces real and imaginary data):

```
>> w=logspace(-1,1,500);          % 500 log-spaced values from 0.1 to 1.0 rad/s

>> den=conv([1 1 0], [1 2]);

>> [ncl,dcl]=cloop(4,den,-1);     % form unity negative feedback system

>> [mag,pha]=bode(ncl,dcl,w);     % calculate closed loop data using w,
                                  % but do not plot

>> max(mag)                       % displays M_p value
```

It is also feasible to use the *ginput* command to pick the peak value from the bode plot (see Appendix 3, or use >> *help ginput*). Again, the accuracy will depend upon the number of points in the plot and, of course, the accuracy in positioning the cursor over the plot. MATLAB can also confirm the gain and phase margins for each case (as in Example 3.12).

Inverse polar plots

This section provides a brief introduction to inverse polar plots. In principle, the inverse polar plot may be used to achieve the same objectives as direct polar plots. The Nyquist stability criterion (see Section 4.3), gain margin, phase margin, steady-state error checks and the various closed-loop frequency response measures all have their counterpart in the inverse plane.

Consider the closed-loop system shown in Figure 3.53, whose overall LTF is given by

$$\frac{Y(s)}{R(s)} = T(s) = \frac{G(s)}{1 + GH(s)} \tag{3.85}$$

The inverse closed-loop transfer function for this system is

$$T^{-1}(s) = G^{-1}(s) + H(s) \tag{3.86}$$

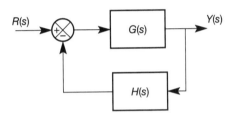

Figure 3.53 A feedback loop with dynamic elements in the feedback path.

For feedback compensation, where the object is to design the dynamic characteristics of $H(s)$ for fixed plant dynamics $G(s)$, it is much easier to work with Equation (3.86) than Equation (3.85). If direct polar plots are used, each change in $H(s)$ requires the rational polynomial $T(s)$ in Equation (3.85) to be evaluated. However, $T^{-1}(s)$ (Equation (3.86)) may be evaluated by using the fixed inverse polar plot $G^{-1}(s)$ of the plant and adding it vectorially to the direct polar plot $H(s)$ of the feedback compensator. In this way, the effect of $H(s)$ on the closed-loop system is seen immediately.

Another system for which it is advantageous to use the inverse notation is one in which a minor feedback loop is used. Figure 3.54 shows $G_1(s)$, a forward path compensator, and $H(s)$, a feedback compensator, for a plant with fixed dynamics $G(s)$. This type of control may have distinct advantages: the system may be easier to build, install and adjust than one with a more complex, single-loop controller. Although the designer will have more work to do, it is often possible to meet performance specifications which would otherwise be unobtainable. An example of the use of minor feedback loops is given in Section 4.5.3.

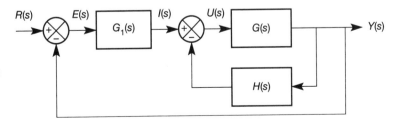

Figure 3.54 A system with an inner feedback loop containing dynamic elements.

An interactive approach to the design of the compensators $G_1(s)$ and $H(s)$ is normally adopted. Typically, the inner loop is designed first: that is, the problem is reduced to the design of a feedback compensator similar to that shown in Figure 3.53. For this stage of the design the inverse notation has a clear advantage. When the dynamics of $H(s)$ are fixed, the dynamics of the feedback compensated plant are fixed and given by

$$\frac{Y(s)}{I(s)} = \frac{G(s)}{1 + G(s)H(s)}$$

Bode and Nichols plots (Section 3.5.3), or a Nyquist plot with M and N circles, can then be used to design the forward path compensator $G_1(s)$.

Figure 3.55 shows a number of inverse polar plots near the origin of the inverse plane. The general shape of the plots depends on the inverse transfer function.

Normally, a system model will have more poles than zeros, and therefore for low frequencies (small values of ω) all but type 0 systems will have loci emanating from the origin of the inverse plane. A type 0 system starts from a point along the $0°$ axis. For a stable closed-loop response, the closer this starting point is to the origin, the lower the steady-state error to a demanded step input. A type 1 system starts along the $90°$ axis, a type 2 system along the $180°$ axis, and so on. If the system's transfer function has no zeros, the phase angle of the inverse locus increases with increasing ω. This produces an inverse locus which is a smooth curve in the anticlockwise direction. When zeros are present, the locus still tends to rotate anticlockwise with increasing ω, but the shape of the curve may be more irregular. For high frequencies (large ω) the (inverse) magnitude values become large. As $\omega \to \infty$ the (inverse) magnitude ratio becomes infinite, and the phase angle becomes $R90°$ for minimum phase systems. Again, R refers to the system rank. Given the system's transfer function, it is therefore possible to use the above argument to predict the general shape of the inverse locus.

Closed-loop performance measures require the inverse equivalent of the M and N circles used with direct polar plots. The closed-loop transfer function of a unity negative feedback system is (see Figure 3.53):

$$\frac{Y(s)}{R(s)} = T(s) = \frac{G(s)}{1 + G(s)} \tag{3.87}$$

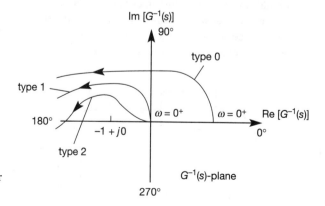

Figure 3.55 The general effect of system type number on the inverse polar (inverse Nyquist) plot.

from which the inverse closed-loop transfer function is

$$T^{-1}(s) = G^{-1}(s) + 1$$

From this equation the $T^{-1}(s)$ locus is obtained by vectorially adding 1 to each point in the $G^{-1}(s)$ locus. Therefore the phase angle of the closed-loop transfer function $T^{-1}(s)$ may be obtained from the open-loop frequency response plot $G^{-1}(j\omega)$, for any value ω, in the following manner. Draw a vector from the $G^{-1}(j\omega)$ locus, from the point with the frequency value ω, to the point $(-1 + j0)$, as in Figure 3.56. The required closed-loop phase angle is the angle this vector makes with the positive real axis. Similarly, a circle of constant magnitude ratio which is centred at the origin of the $T^{-1}(s)$-plane is mapped onto a circle of identical radius in the $G^{-1}(s)$-plane, but centred at the $(-1 + j0)$ point.

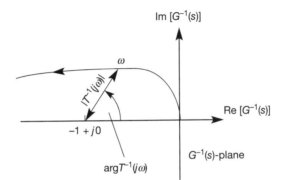

Figure 3.56 Closed-loop information from an open-loop inverse polar (inverse Nyquist) plot.

The above analysis indicates that lines of constant magnitude ratio in the closed-loop $T(s)$-plane map into circles centred on the $(-1 + j0)$ point in the $G^{-1}(s)$-plane. The inverse of the radius of the circle in the $G^{-1}(s)$-plane gives the closed-loop magnitude ratio. Also, lines of constant phase angle in the closed-loop $T(s)$-plane map onto radial lines centred on the $(-1 + j0)$ point in the $G^{-1}(s)$-plane. In this mapping, the angle of any radial line undergoes a sign change.

Example 3.14 *Analysis by an inverse polar plot*

Draw the inverse polar plot of

$$G(s) = \frac{1}{s^4 + 6s^3 + 11s^2 + 6s}$$

and determine the gain margin, phase margin and closed-loop bandwidth. Also, comment on the M_p value.

In inverse notation,

$$G^{-1}(j\omega) = \omega^2(\omega^2 - 11) - 6j\omega(\omega^2 - 1)$$

which is readily plotted on the $G^{-1}(s)$-plane, as shown in Figure 3.57.

The gain margin is the magnitude of $G^{-1}(j\omega)$ when the imaginary part of $G^{-1}(j\omega)$ is zero. This occurs when $\omega = 1$ and the gain margin is 10 (see Figure 3.57).

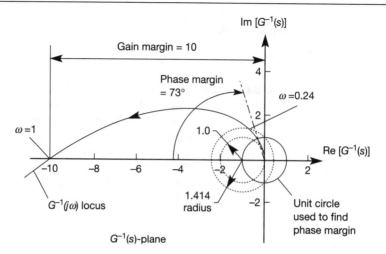

Figure 3.57 Inverse polar (inverse Nyquist) plot for Example 3.14.

Also from Figure 3.57 the phase margin is measured to be 73°.

Since the open-loop system is type 1 with no right-half s-plane poles, the gain and phase margins are sufficient to indicate closed-loop stability. The bandwidth is determined by drawing a circle of radius 1/0.707 (or 1.414, or $\sqrt{2}$) centred at the $(-1 + j0)$ point, as shown. This circle crosses the $G^{-1}(j\omega)$ locus at $\omega = 0.24$ rad s^{-1}.

The value of M_p is determined by drawing the circle of largest diameter centred on the $(-1 + j0)$ point which just touches the $G^{-1}(j\omega)$ locus without crossing it. However, the phase margin of 73° indicates an equivalent damping ratio of approximately 0.73, and therefore the peak magnification is likely to be at the zero frequency point. As expected, Figure 3.57 shows that the largest circle with centre $(-1 + j0)$ which just touches the locus has unit radius.

Example 3.15 *Examples 3.12 and 3.13 repeated using the inverse polar plot*

Draw the inverse polar plots for a system having the open-loop transfer function

$$G(s)K(s) = \frac{K}{s(s+1)(s+2)}$$

where K can take the values 4, 1.5 and 0.4. This is the system analysed in Examples 3.12 and 3.13 using Bode and direct polar plots respectively.

The inverse polar plots are shown in Figure 3.58. The circles centred at $(-1, 0)$ have radii of 1.0, 0.7 and 0.18 and each just touches the $K = 0.4$ locus, the $K = 1.5$ locus, and the $K = 4.0$ locus respectively. Taking the reciprocal of the radius of a particular circle gives the system's M_p value for the indicated gain. The closed-loop phase shift at the M_p value may be measured directly from the inverse polar plot (see Figure 3.56) and the frequency ω_p is found from the frequency response data.

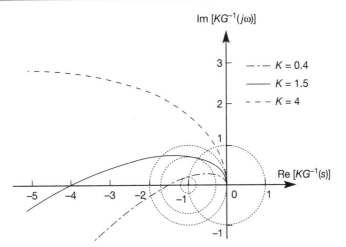

Figure 3.58 Inverse polar (inverse Nyquist) plots for the variable-gain system of Example 3.15.

Using MATLAB for inverse Nyquist plots is easiest if the MVFDTB is available, when the *finv* command can be used. However, assuming this is not the case, but that direct Nyquist or Bode data are present, the inverse frequency response data can easily be generated. For example, in MATLAB any frequency point on the direct Nyquist plot is given in rectangular coordinates by a complex number of the form $x + jy$. The corresponding point on the inverse plot is therefore:

$$\frac{1}{x + jy} = \frac{x}{x^2 + y^2} - j \frac{y}{x^2 + y^2}$$

To calculate individual frequency points, rather than doing normal vector algebra on the entire data set, the dot operator is used to force MATLAB to do point-by-point calculations. Alternatively, the Bode magnitude and phase data can be used. The m-files *fig3_57.m* and *fig3_58.m* on the accompanying disk use the Nyquist and Bode approaches respectively, so as to illustrate both. The commands adding the circles only work if the MVFD toolbox is installed, so they are initially 'commented-out'. Remove the per cent signs to make them active if the MVFDTB is available. In each case, the circles were arrived at by using a larger plot to show the area of interest (use *help axis* to see how this is done), and trying circles of varying radius to find the best.

3.5.3 Rectangular coordinate plot of magnitude and phase, or Nichols plot

In this plot, the y-axis represents the magnitude expressed in decibels, and the x-axis the phase angle in degrees. Linear graph paper could be used to produce a Nichols plot, but it is more common to use a Nichols chart (both named after Nathaniel Burgess Nichols, b. 1914).

The Nichols chart consists of linear graph paper, specially prepared by the addition of superimposed closed-loop magnitude and phase contours (see Figure 3.59, which is simply produced by the MATLAB command *ngrid('new')*). By plotting the *open-loop* magnitude and phase locus using the linear axes of this chart, the system's *closed-loop* frequency response characteristics may be determined from these superimposed contours. For example, if an open-loop system has a magnitude ratio of −8 dB and a phase shift of −80° at a given frequency, and these values are plotted on the linear axes labelled 'open loop gain' and 'open loop phase'

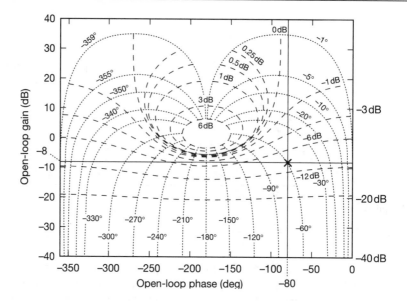

Figure 3.59 A Nichols chart.

in Figure 3.59, the point marked '×' results. The closed-loop contours on the Nichols chart indicate that, at the same frequency, the unity feedback, closed-loop system would have a magnitude of about $-10\,\mathrm{dB}$ (between the $-6\,\mathrm{dB}$ and $-12\,\mathrm{dB}$ contours) and a phase shift of $-60°$.

In principle, a Nichols chart plot may be used for adjusting a system's open-loop gain and phase margins, as well as modifying its closed-loop performance specifications. In practice, Bode plots are used to shape the open-loop response and, since a computer is invariably used to produce these plots, it is more convenient to use computer generated closed-loop frequency response data to check the closed-loop characteristics.

A Nichols chart is essentially a transformation of the M and N circles on the polar plot (see Section 3.5.2) into non-circular M and N contours on a plot of dB magnitude versus phase, in rectangular coordinates. A plot of the open-loop frequency response function $G(j\omega)$ on a Nichols chart yields the gain and phase margins directly (see Figure 3.60).

Information about the gain and phase margins, together with the various closed-loop specifications, is summarized in Figure 3.60. In this figure, M_p, the peak maximum magnitude ratio, is obtained by finding the largest M contour which touches, but does not cross, the $G(j\omega)$ locus. The frequency at M_p, namely ω_p, cannot be read directly, but is usually readily found by interpolating between the $G(j\omega)$ data points. The bandwidth of the closed-loop system is similarly found from the intersection of the $G(j\omega)$ locus with the $-3\,\mathrm{dB}$ M contour.

A change in gain does not alter the shape of the $G(j\omega)$ locus on the Nichols chart, but shifts it vertically. Also, since the magnitude is expressed in dB form, it is common practice to represent the system's open-loop transfer function $G(j\omega)$ in Bode form (see Equation (3.69)). In this form, the effect of a design change such as the inclusion or modification of a compensator, is easily accommodated. The basic plot remains the same, and the modification becomes a problem in graphical

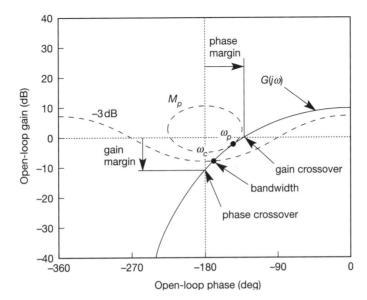

Figure 3.60 Performance and stability criteria on the Nichols chart.

addition. Unlike Bode plots, the Nichols chart does not allow asymptotic approximations to be made. Therefore data points must be calculated directly or transferred from the Bode to the Nichols chart.

Example 3.16 *The system of Example 3.12 on the Nichols chart*

Produce Nichols plots for the system

$$G(s)K(s) = \frac{K}{s(s+1)(s+2)}$$

with $K = 4$, 1.5 and 0.4.

In the Nichols plots of Figure 3.61, the chain dotted line is the locus of frequency points for $K = 4$, the solid line for $K = 1.5$ and the dashed line for $K = 0.4$. With only a monochromatic presentation, computer generated Nichols plots can be confusing. However, by tracing the solid line (the locus of points for $K = 1.5$) the $G(j\omega)K(j\omega)$ locus appears to touch (but not cross) the 3 dB M-circle, making M_p 3 dB (or 1.41) at which point the closed-loop phase angle is about $-80°$. The remaining performance measures can be found as indicated in Figure 3.60. The m-file *fig3_61.m* on the accompanying disk will generate Figure 3.61.

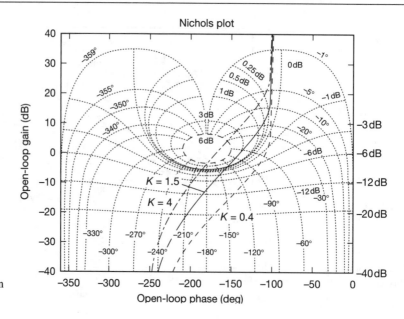

Figure 3.61 Nichols plots for the variable-gain system of Example 3.16.

3.6 Responses of discrete-time models

At this stage it is appropriate to re-examine the ARMA-type difference equation model of Section 2.8.1, and the discrete-time state-space model of Section 2.8.2.

3.6.1 ARMA-type difference equation models

Section 2.8.1 introduced this form of model with the equation:

$$y_n = A_1 y_{n-1} + A_2 y_{n-2} + A_3 y_{n-3} + \cdots$$
$$+ B_0 u_n + B_1 u_{n-1} + B_2 u_{n-2} + \cdots$$

where the A and B values are constants, y is the output and u is the input. As before (see Figure 2.52), the nth output (or input) is the latest one, the $(n-1)$th is the previous one, and so on. How such models work will be demonstrated by examples.

Example 3.17 Investigate the step response of a system modelled by a difference equation

Find the unit step response of the system modelled by the difference equation:

$$y_n = 0.9 y_{n-1} + 0.1 u_n$$

where y starts from zero at step 0 and the sampling interval is 0.1 s.

Since the specified input is a unit step, all the u_n values will be equal to unity and the following iterative approach may be used:

Set *last_y* equal to zero and n equal to 1

Loop: Calculate y as $0.9 \times last\_y + 0.1 \times u$ (NB: u is 1)

Store, graph or output y and t ($t = n \times$ the sampling interval)

Make *last_y* equal to the new value of y

Add 1 to n

Go to *Loop* (unless enough of the response has been calculated!)

This algorithm can be coded as an m-file in MATLAB (see Appendix 3) as follows:

```
last_y = 0;
n = 0;
ts = 0.1;          % Setting the sample interval equal to 0.1 second
y(1) = last_y;     % First value of response (initial condition)
t(1) = 0;          % Initial value of time
for n = 2 : 101    % This will iterate through 100 time steps
    y(n) = 0.9*last_y + 0.1;   % Storing the y values in the form of an array
    t(n) = n*ts;               % Doing likewise with the time values
    last_y = y(n);             % ready for the next step
end
plot(t,y,'x'),grid
```

Note that MATLAB is not designed to work efficiently when programmed using *for* loops to step array indices (there are tricks to make execution of such 'non-vectorized' MATLAB code more efficient, but they are beyond the scope of this text). This usage of MATLAB has been included to illustrate how such simulation might be performed using a conventional technical computing language. Also, since many languages cannot cope with element (0) in an array, the initial conditions are stored in element (1) – so array element numbers are always one greater than the time step number whose data they contain.

The m-file (above) produced the output shown in Figure 3.62. The 'x' characters are used to show the output samples and indicate the discrete-time nature of the response (without this, the *plot* command would draw lines joining up the points as a smooth curve). Note that the output is the digitized version of a continuous-time first-order lag. The interconversion is described in Chapter 4.

An alternative is to use the MATLAB *filter* command which is also defined using a difference equation. The required commands are:

```
≫ u = ones(100,1);
≫ y = filter(0.1, [1 −0.9], u);
≫ plot([0;y], 'x'),grid
```

Notes:
- The *ones* command sets up the unit step input u as a column of 100 unity elements (one for each of the required 100 time steps).

- The *filter* command applies the filter (that is, the system represented by the difference equation model) to the input data stream in u, so as to generate the corresponding output data stream in y. The filter (model) coefficients are specified by rearranging the model in the form:

$$y_n - 0.9y_{n-1} = 0.1u_n$$

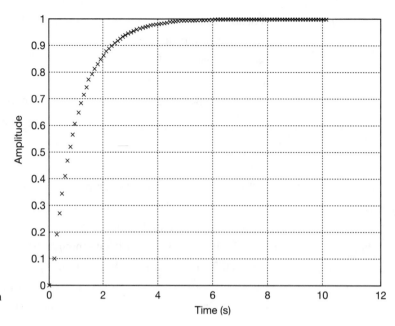

Figure 3.62 Discrete-time step response for the system of Example 3.17.

The first argument in the *filter* command is then a vector containing the coefficients on the RHS of the equation (in this case, just one, namely 0.1), and the second argument is a vector containing the coefficients on the LHS (that is, [1 −0.9]).

- The *plot* command needs to add an extra zero row to the top of the y vector, to represent the initial condition on y (because the *filter* command output begins with time step 1). The resulting plot is similar to Figure 3.62, except that the horizontal axis is calibrated in sample numbers, not time values, and the initial value ($t = 0$) now corresponds to sample number 1.
- A matching 101 element time vector could be set up (for example, using $\gg t = [0{:}0.1{:}10];$) and the command $\gg \text{plot}(t, [0; y], 'x')$, grid would then duplicate Figure 3.62.
- The best way to use MATLAB directly for this example would be to use z-transforms (introduced in Chapter 4). Then the *dstep* command could be used, and the output plot would appear as a sampled-and-held (or 'staircase') type of waveform.

The next example will show that higher-order systems can also be represented in discrete form.

Example 3.18 *A higher-order discrete-time step response*

Investigate the unit step response of the system modelled by the equation:

$$y_n = 1.6y_{n-1} - 0.8y_{n-2} + 0.11u_{n-1} + 0.09u_{n-2}$$

The sampling interval is 0.1 s.

The procedure is based on that given in Example 3.17:

Set *last_y* and *last1_y* equal to zero

Set *last_u* and *last1_u* equal to zero

Set *n* equal to 1

Loop: Calculate *y* as

$$1.6 \times last\_y - 0.8 \times last1\_y + 0.11 \times last\_u + 0.09 \times last1\_u$$

Store, graph or output *y* and *t* (*t* = *n* × the sampling interval)

Make *last1_y* equal to the new value of *last_y*

Make *last1_y* equal to the new value of *y*

Make *last1_u* equal to the new value of *last_u*

Make *last_u* equal to the new value of *u*

Add 1 to *n*

Go to *Loop* (unless enough of the response has been calculated)

If desired (although it is not a recommended usage) MATLAB can be used to code the above example directly:

```
% Demo for second-order discrete example
% Initializing
last1_y = 0;   last_y = 0;   last1_u = 0;   last_u = 0;
u = 1;         % Unit step applied at t = 0
y(1) = 0;      % Output y starts at zero
for n = 2 : 101
    y(n) = 1.6*last_y - 0.8*last1_y + 0.11*last_u + 0.09*last1_u;
    last1_y = last_y;   last_y = y(n);
    last1_u = last_u;   last_u = u;
end;
t = 0.1*[0:100];   % 101 time values for sampling interval = 0.1 s
plot(t,y,'x'); grid
```

The output appears as Figure 3.63.

More elegantly, the *filter* command introduced in Example 3.17 could be used. Note that the denominator and numerator both have the same number of terms, which has been achieved by including the zero coefficient of u_n. The significance of this is explained in Section 3.7.

```
≫ u = ones(100,1);
≫ y = filter([0  0.11  0.09], [1  −1.6  0.8], u);
≫ plot([0;y], 'x'); grid
```

Discrete-time models of this type are next studied in detail in Section 5.7 of Chapter 5.

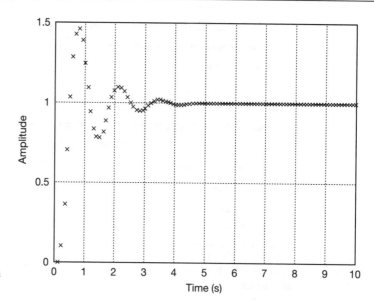

Figure 3.63 Discrete-time step response for the system of Example 3.18.

3.6.2 Discrete-time state-space models

Equation (3.12), repeated below, gives the full solution for the time responses of the state variables in a state-space model.

$$x(t) = \Phi(t)x(0) + \int_0^t \Phi(t - \tau).B.u(\tau).d\tau \tag{3.12}$$

Application of the linear output equation (Equation (2.44)) would then give the output responses. However, direct solution of the convolution integral in Equation (3.12) can be difficult, and it is usual to convert to the discrete-time version of the equation, outlined in Section 2.8. This is how digital computer packages such as MATLAB evaluate their time responses – even when it appears that continuous-time models are used, giving continuous-time results.

For computer solution of Equation (3.12), it would be convenient if a discrete-time, recursive expression existed. That is, if the value of x at a sampling instant k (written as x_k) could be found as a simple function of the values from the previous sampling instant: $x_k = f(x_{k-1}, u_{k-1})$. So long as such a scheme is properly initialized (that is, at $k = 0$), the solution is then very simply found in a step-by-step manner.

In seeking to obtain a discrete-time version of Equation (3.12), assume that the input and state vectors (x and u) will be sampled-and-held at the sample intervals $k_0, k_1, k_2, \ldots$. This gives rise to sampled signals $x_0, x_1, \ldots$ and $u_0, u_1, \ldots$ which will be regarded as remaining *constant* throughout each sampling period as illustrated in Figure 2.51 (which was drawn for a signal y, but x or u could equally well be substituted for y). This means that information is lost between sampling instants. Therefore, as usual in such systems, the samples must be taken sufficiently often to capture all significant events.

It can be assumed that the values at the kth sampling instant are known, and therefore can be used to calculate the values which will exist at the $(k + 1)$th instant.

This assumption is valid since, if the initial conditions (at $k = 0$) are known, together with the recursive, discrete-time expression, then all future values follow.

It is convenient to regard the response of Equation (3.12) as being the sum of two components:

- The *free response* due to the initial conditions only (the first term on the RHS)
- The *forced response* due to the input signals only (the convolution term on the RHS).

For the sampled signal in the interval t_k to t_{k+1}, the state vector at the kth instant (x_k) becomes a fixed initial condition (see Figure 2.51, reading y_k as x_k). If the sampling interval (that is, $t_{k+1} - t_k$) is h seconds, then that part of the solution corresponding to the *free response* (Equation (3.12) with zero input) is seen to be:

$$x_{k+1_{free}} = e^{Ah} x_k = \Phi(h) x_k \tag{3.88}$$

Like the state vector, the input vector is also sampled-and-held (that is, analogously to Figure 2.51), and therefore the input vector u_k is a *constant* over the same interval. This is the key to the solution, since it permits 'de-convolution' of the integral in the *forced response* part of the solution. Hence, ignoring the initial condition term (the 'free response', above), and taking the (now constant) input outside the integral, yields:

$$x_{k+1_{forced}} = \int_0^h e^{A(h-\tau)} \, d\tau \, Bu_k = -A^{-1}[e^{A(h-\tau)}]_0^h \, Bu_k$$

$$= -A^{-1}[I - e^{Ah}] Bu_k = \Delta(h) u_k \tag{3.89}$$

where:

$$\Delta(h) = A^{-1}[e^{Ah} - I]B = A^{-1}[\Phi(h) - I]B$$

Combining Equations (3.88) and (3.89) gives the general discrete-time state equation:

$$x_{k+1} = x_{k+1_{free}} + x_{k+1_{forced}}$$

or

$$x_{k+1} = \Phi(h) x_k + \Delta(h) u_k \tag{3.90}$$

where:

$$\Phi(h) = e^{Ah} = I + Ah + \frac{A^2 h^2}{2!} + \frac{A^3 h^3}{3!} + \cdots$$

and

$$\Delta(h) = A^{-1}[\Phi(h) - I]B = \left[I.h + \frac{Ah^2}{2!} + \frac{A^2 h^3}{3!} + \cdots \right].B$$

Thus, the matrix exponentials can be evaluated using the power series expansion of a general exponential function.

For completeness, the corresponding discrete-time output equation is:

$$y_{k+1} = Cx_{k+1} + Du_{k+1} \tag{3.91}$$

Digital computer simulation

The discrete-time state-space representation of a system (Equation (3.90)) permits the implementation of digital control schemes based on state-space methods. However, it is also suitable for digital simulation of the system.

For stationary systems, provided that the sampling interval h is constant, $\Phi(h)$ and $\Delta(h)$ will be constant matrices. This means that, for such systems, they need only be calculated once.

The procedure is as follows. In the expressions for $\Phi(h)$ and $\Delta(h)$ following Equation (3.90), begin with the identity matrix as the first term in $\Phi(h)$. Starting with $i = 1$, it will then be found that the ith term in $\Delta(h)$ is the ith term in $\Phi(h)$ multiplied by h/i, and the $(i + 1)$th term in $\Phi(h)$ is then the ith term in $\Delta(h)$ multiplied by A. Increment i and continue until sufficient accuracy is reached. Finally, post-multiply $\Delta(h)$ by B at the end of the procedure (this last step is easily forgotten)!

An outline algorithm for performing a digital simulation of the system

$$\dot{x} = Ax + Bu, \qquad y = Cx + Du$$

would then be:

OBTAIN A, B, C and D (the system matrices)

        initial state vector x_0

        initial input vector u_0

SELECT sampling interval h, small enough to cover all happenings of interest

        duration of simulation, T_{end}

CALCULATE $\Phi(h)$ and $\Delta(h)$ to sufficient accuracy

$k = 0$

DO UNTIL $kh = T_{end}$

        EVALUATE x_{k+1} (Equation (3.90))

        EVALUATE y_{k+1} (Equation (3.91)) and STORE or OUTPUT

        OBTAIN   input vector for next time step u_{k+1}

        $k = k + 1$

END DO

PLOT or OUTPUT required results

Sylvester's expansion theorem

The 'power series' expansion method used above for calculating $\Phi(h)$ and $\Delta(h)$, although convergent, may converge only slowly. If a computer is being used to estimate these values by summing terms, then a large number of terms will be required for models of some systems. Typically other methods, leading to closed-form solutions, are required. One such is based on the Sylvester expansion theorem. The result is stated without proof.

If A is an $n \times n$ matrix, and has eigenvalues (see Appendix A1.2) λ_1, $\lambda_2, \ldots, \lambda_n$, then:

$$e^{Ah} = \sum_{i=1}^{n} [e^{\lambda_i h} Z_i]$$

where:

$$Z_i = \frac{\prod\limits_{j=1, j \neq i}^{n} (A - \lambda_j I)}{\prod\limits_{j=1, j \neq i}^{n} (\lambda_i - \lambda_j)}$$

Computer-assisted analysis with MATLAB

Using MATLAB (Appendix 3), the *step* or *lsim* commands will directly perform the above procedure. For the state-space system defined by the continuous-time system matrices A, B, C and D, type

$\gg$ *step(A,B,C,D,i)*   for the step response from the ith input (i is omitted for SISO systems)

If a specific timescale is required, or an input other than a unit step is to be applied, this is also possible. Use the *help step* and *help lsim* commands respectively, for more information.

The command *dstep* works for discrete-time models. However, when using MATLAB's *help* system, note that the help system does not distinguish between A and B for continuous-time models, and Φ and Δ for discrete-time models (all are called A and B).

To obtain a discrete-time model, from a continuous-time one, type:

$\gg$ *[phi,delta]* = *c2d(A,B,h)*   (*d2c* does the inverse)

3.7 Time delays (or transport lags)

A phenomenon often encountered in process equipment is that of transportation lag or time delay. For example, if a fluid is flowing with a velocity v m s^{-1} in a pipe of length L m, the time taken for any individual element of fluid to flow from the entrance to the exit of the pipe is L/v s. This time, usually given the symbol T or τ, represents a delay between an action occurring and its observed response. The real shift theorem, see Table 2.8, may be used to show that for such a process the open-loop system Laplace transfer model is

$$G(s)K(s)e^{-s\tau}$$

There are two or three ways of dealing with time delays in modelling and design. One is to use a *Padé approximation*, which replaces the term $e^{-s\tau}$ with a rational polynomial in s. This method is described in Section 8.6.2. This is seldom accurate, particularly when using the open-loop model to find the closed-loop response, but does give an open-loop transfer function which consists purely of poles and zeros.

The second and more accurate method is to perform all the design work within the Laplace domain. As previously noted, the Laplace transform of a time delay is given by $e^{-s\tau}$, and has the corresponding frequency response function $e^{-j\omega\tau}$. This has a magnitude of unity for all frequencies and a phase angle of $-\omega\tau$ radians, or $-\omega\tau\,360/2\pi$ degrees.

A time delay does not affect a system's magnitude since its contribution is 1 (that is, 0 dB), but it does affect the phase plot (by adding the negative angle just described). The effect on the Bode plot of adding a time delay to a transfer function is that there is no change in the Bode magnitude plot, but the Bode phase plot decreases rapidly towards minus infinity as the frequency increases (see the example in Section 8.5.2). In a direct polar (Nyquist) plot, the dominant feature is that the $G(j\omega)K(j\omega)$ locus spirals into the origin of the $G(s)K(s)$ plane, see Figure 3.64. In Figure 3.64 the original system is shown by the solid line, and it can be seen from the gain and phase margin measures that this system would be stable. The same system, except for the addition of a 1 s time delay, is also shown in Figure 3.64 by the dotted line. An examination of the gain and phase margins for the system with the time delay indicates that the closed-loop response would be unstable. In general, time delays will tend to have a destabilizing effect on a system due to the greatly increased phase lag (but not always a completely *un*stabilizing effect).

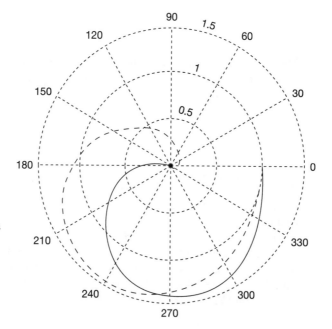

Figure 3.64 The effect of a pure time delay (transport lag) on the direct polar (Nyquist) plot for a type 0 system. The dashed line is for the system with the delay.

Time delays in state-space systems

It is not straightforward to include a pure time delay in a continuous-time state-space model. This is because terms like $e^{-s\tau}$ cannot be forced into the standard state-space form. The available methods are included in Section 8.6.1.

Time delays in discrete-time models

Provided that the sampling interval for a discrete-time model is significantly shorter than the time delay, it is easily included by delaying the signal for a number of sampling intervals equivalent to the length of the time delay. For example, if the input signal to a plant is effectively delayed by 500 ms before having any noticeable effect, and if the sampling interval for a discrete-time model of the plant is chosen to be 50 ms, then the input signal must be delayed by ten time steps. If the model at the start of Section 3.6.1 represents the delay-free system:

$$y_n = A_1\, y_{n-1} + A_2\, y_{n-2} + A_3\, y_{n-3} + \cdots$$
$$+ B_0 u_n + B_1 u_{n-1} + B_2 u_{n-2} + \cdots$$

then, to include a 10-step delay on the input, we rewrite as:

$$y_n = A_1\, y_{n-1} + A_2\, y_{n-2} + A_3 y_{n-3} + \cdots$$
$$+ B_0 u_{n-10} + B_1 u_{n-11} + B_2 u_{n-12} + \cdots$$

The effect of this is that the input will not start to have any effect until the 10-time-step delay has elapsed, when u_{n-10} will first appear in the equation.

Referring back to Example 3.18, note that the modelled system has a one time-step delay on its input (because there is no u_n term on the RHS). The effect of this is seen in the response (Figure 3.63), which displays a one-step delay before the output begins to respond.

In a discrete-time state-space model (Equation (3.90)), the same methodology applies. MATLAB can handle some time delays in discrete-time state-space models using the *c2dt* command. For time delays that do not fit this command, custom simulations may be written along the lines suggested in the algorithm of Section 3.6.2. Alternatively SIMULINK (Appendix 4), if available, would prove an easier solution.

3.8   *Non-minimum-phase transfer functions*

It has been implicitly assumed in all the frequency response analysis performed, that the transfer function is minimum phase. That is, the system's transfer function has all its poles and zeros contained in the open left-hand portion of the s-plane. When a transfer function has a pole or zero in the right-hand half of the s-plane, that pole or zero is said to be non-minimum phase. Consider a polynomial in s which contains two roots, one at $s = +x$ and the other at $s = -x$. The first root is non-minimum phase and the second is minimum phase. To find the magnitude and phase contribution from each root set $s = j\omega$, so that the terms in the polynomials become $j\omega - x$ and $j\omega + x$ respectively.

Representing the minimum-phase root in polar form gives

$$(j\omega + x) = \sqrt{(\omega^2 + x^2)}\, e^{j\tan^{-1}(\omega/x)} \tag{3.92}$$

and the non-minimum-phase term

$$(j\omega - x) = \sqrt{(\omega^2 + x^2)}\, e^{j\tan^{-1}(-\omega/x)} \tag{3.93}$$

Note that the magnitude contribution from each term is the same.

For the minimum-phase element, Equation (3.92), the phase is found in the normal way. However, the non-minimum-phase element, Equation (3.93), requires the inverse tangent of a negative number, which is found to be

$$\tan^{-1}\left(-\frac{\omega}{x}\right) = 180° - \tan^{-1}\left(\frac{\omega}{x}\right) \tag{3.94}$$

Therefore, as the frequency increases, the change of phase introduced by the non-minimum-phase element is $-90°$, whereas the change in phase for the minimum-phase element is $+90°$.

Example 3.19 *Minimum-phase and non-minimum-phase transfer functions*

Analyse the closed-loop responses of the systems with the following open-loop transfer functions, for $K = 1$ in each case:

$$G_1(s) = \frac{K(s+1)}{(s+2)(s+3)} \quad \text{and} \quad G_2(s) = \frac{K(s-1)}{(s+2)(s+3)}$$

Both systems are type 0 and therefore there will be a steady-state error to a unit step input.

System $G_1(s)$

Since the phase never reaches $-180°$ or the gain $0\,\text{dB}$ (see Figure 3.65), the closed-loop response must be stable for $K = 1$. Indeed, the closed-loop system response will be stable for all $K > 0$ (the m-file *fg3_6570.m* on the accompanying disk plots all the responses for this example).

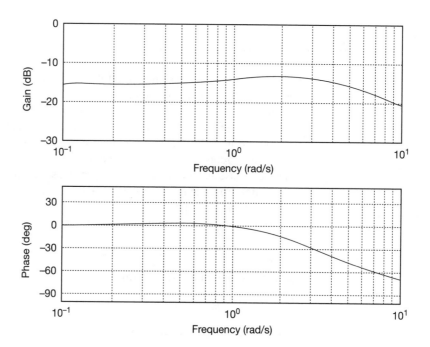

Figure 3.65 Open-loop Bode plot for the system $\dfrac{s+1}{(s+2)(s+3)}$.

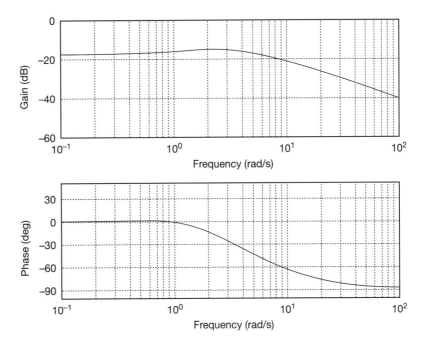

Figure 3.66 Closed-loop Bode plot for the system

$$\frac{s+1}{(s+2)(s+3)}.$$

The closed-loop Bode plot (Figure 3.66) is similar to that of a heavily damped second-order system, see Figure 3.39. At low frequencies the gain plot is 16.9 dB down from the 0 dB line and hence the closed-loop system gain is 1/7 ($= -16.9$ dB). The response of the system to a unit step input is shown in Figure 3.67.

System $G_2(s)$

The open-loop Bode plots for this system are shown in Figure 3.68. By comparing Figures 3.65 and 3.68 it can be seen that the gain plots for both systems are identical. However, the non-minimum-phase zero produces a phase plot which, at low frequencies, approaches 180° (see Equation (3.94)), and at high frequencies approaches $-90°$. Since the system is non-minimum phase, the gain and phase margins cannot be used to determine closed-loop stability. Using Routh's stability criterion, Section 3.3.3, the closed-loop system is stable in the range $0 < K < 6$.

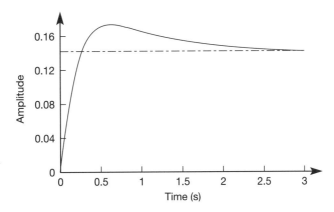

Figure 3.67 Closed-loop step response for the system

$$\frac{s+1}{(s+2)(s+3)}.$$

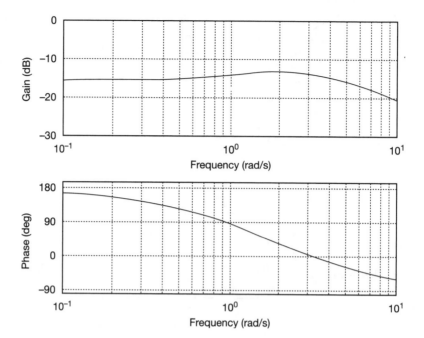

Figure 3.68 Open-loop
Bode plot for the system

$$\frac{s-1}{(s+2)(s+3)}.$$

The closed-loop Bode plot, Figure 3.69, shows a gain plot which is 14 dB down (1/5), and appears non-oscillatory. The closed-loop phase plot again shows that system is non-minimum phase, but gives no indication of the step response characteristics shown in Figure 3.70. In this figure, the final steady-state output to a positive unit step is −0.2. This could have been determined from the closed-loop transfer function and the final value theorem. What is surprising is that the system initially responds in a positive direction and away from its final steady-state value.

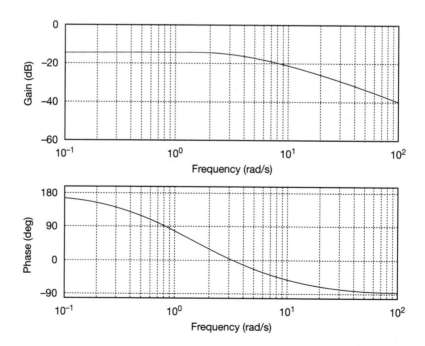

Figure 3.69 Closed-loop
Bode plot for the system

$$\frac{s-1}{(s+2)(s+3)}.$$

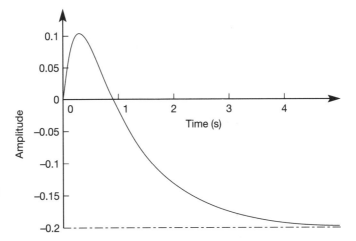

Figure 3.70 Closed-loop step response for the system $\dfrac{s-1}{(s+2)(s+3)}$.

The kind of behaviour exhibited by the non-minimum-phase step response makes such systems rather hard to control, because the controller has to 'know' that the system will initially move in the wrong direction. There are certainly real systems, in need of control, which exhibit such behaviour. One example is the roll dynamics of a warship as it is turned at speed. Initially it rolls into the turn, then rolls out of it. A second example is the control of water level in certain boiler drum systems. If the liquid is boiling fiercely, and cold liquid is added to raise the level in the boiler, the apparent level will initially drop as the cold liquid quenches the rising bubbles in the boiling liquid.

3.9 Simple system identification

In this chapter, the responses of various system models to various types of input signal have been investigated. In particular, step inputs have been used with time-domain models (differential equation models, state-space models and discrete-time models) and sinusoidal inputs, at various values of frequency, with frequency-domain (LTF) models – to produce Bode plots, for example.

Some of the methods used to predict a model's response to these input signals can be used in reverse. That is to say, step response tests or frequency response tests on an existing plant may be used to generate a model of the plant. This is the process of *system identification.*

In this section, it will be assumed that the system to be identified is extremely simple, and can be modelled as being either a first- or second-order system with no zeros in the transfer function. The more generally applicable approach of Bode decomposition, together with methods of handling more complicated systems, will be dealt with in Chapter 8.

It is very important to note that the methods in this section are *only* applicable to the limited class of systems described above. They are included as an introduction to the more complex methods studied later. If faced with an open-loop step response which looks like that of a first-order system (see, for example, Figure 3.16), the present section will simply assume it to be the response of a first-order lag, and derive an appropriate model. The resulting model will therefore predict that the modelled system would be stable in a unity negative feedback closed-loop, for any positive value of forward path gain.

However, a system with a third-order denominator but a second-order numerator (that is, having two finite zeros) could generate a step response looking very similar to Figure 3.16. In this case, although the same first-order model might be derived, the real system could be unstable in a closed-loop arrangement, with disastrous consequences.

These introductory methods are therefore more for academic interest and practice of the fitting techniques, than for practical application.

3.9.1 Obtaining a system model by experiment

In Chapter 2, examples were given explaining how a system can be mathematically modelled by applying appropriate engineering principles (such as Ohm's law and Newton's laws of motion) to its lumped-parameter model. Although lumped-parameter modelling is widely used, considerable skill is required. It is rare, for example, that the data about the individual system components are precisely known, so assumptions have to be made regarding particular 'lumpings', and especially about parameter values. It is also the case that too detailed a model would be impractical for control design purposes, and judgement is again needed in deciding which elements are significant.

For these reasons, if the system exists (as opposed to a system under design), it is desirable to perform some appropriate experiments on the system, from which a model could be obtained. The experimental (or 'identification') approach is even more vital in cases where the system is so complicated (or impenetrable) that it is impracticable to obtain a lumped-parameter model.

System identification normally has two stages; the first is to decide on an overall structure for the model (for example, a transfer function of appropriate order in numerator and denominator) and the second is to determine appropriate model parameters (*parameter estimation*).

When considering model structure, both the 'kind' and 'size' of the selected model are important. In this chapter, only models of the Laplace transfer function 'kind' are considered. The 'size' of the model then becomes important. For LTF models, that is the number of terms to be included in the numerator and denominator. In practice, at least a third-order denominator is normally required or, if a lower-order denominator is used, a time delay would usually be included within the model.

The system's response is a function of the plant dynamics, the applied test input and any imposed initial conditions. In system identification, unless otherwise stated, it will be assumed that the system is stable, and is operating under steady-state conditions (all transients due to the initial conditions having decayed to zero).

To simplify the problem further, only well-defined forcing inputs will be used. The types of input normally used in system identification are:

(1) A steady input ('static test').

(2) A step input.

(3) An impulse input.

(4) A sinewave input ('frequency-response test').

(5) A random input – normally a pseudo-random binary sequence (PRBS).

(6) A general input – the one naturally arising in the normal operation of the system.

Possibilities (5) and (6) are discussed in Chapters 8 and 11 respectively. The approaches of (3) and (4) will be introduced here and developed further in Chapter 8.

3.9.2 The static test

In its simplest form, this test simply consists of applying a steady input and observing the output. It may be that the output does not settle, in which case the system may be assumed to contain an integrator, that is, to be of type 1 or higher (unless it is unstable for other reasons). It is normally advisable to perform the static test over a range of steady input values; ideally increasing from zero to a positive maximum value, then reducing back to zero and onwards to a negative maximum value, and back to zero again. Such a procedure will detect many types of nonlinearity. For example, a common nonlinear characteristic takes the form of hysteresis (see Figure 3.71), which will be detected by this approach. The corresponding response plots for a linear system and for a nonlinear one which does not display hysteresis are included for comparison. Nonlinear systems are considered in Chapter 14.

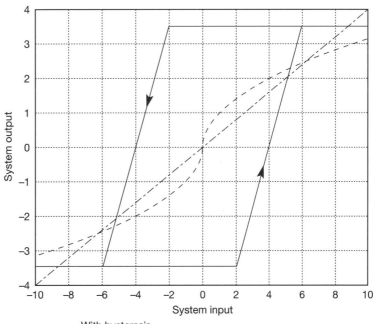

Figure 3.71 Static test characteristics for various systems.

——— With hysteresis

– – – Nonlinear but no hysteresis

—·— Linear

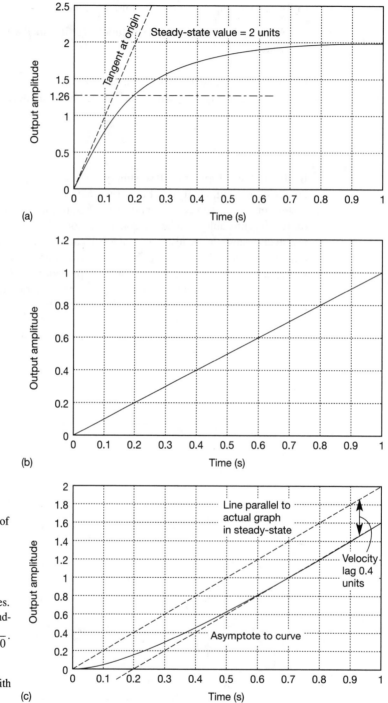

Figure 3.72 A selection of unit step test responses.
(a) A simple lag $\dfrac{10}{s+5}$.
(b) A pure integrator $1/s$.
(c) The elements of (a) and (b) connected in series.
(d) An overdamped second-order system $\dfrac{80}{s^2 + 13s + 40}$.
(e) An underdamped second-order system.
(f) A first-order system with a transport lag.

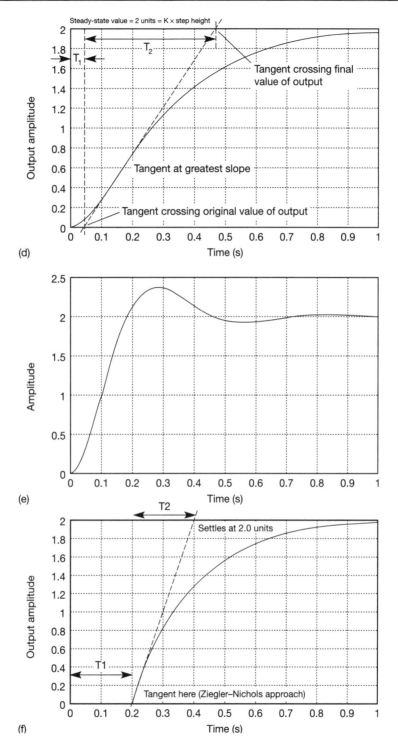

Figure 3.72 (*Continued*)

3.9.3   The step test

This differs from the static test in that the constant input is suddenly applied and the behaviour of the system output is recorded as time progresses. Figure 3.72 shows some possible responses to a unit step input. The solid lines are the recorded responses, while the other lines are constructions to be described in the following example. Subject to the restrictions imposed in the introductory paragraphs of Section 3.9, Figure 3.72(a) is deducible to be that of a simple first-order lag. Its shape is that of the expression $K(1 - e^{-at})$ (cf. Equation (3.18b)) in that it settles at a value K without overshoot, that the initial slope of the graph is aK, and that it has reached a value of approximately 63 per cent of the steady-state value after one time constant (a time of $1/a$, equivalent to τ in Equation (3.18b)). (If the input step were of height h instead of being a unit step, the steady-state output would be Kh and the initial slope would be aKh.) One approach is to determine the value of K from the value at which the response settles; to find the value of a from the time taken to reach 63 per cent of the steady value (remembering that a is the reciprocal of that time); and finally to use the initial slope as a check on the relationship of a and K.

Figure 3.72(b), a straight line through the origin, is the step response of an integrator, while Figure 3.72(c) is that of Figures 3.72(a) and (b) in cascade. In each of Figures 3.72(b) and (c) it is clear that, because of the integration, the response is not settling. A simplified method for producing an approximate model for systems like that of Figure 3.72(c) was devised by Ziegler and Nichols (1942) and will be described in the following example. It is closely allied to the PID tuning method outlined in Section 3.3.4.

Again subject to the restrictions imposed in the introductory paragraphs of Section 3.9, Figure 3.72(d) is the step response of a well-damped second-order system; it superficially resembles Figure 3.72(a) but its initial slope is zero. The latter feature is indicative of a system with at least two more poles than zeros. The exact transfer function of the system of Figure 3.72(d) is difficult to find without the use of computer curve-fitting, but another method devised by Ziegler and Nichols (1942) allows the response to be approximated by that of a first-order lag plus a pure time delay (that is, a transport lag). This approximate method is explained in the examples.

Subject to the restrictions imposed in the introductory paragraphs of Section 3.9, Figure 3.72(e) is the response of an underdamped second-order system; the underdamping produces the typical overshoot in response to a step input. The transfer function is easier to determine from the step response than for the overdamped case; the approach is also demonstrated in the examples. Figure 3.72(f) is the response of a system which is identical to that of Figure 3.72(a) except for a transport lag of τ seconds. That is, there is no response until a time τ has elapsed (the response being otherwise unaltered).

The procedures for determining the transfer functions from the responses are explained by example.

Example 3.20 **Obtaining the transfer functions of systems from unit step responses**

Obtain the transfer functions of the systems whose unit step responses are given in Figure 3.72(a) to (f). Note that the various dotted lines represent the geometrical constructions to be described in the example. Initially, following the plant tests, only the solid curves would be present.

Figure 3.72(a)

In this figure, note that the response settles to a steady value (showing that the system is of Type 0) and that it has a non-zero initial slope. This means that it has one more pole than zero but, in accordance with the simplifying restrictions which apply to this section, it will be assumed to be of first order. The general form of a first-order transfer function is $K/(1 + Ts)$ or $aK/(s + a)$ where $a = 1/T$.

   Further, note that an input of 1 unit gives a steady output of 2 units, so the static gain K is $2/1 = 2$ units of output per unit of input. The output reaches 63 per cent of that steady output in 0.2 s, so the time constant T is 0.2 s.

   The assumed transfer function is therefore $2/(1 + 0.2s)$ or $10/(s + 5)$. The values of K and T can be checked by drawing the tangent to the curve at zero time, which crosses the steady-state value line (output $= 2$ units) at approximately 0.2 s, confirming that the assumed first-order model is valid.

Figure 3.72(b)

This response is a straight line through the origin. As the integral of a constant K with respect to time is simply Kt plus the usual constant of integration, this system may be deduced to be an integrator of transfer function K/s. The value of K is the slope of the graph, which by inspection is $1/1 = 1$ unit of output per second, so the transfer function is $1/s$.

Figure 3.72(c)

This response is produced by a plant which is a cascade combination of those of Figures 3.72(a) and (b), though possibly with a different time constant T and overall gain K. Its transfer function is therefore assumed to be:

$$\frac{K}{s(1 + sT)}$$

The problem is that of finding K and T. For such a system, the response to a unit step $1/s$ will be:

$$Y(s) = \frac{K}{s^2(1 + sT)}$$

which is divisible into partial fractions (Appendix 2) as:

$$Y(s) = K\left(\frac{1}{s^2} - \frac{T}{s} + \frac{T^2}{1 + sT}\right)$$

Taking inverse-Laplace-transforms gives:

$$y(t) = K(t - T + Te^{-t/T})$$

This has the following properties. Both $y(t)$ and its gradient are zero when t is zero (like the response of Figure 3.72(c)) and, when t is much greater than T, it tends to $K(t - T)$, that is, its value is KT below that of a graph of a quantity equal to Kt alone. K is therefore the gradient of the response when t is large, namely $1.6/(1 - 0.2) = 2$ units/second. The value of KT is the 'velocity lag' shown in the figure, which proves to be 0.4 unit, so $T = KT/K = 0.2$ unit, and the transfer function becomes:

$$\frac{2}{s(1 + 0.2s)}$$

The first of the Ziegler–Nichols methods mentioned earlier can be applied to Figure 3.72(c). They postulated that, after the initial period of decay of the exponential, the response is the same as that of an integrator K/s with a transport lag equal to the time value at which the asymptote shown in the figure crosses the line of zero output. In this case, the approximate transfer function would be:

$$\frac{2e^{-0.2s}}{s}$$

Despite the approximate nature of such a result, it is often more accurate than an assumed second-order model for controller design purposes using LTF methods. However, if state-space methods were to be used, the transport delay would generally be unacceptable.

Figure 3.72(d)

This response differs from those investigated so far, in that its initial slope is zero. That feature shows the system to have at least two more poles than zeros. There are two methods of finding a transfer function from a response of this type. The first is another version of the Ziegler–Nichols approximation and it assumes the system to be a first-order lag with a transport lag. The static gain K is found in the usual way as the steady-state output divided by the steady-state input (in this case, $2/1 = 2$ units of output per unit of input). A tangent is then drawn on the graph at the steepest gradient and is produced as necessary to intersect both the initial output and steady-state output levels. The value of time where the tangent crosses the initial output is taken to be the transport lag T_1, while the difference between that time and the time at which the tangent crosses the steady-state output is taken to be the time constant T_2 of the first-order lag. The transfer function will then be

$$\frac{Ke^{-sT_1}}{1 + sT_2}$$

The other method assumes the response to be of an order determined by simple analysis of the plant components (but for the purposes of this section a second-order plant is assumed), and to fit, with computer assistance, a transfer function of the following form:

$$\frac{K}{(1 + sT_1)(1 + sT_2)}$$

Although the static gain will be the same in both cases, in general, none of the lags will coincide. If it is suspected that the system is of higher than second order, the Ziegler–Nichols approach generally gives the better model for design purposes. A least-squares approach to computer curve-fitting is demonstrated in Example 8.2 in Chapter 8.

Figure 3.72(e)

This figure is again significantly different from the others, in that it displays overshoot. Paradoxically, the overshoot makes it easier to determine a transfer function, provided that the system is second-order (again in line with the assumptions made for this section). Again two methods are available. In each case, a transfer function of the form of Equation (3.19) will be assumed (but in general there would be more poles and zeros, as discussed before, which would then prevent this analysis):

$$\frac{K\omega_n^2}{s^2 + 2\zeta\omega_n s + \omega_n^2} \tag{3.95}$$

As noted following Equation (3.23), the decay constant is $\zeta\omega_n$. The damped natural frequency ω_d is given by Equation (3.21) as $\omega_d = \omega_n \sqrt{(1 - \zeta^2)}$.

The damped natural frequency is found from Figure 3.72(e) as

$$\omega_d = \pi/(\text{half-cycle time}) = 11.2 \text{ rad s}^{-1}$$

The decay of the peak height is caused by the multiplying exponential $(e^{-\zeta\omega_n t})$. The height of the first overshoot is about 0.36 unit and the magnitude of the first 'undershoot' is about 0.06 unit. The time between them is about 0.28 s (the overshoot and undershoot can be used in combination because the decay envelopes are symmetrical about the steady-state value – see Figure 3.18). Therefore, it may be deduced that $0.06/0.36 = e^{-a0.28}$ (where $a = \zeta\omega_n$). Taking natural logs of each side, $\ln(0.1667) = -0.28a$. This gives $0.28a = 1.79$, or $a = 6.4 = \zeta\omega_n$.

The numerator gain term K in Equation (3.95) is the static (d.c.) gain, which is seen from the graph to be 2.0.

To obtain the value of ω_n^2, square the formula for ω_d, to obtain:

$$\omega_d^2 = \omega_n^2 - (\zeta\omega_n)^2$$

Since $\omega_d = 11.2 \text{ rad s}^{-1}$ and $\zeta\omega_n = 6.4 \text{ rad s}^{-1}$, then $\omega_n^2 = 166.4$. The resulting transfer function is therefore:

$$\frac{K\omega_n^2}{s^2 + 2\zeta\omega_n s + \omega_n^2} = \frac{332.8}{s^2 + 12.8s + 166.4}$$

The alternative method uses the standard second-order step response graphs in Figure 3.19(a). The damping ratio ζ is determined directly from the percentage overshoot. For Figure 3.72(e), this is $(0.36/2) \times 100 = 18$ per cent. From the standard graph (Figure 3.19(a)) the curve corresponding to such a percentage overshoot gives a ζ value of about 0.48.

The undamped natural frequency ω_n is found by examining the duration of some easily determined feature (for example, the first positive half-cycle) on the actual and standard graphs. On the actual one (Figure 3.72(e)), the duration is 0.28 s; on the standard one (Figure 3.19(a)), the duration is about $(6.1 - 2.4) = 3.7$ units for $\zeta = 0.5$, or $(5.1 - 2.2) = 2.9$ units for $\zeta = 0.4$. Interpolating, $\zeta = 0.8 \times 3.7 + 0.2 \times 2.9 = 3.54$ units.

These 'units' on the standard graph are actually units of $\omega_n t$, so a value of 0.28 s for t corresponds to 3.54 for $\omega_n t$. The value of ω_n is therefore $3.54/0.28 = 12.6$ rad s^{-1}. The transfer function is therefore

$$\frac{2(12.6^2)}{s^2 + 2(0.48)(12.6)s + 12.6^2} = \frac{317.5}{s^2 + 12.1s + 158.8}$$

It is reassuring that both methods, involving, as they do, taking measurements from fairly small-scale graphs, have given similar answers.

Figure 3.72(f)

This response is identical to Figure 3.72(a) except that the response is delayed in time by 0.2 s. As previously, this delay is represented in the transfer function by $e^{-s\tau}$, where τ is the transport lag (Section 3.7). The overall transfer function of the system is therefore $10e^{-0.2s}/(s+5)$.

3.9.4 The impulse-response test

The impulse response has the property of being the inverse Laplace transform of the system transfer function. In addition, by means of convolution (Section 8.2), the response to a general input can be determined from the impulse response by analytical or numerical integration. The transfer function could be determined by recording the unit impulse response, fitting a time function to it, and Laplace-transforming the time function from tables. The main problem with such an approach is the generation of an impulse of sufficient energy (since the 'strength' of an impulse is equal to the product of its height and its duration – and a true impulse is of zero duration). A subsidiary problem is the actual fitting of the curve, though that operation can be performed with computer assistance. The method is discussed in Chapter 8.

3.9.5 The frequency response test

The principle of this test is simple – a sinusoidal signal of varying frequency is applied to the system input and the corresponding system output is measured in terms of both magnitude and phase relative to the input. The system gain is then calculated at each frequency as the ratio of output magnitude divided by input magnitude. The results are normally plotted either as a Nyquist diagram or (more usually for this purpose) as a Bode diagram (Section 3.5.1). The full procedure for determining the transfer function from a Bode diagram is given in Section 8.5. The following description refers to the simpler approach appropriate to minimum-phase systems having no time delays and being of no higher than second order, in accord with the restrictions noted in the early paragraphs of Section 3.9.

The method is as follows.

(1) Draw the asymptotes on the magnitude Bode plot. It is important to remember that their gradients can only be zero or an integral multiple of 20 dB/decade (positive or negative). Where the asymptote gradient changes by only 20 dB/decade, it is often helpful in placing the asymptote to note that the actual graph should be approximately 3 dB inside the 'corner' of the asymptotes. Note that if

two such 'break frequencies' are relatively close together, their responses will add to one another, and this simple rule will no longer apply (Table 3.3 then gives the required corrections for first-order terms).

(2) Examine the gradient at low frequency. If it is zero, that is, the graph is level, the system is of Type 0 with no pure integration or differentiation content. If it has a negative slope of 20 dB/decade, the system will have a transfer function of the form:

$$\frac{K}{s(1+sT)}$$

Note that, if the low-frequency gradient is zero, the value of K (the static gain) can be determined as $K = 10^{(G/20)}$, where G is the low-frequency gain in decibels (G is $20 \log_{10} K$). The determination of K is easier, and more accurate, if the experimentally determined value of the gain is used before calculating the decibel equivalent. The method of determining K if the low-frequency slope is not zero is explained under point (5) below.

(3) Find the frequencies at which the gradient of the asymptotic diagram changes. For the purpose of the following explanation, let a particular change of gradient occur at a frequency ω_c rad s^{-1}.

(a) If the gradient becomes more negative by 20 dB/decade, there is a $(1 + s/\omega_c)$ term in the denominator of the transfer function. Note that this corresponds to the $(1 + s\tau)$ in the denominator of Equation (3.18a), since the time constant τ seconds is the reciprocal of the 'corner frequency' in rad s^{-1}.

(b) If the gradient becomes more positive by 20 dB/decade, there is a $(1 + s/\omega_c)$ term in the numerator of the transfer function.

(c) If the gradient becomes more negative by 40 dB/decade, there is a $(s^2 + 2\zeta\omega_c s + \omega_c^2)$ term, representing underdamped second-order behaviour, in the denominator of the transfer function.

(d) If the gradient becomes more positive by 40 dB/decade, there is a $(s^2 + 2\zeta\omega_c s + \omega_c^2)$ term in the numerator of the transfer function.

(4) If there was an underdamped second-order term in the numerator or denominator, the damping ratio ζ may be found by reference to standard graphs of the behaviour of such systems (Figure 3.39). The procedure is to examine the maximum rise of the actual curve above the asymptote (or the fall below it if the term is a numerator one), and the 'corner error' (the difference between the decibels at the intersection of the asymptotes and the actual decibels at the same frequency), and to compare this error with those from the standard curves: see Example 3.21 below.

(5) When the low-frequency gradient is not zero, the procedure for finding K (which is now not the static gain) is to select a frequency low enough so that the transfer function has reduced to K/s^n (with allowance for the ω_n^2 terms if underdamped second-order behaviour is present), or $K/(j\omega)^n$ in frequency

response form. At such a frequency, the graph should coincide with the low-frequency asymptote. At this low frequency ω, the gain in decibels is $20 \log_{10}(K/\omega^2)$ and again K can be obtained.

(6) The phase graph is vital in determining the presence of non-minimum-phase zeros or poles (terms like $(1 - Ts)$ rather than $(1 + Ts)$ in the transfer function) and time delays, or transport lags. The methods required to deal with systems having such effects are explained in Chapter 8. For the moment, the phase graph gives a check on the result by confirming the system order (a system of order n with no zeros will have a maximum phase lag of $n90°$) and the 'corner frequency' (at this frequency, the phase shift for a first- or second-order system will have completed half its final change).

These methods will be demonstrated by a series of examples.

Example 3.21 *Obtaining the transfer functions of systems from frequency responses*

Obtain the transfer functions of the systems whose frequency responses are shown in Figures 3.73(a) to (e). Note that the addition of the various asymptotes is described during the example. Initially, following the plant tests, only the solid curves would be present. In line with the rest of this section, only simple first- or second-order systems are assumed, whereas plots such as these might generally have been produced by more complex systems having zeros in their transfer functions, and more than one or two poles, with the zeros fairly close to the poles in the s-plane.

Figure 3.73(a)

In the magnitude (dB) part of the figure, the low-frequency gain is constant at 6 dB, so the static gain is given by $K = \text{antilog}_{10}(6/20) = 10^{6/20} = 2.0$. There is only one 'break frequency' of about 5.1 rad s^{-1}, at which the gradient of the graph changes from zero to -20 dB/decade. This indicates the presence of a term $(1 + s/5.1)$ in the denominator of the transfer function. At the corner frequency, the difference between the actual graph and the asymptote corner is approximately 3 dB. As the gradient never increases, or becomes *less negative*, with increasing frequency, there are no numerator zeros, hence the transfer function is:

$$\frac{2}{1 + \dfrac{s}{5.1}} = \frac{2}{1 + 0.2s}$$

The phase graph offers confirmation that the system is first order; the maximum phase shift is $-90°$, and the phase angle is $-45°$ at 5.1 rad s^{-1}, in agreement with the above 'corner frequency'.

Figure 3.73(b)

The magnitude ratio (gain in dB) plot has no changes in gradient, but it does slope downwards at all frequencies at -20 dB/decade. The corresponding phase graph is a constant $-90°$. From rule (2) it may be deduced that the system is an integrator.

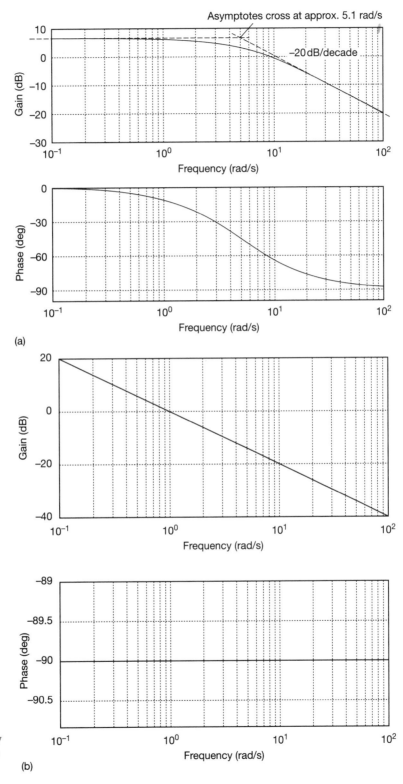

lag

Figure 3.73 Five frequency response plots for Example 3.21.

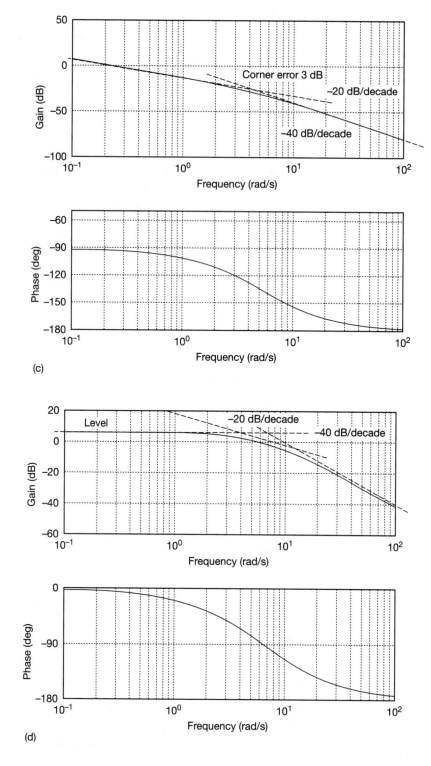

Figure 3.73 (*Continued*)   (d)

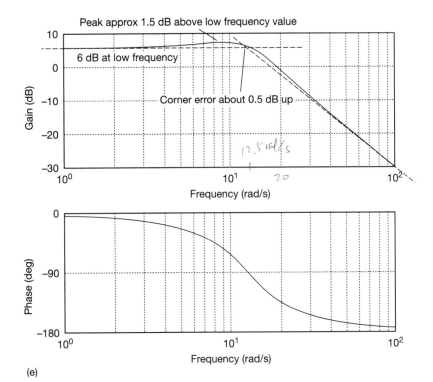

Figure 3.73 (*Continued*)   (e)

The denominator therefore contains a single s. The numerator only contains a constant term K. Selecting the frequency $\omega = 1$ rad s^{-1}, the modulus of the transfer function is

$$\left|\frac{K}{s}\right| = \left|\frac{K}{j\omega}\right| = \frac{K}{\omega} = K$$

The decibel magnification at 1 rad s^{-1} is zero, so $20 \log_{10} K = 0$, giving $K = 1$. The transfer function is therefore $1/s$.

Figure 3.73(c)

In this figure, the gradient of the magnitude graph starts at -20 dB/decade and increases to -40 dB/decade at approximately $\omega = 4.8$ rad s^{-1}. Again, applying rule (2) indicates a transfer function of form:

$$\frac{K}{s(1 + sT)}$$

in which the s is from the initial -20 dB/decade and the $(1 + sT)$ from the increase to -40 dB/decade. The value of T can be found from the corner frequency, so $T = 1/4.8 = 0.21$ s. To find K, again examine the gain in decibels at a low frequency; 0.1 rad s^{-1} is suitable. At this frequency:

$$\frac{K}{s(1 + sT)} = \frac{K}{j\omega(1 + j\omega T)} = \frac{K}{j0.1(1 + j0.021)} \approx \frac{K}{j0.1}$$

The modulus is therefore $10K$, or $20 \log_{10}(10K)$ dB. From the gain (dB) graph, the modulus is 6 dB when $\omega = 0.1$, so $10K = 2$ and $K = 0.2$. The complete transfer function is therefore

$$\frac{0.2}{s(1 + 0.21s)}$$

Figure 3.73(d)

This magnitude plot is level at low frequency, so there are no s terms multiplying the whole of the numerator (pure differentiations) or denominator (pure integrations). At high frequency, it slopes down at -40 dB/decade, suggesting a second-order denominator system (subject to the restrictions given in the early paragraphs of Section 3.9). At no point does the graph go upwards or begin to descend less steeply, so the numerator will contain only a constant. The denominator will therefore contain terms up to and including s^2. The 'corner error' is about 7 dB, so comparison with the standard second-order responses (Figure 3.39(a)) suggests that the system is overdamped (that is, $\zeta > 1$) and consists of a cascaded pair of first-order lags. This requires a -20 dB/decade asymptote to find the break frequencies. A clue as to the position of this further asymptote may be gained by remembering that the 'corner errors' will be approximately 3 dB (the true curve being below the asymptote corner). In fact, the errors will be marginally over 3 dB in this case, as the error for one 'corner' will still be having a slight effect at the corner frequency of the other (see Table 3.3). This illustrates that the asymptotes are *not* necessarily *tangents* to the curve. The asymptote is therefore drawn, giving corner frequencies of 3.3 rad s^{-1} and 12 rad s^{-1}, corresponding to time constants of $1/3.3 = 0.3$ s and $1/12 = 0.083$ s. The level graph at low frequency determines the static gain; $20 \log_{10} K = 6$ dB, so $K = 10^{6/20} = 2.0$. The complete transfer function will be

$$\frac{2}{(1 + 0.3s)(1 + 0.083s)}$$

Figure 3.73(e)

In this final response, it is again clear that the decibel graph is level at low frequency and falling at -40 dB/decade at high frequency. The difference this time is that the asymptote corner is perhaps 0.5 dB below the actual plot. At a slightly lower frequency, the graph peaks even higher above the low-frequency value – by about 2 dB.

The corner frequency (12.5 rad s^{-1} approximately) gives ω_n, and ζ is found by inspecting the standard frequency-response curves (Figure 3.39(a)). In these curves, the corner frequency is normalized to 1 rad s^{-1} and, in the case of the decibel graphs, the static gain is normalized to unity (that is, 0 dB). For this example, the important features of the decibel graph are the peak height and the corner error; the graph which most closely resembles Figure 3.73(e) in those respects being that for $\zeta = 0.5$. Finally, the low-frequency gain is 6 dB, so the static gain is $K = 10^{6/20} = 2.0$.

The general form of the transfer function for an underdamped second-order system, and the version with the identified parameters inserted are:

$$\frac{K\omega_n^2}{s^2 + 2\zeta\omega_n s + \omega_n^2} = \frac{2(12.5^2)}{s^2 + 2(0.5)(12.5)s + 12.5^2} = \frac{312.5}{s^2 + 12.5s + 156.25}$$

3.10 Conclusions

This chapter has introduced many of the standard responses expected from linear system models. Standard input signals have been specified and used in testing models (they can similarly be used for testing real plants). Most notable amongst these are the step input and the steady-state sinusoidal input. In response to these inputs, the performance of a system may be viewed either in the time domain (as a step response plot, for example) or in the frequency domain (as a Bode plot, polar (Nyquist) plot, inverse polar plot or Nichols plot, for example).

It was shown that the poles of a Laplace transfer function model are identical to the eigenvalues of the plant matrix (that is, the A matrix) of the equivalent state-space model, and that these are responsible for the dynamic behaviour of the system, and its stability. Methods of assessing the stability of linear systems were also examined.

The time and frequency responses of standard first- and second-order models have been investigated, and used to predict the behaviour of more complex systems. In this context, both the frequency and time response of a system have been linked to its pole (or eigenvalue) locations in the s-plane.

The responses of various model types, including continuous-time models (such as the state-space model), Laplace transfer function models and discrete-time models (both in the form of discrete-time series, and the discrete-time state-space model), have been derived.

Finally, some of the techniques for calculating model responses have been used to obtain plant models by analysis of test results (either step tests or frequency response tests) – a process known as system identification.

This background is sufficient for a study of control system design techniques.

3.11 Problems

3.1 Derive the time response (to a unit step input) given for the initial system in Example 3.2.

3.2 A number of pole–zero plots are shown in Figure P3.2 with the gain K indicated. As usual, K is the multiplier of the pole–zero function. For each plot:

(a) Find the transfer function $G(s)$.
(b) Determine the inverse Laplace transform.
(c) Sketch the general shape of the transient response using a dominant pole analysis.
(d) Check the response found in (c) using the MATLAB *impulse* function in the CSTB.

3.3 Evaluate the following function $G(s)$ at the s-plane locations $(0, 0)$, $(0, -4)$ and $(-2, 2)$:

$$G(s) = \frac{6(s+1)}{(s+2)(s+3)}$$

3.4 Using the normal rules of complex algebra evaluate the functions $G_1(s)$ and $G_2(s)$, given below, for any positive value of ω on the imaginary s-plane axis.

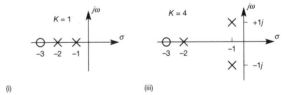

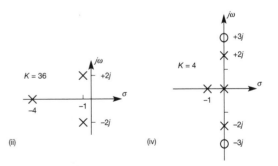

Figure P3.2 Pole–zero maps for Problem 3.2.

Plot the resulting loci $G_1(j\omega)$ and $G_2(j\omega)$ on Argand diagrams. (The Argand diagrams are called the $G_1(s)$- and $G_2(s)$-planes respectively, and the resulting plots are called polar plots.)

Phase lead $\quad G_1(s) = \dfrac{1 + 10s}{1 + s}$

Phase lag $\quad G_2(s) = \dfrac{1 + 0.1s}{1 + s}$

3.5 From the following time-domain performance specifications, establish s-plane performance criteria. In each case use these criteria to sketch the response to a unit step. Assume a second-order system with unity gain.

(a) A 100 per cent rise time of 0.5 s and a peak time of 0.7 s. What would be the peak overshoot and steady-state error for this system?
(b) A subsidence ratio of 3:1 and a 5 per cent settling time of 9 s. What is the peak overshoot for this system?
(c) A peak overshoot of 30 per cent and a 5 per cent settling time of 9 s. What is the subsidence ratio?

3.6 A system has an open loop transfer function given by

$$G(s) = \frac{25}{s(s + 2)}$$

To control this system two alternative closed-loop designs have been proposed as shown in Figure P3.6. Design A is a simple proportional controller and design B is a proportional plus derivative controller. The variable K in both designs is to be set so that the closed-loop damping ratio is 0.5. For both designs determine:

(a) the rise time,
(b) the peak time,
(c) the 5 per cent settling time and

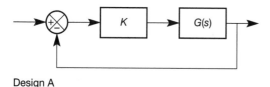

Design A

Design B

Figure P3.6 Alternative arrangements for Problem 3.6.

(d) the percentage overshoot.

Comment on the two designs and explain the differences and similarities between the two responses.

3.7 Indicate the root distribution and hence the stability of the following characteristic equations, using Routh's array (check with MATLAB's *roots* command if available):

(a) $s^3 + 2s^2 + 3s + 1 = 0$
(b) $s^4 + 2s^3 + s^2 + 2s = 0$
(c) $s^5 + 2s^4 + 3s^3 + 6s^2 + 2s + 1 = 0$
(d) $s^5 + s^4 + 5s^3 + 5s^2 + 4s + 4 = 0$
(e) $s^6 + s^5 + 3s^4 + 3s^3 + 2s + 1 = 0$
(f) $s^{10} + s^9 + 2s^8 + 2s^7 + s^6 +$
$\quad 2s^5 + 6s^4 + 7s^3 + 10s^2 + 6s + 4 = 0$

3.8 A system's characteristic equation may be written in the form

$$D(s) = a_0 s^n + a_1 s^{n-1} + a_2 s^{n-2} + \cdots + a_{n-1}s + a_n$$
$$= 0$$

If all the coefficients a_i are positive and exist, show, using Routh's criterion, that:

(a) A second-order system is always stable.
(b) A third-order system is stable if $a_1 a_2 - a_0 a_3 > 0$.

What is the stability condition for a system of fourth order?

3.9 The block diagram shown in Figure P3.9 consists of an inner loop with feedback gain K_T and an outer loop with forward path gain K_B. Find the gain K_T which makes the inner loop critically damped. (For a second-order system to be critically damped, requires that the characteristic equation $s^2 + 2\zeta\omega_n s + \omega_n^2 = 0$ has a damping ratio $\zeta = 1$.) Using this value of K_T, determine:

(a) The range of gain K_B for which the closed-loop system is stable.
(b) The K_B which results in a marginally stable sinusoidal response. What is the frequency of this response?

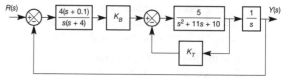

Figure P3.9 System block diagram for Problem 3.9.

3.10 A process control system with unity negative feedback has plant dynamics given by

$$G_p(s) = \frac{4}{(s^2 + 8s + 80)(s + 1)}$$

and incorporates a three-term controller with the idealized transfer function

$$G_c(s) = 20\left(1 + \frac{1}{T_i s} + T_d s\right)$$

Establish the values of T_i and T_d that will ensure the system's closed-loop stability.

3.11 Use the Ziegler–Nichols rules to design a three-term controller for a plant model having an open-loop transfer function given by

$$G(s) = \frac{1}{s(s+1)(s+2)(s+3)}$$

Show that the resulting closed-loop system is stable.

3.12 Determine the amplitude magnification $M(\omega)$ and the phase shift $\phi(\omega)$ for the following open-loop transfer functions:

(a) $G(s) = \dfrac{1}{s(s-1)}$

(b) $G(s) = \dfrac{K}{s(\delta_1 s + 1)(\delta_2 s + 1)}$

(c) $G(s) = \dfrac{10(1+2s)}{s(1+0.02s)(1+0.5s)(1+s)}$

3.13 Use a graphical technique to evaluate $M(\omega)$ and $\phi(\omega)$ at $\omega = 2$ rad s^{-1} for the transfer function

$$G(s) = \frac{0.2s + 1}{(s+1)(0.5s+1)(0.25s+1)}$$

Check your solution by direct calculation of $M(\omega)$ and $\phi(\omega)$.

3.14 For the transfer function

$$G(s) = \frac{10}{s(1+0.25s)(1+0.5s)(1+s)}$$

plot the polar, inverse polar and Bode diagrams. Also produce a plot of dB magnitude versus phase (a Nichols plot). From each of these plots, find the gain and phase margins. If the transfer function gain were reduced from 10 to 1, what would be the new gain and phase margins?

3.15 Repeat Problem 3.14 for the case in which the transfer function $G(s)$ also contains a one-second time delay (transport lag), such that $G(s)$ becomes:

$$G(s) = \frac{10e^{-s}}{s(1+0.25s)(1+0.5s)(1+s)}$$

3.16 Use MATLAB (or some other high-level language or package) to investigate the step response of each of the following system models:

(a) $y_n = 0.8y_{n-1} + 0.4u_{n-1}$
(b) $y_n = 1.2y_{n-1} - 0.5y_{n-2} + 0.14u_{n-1} + 0.1u_{n-2}$

3.17 For the system represented by the following state-space model:

$$A = \begin{bmatrix} 0 & 1 \\ -4 & -0.8 \end{bmatrix}, \quad b = \begin{bmatrix} 0 \\ 1 \end{bmatrix}, \quad c = \begin{bmatrix} 8 & 0 \end{bmatrix},$$

$$d = 0$$

(a) Obtain a discrete-time state-space model, using a sampling period of $h = 0.4$ s. Each term in the model should be accurate to about ± 1 per cent. With access to MATLAB and the Control Systems Toolbox (CSTB), enter A, b and h, and check your result with the command:

  [phi,del]=c2d(a,b,h)

(b) Evaluate, using hand calculation, the first 10 values of the unit step response (at the output), beginning from zero initial conditions. Do this on paper. Again, with MATLAB and the CSTB, the result can be checked using the command in part (a), entering c and d also, and then issuing the command:

  dstep(phi,del,c,d)

(c) Compare the peak output value, and the time at which it occurs, with the equivalent continuous-time solution – convert the original state-space model to a Laplace transfer function model and use the standard response curves (Figure 3.19) or formulae (Equations (3.27), (3.28), and so on).

(d) Comment on the choice of sampling interval.

3.18 The discrete-time state-space model of a system is given by:

$$\Phi = \begin{bmatrix} 0.5 & 1.0 & -0.8 \\ 0.3 & 0.8 & -0.1 \\ 0 & 0.3 & 1.0 \end{bmatrix}, \quad \Delta = \begin{bmatrix} 0.26 \\ 0.05 \\ 0.01 \end{bmatrix},$$

$$c = \begin{bmatrix} 0 & 0 & 1 \end{bmatrix}, \quad d = 0$$

The sampling period is $h = 1/3$ s.

(a) Evaluate the first 15 values of the output response to a unit impulse applied with zero initial conditions (a unit impulse lasts for one sample period in discrete time, and is of such a height that it has unit area).

(b) Estimate the following quantities in the output response:
(i)   Time to the first peak
(ii) Value of the first peak
(iii) Time to the second zero crossing (following the first 'undershoot').

(c) Comment on the likely suitability of the sampling period for this system.

(d) With MATLAB and the CSTB, convert the system to a continuous-time equivalent model by entering the values of *phi*, *del*, *c*, *d* and *h*, and using the command *[a,b]=d2c(phi,del,h)*. The discrete-time response should be compared with the response of the equivalent continuous-time system, using the command *impulse(a,b,c,d)*.

Note: When using the MATLAB command *dimpulse (phi,del,c,d,1,15)* to duplicate the discrete-time impulse response (for example), it is important to note that MATLAB applies a unit-*height* impulse during the first sample, because that is what a sampled-data system would acquire. For the response to a unit-*area* impulse, multiply the results by a factor of 3. This can be done, for example, by the commands: *[y,x]=dimpulse (phi,del,c,d,1,15); plot(3*y)*. To see the correct time axis, use *t=[0:h:(length(y)–1)]*h; plot(t,3*y)*.

3.19 Obtain the transfer functions of the systems whose step responses are given in Figure P3.19(a) to (e). The magnitude of the step input is given in each case. Where appropriate, the Ziegler–Nichols approximation may be used to obtain the approximate transfer function.

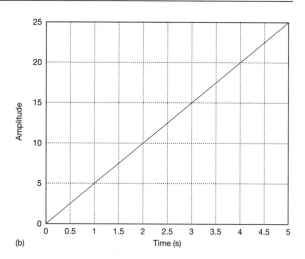

(b)

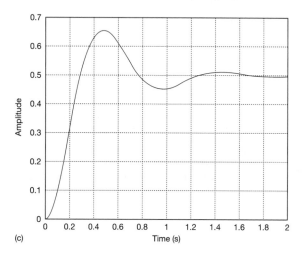

(c)

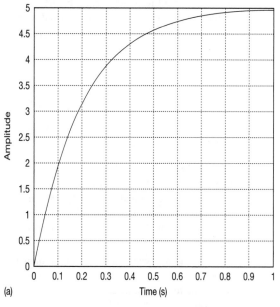

(a)

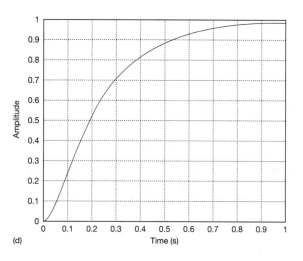

(d)

Figure P3.19 (a) Figure for Problem 3.19(a). Input step = 2 units. (b) Figure for Problem 3.19(b). Input step = 5 units. (c) Figure for Problem 3.19(c). Input step = 0.2 unit. (d) Figure for Problem 3.19(d). Input step = 0.5 unit. (e) Figure for Problem 3.19(e). Input step = 1 unit.

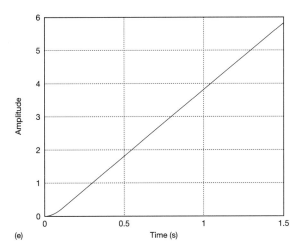

(e)

Figure P3.19 (*Continued*)

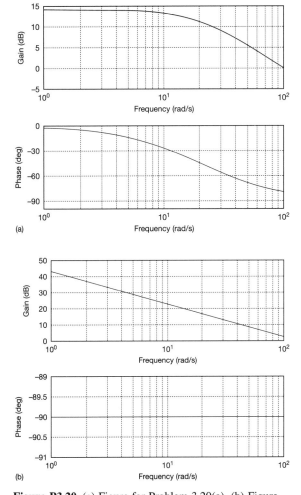

(a)

(b)

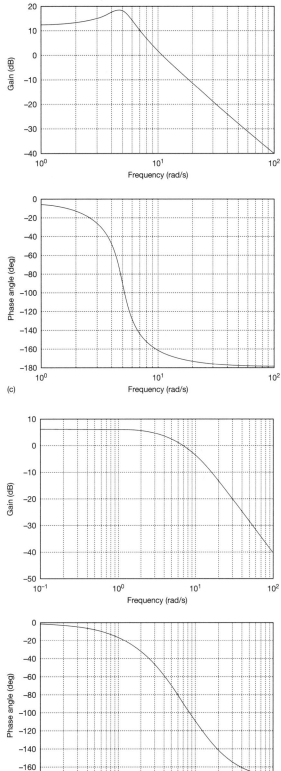

(c)

(d)

Figure P3.20 (a) Figure for Problem 3.20(a). (b) Figure for Problem 3.20(b). (c) Figure for Problem 3.20(c). (d) Figure for Problem 3.20(d). (e) Figure for Problem 3.20(e).

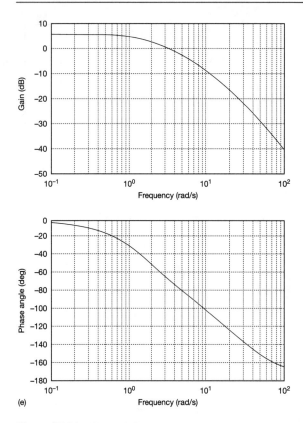

(e)

Figure P3.20 (*Continued*)

3.20 Obtain the transfer functions of the systems whose frequency responses are given in Figure P3.20(a)–(e). It may be assumed that none of these systems is of higher than second order.

4 Single-input-single-output (SISO) design

This chapter covers:

- the Nyquist criterion for investigating a system's stability
- the root locus method for examining the effects of parameter variation
- three-term (PID) controllers
- the design of lead- and lag-type compensators
- the basis of analog controller implementation
- the Smith predictor for use in systems with time delays.

NEW MATHEMATICS FOR
THIS CHAPTER

There are no mathematical techniques in this chapter which have not been used in previous chapters.

4.1   Preview

Previous chapters have examined the modelling of systems and the model's response. Stability and performance characteristics have also been examined in terms of the model's s-plane pole and zero locations. Provided all the poles and zeros are suitably located, an acceptable dynamic response to some forcing input or disturbance may be inferred. However, a problem exists when such criteria are not met.

This chapter expands on the design problem introduced in Section 3.4. It will show how a controller can be designed so that a system's closed-loop response dynamics can be changed. In particular it will consider the determination of closed-loop stability from the system's open-loop frequency response. It will then look at the root locus method and show how the position of the closed-loop system poles can be moved within the s-plane. The use of root locus diagrams and frequency response plots in control system design is then developed. To close the chapter, a number of other simple design techniques are considered. State-space design methods and digital control are covered in the next chapter.

4.2   Preliminaries

Consider the feedback configuration shown in Figure 4.1 and compare it with Figure 3.32 (Section 3.4). It will be noted that an element $H(s)$ has been introduced into the feedback path. This element could represent either transducer dynamics or some form of feedback compensation.

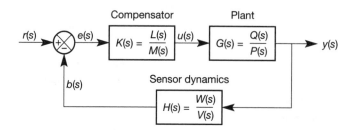

Figure 4.1 A general feedback controller, showing rational polynomials.

The open-loop transfer function of the system of Figure 4.1 is:

$$H(s)G(s)K(s) \tag{4.1}$$

and the closed-loop transfer function between $y(s)$ and $r(s)$ is given by

$$T(s) = \frac{G(s)K(s)}{1 + H(s)G(s)K(s)} \tag{4.2}$$

For this system the return difference equation (or closed-loop characteristic equation) is:

$$1 + H(s)G(s)K(s) = 0 \tag{4.3}$$

On substitution for $H(s)G(s)K(s)$ from Figure 4.1, this becomes

$$\frac{V(s)P(s)M(s) + W(s)Q(s)L(s)}{V(s)P(s)M(s)} = 0$$

or

$$V(s)P(s)M(s) + W(s)Q(s)L(s) = 0 \tag{4.4}$$

and the roots (or zeros) of Equation (4.4) are the poles of the closed-loop system given in Equation (4.2). It may be noted that the closed-loop poles are a function of the open-loop poles and zeros. Further, since the plant transfer function $G(s)$ is normally fixed, the design problem is to select a $K(s)$ and/or an $H(s)$ which guarantee that the closed-loop poles of $T(s)$ are stable (all contained within the open left-hand half of the s-plane) and then to ensure satisfactory performance. This is achieved by shaping the system's frequency response curve, or by placing the dominant closed-loop poles in preferred s-plane positions.

Substituting for $H(s)G(s)K(s)$ into the closed-loop transfer function of Equation (4.2) gives

$$T(s) = \frac{V(s)Q(s)L(s)}{V(s)P(s)M(s) + W(s)Q(s)L(s)} \tag{4.5}$$

from which it is seen that the closed-loop zeros are the zeros of the plant and forward path controller (zeros of $G(s)K(s)$) and the poles of the feedback element $H(s)$.

Equation (4.5) relates $y(s)$ to $r(s)$, but other signals in Figure 4.1 could be chosen as the input and output. The return difference equation (that is, the system closed-loop characteristic equation) remains the same regardless of the selected input–output pair, but the closed-loop numerator is a function of the selected input and output. That is, the closed-loop system dynamics (defined by the closed-loop poles) are independent of the chosen signals, but for the same input, the nature of the response will appear to change with the output. For example, the relationship between the error signal, $e(s)$, and input, $r(s)$, in Figure 4.1 is given by

$$\frac{e(s)}{r(s)} = \frac{1}{1 + H(s)G(s)K(s)} \tag{4.6}$$

and its closed-loop zeros will be found to be the roots of

$$V(s)P(s)M(s) = 0$$

which are the system's open-loop poles. Perhaps the simplest way to visualize the change in output response is to note that Equations (4.2) and (4.6) have the same characteristic equation and therefore the dynamics of the two responses will be the same. However, for a unit step input, the requirement that $y(t)$ becomes close to unity with increasing time requires that $e(t)$ approaches zero. Clearly, $y(t)$ and $e(t)$ will have the same dynamics but the zeros make the two responses very different.

Having examined the effect of pole position on performance (Section 3.2.2) attention will now be given to the influence of a zero on a system's response. Consider the system

$$G(s) = \frac{18}{(s+6)(s^2 + 2s + 3)} \tag{4.7}$$

which has poles at $s = -6$ and $s = -1 \pm j1.4142$. The step response for this system is shown by the solid line in Figure 4.2. Assume a zero is added to the system, such that Equation (4.7) becomes

$$G(s) = \left(\frac{18}{a}\right)\frac{(s+a)}{(s+6)(s^2 + 2s + 3)} \tag{4.8}$$

Note that the gain in Equation (4.8) has been adjusted so that the final steady-state output to a unit step input will still be unity. Step response plots for zeros at $s = -1$, $s = -2$ and $s = -0.5$ are also shown in Figure 4.2. Readers with access to MATLAB (Appendix 3) can use the m-file *fig4_234.m* on the accompanying disk to produce this figure, and test any other values they add.

When compared with the response of the original system, the added zero reduces the rise time, reduces the peak time, but increases the overshoot. The closer the zero is to the imaginary axis the more pronounced the effects. Zeros in the right-hand portion of the s-plane are not considered from the viewpoint of the effects they might have in a compensator since, as shown in Section 3.8, these can produce some very peculiar responses. They are therefore unlikely to be introduced into a system (via a compensator design) deliberately.

The corresponding Bode plots, Figure 4.3, and polar plots, Figure 4.4 (which

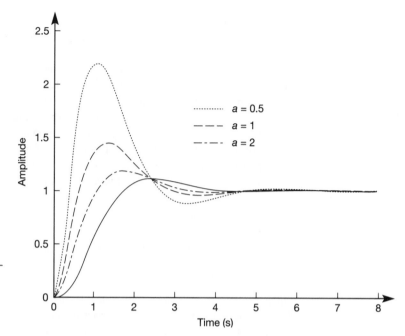

Figure 4.2 The effect on the step response of a third-order system of adding a zero $(s + a)$ (while maintaining steady-state gain).

can also be produced by the MATLAB m-file *fig4_234.m* on the accompanying disk), show that as the zero becomes more positive the peak magnification and bandwidth increase. Introducing a zero reduces the system's rank and the phase shift at large frequencies tends to $-180°$ rather than the $-270°$ phase shift, produced by the system of Equation (4.7). The effect on the polar plot is to rotate the $G(j\omega)$ loci anticlockwise; the closer the zero is to the right-hand portion of the s-plane the greater the rotation.

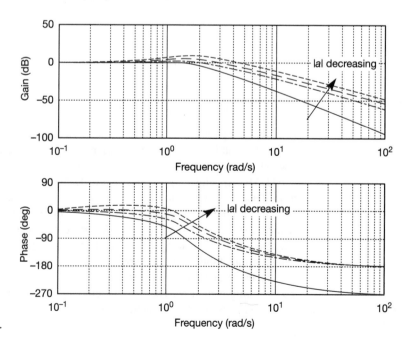

Figure 4.3 Bode plots corresponding to Figure 4.2.

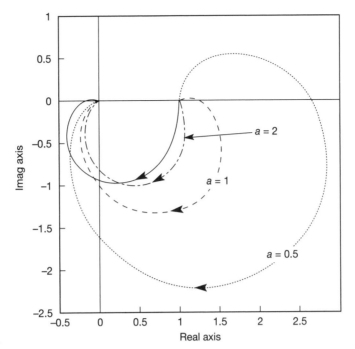

Figure 4.4 Direct polar (Nyquist) plots corresponding to Figure 4.2.

For comparison purposes, assume that a pole is added to the system such that

$$G(s) = \frac{18a}{(s+6)(s^2+2s+3)(s+a)} \qquad (4.9)$$

and that the introduced poles can have the value $s = -2$, $s = -1$ and $s = -0.5$.

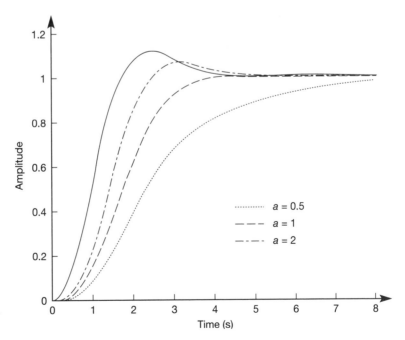

Figure 4.5 Step responses of the same system as Figure 4.2, but adding a pole rather than a zero.

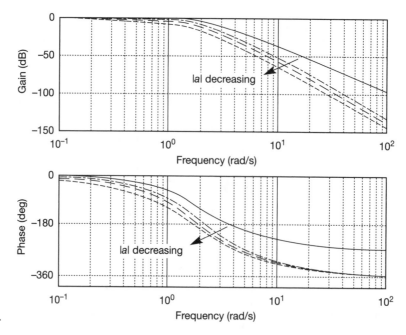

Figure 4.6 Bode plots corresponding to Figure 4.5.

The step responses for the systems described by Equations (4.7) and (4.9) are shown in Figure 4.5. Bode and polar plots for the responses are shown in Figures 4.6 and 4.7 (the m-file *fig4_567.m* on the accompanying disk will produce these).

In the step response plots of Figure 4.5, the pole reduces the oscillations in the system and, as it becomes more dominant, makes the response more sluggish. The

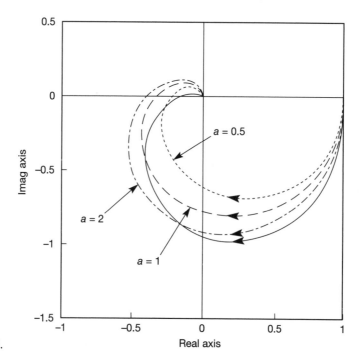

Figure 4.7 Direct polar (Nyquist) plots corresponding to Figure 4.5.

rank of the system is also increased and at large frequencies the phase shift becomes $-360°$. This tends to twist the polar plots clockwise, as shown in Figure 4.7.

It should be evident that having zeros or poles near to the open-loop system's dominant poles can dramatically alter the shape of the open-loop frequency response curve. Such changes will be reflected in the closed-loop frequency response and hence in the closed-loop time-domain response. Adding open-loop zeros and poles through the forward path compensator, $K(s)$, or by the feed-back element, $H(s)$, is therefore one of the foundation stones of control system design.

4.3 The Nyquist stability criterion

A control system design problem exists when the performance of a plant or process is deemed unsatisfactory (for whatever reason). The starting point is the open-loop plant and the requirements are closed-loop stability and performance. Stability requires that all the system's closed-loop poles are contained within the left-half portion of the s-plane and performance requires that the mix of dominant poles and zeros produces a time response with desirable properties.

In terms of Figure 4.1, the stability requirement is that the values of s which make the system's closed-loop characteristic equation (the return difference equation) zero must all have negative real parts. Setting the characteristic equation equal to $F(s)$, then from Figure 4.1 and Equation (4.3):

$$F(s) = 1 + H(s)G(s)K(s) \tag{4.10}$$

or

$$F(s) = \frac{V(s)P(s)M(s) + W(s)Q(s)L(s)}{V(s)P(s)M(s)} \tag{4.11}$$

Comparing Equation (4.11) with the system's open-loop transfer function (see Equation (4.1)):

$$H(s)G(s)K(s) = \frac{W(s)Q(s)L(s)}{V(s)P(s)M(s)} \tag{4.12}$$

reveals the following properties:

- The poles of $F(s)$ are the poles of the open-loop transfer function (Equations (4.11) and (4.12)).
- The zeros of $F(s)$ are functions of the open-loop poles and zeros and of any fixed gain within the controller $K(s)$, feed-back element $H(s)$ and plant $G(s)$.
- Since the zeros of $F(s)$ are also the system's closed-loop poles, for closed-loop stability all the zeros of $F(s)$ must lie in the left-half s-plane.

Nyquist's stability criterion establishes the number of zeros of $F(s)$ which are located in the right-half s-plane. Clearly, for closed-loop stability this number should be zero.

The next few pages introduce the mathematical basis of the Nyquist criterion, but this need not be understood fully in order to make use of the resulting procedures. Nevertheless, an understanding of *why* the criterion works is definitely helpful. The basic idea is that a curve ('contour') drawn in the *s*-plane can be regarded as joining a series of points, each of which is a value of *s* (and will therefore be a complex number). As we travel along such a curve, each value of *s* through which the curve passes can be substituted into any function of *s* which is of interest (for example, $F(s)$, the closed-loop characteristic equation of a system). The resulting values of the function of *s* will therefore trace out their own contour in the new function plane (this is known as a 'mapping' of the original contour). Of course, a curve can be regarded as consisting of an infinite number of points, so this mapping would be rather long-winded by hand. There are analytical techniques to help with it, which are explained below but, in general, it need not be done at all, as computer packages such as MATLAB (Appendix 3) can do it all in a single command. The purpose of the following analysis is to illustrate the concepts. Beginning with an analysis of how poles and zeros of the characteristic equation relate to such *s*-plane mappings, it will eventually be shown that if a closed contour is chosen in the *s*-plane which contains everything in the right-hand half of the plane, then the mapping of the open-loop transfer function of a system as this contour is traversed will provide a plot (a Nyquist plot) from which the required stability information can be determined.

Consider the closed-loop characteristic equation, Equation (4.11), which may be written as

$$F(s) = \frac{K(s + a_1)(s + a_2) \cdots (s + a_m)}{(s + b_1)(s + b_2) \cdots (s + b_n)} \qquad (4.13)$$

where $n \geq m$ and all the *a*'s and *b*'s are constants. The polar form of Equation (4.13) is

$$F(s) = |F(s)|e^{j \arg F(s)}$$

Let the poles and zeros of $F(s)$ be plotted in the *s*-plane. Let Γ_s be any closed contour enclosing some of the poles and zeros of $F(s)$ (see Figure 4.8). Consider the zero at $-a_2$, which is representative of any zero, real or complex, which lies outside the contour Γ_s. To determine the contribution of this zero to $F(s)$ at any point *s* on the contour Γ_s, draw a vector from the zero to the required point on Γ_s.

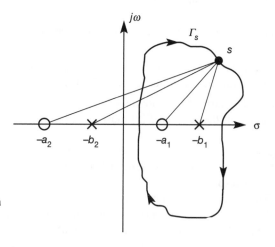

Figure 4.8 Distances from poles and zeros to a point on a closed contour in the *s*-plane.

The modulus contribution to $F(s)$ is given by the length of the vector, and the argument contribution is the angle the vector makes with the positive σ-axis. As usual, the normal mathematical convention is used in which a positive angle indicates that the vector is rotated anticlockwise from the positive real axis. Equation (4.13) indicates that the modulus contribution is $|s + a_2|$ and the argument contribution is $+\arg(s + a_2)$, evaluated at the point s on Γ_s. If the vector now traverses once around the contour, the net contribution to the argument of $F(s)$ must be zero; the vector oscillates about its starting position and its net displacement is zero degrees. Similarly, any pole which is outside the contour, for example $-b_2$, makes a net contribution to the argument of $F(s)$ of zero degrees every time the vector traverses once around Γ_s.

The only other possibilities that need to be considered are zeros or poles of $F(s)$ *inside* the contour. The zero at $-a_1$ is representative of such zeros and, as shown in Figure 4.8, a_1 must be negative since it lies in the right half of the s-plane. However, the argument's contribution to $F(s)$ is independent of the sign of a_1, provided a_1 is within the contour. As the vector from a_1 traverses once around the contour in the clockwise direction, the net contribution to the argument of $F(s)$ is $-360°$. For each such clockwise circuit of the contour the vector is displaced from its starting point by $-360°$.

A vector from any pole within the contour, say $-b_1$, will also be displaced by $-360°$ for each clockwise circuit of the contour. However, since for a pole

$$\arg\left(\frac{1}{s + b_1}\right) = -\arg(s + b_1)$$

the net contribution to the argument of $F(s)$ is $+360°$.

A more formal statement of the above reasoning would conclude that the net angular rotation of the $F(s)$ locus about the origin of the $F(s)$-plane is some multiple of $\pm360°$. Note that it may be shown that the mapping of the closed contour Γ_s onto the $F(s)$-plane produces a closed contour. Consequently,

$$\arg F(s) = 2\pi N = 2\pi(P^* - Z^*)$$

where

> N = net number of anticlockwise encirclements of the $F(s)$ locus about the origin of the $F(s)$ plane
>
> Z^* = number of zeros of $F(s)$ enclosed by Γ_s
>
> P^* = number of poles of $F(s)$ enclosed by Γ_s
>
> and Γ_s is traversed in the clockwise direction.

Generalizing this idea, let Γ_s be a contour enclosing the whole of the right half of the s-plane, but avoiding poles on the $j\omega$ axis, as below. This is the Nyquist contour – see Figure 4.9. For obvious reasons it is often called the 'D-contour'. The vertical of the 'D' travels along the imaginary axis, and poles on the imaginary axis are avoided by introducing a semicircular arc of infinitesimally small radius ρ, as shown. The large semicircle of the 'D' has a radius R which is infinitely large, and therefore must enclose all the right-half s-plane poles and zeros of $F(s)$.

The Nyquist stability criterion may now be stated as

$$Z = P - N \tag{4.14}$$

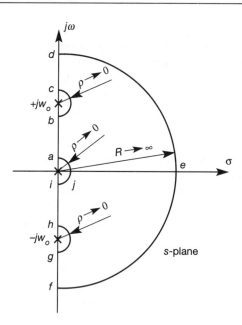

Figure 4.9 The Nyquist 'D' contour.

where

Z = number of zeros of $F(s)$ in the right half of the s-plane. Since $F(s)$ is the closed-loop characteristic equation, Z gives the number of closed-loop poles in the right half of the s-plane. For closed-loop stability Z must be zero.

P = number of poles of $F(s)$ in the right half of the s-plane, and is equal to the number of poles of $G(s)H(s)$ in the right half of the s-plane. P may be obtained directly or from Routh's array (Section 3.3.3). Note that for an open-loop stable system $P = 0$.

N = net number of encirclements by the $F(s)$ locus about the origin of the $F(s)$-plane. Anticlockwise encirclements are positive, and clockwise encirclements negative.

In applying the Nyquist stability criterion, the object is to find Z from a knowledge of P and N. P, the number of open-loop right-half s-plane poles, may be found directly from the open-loop transfer function. As N is the net number of encirclements of the $F(s)$ locus about the origin of the $F(s)$-plane, it may be found by mapping $F(s)$ as s traverses around the Nyquist D-contour. However, it is not necessary to plot the $F(s)$ locus, since the origin of the $F(s)$ plane is

$$F(s) = 0 + j0$$

and, by equating with Equation (4.10), it is evident that an equivalent point exists in the $H(s)G(s)K(s)$-plane, namely

$$H(s)G(s)K(s) = -1 + j0$$

The normal procedure is to plot the open-loop transfer function $H(s)G(s)K(s)$ as s traverses the Nyquist contour. N is then determined from the net number of encirclements of the $H(s)G(s)K(s)$ locus around the critical point $(-1 + j0)$ in the $H(s)G(s)K(s)$-plane.

Closed-loop stability is therefore determined from the system's open-loop transfer function. If the closed-loop system is stable, the open-loop gain and phase margin measures indicate the degree of stability and the adjustments required to improve stability. Since a system's frequency response deals with time delays exactly, the method may be used to test the stability of systems containing such elements.

4.3.1 The Nyquist contour

Figure 4.9 shows a general Nyquist contour which encloses the whole of the right half of the s-plane. The contour has been deformed to avoid the real open-loop pole at the origin and the pair of complex conjugate poles on the imaginary axis. In plotting the open-loop transfer function locus $H(s)G(s)K(s)$, appropriate values of s must be used as s traverses round the Nyquist contour. For example, consider the path ab shown in Figure 4.9. At any point on this path $s = j\omega$, and therefore $H(j\omega)G(j\omega)K(j\omega)$ must be evaluated. This portion of the Nyquist plot is therefore the polar plot of the open-loop frequency response for all frequency values between points a and b.

The various paths along the Nyquist contour and their mathematical equations and range of validity are summarized in Table 4.1.

Table 4.1 Mathematical equations for the Nyquist contour of Figure 4.9.

Path	Equation	Range of validity
ab	$s = j\omega$	$0 < \omega < \omega_0$
bc	$s = \lim_{\rho \to 0} (j\omega_0 + \rho e^{j\theta})$	$-90° \leq \theta \leq 90°$
cd	$s = j\omega$	$\omega_0 < \omega < \infty$
def	$s = \lim_{R \to \infty} \mathrm{Re}^{j\theta}$	$+90° \geq \theta \geq -90°$
fg	$s = j\omega$	$-\infty < \omega < -\omega_0$
gh	$s = \lim_{\rho \to 0} (-j\omega_0 + \rho e^{j\theta})$	$-90° \leq \theta \leq 90°$
hi	$s = j\omega$	$-\omega_0 < \omega < 0$
ija	$s = \lim_{\rho \to 0} \rho e^{j\theta}$	$-90° \leq \theta \leq 90°$

4.3.2 Nyquist stability and inverse polar plots

Applying the Nyquist stability criterion to the inverse transfer function (see Section 3.5.2) gives similar results to using direct polar plots in that the equation

$$Z = P - N$$

still holds (see Equation (4.14)). Again, Z is the number of zeros of the closed-loop characteristic equation in the right half of the s-plane, and for closed-loop stability Z must be zero. Likewise, N is the net number of encirclements of $[H(s)G(s)K(s)]^{-1}$ locus about the $(-1 + j0)$ point in the $[H(s)G(s)K(s)]^{-1}$ plane, anticlockwise encirclements being positive. However, P is now the number of zeros of the open-loop transfer function $H(s)G(s)$ enclosed by the Nyquist contour, and not the number of poles in the right half of the s-plane, as with direct polar plots.

4.3.3 The left-hand rule

Using the Nyquist stability criterion, it is possible to show that for open-loop stable systems of type 0, 1 or 2, only the $H(j\omega)G(j\omega)K(j\omega)$ locus need be plotted in order to determine closed-loop stability. Systems with complex open-loop poles on the imaginary axis are excluded and in these cases (and for open-loop unstable systems) the full criterion must be applied.

Parts of three possible $H(j\omega)G(j\omega)K(j\omega)$ loci are shown in Figure 4.10. In this figure the direction of increasing ω is indicated by means of arrows on the loci. It is also assumed that each locus is obtained from an open-loop stable system of type 0, 1 or 2.

The left-hand rule states that if the $(-1+j0)$ point lies to the left of the $H(j\omega)G(j\omega)K(j\omega)$ locus, then the system is closed-loop stable. In Figure 4.10(a), an observer looking along the locus in the direction of increasing ω would place the

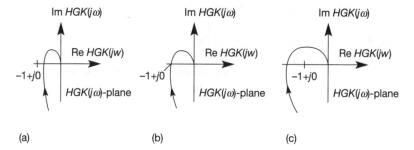

Figure 4.10 Illustrating the 'left-hand rule' for direct Nyquist plots.

(a) (b) (c)

$(-1+j0)$ point to the left of the locus, and therefore the closed-loop system is stable. In Figure 4.10(b), the locus passes through the $(-1+j0)$ point and the system is marginally stable; the loop gain is unity when the phase shift is $-180°$. The system in Figure 4.10(c) is closed-loop unstable since the $(-1+j0)$ point lies to the right of the locus.

Proof of the left-hand rule is left as an exercise for the reader.

Example 4.1 Nyquist's stability criterion applied to an open-loop-stable fourth-order system

A negative feedback closed-loop system has an open-loop transfer function given by

$$G(s) = \frac{K}{(s+3)^3(s+1)} \tag{4.15}$$

Use the Nyquist stability criterion to determine the range of gains K for which the closed-loop system is asymptotically stable.

Since the open-loop transfer function has no poles on the imaginary axis, the Nyquist contour takes the form shown in Figure 4.11. Now, $G(s)$ has none of its poles enclosed by this

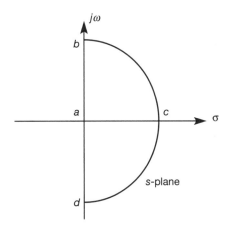

Figure 4.11 The Nyquist 'D' contour for Example 4.1.

contour, so P (Equation (4.14)) equals zero. For stability Z must be zero, and therefore N must be zero. To determine N, the $G(s)$ locus must be plotted as s travels round the Nyquist contour, and this is carried out as follows.

For path ab

Along the path ab, $s = j\omega$ for $0 \le \omega \le +\infty$ (note that this range corresponds to normal, positive frequencies), so

$$G(j\omega) = \frac{K}{(j\omega + 3)^3 (j\omega + 1)}$$

At this point it is useful to fix K at some value. A suitable choice is $K = 27$, which makes the static gain of the system unity. Consequently,

$$|G(j\omega)| = \frac{27}{(\omega^2 + 3^2)^{3/2}(\omega^2 + 1^2)^{1/2}}$$

and

$$\arg G(j\omega) = 0 - 3\tan^{-1}(\omega/3) - \tan^{-1}(\omega)$$

Note that K affects only $|G(j\omega)|$, and not $\arg G(j\omega)$.

A table of frequency response points, Table 4.2, may now be calculated from which to draw the polar plot. These points may be found by solving the modulus and argument

Table 4.2 Frequency response points for the path ab in Example 4.1.

ω	$|G(j\omega)|$	$\arg G(j\omega)$
0	1.0	$0°$
0.5	0.859	$-55°$
1.0	0.604	$-100°$
1.5	0.397	$-136°$
2.0	0.257	$-165°$
3.0	0.112	$-207°$
∞	0	$-360°$

equations at different frequencies, or by using appropriate software (MATLAB can be used as outlined in Appendix 3 – it is used below).

For path bcd

This section of the Nyquist contour (see Table 4.1 and Figure 4.9) is described by

$$s = \lim_{R \to \infty} Re^{j\theta}, \quad +90° \geq \theta \geq -90°$$

Now, from Equation (14.15), for large values of s,

$$G(s) = \frac{K}{s^4 + 10s^3 + 36s^2 + 54s + 27} \approx \frac{K}{s^4}$$

So, for large values of R

$$\lim_{R \to \infty} G(Re^{j\theta}) = \lim_{R \to \infty} \left(\frac{K}{R^4 e^{j4\theta}} \right) = 0e^{-j4\theta}$$

The infinite semicircular arc in the s-plane therefore maps onto a point of zero radius in the $G(s)$-plane (because the modulus is zero). The argument (involving -4θ) shows that the original range of $+90° \geq \theta \geq -90°$ maps to $-360° \leq \theta \leq +360°$. Therefore the $G(s)$ locus approaches the origin of the $G(s)$-plane at $-360°$, makes two complete anticlockwise revolutions (albeit at zero radius in this case) and leaves the origin at $+360°$. Note that the inequality signs specify the direction of the rotation (see Problem 4.1).

For path da

Over this path $s = -j\omega$, and its evaluation produces the mirror image of the $G(j\omega)$ locus along the real axis.

The complete $G(s)$ locus may now be plotted. This is shown in Figure 4.12. At $180°$ the $G(s)$ locus has a magnitude of 0.195, so it does not enclose the $(-1 + j0)$ point, and therefore $N = 0$. From Equation (4.14),

$$Z = P - N = 0$$

and the system is stable. If the system is to become marginally stable, then the gain K must be increased until the $G(s)$ locus passes through the critical point $(-1 + j0)$. Reference to Figure 3.45 shows that the gain margin for this example is $1/0.195 \approx 5.13$. The gain may therefore be increased from its present value ($K = 27$) by this factor to obtain marginal stability. Thus the system is asymptotically stable in the closed-loop for $0 \leq K < 139$.

To obtain the plot using MATLAB (Appendix 3) use the commands:

$\gg$ num = 27;

$\gg$ den = conv([1 3], conv([1 3], conv([1 3], [1 1])));

$\gg$ nyquist(num, den)

The *ginput* command can then be used to confirm the value of 0.195 for the negative real axis crossing. Alternatively, the command *[g,p,wg,wp] = margin(num,den)* will give the gain and phase margins directly. Using *num = num*g;* and repeating the rest of the

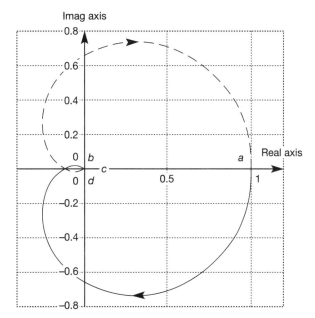

Figure 4.12 Direct Nyquist plot for Example 4.1.

commands will then confirm the condition for limiting stability. To investigate the closed-loop system for stability and performance, use:

》[numc, denc] = cloop(num, den, −1); % apply unity negative feedback

》roots(denc)   % check the closed-loop poles

》step(numc, denc), grid   % closed-loop step response

Example 4.2 *Nyquist's stability criterion applied to an open-loop-unstable second-order system*

Use the Nyquist stability criterion to test the closed-loop stability of the unstable open-loop system

$$G(s) = \frac{1}{s(s-1)}$$

The open-loop system has one pole on the imaginary s-plane axis at $s = 0$, which means that the Nyquist contour must be deformed as shown in Figure 4.13. There is one pole at $s = 1$, which is inside this contour, and therefore $P = 1$ (see Equation (4.14)). The Nyquist contour is mapped onto the $G(s)$ plane as follows.

For path ab

Along this path $s = j\omega$ for $0 < \omega \le +\infty$. Therefore

$$G(j\omega) = \frac{1}{j\omega(j\omega - 1)}$$

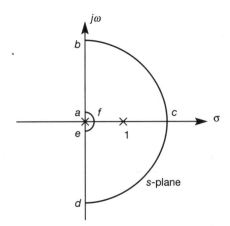

Figure 4.13 The Nyquist
'D' contour for Example 4.2.

from which

$$|G(j\omega)| = \frac{1}{\omega\sqrt{(\omega^2 + 1)}}$$

and

$$\arg G(j\omega) = -[\tan^{-1}(\infty) + \tan^{-1}(-\omega)]$$

$$= -\left[\frac{\pi}{2} + \pi - \tan^{-1}(\omega)\right] = -\frac{3\pi}{2} + \tan^{-1}(\omega) = +\frac{\pi}{2} + \tan^{-1}(\omega)$$

Note that $\tan^{-1}(-\omega)$ has been rewritten as $\pi - \tan^{-1}(\omega)$ (as in Equation (3.94)).

A scale plot of the $G(j\omega)$ locus is inappropriate since $|G(j\omega)| \to \infty$ as $\omega \to 0$. Typically, points close to the origin are plotted to scale, while points away from the origin indicate the general direction.

For path bcd

On this path $s = \lim_{R \to \infty} Re^{j\theta}$ for $+90° \geq \theta \geq -90°$. For large values of s

$$G(s) = \frac{1}{s^2 - s} \approx \frac{1}{s^2}$$

so

$$\lim_{R \to \infty} G(Re^{j\theta}) = \frac{1}{\infty} e^{-j2\theta}$$

The modulus of this expression is zero, and the argument (by similar reasoning to that in Example 4.1) shows that the original range of θ maps to $-180° \leq \theta \leq +180°$.

For path de

This path is the mirror image of the path ab with respect to the real $G(s)$-plane axis.

For path efa

Here

$$s = \lim_{\rho \to 0} \rho e^{j\theta}, \qquad -90° \le \theta \le +90°$$

or, when s is negative,

$$(-s) = \lim_{\rho \to 0} \rho e^{-j\theta}, \qquad -90° \le \theta \le +90°$$

For small values of s

$$G(s) \approx \frac{1}{-s}$$

so

$$\lim_{\rho \to 0} G(\rho e^{-j\theta}) = \lim_{\rho \to 0} \left(\frac{1}{\rho} e^{+j\theta} \right)$$

The modulus of this expression is infinite, and the argument is valid for $-90° \ge \theta \ge +90°$.

The complete $G(s)$ locus may now be plotted. This is shown in Figure 4.14. In this sketch there is one clockwise encirclement of the $(-1 + j0)$ point, and hence $N = -1$. From the Nyquist stability formula

$$Z = P - N = 1 - (-1) = 2$$

The closed-loop system is clearly unstable since there are two closed-loop poles with positive real parts.

Using the MATLAB commands given in Example 4.1, with appropriate numerator and denominator data, gives the required polar plot. To see the detail of Figure 4.14, it is

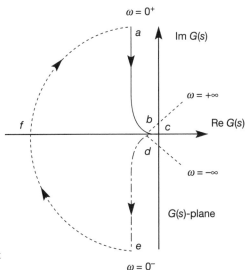

Figure 4.14 Direct Nyquist plot for Example 4.2.

necessary to restrict the frequency range by the commands:

$\gg$ w = logspace (−1,1,100); % 100 log.-spaced frequency values from 0.1 to
% 10 rad s^{-1}

$\gg$ nyquist(num,den,w)

Example 4.3 *Nyquist diagram for a more complicated system*

Plot the Nyquist diagram for the system having the open-loop transfer function

$$G(s) = \frac{2(s+0.1)(s+0.6)(s^2+s+1)}{s^3(s-0.2)(s+1)}$$

and hence determine the stability of the closed-loop system.

The Nyquist contour must avoid the multiple-open loop poles at the origin of the *s*-plane, and therefore takes the form shown in Figure 4.15. There is one open-loop pole within the Nyquist contour at $s = 0.2$, and therefore $P = 1$.

For path ab

Here $s = j\omega$ for $0 < \omega \leq +\infty$, so

$$G(j\omega) = \frac{2(j\omega+0.1)(j\omega+0.6)[j\omega+(1-\omega^2)]}{(j\omega)^3(j\omega-0.2)(j\omega+1)}$$

from which

$$|G(j\omega)| = \frac{2\sqrt{(\omega^2+0.1^2)}\sqrt{(\omega^2+0.6^2)}\sqrt{[\omega^2+(1-\omega^2)^2]}}{\omega^3\sqrt{(\omega^2+0.2^2)}\sqrt{(\omega^2+1)}}$$

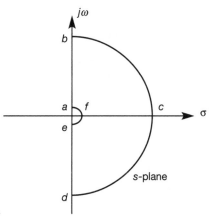

Figure 4.15 The Nyquist 'D' contour for Example 4.3.

and

$$\arg G(j\omega) = \tan^{-1}\left(\frac{\omega}{0.1}\right) + \tan^{-1}\left(\frac{\omega}{0.6}\right) + \tan^{-1}\left(\frac{\omega}{1-\omega^2}\right)$$
$$- 3\tan^{-1}\left(\frac{\omega}{0}\right) - \tan^{-1}\left(\frac{\omega}{-0.2}\right) - \tan^{-1}(\omega)$$

Remember that the term $1/s^n$ gives an argument contribution of $-n\tan^{-1}(\omega/0)$, or $-n90°$.

Now for $\omega < 1$ the term $\omega/(1-\omega^2)$ is positive, so

$$\arg G(j\omega) = \tan^{-1}(10\omega) + \tan^{-1}(1.67\omega) + \tan^{-1}\left(\frac{\omega}{1-\omega^2}\right)$$
$$- \frac{3\pi}{2} - [\pi - \tan^{-1}(5\omega)] - \tan^{-1}(\omega)$$

For $\omega > 1$ the term $\omega/(1-\omega^2)$ is negative, and

$$\arg G(j\omega) = \tan^{-1}(10\omega) + \tan^{-1}(1.67\omega) + \left[\pi - \tan^{-1}\left(\frac{\omega}{\omega^2-1}\right)\right]$$
$$- \frac{3\pi}{2} - [\pi - \tan^{-1}(5\omega)] - \tan^{-1}(\omega)$$

To assist in sketching this locus a minimal table of points is drawn up (see Table 4.3).

Table 4.3 Frequency response points for the path *ab* in Example 4.3.

ω	$|G(j\omega)|$	$\arg G(j\omega)$
0	∞	$-90°$
0.1	761.8	$-8.9°$
0.2	120.2	$+37.3°$
0.5	9.54	$+105.6°$
1.2	1.2	$+199.2°$
∞	0	$+270°$

For path bdc

The required transformation expression is

$$s = \lim_{R\to\infty} Re^{j\theta} \qquad \text{for } +90° \geq \theta \geq -90°$$

For large values of s, $G(s) \to 1/s$, so

$$\lim_{r\to\infty} G(Re^{j\theta}) = 0e^{-j\theta}$$

For path de

This is the mirror image with respect to the real $G(s)$-axis of the $G(j\omega)$ locus.

For path efa

For small values of s, $G(s)$ may be approximated by

$$G(s) \approx 1/(-s^3)$$

For $-s$ the Nyquist path in the s-plane is described by the expression

$$(-s) = \rho e^{-j\theta}, \qquad -90° \leq \theta \leq +90°$$

Therefore

$$\lim_{\rho \to 0} G(\rho e^{-j\theta}) = \lim_{\rho \to 0} \left(\frac{1}{\rho^3 e^{-j3\theta}} \right) = \infty e^{j3\theta}$$

and the argument of this expression is valid for $-90° \geq \theta \geq +90°$. This means that the infinite-radius portion rotates through a total of 540° (that is, 3 × the stated 180° range), or one and one half revolutions. The starting point ($\omega = 0^-$) is $3 \times (-90°) = -270°$, measured from the positive real axis, and the finishing point ($\omega = 0^+$) is $3 \times (+90°) = +270°$, again measured from the positive real axis. Note that the directions of the inequalities have reversed due to the inversion of exponent, and hence the direction of rotation must be from $-270°$ through one and a half clockwise revolutions to $+270°$.

The Nyquist plot of this system is shown in Figure 4.16. There are two anticlockwise encirclements of the $(-1 + j0)$ point, the two inner circles of the Nyquist plot, and one clockwise encirclement of infinite radius from e to a. The net number of anticlockwise encirclements is therefore plus one (two anticlockwise minus one clockwise) so $N = 1$. From the Nyquist stability formula,

$$Z = P - N = 1 - 1 = 0$$

Hence it may be concluded that the closed-loop system is stable.

MATLAB (Appendix 3) can be used to produce the plot very simply, as in the m-file *fig4_16.m* on the accompanying disk.

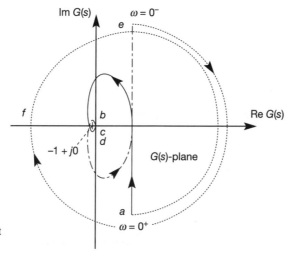

Figure 4.16 Direct Nyquist plot for Example 4.3.

If only stability information is required, it is much easier to investigate the closed-loop characteristic equation directly by using a package such as MATLAB to find the closed-loop poles (as in Example 3.3(b)). If such a package is unavailable then, for low-order systems with no time delays the use of Routh's stability criterion (see Section 3.3.3) may be easier. However, in all other cases, the ease with which polar plots can be produced and the additional information they provide make the Nyquist stability criterion a far superior method. To validate the truth of these statements, it would be a useful exercise to investigate the above examples using Routh's stability criterion.

4.4   The root locus method

The root locus technique provides a design method based on the system's open-loop transfer function, which will give the closed-loop pole positions for all possible changes in a single variable, normally the loop gain. These are plotted on a diagram (the 'root locus' diagram), from which the value of open-loop gain (for example) which will give the most appropriate closed-loop performance can then be chosen.

It has been previously shown that the system's closed-loop zeros, between any two points in the loop, will be a subset of the open-loop zeros and the poles of the feedback element $H(s)$ (see Section 4.2). For these reasons, the root locus method can be used to find all possible closed-loop transfer functions for variations in a single variable. As usual, MATLAB (Appendix 3) or a similar package will normally be used in such investigations. The following descriptions of how to follow the method by hand are provided for use in situations where computer assistance is unavailable, and also to provide some background information.

To introduce the root locus method, assume that in Figure 4.1 $K(s) = K$ and can take any value between $\pm\infty$, $G(s) = s/(s^2 + 1)$ and $H(s) = 1$. For this system the open-loop transfer function (see Equation (4.1)) is

$$K(s)G(s)H(s) = \frac{Ks}{s^2 + 1}$$

and the closed-loop transfer function (see Equation (4.2)) is

$$T(s) = \frac{Ks}{s^2 + Ks + 1}$$

Notice that both the open and closed-loop systems have a zero at $s = 0$. The closed-loop poles are at

$$s = -\frac{1}{2}K \pm \sqrt{\left(\frac{1}{4}K^2 - 1\right)} \tag{4.16}$$

For values of K between 0 and 2 the poles are complex and for K greater than 2 the poles are real. For $K = 2$ there is a double pole at $s = -1$. For any negative value of gain the response is unstable.

Consider the following cases:

(1) $K = -1$. There are two complex poles, at $s = +0.5 \pm j0.866$, and the time response for a unit impulse is given by

$$y(t) = -e^{+0.5t}(\cos 0.866t - 0.577 \sin 0.866t)$$

The positive real part of the poles produces a positive exponential, and consequently the system is unstable.

(2) $K = 0$. This essentially breaks the loop (see Figure 4.1) since at $K = 0$ a change in $E(s)$ has no effect on $U(s)$. However, an impulse response may be obtained for the open-loop system and is given by

$$y(t) = \cos t$$

(3) $K = 1$. There are now two complex poles, at $s = -0.5 \pm j0.866$, and the corresponding impulse response is given by

$$y(t) = e^{-0.5t}(\cos 0.866t - 0.577 \sin 0.866t)$$

The response is now stable, as indicated by the negative exponential.

(4) $K = 2$. There is a double pole at $s = -1$ and the impulse response is given by

$$y(t) = 2(e^{-t} - te^{-t})$$

(5) $K = 4$. This gives poles at $s = -3.732$ and $s = -0.268$ and the closed-loop impulse response is

$$y(t) = 4(-0.0774e^{-0.268t} + 1.0774e^{-3.732t})$$

Note that the effect of the dominant pole is reduced because of its proximity to the closed-loop zero.

Since a simple feedback loop affects only the closed-loop pole positions, it is useful to plot the pole trajectories, or root loci, for all possible changes in the loop gain K. Figure 4.17 shows the closed-loop pole plot, or root locus plot, for the system. The poles and zeros of the open-loop system are indicated by crosses and enclosing circles respectively. The arrow on each locus indicates the direction of increasing loop gain.

It is evident from Figure 4.17 and Equation (4.16) that when K is positive the closed-loop poles will be negative and the system will be stable. When $K = 0$ the system is marginally stable and when $K < 0$ it is unstable. A negative value of K has the same effect as positive feedback and therefore only positive values of K are normally considered (MATLAB, for example, only produces the plot for positive K by default – see below). Looking at Figure 4.17 it is evident that for the fastest rate of decay (when the dominant pole has its most negative value) $K = 2$, and the system has 2 real poles at $s = -1$. For $0 < K < 2$ the poles are complex, and will always have a negative real part greater than -1. For $K > 2$, one pole tends to the

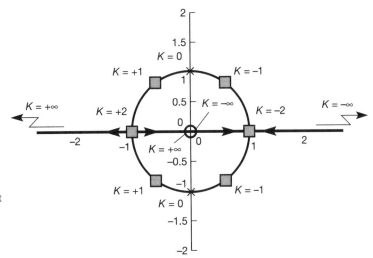

Figure 4.17 Root locus plot for positive and negative values of K in

$$KGH(s) = \frac{Ks}{s^2 + 1}.$$

origin and the other to minus infinity and consequently the system's response takes longer to decay.

Production of the root locus using MATLAB (Appendix 3) is very easy, requiring only the following commands (also, try the m-file *fg4_1718.m* on the accompanying disk).

$\gg$ num $= [1 \ 0]$; % This is Ks with $K = 1$. MATLAB will sweep K
   % automatically.

$\gg$ den $= [1 \ 0 \ 1]$; % note the zero coefficient of s

$\gg$ rlocus(num,den),grid

To pick points off the locus (in order to find the K value, for example) MATLAB has the command *[k,p] = rlocfind* which puts a crosshair on the plot, waits for the click of the mouse button over the required point and then returns the value of K at the selected point and the pole locations for that value of K.

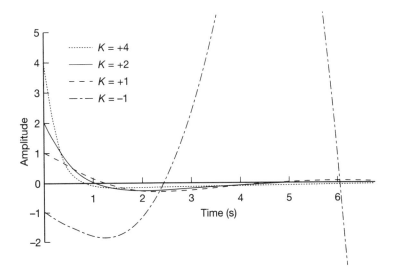

Figure 4.18 Closed-loop impulse responses for several values of K from Figure 4.17.

It is interesting that with this system, faster transient responses are obtained when K is not equal to 2 – see Figure 4.18. For $K < 2$ the damping is reduced and the system responds as expected. However, if the open-loop zero occurs between the forcing input and the measured output, the closed-loop pole–zero map will have a zero at the origin. For large values of K this zero will give approximate closed-loop pole–zero cancellation. For the impulse responses shown in Figure 4.18 (also produced by the m-file *fg4_1718.m* on the accompanying disk), notice that an increase in gain K causes a very fast initial measured response, which quickly approaches its final steady-state value, but takes a considerable amount of time to settle at that value. However, if the open-loop zero occurs in some other part of the system, the measured closed-loop response will become progressively more sluggish as K is increased.

4.4.1 Rules for constructing root locus plots

Normally, root locus plots will be produced by computer, as in the MATLAB example above. However, if such a package is unavailable, it is possible to determine the salient features of root loci by applying simple rules. These rules obviate the necessity to solve the characteristic equation, and thus make the method applicable to quite complicated systems.

The rules, often known as Evans' rules (after Walter R. Evans, whose original paper on the subject of root loci appeared in 1948), are stated below for the closed-loop single-input-single-output system shown in Figure 4.1. To simplify the presentation it will be assumed that the open-loop system's transfer function is $HGK(s)$ where, from Equation (4.1)

$$HGK(s) = H(s)G(s)K(s)$$

and the return difference is set equal to zero to give the closed-loop characteristic equation:

$$F(s) = 1 + H(s)G(s)K(s) = 1 + HGK(s) = 0 \qquad (4.17)$$

The purpose of these rules is to find the roots of $F(s)$, given $HGK(s)$, for all positive values of some constant K in $HGK(s)$.

That is, assuming the system's open loop transfer function has m zeros and n poles with $n > m$, then

$$HGK(s) = K \frac{\displaystyle\prod_{i=1}^{m}(s + z_i)}{\displaystyle\prod_{i=1}^{n}(s + p_i)} \qquad (4.18)$$

(1) *Number of root loci. The number of root loci is equal to the order of the closed-loop characteristic equation.*

For engineering systems $HGK(s)$ will be rational, and the closed-loop characteristic function $F(s)$ must therefore be of the same order as the denominator of $HGK(s)$; that is, of order n. Since $F(s)$ is an nth-order polynomial it will have n roots, each with its own locus.

(2) *Symmetry of loci. The root loci of a characteristic equation having real coefficients are symmetrical with respect to the real axis.*

This follows from the fact that the complex roots of a real characteristic equation can only occur in conjugate pairs.

(3) *Poles of HGK(s). Each pole of HGK(s) lies on a root locus and corresponds to $K = 0$.*

This follows directly from Equations (4.17) and (4.18) as follows. Substitute Equation (4.18) into Equation (4.17) and rearrange to make K the subject. Then substitute $s = -p_i$, where p_i, is the value of one of the poles.

(4) *Zeros of HGK(s). Each zero of HGK(s) lies on a root locus and corresponds to $K = \infty$.*

Again, this follows directly from Equations (4.17) and (4.18), but this time the subject becomes $1/K$, and s is set equal to $-z_i$. If there are r more poles than zeros, then r of the loci will become infinite as $K \to \infty$. These loci are dealt with in rules (5) and (6).

(5) *Asymptotes of root loci. If HGK(s) has r more poles than zeros, the root loci are asymptotic to r straight lines making angles*

$$\frac{(2i + 1)\pi}{r}, \qquad i = 0, 1, 2, \ldots, r - 1$$

with the real axis. The root loci approach asymptotes when $K \to \infty$.

(6) *Point of intersection of asymptotes. Asymptotes intersect on the real axis at a point with abscissa*

$$\sigma_0 = \frac{1}{r} \left(\sum_{j=1}^{n} p_j - \sum_{j=1}^{m} z_j \right)$$

where the p_j and z_j are respectively the poles and zeros of HGK(s).

(7) *Root loci on the real axis. If HGK(s) has one or more real poles or zeros, the segment of the real axis having an odd number of real poles and zeros to its right will be occupied by a root locus.*

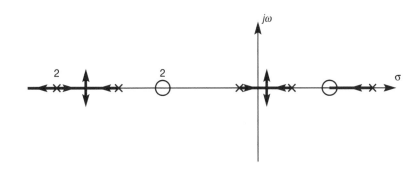

Figure 4.19 Illustrating the existence of root loci on the real axis.

This rule is best demonstrated by means of the example shown in Figure 4.19, in which $HGK(s)$ has six poles and three zeros on the real axis.

(8) *Singular points. These indicate multiple characteristic roots (intersections of loci), and occur at those values of s which satisfy $dK/ds = 0$.*

This rule is probably best demonstrated by an example. Consider a system having the closed-loop characteristic equation

$$s^2 + Ks + 1 = 0$$

as used in the example of Section 4.4. The rule states that there will be singular points when $dK/ds = 0$, and since

$$K = \frac{-(s^2 + 1)}{s}$$

differentiating K with respect to s and setting $dK/ds = 0$ indicates that

$$s^2 - 1 = 0$$

Therefore there are two singular points, at $s = +1$ and $s = -1$. Substituting both values of s back into the characteristic equation indicates that when $s = +1, K = -2$ and when $s = -1, K = +2$. Since only the rules for positive values of K are being considered, the singular point at $s = +1$ would not appear on the root locus plot.

Typically, not all the singular points found using this rule will be for positive values of K; some will be associated with zero or negative values. When producing the root locus plot it is usually obvious which of the singular points are required. In the few cases where it is not clear, the associated value of s may be substituted back into the closed-loop characteristic equation, and the sign of K determined.

(9) *Intersection of root loci with the imaginary axis. The intersections of root loci with the imaginary axis can be determined by calculating the values of K which result in the existence of imaginary characteristic roots.*

These values of K, together with the corresponding imaginary roots, can be found from Routh's array using the method described in Section 3.3.3.

(10) *Slopes of root locus at complex poles and zeros of $HGK(s)$. The slope of a root locus at a complex pole or zero of $HGK(s)$ can be found at a point in the neighbourhood of the pole or zero.*

This technique can be illustrated by considering the complex pole p_1 shown in Figure 4.20, where Ω is the unknown slope of the locus at p_1. The arguments of the complex numbers, represented by vectors drawn from the other poles p_2, p_3 and p_4 and the zero z_1 to a point on the root locus near p_1, obviously differ very little from the angles ϕ_2, ϕ_3, ϕ_4 and θ_1.

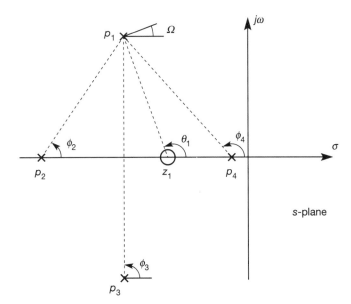

Figure 4.20 Diagram for evaluating the slope of a root locus at a complex pole using the angle criterion.

Now, for some point s on a root locus,

$$\sum_{j=1}^{m} \arg(s - z_j) - \sum_{j=1}^{n} \arg(s - p_j) = (2i + 1)180° \qquad (4.19)$$

where $\arg(s - z_j)$ and $\arg(s - p_j)$ are the angles of the vectors from the zeros and poles of $HGK(s)$ to the point s, and i is some real integer. Equation (4.19) states that the sum of the arguments of all the vectors drawn from s to all the zeros, minus the sum of the arguments of all the vectors drawn from s to all the poles, is given by an odd integer multiple of $180°$ for all points on a root locus.

Therefore it follows that

$$\theta_1 - (\phi_2 + \phi_3 + \phi_4 + \Omega) = (2i + 1)180°$$

where i is an appropriate integer. This equation can be solved for Ω since the angles ϕ_2, ϕ_3, ϕ_4 and θ_1 are easily measured. Equation (4.19) is often called the *angle criterion*.

(11) *Calculation of K on the root locus. The absolute magnitude of the value of K corresponding to any point s_0 on a root locus can be found by measuring the lengths of the vectors drawn to s_0 from the poles and zeros of HGK(s), and then evaluating*

$$|K| = \frac{\displaystyle\prod_{j=1}^{n} |s_0 - p_j|}{\displaystyle\prod_{j=1}^{m} |s_0 - z_j|} \qquad (4.20)$$

This rule states that

$$K = \frac{\text{Product of lengths of vectors from open-loop poles to } s_0}{\text{Product of lengths of vectors from open-loop zeros to } s_0}$$

If the open-loop system has no zeros, then

$$K = \text{Product of lengths of vectors from open-loop poles to } s_0$$

In applying this rule, it is assumed that, on multiplying out all the numerator and denominator factors, $HGK(s)$ will be of the form

$$M(s) = \frac{s^m + b_1 s^{m-1} + \cdots + b_{m-1}s + b_m}{s^n + a_1 s^{n-1} + \cdots + a_{n-1}s + a_n}$$

The coefficients b_0 and a_0 normally associated with s^m and s^n, respectively, must be unity. If these terms are not unity, then they must be made unity by dividing through the appropriate polynomial and adjusting K accordingly. Equation (4.20) is often called the *magnitude criterion*.

Example 4.4 *A root locus analysis of a fourth-order system*

A system has the open-loop transfer function

$$\frac{K(s^2 + 1.5s + 1.5625)}{(s - 0.75)(s + 0.25)(s + 1.25)(s + 2.25)}, \quad K > 0$$

Use Evans' rules to plot the root locus diagram for this system given that, for positive values of K, there are only four singular points, at $s = +0.26$, $s = -1.76$ and $s = -0.75 \pm j1.74$. Determine the values of gain for a stable closed-loop system.

By inspection, the denominator and numerator polynomials are of the correct form. The coefficient associated with s^2 in the numerator is unity, and the coefficient associated with s^4 in the denominator is also unity.

Evans' rules as given in Section 4.4.1 are applied to yield the following results:

(1) Number of roots $= 4$. Therefore the root locus plot will have four branches.

(2) The loci are symmetrical about the real axis.

(3) When $K = 0$, loci start from poles at $s = 0.75$, $s = -0.25$, $s = -1.25$ and $s = -2.25$.

(4) When $K = \infty$ loci terminate at infinity and at the finite zeros $s = -0.75 \pm j1$.

(5) Asymptotes: $HGK(s)$ has two more poles than zeros, so $r = 2$ and two loci approach asymptotes with angular slopes given by

$$\left(\frac{2i + 1}{r}\right)\pi, \quad i = 0 \text{ or } 1$$

that is, at $\pi/2$ and $3\pi/2$ (90° and 270°).

(6) Asymptotes for the loci extending to infinity intersect on the real axis at

$$\sigma_0 = \frac{1}{r}\left(\sum p - \sum z\right)$$

$$= \frac{(+0.75 - 0.25 - 1.25 - 2.25) - (-0.75 + j - 0.75 - j)}{2}$$

$$= -0.75$$

(7) Loci are on the real axis between $s = 0.75$ and $s = -0.25$, and also between $s = -1.25$ and $s = -2.25$.

(8) The singular points are given. (It would be a useful exercise to show that these points are singular points, and to find the associated closed loop gains K.)

(9) To determine the points of intersection of the loci with the imaginary axis using Routh's array, the closed-loop characteristic equation must be found. It is:

$$K(s^2 + 1.5s + 1.5626) + (s^4 + 3s^3 + 0.875s^2 - 2.063s - 0.527) = 0$$

or

$$s^4 + 3s^3 + (0.875 + K)s^2 + (1.5K - 2.063)s + (1.5626K - 0.527) = 0$$

The Routh array for this equation is then

Row				
0	s^4	1	$(0.875 + K)$	$(1.5626K - 0.527)$
1	s^3	3	$(1.5K - 2.063)$	
2	s^2	$\dfrac{3(0.875 + K) - (1.5K - 2.063)}{3}$	$(1.5626K - 0.527)$	

To be on the imaginary axis, row 3 must be an all-zero row – that is, the first and only element must be zero:

$$\left[\frac{3(0.875 + K) - (1.5K - 2.063)}{3}\right](1.5K - 2.063) - 3(1.563K - 0.527) = 0$$

which may be rearranged to give

$$0.75K^2 - 3.377K - 1.634 = 0$$

from which

$$K = 2.251 \pm 2.694$$

Since only positive values of K are required, $K = 4.945$.

Note that this analysis also indicates that the closed-loop system is asymptotically stable for all values of K greater than 4.945.

Letting $s = j\omega$ and equating the resulting imaginary part of the characteristic equation to zero gives

$$-3\omega^3 + (1.5K - 2.063)\omega = 0$$

Therefore

$$\omega = \sqrt{\left(\frac{1.5 \times 4.945 - 2.063}{3}\right)} = 1.34$$

(10) This rule will be used to find the slope of the locus at the complex zero, $s = -0.75 + j$. Let Ω be the unknown slope of the locus at the zero $s = -0.75 + j$ (see Figure 4.21). From the angle criterion (Equation (4.19)),

$$(90° + \Omega) - [(180° - 34°) + 34° + (180° - 63°) + 63°] = (2i + 1)180°$$

where i is some real integer. If i is set equal to -1, then $\Omega = +90°$, which is the required answer. With $i = 0$, $\Omega = 450°$ and, since this is a 360° rotation plus a 90° rotation, the results are identical. Indeed, any value of the integer i will give the correct result once any superfluous 360° rotations have been removed.

From the above rules the root locus plot may now be produced, and is shown in Figure 4.22. Rule (11) could be used to determine the loop gain at any point on the diagram.

Figure 4.22 can be produced using MATLAB (Appendix 3) simply by using the following commands:

```
》num = [1  1.5  1.5625];

》den = conv([1  −0.75],conv([1  0.25],conv([1  1.25],[1  2.25])));

》rlocus(num,den),grid
```

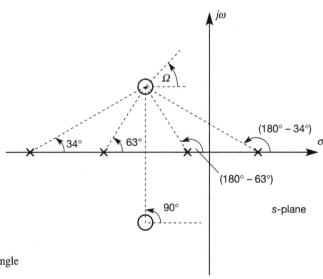

Figure 4.21 Use of the angle criterion in Example 4.4.

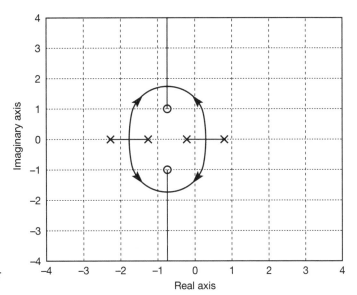

Figure 4.22 Root locus for Example 4.4.

The following command can then be used to find the value of K at any point on a locus:

$\gg$ [k, p] = rlocfind(num,den)

This command produces a 'crosshair' which can be placed at any required point on a locus. Clicking the appropriate mouse button then selects the point and returns the corresponding value of gain in k and the closed-loop pole locations in p.

Example 4.5 *Gain selection using the root locus approach*

A closed-loop, negative feedback control system has an open-loop transfer function given by

$$HGK(s) = \frac{K(s+1)}{s^2(s+9)}$$

Plot the root locus diagram, and hence determine the gain K which will give the closed-loop system its maximum value of relative stability.

Why would this system have good tracking characteristics?
The closed-loop characteristic equation is

$$1 + HGK(s) = 0$$

or

$$s^3 + 9s^2 + Ks + K = 0$$

Evans' rules yield the following results:

(1) The characteristic equation is of degree three, and hence has three roots. Therefore, there will be three root loci.

(2) The characteristic equation has real coefficients, so the loci must be symmetrical with respect to the real axis.

(3) When $K = 0$, there are loci at points $s = 0$ (a double root and therefore also a singular point) and $s = -9$.

(4) When $K = \infty$, there are loci at $s = -1$ and infinity.

(5) The asymptotes have angular slopes

$$\frac{1}{2}(2i + 1)\pi, \qquad i = 0 \text{ or } 1$$

that is, at $\pi/2$ and $3\pi/2$ (90° and 270°).

(6) The asymptotes intersect the real axis at

$$\sigma_0 = \frac{1}{r}\left(\sum p - \sum z\right) = \frac{(0 + 0 - 9) - (-1)}{2} = -4$$

(7) Loci are on the real axis between -1 and -9.

(8) Break points occur when $dK/ds = 0$. The closed-loop characteristic equation may be solved for K to give

$$K = \frac{-(s^3 + 9s^2)}{s + 1}$$

Differentiating K with respect to s and setting $dK/ds = 0$ yields

$$(s^3 + 9s^2) - (s + 1)(3s^2 + 18s) = 0$$

On expansion the above equation becomes

$$-2s^3 - 12s^2 - 18s = 0$$

which may be solved directly to give the singular points $s = -3, -3$ and 0. Substitution of any of these values of s back into the closed-loop characteristic equation indicates that all are associated with positive values of K. Note also that the analysis has indicated that the double open-loop pole at $s = 0$ is a singular point.

(9) The axis crossing points are determined using the Routh array:

Row			
0	s^3	1	K
1	s^2	9	K
2	s	$8K/9$	
3	s^0	K	

Since there are no sign changes in the first column of the array, it may be concluded that the loci do not cross the imaginary axis (so the closed-loop system will be stable) for any positive K. (Note that when using the Routh array for stability assessment, it is permissible to multiply a row by any positive constant. Row 2 could therefore have been written as K.)

(10) This rule is not required.

The root locus plot for this system may now be drawn, and is shown in Figure 4.23. In this figure, the two loci breaking from the double pole at $s = 0$ coalesce on the real axis at $s = -3$, together with the real pole emanating from $s = -9$. Two of the three loci break from this point, and become infinite, with asymptotes passing through the $s = -4$ point on the real axis. The third locus breaks from the $s = -3$ point, and moves with increasing K along the real axis until, at $K = \infty$, it reaches the zero at $s = -1$. From the root locus plot the system will have maximum stability when $s = -3$; for at any other point on the plot at least one closed-loop pole would have a more positive real part.

The loop gain which gives the maximum value of relative stability is found from rule (11) to be

$$K = \frac{|-3||-3||6|}{|2|} = 27$$

The tracking properties of this system are most easily evaluated by using the final value theorem to find the closed-loop transfer function between the forcing input and the error signal. Applying the final value theorem for various forcing inputs shows that the system will track both a step and a ramp input, with zero steady-state error.

To produce Figure 4.23 using MATLAB requires the following commands:

$\gg$ num $= [1 \ 1]$;

$\gg$ den $= [1 \ 9 \ 0 \ 0]$; % note the zero coefficients of s^1 and s^0.

$\gg$ rlocus(num,den),grid

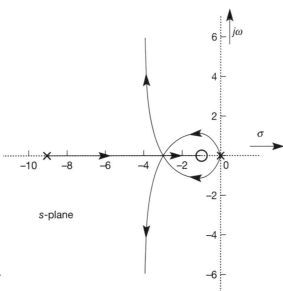

Figure 4.23 Root locus for Example 4.5.

Note that the values of K chosen by MATLAB make the plot cover a larger area than Figure 4.23, so some detail is lost. To obtain Figure 4.23, the command *axis([−10 0 −6 6])* should be used. Note that in versions of MATLAB earlier than v4.0, this command must be issued *before* the *rlocus* command.

As in the previous example, the command:

$$\gg[\text{k, p}] = \text{rlocfind(num,den)}$$

can be used to 'pick' points off the plot.

4.4.2 Root locus and PID controllers

In Chapter 3 the notion of the PID (proportional plus integral plus derivative, or 'three-term') controller was introduced and the empirical Ziegler–Nichols methods were used to establish settings for PID controllers (PID controllers are properly discussed in Section 4.5.2). For some applications, the time responses resulting from the Ziegler–Nichols settings tend to be underdamped, and the methods provide no indication of how the performance may be modified. Here, we show how the root locus method can be used to select controller settings, and how the introduction of control elements within a closed-loop system affects performance.

Consider again the system used in Example 3.10 (Section 3.3.4), in which the plant transfer function is given by

$$G(s) = \frac{6}{(s+1)(s+2)(s+3)}$$

and the PID controller's idealized transfer function is as follows (this was introduced in Chapter 3, and is properly discussed in Section 4.5.2):

$$G_c(s) = K\left(1 + T_d s + \frac{1}{T_i s}\right)$$

Under proportional control, $T_d = 0$ and T_i has to be infinite. The controller therefore becomes simply $G_c(s) = K$ and, for the closed-loop system, the characteristic equation is then:

$$1 + G(s)K = 1 + \frac{6K}{s^3 + 6s^2 + 11s + 6} = 0$$

The root locus plot is shown in Figure 4.24(a) (the MATLAB m-file *fig4_24.m* on the accompanying disk generates all the parts of this figure). In this plot K_c, the critical gain, and K_{ZN}, the Ziegler–Nichols gain (from Example 3.10), are indicated by their corresponding dominant closed-loop pole positions. Clearly, the response dynamics would be made less oscillatory by reducing the controller gain K to 0.6. However, this increases the offset (that is, the steady-state error) caused by a demand step change in the reference input (see Figure 4.24(b)).

Under proportional plus integral (PI) control, the controller is given by:

$$G_c(s) = K\left(1 + \frac{1}{T_i s}\right) = \frac{K}{T_i s}(1 + T_i s)$$

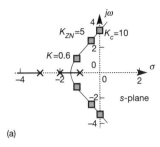

(a)

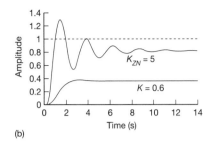

(b)

Figure 4.24 Relationship between the root locus and PID controller tuning for a third-order system. (a) Root locus for proportional control only. (b) Step responses for proportional control only. (c) Root locus for varying K in proportional plus integral control. (d) Step responses for varying K in proportional plus integral control. (e) Root locus for varying T_i in proportional plus integral control. (f) Root locus for varying K in proportional plus integral plus derivative control.

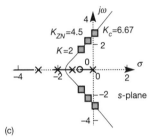

(c)

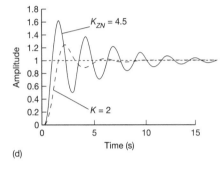

(d)

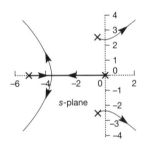

(e)

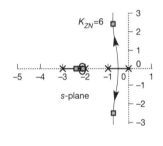

(f)

The closed-loop characteristic equation is $1 + G(s)G_c(s) = 0$, that is:

$$1 + \frac{6K(1 + T_i s)}{T_i s(s^3 + 6s^2 + 11s + 6)} = 0 \tag{4.21}$$

Since the root locus method can only cope with adjustments to one variable at a time, either K or T_i must be fixed. If T_i is fixed at its recommended Ziegler–Nichols value of 1.57 (Example 3.10), then

$$1 + \frac{K(6s + 3.82)}{s(s^3 + 6s^2 + 11s + 6)} = 0$$

and the corresponding root locus plot is as shown in Figure 4.24(c). Note that the PI controller introduces a pole at the origin of the s-plane, and a zero at $s = -0.64$. The controller pole eliminates the offset (steady-state error) for a demanded step change in input. Provided the controller gain is high enough, the zero will give approximate closed-loop pole–zero cancellation and make the complex poles dominant. Again, the Ziegler–Nichols gain K_{ZN} and the gain K_c which makes the closed-loop system marginally stable are indicated on the root locus plot.

From this plot, it would appear that the response could again be made less oscillatory by reducing the gain K to 2, as shown. Further reductions in gain will move the real pole away from the zero, and the complex poles closer to the zero, and hence reduce the effective dominance of the complex poles. The step responses for the system with both the Ziegler–Nichols settings and the reduced gain setting of $K = 2$ are shown in Figure 4.24(d).

To see the effect of adjustments to T_i, the characteristic equation must be modified so that it is of the following form, in which K_1 will be chosen to allow for variation of T_i:

$$1 + K_1 G(s) = 0$$

Again, the controller gain K must be fixed and, as previously, its Ziegler–Nichols value is used. Hence, from Equation (4.21) and with K set at 4.5 (Example 3.10), it is easily shown that

$$1 + \frac{27/T_i}{s(s^3 + 6s^2 + 11s + 33)} = 0$$

If K_1 is now set equal to $27/T_i$, the above closed-loop characteristic equation is of the required form. The root locus plot for $0 \leq K_1 \leq \infty$ is shown in Figure 4.24(e), in which the arrows on the root loci indicate decreasing values of T_i. When $T_i = 1.57$, its Ziegler–Nichols value, the loci on this plot coincide with the points on the loci of Figure 4.24(c), for which $K_{ZN} = 4.5$. Also, if $T_i < 1$ (with $K = 4.5$), the system becomes unstable (because two loci, that is, two poles of the closed-loop transfer function, move into the right half of the s-plane).

Using full PID control, there are three terms which could be adjusted, and therefore two must be fixed during any given root-locus analysis. For convenience T_i and T_d are set at their Ziegler–Nichols values of 0.947 and 0.237 respectively (Example 3.10), and the root locus plot for variations in K is drawn. This is shown in Figure 4.24(f), from which the closed-loop system is seen to be stable for all values of loop gain K. With the proposed Ziegler–Nichols value, $K_{ZN} = 6$ (Example 3.10), there are four closed-loop poles, in the positions shown. Further, since the open-loop and closed-loop zeros between the forcing input and measured output are the same, there is approximate pole–zero cancellation and the dominant poles are at $s = -0.85 \pm j2.67$. By comparing these closed-loop pole positions with those obtained from the P and PI controllers (Figures 4.24(a) and 4.24(c)), this control system is seen to have the better relative stability and damping ratio, while the damped frequencies of all three systems are similar. A plot of the Ziegler–Nichols responses to a step input for P, PI and PID control is shown in Figure 3.29.

4.4.3 Comments on the root locus design procedure

The root locus design procedure is clearly an iterative one, and the more design variables there are, the greater the required number of root locus plots. For three or more variables the method (particularly for hand calculation) becomes unwieldy as a design technique. It is still useful, however, as a test of the system's sensitivity to changes in variables and loop integrity (loop integrity is concerned with the stability and response of a system in the face of some failure: when a loop breaks, how will the system respond?).

Since the root locus method gives the closed-loop pole positions, it provides the designer with considerable insight into a system's stability, performance and response characteristics. With experience, it is possible to sketch the general shape of a root locus diagram and hence show what type of controller or compensator network is required to meet a particular design criterion.

The method's main limitations are its inability to deal with more than one variable at a time, and the difficulty of dealing with time delays (transport lags). Time delays *can* be included, but produce loci which exhibit repeated asymptotic behaviour as described in D'Azzo and Houpis (1995), for example.

4.5 Controller and compensator design

The normal procedure for designing a controller for a single-input-single-output system is an iterative one. Often the nature of the plant determines the structure of the controller, and the problem becomes one of selecting and tuning suitable compensating networks in order to meet the required performance specifications. Sometimes the controller's structure is not immediately evident; sometimes a preferred configuration proves impossible to implement. Such cases test the ingenuity of the designer. Also, the final control system design is invariably a compromise, not only between the various performance specifications but also between the feasible and the economically justifiable. There are no universal solutions. However, there are some simple controller structures and compensating elements which have proved useful, in particular the PID controller and compensating networks consisting of lead and lag elements. Since lag and lead elements have properties similar to the integral and derivative terms in a PID controller respectively, it is useful to examine these properties briefly.

In Section 4.4 it was shown that by adjusting a closed-loop system's proportional gain, K, it is possible to move the closed-loop poles within the s-plane. The system's root locus plot, which indicates all possible closed-loop pole positions for all positive values of K, is therefore a valuable design tool. However, for an open-loop type 0 plant there will be a steady-state error to a step input (see Section 3.4.1) and this might be a problem. The magnitude of the error can be reduced by increasing the proportional gain K but, for many systems, the root locus plot will indicate that large values of gain produce instability. For this reason a purely proportional controller tends to be used with type 1 plant or in stable plants which have an open-loop pole very close to the imaginary axis.

Integral action increases the system's type number by one and a lag compensator introduces a stable pole close to the imaginary axis. For a type 0 plant this eliminates or (in the case of the lag compensator) reduces the steady-state error produced by a step input. Proportional control can then be used to achieve the desired performance. The downside of integral action is that the introduced pole tends to make the closed-loop response more sluggish.

Derivative action reduces the system type number by one and a lead compensator introduces a minimum phase zero close to the imaginary axis. Since a derivative action controller produces an output equal to the time rate of change of the input, the error signal must be changing for there to be any controller output.

This means that derivative action has no effect at steady-state, and is therefore always used in combination with a P or PI controller. Type 2 plants are invariably controlled using a PD controller. When D action is used alongside PI action to give full PID control, it will tend to speed up the closed-loop system response. The downside is that D action increases the system's bandwidth, making it more susceptible to noise.

4.5.1   Controller design concepts

A control system consists of a plant with its actuators and sensors, and a controller. The actuator normally takes a low-energy signal, and transforms and amplifies it in order to produce a corresponding action which is applied to the plant's manipulable input. A sensor typically takes a high-energy signal from the plant's measurable output, and transforms it – ideally linearly – into an equivalent low-energy signal. The controller includes all the compensators, comparators, set points and paths required to complete the system.

In this section, rules of thumb are given for designing the controller and the control system configuration. These are not rigorous, and not applicable in every case, but are intended to provide some insight into the SISO design problem.

Some of the more common design problems concern the control of flow, level, temperature and pressure. Experience shows that satisfactory control may be achieved by using simple PID controllers.

Flow

Flow loops are notoriously noisy, and this precludes derivative action. The controller gain is rarely greater than one, and consequently, unless integral action is included, there is significant steady-state error.

Level

In most level control loops the actual liquid level is relatively unimportant, provided it is between some maximum and minimum value. The transfer function model of a level system will often be of type 1, so satisfactory control can be achieved with a simple proportional controller.

Temperature

In temperature loops, thermal lags (caused by heat transfer) and sensor lags make the loops relatively slow and noise-free, so full PID controllers are often used.

Pressure

Pressure loops can vary considerably in their dynamic characteristics, from very fast loops almost like flow to slow averaging loops almost like level. Therefore, depending on the nature of the loop, a PI or P controller is used.

Returning to the more general discussion, compensators modifying low-energy signals are normally cheaper and easier to adjust than those for high-energy signals. If, for some reason, a forward path compensator has to be combined with the actuator, then it may be more economical to use feedback path compensation. In general though, forward path compensation provides the preferred configuration.

At this point it is worth noting that control is an important aspect of any plant design. If control is likely to be required, dynamic considerations should be taken into account in the plant design. Various hardware configurations may be possible and some, for example those that avoid interactions, are easier to control than others. Allowance should be made in sizing and selecting equipment so that it can handle operational transients. Also, allowance should be made for the inclusion and optimal location of sensors and actuators so that they operate effectively.

In developing the controller, the following design points should be borne in mind:

(1) Feedback paths should be designed so as to avoid time delays and lags. In practice this means that a sensor should be placed as close as possible to the source of the variable it is intended to control. For example, if the temperature of the steam from a boiler is required, then ideally the sensor should be in contact with the steam and located close to the boiler steam outlet point. Measurements at a point on the steam pipe away from the boiler would introduce a time delay, and if the sensor is not in contact with the steam a thermal lag would also be introduced into the feedback path.

Sometimes, it is not possible to measure the controlled variable at the required point. For example, in a metal rolling mill, it is impossible to measure the thickness of the product until it has emerged from the workrolls (and the separation of the rolls is not a useful measure, due to significant bending and flattening effects). The resulting time delay as the metal moves from the mill to the sensor means that such a measurement can only be used to compensate for slowly varying changes (such as thermal expansion of the rolls and roll wear), and not for transient changes of thickness (which will have 'gone' from the mill before the measurement is made). This means that more complex schemes have to be used to control rolling mill output thickness. If the sensor could be placed at the point where the measurement is required, things would be simpler.

(2) Large, frequent measurable disturbances should be compensated for by feed-forward control. Consider the single-loop control system shown in Figure 4.25(a). The controller consists of the forward path compensator $G_c(s)$ and the comparator. Also indicated are the actuator dynamics $G_1(s)$, the plant dynamics $G(s)$, the sensor dynamics $H(s)$ and the disturbance input $D(s)$. Two examples are now given of systems that this model might represent.

The model could represent a valve-actuated system with LTF $G_1(s)$, controlling the flow of some liquid into the plant whose LTF is $G(s)$. $D(s)$ would then represent temperature disturbances (variations) affecting the viscosity, and hence the flow rate, of the fluid, and $G_2(s)$ would be a transfer function converting temperature changes into equivalent flow changes.

In another example, again taken from the metal rolling field, $G_1(s)$ could represent the dynamics of the 'screwdown' system setting the roll gap, $G(s)$ would be the dynamics of the propagation of thickness variations through the mill, $D(s)$ would represent a disturbance in metal thickness entering the mill, and $G_2(s)$ would represent the effect of the incoming disturbance on the mill rolls, including a transport delay to represent the propagation of incoming thickness disturbances from the point of measurement to the roll gap.

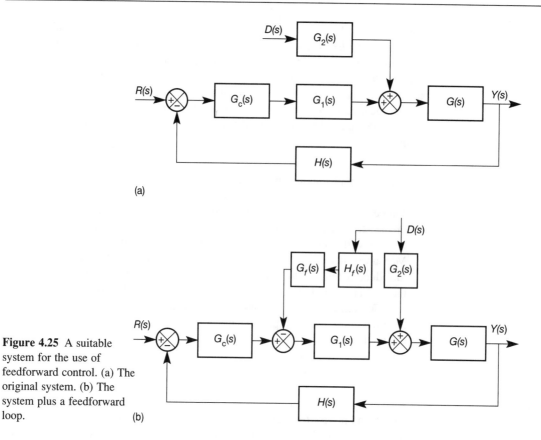

(a)

(b)

Figure 4.25 A suitable system for the use of feedforward control. (a) The original system. (b) The system plus a feedforward loop.

Figure 4.25(b) shows the same system with feed-forward control. For the fluidic example given above, $H_f(s)$ would be a temperature sensor and $G_f(s)$ a compensating network.

For the metal rolling example, $H_f(s)$ would be a thickness sensor at the incoming side of the mill and $G_f(s)$ would be a compensator which generated the necessary control signal to remove the disturbance, and *delayed its application* until the moment the disturbance would arrive in the mill. The necessity for an accurately calculated delay is one difficulty with such schemes – if it is wrong, then the wrong piece of material will be 'corrected'. Nevertheless, this method is routinely used in modern rolling mills.

In either case, for removal of the effect of the disturbance from the input of $G(s)$, it is clear from Figure 4.25(b) that $G_1(s)G_f(s)H_f(s)$ must be made the same as $G_2(s)$, so that:

$$G_f(s) = \frac{G_2(s)}{G_1(s)H_f(s)}$$

(3) Where possible, minor disturbances should be eliminated by introducing a cascade controller (see Figure 4.26). If, in the fluidic example given, the temperature changes were small but still produced flow disturbances, they could be compensated for by placing a cascade controller around the valve

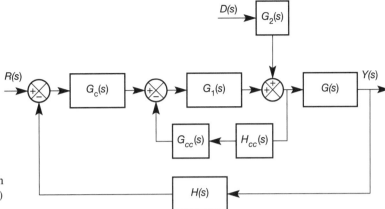

Figure 4.26 A system with an inner ('cascade control') feedback loop.

$G_1(s)$, as shown in Figure 4.26 (note that a 'standard' forward path controller is sometimes called a cascade controller – the context dictates which is intended). Here $H_{cc}(s)$ represents the flow sensor dynamics, and $G_{cc}(s)$ the cascade controller compensator. With this scheme the master controller is providing the setpoint value for the slave cascade control system.

Such arrangements are also often found in complex electrical drive systems. Again using a metal rolling mill as an example, if the mill has several stands, it is necessary to control the tension between them, so as to avoid damaging the product (and the mill). This tension is typically controlled by slight relative variation of adjacent stand speeds. The tension control loop typically provides a set point value to the speed control loop; and the speed control loop in turn provides a set point to the motor current control loop. Sometimes, four or five loops will be found cascaded in this way in such drive systems. In order to avoid severe problems of interaction between the loops, it is normal practice to arrange for the inner loop to have the fastest dynamics, and the outer one the slowest.

(4) Multiloop systems in which the operation of an external loop depends on the operation of an inner loop should be avoided. This is not always possible, as for example in cascade control for disturbance rejection. However, as a general rule loops should be independent.

(5) The control structure should be such that the failure of a loop, or any element in the loop, does not tend to make the overall system unstable.

(6) The manipulated variable should not be saturated. Aircraft control surfaces can deflect only by a certain amount; fluid flows are fixed between certain limits; motors have maximum torques. Once a loop saturates, the system becomes nonlinear and there is often a significant recovery time before the loop becomes fully operative again. The saturation values provide design constraints.

(7) A proposed design should be checked to ensure that it is effective during start-up and shutdown conditions, and not just during normal operation.

4.5.2   PID controllers

Control systems which have a forward path compensator consisting of some PID combination have proved suitable for most industrial applications. Since maintenance and failure problems associated with non-standard equipment can prove costly, it is normal practice to use a commercially available PID controller whenever possible (hundreds of such devices are readily available worldwide). The PID control concepts introduced in Section 1.3.5 and Example 1.6 will now be expanded.

The first of the two standard PID arrangements to be presented is the most convenient mathematically, in that the three terms are completely non-interacting. Consider Figure 4.1, in which the compensator $K(s)$ represents the three-term controller, and the sensor dynamics are assumed negligible ($H(s) = 1$).

The *proportional* term of the controller produces a control signal proportional to the error in the system, so that $u(t) = K_p e(t)$ or $U(s) = K_p E(s)$. It is simply a forward path gain, the effects of which have been considered previously (see Example 4.1, or the whole of Section 4.4, for example). Typically, given a step change in setpoint, low values of K_p give rise to stable responses, but large steady-state errors. Higher values of K_p give better steady-state performance, but worse transient performance (longer settling times, more overshoot and so on). If K_p is made too high, then instability will result (as seen in the root locus plots in Figures 4.24(a), (c) and (e), for example).

If steady-state error is a problem, then instead of increasing K_p, the controller's *integral* term is activated. Here, the control signal generated is proportional to the integral of the error signal. That is

$$u(t) = K_i \int e(t)dt \quad \text{or} \quad U(s) + \frac{K_i}{s} E(s)$$

where K_i is the integral gain.

The integral term gives a controller output which is a ramp, when fed with a constant non-zero input. When the controller input is zero, its output is held steady at its existing value. In a stable closed-loop control system containing integral action, the steady-state error (controller input) is normally zero, the output of the integral term then being held steady at whatever value of $u(t)$ is needed to maintain the plant output $y(t)$ equal to the reference input $r(t)$, thus maintaining the zero error condition.

Although the steady-state error may be reduced to zero by this controller, such performance is achieved at the expense of stability. This is because the integral term increases oscillation and settling time by introducing an extra $90°$ phase lag at all frequencies, thus reducing the phase margin. In some systems, this is not serious. In others, assuming a combined proportional plus integral controller, it is possible to regain satisfactory operation by reducing the value of proportional gain, thus reducing oscillation. This is acceptable, since one reason for using a high value of K_p is to reduce steady-state error, which should now be eliminated by the integral term. It can be seen from the Ziegler–Nichols tuning rules (introduced in Section 3.3.4) that the proportional gain is set 10 per cent lower in a PI controller than in a purely proportional controller for the same plant.

If it is not possible to reduce oscillation sufficiently using a PI controller, then the third term can be added. This term gives a control signal proportional to the time derivative (rate of change) of the error signal. Thus,

$$u(t) = K_d \frac{de(t)}{dt} \quad \text{or} \quad U(s) = sK_d E(s)$$

where K_d is the derivative gain. Since the output signal from this third term responds only to the rate of change of error, it has no effect upon steady-state operation (when rate of change is zero).

During a transient, the 90° phase lead introduced by the zero at the origin of the s-plane increases the system's phase margin and hence increases the damping (thus reducing the oscillations) of the closed-loop system. This increase in damping allows higher values of K_p and K_i to be used than would otherwise be the case. Again, typical increases can be seen by examining the appropriate Ziegler–Nichols rules. (Note that derivative action must be used with care in 'noisy' environments, as the response to the high rates of change of the noisy input signal causes problems.)

The foregoing description leads to a complete three-term controller whose transfer function is:

$$G_c(s) = \frac{U(s)}{E(s)} = K_p + \frac{K_i}{s} + sK_d \qquad (4.22a)$$

where

K_p = proportional gain, selected for adequate rise time

K_i = integral gain (units of gain per second), selected for steady-state accuracy without making performance unacceptably poor

K_d = derivative gain (units of gain × seconds), selected to overcome excessive oscillation or too long a settling time

Often, although less convenient in terms of the interactions created, the controller is rewritten as:

$$G_c(s) = \frac{U(s)}{E(s)} = K_p \left(1 + \frac{1}{sT_i} + sT_d \right) \qquad (4.22b)$$

where (when compared with Equation (4.22a)):

T_i = integral action time (or 'reset time') (seconds) = K_p / K_i

T_d = derivative action time (or 'rate time') (seconds) = K_d / K_p

Further, the gain is often expressed as a 'proportional band', PB, where

$$\text{PB} = (1/K_p) \times 100 \text{ per cent}$$

PB is the percentage of the controller's input range which will cause a change of 100 per cent in its output range. For example, if $K_p = 5$, then the PB = 20 per cent. Under these conditions an error signal change of 20 per cent of the controller's input range would cause the controller output to change by 100 per cent of its range (that is, by the fullest possible amount). If the gain K_p was low ($K_p = 1$, say)

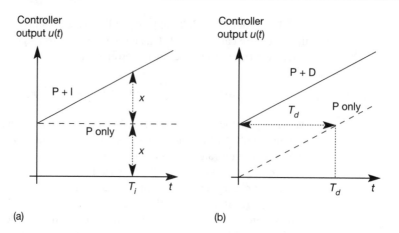

Figure 4.27 Physical meaning of integral and derivative action times.

(a)

(b)

then the proportional band would be high (100 per cent in this case). With a proportional band of 100 per cent, an input error change of 20 per cent of the scale (3.2 mA in the case of a common analog electronic 4 to 20 mA system, for example) would cause the controller output to change by only 20 per cent of full scale.

T_i is the time required for a PI controller to duplicate the effect of the P term acting on a step input (see Figure 4.27(a)).

T_d is the time by which the output of a PD controller (to a *ramp* input) is advanced compared with the P term acting alone (Figure 4.27(b)).

In practice, the proportional channel will have a bias. For proportional control, the actuation signal $U(s)$ is related to the error signal $E(s)$ by the equation

$$U(s) = K_p E(s) + b$$

where b is the controller output bias.

Figure 4.28 shows the closed-loop block diagram of such a proportional controller. In this system the bias provides an offset for the zero error point. With no bias signal and an open-loop system transfer function of type 0, $E(s)$ can be zero in the steady state only if $R(s)$ and $Y(s)$ are zero. However, even if the bias is not zero, the zero steady state $E(s)$ point still requires $R(s)$ and $Y(s)$ to be equal (assuming $H(s) = 1$), but they will not be identically equal to zero. The bias therefore shifts the control system's operating point. Normally, the selected operating point is somewhere in the actuator's mid-range of operation, so that it can respond to both positive and negative error signals.

Ideal derivative action cannot be achieved in practice, so a practical PID

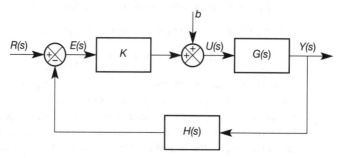

Figure 4.28 A proportional controller with output bias.

controller approximates derivative action via a lead compensator, whose action may be described by

$$T_d s \approx \frac{T_d s + 1}{\alpha T_d s + 1}$$

where typically, for a commercial device, α is fixed somewhere between 1/6 and 1/20. The lead compensator is placed in series with a PI controller. The commercial PID controller transfer function is therefore the product of those of a PI controller and a lead compensator described by the equation:

$$G_c(s) = K\left(\frac{T_i s + 1}{T_i s}\right)\left(\frac{T_d s + 1}{\alpha T_d s + 1}\right) + b \qquad (4.23)$$

For further details on lead compensation, see Section 4.5.4.

The reason why a pure derivative control element is impossible can be seen by visualizing the frequency response of any real system. As frequency increases to a very high value, the system will cease to respond (that is, its gain will become zero). Even atoms cannot vibrate at infinite frequency. The corollary of this statement is that the Bode magnitude plot of any real system must have a high frequency roll-off of at least -20 dB per decade. It will usually be more than this. It cannot be flat, nor can it rise at 20 dB per decade (which the Bode plot of a pure differentiator would do). Therefore, the model of any system should have more poles than finite zeros.

The model sK used to represent the ideal differentiator would only be valid over an extremely restricted frequency range. If a crude differentiator was built using an operational amplifier, having its inverting input connected to the outside world by a series capacitor (C farads), and having a resistive feedback path (R ohms) back to its inverting input from its output, the transfer function may be derived on paper as $-sK$, where $K = CR$ (such an arrangement would appear as in Figure 4.45(a) in Section 4.7, below, with R_1 and C_2 omitted). In practice, at high frequencies, the resistive effects in a real capacitor and the capacitive and inductive effects around the amplifier and feedback resistor would introduce some significant poles, causing the Bode magnitude plot first to level off, and then to fall.

Many CACSD packages (including MATLAB) will not accept a transfer function whose numerator is of higher order than its denominator, because such a system is physically unrealizable. Also, from a mathematical viewpoint, such packages do most of their analysis (transparently to the user) using state-space models. In Section 2.5.1 it was seen that it is impossible to obtain such a model if the corresponding transfer function's numerator is of higher order than its denominator.

PID controller tuning

There have been many techniques proposed for the time-domain tuning of PID controllers, but in practice the Ziegler and Nichols methods seem to have become the recognized standards in the process industries. Using their settings produces an underdamped response (see Section 3.3.4). For process control such responses are desirable, since the oscillations tend to average out, and the overshoot helps to eliminate the effects of stiction and backlash in the valves and actuators ('stiction' is static friction, the initial resistance to motion of contacting surfaces). However, with servo control systems, oscillations are not usually desirable and additional damping will therefore be required. In most cases this can be achieved by reducing the controller's proportional gain during online tuning.

The effect on a system's root locus plot of introducing a PID controller has been examined in Section 4.4.2. Proportional control is ideally suited to the root locus design method. With PI control the design procedure is normally an iterative one. However, a PI controller introduces a pole at the origin of the s-plane and a zero at $-1/T_i$ and this requires the selection of two parameters: T_i, which defines the position of the zero in the open- and closed-loop transfer functions, and K, which locates the closed-loop pole positions. The pole at the origin of the s-plane slows down the response, so T_i is normally chosen so that its associated zero is close to the pole, so as to speed up the response. If derivative action is included in the controller, then a further pole–zero combination is introduced, the relative positions of which depend on the selection of α (see Equation (4.23)). In general, integral action is introduced to increase the system's type number and hence improve its steady-state tracking capabilities. Since most systems contain some noise, derivative action is often not used; however, when it can be used (see Section 4.5.1) the effect is to increase the system's speed of response.

For selecting the settings for a PID controller using frequency-domain techniques, design guidelines have been developed. Typically, a proportional controller should satisfy a given gain margin criterion. For example, the Ziegler–Nichols setting for a proportional controller is based on a gain margin of 2. Checks are then carried out on the phase margin and M_p value to ensure satisfactory performance. With a PI controller the proportional gain would be adjusted to meet the M_p specification (this is effectively defining the closed-loop damping ratio). The introduction of integral action reduces the system's bandwidth, hence reducing the response speed of the closed-loop system. The zero introduced by the PI controller is then used to increase the bandwidth, therefore reducing any loss of response speed. One way of achieving this is to set the integral action time at

$$\frac{1}{T_i} = \frac{\omega_p}{5} \tag{4.24}$$

where ω_p is the resonant frequency (the closed-loop frequency corresponding to M_p). This places the zero in a position where it removes most of the integration phase angle lag at the point where the resonant peak occurs. The introduction of this combined PI action changes the closed-loop response, and consequently the proportional term must be readjusted to meet the bandwidth specification. If at this stage ω_p is essentially unaltered, the design is complete; otherwise the integral action time must be recalculated from Equation (4.24) and the process repeated.

The frequency-domain design of a full PID controller gives three adjustable parameters. A typical procedure for selecting them is first to design a PI controller. The derivative action time T_d is then chosen so that the lead compensator network gives its maximum phase angle advance (see Figure 4.34, below) at the $-180°$ point on the PI-controlled open-loop frequency response curve. Again, the closed-loop frequency response changes with the introduction of new terms, and the various design stages must be repeated until there is no appreciable change.

Finally, note that it is possible to arrange for a PID controller to tune itself in response to changing plant conditions. Such controllers are discussed in Chapter 11.

4.5.3 Velocity feedback (or rate feedback)

Derivative action has the attraction that it improves a system's rise time without producing an over-large overshoot or excessive oscillations. Despite this, it finds little practical application because of its intolerance to noise. There is also another well-known problem with derivative action, known as 'setpoint kick'. This occurs when the setpoint to a control loop is suddenly changed to a new value (thus approximating a step change in demand). Since the feedback signal cannot change instantaneously, such a step is passed directly to the error signal. A derivative controller term acting on an approximation to a step change in error will immediately cause an impulsive-type response to the maximum (that is, saturated) controller output. Although such an impulse in the control signal will only occur once per demand change, it is nevertheless undesirable from the viewpoints of linearity and plant actuator wear.

With some systems these problems can be avoided by using the derivative of the output signal, rather than the error, and summing it with the output of a PI controller. Writing the error signal as $e(t) = r(t) - y(t)$ (cf. Figure 4.1 with $H(s) = 1$), it can be seen that the derivative of the error signal is simply the inverse of that of the output, so long as the setpoint is not changing (so the derivative of $r(t) = 0$). In some systems it is also possible to measure the derivative of the output variable directly and so avoid the problems of noise. For historical reasons, the direct measurement of the derivative of an output for control purposes is known as velocity feedback, or rate feedback.

A classic example of direct derivative measurement is the use of a tachogenerator to measure the angular velocity of rotational position control machinery. Figure 4.29 shows the block diagram of such a system incorporating velocity feedback. The block $K_v s$ represents the tachogenerator, so its output is proportional to the derivative of $Y(s)$. Assuming $Y(s)$ to be an angular position, the output of this block is therefore directly proportional to angular velocity. The closed-loop transfer function for this system is given by:

$$\frac{Y(s)}{R(s)} = \frac{G(s)G_c(s)}{1 + G(s)K_v s + G(s)G_c(s)}$$

It is left as an exercise for the reader to show that, if $G_c(s)$ is a PI controller, then the closed-loop characteristic equation from the above LTF is identical to that of the system under ideal PID control. This is found to be true if $K_v = KT_d$, where K and T_d are the gain and derivative action time of an ideal PID controller (see Equation (4.22b)).

Although velocity feedback produces the same characteristic equation as a

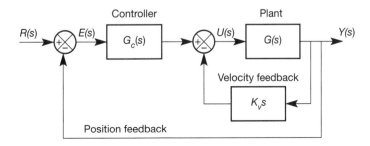

Figure 4.29 A velocity feedback loop.

PID controller, PID action introduces a zero into the closed-loop system, whereas velocity feedback does not (see Example 4.7).

4.5.4 Simple compensators

When it is not possible to meet the desired performance specifications simply by adjusting the system's closed-loop gain, dynamic compensation is required. The simplest form of dynamic compensation is provided by the lag, lead and lag-lead compensators. With these devices, either singly or in combination, it is normally possible to adjust the system's response sufficiently to achieve the required performance. In this sense, the PID controller and its various derivatives are specific combinations of lead and lag compensators.

The following subsections consider the design of lag, lead and lag-lead compensators using root loci and frequency domain methods.

Lag compensation

A stable closed-loop design which has a satisfactory transient response, but too large a steady-state error, can often be improved by means of lag compensation. The transfer function of a lag compensator is

$$G_c(s) = \frac{1 + Ts}{1 + \alpha Ts}, \qquad \alpha > 1$$

The compensator's pole–zero map is shown in Figure 4.30. Since the pole is located at $s = -1/\alpha T$, this compensator is a more general form of PI controller. A PI controller has its pole at the origin of the s-plane, corresponding with αT being very large.

Using the root locus design technique, the pole of the compensator is placed close to the origin of the s-plane in order to improve the steady-state error. Since the response is deemed to be satisfactory, the original shape of the root locus plot should remain unchanged. The change brought about by compensation can be minimized by placing the zero close to the pole. A useful rule of thumb is that the angle contributed by the compensator at the original dominant closed-loop poles should be less than $5°$.

The various frequency design methods enable closed-loop performance to be described in terms of M_p, ω_p and the static error coefficients. The usual design procedure is to fix M_p and then determine the corresponding frequency ω_p and loop gain K. Fixing M_p essentially defines the damping ratio, and from ω_p is found the undamped natural frequency ω_n. The value of ω_n, together with the damping ratio, determines the response settling time.

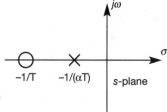

Figure 4.30 Pole–zero map of a lag compensator.

The frequency response function of the lag compensator is

$$G_c(j\omega) = \frac{1 + j\omega T}{1 + \alpha j\omega T}$$

Bode plots for this compensator are shown in Figure 4.31 for $\alpha = 2$, 5 and 10. These can be added to the Bode plots of an uncompensated open-loop system in the manner illustrated in Example 3.11. Phase lag compensation is normally applied to the very low frequency end of the open-loop frequency response. If the object is to reduce the steady-state error of a type 0 system, then α is selected to give the required increase in gain (that is, the forward path gain is adjusted so that the right-hand end of the magnitude plot in Figure 4.31 coincides with the mid-frequency gain of the uncompensated system, thus allowing low-frequency gain to rise). Typically, for the same gain and phase margins as the uncompensated system, the loop gain of the compensated system has to be increased by α. There are practical limits on the magnitude of α, and normally α would not be greater than 10. Once α is selected, T is chosen so that the corner frequency of the zero is well away from the critical frequency range of the system (otherwise the extra phase lag introduced by the compensator may destabilize the system). An acceptable value is

$$\frac{1}{T} = \frac{\omega_p}{10}$$

The above design procedure is intended only as a guide; in an actual study adjustments to these values are likely to be required before an acceptable solution is reached.

Figure 4.32 shows the effect of lag compensation on a type 0 system. In this figure the gain of the compensated system has been increased so that it has the same

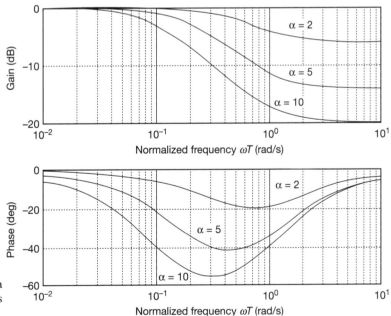

Figure 4.31 Bode plots of a lag compensator for various values of α.

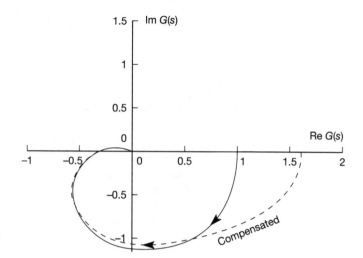

Figure 4.32 The effect of adding lag compensation to a type 0 system.

gain and phase margins as the original system. As a corollary, if the steady-state error of the original design were acceptable, no increase in loop gain would be required and the gain margin of the compensated system would be increased. The increase in gain margin would be accompanied by an increase in phase margin.

In general, the effects of lag compensation are as follows:

- The bandwidth is reduced.

- The predominant time constant of the system is usually increased, producing a more sluggish system, as lag compensation introduces a pole close to the imaginary axis.

- For a given relative stability the steady-state error is improved, and the system has better tracking capabilities.

- For a given steady-state error the gain margin is improved.

The MATLAB Multivariable Frequency Domain Toolbox (Ford *et al.*, 1990) has a *phlag* command for designing such compensators. The *phlead* command can also be used. The difference is that *phlag* works in terms of a specified dB gain change, whereas *phlead* works in terms of a specified maximum phase change.

Lead compensation
Lead compensation is normally used to improve the speed of response of stable systems of type 1 and higher. It can also be used to advantage in systems with time delays.

The compensator transfer function is given by

$$G_c(s) = \alpha \frac{1 + Ts}{1 + \alpha Ts} \qquad \alpha < 1$$

or

$$G_c(s) = \frac{s + 1/T}{s + 1/(\alpha T)} \qquad \alpha < 1 \tag{4.25}$$

There are practical limits to the amount by which the magnitude of α may be decreased, and typically α is not less than 0.1.

The inclusion of a lead compensator can have a pronounced effect on a system's root locus plot. For this reason the root locus design procedure in which the compensator's pole–zero positions are fixed (see Figure 4.33) is normally one of

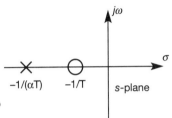

Figure 4.33 Pole–zero map of a lead compensator.

trial and error. However, an approach which can be used to good effect on systems of type 1 or higher is to choose the zero position so that it cancels the most positive (that is, the most dominant) real pole of the open-loop system. (It is assumed that poles at the origin of the system are excluded, and that all other open-loop poles and zeros are in the left-half s-plane.) The frequency response function of the lead compensator is

$$G_c(j\omega) = \alpha \frac{1 + j\omega T}{1 + \alpha j\omega T}$$

and its Bode plots are shown in Figure 4.34 for $\alpha = 0.1$, 0.2 and 0.5. Again, such compensator Bode plots can be added to those of the open-loop system.

In the frequency domain, lead compensation is commonly used to improve a system's phase margin. Improvements in phase margin are accompanied by increases in bandwidth and ω_p, and hence also by an increase in the closed-loop transient response speed. One method of selecting the parameters of the lead compensator is first to choose the desired phase margin, β. The frequency ω_B is then found at which the phase angle of the open-loop plant $GH(s)$ is given by

$$\arg GH(s) = -(180° - \beta + r\delta)$$

where

β = desired phase margin

r = system type number

δ = maximum phase advance of the selected compensator

It may be shown that

$$\delta = \tan^{-1} \frac{1 - \alpha}{2\sqrt{\alpha}} \quad \text{and occurs at a frequency } \omega = \frac{1}{T\sqrt{\alpha}}$$

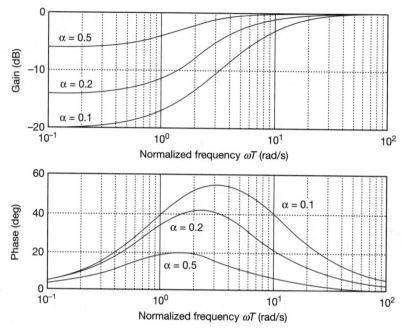

Figure 4.34 Bode plots of a lead compensator for various values of α.

If T is now chosen such that

$$T = \frac{1}{\omega_B \sqrt{\alpha}}$$

then the frequency at which the maximum compensator phase lead occurs will coincide with ω_B. If the gain of the compensated system is adjusted to give a magnitude ratio of 0 dB at ω_B, the system will be found to have the desired phase margin β.

The effect of a phase advance compensator on the polar plots of type 1 and 2 systems is shown in Figure 4.35.

In general, the effects of lead compensation are as follows:

- The bandwidth is increased. This could be a problem if there is significant noise.

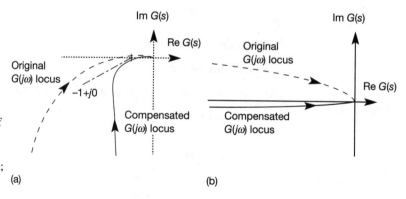

Figure 4.35 The effects of a lead compensator on the polar plots of type 1 and 2 systems. (a) Type 1 system; (b) type 2 system.

- The predominant time constant of the system tends to be reduced, producing faster response times.

- A system's gain and/or phase margin is increased. When increases in phase margin of more than 55° are required (corresponding to $\alpha = 0.1$), a number of lead compensators is cascaded together.

The MATLAB Multivariable Frequency Domain Toolbox (Ford *et al.*, 1990) has a *phlead* command for designing such compensators (*phlag* can also be used as noted above).

Lag-lead compensators

If both the transient and steady-state responses of a system are unsatisfactory, it is usually more economical to use a combined lag-lead compensator instead of individual lag and lead elements. The transfer function of this type of compensator is

$$G_c(s) = \frac{(1 + T_1 s)(1 + T_2 s)}{(1 + \alpha T_1 s)(1 + T_2 s / \alpha)}, \qquad \alpha > 1, \qquad T_1 > T_2$$

Figure 4.36 shows the compensator's pole–zero plot, and Figure 4.37 a sketch of its Bode plots.

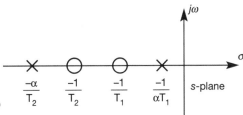

Figure 4.36 Pole–zero map of a lag-lead compensator.

Design techniques for the lag-lead compensator combine the methods previously described for the lag and lead elements.

In general, a lag-lead compensator has all the advantages of both lag and lead compensation, and very few of their usually undesirable characteristics. The use of lag-lead compensators therefore makes it possible to meet many system specifications without incurring the penalties of excessive bandwidth or over-sluggish response.

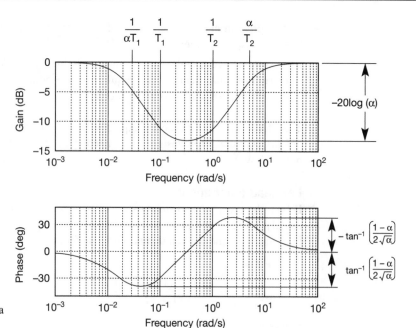

Figure 4.37 Bode plot of a lag-lead compensator.

Example 4.6 *Lead compensation of the antenna positioner*

In Section 2.7, Figures 2.47 and 2.48, a simple model was derived for an antenna azimuth control system. The open loop LTF was found to be:

$$\frac{Y(s)}{U(s)} = \frac{5}{(s+5)(s+1)s}$$

Design a closed-loop controller to give the system zero steady-state error to a step input, a settling time of better than about 5 seconds to within 2 per cent of the steady output and as little overshoot as possible.

As an initial investigation, the root locus plot appears as Figure 4.38(a) and the closed-loop step response with unity negative feedback as Figure 4.38(b) (the open-loop step response does not reveal much, because this is a type 1 system, so the output simply ramps after the initial transient period). Figure 4.38 could have been produced manually, using the rules in Section 4.4.1. In this instance MATLAB was used, and the appropriate m-file will be found as *fig4_38.m* on the accompanying disk (it generates some on-screen instructions).

From Figure 4.38(a) (or by running the MATLAB m-file), it is found that the gain for critical stability is $K = 6$ (note that this is *extra* gain, so the LTF numerator would become 30 for this condition). This is easily verified by a Routh test (Section 3.3).

However, Figure 4.38(b) shows that, rather than needing to increase the gain, it must be *reduced*. This is because there is already a large amount of overshoot (about 20 per cent) and the settling time is too long (about 11 seconds). Increasing the gain would only make these parameters worse.

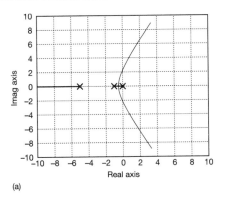
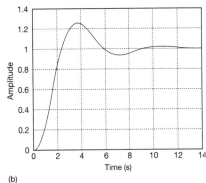

Figure 4.38 Uncompensated antenna-positioner system, Example 4.6. (a) Root locus plot. (b) Closed-loop step response (with UNF).

(a)

(b)

There is no steady-state error problem with this system, so an experiment can be performed to investigate the effect of simply reducing the gain. Applying the magnitude criterion to the root locus plot at the real axis breakaway point (corresponding to unity damping ratio and hence zero overshoot) would give the value $K = 0.2257$ (this value was obtained by running the MATLAB m-file in this case). This implies a corrected numerator term in the open-loop LTF of $5 \times 0.2257 = 1.1285$. However, if this value is used, and the closed-loop response is recalculated, it is found that there is indeed zero overshoot, but the settling time extends to about 15 seconds, which is too long. In fact, there is no reduced value of gain capable of meeting the specification. A proportional-only controller will therefore not suffice and so a dynamic compensator is required.

A lag compensator is used when the transient response is satisfactory, but there is a steady-state error. Neither of these is the case here, so a lag compensator is inappropriate. For the same reasons, a PI controller (which is a special case of a lag compensator) is inappropriate. What is required is a PD controller, to damp out the oscillations and reduce settling time. A real PD controller is equivalent to a lead compensator, so that is the appropriate compensator to try for this system.

The design 'rule of thumb' suggests that the compensator zero should cancel the most dominant non-zero system pole. In this case, this is the pole at $s = -1$, so the lead compensator will be:

$$\frac{(s+1)}{(s+1/\alpha)}, \qquad \text{where} \quad \alpha < 1 \quad \text{(Equation (4.25) with } T = 1\text{)}$$

The controlled system is now as shown in Figure 4.39 and it remains only to choose the value of α. This was carried out by trial and error using the MATLAB m-file, and a value of $\alpha = 0.5$ gave the response of Figure 4.40.

Note that this compensator is only just capable of meeting the specification. By setting $\alpha = 0.56$, a settling time of just under 4 seconds can be achieved. With faster responses, the overshoot becomes too large and the settling time too long.

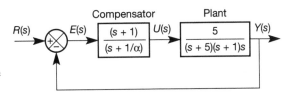

Figure 4.39 Lead compensator applied to the antenna positioner.

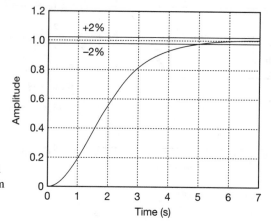

Figure 4.40 Compensated step response of the system of Figure 4.39, with $\alpha = 0.5$.

Example 4.7 *Velocity feedback control of the antenna positioner*

Since the velocity signal appears in the antenna-positioner model (Figure 2.48) and could presumably be made available to the controller by use of a tachogenerator, it would be particularly easy to try to control this system using velocity feedback. The arrangement is shown in Figure 4.41.

A root locus plot can be used to investigate the behaviour as K_v varies. For this purpose, a closed-loop characteristic equation is needed of the form:

$$1 + K_v G(s) = 0$$

From Figure 4.41, the overall forward path transfer function can be seen to be:

$$\frac{Y(s)}{E(s)} = \frac{1}{s} \left[\frac{5}{(s+5)(s+1) + 5K_v} \right]$$

With unity negative feedback, the resulting closed-loop characteristic equation is:

$$1 + \frac{1}{s} \left[\frac{5}{(s+5)(s+1) + 5K_v} \right] = 0$$

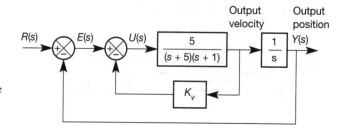

Figure 4.41 Velocity feedback applied to the antenna positioner (Example 4.7).

and may be rearranged to give the required form:

$$s(s+5)(s+1) + 5K_v s + 5 = 0,$$

$$\text{so} \quad 1 + \frac{5K_v s}{s(s+5)(s+1) + 5} = 1 + \frac{5K_v s}{s^3 + 6s^2 + 5s + 5} = 0$$

Again using MATLAB, the resulting root locus plot for varying K_v appears as Figure 4.42, which can be generated using the m-file *fig4_42.m* on the accompanying disk.

This root locus shows that the system will remain stable for any positive value of K_v. However, the greatest relative stability will occur at the point indicated on the plot, where $K_v \approx 0.98$. With this value of K_v, there will be no overshoot in the closed-loop step response. This analysis is also carried out by *fig4_42.m*.

However, the settling time definition allows about 2 per cent overshoot, so the value of K_v can be slightly reduced to give a faster response. After a little experimentation, a value of $K_v = 0.85$ gives a response similar to that of Figure 4.40, but with a settling time of about 3.7 seconds (given modelling inaccuracies, predictions of this slight an improvement may bear little resemblance to what would happen in practice).

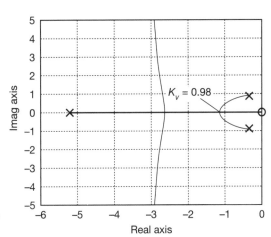

Figure 4.42 Root locus for varying K_v in Figure 4.41.

4.6 The Smith predictor

So far the problem of controlling plants having significant time delays has not been considered. In Chapter 3 (Section 3.7), it was shown that the time delay in a system would add to the phase lag at a given frequency without altering the magnitude. This increase in the phase lag reduces the system's stability margins and can make the system difficult to control. A potential solution for some plants is to adopt a digital control strategy (see Section 7.7).

In practice unfortunately, there is little to choose between the performance of digital and continuous controllers for plant having a significant time delay. Indeed, for disturbance inputs, unless the digital controller samples very rapidly indeed, it can be worse than a continuous controller; the reason being that the sampled system takes longer to detect the disturbance. Under these circumstances one often very successful solution is to use a Smith predictor (named after O. J. M. Smith, who published the method in 1958).

Consider a plant having the transfer function

$$G_p(s) = G(s)e^{-\tau s}$$

The Smith predictor attempts to remove the effect of the τ second time delay (represented by the LTF $e^{-\tau s}$) from the closed-loop control system, so that the controller can be designed as if there were no time delay present. To achieve this, Smith proposed the control scheme shown in Figure 4.43. Ignoring the dashed lines, what happens in Figure 4.43 is that a delay-free model of the plant is used to

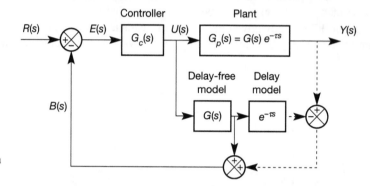

Figure 4.43 The Smith predictor arrangement.

generate the output signal which would exist if the delay were absent (assuming a good model). This delay-free signal is then used in the usual feedback loop (via $B(s)$), instead of the plant output. To help to account for errors in the delay-free model, the delay itself is also modelled, and used to generate what should be a model of the actual plant output, including the delay effect. The dashed lines show how this is then compared with the actual output, $Y(s)$, so that a *modelling error* is also fed back into the control loop via $B(s)$. In this way, the effects of errors in the model of $G(s)$ are reduced.

Figure 4.44 shows an alternative representation of Figure 4.43, in which the various feedback elements have been combined by block diagram reduction, to

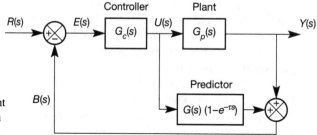

Figure 4.44 An equivalent arrangement for the Smith predictor.

show a *predictor* block, which effectively predicts the system output. Analysing this structure by further block diagram algebra shows that:

$$Y(s) = G_p(s)U(s)$$

and

$$B(s) = G(s)(1 - e^{-\tau s})U(s) + G_p(s)U(s) = G(s)U(s) \tag{4.26}$$

Eliminating $U(s)$ from Equation (4.26)

$$B(s) = \frac{G(s)}{G_p(s)} Y(s)$$

or

$$B(s) = e^{+\tau s}Y(s)$$

That is, the predictor block predicts the effect of the manipulated variable $U(s)$ on the plant output $Y(s)$ and adjusts the feedback signal accordingly.

Further examination of Figure 4.44 allows the stability requirements of the scheme to be assessed. In fact, Figure 4.44 turns out to be similar to an arrangement known as the 'internal model control' (IMC) scheme, examined in Section 13.3. Although it can be analysed by the approach of Chapter 13, block diagram reduction will serve for the present. First, imagine moving the $G_p(s)$ block to the right of the feedback take-off point in Figure 4.44, and adding a copy of it into the vertical feedback path to compensate (as in entry 3 of Table 2.11). This gives two parallel feedback paths, feeding back $U(s)$, whose transfer functions can simply be added to give $G_p(s) + G(s)(1 - e^{-\tau s})$. Now, from the original definitions, $G(s)e^{-\tau s} = G_p(s)$. Therefore, this block becomes simply $G(s)$. The overall transfer function is then easily seen to be given by:

$$\frac{Y(s)}{R(s)} = \frac{G_c(s)}{1 + G_c(s)G(s)} G_p(s) \tag{4.27}$$

Making the following substitution:

$$K(s) = \frac{G_c(s)}{1 + G(s)G_c(s)}$$

the closed-loop transfer function becomes:

$$\frac{Y(s)}{R(s)} = K(s)G_p(s) \tag{4.28}$$

Noting that $G_p(s)$ does not appear in $K(s)$, Equation (4.28) shows that for the closed-loop system to be stable both $K(s)$ and $G_p(s)$ must be stable. Therefore a Smith predictor can only be used with open-loop stable plant. In addition, Equation (4.27) shows that in terms of stability (pole locations) the Smith predictor effectively removes the time delay.

Therefore, the stability of the closed-loop system is identical to that obtained if the plant's time delay had originally been ignored. However, the closed-loop transfer function contains a time delay in its numerator (since $G_p(s) = G(s)e^{-\tau s}$ in Equation (4.27)) and this will modify the response of the system compared with that of the time delay free plant. Like all control system design methods this technique depends upon an accurate plant model (particularly, an accurate model of the delay).

4.7   Some notes on controller implementation

Compensators, and the controllers which contain them, are often implemented digitally, see Section 5.8 and Chapter 7. In this section, the basis of analog implementation will be considered.

In Chapter 2 (Figure 2.15 in Example 2.4 of Section 2.4.2) an electrical lead compensator was constructed from passive components. Similarly, Figure 2.20 in Example 2.7 (Section 2.4.3) showed a mechanical lead compensator constructed from springs and dashpots. Rather than considering all the various realizations, a single example is selected and used to illustrate some important aspects of implementation.

Consider the electrical lead compensator analysed in Chapter 2, which had the differential equation model (see Equation (2.16)):

$$R_1 R_2 C \frac{d}{dt} v_o(t) + (R_1 + R_2)v_o(t) = R_1 R_2 C \frac{d}{dt} v_i(t) + R_2 v_i(t)$$

Taking Laplace transforms with zero initial conditions and gathering terms together gives:

$$(R_1 R_2 Cs + R_1 + R_2)V_o(s) = (R_1 R_2 Cs + R_2)V_i(s)$$

or

$$\frac{V_o(s)}{V_i(s)} = \frac{R_1 R_2 Cs + R_2}{R_1 R_2 Cs + R_1 + R_2} = \frac{s + 1/T}{s + 1/(\alpha T)} \qquad (4.29)$$

where

$$T = R_1 C$$

and

$$\alpha = \frac{R_2}{R_1 + R_2}$$

This result is in agreement with Equation (4.25), confirming that the circuit is a lead compensator.

If an electrical compensator is to be used, then presumably the plant to be compensated has electrical input and output signals. The electrical input signal requirements (in terms of voltage range and current capacity) must be matched by the compensator. Also, the lead compensator has an *insertion loss* factor of α (the d.c. gain of the LTF of Equation (4.29) can be seen to be α, and $\alpha < 1$). Some amplification will therefore usually be necessary, both to recover the lost gain due to inserting the compensator, and to boost the power to a sufficient level to drive the plant input.

It is normal to combine the compensator of Figure 2.15 with an *operational amplifier*, in order to recover the insertion loss (an operational amplifier is an extremely common electronic 'building block' whose principal features are a very high open-loop gain, very high input impedance and very low output impedance). Simply following the circuit of Figure 2.15 with a high input impedance amplifier (so as not to load R_2) is certainly an option. However, the operational amplifier circuit of Figure 4.45(a) is a neater way of solving the problem, and has the added

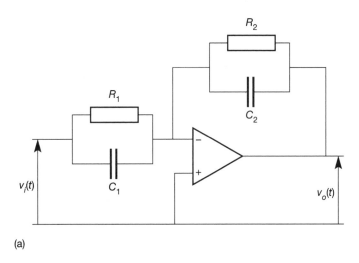

(a)

Figure 4.45 Analog electronic lead or lag compensator. (a) Operational amplifier circuit arrangement; (b) equivalent arrangement for LTF evaluation.

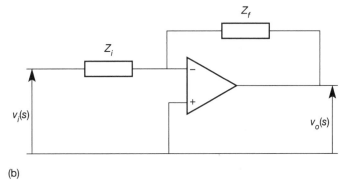

(b)

advantage that it can actually be either a phase lead or a phase lag compensator, depending upon the component choices. In either case, the transfer function of such an operational amplifier circuit is obtained by viewing it as in Figure 4.45(b), and writing the LTF as

$$\frac{V_o(s)}{V_i(s)} = -\frac{Z_f(s)}{Z_i(s)} \tag{4.30}$$

where $Z_f(s)$ and $Z_i(s)$ are the complex (Laplace transformed) impedances of the feedback and input networks, respectively. This result follows directly from the basic characteristics of operational amplifiers.

For each of the parallel RC combinations in Figure 4.45(a), the combined impedance is found by putting the resistance in parallel with the capacitive reactance. This reactance is $X_c = 1/(j\omega C)$ which, in the Laplace domain, becomes $1/(sC)$.

Thus, by the normal rules for calculating parallel impedances:

$$\frac{1}{\text{parallel impedance}} = \frac{1}{Z} = \frac{1}{R} + \frac{1}{X_c} = \frac{1}{R} + sC = \frac{1 + sCR}{R}$$

or

$$Z = \frac{R}{1 + sCR} \tag{4.31}$$

Using the component designations from Figure 4.45(a) in Equation (4.31) and substituting the results into Equation (4.30) gives the compensator LTF as

$$\frac{V_o(s)}{V_i(s)} = -\frac{R_2/(1 + sC_2R_2)}{R_1/(1 + sC_1R_1)} = -\frac{R_2}{R_1} \frac{(1 + sC_1R_1)}{(1 + sC_2R_2)} \tag{4.32}$$

Equation (4.32) can be compared with the LTFs of both the lag and lead compensators given earlier, and made equivalent to either by careful choice of the four component values. For a lead compensator, it is possible to make Equation (4.32) look like Equation (4.25) by rearranging it as follows:

$$\frac{V_o(s)}{V_i(s)} = -\frac{C_1}{C_2} \frac{[s + 1/(C_1R_1)]}{[s + 1/(C_2R_2)]} \tag{4.33}$$

Comparison with Equation (4.25) then shows that $T = C_1R_1$ and $\alpha = (C_2R_2)/(C_1R_1)$. Furthermore, using the final value theorem for the response to a step input in Equations (4.32) or (4.33) shows that the magnitude of the d.c. gain of the compensator will be R_2/R_1 (which could also be seen directly from Figure 4.45(a), since the capacitors have no effect at steady-state d.c.). The requirement that α must be less than unity for a lead compensator is also catered for by appropriate component selection.

Note also the following points:

- Unless the plant actuation system has built-in power amplification, it is likely that more power will be required to drive the plant input than a standard operational amplifier can provide (that is, more than a few mW). There are then two options. A power operational amplifier can be used. These are readily available with output ratings of up to ±30 V at a few amperes – but whereas standard operational amplifiers are extremely cheap, good high power versions are relatively costly. Alternatively, an extra power amplifier can simply be inserted between the compensator and the plant input.

- Note that there is a minus sign in the compensator transfer function. This cannot be ignored, as it produces 180° of phase shift. It is considered in the following example.

- From the point of view of good electronic design, the direct connection from the non-inverting input of the operational amplifier, down to the 0 V common rail (that is, down to the bottom line in Figure 4.45(a)), should strictly be replaced by a resistor having a value equal to the parallel combination of R_1 and R_2. The purpose of this is to minimize offset effects due to imperfections in the operational amplifier.

- Not shown in Figure 4.45 are the power supplies for the operational amplifier.

These are often omitted from such circuit diagrams, on the understanding that ± 15 V (or some other suitable voltage) supplies are connected to the two appropriate pins of the amplifier chip. There is no zero volts supply to standard operational amplifiers, except for deliberate referencing of one input to a common 0 V line, as in the case of the non-inverting input in Figure 4.45(a).

Example 4.8 *An analog implementation of a lead compensator for the antenna positioner*

In Example 4.6, a lead compensator for the antenna positioner was designed. It had values of $T = 1$ and $\alpha = 0.5$. Now assume (for illustration purposes) that the input range to the drive system of the antenna positioner is ± 10 V, and that the actual antenna position is measured by a $360°$ potentiometric transducer, connected across a ± 15 V supply. Design an analog electronic system to implement the controller.

For the sake of this example, assume that the antenna can be rotated through a full $360°$. This would probably not be the case in practice for two reasons:

(1) A $360°$ measurement potentiometer must have a slight gap in its track, in order to allow the two 'ends' of the power supply to be connected (see Figure 4.46). Further, the wiper cannot be allowed to come to rest in this gap, otherwise control will be lost. Note that helical potentiometers, allowing several revolutions, are available.

(2) If complete rotation were allowed, there would be the possibility of more than one revolution in any given direction, thus making reliable electrical connections costly.

Next assume that the middle of the measurement potentiometer track is arranged to coincide with a $0°$ datum position, so that the potentiometer output will be zero volts at this position as shown in Figure 4.46. The total possible output position range is then (just less than) $\pm 180°$, and corresponds to the ± 15 V of the transducer power supply.

The next assumption is that the setpoint will be generated using a similar $\pm 180°$ potentiometer, connected to the same ± 15 V supply. Note that the following detailed design would be different if the setpoint potentiometer were to be fed from a ± 10 V supply so as to match the drive system input. Note also that the setpoint potentiometer and the sensor potentiometer must be good quality servo potentiometers with 'quiet' tracks.

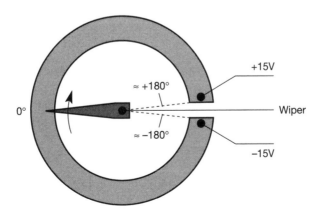

Figure 4.46 Sensor potentiometer for Example 4.8.

This is because the lead compensator (being roughly equivalent to proportional plus derivative action) will tend to amplify electrical noise (see Figure 4.45(a)).

To implement the comparator, another operational amplifier can be used. A sign inversion will be required in the comparator operational amplifier, to cancel out that in the LTF of the compensator (the minus sign in Equation (4.33)). The error signal should be $R - Y$, but the inversion of this is $Y - R$, which is the error signal which must be formed. This can then be connected to the compensator of Figure 4.45(a) to complete the arrangement of Figure 4.39.

Note that if the setpoint potentiometer is at $0°$ (the middle of its track, where $R = 0$ V), then the error signal $(Y - R)$ can swing between ± 15 V (the range of Y). However, if the setpoint is at one end of its track (very near to -15 V say) then the error signal $(Y - R)$ can swing between $+30$ V and 0 V as Y changes between ± 15 V. Similarly, for R very near to $+15$ V, the error could swing between 0 V and -30 V. The error signal can therefore apparently cover twice the range of the setpoint signal. While this is theoretically true, it actually represents an error range of $\pm 360°$, or two complete revolutions. Such errors are impossible, given the assumptions, and it may be taken that the error signal also has a working range of ± 15 V (which corresponds to one complete revolution, so real error signals should be much smaller).

The only difficulty which might be encountered with this assumption is if sudden changes of setpoint can be made of more than $180°$ magnitude. Such a change would cause saturation of the error signal. In these cases, large changes must be *rate limited* so that they are applied as a gradual ramp, rather than a sudden step, thus allowing the output to 'keep up' with the setpoint, and keep the error magnitude relatively small. Adding such a rate limiter between the setpoint potentiometer and the comparator would be equivalent to adding the *pre-filter* block P in Figure 1.5.

It only remains to calculate the component values for the compensator. Recall that the compensator's d.c. gain is set by R_2/R_1. In this example, the requirement is to match the ± 15 V error signal to the ± 10 V plant input, so $R_2/R_1 = 2/3$.

The value of T is 1 second, so $C_1 R_1 = 1$.

The value of α is 0.5, so $0.5 = C_2 R_2/C_1 R_1 = C_2 R_2$.

These can be summarized (in terms of numerical magnitude) as:

$$R_1 = 1/C_1 \qquad R_2 = 2R_1/3 \qquad C_2 = 1/(2R_2)$$

Using these relationships can give impractical component values as C_1 has to be large to keep the resistor values manageably small. The problem with a large C_1 is that it must be a bipolar type. These tend to be expensive and have poor tolerance at large values. The problem with very large R values is that they tend also to have poor tolerance, and to be noise-prone. However, a compromise is to choose C_1 as 0.5 μF (still relatively 'large' for a bipolar capacitor), which gives $R_1 = 2$ MΩ (a 'large' resistance, but probably usable), $R_2 = 1.33$ MΩ and $C_2 = 0.375$ μF.

Using these components, the final system is as shown in Figure 4.47. Notice that in order to obtain the error signal $Y - R$ (as required above) from the comparator, we have fed $+R$ and $-Y$ to an inverting summing amplifier. The signal $-Y$ has been obtained by the simple expedient of reversing the transducer supply connections compared with Figure 4.46.

The implementation of a lag compensator follows the steps given in Example 4.8. A lead-lag compensator can be obtained by cascading two separate compensators

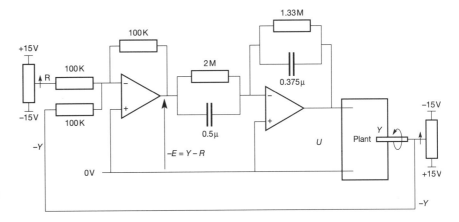

Figure 4.47 Analog implementation of Example 4.8.

(although there are specific lead-lag operational amplifier circuits which achieve the same thing more efficiently).

Similarly, there are special operational amplifier arrangements for three-term (PID) controllers. However, it is easy enough to build the three terms (or two if only PI or PD action is required) separately, each fed from the error signal, and sum their outputs using a summing amplifier (like the one used as the comparator in Figure 4.47). This gives the PID arrangement of Equation (4.22a), in which each term is independently adjustable. Each term takes the form of Figure 4.45(b). For the proportional term, Z_f and Z_i are both resistors, and $K_p = R_f/R_i$ (volts per error volt). For an integral term, Z_f is a capacitor, Z_i is a resistor and $K_i = 1/(RC)$ (volts per second per error volt). For an approximate derivative term, Z_f is a resistor, Z_i is a capacitor and $K_d = RC$ (volt seconds per error volt).

Note that each term will have a negative gain, due to the inversion in the operational amplifier, but that the inversion in the following summing amplifier will correct this. The comparator in front of such a PID controller will therefore *not* need to form the inverse of the error, but must form the error 'proper'. Note also that unless extremely high quality components are used, and good electronic design practice is used when designing the circuit layout, the integral term is likely to 'drift', and the derivative term to be extremely approximate (it will be approximate even with the best possible techniques, for reasons discussed earlier).

A PID controller circuit built along the lines indicated may be fine for laboratory trials, or for use on a pilot plant. However, in an implementation for commercial use there are many other considerations. There is the problem of providing an offset to the controller output, mentioned previously. There is also the problem of obtaining a 'bumpless transfer' from manual control to automatic control, when the PID controller is first switched into the loop. This involves priming the integral term so that the initial controller output matches the manual setpoint at the time of switching on the controller. This, and other similar matters mean that, for most applications, a proprietary three-term controller offers the best solution. (Note that such issues are discussed in greater detail with respect to a digital implementation in Section 9.7.)

4.8 Conclusions

This chapter dealt with several aspects of single-input-single-output control system design. Nyquist's stability criterion was covered, and used as the basis for assessing absolute and relative stability of system models. The root locus technique was then introduced. It was demonstrated that the method can be applied to show the variation of closed-loop poles with any given system parameter. This allows it to be used for three-term (PID) controller tuning (for example), where one parameter at a time can be varied and its effect on the closed-loop poles noted.

The PID controller was studied in some detail. Other common SISO compensators, such as lag and lead compensators and the use of velocity feedback, were also introduced, and demonstrated by worked examples. Some aspects of analog controller implementation have also been discussed.

Finally, the Smith predictor was described as a means of controlling systems with time delays.

4.9 Problems

4.1 On separate Argand diagrams, draw the locus of the functions $f_1(\theta) = e^{j\theta}$ for $+90° \geq \theta \geq -90°$ and $f_2(\theta) = e^{j\theta}$ for $-90° \leq \theta \leq +90°$.

Find the inverse functions $f_1(\theta)^{-1}$ and $f_2(\theta)^{-1}$ and plot the inverse functions on separate Argand diagrams. How would these functions be represented in polar coordinates?

Hint: The function $f_1(\theta)$ matches the large semicircular section of the Nyquist D-contour with R set equal to 1, see Section 4.3.1. To find $f_1(\theta)^{-1}$, select a number of θ values (say $\theta = +90°$, $+45°$, $0°$, $-45°$ and $-90°$). For each θ value find the corresponding rectangular coordinate $(a_1 + jb_1)$ and, using the normal rules of complex algebra, find $(a_1 + jb_1)^{-1}$. These points may be mapped into the $f_1(\theta)^{-1}$-plane.

4.2 Produce the root locus plots for closed-loop control systems with unity negative feedback, and having the open-loop transfer functions given below. In each case the roots of the characteristic equation may be solved directly, and positive values of K assumed.

(a) $\dfrac{K}{s(s+2)}$; (b) $\dfrac{K}{s^2+4s+3}$; (c) $\dfrac{K(s+1)}{s^2+2s+1}$

4.3 A feedback system has the open-loop transfer function
$$GH(s) = \frac{K(s+2)}{s^2-2s+2}$$

Use Evans' rules to plot the root locus diagram for the closed-loop system. Find the gain K that will give the closed-loop system a damping ratio of 0.707. With this value of gain, what would be the system's 5 per cent settling time?

4.4 For the following closed-loop characteristic equations, plot the root locus diagrams for positive values of K:

(a) $s^3 + (2+K)s^2 + (4-2K)s + K = 0$

(b) $s^3 + (5+K)s^2 + (6+K)s + 2K = 0$

(c) $s^4 + Ks^3 + (7K-5)s^2 + 12Ks + 4 = 0$

4.5 A control system with unity negative feedback has an open-loop transfer function given by
$$GH(s) = \frac{K}{s(s+3)(s+6)}$$

(a) Show that the points $s_1 = -0.55 + j3.0$ and $s_2 = 1.1 + j1.4$ are, to a sufficient degree of accuracy, on the root locus. In each case give the corresponding value of K.

(b) Determine the gain K that would give the dominant closed-loop poles a real value of -0.8.

(c) With the gain calculated in part (b), what is the closed-loop transfer function?

(d) Use the graphical method to find the response of this closed-loop system to a unit step input.

4.6 The open-loop transfer function of a closed-loop control system with unity negative feedback is
$$G(s) = \frac{K}{s(s+3)(s^2+6s+64)}$$

Plot the root locus diagram for this system, and hence determine the closed-loop gain that gives an effective damping ratio of 0.707.

4.7   The open-loop transfer function of a closed-loop control system with unity negative feedback is given by

$$G(s) = \frac{90s + 9}{800s^2(5s + 1)}$$

Construct the root locus diagram for this system and hence determine the closed-loop poles.

4.8   A unity negative feedback system has an open-loop transfer function given by

$$G(s) = \frac{K(1 + 5s)}{s(1 + 10s)(1 + s)^2}$$

Draw the Bode diagrams and determine the loop gain K required for a phase margin of 20°. What is the gain margin?

4.9   For the system of Problem 4.8, set the gain K to 1.52 and find the closed-loop transfer function. Draw the closed-loop Bode plot and determine M_p, ω_p, the bandwidth and the cut-off rate.

4.10 Use the Nichols chart shown in Figure 3.59 to obtain the closed-loop Bode plots from the open-loop Bode plots of Problem 4.8.

4.11 A lag compensator (see Section 4.5.4) with the transfer function

$$G_c(s) = \frac{1 + 10s}{1 + 50s}$$

is added to the system of Problem 4.8. Use Bode diagrams to find the reduction in steady-state error following a ramp change in the reference input, assuming the phase margin of 20° is maintained.

4.12 A lead compensator (see Section 4.5.4) with transfer function

$$G_c(s) = \frac{s + 0.1}{s + 0.5}$$

is added to the system of Problem 4.8. Use Bode diagrams to find:

(a) the new phase margin ($K = 1.52$);
(b) the increase in phase margin frequency and the gain K required if the phase margin is returned to 20°.

4.13 Plot the Nyquist diagram for a plant having an unstable open-loop transfer function given by

$$G_c(s) = \frac{K(s + 0.4)}{s(s^2 + 2s - 1)}$$

Determine the range of gain K for which a closed-loop control system with unity negative feedback which incorporated this plant would be stable.

4.14 An integral controller having the transfer function

$$G_c(s) = K/s$$

is placed in the forward path of a control system having unity negative feedback and an open-loop transfer function

$$G(s) = \frac{1}{(s + 1)(2s + 1)}$$

For this system:

(a) determine the value of K that will give a gain margin of 2.5;
(b) using this value of K, plot the full Nyquist diagram and hence prove that the closed-loop system is stable;
(c) find the system's phase margin.

4.15 A closed-loop control system with negative feedback has an open-loop transfer function given by:

$$G(s) = \frac{K}{(s + 3)^3(s + 1)}$$

From the polar plot, determine:

(a) the gain K that will make the closed-loop system marginally stable;
(b) the gain margin when the gain K is adjusted to give a phase margin of 45°;
(c) the phase margin when the gain K is adjusted to give a gain margin of 3.

4.16 Figure P4.16 shows an M circle in the $G(s)$-plane. As usual, the circle has a radius $M/(M^2 - 1)$ and is centred at $M^2/(M^2 - 1)$ (see Section 3.5.2). A line Oa is drawn which is tangent to the M circle at a and passes through the origin of the $G(s)$-plane, as shown. Prove that:

(a) the angle ψ is such that $\sin \psi = 1/M$;
(b) a line perpendicular to the real axis which passes through the point of tangency intersects the real axis at -1. This is the line ab shown in Figure P4.16.

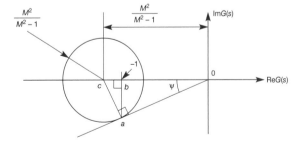

Figure P4.16 M-circle construction.

4.17 The relationships proved in Problem 4.16 are used to design control systems having a given closed-loop peak magnification M_p. The procedure used is first to draw a line from the origin of the $G(s)$-plane at angle ψ, where $\sin \psi = 1/M_p$. By trial and error a circle is found which is centred on the real axis and is tangent to both the $G(j\omega)$ locus and the line at angle ψ. From the tangency point of the line and circle, a line is drawn perpendicular to the real axis. Let this line intersect the real axis at $-T$. Now, if the closed-loop peak magnification is to be M_p, then the loop gain must be adjusted by an amount x, such that $Tx = 1$. Hence to produce the required closed-loop peak magnification, the loop gain must be adjusted by $1/T$.

Construct a polar plot for the system

$$G(s) = \frac{K}{s(1+0.5s)(6+s)}$$

and determine the gain K that will give the closed-loop system a value of $M_p = 1.4$. With this gain, what are the system's gain and phase margins?

4.18 A frequency test on an open-loop system produced the following results:

ω, rad s^{-1}	2.5	3	3.5	4	4.5
Re $G(j\omega)$	-2	-1.75	-1.5	-1.25	-1.0
Im $G(j\omega)$	-2.5	-1.67	-1.17	-0.8	-0.5

Show that M_p has a value of 2.3 when $\omega = 4.7$.

Determine the change in system gain required to give an M_p value of 1.5. What is the new resonant frequency? What is the new damped frequency?

4.19 Construct an inverse polar plot for the system

$$G(s) = \frac{K}{s(1+0.5s)(6+s)}$$

and determine the gain K that will give the closed-loop system a value of $M_p = 1.5$. With this gain, what are the system's gain and phase margins? Also determine the closed-loop bandwidth.

4.20 A closed-loop control system with negative feedback has the open-loop transfer function

$$G(s) = \frac{K}{s^2(s+10)}$$

The closed-loop system is to be stabilized by means of a forward path compensator. The required closed-loop performance specifications are a 5 per cent settling time of 5 seconds, and a subsidence ratio of about $5:1$.

Select an appropriate compensator, and use the root locus method to find suitable values for its parameters.

4.21 Repeat Problem 4.20 using frequency domain methods.

5 A first look at state-space and digital control

5.1 Preview

In Chapter 4, the root locus and frequency-domain approaches to control system design were studied, but time-domain techniques were not considered. There are two reasons for this.

First, Chapter 4 concentrated on simple single-input-single-output (SISO) systems, and these are usually treated by techniques based on frequency response methods. In this chapter, state-space techniques are introduced. These are time-domain methods which also work with SISO systems, but are equally applicable to multivariable systems (that is, systems with multiple inputs and/or outputs – MIMO systems). The techniques of Chapter 4 are not applicable to such systems without severe modification (as explained in Chapter 10).

Secondly, it is normal to implement state-space-based controllers digitally, so these time-domain methods were saved for this chapter, where we also take a first look at digital control, using the z-transform.

Topics covered in this chapter include:

- control of the dynamic behaviour of a system by feeding back simple multiples of the state variables to the input
- descriptions of tests which can be carried out on state-space models to determine whether or not a desired controller can actually be built
- consideration of spare 'degrees of freedom' in state-space designs, which might profitably be used for various purposes
- how the eigenvectors of a system contribute to its time response
- the use of the z-transform to represent discrete-time systems in a transfer function form
- the design of computer control schemes which work by 'digitizing' continuous-time designs
- the use of discrete-time models in digital computer simulations of control loops.

NEW MATHEMATICS FOR THIS CHAPTER

Several aspects of matrix algebra, described in Appendix 1, are used for the first time. Readers unfamiliar with matrix algebra should read Appendix 1 at this point.

The z-transform is introduced for discrete-time systems by making various substitutions for s in the Laplace transfer function (the 'true' z-transform is introduced in Chapter 7).

5.2 The time-domain (state-space) approach to feedback compensation

In previous chapters, the possibility of altering a system's performance by putting a compensator in the feedback path, rather than the forward path, was considered. For example, Section 3.5.2 introduced the inverse polar (Nyquist) plot and Section 4.5.3 looked briefly at velocity feedback.

In time-domain controller designs, the use of feedback compensators is common. In fact, it is the basis for most state-space control designs, which take the state variables of the system and feed them back to the input through a suitable matrix (which effectively plays the part of a feedback compensator).

Although the reasons for doing this will appear different from those given for designing frequency-domain feedback compensators, there can be distinct similarities between the two. For example, consider the frequency-domain model of a position control system, including a velocity feedback loop, as shown in Figure 4.29.

Here, the performance of a proportional plus derivative type of controller is achieved, but without the need to differentiate the output signal explicitly. By avoiding signal differentiation, noise problems are reduced, since the 'velocity' signal is measured directly. Figure 4.29 shows that signals proportional to output *position* and output *velocity* are both fed back.

Now reconsider Figure 4.29 from a state-space viewpoint, and let *position* and *velocity* be chosen as state variables for the system. The two feedback loops are therefore feeding back to the input, signals representing the state variables – hence, it is effectively a *state variable feedback* (or SVF) scheme.

In essence, velocity feedback and SVF improve a system's closed-loop performance by feedback compensation using multiple (appropriate) measurements. This chapter sets out the basis for the design of feedback control systems using state-space models and SVF. Before studying how SVF systems work, and how they can be designed, there are some tests which can be carried out on state-space models to inform this design process. Some of these tests are described in the next section, before studying SVF itself in Section 5.4.

5.3 Controllability and observability

In the frequency-domain design techniques studied in Chapter 4, a controller structure was chosen (typically one using a forward path compensator cascaded with the plant, and an overall unity negative feedback loop) and then the compensator contents were designed to achieve the required results. The design was carried out using considerations such as stability margins on frequency response plots, or root locations on root locus plots. Such design procedures can often be iterative – iterating around the 'change-the-design-and-check-the-new-frequency-response' procedure (for example) a number of times, until the result is deemed acceptable. In such an approach, the possibility seldom occurs to the designer that the system may not be able to be controlled at all; it is just a matter of finding out how much performance improvement the controller can be made to provide.

The design of SVF systems (and of most other systems using state-space models) is very different in approach. As will be revealed later, it is a *synthesis* approach, in which the desired closed-loop performance is specified in advance (perhaps by specifying the required set of closed-loop poles, as we shall see). The information about the required performance, together with the state-space model of the open-loop plant, is fed into an algorithm (normally mechanized in a computer program, although Section 5.4 contains 'by-hand' examples to illustrate how the techniques work). The algorithm then produces the details of the required controller (*synthesized* from the information provided). It is possible for certain combinations of poles and zeros within the open-loop plant to prevent the controller design algorithm from working properly. Therefore, in state-space work, it becomes appropriate to ask, in advance, whether or not a certain design is *possible*. This section introduces some of these ideas.

The concept of *controllability* will be needed immediately in Section 5.4 – it tells the designer whether or not a SVF system can be synthesized to give the required performance. The concept of *observability* is not needed until Chapter 9, but is so closely linked to controllability that it is also introduced here (for interest, note that Chapter 9 is concerned with the problem of estimating the values of internal states of a system which cannot be directly measured, and the observability test tells the designer whether or not this can be done).

Consider the system shown in Figure 5.1(a), with input $U(s)$ and output $Y(s)$. Using the block diagram reduction techniques described in Section 2.6, the Laplace transfer function (LTF) model relating the output to the input is found to be:

$$\frac{Y(s)}{U(s)} = \frac{s-1}{s+4}$$

Also indicated in Figure 5.1(a) are the Laplace transforms of two state variables. Describing the same system using these state variables in the time domain (namely $x_1(t)$ and $x_2(t)$) leads to a state-space model of the usual form:

$$\dot{x} = Ax + bu \quad \text{and} \quad y = cx + du$$

or, in full (see Example 2.28 in Section 2.7 if revision is needed of the technique for obtaining such a model):

$$\begin{bmatrix} \dot{x}_1(t) \\ \dot{x}_2(t) \end{bmatrix} = \begin{bmatrix} -3 & -4 \\ -1 & 0 \end{bmatrix} \begin{bmatrix} x_1(t) \\ x_2(t) \end{bmatrix} + \begin{bmatrix} 4 \\ 1 \end{bmatrix} u(t)$$

and

$$y(t) = \begin{bmatrix} -1 & -1 \end{bmatrix} \begin{bmatrix} x_1(t) \\ x_2(t) \end{bmatrix} + 1u(t)$$

Now the LTF model above suggests that the system is stable, with a single pole at $s = -4$. However, it is clear from Figure 5.1(a) that the system should be second order (there are two poles in the system). The eigenvalues of the A matrix in the state-space model (see Section 3.2.1) are $\lambda_1 = -4$ and $\lambda_2 = +1$, indicating that the system does, indeed, have *two* poles, and that one of them is unstable.

What has happened here is that a pole and zero at the same location ($s = +1$) were unwittingly cancelled out during the derivation of the LTF model, whereas the state-space model preserved all the internal behaviour of the system. In frequency-domain control, such cancellations are often made in deriving LTF models; normally in state-space derivations, they are not (occasionally, a 'minimal

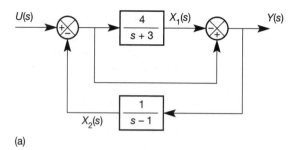

(a)

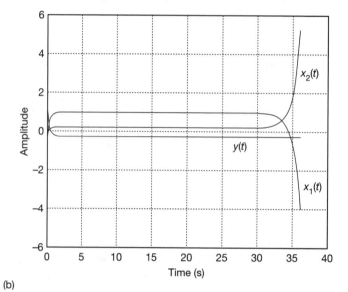

Figure 5.1 A plant for discussion of controllability and observability. (a) Block diagram. (b) Unit step responses.

(b)

realization' is used, in which such cancelling poles and zeros are deliberately removed from the state-space model, but not in this text).

Using control terminology, the unstable mode of response due to the pole at $s = +1$ (that is, e^{+1t}) is said to be *unobservable* at the output $Y(s)$. If the pole were observable, it would appear in the LTF model. The problem is that even though the LTF model makes the system appear to be stable at the output, the unobserved (and unstable) pole at $s = +1$ would cause system failure.

Figure 5.1(b) shows the unit step responses of the state variables and the output. It can be seen that the output appears stable, as predicted by the LTF model. However, it can also be seen that the internal states both behave in an unstable manner, and the system is clearly going to fail within 40 seconds.

The closely related concepts of *observability* and *controllability* describe the effects of identical, but uncancelled, pole–zero pairs in state-space models. Observability was introduced above. Controllability is concerned with the question, 'is it possible to use the input $u(t)$ in some closed-loop scheme which will move the open-loop poles to more favourable locations?'. In the present case (Figure 5.1), is it possible to design a controller to move the unstable open-loop pole at $s = +1$ into the left-hand half of the s-plane, thus stabilizing the system?

Note that the observability of a particular pole is unimportant if information regarding that pole is not required for control purposes (but if the pole is unstable, as above, then the designer must be aware of its existence); while controllability is not important if the open-loop system pole under consideration is already in an acceptable s-plane location.

5.3.1 Controllability

Definitions

A mathematically based definition is that a system is *completely state controllable* if it is possible to cause the state vector to move from any initial value, to any other value, in a finite time.

An alternative (more pragmatic) view of controllability is that a system is completely state controllable if it is possible to move all of its open-loop poles, by state variable feedback, to any arbitrary closed-loop locations (except that complex conjugate pairs of poles must be moved as conjugate pairs).

Note that the two definitions are equivalent, and that the loose term 'controllable' will be used rather than 'completely state controllable' from now on.

Comments

The word 'other' in the first definition is carefully chosen. It does *not* say, 'any *final* value', and the 'other' value may only be transiently achievable. Problem 5.5 illustrates this.

Another point is that although a system may be shown to be theoretically controllable (according to the above definitions), nevertheless, it may not be possible in practice. The reason is that a completely unconstrained system input vector is assumed by the definitions. Thus, the plant input is assumed to be able to take arbitrarily large or small values, and to be capable of arbitrarily large rates of change. These factors are always limited in practice.

Finally, the techniques to be presented for assessing the controllability of a system, even if it is proved to be controllable, give no clue as to how such control may be achieved in practice.

Controllability test

One method of testing the controllability of the state-space model

$$\dot{x} = Ax + Bu, \quad y = Cx + Du$$

involves finding the rank of the following partitioned matrix (see Appendix 1) made up of combinations of the A and B matrices:

$$[B \quad AB \quad A^2B \quad \cdots \quad A^{n-1}B] \tag{5.1}$$

This matrix will often be rectangular, having more columns than rows, and Appendix 1 covers this case. If the test matrix is of full rank (that is, of rank equal to the number of rows in B), then the system is completely state controllable. If it is not of full rank, then the system is only partially state controllable – that is, only a subset of elements of the state vector (or the positions of a subset of the poles) can be adjusted from the input. The *rank deficiency* of the test matrix is equal to the number of uncontrollable modes in the system.

The proof of this test's validity is not given here (see, for example, Friedland (1987)). Computer-aided control system design packages (such as MATLAB) can do the test very easily.

Example 5.1 *Controllability of the antenna-positioning system*

In Example 2.28 in Section 2.7, the following state-space model of an antenna-positioning system was derived (see Figure 2.48(b) and Equations (2.103) and (2.104)):

$$A = \begin{bmatrix} 0 & 1 & 0 \\ 0 & -1 & 1 \\ 0 & 0 & -5 \end{bmatrix}, \quad b = \begin{bmatrix} 0 \\ 0 \\ 5 \end{bmatrix} \quad \text{and} \quad c = \begin{bmatrix} 1 & 0 & 0 \end{bmatrix}$$

Although the design of SVF schemes is not discussed until Section 5.4, imagine that it is desired to use such a scheme to obtain a closed-loop system, whose performance is determined by a set of closed-loop poles at completely different locations from the open-loop pole set. Would such a design be possible? In other words, is the system controllable? For this system:

$$Ab = \begin{bmatrix} 0 \\ 5 \\ -25 \end{bmatrix} \quad \text{and} \quad A^2b = A[Ab] = \begin{bmatrix} 5 \\ -30 \\ 125 \end{bmatrix}$$

To obtain A^2b (for example) by hand, take the existing vector Ab and pre-multiply it by A, as indicated above. In this way, all the columns of the test matrix can be obtained (whatever the order of the system) without raising A to any power higher than unity.

The controllability test matrix is therefore:

$$[b \quad Ab \quad A^2b] = \begin{bmatrix} 0 & 0 & 5 \\ 0 & 5 & -30 \\ 5 & -25 & 125 \end{bmatrix}$$

This has a non-zero determinant and is therefore of full rank (three), so the system is completely state controllable (the actual SVF design is done later, in Example 5.4).

To carry out the rank test using MATLAB (Appendix 3), the following commands may be used:

```
≫ a = [0 1 0; 0 −1 1; 0 0 −5];   % enters the A matrix
≫ b = [0; 0; 5];                 % enters the b vector
≫ test = ctrb(a,b);              % forms the test matrix
≫ rank(test)
```

The *rank* command with no semicolon will display the result, namely the rank of the test matrix, from which the controllability is determined. The last two commands can be merged:

$$\gg \text{rank(ctrb(a,b))}$$

Some further comments on controllability follow Example 5.2.

5.3.2 *Observability*

Definition

A system is *completely observable* if it is possible to reconstruct the state-vector completely from measurements made at the system's output, y (as mentioned earlier, it will be necessary to do precisely this, in Chapter 9).

Observability test

A rank test may also be used to discover whether a system is observable in the above sense. In mathematical language, observability is the *dual* of controllability. This means that the observability test matrix can be generated from the controllability one by using A^{T} instead of A, C^{T} instead of B, and then transposing the result. Sometimes, since transposing a matrix does not affect its rank, the last step is not carried out, giving the following test matrix:

$$[C^{\text{T}} \quad A^{\text{T}}C^{\text{T}} \quad (A^{\text{T}})^2 C^{\text{T}} \quad \cdots \quad (A^{\text{T}})^{n-1} C^{\text{T}}] \tag{5.2}$$

If this matrix is of full rank (that is, the rank is equal to the number of columns in C), then the system is completely observable (so that the values of all the states can be found from information available at the system's output). If it is not of full rank, then the system is only partially observable (meaning that some, but not all, of the system's state information can be obtained from output measurements). The rank deficiency of the test matrix is equal to the number of unobservable modes in the system.

Although Equation (5.2) is easily remembered, since it is similar in form to Equation (5.1), it is often easier to work with the transpose if calculations are being made by hand. Thus, an equivalent observability test is to check the rank of the matrix:

$$\begin{bmatrix} C \\ CA \\ CA^2 \\ \vdots \\ CA^{n-1} \end{bmatrix} \tag{5.3}$$

Example 5.2 *An observability example*

At the beginning of Section 5.3, an introductory example of an unobservable system was given (see Figure 5.1(a)). The observability rank test should be able to predict this unobservability.

From above, the state-space model corresponding to Figure 5.1(a) is:

$$A = \begin{bmatrix} -3 & -4 \\ -1 & 0 \end{bmatrix}, \quad b = \begin{bmatrix} 4 \\ 1 \end{bmatrix}, \quad c = \begin{bmatrix} -1 & -1 \end{bmatrix}, \quad d = 1$$

To form the observability matrix using Equation (5.3) also requires $cA = \begin{bmatrix} 4 & 4 \end{bmatrix}$. The test matrix is then given by:

$$\begin{bmatrix} c \\ cA \end{bmatrix} = \begin{bmatrix} -1 & -1 \\ 4 & 4 \end{bmatrix}$$

This matrix is *not* of full rank, since its determinant is zero. The system is therefore not fully observable. The actual rank of the matrix is one, so the rank deficiency is $2 - 1 = 1$, and there is therefore one unobservable mode.

Using MATLAB to test the observability requires the A and c matrices to be entered in the same way as were A and b in the previous example, and the command:

≫ rank(obsv(a,c))

These tests have confirmed that the system is not fully observable, and the earlier reason suggested for this (that there was a pole–zero cancellation in the LTF model) can now be checked by deriving the LTF from the state-space model.

The LTF model of the system can be found either by using Equation (2.97), or by using the MATLAB command $[num, den] = ss2tf(a, b, c, d)$. The further commands *roots(num)* and *roots(den)* would then give the factors in Equation (5.4), below.

Using the approach of Equation (2.97), the LTF model is given by:

$$\frac{Y(s)}{U(s)} = c[sI - A]^{-1}b + d = \begin{bmatrix} -1 & -1 \end{bmatrix} \begin{bmatrix} s+3 & 4 \\ 1 & s \end{bmatrix}^{-1} \begin{bmatrix} 4 \\ 1 \end{bmatrix} + 1$$

$$= \frac{\begin{bmatrix} -1 & -1 \end{bmatrix}}{s^2 + 3s - 4} \begin{bmatrix} s & -4 \\ -1 & s+3 \end{bmatrix} \begin{bmatrix} 4 \\ 1 \end{bmatrix} + 1$$

$$= \frac{\begin{bmatrix} -1 & -1 \end{bmatrix}}{s^2 + 3s - 4} \begin{bmatrix} 4s - 4 \\ s - 1 \end{bmatrix} + 1 = \frac{-4s + 4 - s + 1}{s^2 + 3s - 4} + 1$$

$$= \frac{s^2 - 2s + 1}{s^2 + 3s - 4} = \frac{(s-1)(s-1)}{(s-1)(s+4)} \tag{5.4}$$

The expected pair of an identical pole and zero is clearly seen, and cancellation of these would yield the original LTF model.

The reason that the plant is not completely observable is that if some disturbance signal excites the mode e^{+t} arising from the denominator term $(s - 1)$, the dynamic effects will be cancelled out by the simultaneous response of the numerator term $(s - 1)$, thus leaving no visible effect at the output. This can be seen happening in Figure 5.1(b). Therefore, by measuring the output, it is not possible to tell that the mode e^{+t} is present in the system.

Incidentally, checking the *controllability* of this plant shows it also to be not completely controllable. The mode e^{+t} therefore cannot be excited from the plant input either. This means that if a disturbance signal affects the mode e^{+t}, there will be no possibility of removing the effects by driving the plant from the input – this would not be an easy plant to control!

In Example 5.2, the system examined was found to contain both uncontrollable and unobservable modes, and these were the same mode in this case. In general, this need not be the case. Imagine a plant LTF model split into two cascaded (that is, series) blocks. In one block is a certain pole (plus other dynamics), and in the other block an identical zero (plus other dynamics which differ from those in the first block). Only if the LTF models of the two blocks are multiplied together will this pole and zero be cancelled.

If the block containing the pole comes *before* that containing the zero, the plant will be controllable (because the mode in question can be driven from the input), but unobservable (because the dynamic response of the zero in the second block will cancel that of the pole, so the associated mode will not be visible at the plant output).

If the block containing the pole comes *after* that containing the zero, the plant will be observable (because the mode in question will contribute directly to the output), but uncontrollable (because the dynamic response of the zero in the first block will cancel that of the pole in the second, so that the pole cannot be driven from the plant input).

Only if the identical pole and zero occur in the same part of the plant will it be simultaneously uncontrollable and unobservable (as was the case in Example 5.2).

This example indicates that controllability and observability are *not* necessarily fundamental attributes of a given overall LTF model of a system (that is, of a system's input–output model), but that they depend upon the particular state-space representation chosen (that is, on the internal structure of the model). Simply by looking at the transfer function of Equation (5.4), no knowledge can be obtained as to how the system is arranged internally. This would only be gained from an appreciation of the real system.

However, any mode cancellations *are* a fundamental attribute of the real plant being modelled. Therefore, if one state-space model of a plant is uncontrollable but fully observable, and a transition is made to a different state-space model (by effectively choosing a different set of state variables) which is fully controllable, then the new model will be found to be unobservable because the mode cancellations will still be present somewhere in the model.

This is another drawback of Laplace transfer function plant models. If identical numerator and denominator terms are cancelled out, the LTF can only represent those subsystems of a plant which are both completely observable and completely controllable (the state-space model corresponding to this LTF model is then called a *minimal realization*, and can be found using the MATLAB *minreal* command).

Finally, note that this also indicates caution when designing frequency-domain controllers which include the concept of pole–zero cancellation. Once again, the apparent order of the final system will be lower than the actual order, with uncontrollable and/or unobservable modes present. However, these modes still actually exist in the physical world – they do not go away when cancelled on paper.

In a real system represented by the model of Figure 5.1(a), all the $(s-1)$ terms in Equation (5.4) physically exist – they are just not seen in an overall LTF model. They could be excited by unmodelled noise or disturbance signals, thus causing behaviour which would be unexpected if only frequency-domain approaches are used.

It is worth noting that mode cancellations are not the only cause of a lack of controllability or observability. For example, repeated eigenvalues (multiple, identical poles) can cause such problems.

5.3.3 Stabilizability

To end this discussion of various tests on state-space models, there is more to be said about systems which are not completely controllable. One very frequent application of state-space methods is to use a feedback system in which the state variables are fed back to the input. One valid reason for doing this is to *stabilize* a system which has some unstable modes in open-loop. For the opening example of Section 5.3 (Figure 5.1), the specification for the design might be, 'move the unstable eigenvalue (pole) at $\lambda = 1$, to $\lambda = -1$' (that is, replace the unstable open-loop mode e^{+t} with the stable closed-loop mode e^{-t}).

If the system is proved to be completely state controllable, then the design can be carried out, by the methods to be described later in this chapter. If the system fails the controllability test, what then?

It is inevitable that any system must have *some* controllable modes, even if only one. This is clear from the rank test, since the test matrix must have a rank of at least one (unless the b vector is empty, implying a system with no inputs which can be driven – and we would not be trying to control such a system!). It is also clear physically – any real system of interest must have an input which can affect at least one of its internal states (again, the b vector would be empty otherwise).

The question then arises as to *which* modes are the uncontrollable ones. In other words, if a pole needs to be moved in order to stabilize the system, is that pole a member of the class of controllable poles, or not? If it is, the system is said to be *stabilizable*, and the design can proceed to move the unstable pole. If the system failed the controllability test, there will be some modes which cannot be affected from the input, but so long as the unstable ones can be moved, the system, although not completely controllable, is stabilizable.

To find out which modes can be moved in this way, and which cannot, a matrix (sometimes called the *sensitivity matrix*, although that term does have other meanings) is formed corresponding to each eigenvalue of the system in turn ($i = 1$ to n for an nth-order system):

$$S(\lambda_i) = [A - \lambda_i I \quad B] \tag{5.5}$$

If $\text{rank}(S(\lambda_i))$ is less than n, then λ_i is an *uncontrollable* mode.

This type of test can also be used to determine which modes of a system are *unobservable*, by using the dual of Equation (5.5) (that is, replacing A with A^T and B with C^T – it is not necessary to transpose the result, as only its rank is of interest).

Example 5.3 *An investigation of stabilizability*

Given a system

$$A = \begin{bmatrix} -1 & 3 \\ 0 & 2 \end{bmatrix}, \quad b = \begin{bmatrix} 1 \\ 1 \end{bmatrix}, \quad c = \begin{bmatrix} 1 & 0 \end{bmatrix},$$

its *stability* can be checked by finding the eigenvalues of the A matrix. In this case, since A is upper-triangular, the eigenvalues are obviously $\lambda_1 = -1$ and $\lambda_2 = +2$. In general the eigenvalues would be found by using the characteristic equation (Section A1.2) or, more likely, by using a computer package – such as the MATLAB command *eig(a)*.

The eigenvalue $\lambda_2 = +2$ implies an unstable mode e^{+2t} in the time response, so it might be decided to try to use a state variable feedback controller (the design of which we discuss below) to move this eigenvalue to some negative value instead.

Checking for *controllability*, it is found that

$$\begin{bmatrix} b & Ab \end{bmatrix} = \begin{bmatrix} 1 & 2 \\ 1 & 2 \end{bmatrix},$$

which is not of full rank, therefore the system is not fully controllable. The rank deficiency is one, so there must be one uncontrollable mode.

The test of Equation (5.5) is then applied to discover which mode is the controllable one. Really, the test need only be applied to the mode it is desired to move, but both modes will be considered here.

$$S(\lambda_1) = S(-1) = \begin{bmatrix} \begin{bmatrix} -1 & 3 \\ 0 & 2 \end{bmatrix} - \begin{bmatrix} -1 & 0 \\ 0 & -1 \end{bmatrix} \begin{bmatrix} 1 \\ 1 \end{bmatrix} \end{bmatrix} = \begin{bmatrix} 0 & 3 & 1 \\ 0 & 3 & 1 \end{bmatrix}$$

and

$$S(\lambda_2) = S(2) = \begin{bmatrix} \begin{bmatrix} -1 & 3 \\ 0 & 2 \end{bmatrix} - \begin{bmatrix} 2 & 0 \\ 0 & 2 \end{bmatrix} \begin{bmatrix} 1 \\ 1 \end{bmatrix} \end{bmatrix} = \begin{bmatrix} -3 & 3 & 1 \\ 0 & 0 & 1 \end{bmatrix}$$

Rank($S(2)$) = 2 (full rank), and rank($S(-1)$) = 1 (rank deficient). Therefore, the mode corresponding with $\lambda_2 = +2$ *is* controllable, while the mode corresponding with $\lambda_1 = -1$ is not. This system is therefore *stabilizable*, because the pole corresponding to the unstable mode can be moved (although the mode for $\lambda_1 = -1$ could not be moved by state-variable feedback, since that mode is the uncontrollable one).

There is no unique MATLAB command for the stabilizability test, but it is easy to construct one. In the code below, the A and b matrices are entered, the eigenvalues of A are saved in a vector (*lambda*), and a *for* loop is used to give the rank of each sensitivity matrix in turn. This could be written in a more elegant and efficient way, but would then be harder for a non-MATLAB-expert to read. Also with this many lines, it would be wise to enter the commands into an m-file for ease of error correction (see Appendix 3). Note that the numerical values in these commands are for the system of Figure 5.1(a). This confirms that that system is *not* stabilizable by state variable feedback:

```
≫ a = [−3 −4; −1 0];   b = [4; 1];

≫ lambda = eig(a)        % no semicolon, so the eigenvalues are displayed
                         % to show the order in which the results will appear

≫ for k = 1:2

   s = [a−lambda(k)*eye(2)  b];   % eye(2) is the MATLAB 2 × 2 identity
                                  % matrix

   rank(s)

   end
```

If a system model proves *not* to be completely controllable, some familiarity with matrix algebra might suggest the use of a similarity transform (Section A1.6) to change to a system model which may be completely controllable – for example to change to the direct-programming type of result in which the companion form is *always* controllable (Section 2.5.1). There is a similarity transform which will, in principle, transform any state-space model to the companion form (see also the description of Ackermann's method, in Section 5.4.7). Unfortunately, it will only work if the system model is controllable already, so the approach is doomed to failure!

The only way it can be achieved generally is to convert the uncontrollable state-space model back to a transfer function model, being careful *not* to cancel any modes in the process, and then go from the LTF model to the companion state-space form by direct programming.

The MATLAB commands below will achieve this:

```
≫ [num,den] = ss2tf(a,b,c,d);

≫ [a1,b1,c1,d1] = tf2ss(num,den);
```

Note, however, that the states of the new model $(a1, b1, c1, d1)$ are related to those of the original model (a, b, c, d) in an unknown manner. Note also that if the original model was uncontrollable, the new one will be unobservable, as discussed earlier.

5.4 State variable feedback

In frequency-domain control, closed-loop systems are typically designed by feeding the plant output back to the input in some way (Chapter 4). The objective is usually to obtain a satisfactory compromise between dynamic (transient) performance and steady-state performance. To give a measure of progress, such indicators as gain and phase margins, rise time, settling time, peak overshoot and steady-state error are used.

To meet such specifications, either the plant itself can be modified (usually not an option), or a controller of some kind can be added (for example, in a cascade arrangement, or in a feedback path).

The mechanism by which such controllers achieve their effects is the introduction of extra poles and zeros into the overall transfer function of the closed-loop (controlled) plant. The most obvious examples are the lead and lag compensators of Section 4.5.4. The locations of such extra poles and zeros are typically not chosen by the designer directly, but result from fixing stability margins based on the inspection of Bode, Nichols or polar plots, or a root locus diagram (all discussed in Chapters 3 and 4). Moreover, if there are undesirable poles in the open-loop system model, they are frequently 'cancelled' by zeros in the compensator which, as has been discussed in this chapter, can lead to unexpected problems.

Sometimes, a pseudo-time-domain design will be done using a root locus diagram, in which a compensator is still designed using frequency-domain transfer functions, but is designed to give a specific required time-domain performance. This would be done by making the two dominant poles of the closed-loop system model (the pair nearest the imaginary axis in the s-plane) have specific locations in the s-plane. These locations would correspond with a certain percentage overshoot and damping ratio in the time response of the closed-loop scheme, and the designer would hope that the other poles and zeros of the system were sufficiently far away to the left in the s-plane so as not to mar the result substantially. Again, such a design would be based on pole–zero cancellations, a typical design rule being to use a lead compensator (if appropriate) whose zero cancels the slowest non-zero open-loop pole of the system (as in Example 4.6).

In time-domain controller design using state-space methods, this choosing of closed-loop pole locations is also a commonly used approach. However, now the desired closed-loop pole locations in the s-plane of *all* the plant poles will be specified. Thus, the aim is to design a controller which effectively moves some (or all) of the open loop poles to the desired closed-loop locations. For obvious reasons, this approach is often called *pole-placement* control.

In order to achieve the desired closed-loop pole locations, a very successful method is to feed back the state vector to the system input in some way, the design of which is the subject of this section. Note that this concept is not entirely alien to frequency-domain control, as was discussed in Section 5.2. However, placing two poles accurately is the best that can usually be achieved by those methods. Any other poles tend to end up in rather random locations.

5.4.1   The basic state variable feedback (SVF) approach

Given the usual state-space representation of a system $\dot{x} = Ax + Bu, y = Cx + Du$, consider the effects of feeding the state vector back to the input. This is done via a feedback matrix K as shown in Figure 5.2, so as to allow any required combination

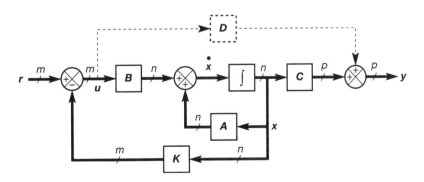

Figure 5.2 State variable feedback – the general arrangement.

of the states to be fed back to each input in the general case. For a system with n states and m inputs, as shown, K must be an $m \times n$ matrix (see Section A1.1.1).

Directly from the figure, it can be seen that with the feedback in place, the closed loop state equation is given by:

$$\dot{x} = Ax + Bu = Ax + B[r - Kx]$$

or

$$\dot{x} = [A - BK]x + Br \tag{5.6}$$

If the D matrix is absent from Figure 5.2, the output equation remains unchanged:

$$y = Cx$$

If the D matrix is present, it can readily be shown that

$$y = [C - DK]x + Dr$$

Assuming, for simplicity, that D is absent, the result is effectively a state-space model of the closed-loop plant, relating r and y, with the same state vector as the open-loop plant, and of the form

$$\dot{x} = A_c x + Br, \quad y = Cx$$

where $A_c = A - BK$ is the closed-loop plant matrix (from Equation (5.6)).

The eigenvalues of A_c are therefore the same as the closed-loop poles of the system. Thus, the elements of K can be chosen to give A_c any required set of closed-loop eigenvalues (poles), *so long as the original system is completely state controllable* (Section 5.3.1).

There is more than one method of choosing the elements of K. Perhaps the easiest to understand is that of making a direct comparison between the required closed-loop characteristic equation (CLCE) and the CLCE which will be obtained from $A_c = A - BK$, and fixing the elements of K to make the two agree. This procedure is illustrated by Example 5.4. Note that this basic method of designing SVF schemes produces *regulators*, which seek to maintain the system at a steady state, and reject disturbances. Their conversion to tracking systems (capable of following varying setpoints) is considered later.

Example 5.4 *A state variable feedback regulator for the antenna positioner*

In Figures 2.47 and 2.48 in Section 2.7, an open-loop system for positioning a communications antenna was introduced. In Example 2.28, a state-space model of the system was obtained, using the physically meaningful state variables shown in Figure 2.48(b). The model appears in Equations (2.103) and (2.104), and is given by:

$$A = \begin{bmatrix} 0 & 1 & 0 \\ 0 & -1 & 1 \\ 0 & 0 & -5 \end{bmatrix}, \quad b = \begin{bmatrix} 0 \\ 0 \\ 5 \end{bmatrix} \quad \text{and} \quad c = \begin{bmatrix} 1 & 0 & 0 \end{bmatrix}$$

Earlier in this chapter (see Example 5.1) this model was shown to be fully controllable. In principle, an SVF controller of the form of Figure 5.2 can therefore be designed to place the closed-loop poles in any desired locations.

Given the dimensions of this model (three states and one input), the 'K matrix' in Figure 5.2 will actually need to be a 1×3 row vector. Writing this as $k = [k_{11} \ k_{12} \ k_{13}]$, the closed-loop plant matrix for use in Equation (5.6) is given by:

$$A_c = A - bk = \begin{bmatrix} 0 & 1 & 0 \\ 0 & -1 & 1 \\ 0 & 0 & -5 \end{bmatrix} - \begin{bmatrix} 0 \\ 0 \\ 5 \end{bmatrix} [k_{11} \ k_{12} \ k_{13}]$$

$$= \begin{bmatrix} 0 & 1 & 0 \\ 0 & -1 & 1 \\ -5k_{11} & -5k_{12} & -5 - 5k_{13} \end{bmatrix}$$

The *actual* closed-loop pole locations will be given by the eigenvalues of this matrix, that is, the roots of the CLCE:

$$|\lambda I - A_c| = \begin{vmatrix} \lambda & -1 & 0 \\ 0 & \lambda + 1 & -1 \\ 5k_{11} & 5k_{12} & \lambda + 5 + 5k_{13} \end{vmatrix}$$

$$= \lambda^3 + (6 + 5k_{13})\lambda^2 + (5 + 5k_{12} + 5k_{13})\lambda + 5k_{11} = 0 \qquad (5.7)$$

Now the *required* CLCE can be specified from the desired closed-loop pole positions and compared with Equation (5.7) to fix the elements of k. This rather begs the question as to where the closed-loop poles should be placed!

A discussion of how to choose a sensible set of closed-loop pole locations in practice is postponed until Section 5.4.8. For the present, a rather arbitrary set of closed-loop pole locations will be specified, designed not particularly to be a sensible choice, but to illustrate various points as the system is reconsidered from time to time in later chapters. The resulting controlled system will have a response with rather more overshoot than would typically be desirable (about 20 per cent, in fact) and a 10–90 per cent rise time designed to be faster than one second.

Bear in mind that the open-loop dynamics contain an element with a one-second time constant, which alone would contribute a rise time of about 2.2 seconds, and if the design attempts to improve the performance too much, very large and fast input changes will be demanded.

For now, the closed-loop poles should be placed at locations in the s-plane given by $\lambda = -1 \pm 2j$ and $\lambda = -10$. Since eigenvalues are the same thing as poles, the required CLCE for the system is therefore:

$$(\lambda + 1 + 2j)(\lambda + 1 - 2j)(\lambda + 10) = \lambda^3 + 12\lambda^2 + 25\lambda + 50 = 0 \qquad (5.8)$$

Comparing the actual CLCE with the required one (Equations (5.7) and (5.8)), the coefficients of λ^3 are equal, so the other coefficients can be compared directly. Doing this, it is found that:

$$6 + 5k_{13} = 12 \qquad \text{therefore } k_{13} = 1.2$$

$$5 + 5k_{13} + 5k_{12} = 25 \qquad \text{therefore } k_{12} = 4 - k_{13} = 2.8$$

$$5k_{11} = 50 \qquad \text{therefore } k_{11} = 10$$

so that $\boldsymbol{k} = [10 \ 2.8 \ 1.2]$.

The feedback signal $\boldsymbol{kx}$ is therefore given by $10x_1 + 2.8x_2 + 1.2x_3$, leading to Figure 5.3, and the unit step response appears in Figure 5.4. These are discussed below. (Section A3.9 describes how MATLAB can produce the plot.)

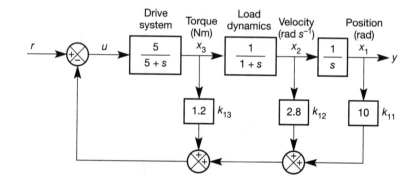

Figure 5.3 The antenna-positioning system with a state variable feedback regulator.

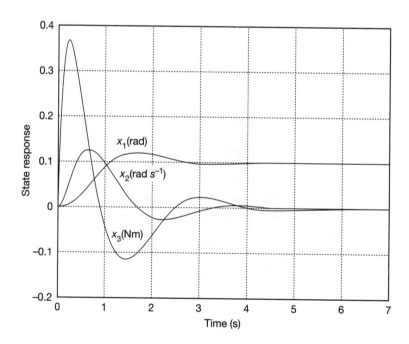

Figure 5.4 Step responses of the arrangement of Figure 5.3.

Although the design is now complete so far as the original specification is concerned, it is instructive to pursue two more aspects of this system before moving on.

Firstly, with a simple SISO system like this one (Figure 5.3), it is easy enough to use normal block diagram reduction methods to check that the closed-loop system is, indeed, the one it was intended to design. If the procedures of Section 2.6.2 are applied to the arrangement of Figure 5.3, the overall transfer function relating the output position y to the reference input r is found to be:

$$\frac{Y}{R}(s) = \frac{5}{s^3 + (6 + 5k_{13})s^2 + 5(1 + k_{12} + k_{13})s + 5k_{11}}$$

$$= \frac{5}{s^3 + 12s^2 + 25s + 50} \tag{5.9}$$

So the CLCE is clearly correct (compare the denominator with Equation (5.8)), and the closed-loop eigenvalues (poles) must therefore be at the desired locations (except for the presence of modelling errors).

5.4.2 Regulator systems – steady-state performance

The second point of interest in Example 5.4 is the steady-state error (s.s.e.) in the system. If this system was expected to follow faithfully the unit step applied at r, it does not seem to perform very well (the output $y = x_1$ in Figure 5.4 settles at only 0.1 unit). This can be confirmed theoretically by applying the final value theorem (Section 2.5.5) to the transfer function of Equation (5.9), when it will be found that the steady-state output following a unit step input is given by

$$\frac{1}{k_{11}} = 0.1 \tag{5.10}$$

This can also be confirmed pragmatically by inspecting Figure 5.3, and noting that the only condition which can produce a steady state is when the integrator input (x_2) is zero (an integrator output will always be ramping unless its input is zero). For this to be the case, x_3 and u must also be zero; therefore $r = 10x_1$ at steady state. If $r = 1$, the output x_1 must therefore be 0.1. This agrees with Equation (5.10).

The reason for this behaviour is that this system is a *regulator*. Regulators are control systems which are designed solely to maintain a plant in a steady state in the face of disturbances which are trying to move it away from that steady state.

Many control systems need to have this regulatory behaviour. For example, many temperature control systems for buildings, ovens, furnaces and other industrial processes are designed to hold the set temperature constant in the face of disturbances. Drive systems in process lines are often required to hold constant speed in the face of load disturbances. Many processes require the maintenance of constant pressures or fluid levels. Sheet material leaving rolling mills usually needs to be produced at constant gauge (thickness) and shape (flatness). Automobile cruise control systems seek to maintain the set speed uphill, or down. Turret-levelling systems for tanks and warships must maintain gun-barrel attitude irrespective of vehicle motion. Inertial navigation platforms must maintain the given heading. Positioning systems for drilling ships and survey vessels must keep station above a given position on the ocean floor. Many control systems in the human body can be modelled as regulators (those for temperature, blood sugar level and light intensity at the retina, to name but three). The list could be almost endless!

The models of such systems used for control system analysis and design are derived *at the required setpoint*, and are studied for *deviations from that setpoint*. For such models, the reference input is therefore zero (because there should be no deviation from the required conditions). For state-space models, it is also arranged that the state vector will have the relationship $x = 0$ at the setpoint. In effect, for a linear system model, the real-world numerical values of the reference input and state vector at the setpoint condition have been subtracted from the block diagram, so as to give the required zero reference and zero state vector at the setpoint (this does not apply to *nonlinear* systems – see Chapter 14). During analysis and design studies, the model then predicts *deviations* of the state vector (and hence the output) from the setpoint values. The setpoint values can be added back in to obtain the real-world values of the output and state vector if necessary.

The simple SVF pole-placement procedure used in Example 5.4 designs regulators. Therefore, the model of Figure 5.3 is assumed to be operating at the required output position. During the modelling process, this required (steady-state) operating position will have been subtracted from x_1, leaving $x_1 = 0$. The reference signal causing the setpoint output will have been subtracted from r, leaving $r = 0$. x_2 and x_3 will similarly be zero. Some texts would place a minus sign on the elements of k, and omit the reference signal and summer altogether, in order to reinforce this point.

Applying a unit step to the system at r therefore disturbs it from its setpoint condition. The regulator attempts to reject this disturbance and return all the state variables to zero, but with the dynamics specified by the desired closed-loop pole locations. In this light, the system behaves quite well. The steady-state output of 0.1 unit shown in Figure 5.4 means that 90 per cent of the disturbance has been rejected from x_1, while all the disturbance is rejected from states x_2 and x_3. The transient behaviour of the system during this disturbance rejection is that specified by the desired pole locations, which were achieved by the feedback system.

If the state-space model of the system of Figure 5.3 is entered into MATLAB (Appendix 3), the linear system simulation command *lsim* can be used to confirm the regulatory action as follows (this is demonstrated by running the m-file *fig5_4.m* on the accompanying disk – its contents can be viewed using any ASCII text editor, or entering *type fig5_4* at the MATLAB prompt):

- Form the closed-loop (SVF) system.
- Set up a reference signal which is zero at every time step.
- Set up a non-zero initial condition state vector.
- Use the *lsim* command to perform the simulation and plot the results.

5.4.3 Tracking systems

In a number of the examples of regulators, given above, it is clear that the setpoint for the system will have to be altered from time to time. The regulatory action must then take place about the new setpoint. If the system is passably linear, and the change of setpoint is relatively infrequent, then the new setpoint values (reference and state vector) can be subtracted from the system model to give a model having zero reference signal and a zero state vector at the new setpoint – the system can then continue to be regarded as a regulator.

However, it will often be the case that the closed-loop system must have zero steady-state error following a step input change (that is, it must exhibit a steady state gain of 1 to a step input). A system which is designed to follow a varying input (reference) signal is called a *tracking* system, or sometimes a *servomechanism*. If a tracking system's reference signal is held constant, it effectively becomes a regulator, so there are not really any fundamental differences between the two types of system, other than the need for a tracker to be able to follow a varying input.

Examples of tracking systems include any of the regulators mentioned before in which a change of setpoint cannot be accommodated by a simple subtraction of values in a linear model; for example, a speed control system in a process line which, in addition to holding a constant speed in the face of disturbances, must start from rest, build up speed according to a predetermined acceleration profile, and slow down in a similar way. Other specific examples include target tracking systems for telescopes, missiles and the like; automatic pilots for aircraft and ships; drive systems which must follow a pre-set trajectory, such as numerically controlled machine tools; and some activity-dependent systems in the human body, such as adrenaline production.

How can the SVF regulator design be altered so that it designs tracking systems? First, a general approach will be presented, and then two other methods will be introduced which are much simpler, but are applicable only to certain classes of systems.

5.4.4 General tracking system design using SVF

The closed-loop regulator designed in Example 5.4 (Figure 5.3 and Equation (5.9)) was of the same order as the original system (both being third order). It is a general property of SVF regulators that they do not increase the system order, whereas dynamic compensators, designed by frequency-domain methods (as in Chapter 4), always increase the system order. This means that SVF has the advantage of simplicity, but the potential disadvantage that the system order may need to be increased if steady-state error (s.s.e.) is to be avoided. This explains why a SVF regulator will not necessarily exhibit zero s.s.e. to a varying input signal.

The requirement for increasing the system order is evident from the frequency domain approaches, where it is often found to be necessary to introduce integral control in order to remove s.s.e. following a step input change (the integral control is added to increase the system type number, but thereby increases the order too).

This gives a pointer as to how s.s.e. might be removed in the SVF arrangement. The approach will be to form the output error and add integral action to remove it, as shown in Figure 5.5. Unlike standard SVF, this does increase the system order due to the extra integrators in the forward path, but hopefully to good effect.

In Figure 5.5, the heart of the system can be seen still to be an SVF regulator with its input at q and its feedback gains in K. The extra components comprise a set of individual integrators (equal in number to the number of plant outputs y, and hence to the number of output reference setpoints in r). These integrators can have zero inputs (that is, zero error between r and y), while providing the non-zero signals at q which are necessary to hold the regulator in the required non-zero state.

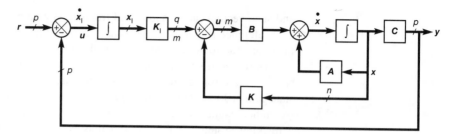

Figure 5.5 SVF tracking system using integral action.

For a system having p outputs, this means that p new integrators will be introduced into the system, and hence p new state variables, which can be called x_I (the 'integral' states). The new $(m \times p)$ matrix K_I is necessary to match the number of these integral states to the number of open-loop plant inputs at u. Also, the new poles due to the integral states will need to be placed in some desirable closed-loop locations, and the elements of K_I give sufficient degrees of freedom to achieve this (that is, the elements are tuneable parameters). In general, the new poles will be placed significantly to the left in the s-plane with respect to the existing ones, so as not to affect substantially the closed-loop dynamics which the regulator has been designed to achieve. However, if they are placed too far to the left, unrealistically large and fast control signals will be called for.

The design process is identical in principle to that employed in Example 5.4. Namely, the block diagram of Figure 5.5 is analysed to find the overall closed-loop state-space model of any system arranged in this form, and then the unknown parameters (the elements of K and K_I) are selected to give the overall closed-loop 'A' matrix the desired set of eigenvalues.

From Equation (5.6), it is known that the SVF system at the heart of Figure 5.5 can be described by the following equation, in which q comes from Figure 5.5, and simply replaces the original r used in deriving Equation (5.6) from Figure 5.2:

$$\dot{x} = [A - BK]x + Bq$$

Now, from Figure 5.5, it is also seen that $q = K_I x_I$ and $\dot{x}_I = r - y = r - Cx$. Combining these equations gives:

$$\dot{x} = [A - BK]x + BK_I x_I \quad \text{and}$$
$$\dot{x}_I = -Cx + r$$

These yield the following overall closed-loop state-space model, where the overall state vector is the original state vector with the new integral states appended to it. Here use is made of *partitioned* matrices, with the partitioning lines omitted for clarity. Such matrices are described in Section A1.3 of Appendix 1.

$$\begin{bmatrix} \dot{x} \\ \dot{x}_I \end{bmatrix} = \begin{bmatrix} A - BK & BK_I \\ -C & 0 \end{bmatrix} \begin{bmatrix} x \\ x_I \end{bmatrix} + \begin{bmatrix} 0 \\ I \end{bmatrix} r \tag{5.11}$$

The eigenvalues of the system matrix of Equation (5.11),

$$\begin{bmatrix} A - BK & BK_I \\ -C & 0 \end{bmatrix}$$

are thus those of the closed-loop (controlled) plant, and can be placed as desired (so long as the system is controllable) by choice of K and K_I. The procedure is illustrated by returning to the example.

Example 5.5 *A state variable feedback tracking system for the antenna positioner*

The antenna-positioning system has only one input and one output (see, for example, Figure 5.3). To arrange it in the form of Figure 5.5, the SVF structure of Figure 5.3 will be required (but with different numerical values for the feedback gains), together with a single unity negative feedback loop feeding back the output to a single reference input, and a single integrator with a scalar gain k_I operating on the resulting error signal. The final system is shown in Figure 5.6, which may be inspected to illustrate the structure, although the numerical values shown for k and k_I are yet to be designed.

The closed-loop plant matrix from Equation (5.11) (with A, b and c as before) is therefore:

$$A_c = \begin{bmatrix} A - bk & bk_I \\ -c & 0 \end{bmatrix} = \left[\begin{array}{ccc|c} 0 & 1 & 0 & 0 \\ 0 & -1 & 1 & 0 \\ -5k_{11} & -5k_{12} & -5 - 5k_{13} & 5k_I \\ \hline -1 & 0 & 0 & 0 \end{array} \right]$$

and so the closed-loop characteristic equation is $|\lambda I - A_c| = 0$, which is easily obtained by expanding the resulting determinant by the last row, and using the previous result from Equation (5.7), as:

$$\lambda^4 + (6 + 5k_{13})\lambda^3 + (5 + 5k_{12} + 5k_{13})\lambda^2 + 5k_{11}\lambda + 5k_I = 0 \qquad (5.12)$$

It is now necessary to compare Equation (5.12) with the *desired* CLCE. Since the same closed-loop pole locations as before are required, the CLCE from Equation (5.8) can be reused, and modified to include the new pole due to the extra integrator. This new pole might be placed at an s-plane location $\lambda = -30$. The required CLCE is therefore:

$$(\lambda + 30)(\lambda^3 + 12\lambda^2 + 25\lambda + 50) = \lambda^4 + 42\lambda^3 + 385\lambda^2 + 800\lambda + 1500 = 0$$

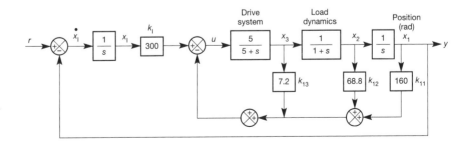

Figure 5.6 The antenna positioning system with a state variable feedback tracker.

and comparing coefficients with those in Equation (5.12) then gives:

$$5k_I = 1500 \qquad\qquad \text{so } k_I = 300$$

$$5k_{11} = 800 \qquad\qquad \text{so } k_{11} = 160$$

$$6 + 5k_{13} = 42 \qquad\qquad \text{so } k_{13} = 7.2$$

$$5 + 5k_{12} + 5k_{13} = 385 \qquad \text{so } k_{12} = 68.8$$

so that $k = [160 \ 68.8 \ 7.2]$ and $k_I = 300$, leading to Figure 5.6.

The resulting unit step response appears in Figure 5.7, and can be seen to exhibit almost the same dynamics as Figure 5.4 (the slight differences are due to the presence of the extra pole arising from the new integrator). However, there is now zero steady-state error to the unit step input, as required.

Note that the MATLAB m-file *fig5_7.m* on the accompanying disk contains a simple modification to allow the plant input signal u to be added to this plot. Doing so indicates that the input signal peaks at a level of about 7 units in response to the unit step reference change. This is within the plant input magnitude saturation limits (if we take one unit to be one volt, and the input range is ± 15 V), but the rate of change is fairly large due to the integral gain of 300 V s^{-1}. Considerations such as these indicate that a unit step is about the most severe change to which this system should be subjected.

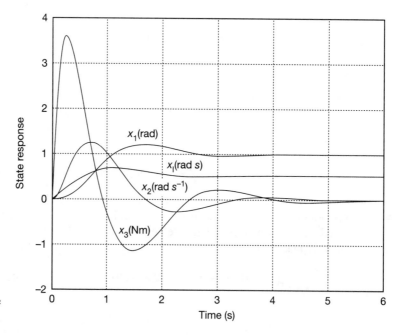

Figure 5.7 Step responses of the arrangement of Figure 5.6.

5.4.5 Tracking system design by gain variation

Equation (5.10) predicted that the steady-state output of the SVF regulator of Figure 5.3, following a unit step input, would be $1/k_{11} = 0.1$. In a simple case such as this, it may be possible to achieve the required output simply by gain variation. Since the output settles at 10 per cent of the input step, the system could apparently be made to behave as a tracking system, and thus produce the traces of Figure 5.7 (except for x_I, which would not be present of course) simply by adding a gain of 10 in series with the reference input r in Figure 5.3 (in other words, in the position of a pre-filter, as shown in Figure 1.5).

This approach would be fine in simulation, but would not be a good idea in practice for two reasons. Firstly, it would potentially place large signal variations on u, which the system may not be able to follow, thus leading to nonlinear responses. Secondly, this gain of 10 would effectively be open-loop. Any error in setting its value, or any variation in its value during use, would be uncorrected by the feedback system, and would therefore result in uncorrectable steady-state errors at the output.

However, the same effect can be obtained by effectively moving the gain inside the feedback loops. Since the steady-state output following a unit step is $1/k_{11} = 0.1$, dividing k_{11} by a factor of 10 ought then to give a steady-state output of unity. However, although the steady-state error depends only upon k_{11}, the closed-loop pole locations are determined by the characteristic equation, which contains every element of $\boldsymbol{k}$. Altering only k_{11} will therefore give zero steady-state error, but will upset the desired closed-loop pole locations.

To maintain these, it is necessary to maintain the gains around each loop at the designed values. A gain of 10 must therefore be added in series with u (that is, in the forward path of Figure 5.3, but *inside* the feedback summer) to compensate for the division by 10 in the feedback path via k_{11}. This extra forward path gain of 10 in turn means that both the gains in the other two feedback loops must be divided by 10. A simple exercise in block diagram manipulation will rapidly show that increasing the forward path gain in Figure 5.3 by a factor of 10, while reducing the three feedback gains by a factor of 10, gives the same overall transfer function as inserting a gain of 10 in series with the reference input, as was originally suggested.

Adding a gain of 10 in the forward path, and using the feedback vector $\boldsymbol{k} = [1 \ 0.28 \ 0.12]$, gives the required responses (those of Figure 5.4, but with the vertical scale multiplied by a factor of 10). These extra gains are inside the feedback loops, and therefore enjoy the benefits of reduced sensitivity to error, common to all negative feedback systems. Also, the final system is still of only third order. However, the maintenance of steady-state error at zero will still not be as reliable as in the 'proper' tracking system of Figure 5.6.

5.4.6 *Tracking system design by use of inherent integration*

The last approach to be considered is quite elegant, but only works for systems which have pure integrators in their forward path in open-loop. Several systems fit this description (but still only a minority of all systems). For example, any system whose output is a linear or angular position derived from a velocity signal, or any system having linear final actuators such as hydraulic cylinders, which effectively integrate flow rate to give position, will be modelled as having a pure integration in their forward path.

In systems such as these, it is sometimes possible to use the built-in integral action to achieve zero steady-state error to a step input, instead of adding extra integrators deliberately. Example 5.6 demonstrates the application of this method to the antenna-positioning system which, due to the output integrator, is a suitable candidate for the method to succeed.

Example 5.6 *A state variable feedback tracker using inherent integral action*

The presence of the integrator in the plant model in Figure 5.3 allows the steady-state condition for the plant to be determined, since the integrator input must then be zero. It has already been noted that this implies $x_2 = x_3 = u = 0$.

From Figure 5.3, the steady-state signal generated at u will thus be simply $r - k_{11}x_1$, and this must therefore be equal to zero. With $k_{11} = 10$ and $r = 1$, this can only happen with $x_1 = 0.1$ (which agrees with the result obtained from the final value theorem in Equation (5.10)).

However, there is another way in which $u = 0$ can be obtained (and hence $x_2 = x_3 = 0$) while maintaining the design values of the feedback gains in k. That is to move the point of application of the reference signal, so that it becomes a reference for the *states* rather than the output. The proposed arrangement is shown in Figure 5.8. Note that reference signals ought also to appear on x_2 and x_3, but they have been omitted, as they should be zero.

Applying a unit step to the reference signal on x_1 in Figure 5.8 now results in the system being driven (with the correct dynamics, since the loop gains are unaltered compared with Figure 5.3) until $x_1 = r$ at steady state – that is, until there is no steady-state error. This works only because the integrator producing x_1 from x_2 can have any desired signal at its output, while its input is steady at zero.

The responses of the system arranged in this manner are indistinguishable from those of Figure 5.4, but with the vertical scales multiplied by 10 as required. The maintenance of zero steady-state error is reliable, as the integral action forces it to happen. The only real drawbacks are firstly, that the plant input at u may again be required to follow very large signals (r is now multiplied by 10 compared with the arrangement of Figure 5.3), and secondly, that the approach only works for the limited class of systems having this type of configuration.

5.4.7 Ackermann's method and the use of MATLAB

The procedure of comparing actual and required closed-loop characteristic equations in order to design feedback gains (as in Example 5.4) is manageable

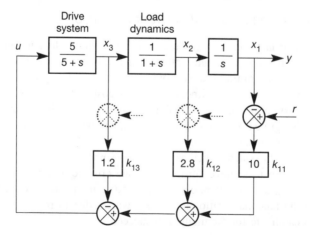

Figure 5.8 Tracking system generation by relocation of the reference input.

by hand for low-order systems, but is not easy for higher-order systems. Nor is it easy to mechanize the procedure in a computer program. In 1972, J. Ackermann published an algorithmic approach to calculating the required feedback matrix for an SVF scheme, using a modified approach. This makes it easy to generate such schemes by use of a computer program.

The method applies only to controllable systems, and was originally developed for single-input-single-output systems only. The method will not be derived here, but will be illustrated in use by hand. In practice, a computer package such as MATLAB would be used to apply the method. For more detail see, for example, Phillips and Harbor (1991). Furuta *et al.* (1988) contains details of an extension of the method to multi-input-multi-output systems. MATLAB has a command *acker* which executes the method for single-input systems (there is a second pole-placement command, *place*, which is numerically more reliable, and also works for multivariable systems and for systems which are not completely controllable).

Ackermann's method is based upon the fact that it is easy to generate the required feedback matrix for a plant whose A matrix is in the controllable canonical companion form (Section 2.5.1). Therefore, assuming the system is fully state controllable, a similarity transform (Appendix 1) is found which puts the plant into companion form, the required feedback matrix is calculated, and the inverse transform back to the original states is finally applied. This is all done in a single procedure, so that the user is not aware of the three distinct steps.

The SISO version of the method could be executed by hand as follows:

- Write down the required CLCE from the desired pole locations.
- Rewrite the required CLCE, substituting the plant matrix (A) for the Laplace operator (s) (or for λ if the CLCE is written in terms of eigenvalues).
- Multiply it out to yield a matrix called $\alpha(A)$.
- Form the controllability test matrix for the plant (Equation (5.1))
$$\mathscr{C} = [\boldsymbol{b} \quad \boldsymbol{Ab} \quad \boldsymbol{A^2b} \quad \cdots \quad \boldsymbol{A^{n-1}b}]$$
- Calculate the required feedback gain vector as
$$\boldsymbol{k} = [0 \quad 0 \quad 0 \quad \cdots \quad 0 \quad 1]\mathscr{C}^{-1}\alpha(A) \tag{5.13}$$

Example 5.7 *State variable feedback regulator design by Ackermann's method*

The design of Example 5.4 is repeated to illustrate the use of Ackermann's method, although this would normally not be done by hand.

The desired closed-loop poles are to be at $\lambda = -1 \pm 2j$ and $\lambda = -10$. The required CLCE for the system is therefore (Equation (5.8)) $\lambda^3 + 12\lambda^2 + 25\lambda + 50 = 0$. The matrices for the state equation are

$$A = \begin{bmatrix} 0 & 1 & 0 \\ 0 & -1 & 1 \\ 0 & 0 & -5 \end{bmatrix} \quad \text{and} \quad b = \begin{bmatrix} 0 \\ 0 \\ 5 \end{bmatrix},$$

so $\alpha(A)$ is:

$$\alpha(A) = \begin{bmatrix} 0 & 1 & 0 \\ 0 & -1 & 1 \\ 0 & 0 & -5 \end{bmatrix}^3 + 12 \begin{bmatrix} 0 & 1 & 0 \\ 0 & -1 & 1 \\ 0 & 0 & -5 \end{bmatrix}^2 + 25 \begin{bmatrix} 0 & 1 & 0 \\ 0 & -1 & 1 \\ 0 & 0 & -5 \end{bmatrix} + 50\boldsymbol{I}$$

$$= \begin{bmatrix} 50 & 14 & 6 \\ 0 & 36 & -16 \\ 0 & 0 & 100 \end{bmatrix}$$

and

$$\mathscr{C} = [\boldsymbol{b} \quad A\boldsymbol{b} \quad A^2\boldsymbol{b}] = \begin{bmatrix} 0 & 0 & 5 \\ 0 & 5 & -30 \\ 5 & -25 & 125 \end{bmatrix} \text{ (from Example 5.1).}$$

The feedback gains are therefore found, using Equation (5.13), to be given by:

$$\boldsymbol{k} = [0 \quad 0 \quad 1]\mathscr{C}^{-1}\alpha(A) = [0 \quad 0 \quad 1] \begin{bmatrix} 0 & 0 & 5 \\ 0 & 5 & -30 \\ 5 & -25 & 125 \end{bmatrix}^{-1} \begin{bmatrix} 50 & 14 & 6 \\ 0 & 36 & -16 \\ 0 & 0 & 100 \end{bmatrix}$$

$$= [10 \quad 2.8 \quad 1.2]$$

which is the same result as was obtained in Example 5.4, leading to the arrangement of Figure 5.4, thus confirming the use of this approach.

Using MATLAB to do the design, which would be far more usual, the commands would be:

> $a = [0 \ 1 \ 0; \ 0 \ -1 \ 1; \ 0 \ 0 \ -5]$;

> $b = [0; \ 0; \ 5]$;

> $p = [-1 + 2j \quad -1 - 2j \quad -10]$; % vector of desired closed-loop pole
% locations

> $k = \text{acker}(a, b, p)$

The *place* command can be used simply by replacing the word 'acker' with the word 'place'.

5.4.8 Choice of locations for the closed-loop poles

The question of where the closed-loop poles ought to be located has not yet been addressed. There are various means of deciding this, but usually a degree of trial and error will be necessary. MATLAB (or other CACSD) simulations of the designed system can easily be persuaded to display the driving inputs to the plant. This is done by defining new system outputs (which will become the same as the inputs it is required to display) and setting appropriate elements of $\boldsymbol{D}$ in the resulting state-space model to unity, thus connecting the inputs of interest to the extra system outputs whose responses can then be displayed. The rows of the C matrix corresponding to the new outputs would be set to zero. The file *fig5_4.m* on the accompanying disk, for example, contains this method for adding the input signal to Figure 5.4. The required modifications can be seen by viewing the file in any ASCII text editor.

The input signals can thus be easily inspected for unrealistic amplitudes or rates of change, and the pole locations adjusted in the light of the guidance below, until matters are satisfactory.

A common approach is to adopt the philosophy of approximating the desired response by that of a second-order model. The response can thus be specified in terms of damping ratio and natural frequency, leading to a second-order transfer function whose characteristic equation will provide the desired pole locations. This only specifies locations for the two dominant closed-loop poles, of course. Usually there will be several more! In order to maintain the performance specified by the dominant poles, any other poles must, by definition, be less dominant. They must therefore be placed as far to the left in the s-plane as possible. However, there are some general rules to bear in mind when doing this, if excessive control actions (which may well cause the plant input to saturate in terms of either amplitude, or rate of change, or both) are to be avoided. A knowledge of how root locations in the s-plane affect transient response is vital (see Section 4.4 on the root locus method).

The most general rule is to move the open-loop poles as little as possible. If an open-loop pole location is acceptable in closed-loop, do not move it. The further a pole is moved, the more control energy will be required. This is because higher gains are required in the feedback matrix to move a pole further, and therefore larger signals are applied at the inputs.

It will also be found that the feedback gains increase as the controllability of a system decreases – a system that is only weakly controllable will require a lot of energy to control it. This is because weak controllability implies a pole rather close to a zero (almost cancelled in the transfer function). From Section 4.4 on the root locus method, it is known that zeros attract poles, so it is always going to take a lot of control effort to separate a pole from a neighbouring zero.

Another useful rule is that if right-half-plane poles are to be moved (in order to stabilize the system), then left-half-plane locations which are a direct reflection in the imaginary axis are often a good starting point.

If the desired pole set contains some relatively fast poles (to the left in the s-plane) and some slow ones (close to the imaginary axis), then the fast ones will require a lot of control input to make the transients die away rapidly, but the slow ones, requiring much lower inputs, will dominate the response. This is not an efficient use of the control inputs. The most efficient expenditure of control energy is achieved if the closed-loop poles are the same distance from the origin in the s-plane (that is, they lie on a semicircle).

If the poles are to be located on a semicircle, the result could be the configuration of a Butterworth filter. If the required radius of the semicircle is ω_n (remembering that circles in the s-plane are contours of constant undamped natural frequency), and if it is required to place n poles, then the pole locations will be given (Friedland, 1987) by the roots of:

$$\left(\frac{s}{\omega_n}\right)^{2n} = (-1)^{n+1} \tag{5.14}$$

where ω_n is chosen to give a high enough rate of response. The roots of this equation imply a set of poles, on a semicircle of radius ω_n, symmetrically spaced about the negative real axis, and with the extreme rightmost ones at angles of $(90/n)$ degrees to the imaginary axis. This means that as the number of poles to be placed (n) increases, this rightmost pair will occur very close to the imaginary axis. For stability, it will then be wise to move them somewhat further to the left in the s-plane than Equation (5.14) would suggest.

Other sets of polynomials also give roots useful as closed-loop pole sets (Franklin *et al.*, 1994).

The methods of optimal control (Chapter 12) will select the closed-loop pole locations in a SVF scheme automatically.

5.5 Multivariable systems – links with other topics

To cope with systems having several inputs and/or outputs, no modifications are needed to any of the methods described in this section (except that Ackermann's method as described above is for single-input systems only – the MATLAB *place* command can be used instead).

The only difference is that when the characteristic equation of the closed-loop plant matrix $[A - BK]$ is compared with the desired closed-loop characteristic equation (as was done in Example 5.4, for instance), a set of homogeneous equations will result. That is to say, there will be more unknowns in the equations (the elements of K) than there are equations to solve for them – leading to an infinite number of possible solutions. This is entirely predictable.

For a single-input nth-order system, there are n poles to be placed and n states to be fed back via k to the single input. k is therefore of size $1 \times n$, and has n unknown elements to be found from the n simultaneous equations which result from the comparison of the two nth-order CLCEs, so the solution is unique. However, if the nth-order system (so comparison of the CLCEs will still generate n simultaneous equations) now has m inputs, K must now feed back the n states to the m inputs, so it must be of size $m \times n$. There will thus be $(m \times n)$ unknowns to fix using only n equations. The same happens when stabilizing an uncontrollable system, as there must be at least one unmovable pole, thus leading, once again, to fewer equations than unknowns.

This excess of unknowns can be used in a number of ways. The most common are:

- Assign arbitrary values to elements of K until only n unknowns remain to be solved for. This can be useful if some elements of K can be set to zero. If it is possible to set all the elements in one column of K to zero, no connection need be made to that state variable, as no feedback is required from it. If the state is unmeasurable, this may avoid the need to build a state estimator (Chapter 9). Similarly, if one row of K can be set to zero, no connection need be made back to the corresponding input. Of course, the practical desirability of deciding not to use any information from one of the state variables, or not to control at an available input, must be assessed before such design decisions are made.

- Use the excess feedback gains to try to control the directions of the closed loop *eigenvectors* as well as the positions of the closed-loop eigenvalues. This gives even more possibilities for shaping the closed-loop time response (as Section 5.6 will show). However, whereas there is complete freedom in choosing the closed-loop pole locations (subject to controllability, moving complex pairs together and not doing anything practically unrealistic), it is not possible to specify closed-loop eigenvector directions at will, since the closed-loop eigenvectors are constrained to lie in a certain sub-space of the state-space. One use of the technique is in systems with uncontrollable poles. Even though an eigenvalue cannot be moved in such a system, its associated eigenvector can probably be altered so as to minimize the contribution of the uncontrollable mode to some particularly sensitive state variable of the system (for example, its output). For more information on entire eigenstructure assignment consult D'Azzo and Houpis (1995).

- Rather than using a pole-placement approach, use some kind of optimization procedure with the flexibility to calculate all the feedback gains, so as to obtain the best possible performance in some predefined sense. Chapter 12 on optimal control pursues this approach.

5.6 Eigenvectors of the plant matrix, and their contribution to the time response

Although space does not permit coverage of entire eigenstructure assignment (see above), the rôle of the eigenvectors of a system in shaping its time response will be described.

Since the eigenvalues of the A matrix are characteristic values of the system model, whatever state vector is chosen, it might be expected that the eigenvectors (Section A1.2) will be, too. It turns out that they are not! A different choice of state variables (leading to a different A matrix) normally leads to a different set of eigenvectors, even though the eigenvalues remain the same. What then is the rôle of the eigenvectors?

In Appendix 1 a geometrical interpretation of eigenvalues and eigenvectors is given, with reference to Figure A1.6. This reinforces the general point that the eigenvectors only specify the *direction* of something (so they are vectors in the sense of physics, as well as the sense of mathematics). Their lengths are not defined.

Example 5.8 **Eigenvectors of the antenna-positioning system**

Using the methods of Appendix 1, find the eigenvectors for the original state-space model of the antenna-positioning system (Equations (2.103) and (2.104)).

The original system A matrix is:

$$A = \begin{bmatrix} 0 & 1 & 0 \\ 0 & -1 & 1 \\ 0 & 0 & -5 \end{bmatrix}$$

and the eigenvalues (by inspection, or from Example 3.1) are $\lambda_1 = 0$, $\lambda_2 = -1$ and $\lambda_3 = -5$.

Now, from the full characteristic equation of the system, $[\lambda I - A]v = 0$ (Equation (A1.6), Appendix 1):

$$[\lambda I - A]v = \left\{ \lambda \begin{bmatrix} 1 & 0 & 0 \\ 0 & 1 & 0 \\ 0 & 0 & 1 \end{bmatrix} - \begin{bmatrix} 0 & 1 & 0 \\ 0 & -1 & 1 \\ 0 & 0 & -5 \end{bmatrix} \right\} \begin{bmatrix} v_{11} \\ v_{21} \\ v_{31} \end{bmatrix}$$

$$= \begin{bmatrix} \lambda & -1 & 0 \\ 0 & \lambda+1 & -1 \\ 0 & 0 & \lambda+5 \end{bmatrix} \begin{bmatrix} v_{11} \\ v_{21} \\ v_{31} \end{bmatrix} = \begin{bmatrix} 0 \\ 0 \\ 0 \end{bmatrix}$$

Expanding these equations:

$$\lambda v_{11} - v_{21} = 0, \quad (\lambda+1)v_{21} - v_{31} = 0 \quad \text{and} \quad (\lambda+5)v_{31} = 0 \qquad (5.15)$$

Each eigenvalue is substituted separately into Equations (5.15), so as to find the corresponding eigenvector. The eigenvectors corresponding to the eigenvalues λ_1, λ_2 and λ_3 are v_1, v_2 and v_3 respectively. The elements of these eigenvectors will be denoted as, for example, $v_{1_{21}}$ to represent the second element of the first eigenvector.

For $\lambda_1 = 0$, Equations (5.15) reveal that $v_{1_{31}} = v_{1_{21}} = 0$, while $v_{1_{11}}$ can have any value. This freedom of choice arises because a *characteristic direction* is being evaluated, and not a magnitude. The scaling of the vector elements makes no difference to its direction. Normally, one element of the eigenvector is set to unity, or the eigenvector is normalized to unit length. In this case, we choose to set $v_{1_{11}} = 1$.

For $\lambda_2 = -1$, Equations (5.15) show that $v_{2_{31}} = 0, v_{2_{21}}$ can be anything, and $v_{2_{11}} = -v_{2_{21}}$. We choose to set $v_{2_{11}} = 1$.

For $\lambda_3 = -5$, Equations (5.15) give $v_{3_{31}}$ can be anything, $v_{3_{21}} = -0.25v_{3_{31}}$ and $v_{3_{11}} = -0.2v_{3_{21}}$. Again, the first element is chosen as unity, $v_{3_{11}} = 1$. These choices give a set of eigenvectors of:

$$v_1 = \begin{bmatrix} 1 \\ 0 \\ 0 \end{bmatrix}, \quad v_2 = \begin{bmatrix} 1 \\ -1 \\ 0 \end{bmatrix} \quad \text{and} \quad v_3 = \begin{bmatrix} 1 \\ -5 \\ 20 \end{bmatrix} \qquad (5.16)$$

These eigenvectors are linearly independent, as they specify independent directions in the state-space. This will be true for the eigenvectors of any system having non-repeated eigenvalues.

If the eigenvectors for one of the other state-space models of the same system are calculated (for example, the direct programming result in Equation (2.99)), again fixing the first element in each equal to unity, the results will be found to differ from Equation (5.16). Different state variables give different characteristic directions.

As usual, packages such as MATLAB take all the hard work out of such calculations. For example, the pair of MATLAB commands:

$$\gg a = [0 \ 1 \ 0; \ 0 \ -1 \ 1; \ 0 \ 0 \ -5];$$

$$\gg [v, e] = eig(a)$$

will return the eigenvalues of A as the diagonal elements of e, and the eigenvectors of A as the columns of v. Note that MATLAB scales the eigenvectors to unit length, so the numbers differ from those of Equation (5.16), although the directions in the state-space are identical – it is purely a scaling change.

5.6.1   Modes of response

The modes of response of a system represented by a state-space model can be specified in terms of the eigenvalues and eigenvectors of the system. This is not surprising, as the eigenvalues are the same as the poles of the system; and the poles have already been shown to determine the modes of response of the system.

To investigate the modes of response, the solution of the state equation is required. A solution via the Laplace transform was derived in Section 3.2.1, where the Laplace domain solution was found to be (Equation (3.13)):

$$X(s) = \boldsymbol{\Phi}(s)X(0) + \boldsymbol{\Phi}(s)\mathbf{B}U(s),$$

$$\text{where}\quad \boldsymbol{\Phi}(s) = [s\mathbf{I} - \mathbf{A}]^{-1} \tag{5.17}$$

Now consider, for simplicity, the unforced system (that is, the system with no input except initial conditions on the states). The solution is then:

$$X(s) = \boldsymbol{\Phi}(s)X(0) \quad \text{or} \quad x(t) = \boldsymbol{\Phi}(t)x(0) \tag{5.18}$$

Example 5.9 **State equation solution for the antenna-positioning system**

For the example system, taking the same state-space model as was used in Example 5.8, the transition matrix is firstly evaluated from Equation (5.17):

$$\boldsymbol{\Phi}(s) = [s\mathbf{I} - \mathbf{A}]^{-1} = \left\{ \begin{bmatrix} s & 0 & 0 \\ 0 & s & 0 \\ 0 & 0 & s \end{bmatrix} - \begin{bmatrix} 0 & 1 & 0 \\ 0 & -1 & 1 \\ 0 & 0 & -5 \end{bmatrix} \right\}^{-1}$$

$$= \begin{bmatrix} s & -1 & 0 \\ 0 & s+1 & -1 \\ 0 & 0 & s+5 \end{bmatrix}^{-1}$$

Using the (adjoint matrix)/(determinant) to find the inverse (Section A1.1.3):

$$\boldsymbol{\Phi}(s) = \frac{1}{s(s+1)(s+5)} \begin{bmatrix} (s+1)(s+5) & 0 & 0 \\ (s+5) & s(s+5) & 0 \\ 1 & s & s(s+1) \end{bmatrix}^{\mathrm{T}}$$

$$= \begin{bmatrix} \dfrac{1}{s} & \dfrac{1}{s(s+1)} & \dfrac{1}{s(s+1)(s+5)} \\[2mm] 0 & \dfrac{1}{(s+1)} & \dfrac{1}{(s+1)(s+5)} \\[2mm] 0 & 0 & \dfrac{1}{(s+5)} \end{bmatrix}$$

The time-domain transition matrix $\boldsymbol{\Phi}(t)$ can be found by taking the inverse Laplace transform of this matrix (that is, the inverse Laplace transform of each element), giving:

$$\boldsymbol{\Phi}(t) = \begin{bmatrix} 1 & 1 - e^{-t} & \frac{1}{5} - \frac{1}{4}e^{-t} + \frac{1}{20}e^{-5t} \\ 0 & e^{-t} & \frac{1}{4}e^{-t} - \frac{1}{4}e^{-5t} \\ 0 & 0 & e^{-5t} \end{bmatrix}$$

Using this result in Equation (5.18), to find the responses to the initial conditions, gives:

$$\begin{bmatrix} x_1 \\ x_2 \\ x_3 \end{bmatrix} = \begin{bmatrix} x_{0_1} + (1 - e^{-t})x_{0_2} + \left(\frac{1}{5} - \frac{1}{4}e^{-t} + \frac{1}{20}e^{-5t}\right)x_{0_3} \\ e^{-t}x_{0_2} + \left(\frac{1}{4}e^{-t} - \frac{1}{4}e^{-5t}\right)x_{0_3} \\ e^{-5t}x_{0_3} \end{bmatrix} \tag{5.19}$$

From results such as Equation (5.19), it is confirmed that the *eigenvalues* give the exponential modes of the system. These appear in the responses of every state variable, in general. The system of Example 5.9 has eigenvalues of $0, -1$ and -5; so terms such as e^{0t} (constants), e^{-1t} and e^{-5t} appear.

The *eigenvectors* specify the distribution of the modes between the states. In Example 5.8, corresponding with the eigenvalue -5, was an eigenvector of $[1 \ -5 \ 20]^T$. This predicts that the mode e^{-5t} will be found 20 times as much in evidence in the x_3 response, as in the x_1 response; and this is verified by the time solution for x at Equation (5.19). Similarly, in the x_2 response, five times as much of the e^{-5t} mode would be expected as in the x_1 response, but with a sign change; and again this is evident in Equation (5.19). The other two eigenvectors successfully predict the distribution of the other two modes in a similar way.

The eigenvectors therefore show the proportion of each mode appearing in each state response. These proportions remain fixed, whatever the excitation of the system, but the actual numbers involved vary with the initial conditions in $x(0)$.

5.6.2 A tighter link between eigenstructure and time response

The type of result in Equation (5.19) can be obtained directly from the eigenvalues and eigenvectors of the system model, removing the necessity to evaluate the transition matrix, or to take inverse Laplace transforms. For systems with distinct (non-repeated) eigenvalues, the procedure is as follows.

(1) From the plant matrix A, obtain the eigenvalues $\lambda_1, \lambda_2, \lambda_3, \ldots, \lambda_n$.

(2) Using the method of Example 5.8 above, calculate the corresponding eigenvectors, $v_1, v_2, v_3, \ldots, v_n$.

(3) Form the *modal matrix* W for the system. This is the matrix whose columns are made up of the eigenvectors, so that $W = [v_1 | v_2 | v_3 | \cdots | v_n]$ (see Section A1.3 of Appendix 1 on partitioned matrices, if necessary).

(4) Determine the inverse of the modal matrix, that is, W^{-1}, and regard it as being made up of a stack of row vectors, such that

$$W^{-1} = \begin{bmatrix} \underline{w_1} \\ \underline{w_2} \\ \underline{w_3} \\ \vdots \\ \underline{w_n} \end{bmatrix}$$

(5) The state response to an initial condition vector $x(0)$, as in Equation (5.19), is then given by the following alternative representation of the transition matrix:

$$x(t) = W \operatorname{diag}[e^{\lambda_1 t}, \ e^{\lambda_2 t}, \dots, \ e^{\lambda_n t}] W^{-1} x(0)$$

$$= \left[\sum_{i=1}^{n} (e^{\lambda_i t} v_i w_i) \right] x(0) \tag{5.20}$$

Example 5.10 *Direct method of state equation solution for the antenna positioner*

From Example 3.1, the system has eigenvalues $\lambda_1 = 0$, $\lambda_2 = -1$ and $\lambda_3 = -5$. From Example 5.8 (Equation (5.16)), the corresponding eigenvectors are known to be:

$$v_1 = \begin{bmatrix} 1 \\ 0 \\ 0 \end{bmatrix}, \quad v_2 = \begin{bmatrix} 1 \\ -1 \\ 0 \end{bmatrix} \quad \text{and} \quad v_3 = \begin{bmatrix} 1 \\ -5 \\ 20 \end{bmatrix}$$

so that the modal matrix is

$$W = \begin{bmatrix} 1 & 1 & 1 \\ 0 & -1 & -5 \\ 0 & 0 & 20 \end{bmatrix}$$

The inverse of this is found to be:

$$W^{-1} = \begin{bmatrix} 1 & 1 & \dfrac{1}{5} \\ 0 & -1 & -\dfrac{1}{4} \\ 0 & 0 & \dfrac{1}{20} \end{bmatrix}$$

the rows of which give:

$$w_1 = \begin{bmatrix} 1 & 1 & \frac{1}{5} \end{bmatrix}, \quad w_2 = \begin{bmatrix} 0 & -1 & -\frac{1}{4} \end{bmatrix} \quad \text{and} \quad w_3 = \begin{bmatrix} 0 & 0 & \frac{1}{20} \end{bmatrix}$$

Direct application of Equation (5.20) then yields the following result which, when multiplied out, is identical to that of Equation (5.19):

$$x(t) = \sum_{i=1}^{3} \left(e^{\lambda_i t} \, v_i \, w_i \right) x(0)$$

$$= \left\{ e^{0t} \begin{bmatrix} 1 \\ 0 \\ 0 \end{bmatrix} \begin{bmatrix} 1 & 1 & \frac{1}{5} \end{bmatrix} + e^{-1t} \begin{bmatrix} 1 \\ -1 \\ 0 \end{bmatrix} \begin{bmatrix} 0 & -1 & -\frac{1}{4} \end{bmatrix} \right.$$

$$\left. + e^{-5t} \begin{bmatrix} 1 \\ -5 \\ 20 \end{bmatrix} \begin{bmatrix} 0 & 0 & \frac{1}{20} \end{bmatrix} \right\} x(0)$$

Problem 5.16 takes these ideas a little further.

5.7 More on discrete-time models and simulation studies

Most of the modelling and controller design work so far has concentrated on continuous-time (or analog) approaches. However, in Section 3.6, the use of discrete-time models was briefly discussed, and the time responses of a couple of such models were generated. Now the derivation of such models will be introduced.

In many cases, 'digital controllers' are actually analog ones implemented by digital methods. Such implementation has a number of advantages: in particular, it is often much easier to adjust compensator parameters in software than by modifying components in a hardware analog controller. Also, although not discussed in this book, it is now normal for large process plants (for example, in the petro-chemical industries) to be controlled by very many individual controllers, whose setpoints are provided by a hierarchy of computers in 'distributed control' or SCADA (supervisory control and data acquisition) schemes, and the digital approach lends itself well to use in such schemes. A balancing disadvantage for simple systems is perhaps that of cost, but the difference is diminishing all the time as microprocessor technology becomes cheaper as well as faster. The major costs of digital control are now often in software development.

The digital implementation of analog control strategies will be explored in Section 5.8. This section continues the investigation of the parallel possibility of simulating the operation of analog components and systems by means of a digital computer. Such simulation is very useful in that it allows the soundness of a controller to be tested without the risk of expensive and dangerous plant damage. Several proprietary packages exist for performing such simulations and

SIMULINK (The Mathworks Inc., 1993c, and Appendix 4) will be used as being representative of these. In addition, MATLAB (Appendix 3) has already been used to simulate systems on several occasions. Every time the *step* command has been issued to generate figures such as Figure 5.7, a digital computer simulation of a continuous-time model has been performed.

The basic principles of simulating an analog system by means of a digital computer were outlined in Section 3.6.1 (Example 3.17). Another simple first-order example will be used in discussing the derivation of the discrete-time model, and different methods of performing the simulation.

Consider a simple lag whose LTF model is

$$\frac{Y(s)}{U(s)} = \frac{5}{1 + 0.5s}$$

Cross-multiplying gives:

$$(1 + 0.5s)Y(s) = 5U(s)$$

Taking inverse Laplace transforms, with zero initial conditions, the time-domain equivalent is:

$$0.5 \frac{dy(t)}{dt} + y(t) = 5u(t)$$

or, dropping the *t*-dependency and rearranging:

$$\frac{dy}{dt} = 10u - 2y \tag{5.21}$$

The procedure is now to solve this differential equation numerically. The simplest approach is by Euler integration whose principle is explained by Figure 5.9. From the figure, at time instant t_{n-1}, the slope of the response is approximately given by:

$$\frac{dy}{dt} \approx \frac{y_n - y_{n-1}}{t_n - t_{n-1}}$$

If the sampling interval $t_n - t_{n-1} = T_s$, then the latest value of y (y_n) is given by:

$$y_n \approx y_{n-1} + T_s \left. \frac{dy}{dt} \right|_{n-1}$$

If the interval T_s is chosen to be short in relation to the rate of change of y, reasonable accuracy is achieved. The following pseudo-code shows how the method would work with this example.

> Set the time *t* equal to zero
> Set a quantity *last_y* equal to the initial value of *y*
> Set, read or input the total simulation time *tmax*

Loop: Read or input the value of *u*

> Calculate *t* as $t + T_s$
> Calculate *y(t)* as *last_y* + $(10u - 2 \times last\_y) \times T_s$ (from Equation (5.21))
> Output and/or graph and/or store *t* and *y(t)*
> Set *last_y* equal to *y(t)*
> End if $t \geq tmax$, else goto *Loop*

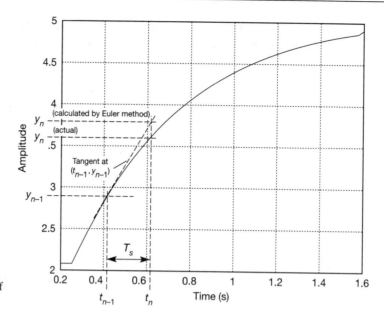

Figure 5.9 The principle of Euler integration.

This algorithm could easily be programmed in a high-level language, or even on a spreadsheet. The responses of Figure 5.10 were plotted using the MATLAB m-file *fig5_10.m* on the accompanying disk (of course, MATLAB is not meant to be used like this – the *step* command would be much better – but it illustrates how the method would be programmed in a general high-level language). It can clearly be seen that integration intervals of 0.05 s or less give a very good representation of the actual system output, but that increasing the integration interval causes an

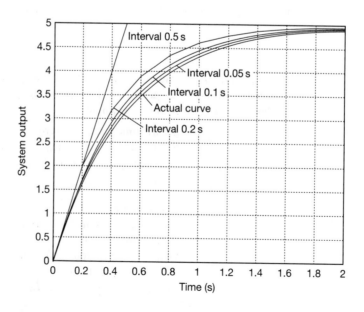

Figure 5.10 Euler integration: system outputs at various integration intervals.

increase in the error. Indeed, when the interval became equal to the time constant (0.5 s), the response moved to the steady-state value in one interval and stayed there!

It is important to note that the different responses are purely a function of the use of the numerical integration algorithm (choice of the integration time step) – the real system does not respond like that at all (except, perhaps, in the case of the 0.05 s sampling interval). This shows that care is needed in using such simulation techniques.

Better algorithms for numerical integration allow the use of longer steplengths for the integration without accuracy loss, resulting in faster algorithm execution. Probably the most widely used method is that of Runge–Kutta, which works as follows in its 'fourth-order' form. Reference should be made to Figure 5.11 during this description.

The procedure starts by calculating dy/dt at the start of the interval, just like the Euler method (that is, at time t_{n-1} in the figure). That value is used to calculate the amount k_1 by which y increases during the time step, using $k_1 = dy/dt \times T_s$. That was the whole of Euler's method – and the first step of Runge–Kutta!

Mathematical common sense suggests that Euler's approach would be more accurate if the average value of dy/dt was known across the interval, rather than just the value at the beginning. Therefore, three further values of dy/dt are used (as described below) to find increases k_2, k_3 and k_4, and a weighted average of those k values is taken to be the actual change in y during the interval.

For determining k_2, the value of y halfway across the interval (at time $t_{n-1} + T_s/2$) is estimated as $y_{n-1} + k_1/2$ and is then used to evaluate dy/dt halfway across the interval from the system equations (such as Equation (5.21)). The value of k_3 is determined similarly except that $y_{n-1} + k_2/2$ is used to work out dy/dt halfway across. Finally, k_4 is calculated using the value of dy/dt at the *end* of the interval (worked out on the basis of $y_{n-1} + k_3$). Finally, y_n is worked out as the weighted average:

$$y_n = y_{n-1} + (k_1 + 2k_2 + 2k_3 + k_4)/6$$

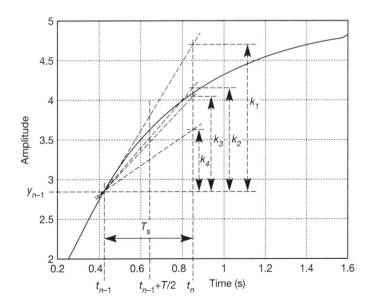

Figure 5.11 The principle of Runge–Kutta integration.

The following algorithm, converted to MATLAB code (again, rather an inefficient use of MATLAB!), was used to produce Figure 5.12 for the example problem (see the file *fig5_12.m* on the accompanying disk).

Set $u = 1$

Set the time t equal to zero

Set a quantity *last_y* equal to the initial value of y

Set, read or input the time step T_s

Set, read or input the total simulation time *tmax*

Loop: Calculate dy/dt as $10u - 2 \times last\_y$ (as in Euler's method)

Calculate k_1 as $T_s \times dy/dt$

Recalculate dy/dt as $10u - 2(last\_y + k_1/2)$

Calculate k_2 as $T_s \times dy/dt$

Recalculate dy/dt as $10u - 2(last\_y + k_2/2)$

Calculate k_3 as $T_s \times dy/dt$

Recalculate dy/dt as $10u - 2(last\_y + k_3)$

Calculate k_4 as $T_s \times dy/dt$

Calculate $y(t)$ as $last\_y + (k_1 + 2k_2 + 2k_3 + k_4)/6$

Set *last_y* equal to $y(t)$

Output and/or graph and/or store t and $y(t)$

Set $t = t + T_s$

Goto *Loop* if $t < tmax$

End

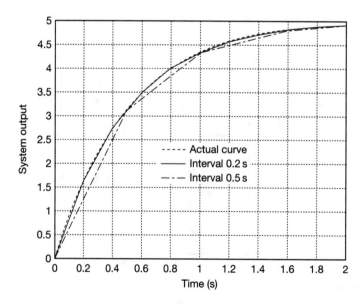

Figure 5.12 System output calculated by Runge–Kutta integration.

Referring to Figure 5.12, it is clear that the calculated results from the simulation are very close to the true values, even when an integration interval equal to the time constant of the first-order system is used. Despite the much greater number of calculations which Runge–Kutta requires at each iteration of the simulation, it does allow an accurate simulation to run very much faster than the Euler method.

Other numerical integration methods exist to deal with the case of systems which have periods of slow change of variables interspersed with episodes of more rapid change. In such cases, it can be advantageous to vary the integration interval – reducing it when the change is rapid and extending it when change is slower – and methods exist to perform the integration with a suitably variable steplength. Typically, Runge–Kutta methods are modified to incorporate steplength control, but there are also other methods such as those due to Adams, Bashforth, Fox, Gear, Moulton and others (see, for example, Conte and de Boor (1972)). Some of these are specialized, such as the Gear algorithm which is specifically aimed at the simulation of 'stiff' systems – namely those with a severe discrepancy between their fastest and slowest dynamics. With some systems, it is found that variable step methods take a long time to optimize their steplength, so that that they save little execution time compared with using Runge–Kutta with a short steplength throughout, but they may nevertheless effect useful savings in the quantity of information which needs to be stored.

In practice, control system simulation studies are best carried out using a specialist simulation package such as SIMULINK. The arrangement shown in Figure 5.13 will be investigated using SIMULINK as a brief example.

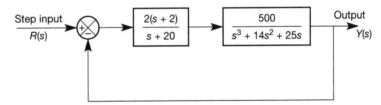

Figure 5.13 Example for SIMULINK simulation.

SIMULINK is used as in Section A4.1.1, noting also that the *discrete* library contains the further elements needed to model digital controllers, while *extras* contains some useful specialist blocks including PID controllers.

Figure 5.14 shows the graphic input for this problem in SIMULINK. It is noteworthy that the step input is applied by default after one second rather than

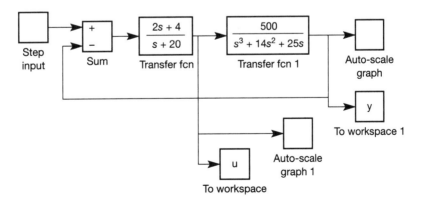

Figure 5.14 SIMULINK arrangement for the lead-compensation example.

initially. This can be overridden, but is often useful in that it allows the state of the system before the application of the step to be confirmed. However, do not be misled into thinking that a one-second transport lag has appeared. The blocks 'y' and 'u' are those used to pass the variable values back to MATLAB, as outlined in Appendix 4; the corresponding values of time t are passed back by typing t in the 'Return Variables' window in the 'Parameters' section of the 'Simulation' menu, in which the integration algorithm, the permitted range of time step used, and the total simulation time are also specifiable. The file *fig5_14.m* on the accompanying disk contains this SIMULINK model. To a novice MATLAB user, its contents will probably not mean very much, as SIMULINK created it automatically from the block diagram input. However, it will run in response to the command *fig5_14* issued at the MATLAB prompt, so long as SIMULINK is installed on your system (just typing the name of the m-file will load SIMULINK automatically and draw the block diagram – click on *Simulation* then *Start*). Note that the version on disk also uses a *multiplexer* ('mux') block from the *connections* library, so as to display both plots on the same graph.

Figure 5.15 shows the controller and plant outputs resulting from running the simulation. Smoother traces could be obtained by taking more samples (reduce the 'Max Step Size' setting in the 'Parameters' section of the 'Simulation' menu, so as not to allow the integration algorithm to take such long time steps). This will also make the 'controller action' look more like Figure 5.15, which was achieved with a maximum integration step size of 10 ms. However, the penalty for doing this is longer execution times.

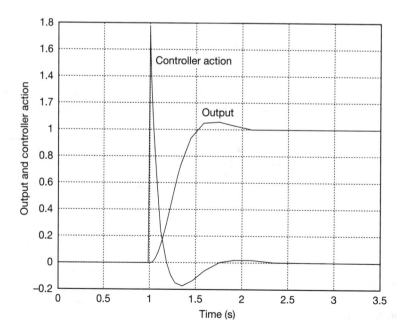

Figure 5.15 Plant step response for the lead-compensation example as plotted by MATLAB using data generated by SIMULINK.

5.8 An introduction to digital controllers

So far we have discussed control by analog methods. Such methods are important not only because they are still used in many areas, but also because many digital controllers are in fact implementing analog control strategies, especially PID (three term). This section will explore how digital controllers operate; how digital algorithms based on an analog controller design may be developed; and the advantages and problems of such 'digitized analog' controllers compared with their analog counterparts.

5.8.1 Components in a digital control loop

A digital controller may be viewed as operating the following cycle of events.

(1) It reads the value of the signal at its input. This may be the error signal in the case of a cascade controller such as a PID, or the digital controller may form the error signal itself by reading the values of setpoint and output and subtracting the latter from the former.

(2) It checks that the read values are within known limits of magnitude and rate-of-change. It may average several values to perform rudimentary noise filtering.

(3) It performs a calculation, based on the input values and on stored values of previous inputs and outputs, to work out what its output should now be.

(4) It outputs that value after checking it against plant limits, and modifying it if necessary.

(5) It pauses for a moment before restarting the cycle by reading its input again.

It is noteworthy that the controller output is maintained at its last value until the next one is calculated, so that a graph of the controller output against time resembles a series of d.c. levels (a 'staircase' type of waveform) rather than a series of impulses. The implications of the latter point are discussed in Chapter 7.

The reason for the pause (item 5 above) is that the time between reading successive inputs, known as the 'sampling interval', must be constant for the controller to operate properly to its specification, but the calculation will not always take the same length of time (in particular, it will be significantly quicker if one or more of the input and output values proves to be zero).

The general form of a digital controller to perform such a series of operations is shown in Figure 5.16, and its block diagram representation in Figure 5.17. The

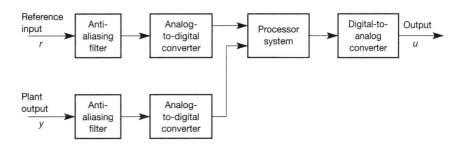

Figure 5.16 General arrangement of a digital controller.

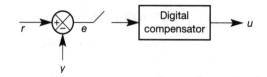

Figure 5.17 Schematic representation of Figure 5.16.

controller will need to include an analog-to-digital converter for each input, a computer (using the word in its most general sense) to perform the calculation, and a digital-to-analog converter for the output.

It will also be necessary to precede the input or inputs by analog low-pass filtering to avoid aliasing. This phenomenon was described in Section 1.3.6 (see Figure 1.15). High-frequency (by electromechanical standards) signals can appear unexpectedly at controller inputs by electromagnetic action if the screening is less than perfect – those caused by the a.c. mains can be particularly prevalent. It is such signals at higher frequencies than those expected in the control loop, that the anti-aliasing filter must eliminate. Otherwise, they would be under-sampled and would appear to the controller as a lower-frequency signal, which might then be indistinguishable from a genuine signal at the frequencies expected in the control loop. Such an analog filter must offer negligible attenuation at the highest natural frequency of the plant, the controller or the resulting closed-loop system, but must offer considerable attenuation above about half the sampling frequency.

The significance of *half* the sampling frequency (the so-called 'Nyquist frequency') is shown by the frequency response graphs of Figure 5.18, in which the responses of an analog first-order lag and its digital equivalent are shown. The response of the digital version is seen to fall to about −26 dB at half the sampling frequency of 100 Hz (or approximately 628 rad s^{-1}) before rising back to 0 dB at the sampling frequency owing to aliasing. The exact interpretation of the 'considerable attenuation' required depends on the likely level of interference and the quality of the screening, but at least 40 dB would be a guide. It is possible

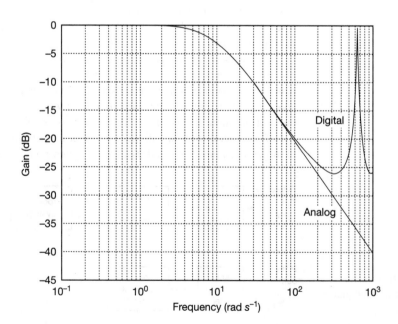

Figure 5.18 Frequency responses of analog and digital first-order lags.

that a similar filter following the D–A converter will be useful in smoothing the discrete steps in the controller action. The plant itself will generally do this smoothing effectively, since not many plants can faithfully follow a 'staircase' waveform, but an output filter might be necessary to prevent excessive plant actuator wear. An example in Section 9.7 develops the implementation aspects of digital control further.

5.8.2 Digitizing analog transfer functions – the z-transform

The conversion of an analog controller transfer function to a digital controller algorithm is generally best done by one of the approximations to the *z-transform*. Just as the Laplace transform maps continuous-time quantities into the complex frequency domain, thus allowing linear LTF models to be written in terms of the variable s, so the z-transform maps LTF models into discrete-time, and allows transfer function models for discrete-time systems to be written in terms of a variable z.

This variable z strictly arises from a discrete-time summation expression, related to the integral definition of the Laplace transform. However, it is usually approximated by various substitutions for s in the LTF, instead. These different possible derivations lead to different possible z-transform models of the same system, each having slightly different properties (or, in some cases, rather more than 'slightly different')!

The variable z is first introduced in the simplest manner possible. That is, a multiplication by z^{-1} is used simply as a notation to represent a *delay* of one time step in a discrete-time signal. Thus, an ARMA model such as the following one, which was used in Section 2.8.1, can be rewritten as shown:

$$y_n = A_1 y_{n-1} + A_2 y_{n-2} + A_3 y_{n-3} + \cdots + B_0 u_n + B_1 u_{n-1}$$
$$+ B_2 u_{n-2} + \cdots$$

transforms to:

$$Y(z) = A_1 z^{-1} Y(z) + A_2 z^{-2} Y(z) + A_3 z^{-3} Y(z) + \cdots + B_0 U(z)$$
$$+ B_1 z^{-1} U(z) + B_2 z^{-2} U(z) + \cdots$$

This can then be rearranged into the z-domain transfer function:

$$\frac{Y(z)}{U(z)} = \frac{B_0 + B_1 z^{-1} + B_2 z^{-2} + \cdots}{1 - A_1 z^{-1} - A_2 z^{-2} - A_3 z^{-3} - \cdots}$$

In obtaining z-transfer functions from LTFs by making substitutions for s, many of the generally used substitutions depend on the property that multiplying by s implies differentiation in the time domain. It has just been demonstrated that multiplying the z-transform by z^{-1} implies an additional delay of one sampling interval. The argument for the simplest of the substitutions suggests that the gradient of the time response of, for example, $e(t)$ at the present time (instant n) is given by

$$\left.\frac{de}{dt}\right|_n = \frac{\text{(latest sample of } e) - \text{(last sample of } e)}{\text{(sampling interval)}} = \frac{(e_n - e_{n-1})}{T_s}$$

In the z-domain, when e_n becomes $E(z)$ (that is, the z-transform of e_n), this will convert to $[E(z) - E(z)z^{-1}]/T_s$, or $E(z)(1 - z^{-1})/T_s$. Multiplying $E(z)$ by $(1 - z^{-1})/T_s$ is therefore equivalent to differentiation, that is, to multiplying $E(s)$ by s. This means that substituting the following expression for s in a LTF:

$$s = \frac{(1 - z^{-1})}{T_s} \tag{5.22}$$

will give an approximately equivalent discrete-time version (in z) of the analog LTF. It is then easy to convert the z transfer function to an algorithm from which a computer program can be written. The procedure for obtaining the algorithm is demonstrated following Example 5.11, below.

A more exact substitution for s (known as the *Tustin transformation*) is

$$s = \frac{2(1 - z^{-1})}{T_s(1 + z^{-1})} \tag{5.23}$$

This more complicated substitution often allows close approximation of the behaviour of the analog controller, using a longer sampling interval (T_s) than would be the case with the simpler substitution for s. The longer sampling interval is beneficial in that either a slower processor can be used to perform the calculation, or a given processor can cope with a faster sampling rate. It is found, however, that the Tustin method is worse for use in a digitized PID controller than is the simpler method; the reason will be revealed later in this chapter.

The 'true' z-transform is not used in this chapter, but is used in Chapter 7, in methods of digital controller design which use the discrete nature of the controller output to advantage. It is discussed in Appendix 5. It gives a more accurate response, but the extra effort required does not produce a commensurate improvement in controller performance when digitizing continuous-time designs.

Yet another method of obtaining a digital equivalent of a particular analog controller is the *matched pole–zero* method. This method relies on the property of the true z-transform that an analog controller pole or zero at $s = -a$ converts to a pole or zero of the digital equivalent at $z = e^{-aT_s}$. Each bracket containing $(s + a)$ is replaced by one containing $(z - e^{-aT_s})$ and an overall gain constant is applied to make the low-frequency gain of the version in z equal to that of the version in s.

These procedures, and the effectiveness of the resulting controllers, will be demonstrated by an example.

Example 5.11 *A digital controller implemented following various different z-transform methods*

A plant modelled by the continuous transfer function

$$\frac{Y(s)}{U(s)} = \frac{10}{s^3 + 7s^2 + 6s}$$

is to be controlled in the closed-loop with unity negative feedback, using a forward path lead compensator of transfer function

$$D(s) = \frac{1.5(s+1)}{(s+3)}$$

The compensator is to be implemented in digital form ($D(s)$ has been used so as to match the notation $D(z)$ used to represent the z-transfer function of a digital compensator). Investigate the performance of implementations having a sampling interval of 0.1 s and being converted into the z-domain by

(1) the simple method $s = (1 - z^{-1})/T_s$;

(2) the Tustin method and

(3) the matched pole–zero method.

Evaluating the three compensator transfer functions:

(1) Using $s = (1 - z^{-1})/T_s$ with $T_s = 0.1$ s gives $s = 10(1 - z^{-1})$ and so

$$D(z) = \frac{1.5[10(1 - z^{-1}) + 1]}{[10(1 - z^{-1}) + 3]} = \frac{(16.5 - 15z^{-1})}{(13 - 10z^{-1})} = \frac{(1.2692 - 1.1538z^{-1})}{(1 - 0.7692z^{-1})}$$

(2) The Tustin substitution for s is

$$s = \frac{2(1 - z^{-1})}{T_s(1 + z^{-1})}$$

(Equation (5.23)) so, with $T_s = 0.1$, the transfer function converts to:

$$D(z) = \frac{1.5\left[\left(\dfrac{2(1 - z^{-1})}{T_s(1 + z^{-1})}\right) + 1\right]}{\left[\left(\dfrac{2(1 - z^{-1})}{T_s(1 + z^{-1})}\right) + 3\right]} = \frac{1.5[20(1 - z^{-1}) + (1 + z^{-1})]}{[20(1 - z^{-1}) + 3(1 + z^{-1})]}$$

$$= \frac{(31.5 - 28.5z^{-1})}{(23 - 17z^{-1})} = \frac{(1.3696 - 1.2931z^{-1})}{(1 - 0.7391z^{-1})}$$

(3) The matched pole–zero method requires the calculation of e^{-aT_s}. For the numerator term, $e^{-1 \times 0.1} = 0.90484$ and, for the denominator term, $e^{-3 \times 0.1} = 0.74082$. The transfer function will therefore be:

$$D(z) = \frac{A(z - 0.90484)}{(z - 0.74082)}$$

where A is a constant chosen to make the low-frequency gain the same as for the continuous version. From Section 2.5.5, the final value theorem shows that the low-frequency gain (following a step input) is determined by setting s equal to zero in the transfer function, giving in this instance $1.5(0 + 1)/(0 + 3) = 0.5$ for the analog implementation. For the digital one, Appendix 5 reminds us that $z = e^{sT_s}$, so setting s to zero makes z equal to unity. Substituting $z = 1$ into the digital compensator transfer function produces $0.3672A$ as the low-frequency gain. Therefore, $0.3672A = 0.5$, giving $A = 1.3618$. The compensator transfer function is therefore

$$D(z) = \frac{1.3618(z - 0.90484)}{(z - 0.74082)} = \frac{(1.3618 - 1.2322z^{-1})}{(1 - 0.74082z^{-1})}$$

The step responses of the controlled systems may be compared by means of an appropriate computer simulation package or by means of the MATLAB m-file *fig5_19.m* on the accompanying disk. Graphs of the step responses are shown in Figure 5.19 and the following table compares them for the maximum overshoot and the time at which it occurs.

Conversion type	Simple	Tustin	Matched p–z
Overshoot, per cent	3.6	2.9	3.0
Peak time, seconds	3.2	3.35	3.35

For comparison, the analog controller would give an overshoot of 1.6 per cent at 3.5 seconds, so all the digital implementations produce some degree of performance degradation.

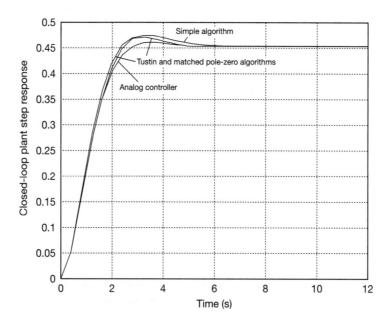

Figure 5.19 Closed-loop step responses for Example 5.11.

In order to produce a digital control system arrangement which will mechanize one of the above transfer functions in z, it will be necessary to produce a suitable program for the processor. A first step in so doing is to convert the transfer function into a discrete-time (difference) equation from which an algorithm, or 'pseudo-code', may be developed. The system resulting from the simple conversion technique will be used as an example.

The compensator transfer function in z is, by definition, equal to the ratio $U(z)/E(z)$ of the z-transforms of the controller action u and the error signal e respectively, thus:

$$D(z) = \frac{U(z)}{E(z)} = \frac{(1.2692 - 1.1538z^{-1})}{(1 - 0.7692z^{-1})}$$

which can be multiplied throughout by $(1 - 0.7692z^{-1})E(z)$ to give:

$$(1 - 0.7692z^{-1})U(z) = (1.2692 - 1.1538z^{-1})E(z)$$

Such equations can be returned easily to the time domain, by recalling that multiplying by z^{-1} represents a delaying of the associated signal by one sampling interval. For example, $E(z)$ represents the latest sample of the error signal and $z^{-1}E(z)$ represents the sample at the previous sampling interval. Returning to the time-domain notation, it can therefore be seen that

$$u_n - 0.7692u_{n-1} = 1.2692e_n - 1.1538e_{n-1}$$

or

$$u_n = 0.7692u_{n-1} + 1.2692e_n - 1.1538e_{n-1} \tag{5.24}$$

This equation enables the controller to calculate the required value of u from the last value of u and the present and previous values of e.

The outline algorithm could then be:

Initialize (see below)

> *Loop*: Reset sampling interval timer
>
> Input e $(= e_n)$
>
> Calculate u_n from Equation (5.24)
>
> Output u_n
>
> Set $u_{n-1} = u_n$
>
> Set $e_{n-1} = e_n$
>
> Wait for end of sampling interval
>
> Goto *Loop*

'Initialize' sets the initial values of u_{n-1} and e_{n-1} to zero, or to their actual values if they are available. The 'Wait for end of sampling interval' is because the calculations are unlikely always to take exactly the same time and, even if they do, that time is unlikely to be the required sampling interval! The 'Set $x_{n-1} = x_n$' lines are included because the value of u (for example) at the present time (u_n) will be u_{n-1} at the next sampling interval. Users of structured languages will doubtless be aghast at the use of the word 'Goto'; it can be readily replaced by a structure of the 'Repeat forever' variety.

The easiest way of ensuring that sampling does take place at equal intervals of time may be to program the 'Loop' routine as an interrupt routine which is called by an appropriate pulse train (often generated by a support chip having a software-settable timing signal generator).

5.8.3 Digital PID controllers

The most common type of controller in industrial use is the three-term or PID controller described in Section 4.5.2. It is now used in digitized form more often than in analog implementation. The digitization can be performed either by one of the substitutions for s already described, or by the following direct method.

The LTF of the ideal PID controller is $K_p[1 + T_d s + 1/(T_i s)]$ from Section 4.5.2 (Equation (4.22b)). Recalling that multiplication by s is equivalent to differentiation, the approximate approach of the '$s = (1 - z^{-1})/T_s$' method is again applicable for the derivative term. Since the time integral of a function is the area beneath its time response graph, the integral term is more easily dealt with by the method of Figure 5.20. The integral of error at the end of sampling interval n is the integral at the end of interval $(n - 1)$ plus the lighter shaded area A which, by geometry, is approximately equal to $(e_n + e_{n-1})T_s/2$ (so long as T_s is short relative to the dynamics of the signal). The problem of the integral in an algorithm can therefore be approached by calculating its latest value as its last one (referred to as the previous integral or *PInt*) plus $(e_n + e_{n-1})T_s/2$. The value of *PInt* will require initialization at step 0, of course (later, an alternative algorithm is presented, which avoids using this 'running sum' approach). The controller action at interval n will therefore be:

$$u_n = K_p \left\{ e_n + \frac{T_d(e_n - e_{n-1})}{T_s} + \frac{1}{T_i}\left[\frac{(e_n + e_{n-1})T_s}{2} + PInt \right] \right\} \quad (5.25)$$

giving, in more manageable form, by algebraic manipulation:

$$u_n = Ae_n + Be_{n-1} + C(PInt) \quad (5.26)$$

where

$$A = K_p(1 + T_d/T_s + 0.5T_s/T_i)$$

$$B = K_p(-T_d/T_s + 0.5T_s/T_i)$$

$$C = K_p/T_i$$

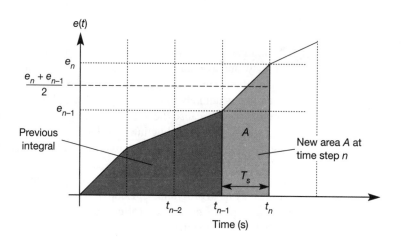

Figure 5.20 Direct method of integral calculation.

It will also be necessary to update the value of the previous integral *PInt* at each sampling interval, by adding on the additional area $0.5(e_n + e_{n-1})T_s$. The algorithm proves to be:

> Initialize: Set initial values of e_{n-1} and *PInt*, and the sample time T_s.

> *Loop*: Reset sampling interval timer
>
> Input the error e_n
>
> Calculate u_n by Equation (5.25) or (5.26)
>
> Set *PInt* equal to $PInt + 0.5(e_n + e_{n-1})T_s$
>
> Set e_{n-1} equal to e_n
>
> Wait for sampling interval duration to reach T_s

> Goto *Loop*

This algorithm does not include provision for detecting and protecting against *integral wind-up*. This is the phenomenon whereby the plant input actuator saturates (for example), so that further increasing the magnitude of the control signal has no effect. Nevertheless, if there is still a non-zero error in the loop, the integrator 'running sum' will continue to add extra area to the integral at every time step. When conditions change, such that the error signal eventually changes sign, the large magnitude which the integral term will have reached prevents the actuator from responding until the integral term has 'unwound'. At best, this introduces long time delays, having the usual destabilizing effect. There is an additional problem in the digital version because, in fixed-point implementation, the accumulating integral term (if unchecked) could overflow its register and cause the controller output u to swing suddenly from full positive to full negative (or vice versa) – clearly any system which actually permits such behaviour has been very poorly designed!

Such wind-up can be protected against in an electronic analog controller by installing Zener diode clamps around the integrator operational amplifier. In the digitized version, a command can be included in the algorithm after the updating of the previous integral, to monitor its value and, if it has exceeded some pre-set percentage of its maximum, to limit it to that value.

Another thing missing from the algorithm is any facility for switching between manual operator control and automatic computer control. This requires steps to be taken to achieve a 'bumpless transfer' between the two. While in manual control, the plant input must be monitored, and the integral term must be kept properly initialized (by setting the value of *PInt*) such that the digital controller is always outputting the same value as the manual operator. When automatic control is switched in, the plant will then see no difference initially.

The previous comments about checking input and output signal levels against magnitude and rate limits should also be borne in mind.

In respect of digital PID control, the above method based on explicit integration and the method of substituting $s = (1 - z^{-1})/T_s$ will give comparable performance to that of an analog controller, if an appropriate value of the sampling interval T_s is chosen. Unlike the case of the lead compensator, however, the Tustin substitution is much less successful.

Example 5.12 *Comparison of an analog and various digitized PID controllers*

A plant of transfer function

$$\frac{Y(s)}{U(s)} = \frac{10}{s^3 + 8s^2 + 17s + 10}$$

is to be controlled by a PID controller having $K_p = 3$, $T_d = 0.333$ s and $T_i = 1.5$ s. Investigate the performance of the closed-loop system both with an analog PID implementation and with the PID digitized by the methods discussed above.

The investigation will be performed by simulation using MATLAB (Appendix 3). Of course, MATLAB will also use a digital simulation of the continuous analog PID controller, so the comparison is not absolutely fair. However, it is assumed that MATLAB's simulation will be sufficiently realistic to allow the comparison to be made.

Firstly, except for the analog version, it is necessary to calculate the parameters of the digitized controller. The criteria for selecting a suitable sampling interval will be discussed shortly but, for now, a sampling interval of 0.1 s will be used.

The PID transfer function is $K_p[1 + T_d s + 1/(T_i s)]$ from Section 4.5.2 (Equation (4.22b)), so substituting the values of K_p, T_d and T_i gives $3[1 + 0.333s + 1/(1.5s)]$ $= 3 + s + 2/s$. It will prove to be simpler to convert to the z-domain if a common denominator of s is used, giving

$$D(s) = \frac{U(s)}{E(s)} = \frac{(s^2 + 3s + 2)}{s}$$

For the simple method, substituting $s = (1 - z^{-1})/T_s = 10(1 - z^{-1})$ gives:

$$D(z) = \frac{U(z)}{E(z)} = \frac{\{[10(1 - z^{-1})]^2 + 3[10(1 - z^{-1})] + 2\}}{10(1 - z^{-1})}$$

$$= \frac{(13.2 - 23z^{-1} + 10z^{-2})}{(1 - z^{-1})}$$

This transfer function is easily describable to MATLAB in precisely the same way as a continuous transfer function. The main pitfall is that the missing coefficient of z^{-2} must be supplied in the denominator (imagine multiplying throughout by z^2 to see why). A suitable MATLAB command would be:

$\gg$numc $= [13.2 \ -23 \ 10];$ denc $= [1 \ -1 \ 0];$

For the explicit-integration method, use of Equation (5.26) produces values of 13.10 for A, -9.90 for B and 2.00 for C. The previously existing 'running sum' integral value (*PInt*) prevents a transfer function approach being used in simulation by MATLAB, because Equation (5.26) cannot be arranged in transfer function form. A custom m-file has been written to determine the step response with this controller.

A Tustin-substitution version will also be derived and investigated to demonstrate the problems encountered with the method when applied to PID controllers. The substitution for s is

$$s = \frac{2(1 - z^{-1})}{T_s(1 + z^{-1})}$$

(Equation (5.23)), so its use, followed by algebraic manipulation, gives the following transfer function in z:

$$D(z) = \frac{U(z)}{E(z)} = \frac{(23.1 - 39.8z^{-1} + 17.1z^{-2})}{(1 - z^{-2})}$$

which is described to MATLAB as:

$\gg$ numc $= [23.1 \ {-}39.8 \ 17.1];$ denc $= [1 \ 0 \ {-}1];$

The closed-loop step response achieved with each of these digitized controllers was plotted by simulation using MATLAB. The m-file is *fig5_21.m* on the accompanying disk. The resulting step response graphs are displayed in Figure 5.21.

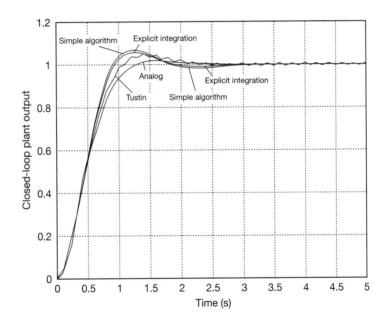

Figure 5.21 Closed-loop step responses for Example 5.12.

The following points are noteworthy in comparing the PID controller results from Example 5.12:

- There is little difference in the performance of the controllers digitized by the simple algorithm and by the explicit-integration method.
- Both of those controllers give a slightly increased step overshoot as compared to the analog implementation (6.7 per cent and 7 per cent in this case as against less than 2 per cent for the analog controller).

- The Tustin version produces a response which is significantly closer to the analog one except for an oscillation at half the sampling frequency. The graph of controller output against time shown in Figure 5.22 shows clearly what is happening!

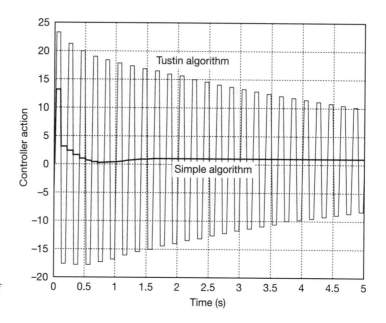

Figure 5.22 Controller action in the PID system of Example 5.12.

The Tustin version has a controller output which is changing value quite violently at every sampling instant. The oscillation is dying away, so the system is not actually unstable, but a heavily oscillatory controller action such as this will cause wear in the controller and plant hardware and an increased level of energy consumption, in addition to any difficulties that may be caused by the actual oscillation in the plant output. The cause of the oscillation is a 'ringing pole' in the controller (see Section 7.6 for an explanation of this phenomenon). For the moment, it will suffice to note that the Tustin substitution is generally unsuitable for digitizing PID controllers.

The velocity algorithm

There is a more commonly used version of the digital PID controller, which operates in an incremental manner, rather than an absolute manner. That is to say that, whereas a mechanization of Equation (5.25) will calculate an absolute positioning signal for the plant actuator every time step, the *velocity algorithm* calculates only the *change* required from the previous position. This has some advantages as follows:

- If the controller fails in certain modes (for example, broken connections or computer failure causing zero control output), then the actuator will simply stay put. Using Equation (5.25) the actuator would move rapidly to zero.

- Many modern plant actuators designed for digital control are incremental in nature. For example, some stepper motor systems require a control increment as an input, rather than an absolute position.

- It will become apparent that there is no need to keep account of the 'running sum' necessary for integral control in Equation (5.25).

- Bumpless transfer from manual control is easier, because when the controller is first turned on, it simply has to calculate the first incremental movement. It does not have to be initialized to give the same absolute position as the manual operator's control, as did Equation (5.25).

To generate the incremental algorithm, Equation (5.25) is effectively evaluated at two successive time steps, and the results are subtracted to leave the incremental change.

Equation (5.25) gave the controller output u_n at step n. To work out the *change* in control signal required at step n compared with step $n - 1$, it is necessary to write down the similar equation one time step earlier (that is, for step $n - 1$), and subtract the two. To be able to do this, Equation (5.25) is initially rewritten so that the 'previous integral' term *PInt* is replaced by a similar term referring to the 'previous integral' which existed one time step *earlier*. Then, when the two results are subtracted, this term will cancel out.

Reference to Figure 5.20 shows that the new area added to the integral at step n is $(e_n + e_{n-1})T_s/2$. By the same reasoning, the 'new' area added at the previous time step $(n - 1)$ would have been $(e_{n-1} + e_{n-2})T_s/2$. *PInt* can therefore be replaced by $PInt_{n-1}$, by writing:

$$PInt = \left[PInt_{n-1} + \frac{(e_{n-1} + e_{n-2})T_s}{2} \right]$$

and then Equation (5.25) becomes:

$$u_n = K_p \left\{ e_n + \frac{T_d(e_n - e_{n-1})}{T_s} \right.$$
$$\left. + \frac{1}{T_i} \left[\frac{(e_n + e_{n-1})T_s}{2} + PInt_{n-1} + \frac{(e_{n-1} + e_{n-2})T_s}{2} \right] \right\} \qquad (5.27)$$

Now, writing Equation (5.25) one time step earlier gives:

$$u_{n-1} = K_p \left\{ e_{n-1} + \frac{T_d(e_{n-1} - e_{n-2})}{T_s} \right.$$
$$\left. + \frac{1}{T_i} \left[\frac{(e_{n-1} + e_{n-2})T_s}{2} + PInt_{n-1} \right] \right\} \qquad (5.28)$$

Subtracting Equation (5.28) from Equation (5.27) then gives:

$$\Delta u_n = u_n - u_{n-1} = K_p \left\{ e_n - e_{n-1} + \frac{T_d(e_n - 2e_{n-1} + e_{n-2})}{T_s} \right.$$
$$\left. + \frac{1}{T_i} \left[\frac{(e_n + e_{n-1})T_s}{2} \right] \right\} \qquad (5.29)$$

Equation (5.29) is the required form, in which Δu_n is the *change* required in the plant actuator position at time step n.

A controller (but not a PID one) including code for integral desaturation and bumpless transfer is developed in Section 9.7.

5.8.4 Sampling interval considerations

So far a figure has been assumed for the sampling interval. In practice, it is necessary to determine a suitable value with regard to the open-loop and intended closed-loop time constants of the system, and to any constraints imposed by the appropriate hardware choice.

Too long a sampling interval will lead to a degradation of the accuracy of the chosen algorithm and often to closed-loop instability, because the sampling process effectively introduces a transport lag of half the sampling interval. Making the interval too short also introduces a problem in that very small changes in controller coefficient values cause a considerable change in the response. This difficulty is most severe if the controller arithmetic is to be performed in fixed-point form (which is likely, to allow the use of a simpler and cheaper processor). The problems will be demonstrated by an example.

Example 5.13 *Performance variation of a digital PID controller with varying sampling interval*

Investigate the performance of the PID controller of Example 5.12, if the simple algorithm is used with sampling intervals of (a) 0.01 s, (b) 0.1 s and (c) 0.5 s.

The first step is to determine the $D(z)$ for the controller in each case.

For $T_s = 0.01$ s, the algebra gives:

$$D(z) = \frac{(103.02 - 203z^{-1} + 100z^{-2})}{(1 - z^{-1})}$$

for $T_s = 0.1$ s, Example 5.12 gave:

$$D(z) = \frac{(13.2 - 23z^{-1} + 10z^{-2})}{(1 - z^{-1})}$$

and for $T_s = 0.5$ s, an analysis gives:

$$D(z) = \frac{(6 - 7z^{-1} + 2z^{-2})}{(1 - z^{-1})}$$

The closed-loop performance was simulated using MATLAB as described in the previous example and the step responses of the three systems are shown in Figure 5.23. The performance of the system with a sampling interval of 0.5 s is clearly unsatisfactory, while that with a sampling interval of 0.01 s appears to be a perfect replica of the analog version (or, at least, MATLAB's digital simulation of it!). Unfortunately, the 0.01 s version proves to have two practical disadvantages. It is only achieved at the cost of a very heavy controller action (an initial peak of over 100 units, as indicated in Figure 5.24) and it is very easily degraded if the coefficients of the digital version have to be rounded (as they do in practice, especially for fixed-point arithmetic implementation).

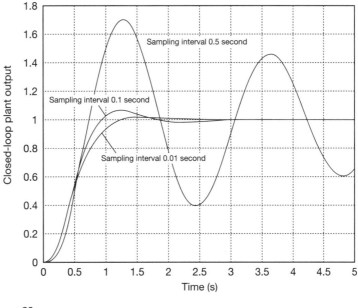

Figure 5.23 PID controller closed-loop responses at varying sampling intervals.

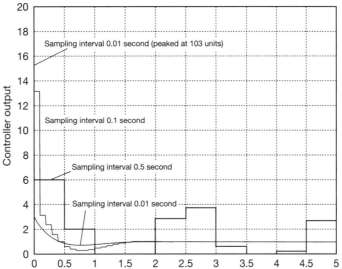

Figure 5.24 Controller action from the PID controllers.

The controller action problem in Example 5.13 is caused by the derivative term of the controller. When a step is applied to the closed-loop system, there is no instantaneous output change, so there is initially no change in the feedback signal. The error signal (the controller input signal) therefore undergoes the same step change as the setpoint signal. In principle, the derivative of that step is an impulse (infinite height, zero width) while the digital controller treats it as having risen in one sampling interval. If the step is of height h, therefore, its rate of rise during the interval in which it is applied is given by h/T_s. That is why shortening the sampling interval increases the initial controller action in the case of a digital PID controller.

It is possible to overcome the problem of heavy controller action by limiting the rate of change of the reference signal to a maximum value at which the controller action is not excessive, and the following algorithm indicates how such *rate-limiting* could be performed. It will be assumed that the maximum permitted rate of change of the reference input r is a quantity *inc* per sampling interval.

The discrete-time controller transfer functions, given by the three versions of $D(z)$ above, can each be written in an equation of the following general form, to evaluate the controller action u (from $D(z) = U(z)/E(z)$).

$$u_n = Au_{n-1} + Be_n + Ce_{n-1} + De_{n-2} \qquad (5.30a)$$

For the implementation pseudo-code, the following variables are used:

$$u = A \times lastu + B \times e + C \times laste + D \times lastle \qquad (5.30b)$$

where (for example) *laste* means 'the value of e at the last sampling instant (instant $n-1$)' and *lastle* means 'the value of e at the last sampling instant but one (instant $n-2$)'.

An algorithm for the implementation of the controller would then be:

> Initialize: Set *lastu*, *laste*, *lastle*, *lastr* all to zero or to their actual previous values if available

> *Loop*: Reset the sampling interval timer to zero
> Input the reference input r and the plant output y
> Calculate dr as $(r - lastr)$
> If $dr > inc$ then set $r = lastr + inc$
> If $dr < -inc$ then set $r = lastr - inc$
> Set $e = r - y$
> Calculate u by Equation (5.30)
> Output u
> Set $lastr = r$
> Set $lastle = laste$
> Set $laste = e$
> Set $lastu = u$
> Wait for sampling interval timer to expire

> Goto *Loop*

The following criteria apply when considering the choice of a suitable sampling interval.

Shannon's sampling theorem requires that, for all the information in a signal to be retained on digitization, the sampling rate must be at least twice as high as the highest frequency contained in the signal. In the context of a control system design, that would mean sampling at a minimum of double the highest cut-off frequency of the system (open- or closed-loop). Unfortunately, sampling at that frequency, it is impossible to produce a satisfactory anti-aliasing filter!

In practice, it is therefore usual to sample at about 10 times the highest cut-off frequency present. The trade-off is between the difficulty of avoiding aliasing (which requires a high sampling frequency) and the performance requirement of the computing device used (which indicates a low sampling frequency). The latter consideration is made worse by the fact that, for a given system, faster sampling requires more computing precision if the algorithm is to work as intended. The following example uses the results of Example 5.13 to make that point.

Example 5.14 *An investigation of accuracy requirements for proper controller operation*

Investigate the requirements on the accuracy of the parameters in the results of Example 5.13, for proper functioning of the controllers.

The controller z-transform for a sampling interval of 0.01 s was found to be:

$$D(z) = \frac{U(z)}{E(z)} = \frac{(103.02 - 203z^{-1} + 100z^{-2})}{(1 - z^{-1})}$$

giving an equation for the controller output u of:

$$u_n = 103.02e_n - 203e_{n-1} + 100e_{n-2} + u_{n-1}$$

This equation presents considerable problems because, if a steady state has been reached (so that the error e is constant), the dependence upon e is only

$$e(103.02 - 203 + 100) = e \times 0.02$$

If 8-bit fixed-point arithmetic is used, only the integers 0 to 255 are allowed or, if negative values are to be included, -127 to $+127$. The overall range of controller coefficients is from -203 to $+103.02$. This means that the coefficients would need to be scaled downwards to fit them into the available range (see below) but, even before such scaling, the 0.02 cannot be distinguished from zero. This will destroy the effect of the integral term altogether.

Sixteen-bit arithmetic allows a number range from $-32\,767$ to $+32\,767$, so the effect of the 0.02 could be retained – though not with its full precision. For example, scaling the coefficients by a factor of 128 to make best use of the number range (while still using a power of 2 – see below) would give $128 \times 0.02 = 2.56$, which must be rounded either to 2 or to 3. Note that this is still not the whole story. It may also be necessary to take into account the fact that a 12-bit A–D converter is being used, for example, with an obvious penalty in accuracy.

If the long sampling interval of 0.5 s is used instead, the transfer function in z is:

$$D(z) = \frac{U(z)}{E(z)} = \frac{(6 - 7z^{-1} + 2z^{-2})}{(1 - z^{-1})}$$

and the equation for the controller output u is:

$$u_n = 6e_n - 7e_{n-1} + 2e_{n-2} + u_{n-1}$$

Now only a number range from -7 to $+6$ in integer steps need be represented, which even 8-bit arithmetic can easily do!

To indicate a little more about how scaling might be used, note that the coefficients in the implementation with a sampling interval of 0.1 s, given in Example 5.12 as 13.2, -23 and 10 respectively, can be scaled for use in an 8-bit fixed-point representation by multiplying them by 4 to give 53 (approximated from 52.8), -92 and 40, which are all within the range -127 to $+127$. The z-transform will then become

$$D(z) = \frac{(53 - 92z^{-1} + 40z^{-2})}{(4 - 4z^{-1})}$$

which will re-convert to the time domain as:

$$4u_n = 53e_n - 92e_{n-1} + 40e_{n-2} + 4u_{n-1}$$

The result will be four times its true value, but a final division by four is easy to achieve by shifting the answer two binary places before it is output. That is why the apparently more accurate approach of multiplying the numerator and denominator of the z-transform by five instead of four (because then the coefficient 13.2 would become an integer value, while 5×23 would still fit the number range) is not particularly good – the final division by five would introduce rounding errors and would also be slow to execute. It is generally therefore better to use powers of two as scaling factors, since these just require shifting operations in the processor. For a much fuller discussion of matters such as these, see Williamson (1991).

5.9 Conclusions

In this chapter a further selection of controller design and implementation techniques has been examined. The foundational state-space approach of state variable feedback was studied, on which most of the more complicated state-space methods of later chapters are based.

It was found that a system can be tested for controllability and, so long as it passes this test, a controller can be designed (in principle) to position the closed-loop poles of the system at any desired s-plane locations. In practice, the control energy requirements and the effects on plant actuator wear, of the selected closed-loop pole locations, must be borne in mind. The allied concept of observability, which is needed later in the text, was also introduced.

Finally on the topic of state variable feedback, it was noted that sometimes there is some flexibility in the design process, which might be used to achieve other things than simply positioning the closed-loop poles. One possibility mentioned was that of affecting the closed-loop eigenvectors in some way and, by way of an introduction to that topic, the part played by the eigenvectors in shaping the time response was investigated.

The other major topic in this chapter was that of digital control. In Chapter 7, purely digital designs are presented, but here the parallel concept of digitizing continuous-time designs was investigated, so as to allow digital computer implementation.

The z-transform was introduced and used to obtain discrete-time transfer functions of various continuous-time compensators. From these transfer functions, algorithms were generated which would allow the compensators to be mechanized. Finally, some of the pitfalls of digital implementation, such as the influence of the sampling period on the results, and the effects of finite wordlength arithmetic in the computer system, were mentioned. In addition, the basics of digital computer simulation were presented, and two commonly used numerical integration techniques were described. Again, the effects on these of the choice of sampling interval were studied.

5.10 Problems

5.1 Figure P5.1 shows a plant comprising a closed-loop, in which the integral of the error in the loop drives a subsystem modelled by a simple first-order lag. For the system shown in Figure P5.1, determine the closed-loop transfer function. By consideration of any mode cancellations (or the lack of them), say whether you would expect the closed-loop system to be controllable and/or observable.

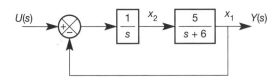

Figure P5.1 A simple closed-loop system.

5.2 For the system of Figure P5.1, obtain a state-space model, using the state variables shown. Check the controllability and observability of this model, and compare the results with those of Problem 5.1.

5.3 A two-input, two-output system is represented by the state-space model:

$$A = \begin{bmatrix} -1 & 1 & 0 \\ 0 & -1 & 0 \\ 0 & 0 & -1 \end{bmatrix}, \quad B = \begin{bmatrix} 0 & 1 \\ 1 & 0 \\ 1 & 0 \end{bmatrix},$$

$$C = \begin{bmatrix} 1 & 0 & 0 \\ 0 & 1 & 0 \end{bmatrix}, \quad D = \begin{bmatrix} 0 & 0 \\ 0 & 0 \end{bmatrix}$$

Test the controllability and observability of the system. Evaluate the various transfer functions in the system, and comment upon the links between these and the results of the controllability and observability tests.

5.4 (a) Investigate the stability of the system represented by the state-space model:

$$A = \begin{bmatrix} -3 & 1 \\ -2 & 1.5 \end{bmatrix}, \quad b = \begin{bmatrix} 1 \\ 4 \end{bmatrix}$$

(b) Show that the system is not completely controllable, but that it is nevertheless possible to design a state-variable feedback scheme to move the unstable pole to the location $s = -4$.

(c) Carry out the design of such a scheme.

(d) Draw a simulation diagram of the system with the feedback in place.

(e) How could the design flexibility in part (c) be used to cope with the situation in which state x_1 was unmeasurable?

5.5 The type of system represented by Figure P5.5 was first proposed as an interesting controllability example by Elgerd (1967). For this system:

(a) Taking the voltages (strictly, potential differences) across the capacitors as both the state variables and the system outputs (three states = three outputs), develop a state-space model of the system.

(b) Find out the conditions under which the system is fully controllable.

(c) If the system is fully controllable, the implication is that the capacitor voltages (the states) can be moved from any arbitrary set of initial voltages to any other arbitrary set of voltages in a finite time. Think of a way in which this might be achieved using the single input voltage in Figure P5.5, and sketch the form of the resulting responses to illustrate how your method achieves its result (no calculated values are required). What would be the practical limitations on the control that could be achieved?

Hint: This *can* be done in principle, and in practice within the limitations mentioned above, without adding anything (such as state variable feedback) to the diagram. However, you will not find the solution in the text – it is up to you to think it out from practical considerations. It is given as an example of the situation in which the mathematics says that the system is controllable, but you might not be able to think how! You may find it easier initially to consider a system having only two of the RC networks present.

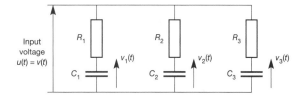

Figure P5.5 An electrical system for controllability investigation.

5.6 Equation (5.5) presented a test matrix for determining the controllability of individual modes of a system. By using the *dual* of this matrix (see the text preceding Equation (5.2) for a definition of the dual), devise a test which will determine the *observability* of individual system modes.

5.7 Apply the test devised in Problem 5.6 to the system of Example 5.2 in Section 5.3.2, and hence confirm the results of that example.

5.8 Consider the system shown in Figure P5.8.
(a) Obtain a state-space model of the system. A selection of possible state variables is indicated.
(b) Develop a state variable feedback controller to place the closed-loop poles at $s = -7$ and $s = -0.6 \pm 2j$. Note that the model of part (a) may contain too many states – a third-order model is required for a third-order system, and if you have a fourth-order one, you should eliminate one of the states before trying the design.
(c) Sketch a modified version of Figure P5.8, with your SVF system in place.
(d) What would be the effect on this design if the feedforward gain in Figure P5.8 were to be changed from 2 to 1? Why?

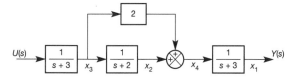

Figure P5.8 An open-loop system for state variable feedback control.

5.9 A system has a transfer function
$$G(s) = \frac{5}{s^3 + 6s^2 + 9s - 50}$$
(a) Obtain a state-space model of the system.
(b) Using state variable feedback, move the unstable pole at $s = 2$ to a stable location. The other two poles should not be moved, and these two poles, together with the moved pole, should lie on a semicircle in the s-plane.

5.10 (a) Using the system of Problem 5.9(a), and the same closed-loop pole set as in Problem 5.9(b), design a tracking system which will have zero steady-state error following a step input. Place the new integrator pole so that it is five times as fast as the fastest pole in the SVF regulator.
(b) Produce a sketch showing how your tracking controller interfaces with the plant, assuming that you can measure all the state variables (problems in Chapter 9 address systems in which this is not the case).
(c) If you have access to MATLAB and the control systems toolbox, build a simulation of the

system and verify its closed-loop eigenvalues and its tracking performance.
Hint: In order to use the MATLAB *feedback* command to feed back the state vector, you will need to define the states as outputs by altering the c and d quantities in the state-space model (because the *feedback* command feeds back outputs, not state variables). The easiest way is to use the *augstate* command, which adds extra outputs to the system model, each one being one of the state variables (type *help augstate* and *help feedback* for details). Also, try inspecting (and, indeed, modifying to suit this problem) the m-files on the accompanying disk.

5.11 Given a system having the following state-space model:
$$A = \begin{bmatrix} -1 & 1 & a_{13} \\ 0 & -2 & 1 \\ 0 & 0 & -3 \end{bmatrix}, \quad b = \begin{bmatrix} 0 \\ 0 \\ 1 \end{bmatrix},$$
$$c = [1 \ 0 \ 0], \quad d = 0$$
(a) Draw its simulation diagram.
(b) Are there any values of a_{13} for which the system would be unstable? (You should be able to answer this part purely by inspection of the state-space model above).
(c) By considering the s-plane pole–zero pattern of the system, determine whether there are any values of a_{13} which would result in an uncontrollable or unobservable mode (there are three values of a_{13} leading to such behaviour – but one of them is impractical).
(d) Confirm the results of part (c) using rank tests.

5.12 The design in this question is similar to previous questions, but the result does not drop out so easily. This is included to illustrate the amount of manual calculation which can be involved, in even a simple design, if computer assistance is not used. The computer-aided solution is easy! The problem involves an unstable system with the following state-space model:
$$A = \begin{bmatrix} 1 & 1 & 0 \\ 0 & -1 & 1 \\ 13 & -1 & -2 \end{bmatrix}, \quad b = \begin{bmatrix} 2 \\ -1 \\ -2 \end{bmatrix},$$
$$c = [1 \ 1 \ 0], \quad d = 0$$
(a) Determine the eigenvalues of the A matrix, confirming the instability.

(b) Design a state variable feedback system to move (only) the unstable pole to $s = -3$ in closed-loop.

(c) Draw a simulation diagram of the resulting scheme.

(d) Show that the resulting closed-loop system will have the required eigenvalues.

(e) If a package such as MATLAB and its control systems toolbox is available, repeat the whole exercise with as much computer assistance as you like.

5.13 (a) Obtain a state-space model for the system

$$\frac{Y}{U}(s) = \frac{2 - s^2}{s^3 + 7s^2 + 6s + 3}$$

(b) Design a state variable feedback regulator (assuming that all your chosen state variables are measurable) to position the closed-loop poles in a Butterworth configuration at an associated frequency of 2 rad s^{-1}.

(c) Calculate the steady-state output of the closed-loop system in response to a unit step disturbance on the new 'reference' input, and comment on its value.

(d) Convert the design to that of a tracking system, placing the integrator pole five times as fast as the fastest closed-loop system pole. Confirm, by calculation, that the new closed-loop steady-state performance is as required.

5.14 (MATLAB and the control systems toolbox, or a similarly featured CACSD package will be necessary to attempt this problem.)

(a) For the regulator system designed in part (b) of Problem 5.13, determine (by simulation) the peak amplitude of the system *input* signal (u) following a unit step disturbance to the closed-loop 'reference' input.

Hint: If using MATLAB, alter the '*C*' and '*D*' quantities in the state-space model so as to define an extra system output which is simply equal to the input (add an extra row of zeros to *C*, and a '1' as an extra row for *D*). The MATLAB *step* commands etc. will then plot its response, as they always plot all outputs.

(b) Using the simulation of part (a) discover, by experimentation, to what value the frequency associated with the Butterworth pole-set design would have to be altered, in order that the plant input signal would not exceed 1 unit following the unit step 'reference' disturbance.

5.15 (a) For the system model with *A* matrix

$$\begin{bmatrix} 2 & 1 \\ -4 & -3 \end{bmatrix},$$

find the eigenvalues and eigenvectors.

(b) Assess the stability of the system.

(c) Find the responses of $x_1(t)$ and $x_2(t)$ due to each of the following sets of initial conditions:

(i) $x_0 = \begin{bmatrix} 3 \\ 0 \end{bmatrix}$ and (ii) $x_0 = \begin{bmatrix} 1 \\ 1 \end{bmatrix}$

Use both the direct method, and the method via the transition matrix and inverse Laplace transform, as a cross-check.

Notice how the eigenvectors always specify the distribution (proportion) of the modes between the state variables.

5.16 (a) Find the eigenvalues of

$$A = \begin{bmatrix} -6 & -3 \\ 4 & 1 \end{bmatrix}$$

(b) Comment on the stability of the system.

(c) Find the eigenvectors, and hence write down the equation of the state response to a general initial condition vector x_0.

(d) Plot the eigenvectors in the state-space (x_1 vs. x_2 plane, in this case).

(e) Choose some initial condition vectors aligned with each eigenvector, and use the result of part (c) to calculate the time responses for each.

You should find that, in every case, only the mode corresponding with the eigenvector to which the initial condition vector is aligned, appears in the response. This is because such initial condition vectors generate zeros which cancel the poles associated with the non-appearing modes.

5.17 A system has eigenvalues $\lambda_1 = -2$, $\lambda_2 = -4$ and $\lambda_3 = -3$. Write out its characteristic equation, and hence the companion form of the *A* matrix for a state-space model.

5.18 Repeat Problem 5.17 for the system with eigenvalues λ_1, $\lambda_2 = -1 \pm 2j$. Use a similarity transform (see Section A1.6) $W^{-1}AW$ (where *A* is the companion form plant matrix and *W* is the corresponding modal matrix), to transform *A* to a diagonal matrix having the eigenvalues λ_1 and λ_2 on the diagonal.

5.19 Write a program in a suitable high-level language (or MATLAB or spreadsheet if you prefer) to obtain the step response of the first-order system $50/(s + 20)$ over a period of 0.2 s. Use (a) Euler and (b) 4th-order Runge–Kutta integration with steplengths of both 0.005 s and 0.2 s in each case. Run your simulations and compare the results.

5.20 Use an appropriate simulation package to investigate each of the following systems.

(a) A plant

$$\frac{Y(s)}{U(s)} = \frac{40}{s^2 + 10s + 20}$$

controlled by a PID controller having $K_c = 2$, $T_d = 0.1$ s and $T_i = 0.5$ s. Can you achieve a faster response without causing more than about 5 per cent step overshoot, by varying the controller parameters? Note that most packages should not allow you to use the PID controller in its ideal form; you will need to make the controller transfer function

$$\frac{U(s)}{E(s)} = \frac{K_c(T_d s^2 + s + 1/T_i)}{s(\Delta s + 1)}$$

where Δ is a small number, in order to make its transfer function 'proper' ($\Delta = 0.02$ worked well). SIMULINK has a PID controller block available (in a 'PID controllers' library within the 'Extras' library).

(b) An inertial accelerometer, consisting of a frame which is fixed to the object whose acceleration is to be measured. Within the frame, a small mass m is suspended by a spring (of stiffness K_s) and a parallel damper (of damping coefficient K_d), so that it can move relative to the frame, in the direction of acceleration.

If the displacement of the frame from some datum position is x_1 (m) and that of the internal mass m from the same datum position is x_2 (m), then the displacement of the accelerometer mass relative to its frame is $(x_2 - x_1)$ (m). This displacement is measured, and is assumed to represent the acceleration

$$\frac{d^2 x_1}{dt^2}$$

according to the transfer function

$$\frac{X_2(s) - X_1(s)}{s^2 X_1(s)} = \frac{m}{ms^2 + K_d s + K_s}$$

Let m be 0.2 kg, K_s be 5 Nm^{-1} and K_d (in N/(ms^{-1})) is yet to be determined. Investigate the accelerometer with a view to determining the circumstances in which it will measure acceleration and not, for example, velocity. A

suitable value for the damping constant K_d should be found by experiment or otherwise. Frequency and/or step response data could be used.

5.21 A plant of transfer function

$$\frac{Y(s)}{U(s)} = \frac{20}{s(s+1)(s+10)}$$

is to be controlled in the closed-loop with a cascade lead compensator of transfer function

$$D(s) = \frac{U(s)}{E(s)} = \frac{10(s+1)}{(s+10)}$$

implemented in digital form. Obtain the transfer function in z for the controller converted using a sampling interval of 0.1 s in conjunction with:
(a) the 'simple' conversion method;
(b) the Tustin conversion method;
(c) the matched pole–zero conversion method.

5.22 In Problem 5.21, what sampling interval would you have recommended on the basis of the 'ten times the closed-loop natural frequency' rule? Would it have been in reasonable agreement with that using the 'five times the highest natural frequency in the loop' rule?

5.23 Repeat Problem 5.21 with sampling intervals of 0.2 s and 0.05 s used with the simple conversion method. Use an appropriate computer simulation program to investigate the performance of the resulting controllers.

5.24 A PID controller having $K_c = 1$, $T_d = 0.2$ s and $T_i = 1$ s is to be implemented in digital form.
(a) What sampling interval would you recommend?
(b) Obtain the transfer function in z for the required digital controller using the 'simple' conversion method.
(c) Use your transfer function to write an algorithm for the controller.

5.25 Obtain an algorithm for the digital PID controller of Problem 5.24 using the explicit-integration conversion method. The algorithm should include rate-limiting for input steps and appropriate measures to combat 'integral wind-up'.

5.26 Use the Tustin substitution to obtain a transfer function in z for the controller of Problem 5.24, and show by computer simulation that an oscillatory controller action results.

6 'On–off' control and practical control devices

6.1   Preview

This chapter can be understood without having studied much of the remainder of the text. A basic familiarity with feedback control concepts and with the PID (three-term) controller is required; and Section 4.7 on controller implementation should be read.

Every other chapter of this text involves the analysis or design of feedback control loops that operate continuously (be they controlled by analog or digital controllers). There is another class of control systems, in which sequences of actions must be performed in order, at the correct times relative to each other, or when certain sets of conditions become true. Anyone who has seen robotic manufacturing systems in operation has witnessed the results of this kind of control.

Such sequential control systems are often ignored in control engineering texts, because the control systems themselves (but not the applications – such as the robots) are perceived to require little analysis, and are seen as 'common sense' (as indicated by the first sentence of this chapter). However, an outline appreciation of this type of control can be extremely useful, not least because the most common device for performing such control (the programmable logic controller) is now able to perform many other kinds of control too (such as PID loops), and is therefore the preferred means of

implementation for many control schemes where a mixture of continuous and sequential control is required.

In this chapter, the topics studied include:

*   the idea of the 'on–off' type of control as opposed to the control of an analog variable
*   some methods by which such control can be achieved, and the relative advantages and drawbacks of each method
*   programmable logic controllers (PLCs) and their application to 'on–off' control
*   practical implementation of the controllers treated in earlier chapters, including the use of PLCs.

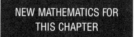

There is no new mathematics in this chapter. The one numerical example does use a difference equation, of the kind introduced in Section 2.8.1. The basic concept of a digital logic signal is also used, whereby a signal can adopt one of two levels: 'logic 1' (representing a condition which is true, or a voltage level of, say, $+15$ V), or 'logic 0' (a false condition, or a voltage level of, say, 0 V). However, the associated logical operations of AND, OR, NOT, and so on are not used in this text.

6.2 'On–off' control

Many practical control problems consist solely of the requirement to switch equipments on and off in the correct sequence, at the right time, or when a particular combination of events occurs. For the purposes of this chapter, such systems will be termed 'sequential' control systems. Even systems which predominantly involve continuous or digital control loops (using, for example, PID control) may also contain such sequential control requirements.

Such a system is exemplified by the simple arrangement of Figure 6.1, the specification for which is as follows. The conveyor is to be started when push-button PB1 is pressed and is to be capable of being stopped by an emergency-stop

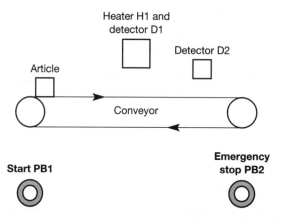

Figure 6.1 A simple conveyor system.

button PB2, which closes a switch when pressed (often, for 'fail-safe' reasons, an emergency-stop button will *open* a circuit, but PB2 in this example closes one). When the article on the conveyor reaches detector D1 (assumed to close a switch when the article reaches it) the heater H1 is to be energized for five seconds. The conveyor is to continue to run until the article reaches detector D2 (also closing a switch when the presence of the article is detected).

The technology that can be used to perform the control includes the following.

- Switches, relays, and timer units. A relay is a switch operated magnetically by a current in a coil; either the switch is open with no current and closes when a current flows ('normally open'; shown in diagrams as N/O) or vice versa ('normally closed'; N/C). Various changeover configurations and latching arrangements are also possible. The arrangement for the 'normally closed' case is shown diagramatically in Figure 6.2, in which the operating current flows in the coil, causing the normally closed switch to open. The coil is normally electrically isolated from the switch contacts. The control approach using switches and electromechanical relays has the advantages of simplicity and direct connection between the components implementing the control logic and the system outputs. The balancing disadvantages are that significant voltages and currents are being handled throughout the control system, which is wasteful of power and causes contact wear on the switches; and that any timer elements are likely to be electronic anyway (although pneumatic timers are available). The

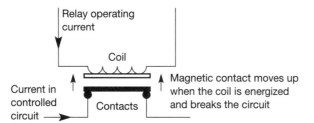

Figure 6.2 An electromechanical relay (normally closed type shown).

method is likely to be the most effective one if the logic is relatively simple in relation to its number of outputs; it might actually be the best in this case, and almost certainly would be were timing functions not required. The circuit principles of the method are replicated in the 'ladder diagram' method of programming PLCs (see later in this chapter).

● Hard-wired electronic logic is likely to prove cheaper than switches and relays if the logic functions required are complicated in relation to the number of inputs and outputs, because logic circuits are now very cheap and consume little power. Again, however, there are balancing disadvantages. The logic circuit outputs are typically $0\,V$ for a logic 0 and either $5\,V$ or $15\,V$ for a logic 1. The current capacity is poor, so circuitry with adequate drive capability must be connected between the logic system outputs and the loads to be controlled. Such circuitry (often based on opto-isolated devices) frequently costs more than the logic itself. It is also necessary to provide an adequate stabilized power supply for the logic. If $5\,V$ logic is used, it can be difficult to achieve sufficient immunity from electrical 'noise' in some circumstances. Finally, the method shares with the 'switches and relays' approach the drawback that any changes require hardware modification to the circuit (except perhaps timer settings, which may be adjustable via thumbwheel or other switches). It is unlikely to offer advantages over switches and relays in an example such as that of Figure 6.1, except in respect of the timer for the heater.

● Fluidic logic systems are also available. These are expensive to install and maintain compared with the other methods, but their major advantage is that of intrinsic safety, having no electrical connections (important for operation in inflammable atmospheres).

● Dedicated microprocessor-based systems offer the advantage over hardwired logic that the logic functions implemented are performed in software so that, provided that the same input and output connections are to be used, amendments can be made simply by downloading a new program or inserting a new ROM (a read-only memory chip) containing the new program. The method therefore has the advantage of flexibility which often makes it the first choice. It shares with electronic logic the drive capability problem and the power-supply requirements, and it tends to be more expensive if the logic functions required are reasonably simple; the more complicated they are, the more the software approach comes into its own. It also shares the interfacing difficulty of hardwired logic, which means that a specialist electronic engineer is likely to be needed to implement the hardware side of such a system. It was for that reason that PLCs were introduced.

● Programmable logic controllers (PLCs) are often now the standard solution where the logic requirements are not ultra-simple. They are, technically, microprocessor-based systems, and therefore have the same flexibility of software implementation of the required control logic. They also have inputs which operate by detecting the state of switches connected across them (making them compatible with switch and relay logic) and outputs which open and close switches which can be connected in external circuits. They can generally be programmed in either of two ways: in a code resembling a computer assembly language, or by ladder diagrams. The latter are simply representations of the logic as if it were implemented by means of switches and relays! The method originated in the American automobile industry (as did the PLC itself), to meet the needs of car assembly lines which had to be regularly changed to produce a different model. The programmable flexibility of the PLC made such changes relatively cheap and easy, while the ability to program the PLC using ladder logic meant that the technicians familiar with switch and relay logic systems could easily perform the programming. No specialist electronics engineers or programmers were necessary. Over the past few years, PLC systems have become much more sophisticated than simple switching and timing devices, being able to handle analog signals and PID control loops too, for example. As a result, PLCs are now one of the most popular choices for rapid and relatively inexpensive implementation of straightforward control systems.

6.3 Ladder logic and PLCs

The term 'ladder logic' describes a diagrammatic form of expression of on–off control situations in which the requirements are drawn as if they are to be implemented in the form of switches, relays, delay elements and associated blocks. It is the most usual method of input of such systems to a programmable logic controller (PLC) and, as mentioned above, it was originally introduced to make PLC programming easier for works electrical staff whose previous experience was with such hardware devices rather than with software. The method is explained via the PLC arrangement of Figure 6.3, in which two user equipments are controlled on an on–off basis in response to three switch inputs connected to terminals 00, 03 and 05. These could be ordinary hand-operated switches, or detectors of an on–off type such as limit switches. The large box represents the PLC itself, so that all components shown inside it are internal to the PLC. Although the PLC is based on

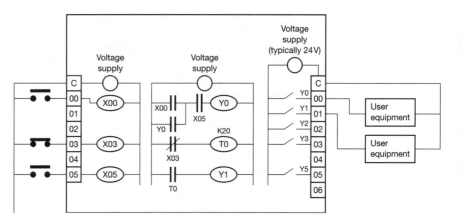

Figure 6.3 A typical simple PLC arrangement.

microprocessor technology, throughout this chapter the terms 'contact', 'coil', 'relay' and so forth will be used as though these elements within the PLC are physical devices. Of course, they do not physically exist, being just parts of the software, but that fact is transparent to the user.

The elements X00, X03 and X05 on the left-hand side of the diagram represent the operating coils of relays which are connected in series with the external switches and a power supply as shown. If a switch is closed the corresponding relay coil is energized. There will be more inputs available on a practical PLC (there can be hundreds); just a few have been shown in order to illustrate the principle. Notice that the external switches connected to X00 and X05 are normally open (N/O) while that on X03 is normally closed (N/C). In practice, some of these switches would be detectors such as limit switches or inductive or capacitive sensors, which would indicate when an object had reached a particular position, or a variable had reached a particular value.

Referring now to the middle circuitry in the PLC, the 'capacitor-type' symbols represent switch contacts which will open or close as if they were relay contacts operated by the appropriate coil. For example, switch X00 is open if coil X00 is not energized and closed if it is. The same principle applies to the other switches except for X03, which is *closed* if coil X03 is not energized and open if it is.

Y0 and Y1 are similar in principle to X00, X03 and X05, but as well as being able to operate switches within the middle block of the PLC, they can also operate those such as Y0 and Y1 on the right of the diagram, which are directly controlling equipments external to the PLC. It is usual for these external switches to have a 24 volt supply to them from within the PLC. If other voltages, or powers beyond the capability of the PLC switches, are to be switched, the currents supplied by the PLC outputs can be used to operate heavier-duty switching devices outside the PLC.

The element T0 operates like the other 'relay coils', except that the operation it controls is delayed by the time specified by the constant (K20 here) beside the coil. Such timers often only delay a closing of the switch (or only an opening if it is normally closed), the opposite operation being undelayed. A more complicated arrangement is needed if the opening operation also requires delaying. For the purposes of this chapter, it will be assumed that energizing a timer coil closes the linked normally open switch after the specified delay, but de-energizing it reopens the switch immediately.

Arrangements of switches and relays of the type contained in the middle block of the PLC are known as 'ladder logic' (from the resemblance of the diagram to a ladder) and the principle is that it works exactly like a similar electric circuit. To program the PLC, a computer-based device is attached and the ladder logic is built up on the computer screen by means of the keyboard and suitable computer software. This software both arranges the screen and converts the ladder logic into the code (resembling computer assembler code) that the PLC actually uses. It is also possible to program the device from a keyboard in its assembler code, but that method is less frequently used. For either method of programming, the programming device is normally detachable from the PLC, partly because of cost (one device can be used for several PLCs) and partly to make it impossible for unauthorized persons to alter the program – a real possibility if it is controlling a manufacturing process whose operator is paid according to output!

To show how a 'ladder diagram' program works, the availability of the elements shown in Figure 6.3 will be assumed. They are based on those of

Mitsubishi-made PLCs, but their range and function has been simplified for easier understanding during a first treatment. The first example of this chapter will explore how the system of Figure 6.1 could be controlled by such a PLC.

Before beginning to solve the example, it is worth noting the portion of the arrangement in Figure 6.3 including X00, X05, and Y0. It is a 'latching' arrangement whereby, assuming that switch X05 is closed, Y0 will be latched on (by its own 'contacts') when switch X00 closes, even if X00 is only closed for an instant. Y0 can subsequently be switched off only when switch X05 opens – even if again it is only for an instant. Such a latching circuit is important if the external switches are of the push-on-release-off type, or if they are detectors which operate only momentarily.

It is noteworthy that although most types of PLC can be programmed either in assembler language or by the ladder-logic method, the ladder logic is much the more common approach in practice. There are also other methods of programming PLCs, as mentioned later.

Example 6.1 *Produce a ladder diagram for PLC control of the system of Figure 6.1*

The following connections will be assumed:

> X0 is the 'start' push-button PB1.
>
> X1 is the 'emergency-stop' push-button PB2.
>
> X2 is the detector D2.
>
> X3 is the detector D1.
>
> Y0 is connected in the conveyor circuit.
>
> Y1 is connected in the heater circuit.

The conveyor is to start when the push-button switch PB1 (X0) is pressed, and is to stop when either detector D2 (X2) or push-button PB2 (X1) produces a switch closure. It would be easier if D2 and PB2 opened a switch (as noted earlier, this could often be the case for an emergency-stop button, such as PB2). This point can easily be solved in the PLC by using the inputs corresponding to PB2 and D2 to operate normally closed contacts within the PLC.

The first rung of the ladder diagram in Figure 6.4 (containing X0, Y0, X1 and X2) uses the 'latching on' idea mentioned above. If D2 and PB2 have not operated, the two normally closed switches will be in the closed position, so momentary operation of PB1 will close the X0 contacts and output Y0 will then be energized and will latch on via its own contacts in parallel with X0, thus running the conveyor. When either PB2 or D2 gives a contact closure (even momentarily), the conveyor will stop. Incidentally, note that for safety purposes, the emergency-stop button (PB2) would be linked directly into the conveyor drive system, and not only via the PLC.

The second rung includes a timer element; it is assumed that a timer T0 is available, which delays the operation of any switch (controlled by the timer contacts) by the number

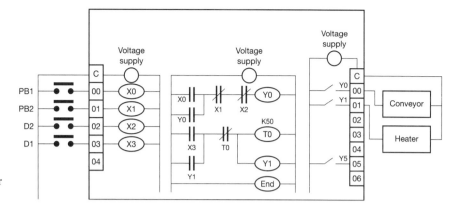

Figure 6.4 PLC implementation of conveyor control.

of tenths of seconds specified by the timer constant (shown as K50 here). It is interesting to note that switch-off is not delayed, so a more complicated circuit is needed here because the switch-on of the heater is to be immediate when D1 closes, but the switch-off is to be delayed. This is achieved by using detector D1 to energize coil X3, latching the circuit on and closing switch Y1 to operate the heater. The timer T0 will not have any effect until $50/10 = 5$ seconds have elapsed, when it will interrupt the circuit to Y1 and switch it off – also 'unlatching' its own input.

For a simple example like this, the PLC may be too expensive in comparison to switches, relays and a timer, but the example serves to illustrate the basic principles of PLC operation.

It is noteworthy that PLCs can also be used to implement digital control of continuous systems, as will be described shortly. For more detail on PLC systems see Webb (1992) and Kalani (1988).

6.4 Practical 'continuous' controller implementation

The word 'continuous' is in inverted commas because a digital controller, strictly speaking, deals with discrete rather than continuous quantities; the notion of 'continuous' in this context is to distinguish between sequential (that is, on–off) control and the control of variables which can adopt any value within a given range.

Controller design discussed elsewhere in the text is centred on obtaining a transfer function for the continuous controller, be it analog or digital. This section will focus briefly on methods of producing a controller with that transfer function. First, the analog electronics approach using operational amplifier (op-amp) circuits will be very briefly described, following on from the ideas introduced in Section 4.7; then the possibilities for digital control in 'continuous' situations, including the use of PLCs. Finally, some of the methods of interfacing the controller to the plant it controls will be described briefly.

It is perfectly possible to make non-electronic analog controllers (pneumatic, or even purely mechanical in some situations), but they are now rare in new equipment (their op-amp successors are increasingly so, too). Figure 6.5 shows an op-amp based PID controller having $K_c = -(R_2/R_1 + C_1/C_2)$, $T_d = -C_1R_2/K_c$, and $T_i = -K_cR_1C_2$. The derivation of these expressions is left as an exercise for the

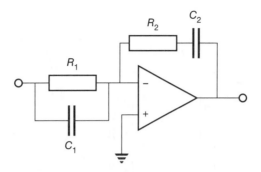

Figure 6.5 Operational-amplifier based controller circuit.

reader (who is reminded that the gain for the inverting op-amp configuration is given by Equation (4.30) in Section 4.7).

Unfortunately, the capacitor directly from the input to the inverting input of the op-amp has low impedance at high frequency and therefore makes the circuit liable to high-frequency noise (a problem with PID control in general). In addition, it can be seen that it is not possible to set the values of all three controller terms independently of each other. An alternative PID arrangement is suggested in Section 4.7, where a lag-lead compensator circuit is also given. Most transfer functions can be generated by circuits such as these, but they suffer from the normal op-amp limitations – especially drift (the slow variation of output voltage with temperature) in slow systems, which many control systems are. They also lose out in comparison with digital controllers in respect of being less flexible (in software, it is possible to change parameters easily, or even to change the complete control strategy) and in being generally unsuitable for installation in digital distributed control networks or SCADA systems (discussed later in this section). They do, however, still have a place in low-cost equipment; in controllers which must operate on very 'fast' plant, for which the sampling times of digital controllers are too long; and in interfacing situations as will shortly be described.

Before leaving op-amps, the question of their connection to the plants to be controlled, and to the transducers measuring the controlled variables, will be considered. Here the op-amp has the advantage over digital systems in that, like the real world of the plant, it is analog. The transducer with its detection system is likely to give an output based on accepted standards: a voltage between 0 and 5 V or 0 and 10 V, or a current between 0 and 20 mA or, more usually, 4–20 mA. The 4–20 mA version has the advantage that it is easy to check if the loop is complete, as the situation of zero current cannot arise during normal operation (the pneumatic 3–15 p.s.i. pressure standard works on the same principle). The voltage possibility offers few problems to the op-amp based controller (it can be easily amplified and/or level-shifted if required – see Figure 6.6, in which these operations are achieved by the variation of R_3 and R_2 respectively) and it can be easily converted by an analog-to-digital converter to give a digital input. The current-based possibilities are a little more complicated to handle, but standard electronic modules to convert either current range to a voltage, or to digitize it directly, exist. Actuators of most types are made to accept inputs of one of the above formats.

There is now a considerable choice of technologies for the implementation of digital controllers. It would be possible in principle to use hardwired logic, but that approach suffers from the disadvantage of inflexibility – it is necessary to modify the hardware configuration to change even a coefficient value. Practically all digital

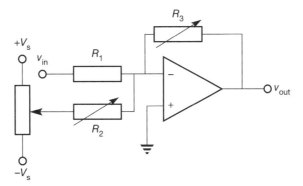

Figure 6.6 A level-shifting circuit.

controllers are therefore based on some form of processor; the following are some of the possibilities.

- A small general-purpose computer such as a PC is easy to buy or make. Appropriate input–output cards can be inserted containing the necessary filters, A–D and D–A converters, and circuits to give the required 4–20 mA current output or the appropriate voltage range. Such a computer is able to act as an operator interface and display and it can be programmed in a high-level language. Its balancing disadvantages are that it is expensive for use simply as a controller because of its keyboard, screen and large memory provision and that its general-purpose processing chip is nevertheless likely to be slow in performing multiplications, which are inherent in the operation of digital controllers. However, the speed of modern PCs means that programs for performing straightforward digital control operations can cycle at up to a few kHz, which is usually more than adequate for the purpose. The situation is different if they have to perform rapid matrix operations (such as those involved in on–line identification for self-tuning control (Chapter 11), or for state-estimation using a Kalman filter (Chapter 9)). For use on plant, PCs in ruggedized, hose-proof cases are available, but at extra cost.

- A dedicated microprocessor-based system will be cheaper than a general-purpose computer, in that the system can be designed to include only the features required for control purposes. These will still include the converters, filters and analog circuits described earlier; the program for implementing the control algorithm will probably be stored in ROM, with parameters normally being user-settable. The setting was traditionally done via a keypad or thumbwheel switches on the unit itself, but it is now common for many controllers to be distributed around a plant, and to be networked to a central computer which sets both the parameters and setpoints, as well as allowing monitoring of plant outputs. Such an arrangement is known as a SCADA system (supervisory control and data acquisition – see Kalani (1988) and Webb (1992)).

   The processors used in the controllers themselves may be general-purpose microprocessor cards (often 16-bit), or – more commonly recently – digital signal processing (DSP) chips. Eight-bit microcontrollers are also employed, especially where most of the task involves sequential on-off control rather than analog control (for which the limited arithmetic capabilities of 8-bit controllers make them rather slow). Originally, DSP chips performed basically fixed-point arithmetic, but devices with floating-point capabilities within their own instruction set are now becoming popular. It is still possible to perform the numerical calculations faster in fixed-point arithmetic, however, and Example 6.2 (below) shows how the coefficients can be determined so that such an approach can be used.

The advantages of the DSP chips over general-purpose ones are that they can perform multiplications as fast as additions (it takes much longer to multiply by shifting and adding as would be usual in a general-purpose processor) and they often have useful hardware features such as timers and interrupts which make it easy to branch to the routine which performs the calculations at equal intervals of time as is necessary for digital controllers. Additionally, their support chips performing the A–D and D–A conversion are not only appropriately fast but often are linked to the main processor via a high-speed serial link, saving pins, and incorporate their own anti-aliasing filter (see Section 1.3.6) whose cut-off frequency can be set in software. Care is needed in this respect as these filters are intended for digital filtering applications and they often have a low-frequency cut-off as well as a high-frequency one. This is not normally appropriate for control systems, where steady-state performance is usually important.

- The capabilities of programmable logic controllers (PLCs) in executing the on–off type of control have already been examined, but the more versatile varieties are capable of performing digital control of analog quantities. In keeping with the idea of a PLC being a directly applicable control device which contains its own means of interfacing with the outside world, the A–D and D–A arrangements are included and work from and to the standard voltage or current ranges (for example, 4–20 mA). One word of caution is to check whether anti-aliasing filtering is included. They are again programmable either in ladder logic or in their own form of assembler code. It must be said that ladder logic is not very convenient for this type of operation, as Example 6.2 will show. However, the recent IEC 1131-3 standard and the use of higher-level PLC programming languages (see, for example, Lewis (1995)) promises to improve matters, as they provide for reusable code blocks and other higher-level 'software engineering' improvements. In Example 6.2, below, the use of ladder logic for analog operations is illustrated.

In practice, industrial control systems normally involve both on–off and analog control, which makes the PLC a particularly strong candidate as the more advanced types can perform both tasks in the one unit, as well as interface with a central SCADA computer for logging, operator display and supervisory control. In a large distributed application, the solution is often to use several PLCs for the on–off control and often a purpose-built microprocessor-based controller unit for each process variable, as this means that less expensive PLC types can be used. A central computer will again communicate with all the local control devices and allow central operator monitoring and operation.

Example 6.2 *To implement a simple controller by computer, DSP and PLC*

Consider a simple controller which has the equation:

$$u = (0.6 \times e) + (0.4 \times last\_e)$$

where u is the controller output, e is the latest sampled value of the error (that is, the controller input) and $last\_e$ is the value of the error one sampling interval ago. Show how this simple controller task can be implemented in floating-point arithmetic by a general-purpose computer; in fixed-point by a device such as a DSP chip; and in ladder logic for a PLC.

The equation is firstly arranged in the form of an algorithm:

> Set $last\_e$ equal to 0
>
> Loop: Input e
>
> Calculate $u = (0.6 \times e) + (0.4 \times last\_e)$
>
> Output u
>
> Set $last\_e$ equal to e
>
> Wait for the next sampling instant
>
> Goto Loop

A computer program written in BASIC (chosen for this demonstration as it is easily understood even by those unfamiliar with it, whereas (for example) 'C' is much less so) might be:

> 10 let laste = 0
>
> 20 let e = input(port address)
>
> 30 let u = 0.6*e + 0.4*laste
>
> 40 output(port address, u)
>
> 50 let laste = e
>
> 60 gosub 600: rem go to subroutine to provide the delay
>
> 70 goto 20

It is assumed that a subroutine to provide a suitable delay will begin at line 600. It could be of the following form:

> 600 let delay% = 500: rem % means it is an integer
>
> 610 let delay% = delay% − 1
>
> 620 if delay% >0 then 610
>
> 630 return

The operation is that the number 500 is reduced by 1 every time the loop 610–620 is executed; the return instruction in 630 is executed when the number has decremented to zero.

It will also be necessary in practice to scale the incoming value of e which is input from the A–D converter, and to perform the inverse scaling on the calculated value of u prior to its output.

It is clear that the high-level language program is a straightforward solution, though a short initial block will also be needed to initialize the input and output ports. The delay subroutine is a somewhat inaccurate method of performing the timing to obtain the appropriate sampling interval, as the time to execute the multiplication instructions will vary slightly; in particular, it will be shorter if a multiplicand is zero. A more accurate way of performing the timing of the sampling would be to write lines 20 onwards (with the delay subroutine call omitted) as a subroutine called by an external timer-driven interrupt.

The fixed-point version of the solution (for use with a DSP chip, for example) has to be modified to suit the situation, in that the multiply instruction will either multiply two 8-bit quantities to give a 16-bit product, or it will multiply two 16-bit quantities to give a 32-bit product. It is necessary both to scale the multipliers appropriately and to do likewise with the quantity to be output. In this case, for simplicity, it will be assumed that the inputs can be between 0 V and 10 V and that those voltages convert to decimal numbers between 0 and 255. It can be noted that multiplying an 8-bit binary number by 256 shifts it eight places to the left, so that 00000000xxxxxxxx multiplied by 256 decimal becomes xxxxxxxx00000000. This suggests a straightforward method of doing the calculation required here. If 0.6 is converted to $0.6 \times 256 = 153.6$ (rounded to 154) and 0.4 to $0.4 \times 256 = 102.4$ (rounded to 102), a procedure similar to the following could be used.

> Clear register B
>
> Loop: Input e and place it in register A
>
> Calculate $u = (154 \times e) + (102 \times \text{contents of B})$; put the answer in registers C and D
>
> Output u from C (the most significant 8 bits of the product)
>
> Set *last_e* equal to e, storing it in register B
>
> Wait for the next sampling instant
>
> Goto Loop

This version required more careful consideration of the scaling than was required for a program in a high-level language. The problem would have been more difficult if (as is often the case) 0–255 actually had to represent a range such as -5 V to $+5$ V. An offset would then have been needed in the program. The likely approach would have been to subtract 128 from the value read in from the A–D converter. The negative value would be stored in the computer and used in the program in '2's complement' form.

Two's complement arithmetic works by representing negative numbers by their 2's complement form, obtained by changing every digit (1 to 0 and 0 to 1) and adding 1 to the result. For example, decimal 27 in 8-bit binary is 0001 1011. In 2's complement, -27 would be 1110 0100 + 1 = 1110 0101. Using this format means that the most significant digit always represents the sign, not a magnitude; thus, for example, an 8-bit number can only represent a number range of -127 to $+127$, or 0–255. The representation has the following advantages:

- Subtraction simply means forming the 2's complement of the number to be subtracted and adding it to the number from which the subtraction was to be done. For example, $48 - 27$, using the 2's complement representation of 27 from above, is

$$0011\ 0000 + 1110\ 0101 = 0001\ 0101\ \text{or } 21\ \text{decimal}.$$

- Multiplication and division still work. If the most significant bit of the result is a 1, then the result is negative, and the 2's complement of the result will give the magnitude.

Examination of the PLC ladder-logic solution confirms that analog control by such an approach is somewhat long-winded! Many PLC types, however, have PID elements provided, so that only the required PID parameters would have to be input. Even without that facility, the PLC implementation does have the balancing benefit that many of the hardware problems have already been solved inside the unit. With reference to the program of Figure 6.7, the following points are noteworthy.

(1) The format of each instruction is that the operation to be performed is indicated to the PLC in an op-code which is entered in a register F670. Parameters for the instruction are entered in registers F671, F672 and so on before the line containing the op-code. The first instruction in the example program (Figure 6.7) clears the register (number 720) in which the quantity *last_e* is to be stored. So the three sub-rungs of the instruction read: 'Constant 0 … into register 720 … put'. The instruction will be carried out whenever the contact labelled M400 closes.

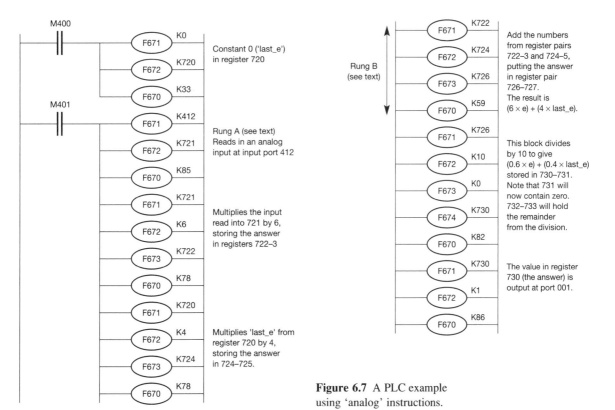

Figure 6.7 A PLC example using 'analog' instructions.

(2) The second instruction (Rung A) will be carried out whenever the contacts M401 close. This will be done by the loop timer. The first three sub-rungs acquire the present value of *e* and can be read: 'Analog input value from port 412 ... into register 721 ... put'.

(3) The input is read in as an 8-bit binary number (which is therefore an integer between 0 and 255) and it is converted to, stored and manipulated as a three-digit BCD (binary coded decimal) number within the PLC. BCD is a common, but not mandatory, number format within programmable logic controllers.

(4) All such numbers are integers, so the 0.6 and 0.4 required in the control calculation cannot be handled directly. The solution is to multiply by 6 and 4 respectively and to divide the answer by 10. The next four sub-rungs of the ladder do the multiplication by 6, and might be read: 'The number in register 721 ... by a constant 6 ... putting the answer in registers 722 and 723 ... multiply'. It is noteworthy that the product of two three-digit BCD numbers will need a six-digit location, or two three-digit locations, to store it. The following four sub-rungs perform a similar function for multiplying *last_e* by 4; and the four sub-rungs after that perform the required addition (labelled 'B' in Figure 6.7).

(5) Division is even more complicated! The division instruction used in the next five sub-rungs after 'B' divides a six-digit number by a specified six-digit constant (10 is only two digits, but the only other available instruction is three-digit by three-digit, which may not be adequate for the dividend). The register specification for the answer (730 in this case) means that the quotient appears in 730 (least significant three digits) and 731 (most significant three) and the remainder likewise in 732 and 733. In the present case, the quotient cannot exceed three digits (why is this?), so register 730 will now contain the answer required. It would be possible to round up to the nearest whole digit by reference to the remainder. If its most significant three digits were 128 or greater, the answer in 730 should be incremented.

(6) A further instruction resembling the instruction to read in the analog input is used to produce the analog output. It will also be necessary to utilize the logic capabilities of the PLC to operate switch M400 once at switch-on, and switch M401 at each sampling instant.

6.5   The real world beyond the controller

This section will discuss how the low-power analog output of the op-amps or D–A converter can be converted to appropriate plant operation. In many practical cases, the direct problems for the control engineer can be solved simply by ensuring that the controller output signal is of the appropriate form – for example, 4–20 mA current – for input to proprietary interfacing circuitry. This section addresses the issues raised by the next stage of the operation.

For a position- or speed-control arrangement, it will be necessary to convert the voltage or current into a force or torque. One method is by an electric motor, which will probably require much more voltage and current than the controller can itself supply (though power op-amps are available which can supply enough current to drive a small motor directly). If it is a d.c. motor, a generally successful approach is to use the controller output to control the duty cycle of a 'chopper' circuit (as indicated in Figure 6.8(a)) to control the armature current and voltage. The chopper is basically a switch which at any instant is either fully closed or fully open, so that

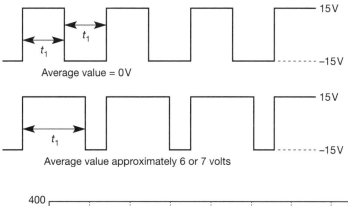

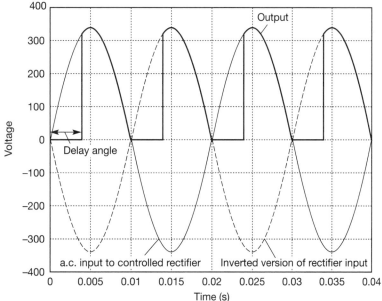

Figure 6.8 (a) The 'chopped' waveform. (b) The controlled-rectifier waveforms.

no power is consumed in the switching devices; it is therefore a very energy-efficient control device. The electronic switching devices in practical choppers do not switch absolutely instantaneously, so practical choppers do consume some power in operation, but it is very much lower than would be the case if an analog current controller such as a power transistor operating in linear mode were to be used. The chopper method is also good in respect of linearity, as the average value of the armature voltage is proportional to the time t_1 in Figure 6.8(a).

Another approach is to use controlled rectification of an a.c. supply (Figure 6.8(b)), in which the average output voltage is varied by varying the delay angle at which the rectifier is allowed to begin to conduct. A disadvantage of this method is that it is not linear – the average voltage depends on the cosine of the delay angle – and so the controller may need to be modified to insert an opposite nonlinearity. These methods of power control are fully discussed in power electronics texts such as Williams (1992). For the purpose of designing the overall control system, it is often possible to treat the interface between the controller and the motor simply as a linear amplifier without dynamics.

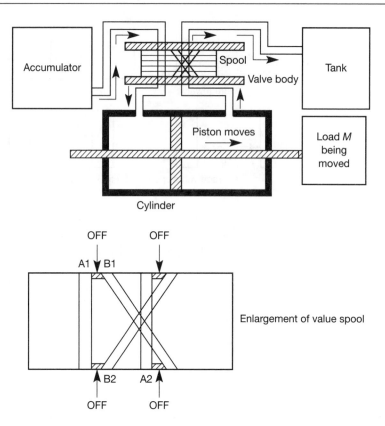

Figure 6.9 A hydraulic system for positioning a mass.

It is also possible to use hydraulic or pneumatic technologies in implementing position and speed control; a commonly used hydraulic arrangement is shown in Figure 6.9. Hydraulic systems are at their best when the movements are quite limited (although they are used in passenger lifts!) but the required forces are very high, which circumstances are often found in rolling and forging operations in the metal-forming industries, and in machinery used in the construction industry, for example.

In Figure 6.9, the accumulator represents the arrangement which maintains a supply of hydraulic fluid at high pressure. When it is required to move the load M to the right, an appropriate current is caused to flow in the operating coil of the spool valve. The spool is moved by magnetic action to allow fluid to flow through channel A1 into the left-hand end of the hydraulic cylinder, which exerts a pressure on the piston, moving it to the right. Its movement will displace fluid from the right-hand end of the cylinder to flow back to the tank via channel A2. If a right-to-left movement is required, the spool-valve current is reversed, moving the spool the other way and allowing fluid to flow into the right-hand end of the cylinder via channel B1 and from the left-hand end of the cylinder via channel B2. The channels are shown diagrammatically to illustrate the principle of operation – they will be more complex in their actual routing in the spool, so as to stay separate from each other.

The spool valve is designed so that the relationship between spool displacement and spool current is reasonably linear, so that the spool can be modelled as a linear system (usually second-order to a reasonable approximation). The hydraulic situation is rather more complicated, with some deadband (the spool movement from the centre 'off' position must exceed a minimum value before any hydraulic fluid flows) and the valve position-flow relationship is usually nonlinear, even when flow does begin.

It is possible for inexpensive spool valves (but not the more expensive ones intended for proportional control in servo systems) to have to move 20 per cent of their travel from the 'off' position before fluid flow begins. It is possible to reduce this by techniques such as applying a small a.c. signal of appropriate amplitude and frequency (known as a dither signal) to the valve spool. Although this greatly reduces stiction (static friction – see Chapter 14), and reduces the effects of the deadband, nevertheless some deadband will remain. Without it, even slight wear on the spool would allow flow into both ends of the cylinder, pressurizing it permanently and causing fluid leakage past the piston seals. Chapter 14 examines the analysis of systems containing such nonlinearities.

The dynamics from the cylinder onwards are dependent on the compressibility of the oil (which changes as it ages and acquires dissolved impurities) and the properties of the load itself.

Many thermal and flow processes also depend on valves which are electrically operated in order to control the flow of a gas or liquid (which would be the fuel or coolant in the case of a thermal process). For a large valve, a two-stage approach is often used; a small electrically operated valve, possibly of the spool type, controls the flow of air from a compressed-air reservoir (or less commonly hydraulic fluid) which operates a larger valve controlling the main flow. The use of hydraulics or pneumatics to provide the power for the system is often more economic than an all-electrical solution, especially if a compressed air supply or equivalent hydraulic system provision is required for other purposes also. The balancing disadvantage is that hydraulic valves normally have significant deadband about their closed position (if they did not, leakage would be likely to occur as noted above).

6.6 Conclusions

This chapter has introduced the idea that not all control systems are continuously-acting, and not all controllers involve feedback loops designed by frequency-domain or state-space methods.

The PLC was introduced as a popular and flexible approach to the control of sequential systems. It was pointed out that modern PLCs can also handle systems involving continuous control loops (of the PID variety, for example). An example of the use of a PLC to handle analog quantities was given.

Finally, some basic aspects of controller implementation, and of the interfacing of controller and plant, were discussed.

6.7 Problems

6.1 Explain the operation of the PLC program shown in Figure 6.3 (the first rung has already been explained).

6.2 Modify the PLC program of Figure 6.4 to make the conveyor return the article to its starting position after a 10-second pause, once it reaches detector D2. (You will need to define an additional detector to ascertain when it has reached the original position, and an additional output for reverse movement of the conveyor.)

6.3 An analog controller is to be implemented digitally by means of a general-purpose computer working in a high-level language. The input and output voltage ranges are -5 V to $+5$ V and the input and output operations are performed by 8-bit converters such that -5 V converts to 1 and $+5$ V converts to 255.

(a) What number does 0 convert to?

(b) If 128 is subtracted to give a binary value with zero in the correct place, and if the 2's complement representation of negative values is used, to what will

(i) -2 V and

(ii) $+2$ V convert?

(c) It is proposed that the coefficients of the controller should be converted to binary on the basis that a coefficient of 1.0 should convert to 64. What should coefficients of $+0.5$ and -0.5 convert to if 2's complement arithmetic is used?

6.4 Write a ladder-logic program to input one of the quantities mentioned in Problem 3 above, and subtract its offset. How could negative values be managed in such a program for multiplication, if 10's-complement arithmetic were used? (10's complement works like 2's complement – see Example 6.2. The number is subtracted from ...00000 to perform the conversion. Thus -27 becomes ...99973, and $48 - 27 = 00048 + 99973 = 00021$. Its properties in respect of multiplication and division correspond to those of the 2's complement form.)

6.5 In a small practical process control application, three process temperatures are to be controlled. Local alarms are to be activated if any temperature goes too high or too low, and data are to be logged by a central SCADA PC. Discuss the technology which could be used to implement the control.

7 'True' digital controllers

7.1 Preview

The reader should be able to understand this chapter if Chapters 1 to 4 and Section 5.8 have been read.

In Section 5.8, the design of digital controllers by producing an analog design and implementing it digitally was explained. That approach is widely used and is very successful in many instances. In particular, it enables the robust capabilities of PID control to be combined with the benefits of digital controllers in respect of precision and ability to communicate with and be supervised by an overall supervisory computer. A balancing disadvantage is that digitization must always produce some degradation of the 'analog' controller dynamics, so that, in an ideal world where operational amplifiers did not drift, had no nonlinearities and no noise, the analog controller would be superior in performance.

An alternative approach is to design the controller from the outset on the basis that it is digital, and to make use of the fact that its output is constant between sampling instants. This chapter will show that, in suitable cases, it is possible to produce higher performance from such a controller than an analog one could achieve.

The analysis is best performed by the z-transform approach using the 'true' z-transform (rather than one of the approximations which were used in Section 5.8). The reason for choosing to use the 'true' z-transform will emerge as the chapter proceeds.

In this chapter, the reader will learn:

- the nature of 'true' digital controllers as opposed to those produced by digitizing analog controllers
- how to produce a z-transform representation of a plant whose Laplace transfer function is known
- how to obtain the transfer functions in the z-domain for controllers based on dead-beat, Kalman and Dahlin algorithms
- how to modify the Dahlin controller to remove the effect of 'ringing poles' in the controller
- the benefits and problems of each type of controller discussed.

> NEW MATHEMATICS FOR THIS CHAPTER
>
> The 'true' z-transform is covered in Appendix 5 and referenced there as required.

7.2 A z-transform representation of a digital control system

Figure 7.1 represents the arrangement of a simple digital control system. The switches represent the sampling action of the control computer as it inputs the value of the error signal e, performs its calculations, and outputs the required value of the controller action u. (In practice, the computer would probably sample the input r and the output y and subtract y from r to give e as the first step of its sums, but the analysis is unaffected as we explain later.)

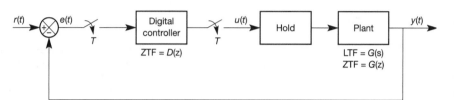

Figure 7.1 A simple digital control system.

One complication is the way in which the output of the computer is used in order to produce the controller action u. The computer is producing a number every sampling interval and outputting it as shown by the shaded impulses in Figure 7.2, but common sense suggests that a signal like the broken or solid lines in Figure 7.2 will be more effective in practice.

The signal represented by the broken lines in Figure 7.2 is produced from the shaded impulses by simply 'catching' the computer output at each sampling instant and holding u constant at that value until another output is produced. The solid-line

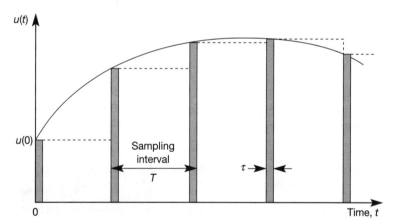

Figure 7.2 Representations of a sampled signal.

signal in Figure 7.2 is a 'smoothed' version of that represented by the broken lines – in other words, it is what we think the signal probably ought to look like in continuous-time, based upon its known values at the sampling instants. Intuitively, the solid-line signal will be the best version to use, but there are three powerful arguments to the contrary:

(1) It is easy in practice simply to 'catch' and hold the output value, but more difficult to smooth it effectively (see also point 3 below).

(2) The plant will itself prove to be a very effective source of smoothing if the sampling rate is reasonably high, so we need not do it (except perhaps as a result of considerations such as reducing actuator wear).

(3) If we smooth, we cannot enjoy the advantages of the 'true' digital control approach, which utilizes the discrete nature of the controller output. One of the problems with 'smoothing' is that this implies filtering of the signal in some way, which would inevitably introduce some phase lag. The 'smoothed' version would therefore occur later in time than shown in Figure 7.2 – we cannot actually generate the signal represented by the solid line in Figure 7.2, but only some degraded version of it.

We call the device which catches and holds the computer output a *zero-order hold*. The 'zero-order' means that it catches only the value of the output and not, for example, its rate of change. If the rate of change were stored and held constant, the hold would be called a first-order hold. The difference in their performance is shown in Figure 7.3. In practice, first- and higher-order hold elements are not used, as negligible advantage would result.

The rules for converting the elements of a diagram such as Figure 7.1 to a z-transform representation are:

- If the block in question has, or behaves as if it has, a sampler both before and after it, its Laplace transfer function should be converted directly to a z transfer function by means of the tables (Table A5.1 in Appendix 5).

- Any blocks which are not separated by a sampler should be combined in Laplace form before conversion.

- When all blocks have been converted to the z-domain, the usual rules of block diagram handling apply (Section 2.6); and the z-transform of the response of the system to any input may be determined by multiplying the overall transfer function in z by the z-transform of the applied input.

- In determining the response, it must be borne in mind that the z-transform only gives the value of the response *at the sampling instants*.

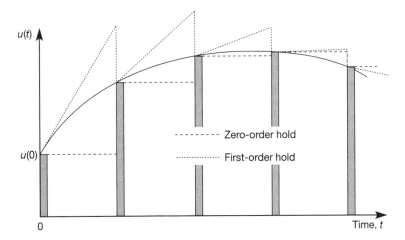

Figure 7.3 Zero-order and first-order holds compared.

Referring to Figure 7.1, we will seek to do the conversion to the z-domain by using the rules above. The 'hold' and the plant are not separated by a sampler (the hold would not work if they were!) so they must be combined in the s-domain before conversion. This means that we must determine the Laplace transfer function (LTF) of the hold. The following derivation refers to Figure 7.2, but considering *only* the samples themselves (the shaded impulses).

The input signal to the hold is a series of pulses, the strength of each of which is equal to its height (that is, the value of u at the peak of the impulse) times the time for which the switch is closed (τ), giving $u\tau$. When a new pulse arrives, the hold must discard its last value and acquire the new one. To acquire the new one, it converts an impulse of strength $u\tau$ to a steady value of u. Integration will convert the impulse into a steady value equal to the strength (area) of the impulse – and in Laplace terms integration means dividing by s. We will also have to divide by the time τ as we wish to hold u, not $u\tau$. The next problem is how to remove the value already stored! If we just integrate, the new value will be added to the old one, but here it is required to replace the old value, so the latter must be subtracted when the new one is stored.

So 'catching' the new value implies a transfer function $1/(s\tau)$. Discarding the old one therefore implies the subtraction of a similar quantity representing the impulse 'caught' one sampling interval (T) previously, and that delay can be represented by e^{-sT} or z^{-1}. The combination leads to a transfer function for the hold of

$$\frac{1}{s\tau} - \frac{e^{-sT}}{s\tau} \quad \text{or} \quad \frac{1 - e^{-sT}}{s\tau}$$

The τ in the denominator appears to be unknown and therefore a problem, but it will divide out in due course with the τ which, as is explained in Appendix 5, is 'hiding' within the z-transform. The practical conversion for the zero-order hold is therefore

$$\frac{1 - e^{-sT}}{s}$$

This is now combined with the $G(s)$ for the plant before the combined transfer function in s is converted to one in z. The procedure is simplified by the fact that the $(1 - e^{-sT})$ just converts to $(1 - z^{-1})$.

Example 7.1 *Converting a system like Figure 7.1 to the z-domain*

Consider a system arranged as in Figure 7.1, with the plant LTF $G(s) = k/s$. For simplicity, at this stage, we shall imagine that the digital controller simply has a gain of unity. Find the z-domain transfer function relating y to r.

$G(s)$ combines with the LTF of the zero-order hold (discussed above) to give

$$k \frac{1 - e^{-sT}}{s^2}$$

Note now that this is of the form $F(s)(1 - e^{-sT}) = F(s) - e^{-sT} F(s)$. The z-transform of the first term will be the z-transform of $F(s)$ $(= F(z)$, say). The z-transform of the second term will also be $F(z)$, but delayed by one sampling period, that is, $z^{-1}F(z)$. The overall z-transform is therefore $(1 - z^{-1})F(z)$. Thus, returning to the example, the numerator term simply transforms to $(1 - z^{-1})$, leaving us to transform the remaining term k/s^2. From the tables, we find this to be

$$\frac{kTz}{(z-1)^2}$$

We also note that $(1 - z^{-1}) = (z - 1)/z$. Multiplying the two terms gives:

$$\frac{z-1}{z} \cdot \frac{kTz}{(z-1)^2} = \frac{kT}{(z-1)}$$

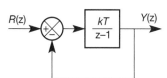

Figure 7.4 A z-domain unity negative feedback system.

The system has therefore already been reduced to that shown in Figure 7.4. It is now possible to apply the '$G/(1 + GH)$ rule' of block diagram reduction (Section 2.6), because all the remaining blocks are effectively separated by samplers. Note that, as the samplers all open and close simultaneously, it will make no difference if the digital controller is arranged so as to form the error signal itself, by inserting further samplers between the reference input and the summer and in the feedback link. There is effectively a sampler at the output, as it has been noted that z-transforming only gives signal values at sampling instants. The closed-loop transfer function in z is therefore

$$\frac{Y(z)}{R(z)} = \frac{kT/(z-1)}{1 + kT/(z-1)} = \frac{kT}{z - 1 + kT} \tag{7.1}$$

As the z-transform of a unit step is $z/(z - 1)$, the z-transform of the step response of the closed-loop system will therefore be given by

$$Y(z) = \frac{z}{z - 1} \times \frac{kT}{z - 1 + kT}$$

which can be divided into partial fractions as

$$Y(z) = \frac{1}{z - 1} - \frac{1 - kT}{z - (1 - kT)}$$

The step response therefore has two parts. The first, of transform $1/(z - 1)$, looks similar to the entry $z/(z - 1)$ in Table A5.1 (Appendix 5), which is the transform of a unit step. However, there is no numerator term in z, so a multiplication by z^{-1} must also therefore be involved, which represents a one-interval delay. The first term can therefore be deduced to be a unit step delayed by one sampling interval.

The second term requires a little more thought to match it with an entry in Table A5.1 (Appendix 5). Since $(1 - kT)$ is a constant, as is e^{-aT}, it resembles a constant times

$$\frac{z}{z - e^{-aT}}$$

which is the z-transform of e^{-aT}. However, once again there is no numerator z term, so the one-interval delay must be present. Also, for stability, e^{-aT} must be positive and less than unity, and we note that $(1 - kT)$ will only be positive if $kT < 1$.

To determine the actual response of the system of Example 7.1, we need to invert the z-transformed result back to the time domain. It is probably easier to do this by the 'difference-equation' method (Appendix 5, Example A5.2). Doing so produces the step response graphs shown in Figure 7.5 (the MATLAB m-file *fig7_5.m* on the accompanying disk produced the plots), in which the following points are noteworthy:

- Only sampling-instant data points are shown; no attempt has been made to join them up because the z-transform only tells us what happens at sampling instants.

- The response depends on the product kT, so increasing the sampling interval has the same effect as increasing the system gain. That observation is strictly true only for this particular $G(s)$, but it is invariably found that increases to T and k are generally similar in their effect.

- From Equation (7.1), we see that the closed-loop system has a *z-plane* pole at $z = (1 - kT)$. This is real for all real values of k and T, and will move in the negative direction as k or T increases. The z-plane stability rule is that all poles must be inside a circle of unit radius centred on the origin for stability (see

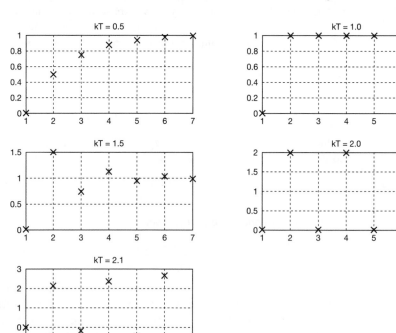

Figure 7.5 Several time-domain responses for the system of Example 7.1.

Section A5.5.1), so it will be crossing the stability boundary when $z = -1 = (1 - kT)$, that is, when $kT = 2$. The responses in Figure 7.5 for $kT = 1.5$, 2.0 and 2.1 confirm that $kT = 2$ is indeed the stability boundary.

- The response for $kT = 1$ is called a *dead-beat* response, in which the response follows the reference signal step with a delay of just one sampling interval. The step response therefore settles in one sampling interval from the application of the step input. Such a response is clearly a very desirable one from the performance point of view, but it is unfortunately unachievable in practice for systems above first order. This point is further discussed later in this chapter.

Unfortunately, most 'plants' are of second order or above, which means that modified methods must be used. The first difficulty in respect of such systems is in their conversion to the *z*-domain, the procedure for which will be explained by two examples.

Example 7.2 Obtaining z-transfer functions for higher-order systems

Produce *z*-transfer functions for systems having each of the following LTFs preceded by a zero-order hold, and assuming a sampling interval of 0.4 s:

$$\text{(a)}\ \ G(s) = \frac{8}{s^2 + 5s + 4} \qquad \text{and} \qquad \text{(b)}\ \ G(s) = \frac{8}{s^2 + 2s + 4}$$

System (a)

The first step is to multiply $G(s)$ by the LTF of the hold

$$\frac{1 - e^{-sT}}{s}$$

to give the overall LTF of the system. We remember that the expression $(1 - e^{-sT})$ simply multiplies the result in z by $(1 - z^{-1})$, so we will leave it out for the moment. The transform in s to be converted is therefore

$$\frac{1}{s} \cdot \frac{8}{s^2 + 5s + 4}, \qquad \text{or} \qquad \frac{8}{s(s^2 + 5s + 4)}$$

Unfortunately, nothing precisely analogous to this expression appears in the table (Table A5.1), so it will have to be split into partial fractions. First, we test the denominator factor $(s^2 + 5s + 4)$ to see if it will factorize further. It does, and can be written as $(s + 1)(s + 4)$. There is also the s term multiplying everything, so we need the following partial fractions:

$$\frac{A}{s} + \frac{B}{s + 1} + \frac{C}{s + 4}$$

These will go over a common denominator as follows:

$$\frac{A(s+1)(s+4) + Bs(s+4) + Cs(s+1)}{s(s+1)(s+4)} = \frac{8}{s(s+1)(s+4)}$$

Comparing the coefficients on each side, we obtain

$$s^2: \quad A + B + C = 0$$

$$s^1: \quad 5A + 4B + C = 0$$

$$s^0: \quad 4A = 8 \quad \text{and so} \quad A = 2$$

Substituting the last result back into the first two and solving for B and C, we obtain:

$$B = -8/3 \quad \text{and} \quad C = 2/3$$

So our LTF becomes

$$\frac{2}{s} - \frac{\frac{8}{3}}{s+1} + \frac{\frac{2}{3}}{s+4}$$

and we convert from the tables, where we find that

$$\frac{1}{s} \rightarrow \frac{z}{z-1}, \qquad \frac{1}{s+1} \rightarrow \frac{z}{z-e^{-0.4}} \qquad \text{and} \qquad \frac{1}{s+4} \rightarrow \frac{z}{z-e^{-1.6}}$$

The resulting expression is

$$\frac{2z}{z-1} - \frac{\frac{8}{3}z}{z-e^{-T}} + \frac{\frac{2}{3}z}{z-e^{-4T}}$$

which we must now multiply by the contribution from the hold element, that is, $(1 - z^{-1}) = (z-1)/z$, giving

$$2 - \frac{\frac{8}{3}(z-1)}{z-e^{-0.4}} + \frac{\frac{2}{3}(z-1)}{z-e^{-1.6}}$$

Now comes the worst aspect of the task! It is necessary to put this expression over a common denominator and calculate all the actual numbers. This leads to the result:

$$\frac{2(z-e^{-0.4})(z-e^{-1.6}) - \frac{8}{3}(z-1)(z-e^{-1.6}) + \frac{2}{3}(z-1)(z-e^{-0.4})}{(z-e^{-0.4})(z-e^{-1.6})}$$

$$= \frac{0.3471z + 0.1792}{z^2 - 0.8722z + 0.1353}$$

It is as well to perform a static check on such results, as they are very liable to algebraic error! The check is most easily done via the frequency response, that is, we put $s = j\omega$ in the Laplace version (and $\omega = 0$) and $z = e^{j\omega T}$ in the z version (and $z = 1$ when $\omega = 0$). The Laplace version gives $8/4 = 2$, whereas the z version gives

$$\frac{0.3471 + 0.1792}{1 - 0.8722 + 0.1353} = \frac{0.5263}{0.2631} = 2.0004$$

The slight discrepancy is caused by numerical rounding errors.

System (b)

The second system, having $G(s) = 8/(s^2 + 2s + 4)$, will now be converted. Again we multiply by

$$\frac{1 - e^{-sT}}{s}$$

for the hold and temporarily omit the $(1 - e^{-sT})$, giving

$$\frac{8}{s(s^2 + 2s + 4)}$$

Again partial fractions will be needed – but of a different type, as it now proves that the denominator quadratic term has complex roots. We must now express the LTF as:

$$\frac{A}{s} + \frac{Bs + C}{s^2 + 2s + 4} = \frac{A(s^2 + 2s + 4) + s(Bs + C)}{s(s^2 + 2s + 4)} = \frac{8}{s(s^2 + 2s + 4)}$$

Comparing the coefficients on each side, we obtain

$$s^2: \quad A + B = 0$$

$$s^1: \quad 2A + C = 0$$

$$s^0: \quad 4A = 8 \quad \text{and so} \quad A = 2$$

Substituting the last result back into the first two, we obtain:

$$B = -2 \quad \text{and} \quad C = -4$$

So we have

$$\frac{2}{s} - \frac{2s + 4}{s^2 + 2s + 4} \tag{7.2}$$

We now know how to convert the $2/s$, but the other term is a 'new' problem! The tables (Appendix 5) give the following clues:

$$\frac{\omega}{(s + a)^2 + \omega^2} \rightarrow \frac{ze^{-aT} \sin \omega T}{z^2 - 2ze^{-aT} \cos \omega T + e^{-2aT}} \quad \text{and}$$

$$\frac{s + a}{(s + a)^2 + \omega^2} \rightarrow \frac{z^2 - ze^{-aT} \cos \omega T}{z^2 - 2ze^{-aT} \cos \omega T + e^{-2aT}}$$

The next step is to sort out what a and ω need to be. The denominator of both the above Laplace expressions multiplies out to $(s^2 + 2as + a^2 + \omega^2)$, which has to correspond to the denominator term $(s^2 + 2s + 4)$ from Equation (7.2). It therefore follows that $2a = 2$, so $a = 1$, and that $a^2 + \omega^2 = 4$, so $\omega = \sqrt{4 - a^2} = \sqrt{3} = 1.732$. So we have the transform pairs:

$$\frac{1.732}{(s + 1)^2 + 3} \rightarrow \frac{ze^{-0.4} \sin(1.732 \times 0.4)}{z^2 - 2ze^{-0.4} \cos(1.732 \times 0.4) + e^{-2 \times 0.4}} \quad \text{and}$$

$$\frac{s + 1}{(s + 1)^2 + 3} \rightarrow \frac{z^2 - ze^{-0.4} \cos(1.732 \times 0.4)}{z^2 - 2ze^{-0.4} \cos(1.732 \times 0.4) + e^{-2 \times 0.4}}$$

What we actually require is

$$\frac{2s+4}{s^2+2s+4}$$

and the procedure is to construct the latter expression from multiples of the two former ones (which is feasible, because the denominators already match).

If we took twice the second transform pair, the numerator would be $(2s+2)$. The term in s is now correct, so we just need to find the remaining 2 from the first transform pair. Its numerator is actually 1.732 as it stands, so we must multiply it by 2/1.732.

So our

$$\frac{2s+4}{s^2+2s+4}$$

becomes

$$\frac{2[z^2 - ze^{-0.4}\cos(1.732 \times 0.4)] + \frac{2}{1.732}[ze^{-0.4}\sin(1.732 \times 0.4)]}{z^2 - 2ze^{-0.4}\cos(1.732 \times 0.4) + e^{-2 \times 0.4}}$$

which works out numerically to

$$\frac{2z^2 - 1.03157z + 0.494361z}{z^2 - 1.03157z + 0.449329} = \frac{2z^2 - 0.537209z}{z^2 - 1.03157z + 0.449329}$$

Now all that remains is to subtract that expression from $2z/(z-1)$ (for the original $2/s$ term in Equation (7.2)) and multiply by $(1 - z^{-1}) = (z-1)/z$ from the hold:

$$\left(\frac{2z}{z-1} - \frac{2z^2 - 0.537209z}{z^2 - 1.03157z + 0.449329}\right)\frac{z-1}{z}$$

Putting it all over a common denominator we obtain:

$$\frac{2(z^2 - 1.03157z + 0.449329) - (z-1)(2z - 0.537209)}{z^2 - 1.03157z + 0.449329}$$

and simplifying gives:

$$\frac{0.474069z + 0.361449}{z^2 - 1.03157z + 0.449329} \tag{7.3}$$

(Static check:

$$\frac{0.474069 + 0.361449}{1 - 1.03157 + 0.449329} = \frac{0.835518}{0.417759} = 2.00$$

which is exactly correct.)

The analysis involved in converting block diagrams to the z-domain is quite laborious algebraically; the reader may be relieved to discover that the subsequent calculations to produce the z-transfer function of the required digital controller are significantly easier! It is also possible to obtain assistance from MATLAB (Appendix 3) and other packages with the s-to-z conversion. There are some examples in Section A5.5.5.

7.3   *Obtaining the z-transfer function of the controller*

Referring to Figure 7.1, in which the digital controller will have a z-transfer function $D(z)$ and the open-loop plant a z-transfer function of $G(z)$, the procedure is as follows:

(1) Specify the required closed-loop unit step response $y(t)$.

(2) Obtain its z-transform $Y(z)$.

(3) Determine the required closed-loop z-transfer function

$$\frac{Y(z)}{R(z)} = G'(z)$$

by dividing $Y(z)$ by the z-transform of a unit step, $z/(z-1)$.

(4) We know from the '$G/(1+GH)$' result that

$$G'(z) = \frac{D(z)G(z)}{1 + D(z)G(z)} \tag{7.4}$$

and we now know $G(z)$ from the analysis in Section 7.2. Rearranging Equation (7.4) gives:

$$G'(z) = D(z)G(z) - G'(z)D(z)G(z) \quad \text{and}$$

$$D(z) = \frac{G'(z)}{[1 - G'(z)]G(z)} \tag{7.5}$$

(5) Ensure that no undesirable oscillation of controller output $u(t)$ or plant output $y(t)$ occurs between sampling intervals (or otherwise in the case of u). The procedure for this check will be explained by subsequent examples.

The procedure sounds rather complicated, but the following specific cases will clarify how to use it in practice.

7.4   *A dead-beat controller*

A dead-beat response was described following Example 7.1 as being one in which an input step (or, in fact, any other input) is faithfully followed with only one sampling interval of delay. It was also stated that such a response cannot be achieved for plants higher than first order. The following analysis will attempt to derive $G(z)$ for such a controller. The numbering refers to the steps outlined in Section 7.3.

(1) The intended response is that of a unit step delayed by one sampling interval, so:

$$Y(z) = \frac{z}{z-1} z^{-1} = \frac{1}{z-1}$$

(2) The closed-loop z-transfer function is therefore

$$G'(z) = \frac{1/(z-1)}{z/(z-1)} = \frac{1}{z}$$

(3) From Equation (7.5),

$$D(z) = \frac{G'(z)}{[1 - G'(z)]G(z)} = \frac{1/z}{[1 - 1/z]G(z)}$$

Example 7.3 *Obtain the z-transfer function D(z) for a dead-beat controller*

This dead-beat controller is to work in the configuration of Figure 7.1, with a plant modelled by the $G(z)$ given in Equation (7.3). Substituting this $G(z)$ into the result of step (3) above, we obtain:

$$\frac{1/z}{(1 - 1/z)\dfrac{0.474069z + 0.361449}{z^2 - 1.03157z + 0.449329}} = \frac{z^2 - 1.03157z + 0.449329}{(z-1)(0.474069z + 0.361449)}$$

$$= \frac{z^2 - 1.03157z + 0.449329}{0.474069z^2 - 0.11262z - 0.361449}$$

Dividing throughout by $0.474069z^2$, so as to obtain a leading denominator term of unity, and only negative powers of z (corresponding to time delays) yields:

$$D(z) = \frac{2.1094 - 2.176z^{-1} + 0.9478z^{-2}}{1 - 0.2376z^{-1} - 0.7624z^{-2}}$$

Unfortunately, it will emerge that the controller derived in Example 7.3 does not actually cause the plant output to settle after one sampling interval. The actual behaviour will be investigated by each of two methods.

7.4.1 *Investigation of the dead-beat result by simulation of the system response*

This can be done by MATLAB (Appendix 3), ACSL (Mitchell and Gauthier Associates, 1987), SIMULINK (Appendix 4), or other appropriate packages. The MATLAB m-file *fig7_6.m* on the accompanying disk produced the response of Figure 7.6 for a unit step input to the closed-loop system (in the MATLAB environment, SIMULINK would actually be better, and is used later on). It will be seen that, if we examine the response at the sampling instants only, it does indeed seem as if the plant has settled. Looking between the samples reveals the true situation! It is the digital control equivalent of a Mr Bailey who, in pre-national power grid days, was in charge of the control room in a power station in a city in Northern England. He was not very expert at the precise control of frequency, but his superiors remained in ignorance because he did manage to achieve almost exactly 50 hertz (the British standard mains frequency) on the hour and half-hour when readings had to be logged. He was known to his colleagues as 'Passing-

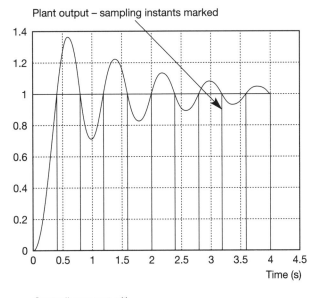

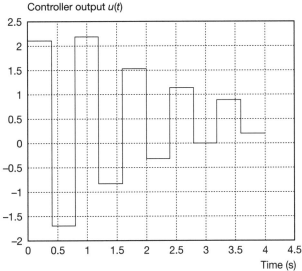

Figure 7.6 Step responses of a closed-loop system with a 'dead-beat' controller.

through-50' Bailey, and the frequency he achieved must have resembled Figure 7.6, though it is not certain that his oscillation was even a decaying one.

7.4.2 Investigation of the result by inspecting the output of the controller

Assuming that the output y has settled, then the loop error e must have done so too. In the case of the dead-beat controller, the error signal e is very simple in form; it is 1 at time $t = 0$ and 0 thereafter. It is therefore possible to investigate the controller output by converting $D(z)$ into difference-equation form.

From Example 7.3, we have

$$D(z) = \frac{U(z)}{E(z)} = \frac{2.1094 - 2.176z^{-1} + 0.9478z^{-2}}{1 - 0.2376z^{-1} - 0.7624z^{-2}}$$

Multiplying out, this gives

$$U(z)(1 - 0.2376z^{-1} - 0.7624z^{-2})$$
$$= E(z)(2.1094 - 2.176z^{-1} + 0.9478z^{-2})$$

Remembering that multiplying by z^{-n} is equivalent to delaying the time-domain quantity by n sampling intervals, we can write:

$$u_n - 0.2376u_{n-1} - 0.7624u_{n-2} = 2.1094e_n - 2.176e_{n-1} + 0.9478e_{n-2}$$

and rearranging to obtain u_n we obtain

$$u_n = 0.2376u_{n-1} + 0.7624u_{n-2} + 2.1094e_n - 2.176e_{n-1} + 0.9478e_{n-2}$$

For the present system, we start with $e_n = 1$ and all the other e and u terms equal to zero. This makes u_n equal to 2.1094. That value becomes u_{n-1} in the next interval, at which time e_{n-1} will hold the initial value of 1, and all the other e terms (including e_n) will be zero. The value of u after one sampling interval is therefore:

$$u_n = 0.2376 \times 2.1094 + 0.7624 \times 0 + 2.1094 \times 0$$
$$- 2.176 \times 1 + 0.9478 \times 0 = -1.6748$$

This is opposite in sign to the last value, so the system has clearly not settled. The following two types of controller were devised in order to keep as many of the advantages of dead-beat control as possible, without the troublesome inter-sample oscillation.

7.5 Kalman's controller algorithm

Kalman reasoned that, while it is impossible to cause a system to settle in one sampling interval if it is above first order, it may be possible to make it settle in a greater, but still finite, number of intervals. His analysis revealed that a number of intervals equal to the order of the system is required, and that the following analysis produces the correct $D(z)$ for a second-order system. Reference is made to quantities in Figure 7.1, which are initially expressed as time series.

The reference input, being a unit step, can be expressed as: $R(z) = 1 + z^{-1} + z^{-2} + z^{-3} + \dots$. The system output for a second-order system should settle in two sampling intervals, so it will start at zero, reach some value α after one sampling interval, and settle at a value of unity after two sampling intervals. Its series in z will therefore be: $Y(z) = 0 + \alpha z^{-1} + z^{-2} + z^{-3} + \dots$.

The controller output needs a little more thought. It must give an output in excess of the steady-state value in the first sampling interval, then 'apply the brakes' in the second one, in order to 'stop' the output in the right place. Subsequently the controller output must settle at the value required to give the required steady-state output – that is, the required output value of 1 divided by the steady-state gain of the plant. The series for the controller output will therefore be:

$$U(z) = \beta + \Gamma z^{-1} + \frac{1}{k}(z^{-2} + z^{-3} + z^{-4} + \cdots) \qquad (7.6)$$

where k is the steady-state gain of the plant. It is now possible in principle to work out the closed-loop transfer function by using $Y(z)/R(z)$, but both appear to be unhelpful infinite series, and we do not yet know what the required value of α is. The first problem is easily resolvable as $R(z)$ is expressible in closed form as $z/(z-1)$, or $1/(1-z^{-1})$. To divide by $R(z)$, therefore, we multiply by $(1-z^{-1})$. If we multiply the series for $Y(z)$ in that way, we obtain

$$Y(z)/R(z) = (\alpha z^{-1} + z^{-2} + z^{-3} + \cdots) - (\alpha z^{-1} + z^{-2} + z^{-3} + \cdots)z^{-1}$$

$$= \alpha z^{-1} + (1-\alpha)z^{-2} \qquad (7.7)$$

(The purists will argue that a term in $z^{-\infty}$ has been ignored, but by that time the plant will have been scrapped and we shall have retired.) We now note that $G(z)$ is equal to $Y(z)/U(z)$. Again, this relation does not look very helpful as both $Y(z)$ and $U(z)$ are in the form of infinite series. This problem can be solved by multiplying both by $(1-z^{-1})$, giving for $Y(z)$:

$$(1-z^{-1})y(z) = (1-z^{-1})(0 + \alpha z^{-1} + z^{-2} + z^{-3} + \cdots)$$

$$= \alpha z^{-1} + (1-\alpha)z^{-2}$$

and for $U(z)$ from Equation (7.6):

$$(1-z^{-1})u(z) = (1-z^{-1})\left[\beta + \Gamma z^{-1} + \frac{1}{k}(z^{-2} + z^{-3} + \cdots)\right]$$

$$= \beta + (\Gamma - \beta)z^{-1} + \left(\frac{1}{k} - \Gamma\right)z^{-2}$$

Therefore:

$$\frac{Y(z)}{U(z)} = G(z) = \frac{\alpha z^{-1} + (1-\alpha)z^{-2}}{\beta + (\Gamma - \beta)z^{-1} + \left(\dfrac{1}{k} - \Gamma\right)z^{-2}} \qquad (7.8)$$

As $G(z)$ is known, it is now possible to determine α, β and Γ by comparing coefficients, and then proceed to determine the required $D(z)$. There is one hidden pitfall, which will be explained by obtaining $D(z)$ for the same arrangement as was used in the dead-beat controller example.

Example 7.4 *Obtain a Kalman controller to be used in place of the dead-beat one in Example 7.3*

This example reconsiders the design of the digital controller $D(z)$ for Example 7.3, but using a Kalman controller in place of the dead-beat one.

Comparing the expression for $G(z)$ in Equation (7.8) with that known from Equation (7.3), we have

$$\frac{0.474069z^{-1} + 0.361449z^{-2}}{1 - 1.03157z^{-1} + 0.449329z^{-2}} = \frac{\alpha z^{-1} + (1-\alpha)z^{-2}}{\beta + (\Gamma - \beta)z^{-1} + \left(\dfrac{1}{k} - \Gamma\right)z^{-2}} \quad (7.9)$$

Direct coefficient comparison does not work because the coefficients in the numerator on the right-hand side (RHS) of Equation (7.9), being α and $(1-\alpha)$, add up to unity, but those on the LHS do not. The transfer function on the LHS has therefore to be scaled by multiplying numerator and denominator by an appropriate factor, so as to make the numerator coefficient sum equal to unity. A little thought will reveal that the required factor is the reciprocal of the present sum of the numerator coefficients. In this instance, that factor is $1/(0.474069 + 0.361449) = 1.1969$ and so Equation (7.9) becomes

$$\frac{0.567z^{-1} + 0.433z^{-2}}{1.197 - 1.235z^{-1} + 0.538z^{-2}} = \frac{\alpha z^{-1} + (1-\alpha)z^{-2}}{\beta + (\Gamma - \beta)z^{-1} + \left(\dfrac{1}{k} - \Gamma\right)z^{-2}}$$

Coefficient comparison now produces:

$\alpha = 0.567$ by inspection

$\beta = 1.197$ by inspection

$(\Gamma - \beta) = -1.235$, so $\Gamma = -1.235 + \beta = -1.235 + 1.197 = -0.038$

Finally, $(1/k - \Gamma) = 0.538$, so $1/k = 0.538 + \Gamma = 0.5$ and $k = 2$. (The value of k should agree with that from our previous static gain check which, encouragingly, it does.)

To find $D(z)$, we remember from Equation (7.7) that the closed-loop transfer function was $G'(z) = \alpha z^{-1} + (1-\alpha)z^{-2}$. We also have the result from Equation (7.5) that

$$D(z) = \frac{G'(z)}{[1 - G'(z)]G(z)}$$

$$D(z) = \frac{0.567z^{-1} + 0.433z^{-2}}{1 - (0.567z^{-1} + 0.433z^{-2})G(z)}$$

$$= \frac{(0.567z^{-1} + 0.433z^{-2})(1.197 - 1.235z^{-1} + 0.538z^{-2})}{(1 - 0.567z^{-1} - 0.433z^{-2})(0.567z^{-1} + 0.433z^{-2})}$$

$$= \frac{1.197 - 1.235z^{-1} + 0.538z^{-2}}{1 - 0.567z^{-1} - 0.433z^{-2}}$$

This controller performs well, as the step responses of Figure 7.7 clearly show (The MATLAB m-file *fig7_7.m* on the accompanying disk produced the figure). The graph of $u(t)$ in Figure 7.7 confirms that the calculated values of β and Γ are indeed the values of $u(t)$ in the first two sampling intervals (there may be a hint of movement at the third sampling instant, due to numerical rounding errors).

Plant output $y(t)$

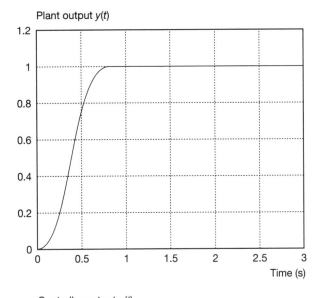

Controller output $u(t)$

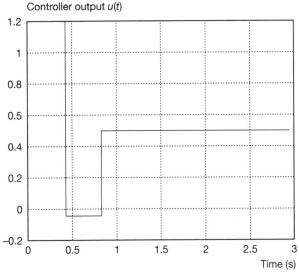

Figure 7.7 Step responses of a closed-loop system with a Kalman controller.

7.6   Dahlin's controller algorithm

Unfortunately, the Kalman controller involves a difficult compromise in respect of the sampling interval, between needing a heavy controller action (if the interval is short) and leading to problems of avoiding aliasing (Section 5.8.1) (if it is too long). An alternative approach is to use a modified dead-beat algorithm which produces a first-order exponential response to an input step, that is, the form of the unit step response is $(1 - e^{-at})$, where a is a constant. In this way, the speed of the response can be adjusted by means of the controller parameters, without changing the sampling interval. This is known as the *Dahlin* approach, and leads readily to the determination of the required $D(z)$ by the usual method.

If $y(t) = 1 - e^{-at}$, then from tables (Appendix 5)

$$Y(z) = \frac{(1 - e^{-aT})z}{(z - e^{-aT})(z - 1)}$$

The closed-loop transfer function $G'(z)$ is readily determined by dividing $Y(z)$ by $R(z)$, the latter being $z/(z - 1)$ again for the unit step. This operation gives

$$G'(z) = \frac{1 - e^{-aT}}{z - e^{-aT}} \tag{7.10}$$

We can then again use

$$D(z) = \frac{G'(z)}{[1 - G'(z)]G(z)}$$

from Equation (7.5).

Example 7.5 *A Dahlin controller with a time constant of 1.5 s for the exponential rise*

This example again uses the plant of Examples 7.3 and 7.4, and develops the controller $D(z)$ using Dahlin's method.

As before,

$$G(z) = \frac{0.474069z^{-1} + 0.361449z^{-2}}{1 - 1.03157z^{-1} + 0.449329z^{-2}}$$

A time constant of 1.5 s gives $a = 1/1.5$. The previously defined sampling period is $T = 0.4$ s so, from Equation (7.10):

$$G'(z) = \frac{1 - e^{-aT}}{z - e^{-aT}} = \frac{1 - e^{-0.4/1.5}}{z - e^{-0.4/1.5}} = \frac{0.23407}{z - 0.76593}$$

Then, from Equation (7.5),

$$D(z) = \frac{0.23407/(z - 0.76593)}{[1 - 0.23407/(z - 0.76593)]G(z)} = \frac{0.23407}{(z - 1)G(z)}$$

$$D(z) = \frac{0.23407(1 - 1.03157z^{-1} + 0.449329z^{-2})}{(z - 1)(0.474069z^{-1} + 0.361449z^{-2})}$$

$$= \frac{0.494 - 0.509z^{-1} + 0.222z^{-2}}{1 - 0.2376z^{-1} - 0.7624z^{-2}}$$

The closed-loop system using the controller designed in Example 7.5 was simulated and the step responses of Figure 7.8 resulted. It will be noted that the output $y(t)$ is broadly of exponential form but there is, as was observed in the dead-beat controller, a decaying oscillation superimposed upon the exponential. Once again, the exponential is exactly followed at the sampling intervals. The controller output $u(t)$ shows the nature of the problem; like the digitized PID controller using the Tustin transform (Section 5.8.3, Example 5.12) the controller has a 'ringing pole'. One way to avoid this is to replace the 'ringing pole' with an equivalent steady-state gain.

Plant output $y(t)$

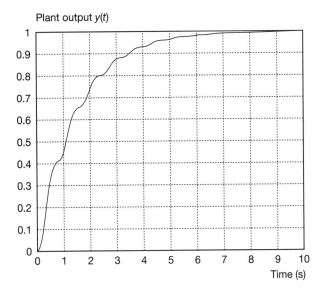

Controller output $u(t)$

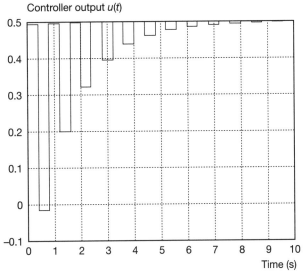

Figure 7.8 Step responses of a closed-loop system with a Dahlin controller.

If we examine the denominator of $D(z)$ in Example 7.5, it can be seen to factorize as

$$(1 - z^{-1})(0.474069 + 0.361449z^{-1})$$

The poles are therefore at $z = 1$ and $z = -0.361449/0.474069 \approx -0.76$. The latter is a real pole to the left of the (z-plane) imaginary axis, and will therefore display ringing (if the closed-loop poles of the entire system comprising $D(z)$, $G(z)$ and the UNF loop are evaluated, it will be found that this pole at $z \approx -0.76$ appears in the closed-loop pole set, and is accompanied by a matching one at $z \approx +0.76$ as discussed in Section A5.5.2). It is therefore probably best to replace the 'ringing pole' by an equivalent static gain (that is, at $z = 1$) of $1/(0.474069 + 0.361449) = 1.197$. This substitution gives

$$D(z) = \frac{1.197(0.494 - 0.509z^{-1} + 0.222z^{-2})}{1 - z^{-1}}$$

$$= \frac{0.591 - 0.609z^{-1} + 0.266z^{-2}}{1 - z^{-1}}$$

This controller was found to give the step-response results shown in Figure 7.9. A slight (about 3 per cent) overshoot has resulted, but the controller action is much less oscillatory. The files *fig7_8.m* and *fig7_9.m* on the accompanying disk will produce these figures.

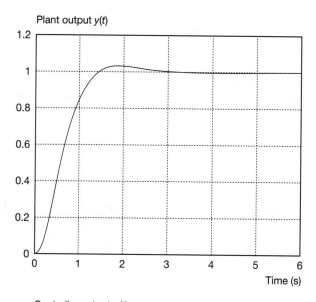

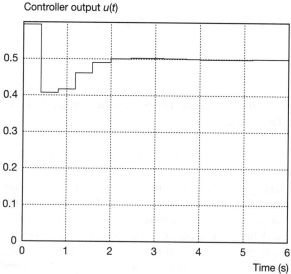

Figure 7.9 As Figure 7.8, but with the 'ringing poles' removed.

7.7 Modification for plants with transport lags

The same analytical approaches are still applicable to situations which are complicated by transport lags. In each case, if $G(z)$ is the open-loop plant z-transfer function with no transport lag, then $z^{-n} G(z)$ will be that for the plant with transport lag, where n is the number of sampling intervals in the transport lag.

Since the plant cannot respond until the transport lag has elapsed, the desired response $Y(z)$ must also be delayed by that time (that is, be multiplied by z^{-n}). A transport lag containing an integer number of sampling intervals is a great help in that respect!

Example 7.6 *Obtain Kalman and Dahlin controllers for a plant with a transport lag*

The plant for this example is 'system (a)' from Example 7.2, but with a transport lag of 0.8 s. It is sampled at 0.4 s intervals. The Dahlin controller is to have a time constant of 2 s. The z-transfer function of the plant *excluding the transport lag* (from Example 7.2) is

$$G(z) = \frac{0.3471z^{-1} + 0.1792z^{-2}}{1 - 0.8722z^{-1} + 0.1353z^{-2}}$$

Note that the figures for this example could be produced by MATLAB m-files similar to those used for the previous figures. However, in this case, the files *fig7_10.m* and *fig7_11.m* on the accompanying disk contain SIMULINK models (see Appendix 4), which were used instead. Readers with SIMULINK installed can simply enter the filename at the MATLAB prompt (SIMULINK will then load automatically) and click on *Simulation* then *Start*.

(a) Design of the Kalman controller

There are two 0.4 s sampling intervals in the transport lag, so the $G(z)$ and the output $Y(z)$ must be multiplied by z^{-2}.

From Equation (7.7), the closed-loop transfer function $G'(z)$ will be:

$$G'(z) = Y(z)/R(z) = z^{-2}(\alpha z^{-1} + (1 - \alpha)z^{-2}) = \alpha z^{-3} + (1 - \alpha)z^{-4} \quad (7.11)$$

Using this result to generate $G(z)$ in Equation (7.8) and comparing with the $G(z)$ above leads to the values of α, β and Γ as follows ($G(z)$ is expressed as 'num$g$'/'den$g$' for use below):

$$G(z) = \frac{\mathrm{num}g(z)}{\mathrm{den}g(z)} = \frac{(0.3471z^{-1} + 0.1792z^{-2})z^{-2}}{1 - 0.8722z^{-1} + 0.1353z^{-2}}$$

$$= \frac{(\alpha z^{-1} + (1 - \alpha)z^{-2})z^{-2}}{\beta + (\Gamma - \beta)z^{-1} + \left(\dfrac{1}{k} - \Gamma\right)z^{-2}}$$

As before, it is necessary to scale the fraction on the LHS by dividing numerator and denominator by $(0.3471 + 0.1792)$ to make the numerator coefficient comparison valid. The division produces the following result:

$$G(z) = \frac{0.6595z^{-3} + 0.3405z^{-4}}{1.9001 - 1.6572z^{-1} + 0.2571z^{-2}} = \frac{\alpha z^{-3} + (1 - \alpha)z^{-4}}{\beta + (\Gamma - \beta)z^{-1} + \left(\frac{1}{k} - \Gamma\right)z^{-2}}$$

By inspection, $\alpha = 0.6595$ and $\beta = 1.9001$. Also $\Gamma - \beta = -1.6572$, so $\Gamma = 1.9001 - 1.6572 = 0.2429$. We also have $1/k - \Gamma = 0.2571$, so $1/k = 0.2571 + 0.2429 = 0.5000$, and $k = 2$, which is precisely in accord with the 'static test' value (setting $z = 1$ in the original $G(z)$).

From Equation (7.5),

$$D(z) = \frac{G'(z)}{[1 - G'(z)]G(z)}$$

and from Equation (7.11),

$$G'(z) = \alpha z^{-3} + (1 - \alpha)z^{-4} = 0.6595z^{-3} + 0.3405z^{-4}$$

Therefore

$$D(z) = \frac{(0.6595z^{-3} + 0.3405z^{-4})[\text{den}g(z)]}{(1 - 0.6595z^{-3} - 0.3405z^{-4})[\text{num}g(z)]}$$

$$= \frac{1.9001 - 1.6572z^{-1} + 0.2571z^{-2}}{1 - 0.6595z^{-3} - 0.3405z^{-4}}$$

When tested by simulation, the system gave the step-test results shown in Figure 7.10. It can clearly be seen that, though the response is clearly a Kalman-type one, it does not begin until the end of the 0.8 s transport lag.

(b) Design of the Dahlin controller

The Dahlin derivation also follows the same pattern as that with no transport lag, but the output $y(t)$ is delayed by 0.8 s, or two sampling intervals. From the analysis leading to Equation (7.10), $Y(z)$ will therefore be

$$\frac{z(1 - e^{-aT})z^{-2}}{(z - 1)(z - e^{-aT})}$$

and the closed-loop transfer function will be:

$$G'(z) = \frac{Y(z)}{\text{step}(z)} = \frac{Y(z)(z - 1)}{z} = \frac{(1 - e^{-aT})z^{-2}}{z - e^{-aT}}$$

We now apply Equation (7.5) again:

$$D(z) = \frac{G'(z)}{[1 - G'(z)]G(z)}$$

Plant output $y(t)$

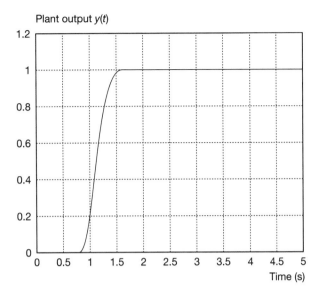

Controller output $u(t)$

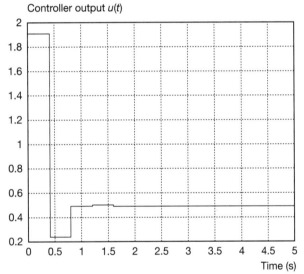

Figure 7.10 Step responses of a closed-loop system with a transport lag, controlled by a Kalman controller.

and, as $e^{-aT} = 0.81873$, a fairly serious bout of algebra produces the result:

$$D(z) = \frac{0.18127(1 - 0.8722z^{-1} + 0.1353z^{-2})}{(0.3471z^{-1} + 0.1792z^{-2})(-0.18127z^{-2} + z - 0.81873)}$$

$$= \frac{0.18127 - 0.1581z^{-1} + 0.02453z^{-2}}{0.3471 - 0.10498z^{-1} - 0.14672z^{-2} - 0.06292z^{-3} - 0.03248z^{-4}}$$

which would finally be divided throughout by 0.3471 to obtain an easier form for implementation. The system using this $D(z)$ was again tested by simulation and produced the step responses of Figure 7.11.

Plant output $y(t)$

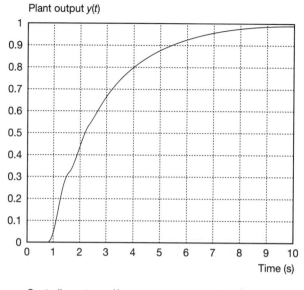

Time (s)

Controller output $u(t)$

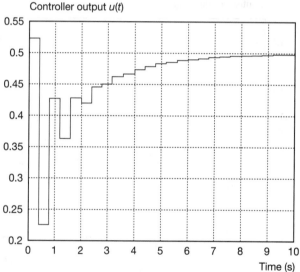

Time (s)

Figure 7.11 Step responses of a closed-loop system with a transport lag, controlled by a Dahlin controller.

7.8 A comparison of the 'true' digital controllers

It is clear from Figure 7.6 that the 'dead-beat' controller is not really dead-beat at all; in fact, its output for second- and higher-order systems at best only settles in infinite time, and the step response displays considerable overshoot and an oscillation which decays only slowly. It may therefore be dismissed as a practical controller type.

The Kalman controller appears to give an excellent step response, settling in only two sampling intervals in the case of second-order systems. It is therefore an attractive option where it can be used, but its use suffers from the following difficulties:

- The settling time is very closely related to the sampling interval. The maximum sampling interval which can be used is restricted by Shannon's sampling theorem and the need to avoid aliasing (see Section 5.8.1) and, even if the usual minimum sampling frequency of five times the highest natural frequency of the closed-loop system, or any element, is used (that is, the longest feasible sampling interval), the required value of the controller output may be unachievably high.

- The Kalman controller performance is less robust in the face of plant parameter changes and nonlinearities than is that of, for example, a digital PID controller. This characteristic is especially pronounced in the case of a variable plant transport lag, though the true digital controllers, including the Kalman controller, are actually better able than 'digitized analog' controllers to cope with a *constant* transport lag.

- The analysis for systems of higher than second order is more difficult (but still far from impossible).

The Dahlin controller, as derived and before the removal of its ringing pole, suffers from the main drawback of the 'dead-beat' controller in having inter-sample oscillations, though they are much less pronounced than those of the dead-beat type. They can be looked upon as arising because the intended step response is that of a first-order system whose rate of change of output becomes non-zero immediately the step is applied. This is not the natural step response of a second-order system! As well as producing the inter-sample oscillations, the Dahlin controller often results in a higher initial controller output than a 'digitized analog' controller of comparable overall performance. The removal of the ringing pole cures the inter-sample oscillation problem, but also causes some degradation of the performance (in Example 7.6, a maximum overshoot of zero became just over 3 per cent, which would admittedly often not be a problem in practice unless the system were for parking a car or for landing a rigid space vehicle on the Moon!).

An additional feature of all the controllers explained in this chapter concerns the trade-off between the required controller gain and the sampling interval – very short sampling intervals result in very high controller gains being required. This point is perhaps intuitively clearer for the Kalman than for the other controllers, as the settling time (after any transport lag has elapsed) is proportional to the sampling interval and, the faster a system is to be driven, the higher will be the signals required to drive it. The reader may wish to experiment with software such as MATLAB to verify this point.

These difficulties with all the 'true' digital controller types have biased most practical control engineers towards the 'digitized analog' approach. This is sometimes unfortunate as especially the Kalman controller can be an excellent choice when circumstances allow its use.

7.9 Conclusions

In this chapter, the principles, derivation and performance of 'true' digital controllers have been treated in some detail. Firstly, the notion of designing the controller from the beginning on the basis of its being a sampled system was explained, and the need to produce a discrete-time plant model via the true z-transform was argued. The derivation of such a model including a zero-order hold, always effectively present in practice, was performed and appropriate worked examples given.

The principle of defining the required step response, obtaining the appropriate closed-loop transfer function in the z-domain, and deriving the controller z-transfer function was explained, and controllers based on the dead-beat, Kalman and Dahlin algorithms were derived for plants with and without transport lags. In the case of the Dahlin controller, the removal of the 'ringing poles' in the controller was discussed. Finally, the performance of the various types of controller was examined and discussed relative both to each other and to controllers based on analog principles.

7.10 Problems

7.1 Produce a z-transform representation of the system shown in Figure P7.1. Use it to determine:
(a) What is the range of values of K which will result in a stable system, if a sampling interval of 0.1 s is used?
(b) What range of values of the sampling interval T_s will result in a stable system if a gain K of 5.0 is used?
(c) For each of the cases (a) and (b), what are the conditions which will produce a dead-beat response to a step input?

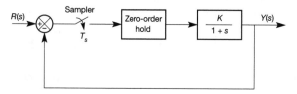

Figure P7.1 The system to be discretized for Problem 7.1.

7.2 Obtain the z-transfer function for each of the following plants in cascade with a zero-order hold, using a sampling interval of 0.1 s in each case.

(a) $\dfrac{50}{s^2 + 9s + 20}$ (b) $\dfrac{50}{s^2 + 4s + 20}$

7.3 For the plant of Problem 7.2(a), obtain the z-transfer function of a controller to achieve closed-loop control with unity negative feedback by each of the following strategies:
(a) Dead-beat
(b) Dahlin (for a response time constant of 0.5 s)
(c) Dahlin (as above but with the ringing pole removed)
(d) Kalman (for a settling time of 0.2 s).
Test their operation by simulation if possible.

7.4 For the controllers of Problem 7.3, determine the initial controller action produced in response to a unit step input to the closed-loop system. Examine the controller actions of the various controllers by simulating the system if possible.

7.5 Consider a plant which has the transfer function of Problem 7.2(b) in addition to a transport lag of 0.2 s. The plant is to be controlled in a unity negative feedback configuration, using each of the following controllers. A sampling interval of 0.1 s is to be used.
(a) A Kalman controller.
(b) A Dahlin controller of exponential time constant 0.5 s.
Obtain z-transfer functions for both controllers and modify that of part (b) to remove any ringing poles.

8 System identification and modelling revisited

8.1   Preview

In Section 3.9, we investigated the use of simple step tests and frequency response tests. The idea was that these could be carried out on existing systems, to aid the process of generating models for them. This chapter builds on that work. There are alternative test signals which can be used in plant identification. One class of such signals is low-level pseudo-noise signals, which can (in principle) be superimposed on the input of a plant during normal operation. That is clearly a potential advantage, and is usually not the case with step or frequency response test signals.

Thus, the emphasis in this chapter is on models generated using correlation techniques, after testing the plant using a pseudo-random binary sequence (PRBS testing). In particular, it proves to be possible to obtain results equivalent to those obtained by frequency response testing, but with much less time spent on plant than that kind of testing demands.

We also mentioned in Section 3.7 that time delays (transport lags) cannot be included accurately in continuous-time state-space models. In this chapter, we extend our general discussion of modelling, by looking at how approximate continuous-time state-space models can be obtained for systems containing time delays.

In this chapter, the reader will learn:

- how a system can be modelled by cross-correlation testing to determine the impulse response
- what types of input are suitable for such a test
- how to generate a pseudo-random binary sequence (PRBS) for use as such an input
- how to perform such a test in practice
- how to obtain the frequency response from the results of such a test
- how to represent time delays as rational functions in the Laplace operator s.

NEW MATHEMATICS FOR THIS CHAPTER

The ideas of correlation functions are introduced and explained as required. The Padé approximation is discussed and applied to the modelling of time delays.

8.2   The principles of correlation testing

The techniques of step, impulse and frequency response testing were explained in Section 3.9, where it was seen that frequency responses plotted in the form of Bode diagrams allowed plant transfer functions to be determined readily by the application of a standard procedure. It even proved to be possible to use curve-fitting methods to obtain an accurate transfer function model from the results of a step or frequency response test, provided that the general form of the transfer function was known (or accurately guessed) beforehand. It therefore appeared that the available approaches constituted a complete 'armoury' of methods which could determine the transfer function of any reasonably linear system. This assumption is not enormously wide of the mark in respect of what the methods are capable of achieving, but all three offer some difficulties in practice.

The *step response*, in view of its sharp change in input level, can produce system output behaviour violent enough to cause system damage (on susceptible plants) or to cause the engineers in charge of the plant to forbid this type of test because of fears of system damage (quite frequently). The application of a small enough step input to overcome this objection often does not work well in the case of systems which have slight stiction (static friction – see Section 14.3.2) but are reasonably linear for bigger inputs. This problem reduces the usefulness of the step-response test in practice, though it is still useful and widely employed on suitable plants.

As was explained in Chapter 3, the feasibility of *impulse response* testing is restricted by the fact that a true voltage impulse (for example) has a finite voltage-time area but zero duration – therefore implying infinite voltage! The practical feasibility of the method therefore depends on the ability of the tester to generate an impulse of sufficient energy to produce an accurately measurable response, while still being of sufficiently short duration that the system response approximates well to that which would be produced by a true impulse. The impulse response of a plant is nevertheless very useful in that it can be very readily converted by simple software into a frequency response. In the time domain, the impulse response is also interesting in itself, as it will give the transfer function directly if a time response function is fitted to it and the Laplace transform taken, though the conversion to the frequency response is likely to prove to be the easier and more systematic method.

The latter comment begs the question 'but why not just measure the *frequency response* directly?', which will now be answered. The difficulty with direct frequency-response determination is the time taken to perform the test. Unless a sweep oscillator is used (generally not an accurate method in any case, especially with regard to phase, as the system does not settle to a steady state at any frequency), a range of frequencies will have to be applied, the system allowed to settle to the steady-state a.c. response at each frequency, and the magnitude and phase of the response measured. This procedure could be very time-consuming on a plant with long time constants. An idea of the relative timescales of obtaining a frequency response by the method to be described in the next two sections, and by the direct method, can be gained from a recent test by the authors on a piece of electro-thermal laboratory equipment having time constants between 0.1 and 1.0 s. The direct method took about 20 minutes to perform, while the time actually collecting results on the equipment by the alternative method was less than one minute! Analysing the results took rather longer, but that operation can be performed 'back at base', thus disrupting plant operation for a much shorter period in a real-world case.

This section will explain how the system impulse response can be obtained by a steady-state method involving the cross-correlation of the system output and input, and using a type of input signal (a pseudo-random binary sequence or PRBS) which does not provoke a violent initial system response and can often be performed while the plant is actually running normally. The impulse response of the plant is directly related to its Laplace transfer function, as the latter is the Laplace transform of the unit impulse response.

The following analysis appears quite involved mathematically, though it is in fact more straightforward than it seems. It may help the reader if it is first explained that the purpose of the analysis is as follows:

A. To show that the graph produced by plotting the output of the linear system of Figure 8.1 against the delay τ is the same as the graph of the impulse response of the plant against time, provided that a suitable input signal is used.

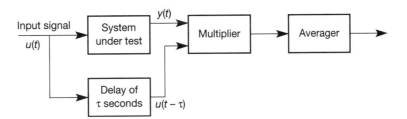

Figure 8.1 Experimental arrangement for cross-correlation testing.

B. To determine what types of input signal will produce the situation of A.

Consider the arrangement of Figure 8.1. This is the test method we shall use, so we shall analyse it to discover its behaviour. At this stage, the input $u(t)$ can be anything. We first obtain the output $y(t)$ by regarding the input $u(t)$ as being composed of a large number of impulses as shown in Figure 8.2. If we determine the response at a time t due to the general shaded impulse shown, then we can evaluate the output y at a time t by combining the responses at time t due to all the impulses.

If the system under test has a response to a unit impulse at time $t = 0$ given by $h(t)$ (and we remember that it is $h(t)$ which we are trying to determine) and the system is assumed to be linear, its response to the shaded impulse applied at time $(t - \lambda)$ is given by:

$$u(t - \lambda) \, d\lambda \times h(\lambda)$$

where

$u(t - \lambda) \, d\lambda$ is the strength (area) of the applied impulse

$h(\lambda)$ is the response of the system to a unit impulse applied a time λ previously

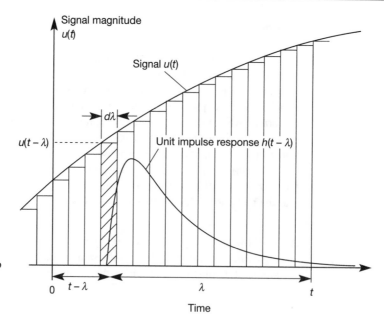

Figure 8.2 The response to
a general input signal,
represented as a series of
impulses.

If we now assume the input $u(t)$ to have been applied since 'time immemorial'
(that is, $t = -\infty$), we can determine the complete response $y(t)$ by letting $d\lambda$ tend
to zero and integrating with respect to λ:

$$y(t) = \int_0^\infty u(t - \lambda)h(\lambda)\ d\lambda \tag{8.1}$$

We now multiply by $u(t - \tau)$ to give the multiplier output.

$$y(t)u(t - \tau) = \left[\int_0^\infty u(t - \lambda)h(\lambda)\ d\lambda \right] u(t - \tau) \tag{8.2}$$

Finally we obtain the averager output. The method is to calculate the area under
the function in Equation (8.2) by integration from time $t = 0$ to time $t = T$, and
then to divide by T to obtain the average. Developing the analysis from Equation
(8.2) in that way, we obtain:

$$\frac{1}{T} \int_0^T y(t)u(t - \tau)\ dt = \frac{1}{T} \int_0^T \left[\int_0^\infty u(t - \lambda)h(\lambda)\ d\lambda \right] u(t - \tau)\ dt \tag{8.3}$$

The left-hand side of this equation represents a common definition of the *cross-
correlation* of y and u and it represents the output of the averager in our test system
of Figure 8.1. The following analysis will attempt to make sense of the right-hand
side of Equation (8.3). In that context, we note two points:

(1) $u(t - \lambda)$ has no τ terms in it and is therefore a constant in integrations with
respect to τ.

(2) $u(t - \tau)$ has no λ terms in it and is therefore a constant in integrations with respect to λ. The $u(t - \tau)$ can therefore go inside the integration with respect to λ, making the right-hand side of Equation (8.3) equal to:

$$\frac{1}{T} \int_0^T \left[\int_0^\infty u(t - \lambda)u(t - \tau)h(\lambda)\,d\lambda \right] dt \qquad (8.4)$$

We can now change the order of integration, giving:

$$\int_0^\infty \left[\frac{1}{T} \int_0^T u(t - \lambda)u(t - \tau)\,dt \right] h(\lambda)\,d\lambda \qquad (8.5)$$

Our attention now focuses on the inner integral within the square brackets. If it was simply:

$$\frac{1}{T} \int_0^T u(t)u(t - \tau)\,dt \qquad (8.6)$$

it would, by the same definition as the cross-correlation of y and u above, be the *autocorrelation* of u. If we could use an input test signal $u(t)$ whose autocorrelation plotted against delay τ happened to have a similar graph to an impulse, then this would have to be (from Equation (8.6)) a 'spike' when $\tau = 0$. However, in respect of the inner integral of Equation (8.5), the spike would be at $\tau = \lambda$, not $\tau = 0$. The version in Equation (8.5) is therefore that in Equation (8.6) but shifted to the right along the τ axis by λ.

This means that the integration with respect to λ in Equation (8.5) will be performed on the product of $h(\lambda)$ and the spike as shown in Figure 8.3, and the integral will be equal to the product of the area of the spike and the value of $h(\lambda)$ where the spike occurs. By varying the value of τ in our experimental arrangement, we can therefore produce corresponding values of $h(\tau)$ times the area of the spike. All we need is a suitable input!

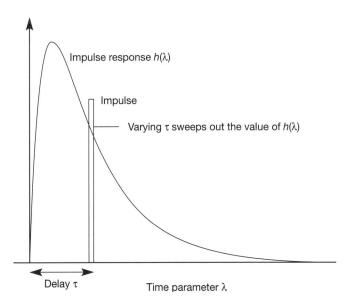

Figure 8.3 The impulse response, and the 'spike' sweeping out the valve.

8.2.1 Selection of a suitable input test signal

The following signals satisfy the autocorrelation requirement.

An impulse

The product of an impulse and a delayed version of itself is only non-zero if the impulses are simultaneous, that is, if the delay is zero. Unfortunately, it is very difficult to generate an impulse of significant power – signals of finite 'area' and zero duration need to be of infinite magnitude! That difficulty also militates against direct impulse-response testing.

White (Gaussian) noise

A true white-noise signal has a mean value of zero, is equally likely to be positive or negative at any instant, and has normally distributed amplitude with probability symmetrical about zero. Its autocorrelation at zero delay (equal to the mean-square value of the signal) is non-zero as, in the averaging over time, we always have $+$ coinciding with $+$, or $-$ coinciding with $-$; the product is therefore always positive or zero.

With a delay, however, the product is equally likely to be positive or negative and observes the same amplitude probability distribution in the positive and negative directions, so its average value is zero. White noise is therefore a good test signal for our purpose, but it does introduce certain problems.

- The above properties only strictly apply to white noise of infinite bandwidth. What can be generated experimentally is an approximation to *bandlimited* white noise, which will give a reasonable approximation to the autocorrelation 'spike'.

- Producing a wide-bandwidth analog delay without attenuation (which would be required for the cross-correlation test setup of Figure 8.1) is quite difficult.

A pseudo-random binary sequence (PRBS)

A PRBS is an effectively random sequence of logic 1 levels and logic 0 levels. For use as a test signal in this respect, logic 1 will be a positive voltage V and logic 0 an equal negative voltage $-V$. It is a very convenient signal to use because it can be generated very easily by means of a serial-input shift register with feedback via an exclusive-OR gate. The signal is not truly random because the sequence repeats itself every $2^n - 1$ bit intervals for an n-bit shift register. It is noteworthy that not all combinations of bits for the feedback connection work; tables of ones which do work are given in Horowitz and Hill (1980) and Godfrey (1993). Figure 8.4 shows one successful possibility for a 9-bit register.

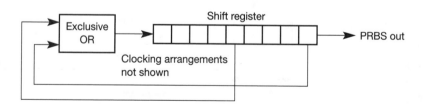

Figure 8.4 A PRBS generator using a nine-bit shift register.

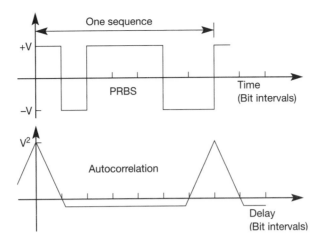

Figure 8.5 A seven-bit
PRBS and its
autocorrelation function.

The autocorrelation properties of a PRBS resemble those of white noise but, unlike white noise, it is very easy to delay. The autocorrelation of a 7-bit PRBS is shown in Figure 8.5, along with the sequence itself. We note that:

- The autocorrelation repeats itself every seven bit intervals of delay. This is to be expected – so does the PRBS!

- A spike at the origin is observed (like white noise) but, unlike white noise, it is triangular with a height V^2 and a base of two bit intervals. Between spikes, the autocorrelation 'sits' on a steady negative level of $-V^2/n$ rather than on zero ($n = 7$ in this case).

8.2.2 The use of a PRBS signal as a test input

If we use such a signal as a test input in the arrangement of Figure 8.1, we are assuming that it has an autocorrelation function which is of the same shape as a time-domain impulse, in order for Equation (8.5) (the averager output) actually to provide the impulse response of the test system (thus leading to a model). Since this is not quite true, we need to consider the differences.

The effect of the triangular autocorrelation function is that the cross-correlation graph obtained in the experiment closely resembles the time impulse response but is not identical to it. The main difference is that the cross-correlation graph 'sits' on a steady negative value rather than on zero. This difficulty can easily be solved by adding the steady negative value to all the averager output readings, or simply redrawing the delay/time axis along the steady negative value (effectively the same thing). Our cross-correlation graph would then be of the same form as the time impulse response.

A remaining difficulty is to decide how strong the 'impulse' (represented by the autocorrelation function of the PRBS) was, as the transfer function (which is often what we would like to determine) is the Laplace transform of the *unit* impulse response. The answer is quite straightforward – the strength of the 'impulse' is the area of the autocorrelation triangle, namely, its height multiplied by one bit interval.

8.3 Practical aspects of correlation testing

In this section, an example will be worked to illustrate the method. The effectiveness of the method in the face of nonlinearities will be discussed, as will the issues relating to the choice of the PRBS length and bit interval.

Example 8.1 *Fitting a Laplace transfer function model to a cross-correlation result*

A control system element was tested by the cross-correlation method using a PRBS input of sequence length 127 bits, bit interval 0.02 s and levels ± 2 V. The resulting input–output cross-correlation is plotted in Figure 8.6. Obtain the transfer function of the system.

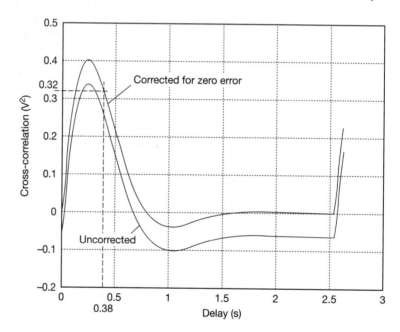

Figure 8.6 Cross-correlograms for Example 8.1.

By inspection, the response appears to be a decaying sinewave. The cross-correlation settles at a value of $-0.064\,\text{V}^2$, so we add that value to all the readings, resulting in the corrected graph in the figure. That graph has zero crossings at delays of 0.77 s and 1.52 s, giving two almost equal half-cycles averaging 0.76 s. This result indicates:

(1) It is a decaying sinewave rather than a cosine wave.

(2) An expression $Ce^{-at}\sin(\omega t)$ can be fitted to the curve and, by fitting it, we can progress towards the transfer function by the following steps:

(3) As the cycle time is $2\pi/\omega$, we have $1.52 = 2\pi/\omega$ and so $\omega = 2\pi/1.52 = 4.13\ \text{rad s}^{-1}$.

(4) The quantity a is now found from the delay between successive peaks, which are 0.76 s

(half a cycle) apart. The height of the first positive peak is $0.405\,V^2$ and the depth of the first negative peak is $(-)0.04\,V^2$ (it is acceptable to consider one positive and one negative peak, since the decay envelope is symmetrical about the zero amplitude axis). So $e^{-0.76a} = 0.04/0.405$, and by taking natural logarithms, $-0.76a = -2.315$, giving $a = 3.05\,\text{s}^{-1}$.

(5) We can now find the value of C from the value of the cross-correlation halfway across the first half-cycle. At this point, $t = 0.38\,\text{s}$ and the measured value from the cross-correlogram is about 0.32. We choose this point because $\sin(\omega t) = 1$ there, so we can omit it from the expression, thus easing the analysis:

$$0.32 = Ce^{-3.05\times0.38} \text{ and so } C \approx 1.02\,V^2$$

We have now fitted an expression to the curve which, viewed in the time domain, gives the system impulse response to an impulse of strength V^2 times the bit interval of the PRBS, which evaluates to $2^2 \times 0.02 = 0.08$ units. The response to a unit impulse is therefore $1/0.08$ times our expression, or $12.75e^{-3.05t}\sin(4.13t)$. The transfer function is the Laplace transform of the unit impulse response and this is readily found using the result that $e^{-at}\sin(\omega t)$ transforms to $\omega/[(s+a)^2 + \omega^2)]$. Our transfer function therefore becomes $12.75 \times 4.13/[(s+3.05)^2 + 4.13^2]$, which simplifies to $52.7/(s^2 + 6.1s + 26.4)$.

The authors will now admit that the cross-correlation results in Example 8.1 were calculated on the basis of a 'plant' of transfer function $50/(s^2 + 6s + 25)$. The differences between that transfer function and the one obtained in the example are as follows.

The static gain was obtained as 1.996, rather than 2.0 (an error of about 0.2 per cent). The undamped natural frequency was obtained as $5.14\,\text{rad}\,\text{s}^{-1}$, whereas it should be 5.0 (an error of about 2.8 per cent). The damping ratio was obtained as 0.58 rather than 0.6 (a 3.3 per cent error). The accuracy of the transfer function determination is acceptable in view of the limitations of graphical measurement and the rounding errors in the above analysis.

If greater accuracy in determining the parameters is required, it is possible to do the fit by a least-squares procedure. At each bit interval of the PRBS, the value of the cross-correlation calculated from the fitted expression is subtracted from that taken from the graph and the difference of the values, known as the error, is squared. The sum of those squares is then calculated. The parameters of the fitted expression are then iteratively adjusted and the new sum of the squares of the errors calculated each time until a set of parameters which minimizes the sum of squares of the errors is found.

Computer assistance is usually needed to accomplish the fitting in a reasonable time! Such an approach to fitting the curve is normally necessary for a situation where the general form of the curve is more obscure, such as the 'double exponential' response typical of overdamped second-order systems (Figure 8.7) where there is no simple accurate method of separating the exponentials.

For higher-order systems where the structure of the model may be unclear, it may be better to use a discrete-time model of the form described in Section 3.6.1:

$$y_n = \sum_{r=1}^{l} A_r y_{n-r} + \sum_{r=0}^{m} B_r u_{n-r} \tag{8.7}$$

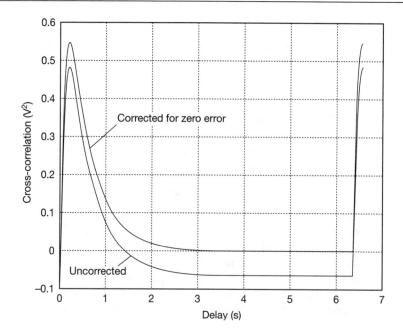

Figure 8.7 Cross-correlograms for an overdamped second-order plant.

With such a model, all we need to decide before fitting the coefficients are the required orders of the model for input and output – the m and l in Equation (8.7). If we are unsure of these and do the fitting successively for higher-order models, the residual least-squares error will fall very substantially when the correct order is reached. The fitting can be done automatically by available computer packages, but caution is required in their use. For example, one of the authors successfully fitted a result manually to a set of test results for which an automated package had simply generated zeros for the A and B coefficients. A computer-aided parameter-fitting example for the system element of Example 8.1 follows, in which MATLAB (Appendix 3) is used to do the computation.

Example 8.2 *A computer-assisted solution to Example 8.1*

Determine the best-fit parameters C, a and ω in Example 8.1.

We will start from the values estimated in Example 8.1: $C = 1.02$, $a = 3.05$, $\omega = 4.13$. The procedure will be to write a MATLAB m-file to determine the sum of squares of the errors, given values of C, a and b; and to use it to find the optimum values. We give the details of the code here so that readers who do not have MATLAB can see how to achieve the same result in another high-level language. The only major difference will be that in the 'other language' it will be necessary to use 'for' loops (or some equivalent construction) to perform the vector multiplications which MATLAB does with a single multiplication operation.

We will assume that the first 127 values on the cross-correlation graph (corresponding to times from 0 to 126 bit intervals of 0.02 s, that is, 0–2.54 s) are entered into a MATLAB row vector called *ruy* (or a suitably-dimensioned array in another language). The m-file will proceed as follows (see Appendix 3 for more details on MATLAB m-files).

```
% First we produce corresponding calculated values from Cexp(−at)sin(wt):

t = 0.02 * [0:126];              % set up the time vector

rr1 = c * exp(−a*t) .* sin(w*t);  % calculated values from model (N.B.
                                  % dot* command to do element-by-
                                  % element multiplications)

% Then we calculate the errors:

err = ruy − rr1;

% Now to square and add – easy with matrices!

ssq = err * err'                 % no semicolon, so we see the
                                 % sum-of-squared-errors
```

The above m-file will now be assumed to have the filename cssq.m.

The following example illustrates the use of the file; the user input to MATLAB is shown following MATLAB prompts (≫), and the output from MATLAB occurs between these. The bracketed comments are those of an author after the experiment.

```
≫ c = 1.02;     % setting the initial parameter values

≫ a = 3.05;

≫ w = 4.13;

≫ cssq          % runs the m-file to produce the sum-of-squared-errors

ssq = .0051

≫ c = 1.03;     % try tuning the value of C

≫ cssq

ssq = .0059              (Worse!)

≫ c = 1.01;

≫ cssq

ssq = .0048              (Improving!)

≫ c = 1.0;

≫ cssq

ssq = .005               (C = 1.01 was best)

≫ c = 1.01; a = 3.04;    % now try tuning a

≫ cssq

ssq = .0049              (Worse!)
```

(Continue in this way, one parameter at a time, until no further improvement is possible by altering any parameter in any direction.)

The above working makes it apparent that obtaining the transfer function directly from the impulse response is rather less straightforward than doing so from the frequency response by asymptotic approximation. Fortunately, given the impulse response at bit intervals of the PRBS test signal, it is easy to obtain the frequency response in magnitude and phase if computer assistance is available.

Suppose the overall time response to a unit impulse is $h(t)$. Section A5.2 shows that the Laplace transform of the impulse response sampled at intervals T (and therefore the system transfer function, since the Laplace transform of a unit impulse is 1) can be given by:

$$\tau[h(0) + e^{-sT}h(T) + e^{-2sT}h(2T) + \cdots]$$

or

$$\tau \sum_{n=0}^{\infty} h(nT)e^{-nsT}$$

Strictly speaking, τ is the width of the impulses in Figure 7.2, as discussed in Appendix 5. However, in the absence of a convincing brief and rigorous argument, it will be left that the right answers result if we take τ to be equal to the sampling interval T in the present context!

The frequency response now results from the substitution of $j\omega$ for s, giving:

$$h(j\omega) = T \sum_{n=0}^{\infty} h(nT)e^{-nj\omega T} \tag{8.7}$$

In practice, the upper limit for n will be the number of bits in the PRBS but, if the length and bit interval have been chosen correctly, the response should have decayed effectively to zero by that time. The MATLAB m-file $eq8\_7.m$ on the accompanying disk will perform the calculations of Equation (8.7) on the series $ruy(t)$ (cross-correlation representing the impulse response) of length 127 bits and obtain the decibel gain dB and the phase angle ag in degrees.

This m-file was used to convert the impulse response from Examples 8.1 and 8.2 to a frequency response, which is shown in Figure 8.8. The next example obtains the transfer function from those graphs and shows that it is effectively the same as that obtained directly from the impulse response. It is noteworthy that the response at the highest frequencies is a little in error because of aliasing (see Section 5.8.1).

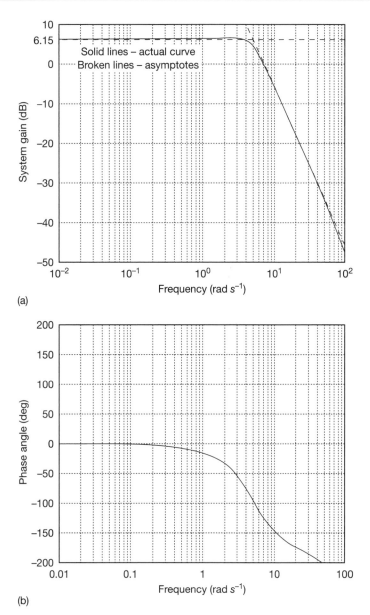

(a)

(b)

Figure 8.8 Bode plots for Example 8.3. (a) The magnitude plot with asymptotes added. (b) The phase plot.

Example 8.3 *Checking the cross-correlation result by Bode identification*

Using the derived graphs of the frequency response, obtain the transfer function and compare it with that derived directly from the impulse response.

Referring to the amplitude graph (Figure 8.8(a)), the asymptotes have been plotted and we deduce:

(1) The gain at low frequency is constant at 6.15 dB, so the system is of type 0.

(2) There is only one high-frequency asymptote and its slope is -40 dB/decade. The 'corner error' is -1.55 dB approximately, suggesting an underdamped second-order system (which is also suggested by the slight peak in the response).

(3) The net transfer function is of the form

$$G(s) = \frac{K\omega_n^2}{s^2 + 2\zeta\omega_n s + \omega_n^2}$$

(we are assuming that the system is second order and that there are no extra poles and zeros so close together that their presence does not greatly affect the plot).

(4) The undamped natural frequency is easily obtained from the asymptote intersection as $\omega_n = 5$ rad s^{-1}.

(5) K is obtained from the low-frequency gain. $20\log_{10} K = 6.15$ dB so $K = \text{antilog}(6.15/20) = 2.03$.

(6) ζ is obtained by reference to standard second-order system response curves (see Figure 3.39(a) in Section 3.5.1) or by noting that the magnitude of the system transfer function at the undamped natural frequency is given by $K/(2\zeta)$. The corner error of -1.55 dB and the d.c. gain of 6.15 dB mean that the gain at the corner frequency (which is the undamped natural frequency) is 4.6 dB, or 1.7. Since $K = 2.03$, this suggests a damping ratio of approximately 0.6.

The overall transfer function is now

$$G(s) = \frac{2.03 \times 5^2}{s^2 + 2 \times 0.6 \times 5s + 5^2} = \frac{50.75}{s^2 + 6s + 25}$$

This answer is accurate, except for the static gain which is 1.5 per cent too high; some error is to be expected in view of the use of the logarithmic scale in obtaining the value.

8.4 Choice of PRBS parameters

The selection of the bit interval and the sequence length for a cross-correlation test using a PRBS follow quite simple criteria. The overriding requirement is that the total time duration of the PRBS must be enough to allow the impulse response of the test system to decay effectively to zero. Within that constraint, it would seem desirable to use a very long sequence in conjunction with a very short bit interval for the PRBS, and that would indeed be the best approach if the test system were perfectly linear. Practical 'plants' seldom are, however, and stiction in particular will cause difficulties if the bit interval is so short that appreciable output change does not occur during it.

A bit interval of about one-fifth of the shortest plant time constant is often the best we can achieve. The sequence length is then simply determined by the impulse-response decay time.

The method has several advantages over frequency-response testing of the traditional type with oscillator, data logger (or even chart or magnetic tape recorder) and oscilloscope. One is that plant noise will be removed by the correlation operation as far as the measurements are concerned; tests have shown that white noise of several times the magnitude of the PRBS input can be added to the plant output without producing more than slight degradation of the results. For the same reason, it is possible in principle to do the test while the plant is in normal use, but plant managers tend to be sceptical! Another advantage is that the time required on plant can be quite short – the response can be logged at the time and taken away for the correlation to be performed offline.

The method is not, however, suitable for systems having more than slight nonlinearities, especially if they are in the form of stiction or backlash (see Chapter 14). This limitation is to be expected in view of the linearity assumptions inherent in the derivation of the principles of the method.

As with the identification of transfer functions from Bode plots, it is obviously necessary to have some knowledge of the plant, before the method can be successfully applied. In this case, some notion of the order of the plant, and the shortest time constant are required. Once again, some simple modelling from physical considerations along the lines suggested in Chapter 2 will be a wise starting point.

8.5 Another look at Bode identification

In Section 3.9.5, the basic principles of identifying a system from its Bode frequency response plots were introduced. However, only simple systems were considered there. In particular, only minimum-phase systems were considered. Now we revisit the process in a little more detail. In this section it will be assumed that frequency response data exist, and were obtained from a stable system.

As we know, for transfer function identification, the reverse procedure to the Bode asymptotic approximation is carried out on the plot of dB magnitude versus logarithmic frequency. It consists of the following steps:

(1) The dB magnitude curve is approximated by a series of piecewise straight lines, or asymptotes. The asymptotes are chosen such that their slope is given by an integer; a unit change in slope is defined as being a gradient change of 20 dB/decade.

(2) The asymptotes are decomposed into several components.

(3) The corner frequencies are noted.

(4) The corresponding transfer function is written in pole–zero form.

8.5.1 A note on non-minimum-phase systems

In using step 4 above, particular attention must be paid to the plot of phase versus frequency response. For a minimum-phase process (one having poles and zeros with negative real parts), the magnitude and phase frequency responses are interdependent: a pole will tend to cause a $-90°$ phase shift, and a zero will tend to cause a $+90°$ phase shift. For a non-minimum-phase process the converse is true; a non-minimum-phase pole will tend to cause a $+90°$ phase shift, and a non-minimum-phase zero will tend to cause a $-90°$ phase shift (see Section 3.8). Since the system is assumed to be stable, all the poles will have negative real parts. However, the sign of the zero must be selected by reference to the curve itself.

8.5.2   A note on time delays (transport lags)

The Bode plots may also be used to determine the presence of a time delay in the process response, since the high-frequency phase plot decreases rapidly to minus infinity. The value of the time delay can be determined by plotting phase against frequency (not log of frequency), and measuring the negative slope of the resulting straight line. This is proved simply by considering the frequency response of a time delay:

$$e^{-s\tau} = e^{-j\omega\tau} = 1 \arg(-\omega\tau)$$

Alternatively, the normal Bode plots against logarithmic frequency axes can be used as follows:

(1) From the high-frequency slope of the magnitude plot, determine the net system order (that is, the multiple of -20 dB/decade slopes present at high frequency).

(2) Evaluate the expected final value of the phase plot at high frequencies, taking into account every pole and zero, including any non-minimum-phase ones, but ignoring the transport lag.

(3) At a specific value of high frequency, compare the actual phase lag with that expected from step 2. The excess phase lag is due to the transport lag. Since this is equal to $-\omega\tau$, where ω is the chosen frequency, the transport lag τ can easily be determined.

If the frequency response is to be calculated from a time domain response using the Fourier transform, any time delay should be removed from the time-domain output before transformation takes place.

Example 8.4 *Transfer function identification from a Bode plot*

Table 8.1 lists the experimentally obtained harmonic responses for a component of a system, following a frequency response test using an input of 0.5 V peak–peak. Find the component's transfer function, explaining clearly how the result is obtained.

A transfer function may be obtained by plotting the data in Bode form and using the asymptotic approximation to determine the corner frequencies, and hence the pole–zero locations. To do this, the data in Table 8.1 must be changed into Bode form, as in Table 8.2. Bode plots may now be produced using standard log-linear graph paper, as shown in Figure 8.9.

Initially the Bode gain plot has a slope of $+20$ dB/decade and the phase plot shows a constant phase shift of $+90°$, so the system has a zero at $s = 0$. Two corner frequencies can be identified at 0.1 and 20 rad s^{-1}, and in both cases the gain plot changes slope by -20 dB/decade accompanied by a fall in phase.

The Bode gain plot has a final slope of -20 dB/decade, but the phase plot does not tend to $-90°$. The final slope of the gain plot suggests that the system has one more pole than zero, and the fact that the phase plot appears not to be settling at all suggests that the component under test contains a transport lag (time delay).

Table 8.1 Plant test results for Example 8.4.

Frequency (rad s^{-1})	Peak–peak output voltage (for 0.5 V in)	Phase shift (degrees)
0.0010	0.008	89.42
0.0023	0.018	88.69
0.0052	0.041	87.02
0.012	0.094	83.24
0.027	0.207	74.89
0.061	0.417	58.38
0.139	0.649	35.26
0.316	0.763	16.46
0.720	0.792	5.44
1.638	0.796	−2.13
3.728	0.786	−11.16
8.483	0.736	−27.17
19.31	0.576	−54.76
43.94	0.331	−90.57
100.0	0.157	−135.9

Table 8.2 Results of Table 8.1 manipulated for plotting Bode diagram.

Frequency (rad s^{-1})	Gain (number)	Gain (dB)	Phase shift (degrees)
0.0010	0.0160	−35.9	89.42
0.0023	0.036	−28.8	88.69
0.0052	0.082	−21.7	87.02
0.012	0.188	−14.52	83.24
0.027	0.414	−7.66	74.89
0.061	0.834	−1.58	58.38
0.139	1.298	2.27	35.26
0.316	1.526	3.67	16.46
0.720	1.584	4.00	5.44
1.638	1.592	4.04	−2.13
3.728	1.572	3.93	−11.16
8.483	1.472	3.36	−27.17
19.31	1.152	1.23	−54.76
43.94	0.662	−3.58	−90.57
100.0	0.314	−10.06	−135.9

The conclusion that the system has one more pole than zero, together with the previous results, suggests a transfer function (minus time delay) of the form

$$G(s) = \frac{Ks}{(1 + s/0.1)(1 + s/20)}$$

However, all we can really say is that the transfer function is *likely* to be minimum phase (with a time delay, yet to be determined). The system could still be non-minimum phase. However, this would require the transfer function to have the same number of right-half

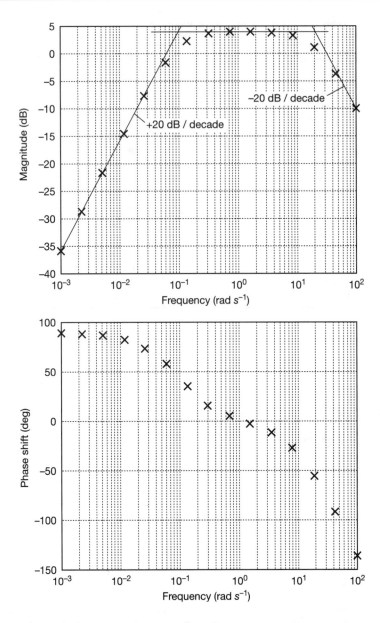

Figure 8.9 Fitting a Laplace transfer function to a Bode plot.

s-plane poles as zeros, which in turn implies that the system is unstable – if this were the case, we should know it!

Also, there may be other minimum-phase poles and zeros present, which are rather close to each other, so that their effect is not visible on our rather crude Bode plots. We shall assume that we have sufficient knowledge of this system (from an initial simple modelling exercise, or from consideration of the number of significant energy storage elements within the component, which predicts the required order of model) to suspect that the above transfer function, together with a time delay, is likely to be the correct result.

We have another problem too, compared with our earlier simple identification exercises, namely that the 'd.c. gain' is not readily apparent for this system. The Bode gain K in the transfer function above may be determined from the dB magnitude expression for the suggested $G(s)$:

$$20 \log_{10}(G(j\omega)) = 20 \log_{10} K + 20 \log_{10}\omega - 20 \log_{10}\sqrt{\left[1 + \left(\frac{\omega}{0.1}\right)^2\right]}$$

$$- 20 \log_{10}\sqrt{\left[1 + \left(\frac{\omega}{20}\right)^2\right]}$$

At a frequency of $\omega = 1$ rad s^{-1} the system gain is about 4 dB, so

$$\frac{4}{20} = \log_{10} K + 0 - 1.0022 - 0.0005$$

or

$$K = 10^{1.203} = 16$$

Consequently, the transfer function so far becomes

$$G(s) = \frac{16s}{(1 + 10s)(1 + 0.05s)}$$

Now we need to consider the transport lag we believe to be present. Since the delay-free LTF has one more pole than zero, and is assumed to be minimum phase, we would expect a high-frequency phase shift of $-90°$. At our highest-frequency point, $\omega = 100$ rad s^{-1}, and the measured phase shift is seen to be $-135.9°$. The excess phase shift therefore seems to be $-45.9°$, and this is therefore assumed to be due to the transport lag.

Before continuing, however, notice that the highest frequency break-point (20 rad s^{-1}) is sufficiently close to 100 rad s^{-1} for the phase lag due to that breakpoint not yet to have reached $-90°$. In other words, we did not test to sufficiently high a frequency.

The effect of this can be compensated for, using the expression for the phase lag of a first-order lag element (Equation (3.76) in Section 3.5.1), namely phase lag $= (-)\tan^{-1}(\omega/\omega_c)$. This tells us that, at $\omega = 100$ rad s^{-1}, the phase lag due to a pole at $\omega_c = 20$ rad s^{-1} is $(-)\tan^{-1} 5 = (-)78.8°$. The excess phase lag due to the transport lag is therefore not $-45.9°$, but $-57.1°$.

Now (from point 3 preceding this example), $-57.1° = 0.997$ rad $= -\omega\tau$ which, since $\omega = 100$ rad s^{-1}, gives $\tau = 0.01$ s.

The final transfer function is therefore:

$$G(s) = \frac{16s\, e^{-0.01s}}{(1 + 10s)(1 + 0.05s)}$$

Note that the MATLAB m-file *fig8_9.m* on the accompanying disk shows how to include the effect of a time delay in a MATLAB Bode plot.

8.6 *Modelling time delays in multivariable systems*

Imagine that we have carried out an identification test on a piece of plant, using either the methods of this chapter, or the simpler methods discussed in Section 3.9. As a result, we shall have acquired information which we can use to form a model of the plant. This model may take various forms. We have discussed differential equation models, Laplace transfer function models, state-space models, discrete-time state-space models, difference equation models and z-transform models; and there are others.

We have noted that many process plants contain pure time delays (transport lags, or 'distance–velocity lags'), and our identification tests may have convinced us that the plant under investigation contains such delays. Since time delays generally destabilize control loops, it is important to be able to include their effects in any plant models used for design purposes.

In Section 3.7 we discovered how to include a pure time delay in a Laplace transfer function model as $Y(s) = e^{-s\tau}U(s)$, where τ is the duration of the delay in seconds (and we identified one in Example 8.4, above). This also gives us a way of handling time delays in multivariable (multi-input-multi-output) systems, because we can include such terms in transfer function matrices and then use the methods of Chapter 10. We also saw how to include time delays in difference equation models in Section 3.7 and in z-transform models in Section 7.7.

Unfortunately, much of the work on multivariable systems is done using state-space models, and we have a problem if we want to use continuous-time state-space methods with systems containing pure time delays. This is because it is impossible directly to convert either the Laplace transfer function model $Y(s) = e^{-s\tau}U(s)$, or the equivalent time-domain model $y(t) = u(t - \tau)$ into the state-space form. There are two basic approaches to overcoming this problem, and we briefly discuss them below.

8.6.1 *Discrete-time state-space models*

The most accurate solution is usually to resort to discrete-time state-space models, in which the time delay can be included by delaying the appropriate signal by a number of sampling intervals equivalent to the delay. However, for accuracy it is necessary to have an integer number of sampling intervals corresponding to the delay. This is precisely analogous to the way we handled time delays using the z-transform, so we need not spend much time on it. It is also likely to increase the model order in the same way.

Example 8.5 *A time delay in a discrete-time state-space model*

Say we represent a system by a second-order discrete-time state-space model:

$$x_{k+1} = \begin{bmatrix} 0.981 & 0.019 \\ -1.909 & 0.904 \end{bmatrix} x_k + \begin{bmatrix} 0.0002 \\ 0.019 \end{bmatrix} u_k \quad \text{and} \quad y_{k+1} = \begin{bmatrix} 1 & 0 \end{bmatrix} x_{k+1}$$

Say that we then discover that this representation is not sufficiently accurate, and that we ought to include a pure time delay of two sampling periods in the output equation. For *simulation studies*, the only modification required would be to change the output equation to $y_{k+1} = [1\ 0]x_{k-1}$ (that is, we form the output at any given time step from the value of x_1 two time steps ago).

However, for use in discrete-time state-space *design methods* (such as those of Chapter 12), we need to keep to the general form $y_{k+1} = Cx_{k+1}$, so that the model remains consistent in time. To achieve this, we might introduce two new state variables x_3 and x_4, such that $x_{3_{k+1}} = x_{1_k}$ and $x_{4_{k+1}} = x_{3_k}$. In this way, x_4 becomes a version of x_1 delayed by two time steps, and can be used in the output equation as required. The modified model is therefore fourth order, and would be:

$$
x_{k+1} = \begin{bmatrix} 0.981 & 0.019 & 0 & 0 \\ -1.909 & 0.904 & 0 & 0 \\ 1 & 0 & 0 & 0 \\ 0 & 0 & 1 & 0 \end{bmatrix} x_k + \begin{bmatrix} 0.0002 \\ 0.019 \\ 0 \\ 0 \end{bmatrix} u_k \text{ and } y_{k+1} = [0\ 0\ 0\ 1]x_{k+1}
$$

If the time delay had been found to be elsewhere in the system, we should proceed in exactly the same manner, introducing new discrete-time states as required, and maintaining the correct state-space form.

8.6.2 Continuous-time state-space models

To include a time delay in a continuous-time state-space model, the usual approach is to use some other function which *approximates* the behaviour of the time delay, and which can be converted to state-space form. Two such approximations are given here, but both have potentially serious disadvantages.

Approximation of a time delay by a series of cascaded first-order lags

Here we simply replace the time delay of τ seconds by a term of the form

$$
e^{-s\tau} \approx \frac{1}{(1 + s\tau/n)^n}
$$

The higher the value of n, the better the approximation – but the higher the order of the resulting model. Figure 8.10 shows some step responses for various values of n.

This model has the advantage that its step response has the correct qualitative form, with no oscillations, but it often needs to be of a high order to succeed. Another difficulty is that, if we make the approximation of sufficiently high an order to be realistic, we shall often end up with numerical problems in the ensuing designs. For example, setting up a 20th-order version of such a transfer function, and asking MATLAB (Appendix 3) for the step response, results in an apparently unstable system. However, the model is not really unstable (after all, it is just a series of first-order terms); rather, the computational methodology is at fault. Note that this is not a poor reflection on MATLAB; the problem is a numerical one and the same type of effect is likely to occur with any package.

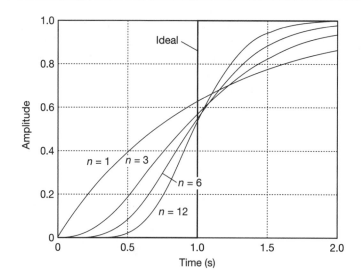

The Padé approximation to a time delay

In this approach, the function we desire to approximate ($e^{-s\tau}$ in this case) is expanded as a power series in the Laplace operator s and then equated to the rational polynomial

$$\frac{a_0 + a_1 s + a_2 s^2 + \cdots}{b_0 + b_1 s + b_2 s^2 + \cdots}$$

proceeding to as high or low an order as we think fit. In general applications, to give models with finite bandwidths, the order of the numerator would be chosen to be lower than that of the denominator. However, they are chosen to be *equal* when approximating pure time delays.

We can expand the time delay as

$$e^{-s\tau} = 1 - s\tau + \frac{s^2\tau^2}{2!} - \frac{s^3\tau^3}{3!} + \cdots$$

and when equated with the general result above, using terms up to order n, the mathematics permits a number of arbitrary choices. This means that you might see different numerical results in different texts. For example, if you do the maths for a second-order approximation, you will discover that the numerator and denominator coefficients of s^2 can be any number we wish to choose, so long as they are identical. The version of the result used in MATLAB is as given below, and Figure 8.11 contains a few example step responses:

$$e^{-s\tau} \approx \frac{2 + \sum_{i=1}^{n} \frac{(-s\tau)^i}{i!}}{2 + \sum_{i=1}^{n} \frac{(s\tau)^i}{i!}}$$

This model has the advantage that its step response can be nearer to the correct form in terms of 'sharpness', but the disadvantage is the obvious pre-step

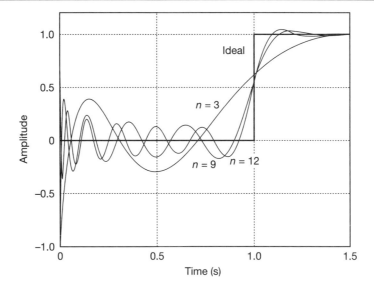

Figure 8.11 Approximating a transport lag by an nth-order Padé approximation.

oscillation, the overshoot and the non-minimum-phase nature of the response. That is to say, the step response begins by moving in the 'wrong' direction, which is very misleading for a controller design based on such a model.

8.7 Conclusions

In this chapter we have revisited some of the areas we introduced earlier, but in greater detail than was appropriate earlier in the text. These areas included the identification of Laplace transfer models from Bode diagrams (where we mentioned the effects of non-minimum-phase terms and transport lags) and the representation of transport lags in state-space models (both continuous- and discrete-time). In the continuous-time domain, we introduced two methods of approximating time delays by rational functions in s.

The main topic of the chapter was the use of PRBS signals for system identification. We discussed the theory behind the method, and also the choice of PRBS parameters and the fitting of models to the resulting correlograms. The fitting was done both 'by hand' and with computer assistance.

8.8 Problems

8.1 A pseudo-random binary sequence has a length of 31 bits, a bit interval of 0.2 s and voltage levels of ±5 V. Sketch the graph of its autocorrelation function against time.

8.2 A pseudo-random binary sequence having a bit interval of 0.02 s and voltage levels of ±2 volts is used to test three plants by the cross-correlation method. The results are shown in Figures P8.2(a) to P8.2(c). Obtain the transfer function for each plant.

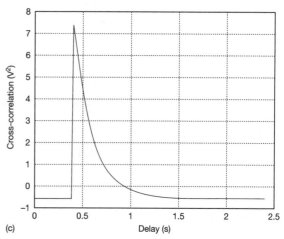

(c)

Figure P8.2 (*Continued*)

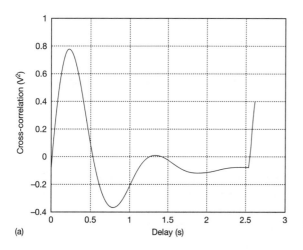

(a)

8.3 The graphs of Figures P8.3(a) and P8.3(b) were obtained by converting the unit impulse response obtained from the graph of Figure P8.2(a) into a frequency response.
   (a) Obtain the transfer function from the frequency response.
   (b) The phase response goes beyond −180 degrees, despite the fact that the system appears from the decibel graph to be second-order. Would you expect this? Why?

8.4 The unit impulse response of a particular plant is given by $5e^{-2t}\sin(5t)$, where t is the time in seconds. You are advised to do this question by using

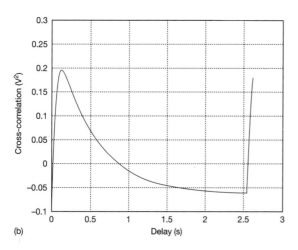

(b)

Figure P8.2 (a) First cross-correlogram for Problem 8.2. (b) Second cross-correlogram for Problem 8.2. (c) Third cross-correlogram for Problem 8.2.

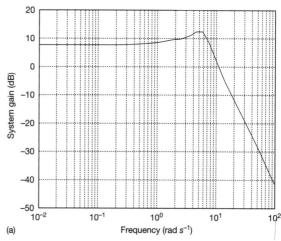

(a)

Figure P8.3 (a) Bode magnitude plot for Problem 8.3. (b) Bode phase plot for Problem 8.3.

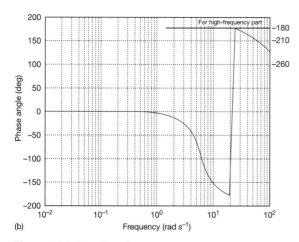

(b)

Figure P8.3 (*Continued*)

MATLAB or a similar package, or by programming in a high-level language.

(a) Set up a sequence of values of t from 0 to 2.52 s in steps of 0.02 s (127 values altogether).

(b) Use the values to generate the frequency response in magnitude and phase and plot it in the form of a Bode diagram.

(c) From the Bode diagram, obtain the Laplace transfer function.

(d) Check the answer by taking the Laplace transform of the impulse response.

8.5 An experiment to determine the transfer function of a particular control system element by the cross-correlation method using a 127-bit PRBS produced the results plotted in Figure P8.5. The vertical axis is calibrated in volt^2. The static gain of the plant was measured separately as 5.0 units.

(a) Determine the transfer function.

(b) What were the amplitude and bit interval of the PRBS signal?

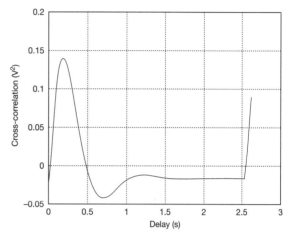

Figure P8.5 Cross-correlogram for Problem 8.5.

9 Observers and state estimation

In this chapter the reader will learn:

- that unmeasurable system states can often be estimated from the system outputs
- that these state estimates can be used in state variable feedback schemes instead of the real states
- that the states can even be estimated in systems in which the states and the measurable signals (outputs) are corrupted by noise
- how to communicate vector–matrix designs to those not trained in state-space methods or matrix algebra
- the rudiments of digital implementation of a scheme containing an observer and a state-variable feedback tracking loop, including bumpless transfer between manual and automatic control, and avoidance of integral wind-up.

9.1   Preview

Readers who have studied Chapters 1 to 5 can understand this chapter. It is concerned with the problem of wishing to measure the state variables of a system in order to feed them back in a state variable feedback (SVF) scheme, for example (see Chapter 5), but finding that some of the state variables are unmeasurable. This chapter provides some methods of estimating such unmeasurable states from measurements of the system's outputs, and covers many practical aspects of using such systems.

The Kalman filter is also studied in this chapter. It takes things one stage further by considering that the plant output signals (from which the state variables are to be estimated) may be corrupted by noise. The Kalman filter is a rather advanced topic, and its study can be omitted if desired, without compromising anything else in the book.

NEW MATHEMATICS FOR THIS CHAPTER

Most of the chapter requires no new mathematics. The matrix algebra has all been used before, especially in Chapter 5; but if a refresher is needed, take some time to read Appendix 1 now, where all the required matrix algebra is presented. Section 9.8 (on Kalman filtering) requires statistical analysis of random signals. This can seem rather daunting on a first reading, but it is all covered in Appendix 6 and, if taken one step at a time, can be assimilated by undergraduate students. Appendix 6 also provides enough mathematics to give a complete derivation of the Kalman filter, if required, which is lacking in most textbooks.

9.2 Introduction

In Section 5.4, we discussed the placement of the closed-loop poles (eigenvalues) of a system by the use of state variable feedback. This required the measurement of the system state variables, so as to be able to feed them back to the plant inputs. We assumed that the state variables will always be available for feedback.

This may well be the case in a simple system model such as that of Figure 9.1, where we choose the state variables to be physically meaningful. However, it is not usually the case, for reasons such as the following:

- There will almost always be blocks in the system which are not adequately represented by models of first order. More than one state variable will need to be assigned to describe such a block (*n* state variables are needed for an *n*th order block). Therefore, we shall not be able to manage with just one state at the block's output (as in Figure 9.1), but will need one or more states 'buried' inside the block somewhere. These will, presumably, *not* be available for measurement.

- Some of the state variables may be physically incapable of being measured. For example, the temperature *inside* a metal ingot may well be a state variable in a model of a reheating furnace. It would probably be impossible to measure this temperature on a routine production basis for use in a continuous temperature control scheme for the furnace.

- While it may be physically possible to measure some state variables, it may be uneconomical to do so. For example, temperatures inside ingots *have* been measured by thermocouples inserted into drillings in the ingot, and sent through the furnace. However, this was done only as a research exercise, for assistance in developing and validating a plant model. It would be prohibitively expensive, in terms of both time and materials, in normal production.

- Plant descriptions often arise naturally as simple input–output transfer functions, following frequency response testing for example (or as transfer function matrices for multivariable systems). When such models are converted to state-space models using techniques such as direct programming (Section 2.5.1), the only state variables to which we would have direct access would be ones assigned to the plant outputs. All other states would be generated by the mathematics, and may well not even correspond to physical variables at all. In this case, they clearly cannot be measured for use in a feedback scheme, as we do not even know where to look for them.

Factors such as these mean that usually we cannot implement a full state variable feedback scheme with the available states (although adequate performance can sometimes be achieved by using only the subset of the states which *are* available).

Fortunately, there is normally a way out of this problem. So long as a system is *observable* in the sense defined in Section 5.3.2, we can estimate the states we

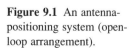

Figure 9.1 An antenna-positioning system (open-loop arrangement).

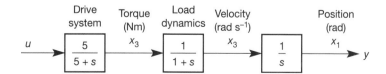

cannot measure, and use the estimates in the feedback scheme instead of the unmeasurable values. A dynamical system which performs such *state estimation* is often called an *observer*. The terms 'state estimator' and 'observer' tend to be used interchangeably (even in this chapter!); but, strictly speaking, they are subtly different kinds of system.

An 'observer' is strictly a system that includes a model of the dynamics of the plant, and assumes that the states of the model mirror the behaviour of those of the plant, usually with some kind of feedback to minimize errors (as we shall see). The plant states are assumed to be entirely deterministic, so that the states of the model are taken to *observe* (or mimic) precisely the behaviour of those of the plant.

A 'state estimator' also contains a model of the plant, but it is not assumed that the model states are automatically identical to the plant states. Usually, the plant states and outputs are taken to be corrupted with random noise, so that we can never be certain of their true values. The state estimator then reconstructs the best possible estimate (in some sense) of the plant states, from whatever noisy measurements can be made. In such an application, the estimator is deliberately being used to filter noisy signals.

9.3   *Simple observers*

There are two ideas that will help to introduce the basic concept of an observer although, as they stand, they are too simplistic for practical applications.

First, assume that we are using a standard, continuous-time, state-space model of the plant, as shown in Figure 9.2. If C is square and of full rank, then C^{-1} exists, and if the D matrix is not present, the output equation $y = Cx$ can be rearranged as:

$$x = C^{-1}y \qquad (9.1)$$

Equation (9.1) indicates that we could obtain the state vector x directly from measurements at the plant outputs y. If the D matrix were to be present in the state-space model, we would obtain $x = C^{-1}[y - Du]$, and we would then additionally use input measurements.

Unfortunately, the situation in which the C matrix for the plant is square and of full rank is very rare (it implies a system which just happens to have a number of independent outputs precisely equal to the number of states), so we cannot actually use this method. However, the method does indicate that it may be possible to extract state estimates by applying the elements of the system model (C and D above) to plant measurements (y and u above). We can extend this notion into the second 'simple-minded' approach that we might adopt.

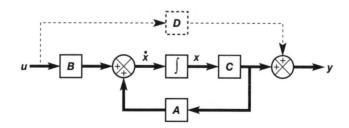

Figure 9.2 The standard state-space model of any linear plant.

This is analogous to the 'Smith predictor' used in SISO systems to compensate for time delays (Section 4.6). In other words, we run a model of the plant in parallel with the plant itself, drive the model from the same input as the plant, and extract the state estimates from the model as shown in Figure 9.3. We use the notation $\hat{x}$ to represent the *estimated value* of x.

In principle, if we initialize the observer of Figure 9.3 with the same initial conditions as the plant, then the estimates in $\hat{x}$ should follow the actual states in x. In practice it will not work because the observer is *open-loop*. Errors are bound to arise in the estimation due to the fact that the A and B matrices constitute only an approximate model of the real plant, and there is no mechanism for correcting these errors in Figure 9.3. In addition, the calculation required to work out the plant initial conditions (that is, working out x_0) each time the observer is first 'switched on', is fairly involved. The initial condition estimates will also therefore be somewhat in error.

The consequence of these weaknesses is that the estimated state vector $\hat{x}$ will diverge more and more from the actual state vector x as time passes, which is of no practical use.

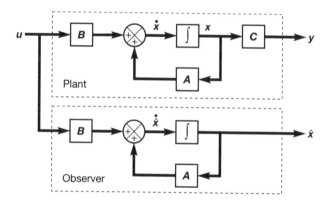

Figure 9.3 A plant with a simple observer to estimate the state vector.

9.4 A better estimator – the full-order observer

The full-order observer to be described here estimates *all* the state variables of the plant to which it is connected. The order of the observer is therefore the same as that of the plant. It is also sometimes called an 'identity observer', or an 'asymptotic state estimator'. The reasons for these other names will become apparent as we progress. Such an observer was first designed by David Luenberger, so it is also known as a 'Luenberger observer'.

The arrangement of Figure 9.3 only makes use of the plant *input*. An improvement is to make use additionally of the plant *output* in some way, to 'close a loop' around the observer, in such a manner as to tend to correct any estimation errors. The proposed arrangement is shown in Figure 9.4. We have effectively added the C matrix to the observer of Figure 9.3, so as to form the estimated plant output $\hat{y}$. This can then be subtracted from the actual output y to form the output estimation error $[y - \hat{y}]$. The error can be fed back into the observer in such a way as to tend to reduce itself to zero, in the normal manner of negative feedback systems.

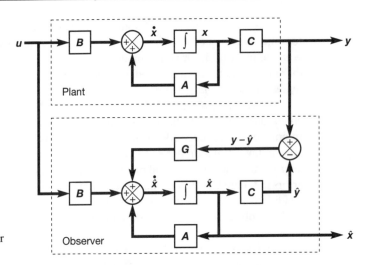

Figure 9.4 Intuitive arrangement of a full-order observer.

The new matrix G is required for two reasons. Firstly, its dimensions match the number of outputs to the number of states, so that the feedback connection is physically possible. If the plant has n states and p outputs, G will be of size $n \times p$. Secondly, the freedom to choose the values of the elements of G provides the means to give the system a suitable dynamic performance, as we shall discover.

9.4.1 Derivation of the design rules for full-order observers

It is possible to derive the design rules for full-order observers directly from the arrangement of Figure 9.4. However, although such an approach has a certain physical appeal (because we can see exactly how the system works from Figure 9.4), we shall use an approach which leads more directly to the design rules.

The signal fed back to the observer input (that is, neglecting the contribution from u for the moment) can be seen, from Figure 9.4, to be $A\hat{x} + G[y - \hat{y}]$. Now $\hat{y} = C\hat{x}$, so we might rewrite this as $[A - GC]\hat{x} + Gy$. If we now define a new $n \times n$ matrix F, such that:

$$F = A - GC \qquad (9.2)$$

we then have a feedback signal $F\hat{x} + Gy$, leading to Figure 9.5 (which we shall use later).

Equation (9.2) is the design equation for the observer. Now we shall show that the system will work, and discover how to assign suitable values to the elements of F (and hence G).

In Figure 9.4, the full observer input is given by $\dot{\hat{x}} = A\hat{x} + G[y - \hat{y}] + Bu$. So we are, as originally suggested, using the error between the actual and estimated plant outputs to drive the observer, in addition to the plant input. This reduces the need for extremely accurate A and B matrices in the plant model, because the feedback action will tend to correct errors caused by deficiencies in these matrices. However, as in all feedback loops, the sensitivity to errors is *reduced*, not *eliminated*, so we should still use the best models we can obtain.

Although the A and B matrices can now be less accurate than before, the C matrix must still be very accurately known. The reason is that the feedback action

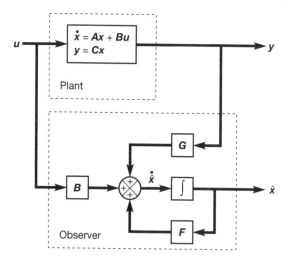

Figure 9.5 Alternative arrangement of the full-order observer.

via G will continue to drive the observer until $y = \hat{y}$, and will then cease. Under this condition, x will be equal to $\hat{x}$ if, and only if, the C matrix in the observer is identical to that in the plant. Fortunately, this often can be the case, as outputs tend to be chosen to be state variables, often giving a C matrix containing only ones and zeros.

If the C matrices of the observer and plant differ, the observer will still settle down with $y = \hat{y}$ (assuming it is asymptotically stable – we ensure this later), but x will then not be equal to $\hat{x}$. The amount of error will depend upon the inaccuracies in C. Thus the observer should work, so long as the C matrix is fairly accurate. The fact that the observer should settle with $\hat{x} = Ix$, where I is the identity matrix, gives rise to the name, 'identity observer'.

As an added bonus, it is no longer necessary to initialize the observer states to have the same values as the plant states at switch-on, which was always going to be a problem. The observer states can have any initial values we wish (usually zero, for convenience), because the feedback action will drive the estimates to the correct levels as time passes (this is why the observer is sometimes called an 'asymptotic state estimator'). We must, however, take steps to ensure that the time it takes for the estimates to settle to the correct values is acceptably short (see below). Similarly, after the observer is initially energized, we must not use the estimates in a control loop until the transients have passed, and the estimates have settled to the correct values.

The dynamic response of the observer will be determined by the eigenvalues of the observer's 'system matrix', namely the F matrix in Figure 9.5. We therefore now have our design procedure (an example follows later):

- Choose the elements of F to place the observer eigenvalues (poles) so as to give the desired dynamic performance from the observer.
- Use Equation (9.2) to derive the elements of G necessary to satisfy the design equation for the overall system.

9.4.2 System analysis with the full-order observer

The design process is straightforward, and has just been stated. However, we can gain more insight into *how* we might choose the observer eigenvalues, and what freedom we have in so doing, by investigating the effects of the observer upon the overall system (that is, the plant + observer combination). This section can safely be omitted if you wish.

Recall from Section 2.7 that the conversion from a state-space model to a Laplace transfer function (or transfer function matrix) is given by $y(s) = C[sI - A]^{-1}Bu(s)$. Since $y(s) = Cx(s)$, the transfer function relating the *state vector* to the input is therefore $x(s) = [sI - A]^{-1}Bu(s)$.

From Figure 9.5, the transfer function relationship between the estimated state vector $\hat{x}$ and the plant input and output vectors is made up of two parts: one via B from u, and the other via G from y, hence:

$$\hat{x}(s) = [sI - F]^{-1}Bu(s) + [sI - F]^{-1}Gy(s)$$

However, for the plant $y(s) = C[sI - A]^{-1}Bu(s)$, therefore,

$$\hat{x}(s) = [sI - F]^{-1}Bu(s) + [sI - F]^{-1}GC[sI - A]^{-1}Bu(s)$$

Taking a factor of $[sI - F]^{-1}$ from the left of the expression, and a factor $Bu(s)$ from the right:

$$\hat{x}(s) = [sI - F]^{-1}\{I + GC[sI - A]^{-1}\}Bu(s)$$

Taking a further factor of $[sI - A]^{-1}$ from the right:

$$\hat{x}(s) = [sI - F]^{-1}\{sI - A + GC\}[sI - A]^{-1}Bu(s)$$

But, from the design equation, Equation (9.2), $-A + GC = -F$, therefore

$$\hat{x}(s) = [sI - F]^{-1}[sI - F][sI - A]^{-1}Bu(s) \tag{9.3}$$

$$= [sI - A]^{-1}Bu(s)$$

$$= x(s) \tag{9.4}$$

What has happened here is that the interaction between the various inputs to the observer has generated some *zeros* (represented by the term $[sI - F]$ in Equation (9.3)) which have cancelled out the observer poles (represented by the term $[sI - F]^{-1}$ in Equation (9.3)). Equation (9.4) shows that this results in an observer that gives us an estimate of the system state for any input vector u.

However, since the interaction generates the zeros at the observer's *input*, our previous work on controllability suggests that the mode cancellations will result in an observer that is *uncontrollable* from the system input (Section 5.3).

This can be verified by forming the overall state-space equations for the plant with the observer, using the equations

$$\dot{x} = Ax + Bu \qquad \text{(the state equation of the plant) and}$$

$$\dot{\hat{x}} = GCx + F\hat{x} + Bu \qquad \text{(the state equation of the observer from Figure 9.5, with } y = Cx)$$

These can be combined to give a partitioned system (see Section A1.3):

$$\begin{bmatrix} \dot{x} \\ \dot{\bar{x}} \end{bmatrix} = \begin{bmatrix} A & 0 \\ \hline GC & F \end{bmatrix} \begin{bmatrix} x \\ \bar{x} \end{bmatrix} + \begin{bmatrix} B \\ B \end{bmatrix} u \tag{9.5}$$

The controllability test matrix for this system (Section 5.3.1) is

$$\begin{bmatrix} B & AB \\ \hline B & [GC + F]B \end{bmatrix}$$

which is not of full rank because $GC + F = A$ (from Equation (9.2)), therefore the system is not completely controllable.

From the partitioned system matrix in Equation (9.5), the eigenvalues of the overall system can be seen to be those of A and those of F (because the system matrix is block triangular – Section A1.3). If the plant (A, B) is controllable (which it usually will be, if we are taking the trouble to design an observer to allow us to feed back its states), then it must be the observer that is not. As a consequence of this, any non-zero initial conditions, or any disturbances entering the system around the observer, may excite the observer modes and lead to uncontrollable responses appearing in the state estimates.

To minimize such effects, the observer poles should be placed far to the left of the system poles in the s-plane. Thus, such transient disturbances will decay very rapidly compared with the dynamics of the plant. Also, since such pole locations imply that the observer poles are much 'faster' than the plant poles, the state estimates will converge very quickly (with respect to the plant dynamics) onto the correct values – another desirable effect.

9.4.3 Guidelines on observer pole locations

The previous section (which you may have omitted) suggested that the observer eigenvalues (poles) ought to be placed far to the left of those of the plant in the s-plane. This ensures stability of the observer, and makes it much 'faster' than the plant. It is directly analogous to the rule used in the design of instrumentation systems, which states that the dynamics of the measuring instrument ought to be significantly faster than those of the quantity it is trying to measure (unless some deliberate filtering effect is required). Here we are trying to 'measure' the plant's state vector, and the observer is our chosen instrument.

In the light of this, it may seem wise to place the observer poles as far to the left in the s-plane as possible. Such an approach has a number of drawbacks, including the fact that the faster the observer is, the more noise it will pass through to the state estimates. Also, the faster the observer, the larger will be the numerical values appearing in the algorithm (as following examples will show). This can lead to computational difficulties.

As a result, it is usual to try placing the observer poles so that they are about five to ten times faster than the fastest plant pole. Simulations will then help to show whether or not the ensuing design is adequate.

If there are several observer poles, it is sometimes decided to place them off the negative real axis, in complex conjugate pairs, such that they exhibit a damping ratio of, say, 0.707. This gives the observer a second-order response with about 4.3 per cent overshoot in the step response, and has the fastest rise time that can be achieved without a resonant peak appearing in the frequency response (the latter would be undesirable for stability reasons).

Example 9.1 *Design of a full-order observer*

Consider part of a system which controls the pressure in a boiler drum. This particular plant can be represented very approximately as a SISO system with the transfer function

$$\frac{Y}{U}(s) = \frac{1}{s^2 + 3s + 4}$$

where $Y(s)$ is the Laplace transform of the drum pressure and $U(s)$ the Laplace transform of the signal to the fuel control system. The poles of this system will be found to be at approximately $s = -1.5 \pm 1.3j$. It may be desired to speed up the dynamic response of the system and reduce the overshoot by using a state variable feedback scheme to move the poles to different locations in the closed loop (perhaps to $s = -3 \pm 2j$). A state-space representation of the LTF model above (using direct programming, by inspection) is:

$$A = \begin{bmatrix} 0 & 1 \\ -4 & -3 \end{bmatrix}, \quad b = \begin{bmatrix} 0 \\ 1 \end{bmatrix}, \quad c = \begin{bmatrix} 1 & 0 \end{bmatrix}$$

Although it is desired to apply a state variable feedback controller to this plant, the state x_2 is not available to be fed back. Since the state-space model was obtained by direct programming, x_1 is at the output (as we see from the c vector), but x_2 is at some unknown, and probably unmeasurable, location inside the plant.

Let us design a full-order observer to estimate the states of this system.

Firstly, we must check that the system is observable. For a second-order system, such as this, the observability test matrix (Section 5.3.2) is

$$\begin{bmatrix} c \\ cA \end{bmatrix} = \begin{bmatrix} 1 & 0 \\ 0 & 1 \end{bmatrix}$$

This is of full rank, so we can proceed with the design.

Looking at Figure 9.5, since the observer is always of the same order as the plant for this kind of design, F must be 2×2. There must therefore be two signals at the observer summer, so in our case the 'G matrix' must actually be a 2×1 vector g, so that it feeds these two signals from the single plant output. We write the contents of F and g as unknowns, and apply the design equation (Equation (9.2)):

$$F = A - gc \Rightarrow \begin{bmatrix} f_{11} & f_{12} \\ f_{21} & f_{22} \end{bmatrix} = \begin{bmatrix} 0 & 1 \\ -4 & -3 \end{bmatrix} - \begin{bmatrix} g_{11} \\ g_{21} \end{bmatrix} \begin{bmatrix} 1 & 0 \end{bmatrix} = \begin{bmatrix} -g_{11} & 1 \\ -4 - g_{21} & -3 \end{bmatrix}$$

From this result we immediately see that

$$f_{12} = 1 \quad \text{and} \quad f_{22} = -3$$

We also see that

$$f_{11} = -g_{11} \quad \text{and} \quad f_{21} = -4 - g_{21} \tag{9.6}$$

Equation (9.6) represents two equations in four unknowns. We use the spare degrees of freedom to allow us to choose the observer eigenvalues (pole positions). The procedure is:

- Decide where we want the observer poles to be located.
- Write down the observer characteristic equation implied by these desired pole locations.
- Write down the actual observer characteristic equation in terms of the unknowns in F.
- Compare the two characteristic equations and solve for the unknown elements of F.
- Use Equation (9.6) to solve for the unknown elements of g.

Where do we want the observer poles to be?

From the denominator of the original transfer function (or the eigenvalues of A), the plant has poles at $s = -1.5 \pm j \sqrt{7/2}$. Therefore, to make the observer ten times 'faster' than the plant, we might choose to place its poles so that they have real parts of $s = -15$. We have two poles to place, so we have three options:

- Make both poles real, but different, at locations on and to the left of $s = -15$.
- Make both poles real and equal at $s = -15$.
- Make the poles a complex conjugate pair having real parts of $s = -15$.

From our knowledge of second-order systems, we know that the first option will give an overdamped observer response. It would be unusual to do this, because we want the observer to follow the plant behaviour as rapidly as possible, so we reject this option.

The second option will give a critically damped observer, having the fastest possible rise time without any overshoot in its step response. This is a sensible possibility.

The third option will give faster rise times in the estimates, but at the expense of some overshoot in their step responses. This may or may not be desirable, depending on the application. In any case, the damping ratio would be kept above 0.707 (that is, the magnitudes of the imaginary parts of the observer pole locations must not be greater than that of the real part), so as to avoid resonance in the observer's frequency response. This value of damping ratio corresponds with a step response overshoot of about 4.3 per cent.

For this example, we shall adopt the option of having both poles real and equal.

The desired characteristic equation of the observer

We have decided to place both observer poles at $s = -15$. In order to achieve this, the observer needs to have a characteristic equation as follows:

$$(s + 15)^2 = s^2 + 30s + 225 = 0 \tag{9.7}$$

The actual characteristic equation of the observer

The observer dynamics are entirely determined by the eigenvalues of the F matrix (compare the observer in Figure 9.5 with the standard state-space model of Figure 9.2). Therefore, the actual characteristic equation will be:

$$|\lambda I - F| = \left| \lambda I - \begin{bmatrix} f_{11} & 1 \\ f_{21} & -3 \end{bmatrix} \right| = \left| \begin{bmatrix} \lambda - f_{11} & -1 \\ -f_{21} & \lambda + 3 \end{bmatrix} \right|$$

$$= \lambda^2 + (3 - f_{11})\lambda - 3f_{11} - f_{21} = 0 \tag{9.8}$$

Comparison of characteristic equations

In order for the actual system to behave in the way we desire, Equations (9.7) and (9.8) must agree. The fact that Equation (9.7) is written in terms of the Laplace operator s, while Equation (9.8) is written in terms of λ, is of no consequence. The former was written in terms of poles, and the latter in terms of eigenvalues, which are the same thing.

Comparing the two equations, the coefficients of s^2 and λ^2 agree, so we can compare the other coefficients directly.

Comparing coefficients of s and λ shows that $30 = 3 - f_{11}$. Hence $f_{11} = -27$.

Comparing the constant terms shows that $225 = -3f_{11} - f_{21}$. Hence, $f_{21} = -144$.

Solving for the elements of g

From Equations (9.6), we now see that $f_{11} = -g_{11}$ so $g_{11} = 27$

and $f_{21} = -4 - g_{21}$ so $g_{21} = 140$

Summing up, the required observer is given by:

$$F = \begin{bmatrix} -27 & 1 \\ -144 & -3 \end{bmatrix}, \quad g = \begin{bmatrix} 27 \\ 140 \end{bmatrix}$$

Drawing the observer, for comparison with later results, we have designed the system of Figure 9.6(a), which employs directly the structure of Figure 9.5. It is a good idea to be

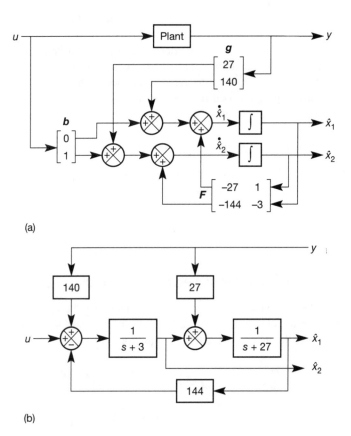

(a)

(b)

Figure 9.6 (a) Full-order observer for Example 9.1. (b) Simplified version of the observer.

able to remove the matrices and vectors from such a diagram, so that the structure can be understood by those who do not know anything of state-space methods. This is easily done using the methods of Section A1.1.1, and then simplifying the direct feedback loops around the integrators, to give the more widely understood representation of Figure 9.6(b) (see Problem 9.1).

Figure 9.7 shows a step response of the full-order observer applied to this system. The system is initialized to a steady-state output of 1 unit, but the observer is initialized with its states at zero. It can be seen that the observer states, as predicted, take a while to home in on the correct values. However, after the first 600 ms or so, the estimated states faithfully predict the real states.

An extra step of 1 unit is applied at the plant input after about 1 s, and the resulting behaviour of the estimated states is indistinguishable from that of the real states on the scales used in the figure. We might expect perfect behaviour from this simulation, because we designed the observer based upon a perfect model of the plant (that is, we use the same model to simulate the plant that we used to design the observer).

Figure 9.7 was obtained by simulating a single state-space model of the entire system of Figure 9.5, with the values of Figure 9.6(a). We derived such a model at Equation (9.5). It is instructive to rerun the simulation, but with an altered plant model, so that the observer model parameters are effectively incorrect.

Computer-aided design

It is actually very easy to design a full-order observer using the *place* command from the MATLAB Control Systems Toolbox (Appendix 3), with no need to do all the preceding mathematics. However, we leave this until the latter part of Example 9.3. For now, the MATLAB m-file that created Figure 9.7 is available on the accompanying disk as *fig9_7.m*.

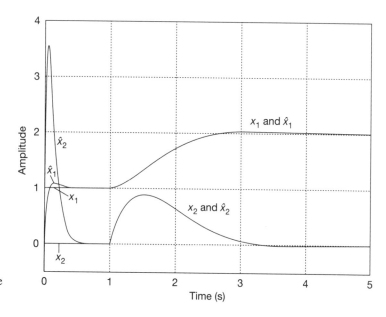

Figure 9.7 Response of the full-order observer.

9.5 Reduced-order observers

The full-order observer of the previous section estimates all the system states. In practice, it is possible to obtain at least some of the state information from the system output directly (since the output equation is $y = Cx + Du$, any plant model having an output and some finite dynamic behaviour must have at least one state which contributes to y). It therefore seems inefficient to estimate every state variable – perhaps we can just estimate a subset of them, thus reducing the required order of the observer. This will make the observer easier to implement (and hence faster, assuming a digital computer implementation is to be used). It will also make the observer more accurate, since we shall effectively use the measured values of some of the states, and only use estimated values of the others.

As an illustration, in Example 9.1, the plant had a c vector of [1 0]. This immediately tells us that the output y is the same as the state x_1 (since $y = cx$). Therefore we do not need to estimate x_1 at all – it is the boiler drum pressure, which can be measured. If we can manage to estimate only x_2 in some way, the observer will be *reduced* in *order* from a 2 × 2 system to a scalar (that is, 1 × 1) system, thus reducing the required computing power significantly (assuming a computer implementation). In addition, since we will have the *measured* version of x_1, and an estimated version only of x_2, the result will be more accurate.

The only times when a full-order observer is used in practice are when the measurements are very noisy, and the observer can do some useful filtering. In such a case, we might prefer to trust the estimates, rather than trying to use the noisy real signals (but we might prefer to use a Kalman filter instead – see Section 9.8).

In general, for a system having n states and p outputs, the C matrix will be of dimension $p \times n$. If p is less than n (fewer outputs than states), and C has rank p, then the p outputs are independent of each other, and will contain sufficient information to deduce the values of p of the n system states (equivalent to solving the simultaneous equations obtained from $y = Cx$).

In such a case, we need only estimate the remaining $(n - p)$ states. This gives the benefits of reduced computation and increased accuracy mentioned above, but at the expense of a more complicated design procedure. As in the case of the full-order observer, we shall derive the appropriate design method and then show, by example, that it is really quite straightforward to apply.

9.5.1 Derivation of the design rules for reduced-order observers

We begin by returning to our original notion that we might be able to obtain the states by solving the output equation to give $x = C^{-1}y$ (Equation (9.1)). As we said, this is usually impossible. However, we can make use of the idea as follows.

Let us define a new vector of signals z related to the system states by the linear transform $z = Tx$, where T is some matrix of our choice (note that this z is nothing to do with the z-transform!). We can combine this equation and the system's output equation in the partitioned arrangement:

$$\begin{bmatrix} y \\ \hline z \end{bmatrix} = \begin{bmatrix} C \\ \hline T \end{bmatrix} x = \Psi x \tag{9.9}$$

If we define T to be of such a size that the partitioned matrix Ψ is square (that is, T must be of size $(n - p) \times n$), and if we are careful about the choice of contents of T, such that the partitioned matrix is also of full rank, then $\boldsymbol{\Psi}^{-1}$ will exist, and we can rearrange Equation (9.9) as

$$x = \begin{bmatrix} \dfrac{C}{T} \end{bmatrix}^{-1} \begin{bmatrix} y \\ z \end{bmatrix} = \boldsymbol{\Psi}^{-1} \begin{bmatrix} y \\ z \end{bmatrix} \tag{9.10}$$

We know C from the plant model, we can measure the plant outputs in y, we shall know the contents of T once we have discovered how best to choose them, so that leaves z as the only unknown on the right-hand side of Equation (9.10). If we can find z, we can calculate the state vector x. In the observer it will again be called $\hat{x}$ as it is still an estimate of the state vector (but only a partial estimate this time, as it will also contain some measured values).

Now, from the size of T defined above, and from the arrangement of Equation (9.10), it is clear that z must be a column vector of length $(n - p)$. We shall use an observer of order $(n - p)$ to produce z for us. Since this observer is of lower order than n, it is a *reduced-order observer*. Of course, we do not yet know how to design the observer.

As in the case of the full-order observer, we shall follow an approach which introduces the design rules, and then show that a system following such rules will work as required.

Figure 9.8 shows a suitably modified form of our previous design (from Figure 9.5). The differences are that the observer is of lower order (as indicated by the numbers of signals shown against the signal paths), the $\boldsymbol{\Psi}^{-1}$ matrix has appeared to create $\hat{x}$ from y and z according to Equation (9.10), and another new matrix, J, has appeared on the observer input.

The J matrix (Figure 9.8) is used instead of B (Figure 9.5) because the observer is of lower order than the plant, and so B would have the wrong number of rows. We shall define J later.

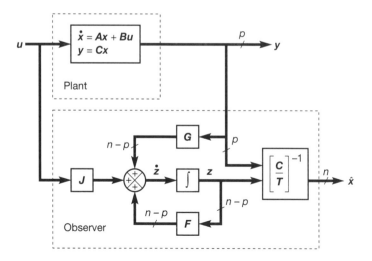

Figure 9.8 The reduced-order observer.

From Figure 9.8, we can write the transfer function to z (in the same way as we did in Section 9.4.2) as:

$$z(s) = [sI - F]^{-1} Ju(s) + [sI - F]^{-1} Gy(s)$$

$$= [sI - F]^{-1} \{Ju(s) + GC[sI - A]^{-1} Bu(s)\} \tag{9.11}$$

In order to be able to extract a factor $Bu(s)$ from the right of this expression, as we did in the derivation of the full-order observer, we introduce the design *definition*:

$$J = TB \tag{9.12}$$

Substituting into Equation (9.11), together with the original definition $z(s) = Tx(s)$, gives:

$$Tx(s) = [sI - F]^{-1} \{T + GC[sI - A]^{-1}\} Bu(s)$$

$$= [sI - F]^{-1} \{T[sI - A] + GC\}[sI - A]^{-1} Bu(s)$$

$$= [sI - F]^{-1} [Ts - TA + GC]x(s) \tag{9.13}$$

To obtain an observer that has the desired effect, we introduce the second design *definition*:

$$FT = TA - GC \tag{9.14}$$

Substituting for $(-TA + GC)$ in Equation (9.13):

$$Tx(s) = [sI - F]^{-1} [Ts - FT]x(s)$$

$$= [sI - F]^{-1} [sI - F] Tx(s) = Tx(s)$$

So, again we find that cancellation of poles by zeros generated at the observer input has led to a consistent result. Since the result is correct, we may use Equations (9.12) and (9.14) as the design equations for such a reduced-order observer.

9.5.2 Design methodology for the reduced-order observer

In the process of carrying out the design, several unknown quantities will arise. Consider, for example, the case of a SISO third-order system. Such a system has one output and three states, so $n = 3$ and $p = 1$ in Figure 9.8. We therefore have to estimate two states. The vector z will contain two elements, and the observer will be second order, so that the F matrix will need to be of size 2×2.

In order to couple the single output into the second-order observer, g will need to be 2×1. The c vector will be of size 1×3 for such a system, and the T matrix needs to square this up. T will therefore have to be of size 2×3, in order that Ψ becomes 3×3.

The contents of F, g and T thus constitute 12 unknowns, even in such a simple system (note that j will be of size 2×1, but is not a new 'unknown' as it depends only upon b and T).

There are usually more unknowns than we have equations to assign them, thus giving some spare degrees of freedom. It is therefore often possible to assign arbitrary values to some of the unknowns. As a starting point in this process, the following tip has been found to give useful results over the years, and can reduce the algebra in many cases.

Since plant outputs are often chosen as state variables, it is often the case that the C matrix for the plant can be partitioned into the form $[I \ 0]$. In such a case, it is usually beneficial to choose the T matrix as $[T' \ I]$ (where T' is a sub-matrix of T, of the required dimensions).

For example, for a third-order plant with two outputs y_1 and y_2, we might well choose $x_1 = y_1$ and $x_2 = y_2$, giving

$$C = \begin{bmatrix} 1 & 0 & 0 \\ 0 & 1 & 0 \end{bmatrix}$$

To square up this matrix, t must be of size 1×3 and, according to the suggestion above, we might choose t to be $[t_{11} \ t_{12} \ 1]$ so that the matrix

$$\begin{bmatrix} C \\ \hline t \end{bmatrix} = \begin{bmatrix} I & 0 \\ \hline t' & 1 \end{bmatrix} = \begin{bmatrix} 1 & 0 & 0 \\ 0 & 1 & 0 \\ \hline t_{11} & t_{12} & 1 \end{bmatrix}$$

is guaranteed to be of full rank and is easily invertible. If the C matrix is not of the stated form, it must simply be borne in mind that the inverse of Ψ must exist.

Further reduction of the number of unknowns is achieved by deciding where the observer eigenvalues should be located. This gives constraints on the values in the F matrix in order to satisfy the required observer characteristic equation.

We then use the design equation (Equation (9.14)) to generate several simultaneous equations which can be solved for the remaining unknown quantities.

Example 9.2 *Design of a reduced-order observer*

We use the same boiler drum pressure control system as in Example 9.1, namely:

$$A = \begin{bmatrix} 0 & 1 \\ -4 & -3 \end{bmatrix}, \quad b = \begin{bmatrix} 0 \\ 1 \end{bmatrix}, \quad c = \begin{bmatrix} 1 & 0 \end{bmatrix}$$

Information about one state must be available from the (single) output. We therefore now seek to estimate only the other state. For this system, with such a simple c matrix, it is clear that x_1 is the measurable state (measurable directly since $y = x_1$ is the drum pressure), while x_2 is the state to be estimated. For systems in general, the outputs will be linear combinations of several states, and it will therefore not be clear which are measurable, and which are to be estimated. This information *can* be determined, but it is unnecessary, because the design method will sort it all out automatically. In any case, the reduced-order observer does not directly estimate the 'missing' state information. Rather, the vector z (Figure 9.8) contains a set of estimated signals which only give the actual state estimates when combined with the measurements y via the matrix Ψ^{-1}.

Firstly, consider the dimensions of the various vectors involved. Reference to Figure 9.8 with $p = 1$ and $n = 2$ (n is the system order, remember) may be helpful:

t is required to square up c, therefore t will be a 1×2 vector.

The observer is to estimate a single state, therefore f will be scalar.

g will couple a single output into the scalar observer, so g will also be scalar.

j will couple a single input into the scalar observer, so j will be scalar too.

The design equation (Equation (9.14)) for this system is therefore $ft = tA - gc$. In terms of known and unknown quantities:

$$f \begin{bmatrix} t_{11} & t_{12} \end{bmatrix} = \begin{bmatrix} t_{11} & t_{12} \end{bmatrix} \begin{bmatrix} 0 & 1 \\ -4 & -3 \end{bmatrix} - g \begin{bmatrix} 1 & 0 \end{bmatrix}$$

and multiplying out gives

$$\begin{bmatrix} ft_{11} & ft_{12} \end{bmatrix} = \begin{bmatrix} -4t_{12} - g & t_{11} - 3t_{12} \end{bmatrix}$$

Therefore, equating elements:

$$ft_{11} = -4t_{12} - g \tag{9.15}$$

and $\quad ft_{12} = t_{11} - 3t_{12} \tag{9.16}$

Here we have two equations in four unknowns. Since c can be partitioned into the form $[I \; 0]$, we follow the guideline above, and choose to set $t_{12} = 1$.

Also, as in the case of the corresponding full-order observer, we place the observer eigenvalue at $s = -15$. Therefore $f = -15$.

Substituting these values into Equations (9.15) and (9.16) we obtain:

$$-15t_{11} = -4 - g \quad \text{and} \quad -15 = t_{11} - 3$$

Solving these equations, $t_{11} = -12$ and $g = -184$. Using the other design equation (Equation (9.12)), we have

$$j = tb = \begin{bmatrix} -12 & 1 \end{bmatrix} \begin{bmatrix} 0 \\ 1 \end{bmatrix} = 1$$

Therefore, the final design is:

$$t = \begin{bmatrix} -12 & 1 \end{bmatrix}, \quad g = -184, \quad f = -15, \quad j = 1$$

and

$$\begin{bmatrix} c \\ t \end{bmatrix}^{-1} = \begin{bmatrix} 1 & 0 \\ -12 & 1 \end{bmatrix}^{-1} = \begin{bmatrix} 1 & 0 \\ 12 & 1 \end{bmatrix}$$

The system therefore appears as shown in Figure 9.9(a). Removing the matrix, as in Example 9.1, gives Figure 9.9(b).

Comparing the solution in Figure 9.9(b) with that in Figure 9.6(b), we see that a significant simplification has been achieved. We also see, as expected, that $\hat{x}_1$ is not an estimate at all, but is the output measurement. The system of Figure 9.9(b) is therefore far preferable to that of Figure 9.6(b), being simpler, easier to construct, faster in execution (for a computer implementation) and more accurate. The only penalty is the more involved design procedure.

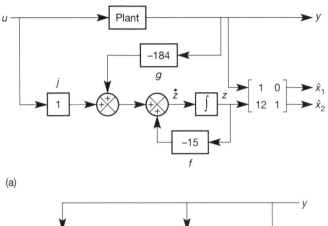

(a)

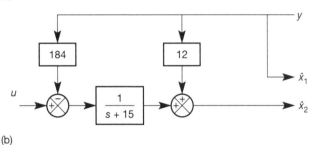

Figure 9.9 (a) Reduced-order observer for Example 9.2. (b) Simplified version of the observer.

(b)

In the same way as we did leading to Equation (9.5), we can form an overall state-space model of the plant plus the observer system. From Figure 9.8, if we choose the state vector for the combined system to be (x stacked over z) we obtain, in the general case:

$$\begin{bmatrix} \dot{x} \\ \dot{z} \end{bmatrix} = \begin{bmatrix} A & 0 \\ \hline GC & F \end{bmatrix} \begin{bmatrix} x \\ z \end{bmatrix} + \begin{bmatrix} B \\ J \end{bmatrix} u$$

Additionally using the matrix

$$\begin{bmatrix} C \\ \hline T \end{bmatrix}^{-1}$$

to form the state estimates ($\hat{x}$) from y and z, a simulation using the values from Figure 9.9(a), and the same input signal as in Example 9.1, gave the responses of Figure 9.10.

Again we see that a transient period is necessary for the observer to home in on the system states, and again we see that after the first 500 ms or so, the state estimates are indistinguishable from the real states. However, in this case we have the added advantage, as expected, that $\hat{x}$ and x_1 are actually the same signal, and there is no difference between them at any time, even during the initial transient period. As in Example 9.1, we have tested the observer on the same plant model as we used to design it. Again, it is useful to rerun the simulation with a mismatched plant model, and see what happens then. Readers who have MATLAB (Appendix 3) can do this by modifying the m-file *fig9_10.m* on the accompanying disk.

The example above, being scalar, was very simple. However, application of the design procedure to higher-order plants is straightforward. We shall not pursue such an example here, as one will occur naturally in the next section.

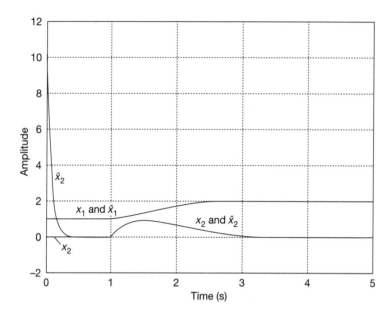

Figure 9.10 Response of
the reduced-order observer.

9.6   *Use of observers in closed-loop systems*

The most common use of observers is to estimate the unmeasurable states of a
system for use in a feedback scheme. It is entirely feasible to design an observer
along the lines suggested in Section 9.4 or 9.5, and then to use the estimates of the
states (that is, the observer outputs) instead of the real states in a state variable
feedback controller design, such as that discussed in Section 5.4, or an optimal
controller (see Chapter 12). If we have designed our observer correctly, so that the
dynamics of the estimates are faster than those of the states, such a system should
work.

9.6.1   *The separation principle*

In discussing the design of observers, we have taken care to make their dynamics
significantly 'faster' than those of the plant they are observing. Therefore, we ought
to be able to treat the state estimates as if they were the actual states, effectively
ignoring the observer dynamics.

It is possible to take this concept further, and *prove* that the observer has no
effect upon the dynamics of the closed-loop scheme (other than transient effects).
As a result, we can design our observer separately from the feedback loops as
suggested above, leading to a simpler design process than would otherwise be the
case. This is called the *separation principle*. We shall see a (non-rigorous) 'proof'
of this as we pursue the design of such a scheme.

Our overall policy will therefore be:

(1) Design the observer to be sufficiently 'fast', so as to give it a transient
performance much faster than that of the plant.

(2) Design the feedback elements so that the closed-loop system has the required dynamic performance (that is, eigenvalues).

9.6.2   Derivation

Our overall aim is to give the closed-loop system of Figure 9.11 a certain dynamic performance relating y and r. We shall express this requirement, as in the case of the SVF schemes discussed in Section 5.4, in the form of a required set of closed-loop eigenvalues. In principle, we therefore need to determine the closed-loop plant matrix resulting from Figure 9.11, so that its eigenvalues can be found and compared with the required ones. In practice, we shall describe an easier way, which we can use instead, but we shall nevertheless form the closed-loop matrix to show what is happening in the closed-loop scheme.

Note also that it is possible to fix the closed-loop *eigenvectors*, and the same comments apply as were made in Section 5.5.

From the summations at the integrator block inputs in Figure 9.11, we can write the following equations (the integrator block for x is buried inside the 'Plant' block, of course; but substitution into the state equation of the signals summing to give u gives the same result):

$$\dot{x} = Ax + Bu = Ax - BK \begin{bmatrix} C \\ \overline{T} \end{bmatrix}^{-1} \begin{bmatrix} y \\ \overline{z} \end{bmatrix} + Br \qquad (9.17\text{a})$$

$$\dot{z} = Fz + Ju + Gy = Fz - JK \begin{bmatrix} C \\ \overline{T} \end{bmatrix}^{-1} \begin{bmatrix} y \\ \overline{z} \end{bmatrix} + Jr + GCx \qquad (9.17\text{b})$$

Now, from Equation (9.10), the quantity

$$\begin{bmatrix} C \\ \overline{T} \end{bmatrix}^{-1} \begin{bmatrix} y \\ \overline{z} \end{bmatrix} = \boldsymbol{\Psi}^{-1} \begin{bmatrix} y \\ \overline{z} \end{bmatrix} = x$$

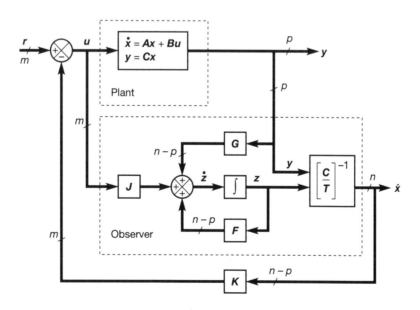

Figure 9.11 Reduced-order observer in a state variable feedback scheme.

Also, from the definition of Equation (9.12), we remember that $J = TB$. Making these substitutions in Equations (9.17) leads to:

$$\dot{x} = [A - BK]x + Br$$

$$\dot{z} = [GC - TBK]x + Fz + TBr$$

or, in partitioned matrix form:

$$\begin{bmatrix} \dot{x} \\ \hline \dot{z} \end{bmatrix} = \begin{bmatrix} A - BK & 0 \\ \hline GC - TBK & F \end{bmatrix} \begin{bmatrix} x \\ \hline z \end{bmatrix} + \begin{bmatrix} B \\ \hline TB \end{bmatrix} r \qquad (9.18)$$

The 2×2 (partitioned) matrix in Equation (9.18) is the 'system' matrix of the overall closed-loop scheme. It is this matrix whose eigenvalues we must find. Since this matrix is block triangular, the eigenvalues are those of the diagonal blocks; namely, those of F (that is, of the observer), and those of element $(1, 1)$ which, from Equation (5.6) in Section 5.4, can be seen to be those of the standard SVF arrangement.

This shows us that the part of the system involving the generation of the vector z (that is, the observer) is dynamically separated from the remainder of the system, and effectively does not form a part of the feedback loop (there is no direct contribution from z into $\dot{x}$ in Equation (9.18) but, of course, if the observer were not present, we could not have generated Equation (9.18) in the first place – it is just a convenient rearrangement of the system). So, as we suggested earlier, we may design the observer first, using the same procedure as in Section 9.5.2, and then design K to give element $(1, 1)$ of the partitioned plant matrix in Equation (9.18) the required eigenvalues.

At this point we illustrate the entire design procedure by an example.

Example 9.3 *Including a reduced-order observer in a feedback regulator*

In Section 5.4.1, we considered the design of a full state feedback scheme for an antenna positioner. The open-loop system is shown in Figure 9.1.

Let us again design a system to place the closed-loop poles at $s = -10$ and $s = -1 \pm j2$. The *required* CLCE (using λ to represent eigenvalues, rather than s to represent poles) is therefore as given by Equation (5.8), repeated here as Equation (9.19):

$$(\lambda + 1 + 2j)(\lambda + 1 - 2j)(\lambda + 10) = \lambda^3 + 12\lambda^2 + 25\lambda + 50 = 0 \qquad (9.19)$$

We shall use the same state-space model as before:

$$A = \begin{bmatrix} 0 & 1 & 0 \\ 0 & -1 & 1 \\ 0 & 0 & -5 \end{bmatrix}, \quad b = \begin{bmatrix} 0 \\ 0 \\ 5 \end{bmatrix}, \quad c = \begin{bmatrix} 1 & 0 & 0 \end{bmatrix} \qquad (9.20)$$

but now we introduce the added complication that it is not possible to make any connections to the plant except at the input and output. In other words, x_2 and x_3 are not measurable.

From the output equation, $y = cx$, we see that $y = x_1$. We therefore have a direct measurement of x_1 for feeding back, but we need to estimate the values of x_2 and x_3. We shall therefore design an observer to estimate two states. Although this will be harder to design than a full-order observer, it will be simpler and quicker to implement, and also more accurate.

We have previously checked the controllability of this system (Example 5.1 in Section 5.3.1). Before proceeding further, we must check its observability (there is no point trying to design an observer for an unobservable system since, by definition, the unobservable states will be those that we wish the observer to estimate).

To form the observability check matrix, we need the following components:

$$A^T = \begin{bmatrix} 0 & 0 & 0 \\ 1 & -1 & 0 \\ 0 & 1 & -5 \end{bmatrix}, \quad c^T = \begin{bmatrix} 1 \\ 0 \\ 0 \end{bmatrix}, \quad A^T c^T = \begin{bmatrix} 0 \\ 1 \\ 0 \end{bmatrix}, \quad (A^T)^2 c^T = \begin{bmatrix} 0 \\ -1 \\ 1 \end{bmatrix}$$

So, the check matrix is:

$$[c^T \quad A^T c^T \quad (A^T)^2 c^T] = \begin{bmatrix} 1 & 0 & 0 \\ 0 & 1 & -1 \\ 0 & 0 & 1 \end{bmatrix}$$

which is clearly of full rank, so the system is fully observable and we are able to proceed.

Until we become really familiar with such designs, it is always good policy to draw a block diagram of the scheme we are about to design, in order to see the required dimensions of all the matrices involved. It has been found, over the years, that students (and practising engineers) make far fewer mistakes if such a step is carried out. Figure 9.12 illustrates this step.

From Figure 9.12, we can determine:

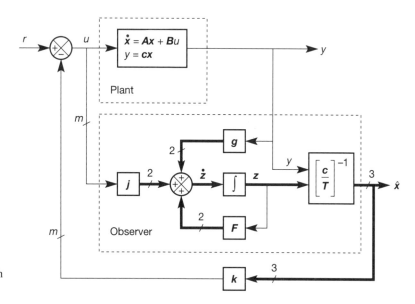

Figure 9.12 Structure for the closed-loop design with reduced-order observer.

- The plant is SISO, so only scalar signals exist at u and y.
- The matrix T must square up the c vector, so T must be of size 2×3. Note that even if we had no intention of implementing the block which generates $\hat{x}$ (Figure 9.11), the T matrix must still be of these dimensions for the rest of the matrix algebra to work.
- The observer must estimate two states, so two signals exist at the observer output and input. Note that the observer outputs are *not* equal to the unmeasurable states, x_2 and x_3, but form the vector z which, together with the plant output y, is used to *generate* x_2 and x_3.
- Bearing in mind that a summation block can only sum signals of the same dimensions, all the other dimensions in Figure 9.12 follow naturally (the narrow signal paths are all scalar).
- Since the dimensions of a matrix are always (output) × (input) (Section A1.1.1), we can now read off all the dimensions from Figure 9.13: g and j are 2×1 vectors, F is a 2×2 matrix, k is a 1×3 vector and T is a 2×3 matrix. Also note that the dimensions of the design equations (Equations (9.12) and (9.14)) are now automatically correct.

Observer design

The c vector for the plant can be partitioned into the general form $[I \,|\, 0]$ (in this case the identity 'matrix' is simply the scalar unity, and the null 'matrix' a 1×2 null vector). We can therefore try the 'rule of thumb' introduced previously, which suggests that we might choose the T matrix to be a suitably dimensioned version of $[T' \,|\, I]$. With our dimensions, this means that we try:

$$T = \begin{bmatrix} t_{11} & 1 & 0 \\ t_{21} & 0 & 1 \end{bmatrix} \tag{9.21}$$

Note that by doing this we have already arbitrarily assigned values to four of our 'unknowns' (that is, to t_{12}, t_{13}, t_{22} and t_{23}). We do this because it will reduce the pain in the following analysis, but it is possible that we may not ultimately have the flexibility to make such assignments. We shall know that this is the case if we eventually end up with impossible assignments, such as $2 = 3$. If this happens, we must return to the start, and keep these variables as unknowns. However, the 'rule of thumb' generally works well, so if we are doing a design on paper it is usually worth taking the risk. We shall see how to use MATLAB to perform a computer-assisted design later on.

Applying the reduced-order observer design equation $FT = TA - gc$ (Equation (9.14)), with A from Equation (9.20) and T from Equation (9.21), we obtain:

$$\begin{bmatrix} f_{11} & f_{12} \\ f_{21} & f_{22} \end{bmatrix} \begin{bmatrix} t_{11} & 1 & 0 \\ t_{21} & 0 & 1 \end{bmatrix} = \begin{bmatrix} t_{11} & 1 & 0 \\ t_{21} & 0 & 1 \end{bmatrix} \begin{bmatrix} 0 & 1 & 0 \\ 0 & -1 & 1 \\ 0 & 0 & -5 \end{bmatrix} - \begin{bmatrix} g_{11} \\ g_{21} \end{bmatrix} [1 \quad 0 \quad 0]$$

which, when multiplied out, gives:

$$\begin{bmatrix} f_{11}t_{11} + f_{12}t_{21} & f_{11} & f_{12} \\ f_{21}t_{11} + f_{22}t_{21} & f_{21} & f_{22} \end{bmatrix} = \begin{bmatrix} -g_{11} & t_{11} - 1 & 1 \\ -g_{21} & t_{21} & -5 \end{bmatrix} \tag{9.22}$$

By comparing elements $(1, 3)$ and $(2, 3)$ on each side of Equation (9.22), we see that

$$f_{12} = 1 \quad \text{and} \quad f_{22} = -5 \tag{9.23}$$

We can assign values to the remaining elements of F by placing the observer poles as required. The open-loop system has a fastest pole at $s = -5$. Note that the closed-loop system has a faster pole at $s = -10$, but it is the open-loop system whose states the observer must track.

In the interests of avoiding very large numerical gains in the final scheme, we shall place the observer poles at $s = -20$. This is only four times as fast as the fastest (open-loop) system pole, and we shall need to test the suitability of this decision by simulation later.

With two observer poles at $s = -20$, the *required* characteristic equation for the observer is therefore $(s + 20)^2 = s^2 + 40s + 400 = 0$.

The *actual* CE of the observer is:

$$|[\lambda I - F]| = \begin{vmatrix} \lambda - f_{11} & -1 \\ -f_{21} & \lambda + 5 \end{vmatrix} = \lambda^2 + (5 - f_{11})\lambda - 5f_{11} - f_{21} = 0$$

Comparing coefficients of s and λ in these two equations, we find that

$$f_{11} = -35 \quad \text{and} \quad f_{21} = -225 \tag{9.24}$$

Substituting these values, and those from Equation (9.23), into Equation (9.22), we now have:

$$\begin{bmatrix} -35t_{11} + t_{21} & -35 & 1 \\ -225t_{11} - 5t_{21} & -225 & -5 \end{bmatrix} = \begin{bmatrix} -g_{11} & t_{11} - 1 & 1 \\ -g_{21} & t_{21} & -5 \end{bmatrix}$$

Comparing elements $(1, 2)$ and $(2, 2)$, we now see that:

$$t_{11} = -34 \quad \text{and} \quad t_{21} = -225 \tag{9.25}$$

Then, substituting these values into elements $(1, 1)$ and $(2, 1)$, we get:

$$g_{11} = -965 \quad \text{and} \quad g_{21} = -8775 \tag{9.26}$$

Finally, using design Equation (9.12),

$$j = Tb = \begin{bmatrix} -34 & 1 & 0 \\ -225 & 0 & 1 \end{bmatrix} \begin{bmatrix} 0 \\ 0 \\ 5 \end{bmatrix} = \begin{bmatrix} 0 \\ 5 \end{bmatrix} \tag{9.27}$$

This completes the observer design. The contents of F, T, g and j are given in Equations (9.21) to (9.27), for use in the arrangement of Figure 9.12, as follows:

$$F = \begin{bmatrix} -35 & 1 \\ -225 & -5 \end{bmatrix}, \quad g = \begin{bmatrix} -965 \\ -8775 \end{bmatrix}, \quad j = \begin{bmatrix} 0 \\ 5 \end{bmatrix}, \quad T = \begin{bmatrix} -34 & 1 & 0 \\ -225 & 0 & 1 \end{bmatrix} \tag{9.28}$$

Feedback design

Having completed the observer design, we now turn our attention to the feedback paths. The closed-loop eigenvalues we *require* lead to the CLCE (Equation (9.19)), repeated here:

$$\lambda^3 + 12\lambda^2 + 25\lambda + 50 = 0$$

We now make use of the fact that we have designed the observer in isolation from the feedback loops.

We want Figure 9.12 to have a certain performance, as determined by its closed-loop eigenvalue set. If the states were available for measurement, we could obtain this simply by designing a standard SVF scheme (Section 5.4). We have *made* the states (or, to be more precise, their estimates) available for measurement, by including the matrix Ψ^{-1} leading to Figure 9.11, and the separation principle says that we can now apply the simple SVF design procedure to give the vector k for feeding back the three estimated states in $\hat{x}$ (for systems with more than one input, K would be a matrix, of course).

We now note that Ψ^{-1} will effectively be in series with the new overall state feedback vector k, so we finally multiply them out to obtain the required feedback configuration.

Adopting this approach, we need to design a SVF vector k, as in Section 5.4, so that the eigenvalues of the closed-loop plant matrix $[A - bk]$ match those given by Equation (9.19) (that is, we ignore entirely the presence of the observer, and assume the state estimates to be correct). Thus:

$$\left| \lambda I - \left\{ \begin{bmatrix} 0 & 1 & 0 \\ 0 & -1 & 1 \\ 0 & 0 & -5 \end{bmatrix} - \begin{bmatrix} 0 \\ 0 \\ 5 \end{bmatrix} [k_{11} \quad k_{12} \quad k_{13}] \right\} \right|$$

$$= \begin{vmatrix} \lambda & -1 & 0 \\ 0 & \lambda+1 & -1 \\ 5k_{11} & 5k_{12} & \lambda+5+5k_{13} \end{vmatrix} = 0$$

or

$$\lambda^3 + (6 + 5k_{13})\lambda^2 + (5 + 5k_{13} + 5k_{12})\lambda + 5k_{11} = 0$$

Comparing coefficients with Equation (9.19), we find that $k = [10 \ 2.8 \ 1.2]$, as in Example 5.4. Now, this feedback vector will be connected to the output of the matrix Ψ^{-1} in Figure 9.12. We have:

$$\left[\frac{c}{T} \right]^{-1} = \begin{bmatrix} 1 & 0 & 0 \\ -34 & 1 & 0 \\ -225 & 0 & 1 \end{bmatrix}^{-1} = \begin{bmatrix} 1 & 0 & 0 \\ 34 & 1 & 0 \\ 225 & 0 & 1 \end{bmatrix}$$

and

$$\hat{x} = \left[\frac{c}{T} \right]^{-1} \begin{bmatrix} y \\ z_{11} \\ z_{21} \end{bmatrix}$$

(from Equation (9.10)). Now, our feedback signal needs to be

$$k\hat{x} = k\left[\frac{c}{T} \right]^{-1} \begin{bmatrix} y \\ z_{11} \\ z_{21} \end{bmatrix}$$

so we obtain:

$$k\hat{x} = \begin{bmatrix} 10 & 2.8 & 1.2 \end{bmatrix} \begin{bmatrix} 1 & 0 & 0 \\ 34 & 1 & 0 \\ 225 & 0 & 1 \end{bmatrix} \begin{bmatrix} y \\ z_{11} \\ z_{21} \end{bmatrix}$$

$$= 10y + 2.8(34y + z_{11}) + 1.2(225y + z_{21})$$

$$= 375.2y + 2.8z_{11} + 1.2z_{21} \tag{9.29}$$

Removing the vectors and matrices

Our final system appears as shown in Figure 9.12, with the quantities of Equations (9.28) replacing the unknowns, and with k and $\boldsymbol{\Psi}^{-1}$ combined, giving the result of Equation (9.29) for the required feedback signal. However, it is useful to redraw the system with the vectors and matrices removed, using the methods of Section A1.1.1. This allows us to communicate our design to an engineer or technician untrained in state-space methods, so that he or she can understand how to implement the structure, even if its derivation is not understood. Such a rearrangement is shown in Figure 9.13.

Simulation studies and computer-aided design

A selection of time responses of this system appears in Figure 9.14. These were obtained by simulating the overall state-space description of the system given in Equation (9.18), using the MATLAB *lsim* command. This produced the traces of the actual states x, which were extracted from the combined state vector (x and z) resulting from the simulation. The estimated states $\hat{x}$ in the figure were then obtained by forming the vector (x_1 stacked over

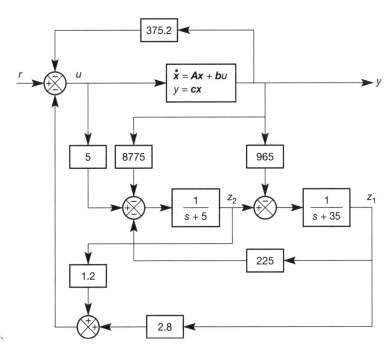

Figure 9.13 The designed closed-loop system drawn without vectors or matrices.

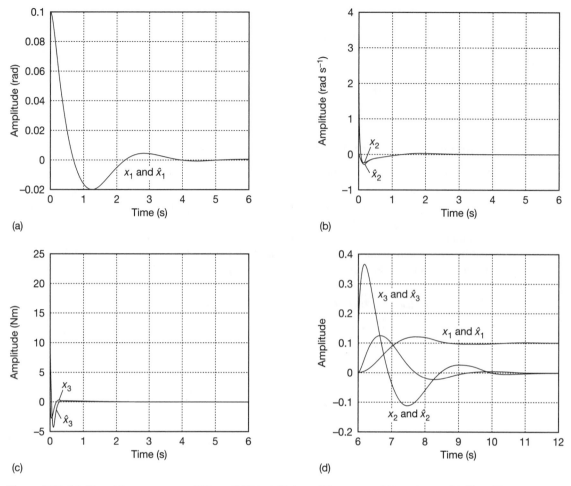

(a)

(b)

(c)

(d)

Figure 9.14 (a) Closed-loop response of Figure 9.13 to initial conditions. x_1 and its estimate. (b) Closed-loop response of Figure 9.13 to initial conditions. x_2 and its estimate. (c) Closed-loop response of Figure 9.13 to initial conditions. x_3 and its estimate. (d) Closed-loop response of Figure 9.13 to a subsequent unit step applied after 6 seconds – all three states and their estimates.

z) and forming the estimates by premultiplying by Ψ^{-1} (Equation (9.10)). The MATLAB m-file is *fig9_14.m* on the accompanying disk.

In the previous examples, we initialized the plant at a steady state, and noticed how the observer converged onto the correct values. In this case, since the plant is closed-loop, we cannot have a steady state until the feedback signals are also correct. In order to show how the estimates converge to the correct values, we deliberately set up an initial non-zero output ($y = x_1 = 0.1$ rad), with all other signals zero, and allow the system to settle for six seconds.

Figures 9.14(a) to (c) show the resulting behaviour. In Figure 9.14(a), we see that the estimate of x_1 is identical to the actual signal at all times. We would expect this, since the reduced-order observer makes use of the fact that x_1 is measurable. The signal is driven to zero as the feedback loops reject the non-zero initial condition.

The other two states (x_2 and x_3) are not measurable, and must be estimated by the observer (by feeding y and z through the matrix $\boldsymbol{\Psi}^{-1}$ as discussed above). Figures 9.14(b) and (c) show the responses of these states and their estimates as the system settles to steady state. Here we begin to see the problems of large gains arising from the observer design. The initial output of 0.1 radian is passed directly back to the input via the gain of 375.2 in Figure 9.13, as -37.52 radians. In the real world, this would saturate the control amplifier at the plant input, and a nonlinear response would result. Nonlinear effects are considered in Chapter 14, while Section 12.6.2 revisits this antenna-positioning problem from an optimal control viewpoint, allowing the plant input to be limited to an appropriate level. Returning to our present example, we see the feedback effect as the very large initial transient in x_3, and the rather smaller one in x_2. We also see that, as in the previous designs, the estimates take some time (about 500 ms) to converge onto the true values.

After waiting six seconds for the plant to reach steady state, a unit step is applied at r (note that desired inputs could be applied at any time after the first 500 ms, once the estimates had converged to the correct values). Figure 9.14(d) shows the results. As discussed in Section 5.4.2, this is a *regulator* design, so it tries to reject the change at r as a disturbance, and return all the states to zero. It does not quite succeed in this, as x_1 settles at 0.1 rad. However, the important point in the present context is that all the state estimates are indistinguishable from the true values on the scale of the plot, so the observer system is clearly working. Also, the responses can be compared with those in Section 5.4.1 (Figure 5.4), which confirms the correct dynamics – so the feedback design is also correct (we convert the system to a *tracking system* in Example 9.4).

Another thing we can investigate is the choice of observer pole locations. However, changing the observer poles implies repeating the entire design, which is not very appealing. It is time to see how a typical computer-assisted design might progress.

Computer-assisted design

Using MATLAB's control systems toolbox, it is actually more convenient to design an optimal regulator (Chapter 12) using a Kalman filter as a state estimator (see Section 9.8), than to perform the simpler design we wish to do. This is because it is not particularly straightforward (although it is possible) to explain how to get MATLAB to design the *reduced*-order observer.

In order to illustrate the kind of thing that *can* be done simply by a CACSD package such as MATLAB we shall revert to a *full*-order observer to estimate all the states, and then use the standard state variable feedback scheme to feed back the estimates. The feedback design is therefore identical to that used with the reduced-order observer in Example 9.3, but the calculated $\boldsymbol{k}$ vector is used directly as the SVF gain vector, since the matrix $\boldsymbol{\Psi}^{-1}$ will not be present in the full-order observer.

A design of this kind is very easy using MATLAB and the Control Systems Toolbox (see Appendix 3). Since we do not rely on the reader having access to MATLAB, we simply present this as one possible CACSD solution. We therefore outline the MATLAB method here, so as not to bore those without access to it. Readers with the appropriate MATLAB setup will find all the commands in the m-file *fig9_15.m* on the accompanying disk (but it will need one line 'commenting out' as noted in the file, in order to produce plots analogous to Figure 9.14, rather than producing Figure 9.15).

The steps in the m-file are:

(1) We specify the open-loop plant state-space model (Equation (9.20)). Note that MATLAB needs the full model – including the zero 'D matrix'.

(2) We design the feedback vector k required to place the closed-loop poles in the desired locations using standard SVF. We shall be feeding back the estimates, rather than the real states, but the design is identical. We use the MATLAB *place* command, which requires only the A and B matrices of the plant, and a vector of desired closed-loop pole locations.

(3) We design the observer. This will be a full-order observer, as designed in Section 9.4, using the design equation $F = A - GC$ (Equation (9.2)), and having the structure of Figure 9.5. Here we can again use the *place* command, making use of the duality which often exists in this area of control engineering (for example, the similarity of the controllability and observability rank tests). If we use the transpose of the A matrix, use C^T instead of B, and transpose the result, the *place* command will give us the gain matrix required to feed into the observer in order to place the observer poles in the desired locations (that is, the G matrix in Figure 9.5). There is one small restriction in the *place* algorithm, namely that we cannot specify more multiple observer poles at the same location than the number of plant outputs. In our case, the plant has only one output, so we can only place one observer pole at -20 on the real axis. In the m-file, we arbitrarily put the others at -19 and -21 to allow comparison with the previous design, although placing them off the real axis with a damping ratio approaching 0.71 would probably be a better idea from a numerical-reliability viewpoint.

(4) The F matrix is constructed using $F = A - gc$ (that is, the design equation, Equation (9.2)).

The design is now complete, and requires simulation to assess its performance. The system we have designed is that of Figure 9.5, plus an SVF vector k feeding back $\hat{x}$ to be subtracted from a reference input r. If we analyse the entire closed-loop system thus formed, we find that the combined closed-loop state equation can be written:

$$\begin{bmatrix} \dot{x} \\ \hline \dot{\hat{x}} \end{bmatrix} = \begin{bmatrix} A & -bk \\ \hline gc & F - bk \end{bmatrix} \begin{bmatrix} x \\ \hat{x} \end{bmatrix} + \begin{bmatrix} b \\ b \end{bmatrix} r = A_c \begin{bmatrix} x \\ \hat{x} \end{bmatrix} + b_c r$$

In MATLAB, we can form the partitioned matrix A_c and vector b_c in this equation in precisely the same way as they are written here, in order to obtain the overall state-space model of the closed-loop scheme.

If the built-in *step* or *impulse* commands are used, only the outputs are plotted by default, so it often makes sense to define every state as an output. The output equation is modified to accomplish this.

To get the same kind of results as in Figure 9.14, we need to specify an initial condition of 0.1 rad for x_1 and zero for everything else; and an input signal at r which is held at zero for about six seconds, before rising abruptly to unity. The *lsim* command will then generate the responses.

The plots produced by this sequence of operations are indistinguishable from those in Figure 9.14(d) between 6 and 12 seconds. In the earlier portion, where the estimates are converging on the correct values, there are differences due to the different design approach, and the fact that x_1 is now also estimated. However, these differences are not large, and the behaviour is qualitatively the same. In fact, this design behaves better, in that no signal magnitude ever exceeds about 3.2 units. It does, however, take more computing time, and is less accurate than the reduced-order design in the face of disturbances and poor plant models.

To investigate the effects of changing the observer pole locations, it is now only necessary to rerun the above sequence of commands with a different vector of required observer pole locations fed to the *place* command which designs the observer. Experimenting with this, we find that the observer can, for this simple and well-behaved system, be made very slow before any differences become apparent in the 6–12 s region (the unit step response).

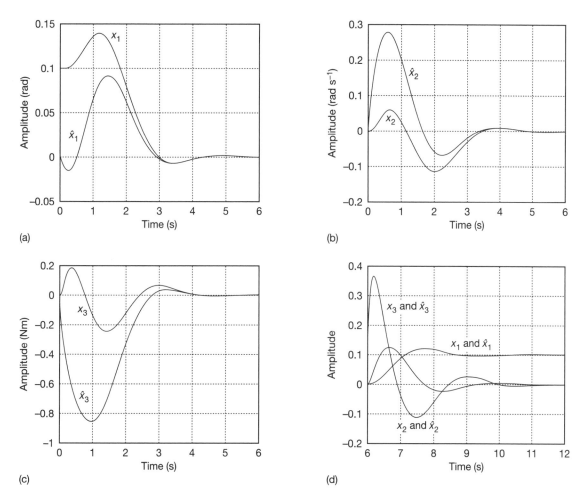

Figure 9.15 (a) Closed-loop response of a slow full-order observer + SVF, to initial conditions. x_1 and its estimate. (b) Closed-loop response of a slow full-order observer + SVF, to initial conditions. x_2 and its estimate. (c) Closed-loop response of a slow full-order observer + SVF, to initial conditions. x_3 and its estimate. (d) Closed-loop response of a slow full-order observer + SVF, to a subsequent unit step applied after six seconds. All three states and their estimates.

Figure 9.15 shows a set of responses with the observer poles selected to be actually *slower* than the fastest open-loop plant pole. The poles used were at -1.4, -1.5 and -1.6. The transient performance as the observer homes in on the correct values can be seen to last much longer than that of Figure 9.14, but the amplitudes are much lower, due to the reduced observer gains. Once the correct values are attained, it is almost impossible to detect any differences between the true and estimated values during the step response (6–12 s). This would not be true for all systems of course; it depends heavily upon the system order and the dynamics.

9.7   Digital implementation of a tracking feedback system including a reduced-order observer

Students (usually) eventually understand the kind of design procedures we have gone through in Example 9.3, but have no idea how to implement the result. We cannot go into great depth here, but in order to illustrate some of the principles involved, and especially to indicate methods of implementation of 'modern-control' schemes, an outline algorithm will be produced for the system we have been using as an example throughout the text.

For digital *simulation*, MATLAB commands could be used to evaluate everything. However, for *implementation*, it will normally be necessary to use some other computer language (Fortran, C, and so on). We shall therefore pursue the numerical parts of the algorithm without reference to any particular language, and we shall assume no language commands of higher level than matrix manipulation commands. If the chosen language cannot directly cope with matrix multiplications and additions, it is usually an easy matter to write some simple routines to allow it to do so (using the rules from Section A1.1, for example). In Fortran this is particularly easy, as the dynamic dimensioning feature allows routines that can cope with any sizes of matrix or vector to be easily written. In other languages, some ingenuity may be necessary to avoid having to write one routine per matrix size.

Direct digital control (DDC) is a wide topic in its own right, and several texts exist which address the issues involved, for example Åström and Wittenmark (1990), Bennett (1994), Bennett and Linkens (1982), Leigh (1992) and Williamson (1991). In addition, control engineering texts are available which are written from a purely digital (or, at least, discrete-time) standpoint, for example Franklin *et al.* (1990), Hostetter (1988), Kuo (1992) and Phillips and Nagle (1990).

The system we designed in Example 9.3, using the reduced-order observer, can be drawn as a simulation diagram containing only summers, integrators and gains. Such a diagram would look like Figure 9.13, with the two first-order blocks each replaced by an integrator with a negative feedback gain. It would therefore be very easy to implement the system using analog electronics (in principle, it could be done with just five or six operational amplifiers, following the ideas introduced in Section 4.7). However, for more complex (that is, higher order) systems, the problems of analog component parameter drift, susceptibility to noise and general lack of flexibility for changes would mean that we would almost certainly prefer a digital implementation, in which the matrix algebra could be done directly.

Computer control would also bring the added benefits of noise immunity, lack of drift, reprogramming flexibility and the capacity to perform other tasks (such as displaying trends and printing logs). These days, the only case where the analog implementation really wins is where the system has very fast dynamics (especially 'stiff' systems containing both very fast and very slow dynamics, which require complicated integration algorithms). In such a case, it may not be possible to justify (or even to find) a sufficiently powerful computer to allow sufficiently fast signal conversion and calculation times. The analog solution, on the other hand, remains a 'real-time' controller, however fast the plant dynamics may be. Nevertheless, for the rest of this section, we shall assume a computer implementation of the general form of Figure 9.16, following the general principles of computer control outlined in Section 5.8.1. In this case, the calculations to generate the observer output, the generation of the feedback signals and the summing of the feedback signals with the reference are all carried out in software in the computer.

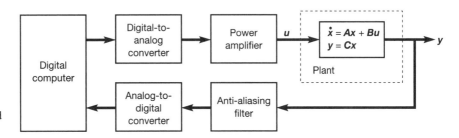

Figure 9.16 Computer control arrangement for reduced-order observer and feedback loops.

An important facet of any computer control scheme is the consideration of *safety measures*, but due to limited space, we shall only indicate a few aspects of programming for safety here.

It should be borne in mind at all times that we should not assume that the plant actuators will always move to the positions demanded by the control computer. All such movements should therefore be checked – effectively closing a monitoring loop around each actuator – and alarm conditions signalled, and action taken, as appropriate.

Similarly, whenever an analog signal is digitized and sampled by the computer, we must consider the likelihood that it will be contaminated by noise, despite anti-aliasing filtering (Section 5.8.1), and that the sample may be taken at a peak (for example) of a noise spike, or that input equipment may fail. It is therefore wise to check all inputs against known magnitude and rate-of-change limits, or even to average a number of readings for each sample.

In all these cases, the computer should raise appropriate alarms and/or print logs if unacceptable conditions are discovered. An orderly return from automatic to manual control may well be carried out, if necessary.

In order to save space, and to avoid obscuring the issues at hand, such matters as these are omitted from the following discussion, although they would be included in a 'real-world' implementation. Two matters to which we shall give more consideration are the following.

9.7.1 Bumpless transfer

The topic of *bumpless transfers* from manual to automatic control (and back again) was introduced in Sections 4.7 and 5.8.3. As a reminder, even when the plant is under manual control, the automatic scheme must track the operating conditions so that, when the switchover to automatic mode is made, the computer takes over precisely where the manual operator leaves off. This can often mean (as in the next example) that the bulk of the calculations must be done irrespective of whether the plant is in manual or automatic mode.

In order to achieve a bumpless transfer back from automatic into manual control, the computer must have some means of updating the human operator's setpoint-setting device, in accordance with any changes it may make to the plant inputs during the automatic phase of operation. For example, if the human operator controls the plant by turning a potentiometer, then the potentiometer may well be of the type having a shaft which passes all the way through it so that, during the automatic phase, the computer can control the plant using the self-same potentiometer, via a stepper motor on the rear shaft behind the control panel.

9.7.2   Integral wind-up

This is another important matter which has been mentioned earlier (Section 5.8.3). Again, by way of a reminder, this happens if elements in the control loop (for example, plant actuators) saturate, making it impossible for steady-state errors to be removed as quickly as they theoretically ought to be, according to a purely linear design.

Assuming that some kind of integral control is present (which is usually the case, so as to remove steady-state errors), the effect of saturated signals is that the integrator always 'sees' a non-zero error at its input. Therefore its output will continuously increase or decrease as a ramp, with the effects noted in Section 5.8.3.

To overcome this effect, *integral desaturation* methods are employed. The simplest method is to specify the maximum and minimum outputs allowed from the controller, and to refuse to update the integral term if doing so would violate these limits (thus duplicating the effects of Zener diode clamping in an analog electronic controller).

9.7.3   The system to be implemented

In Example 9.3, we designed a SVF scheme to place the closed-loop poles of a third-order system, given that we were only able to make measurements at the system output. We therefore designed a reduced-order observer to estimate the missing state information. The resulting solution appeared in Figure 9.13.

Recall that the system of Figure 9.13 is a *regulator*. We did not bother to convert it to a tracking system in Example 9.3 because, at that time, we were only interested in observer design. Now, we shall want the output y to track the reference input r with zero steady-state error following a step change. We know it does not do this at present (see the responses in Figure 9.14(d), where the unit step response settles at only 0.1 unit). What is required is some additional integral action of the type introduced in Section 5.4.4. Example 9.4 considers this conversion to a tracking system.

Example 9.4 *Including the reduced-order observer in a tracking system*

Again, we can effectively use the approach of ignoring the presence of the observer, and incorporating the matrix $\left[\frac{c}{T}\right]^{-1}$ into the feedback path, so as not to need to evaluate it 'online'. In Section 5.4.4 (Example 5.5), we designed a tracking scheme for the system of Example 9.3 (but without the observer) which involved a forward path integrator with a gain of 300, and state variable feedback gains of $k = [160\ 68.8\ 7.2]$ (Figure 5.6). To apply this to our present design (that is, including the observer) it is only necessary to include the matrix $\left[\frac{c}{T}\right]^{-1}$ into the previously designed feedback path, as we did leading to Equation (9.29) for the straightforward SVF design. Doing this, we obtain the following values, leading to the arrangement of Figure 9.17:

$$k\hat{x} = k\left[\frac{c}{T}\right]^{-1}\left[\frac{y}{z}\right] = [160\quad 68.8\quad 7.2]\begin{bmatrix} 1 & 0 & 0 \\ 34 & 1 & 0 \\ 225 & 0 & 1 \end{bmatrix}\begin{bmatrix} y \\ z_{11} \\ z_{21} \end{bmatrix}$$

$$= 160y + 68.8(34y + z_{11}) + 7.2(225y + z_{21}) = 4119.2y + 68.8z_{11} + 7.2z_{21}$$

To seven decimal places, MATLAB reports the closed-loop eigenvalues of the system of Figure 9.17 to be $-1 + 2j, -1 - 2j, -10, -20, -20$ and -30. The first three are the designed closed-loop eigenvalues of the closed-loop plant, so we have succeeded in placing these correctly. The fourth and fifth are the designed eigenvalues of the observer (fast compared with the plant) and the sixth is the designed value for the pole location due

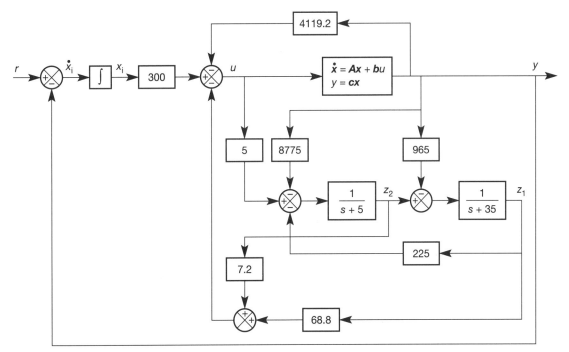

Figure 9.17 Figure 9.13 converted to a tracking system.

to the extra state introduced by the integral controller. So long as the observer has had time to attain the correct state estimates, the simulated step response of the system of Figure 9.17 is so close to that of Figure 5.7 that the plot is not worth repeating here (there is a MATLAB m-file *fig9_17x.m* on the accompanying disk that will do it).

We now wish to implement the system of Figure 9.17.

9.7.4 Choice of sampling rate

The control computer must sample all the appropriate signals, make its calculations and then send the calculated control signal u to the plant. Only then can it take the next sample of the signals. This process will be repeated indefinitely, for as long as the computer is required to control the plant.

The first question we must ask ourselves is, 'how fast must we sample the signals and execute the control loop?'. The importance of this question can be viewed in two ways. Firstly, we can use the answer to help us to decide upon the type of computer required (in terms of the minimum required processing speed). Secondly, once a computer has been specified, we shall then know how much time remains in which the computer can do other things (such as logging values, or carrying out other control functions).

The analysis required to obtain an accurate figure for sampling rate is fairly complicated, and Leigh (1992) gives a good introduction to the problem. In cases where computer speed is not a problem, it is often reasonably safe to sample at ten times the rate suggested by the fastest system pole (eigenvalue) – that is, choose a sampling time of 10 per cent of the fastest system time constant. However, the sampling rate chosen by such an unthinking approach should always be tested in simulation studies. For example, if it is too fast, then a more expensive computer and plant actuators may have to be purchased than are strictly necessary. Also, actuator wear will be more rapid, due to more movements per second. On the other hand, if one of the aims of the closed-loop system is to provide a large increase in speed of response, then 10 per cent of the fastest open-loop time constant may turn out not to be fast enough to obtain the best possible performance.

Example 9.5 *Sampling intervals for the antenna-positioner example system*

Inspection of Figure 9.1 indicates that the fastest time constant of the open-loop plant in our example is 0.2 s. We therefore aim to update the control signals to the plant every 20 ms, say.

In Example 9.3, we placed the observer's poles at $s = -20$, implying observer time constants of 50 ms. The observer might be regarded as 'not really physically existing', because it is only implemented as code inside the computer program. Nevertheless, it should be capable of exhibiting signals changing with time constants of 50 ms. We must therefore resist the temptation to evaluate its algorithm just when a signal is required by the plant (that is, every 20 ms), and instead we must 'sample it' (that is, perform its calculations) say every 5 ms (that is, four times per plant control period). This will hopefully give reasonable initialization and state-estimate-tracking performance, but see the final comment below.

The final dynamic element which the controller must contain is the integrator which produces the input-tracking action. With a step input, the output of an integrator is a linear ramp, so the sampling period is immaterial. However, once the integrator becomes part of a closed-loop scheme, its behaviour changes to a first-order type response. The integrator has an eigenvalue of -30 associated with it, implying a time constant of 33 ms, and a sampling period of 3.3 ms. For convenience, we want the sampling period to be some sub-multiple of the others, so we could choose 2.5 ms. However, in the interests of cheaper computing, we shall try fixing this sampling interval at 5 ms, the same as that of the observer.

We might find that sampling at 5 ms is not quite fast enough (because the observer apparently ought to be sampled at between 2 ms and 4 ms intervals).

9.7.5 Other considerations

Computer wordlength

Sometimes, this is decided by default, in that if a 'fast' machine is required in order to carry out the high-level vector–matrix processing within the required sample period, it is likely that it will be of the 16-bit or 32-bit variety as a matter of course. Nevertheless, we shall probably still have a choice as to the wordlength of the associated analog-to-digital and digital-to-analog converters.

The wordlength is chosen on the basis of the accuracy requirements. Bear in mind, though, that roundoff errors will make the requirement more stringent than may be suggested simply by the accuracy required in plant measurements. At 8-bit accuracy, the signal resolution is about 0.4%. At 12-bit and 16-bit accuracies, the signal resolution is about 0.025% and 0.0015% respectively. Williamson (1991) treats the subject fully. See also the comments in Section 5.8.4.

Manual/automatic mode change-over (bumpless transfer)

In manual control mode, the plant will be in the open-loop configuration of Figure 9.1. The operator's setpoint (reference) value will be applied directly to the input u and no feedback loops will be closed (except via the operator observing (*sic*) the plant behaviour and altering the setpoint in an appropriate manner).

However, as mentioned previously, the computer will still need to track the value of reference setting required for a bumpless transfer into automatic control mode. It does this by reading u and y (via the arrangement of Figure 9.16), and calculating all the feedback signals shown in Figure 9.17, just as if it was controlling the plant. The only difference from Figure 9.17 is that the control loop is broken at the point where the symbol u appears, because the manual operator's input is being fed in there, not the computer control signal. In addition, the computer's reference signal r is not necessarily the same as the manual operator's setpoint, but is set equal to y at each time step, so that the tracking integrator *input $\dot{x}_i$* remains at zero.

During manual control, we shall also (re)initialize the setpoint-tracking integrator's *output x_i* at each sampling instant, so that this signal (multiplied by the gain of 300), when algebraically summed with the two feedback signals which constitute kx, causes the computer controller output (which will be applied at the point u when in automatic mode) to be the same as the measured value of u as supplied by the human operator. That is to say, at each time step in manual mode, x_i is initialized to the value

$$x_i = \frac{u + 4119.2y + 68.8z_1 + 7.2z_2}{300}$$

where u is the control signal being applied by the human operator. In this way, when automatic control mode is selected, the plant will initially see no difference whatsoever in the value applied at u.

If the plant configuration was such that there was no integrator between r and u, then the initialization would involve calculating a suitable value of r at each time step, so as to match the computed value of u to the humanly supplied one. Since the integrator is present, its output can be initialized to any desired steady value when its input is zero.

These things will become clearer with reference to Figure 9.18, later.

Interlocking

In practice, the arrangements will be more complex. There will probably be indicators to show the operator whether control is in automatic or manual mode. There will also be a 'computer healthy' or 'control available' indicator which the computer will extinguish under initialization or fault conditions, and which will be interlocked with the automatic/manual selection hardware (push-buttons, relays, PLC outputs and so on). Other interlocking hardware will also be present to ensure that: (a) automatic mode cannot be selected unless all permissive signals are present (for example, 'computer healthy', 'lubrication on', and so on) and (b) the operator can *always* regain manual mode *irrespective of the state of the computer*.

Watchdog timer

A further fail-safe technique is the use of a 'watchdog timer'. This is simply a hardware monostable circuit which will time-out (thereby raising an alarm, selecting manual mode, or whatever else may be appropriate) unless it is periodically reset by the computer software. Thus, inside the main control loop will be an instruction to send a signal to reset the watchdog timer; and unless this is faithfully executed (say) every 100 ms, the timer will time-out and raise the alarm.

Updating the operator's setpoint (reference) setting while in automatic mode

We have mentioned that this needs to be done in order to achieve a bumpless transfer from automatic back into manual mode. Again, there are several possible approaches.

One common approach is to work out the *change* which the computer controller applies to the plant control variable u at each sampling interval, and to update the operator's setpoint generator (for example, potentiometer) by that amount.

This *incremental* approach is simple to implement, but has the disadvantage that if the operator moves the setting while in automatic mode, the correct setting cannot be regained. Such an approach is therefore only permissible if it is not possible for the operator to do this (for example, the computer moves the potentiometer via a high-torque stepper motor which cannot be rotated by hand).

A far better method is an *absolute* approach, in which the operator's setpoint setting is stored by the control computer at the instant when automatic mode is selected, and the correct setting relative to that single value is then sent to the setpoint generator every time the plant control input is updated. Thus, even if the operator moves the setpoint potentiometer, the computer controller will return it to the correct position at the next control instant.

This effectively means that as well as driving the plant input to the value u at each control instant, the operator's setpoint generating device (potentiometer, say) is set to correspond with the value u at the same time. The reader may therefore wonder why we bother to store the value of the operator's potentiometer at the instant of selecting automatic control, and then add to this value the changes made by the computer. Why not just send the actual value of u to the potentiometer at each control instant? The reason is that the method suggested overcomes open-loop calibration errors (including human errors such as the control potentiometer having been fixed to its shaft at the wrong angle!).

9.7.6 The observer

Figure 9.17 contains a continuous-time observer. This is more obvious as the state-space model in Figure 9.12, on which Figure 9.17 is based (the outer setpoint-tracking loop being the only addition in Figure 9.17). The observer state equation is:

$$\dot{z} = Fz + ju + gy$$

which has the usual convoluted solution for z (see, for example, Equation (3.11) in Section 3.2.1, which would be written in terms of z rather than x, and would have a second input term for gy – note again that this z has nothing to do with the z-transform!).

For digital implementation purposes, it is far preferable to work with a discrete-time version of the equation. Such a version was developed in Section 3.6.2, where its digital computer simulation was also discussed. The only real difference between the state equation solved in Section 3.6.2 and the observer equation here, is that the observer equation effectively has *two* separate inputs u and y. The only effect this has upon the previous result is that a separate Δ term is included for each input. The discrete-time version of the observer is therefore given by:

$$z_{k+1} = \Phi_{obs}z_k + \Delta_u u_k + \Delta_y y_k$$

where

k = sample number

h_{obs} = sampling period for the observer

Φ_{obs} = discrete-time observer 'system' matrix, $= e^{Fh_{obs}}$

Δ_u = discrete-time observer 'input' matrix from u,

$\quad = F^{-1}[e^{Fh_{obs}} - I]j$

Δ_y = discrete-time observer 'input' matrix from y,

$\quad = F^{-1}[e^{Fh_{obs}} - I]g$

Recall from Section 3.6.2 that there are various means of evaluating $\boldsymbol{\Phi}_{obs}$, and the two Δ matrices (actually, vectors in this case). However, in our case, since h_{obs} is very small (5 ms), the first five terms of the matrix power series expansions given in Chapter 3 (following Equation (3.91)) will probably suffice.

9.7.7   The setpoint-tracking integrator

In Figure 9.17 this appears as a simple integrator calculating x_i from $\dot{x}_i$. Its input is $(r - y)$. The discrete-time equivalent of such a scalar pure integrator is:

$$x_{i_{k+1}} = x_{i_k} + h(r_k - y_k)$$

where

$$k = \text{sample number}$$

$$h = \text{sampling period for the integrator}$$

9.7.8   The control software

We can now consider the implementation of Figure 9.17 (or its state-space equivalent), bearing in mind the deliberate omissions mentioned previously. Figure 9.18 shows an outline flowchart of the scheme. Note that it is necessary to read values from the plant (including the status of the auto/manual switch), to send values to the plant and to be able to time events. Any computer designed for process control applications will be able to do these things (indeed, most PCs can do them if equipped with suitable I/O cards and software). The details of these actions are omitted as they are machine- and language-specific.

A further point is that a software flag would be included in order to force the program flow to proceed at least once down the manual control route, before allowing the automatic route to be selected. This is to ensure correct initialization of the internal computer reference r and the integrator output x_i. It also gives the observer a minimum of eight samples (four down the manual route, and four down the automatic route) to acquire the correct state estimates. If an immediate entry to automatic mode were allowed, the number would be halved.

Once under computer control, it should also be noted that any desired setpoint changes can simply be summed with the internal reference value r. Integral desaturation is achieved by testing the new value of x_i generated in the lower right-hand box in Figure 9.18. If its value exceeds the set limit (either positively or negatively), then the limiting value is used instead. This mimics the action of Zener diode clamps in an analog controller, and prevents the integrator from ramping into unrealistic territory.

9.7.9   Simulation

The flowchart of Figure 9.18 forms an excellent basis not only for control of the plant, but also for digital simulation of the entire scheme in the language of the reader's choice. This can be used to check on the effects of varying sampling rates and so on. To convert the flowchart to a full simulation, rather than a plant controller, the following steps are necessary:

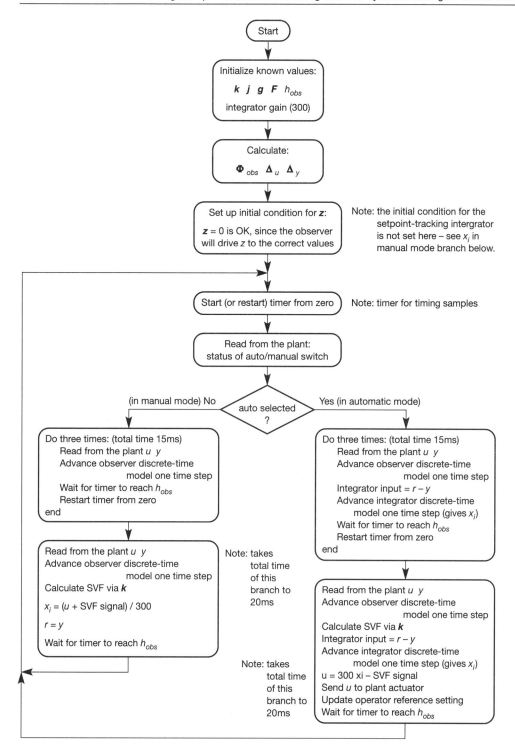

Figure 9.18 An outline flowchart for digital implementation of the system of Figure 9.17.

- Rather than reading values from the plant, provide sets of sampled input data to give the required values of u at each sampling instant under simulated manual control, the state of the auto/manual selector switch at each sampling instant, and any disturbances or setpoint changes to be made while under computer control.

- At the point in the flowchart where the discrete-time matrices for the observer are calculated, calculate those for a discrete-time description of the plant, too.

- Where the initial conditions are set up, also set up the initial conditions for the plant.

- Use the known state-space description of the plant to provide the signal y (from the input signal stream defined for u) wherever the flowchart asks for y to be read from the plant.

- Ignore all the timer commands.

- At some suitable point, store or plot any required values against time (time $= h_{obs} \times$ number of observer evaluations carried out).

Example 9.6 *Digital simulation of the tracking system plus reduced-order observer*

The results of running a simulation, programmed in MATLAB according to the instructions above, appear in Figure 9.19. The appropriate m-file is *fig9_19.m* on the accompanying disk. The code in this file is an incredibly inefficient way to use MATLAB, but it has been done to make the coding as much like any other high-level language as possible. None of MATLAB's high-level commands (*c2d*, *lsim*, and so on) are used. Only the equations from Sections 9.7.6 and 9.7.7 are used, together with the flowchart of Figure 9.18. Even the discrete-time plant model was obtained by using five terms of the power series expansions. The simulation should therefore give results which could be duplicated using any other programming language.

For this simulation, a manual control input of an arbitrary form was applied, and the system was allowed to settle for one and a half seconds. Then automatic control was introduced, and a setpoint change of 1 unit was superimposed on the internal reference r after waiting half a second (the delay is present to show that the transfer to automatic was bumpless).

In order to prove the bumpless transfer, it was necessary to have the open-loop (manually controlled) plant in a steady state at the instant automatic mode was selected. Of course, the system will work perfectly well, whatever state it is in when automatic mode is selected; but if conditions are not steady, the fact that the transfer is bumpless (which we wish to confirm) will be masked by the natural transient response of the plant. The presence of the integrator in the open-loop plant means that a steady state in manual control can only be achieved following impulsive-type inputs.

Although it cannot easily be deduced from Figure 9.19(a) (but it can be from reading the m-file), the manual part of the control input takes the form of two short positive impulses, with one negative one in between them. These are applied over the first few sampling intervals (by a very fast human operator!) and are very carefully chosen to minimize the settling time of the plant (the natural settling time, as we know, is about six seconds, so this approach was taken to keep the number of samples as small as possible, purely to minimize data storage and wasted space on the time axis of the plots).

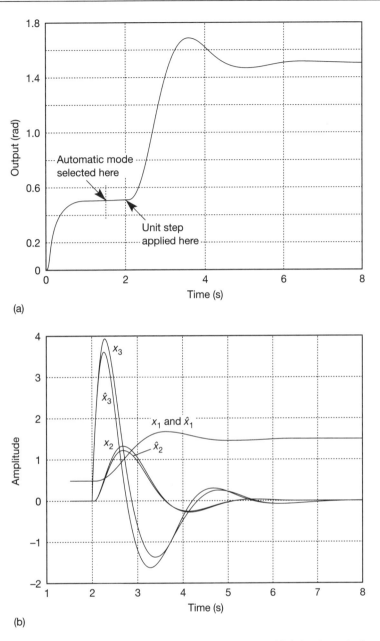

Figure 9.19 (a) A digital simulation of the system of Figure 9.17 and the flowchart of Figure 9.18. (b) The states and estimates for the latter part of Figure 9.19(a).

Figure 9.19(a) shows the resulting response of the plant output (which is also x_1). It can be seen that there is no transient disturbance ('bump') as automatic control is selected at 1.5 s, and that the subsequent unit step performance under computer control (after 2 s) is substantially correct (compare with Figure 5.7 for example). The differences are slight, and are due to the fact that this simulation is carried out in discrete time (implicitly assuming zero-order holds on the signals), while Figure 5.7 was obtained from a continuous-time state-space model of the overall system, using MATLAB's continuous-time simulation commands. Also, Figure 5.7 had no observer, so comparison with Figure 9.14(d) is another possibility (but that system had no tracker, so the amplitude scale is a factor of 10 'out', and the dynamics are slightly different due to the lack of the tracking integrator).

Figure 9.19(b) shows the responses of the states and their estimates during the unit step response. The behaviour is slightly modified compared with Figure 9.14, for the reasons just given. However, the accuracy of the output in Figure 9.19(a) indicates that they are probably good enough for practical purposes. In order to get much better agreement between the states and their estimates, it is necessary to decrease the observer (and tracking integrator) sampling period to about 1 ms (while still updating the plant control input every 20 ms). The MATLAB m-file contains comments indicating how to try this.

9.8 The Kalman filter

Advanced section

This section contains material which is qualitatively different from that in the rest of the text, in that it deals with random (or *stochastic*) signals. If desired, this section can be omitted without affecting the reader's understanding of anything else in the book.

So far, we have assumed *deterministic* conditions, by which we mean that so long as we can initialize our system models correctly, and so long as we know the applied inputs, we can always predict the outputs at any given time. This is not so in stochastic systems, where the presence of random fluctuations (noise) makes it impossible to predict with *certainty* what the values of any of the signals in the system will be at any given time (in much the same way as it is impossible to predict, with certainty, the precise arrival time of a bus, the behaviour of the stock market, or next week's weather).

To handle this uncertainty, we need some rather different mathematical tools, based on statistical ideas and probability. Most general textbooks leave these out, but we shall not. To try to avoid too much detail, we give the background and outline of the Kalman filter here, with enough detail to be able to design and use one, but we relegate the more detailed mathematical parts to Appendix 6. It is not necessary to understand the whole derivation, but it can give a better insight into the filter's operation. The framework of our derivation follows that of Healey (1979), but with the gaps filled in. For a more specialist approach consult Anderson and Moore (1979) or Bozic (1979).

9.8.1 Introduction – the filtering problem

The observers we have designed to date have all provided nice clean traces of the state estimates (because they used deterministic models). In practice, it is possible that the plant input and output signals, which feed the observer, will be so heavily contaminated with noise that our observers will not work very well. This might be electrical noise in the form of switching transients or other signals picked up by cabling, or it might be an unavoidable physical effect, such as the effects of wave motion on measurements of ship position (in positioning systems for drilling and survey vessels, for example – Grimble *et al.* (1979)).

The observers we have designed estimate the values of unavailable signals. If the signals used to form these estimates are themselves apparently too noisy to read properly, we need a more sophisticated solution. We are effectively trying to estimate something we do not know, from something whose value is uncertain!

The Kalman filter (KF) is a rather clever device with a number of applications. One application is the extraction of state estimates from noisy signals, and it is this possibility that interests us here. In order to work, the KF needs to know something about the statistics of the noise present in the signals. The question of how we might provide this information in real situations is left until later.

9.8.2 Derivation of the Kalman–Bucy filter

Although the KF is normally referred to as a 'Kalman Filter', it was actually developed by both Kalman and Bucy. As we mentioned earlier, we shall give sufficient information here to understand the structure and use of the KF, but the hard sums are relegated to Appendix 6. We choose to pursue the derivation in discrete time, as it is perhaps easier to follow than the continuous-time version. Also, since the result will normally be implemented digitally, it is, in any case, the most convenient form.

Consider the discrete-time system (see Section 3.6.2) represented by the usual state and output equations (the D matrix is assumed absent, for simplicity):

$$x_{k+1} = \Phi x_k + \Delta u_k$$

$$y_{k+1} = C x_{k+1}$$

The model of this system is shown in Figure 9.20, where the system is corrupted by an additive vector of random noise signals $w(t)$ representing all system disturbances, modelling errors and so on; and the output is corrupted by another vector of random signals $v(t)$, representing measurement noise, discretization errors, and so on. There can be any number of disturbances contributing to $w(t)$, so we need a matrix (Γ) to couple these into the system. We assume one disturbance in $v(t)$ per system output. Figure 9.20 gives rise to the modified model:

$$x_{k+1} = \Phi x_k + \Delta u_k + \Gamma w_k \qquad (9.30)$$

$$z_{k+1} = C x_{k+1} + v_{k+1} \qquad (9.31)$$

Assuming that we have access only to the input and measurement signals (u and z), we now want to try to extract a meaningful estimate of x (namely $\hat{x}$). This could

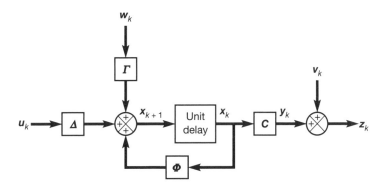

Figure 9.20 A discrete-time system model, subjected to random disturbances.

then be used to represent the states in a feedback control scheme, for example, as we have done before.

Problem statement

Given all information up to an instant $k + 1$, we will know the following:

$$u_0, u_1, u_2, \ldots, u_k, u_{k+1} \quad \text{and} \quad z_0, z_1, z_2, \ldots, z_k, z_{k+1}$$

From this information, we wish to estimate x_{k+1}. Remember that this estimate will be a random variable, and that we are trying to extract a suitable signal. The observers discussed previously are therefore inappropriate.

Estimate notation

Let $\hat{x}_{i|j} = E[x_{i|j}]$ represent the 'best' (in some sense) estimate of x at step i, given information up to and including step j. Our task is therefore to find the value of $\hat{x}_{k+1|k+1}$. See Appendix 6 for the definition of the 'expected value' operator $E[\cdot]$ (it is basically the mean value in the present context).

Development

For efficient computer implementation, we shall develop a *recursive* scheme. We can therefore assume the previous estimate $\hat{x}_{k|k}$ to be available, because we shall provide the initial value $\hat{x}_{0|0}$, and application of the recursion will then generate any required future value. We also assume the noise signals to be of zero mean value, that is, $E[w_k] = E[v_k] = \mathbf{0}$ (if this is not the case, the means can be subtracted and the model adjusted accordingly).

The best estimate that Equation (9.30) can give us (using the mean value of w_k) is then:

$$\hat{x}_{k+1|k} = \Phi\hat{x}_{k|k} + \Delta u_k \tag{9.32}$$

and we therefore obtain a one-step-ahead *predicted measurement* from Equation (9.31):

$$\hat{z}_{k+1} = C\hat{x}_{k+1|k} = C(\Phi\hat{x}_{k|k} + \Delta u_k)$$

At time step $(k + 1)$, we can measure z_{k+1} from the plant output, and hence find the *prediction error* $\tilde{z}$ as:

$$\tilde{z}_{k+1} = z_{k+1} - \hat{z}_{k+1}$$

One way of improving the state estimate given by Equation (9.32) would now be to add some proportion of this prediction error to each element of the state vector, in such a way as to try to drive the prediction error to zero; for example, $\hat{x}_{k+1|k+1} = \hat{x}_{k+1|k} + K\tilde{z}_{k+1}$. Using Equation (9.32):

$$\hat{x}_{k+1|k+1} = \Phi\hat{x}_{k|k} + \Delta u_k + K\tilde{z}_{k+1} \tag{9.33}$$

Notes

(1) The matrix K determines the distribution of $\tilde{z}$ between the states, and is usually called the filter gain matrix, the Kalman gain matrix, or simply the Kalman gain.

(2) We have $\hat{x}_{k+1|k+1}$ on the LHS of Equation (9.33), because we are now using information up to instant $k+1$. The new information has been introduced by the term $\tilde{z}_{k+1}$, and so the sequence $\tilde{z}_1, \tilde{z}_2, \ldots, \tilde{z}_k, \tilde{z}_{k+1}$ is often called the *innovations* sequence.

Equations (9.32) and (9.33) now constitute a 'predictor–corrector' system (similar to those found in many numerical integration algorithms) which attempts to drive $\tilde{z}$ to zero as time passes.

Substituting the definitions of $\tilde{z}_{k+1}$ and then $\hat{z}_{k+1}$ into Equation (9.33), we obtain:

$$\hat{x}_{k+1|k+1} = \boldsymbol{\Phi}\hat{x}_{k|k} + \boldsymbol{\Delta}u_k + K[z_{k+1} - C(\boldsymbol{\Phi}\hat{x}_{k|k} + \boldsymbol{\Delta}u_k)]$$

or

$$\hat{x}_{k+1|k+1} = [I - KC][\boldsymbol{\Phi}\hat{x}_{k|k} + \boldsymbol{\Delta}u_k] + Kz_{k+1} \tag{9.34}$$

Equation (9.34) now represents a recursive estimator, in which the new estimate $\hat{x}_{k+1|k+1}$ depends only upon the current output z_{k+1}, the previous estimate $\hat{x}_{k|k}$ and the previous input u_k.

The choice of the gain matrix K determines the filter's performance. We shall consider two choices: one (very briefly) leading to the Luenberger observer, and the other (in full) to the Kalman filter.

The Luenberger observer

If the noise signals are zero for all k (that is, the system model is entirely deterministic), then Equation (9.34) gives the Luenberger observer. In this case $\hat{x} \approx x$ at every step k, so long as the filter gain matrix K is chosen such that the eigenvalues of $[I - KC]\boldsymbol{\Phi}$ are faster than those of the plant $\boldsymbol{\Phi}$. This is precisely analogous to our previous approach to observer design in continuous time, and the resulting system, structured according to Equation (9.34), appears in Figure 9.21.

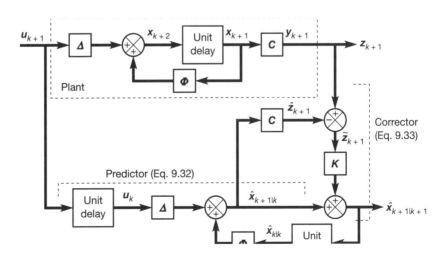

Figure 9.21 The Luenberger Observer (deterministic case) or Kalman filter (stochastic case).

The Kalman filter

Since it is defined by the same equation (Equation (9.34)), the KF is identical in structure to the Luenberger observer (Figure 9.21). However, we now consider the noise signals to be present. Thus, the signal z, and all the various estimates, become *random* variables. The difference between the schemes is in the way we choose the contents of the filter gain matrix K.

We now define the *estimation error* as:

$$\tilde{x}_{k+1} = x_{k+1} - \hat{x}_{k+1|k+1} \tag{9.35}$$

which, from Equation (9.34), can be seen to be a function of K. We also define the *covariance matrix* of the estimation error (see Appendix 6 for more on covariance) as:

$$P_k = \mathrm{E}[\tilde{x}_k \tilde{x}_k^{\mathrm{T}}]$$

The design of the KF follows from a choice of the Kalman gain matrix K, such that the covariance matrix P_k is minimized. The KF is therefore a *minimum variance estimator*.

The derivation of K to achieve this is not exactly straightforward, but is given in full in Appendix 6, as the authors are unaware of any other textbook which contains the derivation with no steps omitted. In Appendix 6, we find that:

if $\hat{x}_{k|k}$ is the minimum-variance estimate of x at instant k,

$\tilde{x}_k$ is the estimation error $(x_k - \hat{x}_{k|k})$,

$P_k = \mathrm{cov}(\tilde{x}_k) = $ (that which is to be minimized),

$Q = \mathrm{cov}(\text{system noise}, w)$ (see below and Appendix 6, Section A6.2.4),

$R = \mathrm{cov}(\text{measurement noise}, v)$ (see below and Appendix 6, Section A6.2.4),

then the KF for the minimum-variance estimate at the next time step (that is, $\hat{x}_{k+1|k+1}$) is given by the recursive scheme:

$$P_k^* = \Phi P_k \Phi^{\mathrm{T}} + \Gamma Q \Gamma^{\mathrm{T}} \tag{9.36}$$

$$K_{k+1} = P_k^* C^{\mathrm{T}} [C P_k^* C^{\mathrm{T}} + R]^{-1} \tag{9.37}$$

$$\hat{x}_{k+1|k+1} = [I - K_{k+1} C][\Phi \hat{x}_{k|k} + \Delta u_k] + K_{k+1} z_{k+1} \tag{9.38}$$

$$P_{k+1} = [I - K_{k+1} C] P_k^* \tag{9.39}$$

such that the Kalman gain matrix K is chosen on the basis of the previous error covariance and the noise statistics, so as to minimize the variance of the next estimate.

A selection of practical points and some examples

We end this chapter with a couple of examples of Kalman filter application. These examples will bring out some of the points which need bearing in mind when using the KF. Before the examples, we list a number of the more important things to look out for, and one or two design guidelines:

- It is most important to remember that although the KF is a clever device, it is not magical! It can only work with the information we give it. Specifically, the KF *does not* evaluate the characteristics of the plant and measurement noise as it operates. It cannot do so, because it has no way of knowing for certain what is really the correct signal and what is really unwanted noise. Therefore, the *only* information the KF uses about the noise signals is contained in the Q and R matrices which we must specify. These matrices (together with the $\varGamma$ matrix) completely determine what the Kalman gain matrix K will do, and what the final estimation error covariance matrix will be. If the characteristics of the noise signals change, the Kalman gain will *not* alter, unless we give the KF a new Q and/or R matrix to tell it about the change. This is discussed further in Example 9.7, below.

- In the light of the last point, the accuracy of the filter's performance clearly depends heavily upon the accuracy of Q and R. As with the other observers we have investigated, it also depends upon having an accurate C matrix for the plant, although A and B need not normally be very accurately known, due to the inherent feedback nature of the system (but poor modelling of A and/or B will lead to poor dynamic performance).

- Equation (9.38) is very similar to Equation (9.34), the difference being that the Kalman gain matrix is non-stationary, but varies from one time step to the next as defined by Equation (9.37). A rigorous implementation of the algorithm, with K changing every step, therefore allows the KF to cope with non-stationary and (to an extent) nonlinear systems.

  For linear systems, however, it can be shown that if the system of equations is iterated, the values in K eventually converge to constants (see the examples below). With some loss of dynamic accuracy, but great savings in computational effort, it is therefore often possible to use a *stationary Kalman filter*, as given by Equation (9.34), with the matrix K being pre-computed and held constant. Such a filter is identical in structure to the Luenberger observer (Figure 9.21), but K will have different contents, as the system has different objectives. MATLAB's control toolbox commands such as *dlqe* design such stationary KFs.

- The covariance matrix for the measurement noise R can often be intelligently estimated from plant knowledge. For example, assuming that the noise signals on the various measurements are uncorrelated (quite usual if they arise from different sensors), R will be diagonal (see Section A1.1.3 for the definition of a diagonal matrix). Each diagonal element is then the variance of the noise in that particular measurement channel.

  Recalling that the variance of a set of samples of a signal is the mean square of their deviations from the mean value, if an r.m.s. noise level (or an accuracy figure, perhaps) is available for the transducer, this can be squared to give the variance (mean square value) for the appropriate diagonal element of R. Discretization errors in analog-to-digital converters can be handled in a similar way.

  Any other plant knowledge available from manufacturers' literature and/or measurements on the plant can be included; the more, the better.

- The covariance matrix for the disturbances to the plant Q is more of a problem. Often, very little real information will be available. Any that is should be used. Although we have assumed in the derivation that the noise vector w contains only white noise, in practice it is supposed to cover all disturbances including steps, spikes, coloured (that is, time-correlated and band-limited) noise, modelling errors and so on.

Again, if the various noise sources comprising the elements of w are assumed to be uncorrelated, Q will be diagonal. Sometimes, the initial diagonal elements of Q have to be set more or less at random, and then tuned in full simulation studies, including simulations of every known disturbance signal, to get the most believable estimates of the states, before trials on plant. However, this can be a very time-consuming exercise for high-order systems with many disturbance signals.

This approach seems like avoiding the issue of setting Q 'properly', and indeed it is; but pragmatically, it is often the only sensible way to proceed. After all, we want a system that works in the real world, not necessarily one in which every mathematical nicety is fulfilled. In simulation studies, uncertain elements of Q might be tuned empirically until fluctuations in the state estimates remain within some pre-specified band.

- The following general characteristics of the KF may also be of help in 'tuning' Q and R.

  If either the elements of Q decrease in magnitude, or the elements of R increase in magnitude, the implication is that there is then relatively more noise on the measured signals (that is, the plant outputs) than the states. The KF therefore assumes that the state estimates generated by the 'predictor' part of the filter are relatively more reliable than the plant output measurements, and reduces the magnitudes of the elements in the Kalman gain matrix accordingly, so that smaller corrections are made to the predicted state estimates. In other words, the KF places more emphasis on the predictions, and less on the measurements.

  Conversely, if either the elements of Q increase in magnitude, or those of R decrease, the implication is that there is then less noise on the measured signals relative to the states. The KF therefore assumes that the state estimates generated by the 'predictor' part of the filter are less reliable, and require more correction. The magnitudes of the elements in the Kalman gain matrix will increase accordingly, so that the KF places more emphasis on the measurements, and less on the estimates from the predictor.

  In all cases, the estimate error covariance matrix P should be symmetric (apart from small numerical errors in off-diagonal terms), and in general, small elements in P indicate that the KF believes the estimates to be reliable. Large elements imply distrust.

- The diagonal elements of Q cannot simply all be set to zero, or else the Kalman gain matrix K will eventually decay to zero, too. This is because the condition $Q = 0$ represents a no-noise condition on the plant (also implying no modelling errors, since the process noise is supposed to include these, too). The KF therefore decides that the most reliable thing to do is to make no corrections to the state estimates (predictor, but no corrector). Once the initial-condition transients have died away, the KF therefore settles with zero gain. The KF then effectively ends up open-loop, and the estimates will deteriorate as time passes, due to the lack of any form of corrective feedback.

  In order to ensure that this does not occur, it is a good idea to choose the elements of Q so that it has no zero diagonal elements (or not many, at least), and to choose $\boldsymbol{\Gamma}$ (see below) so that the pair of matrices $(\boldsymbol{\Phi}, \boldsymbol{\Gamma})$ forms a controllable system. Simulation studies must be carried out to confirm correct operation if these conditions are stretched.

- The matrix $\boldsymbol{\Gamma}$ which defines the coupling of the process noise into the states may also be difficult to define in practice. However, a knowledge of which noise sources affect which states can often be gained by thinking about the practical system.

  As a silly, but simple example, consider a system for transporting a fluid in the open air, having two states: x_1 representing the torque driving the system, and x_2 related to the mass in the system. If we considered wind and rain as disturbances w_1 and w_2 respectively, we would expect the wind to affect the torque and have zero effect on mass, while the rain would have some effect upon the mass (probably a rather small effect), and possibly some very minor effect on torque depending upon the configuration of the system. The $\boldsymbol{\Gamma}$ matrix for such a system might therefore be of the form

$$\boldsymbol{\Gamma} = \begin{bmatrix} 1 & 10^{-6} \\ 0 & 0.001 \end{bmatrix}$$

  (the actual numbers, of course, depending entirely upon the application).

- The 'best estimate' vector requires initialization as $\hat{x}_{0|0}$. A sensible procedure is to see if any elements of x appear in z (that is, the corresponding row of the C matrix contains a number in the appropriate diagonal position only); and, if so, to use a first set of measurements z to initialize these elements of $\hat{x}_{0|0}$, the remainder being set to zero (unless other elements can be sensibly estimated from other plant knowledge or tests).

- The estimation error covariance matrix also requires initialization as $\boldsymbol{P}_0$. It is usual to assume poor initial estimates, and to set $\boldsymbol{P}_0$ to some large diagonal matrix accordingly.

  The only alternative would be to initialize $\boldsymbol{P}_0^*$ directly (Equation (9.36)). This is the version of $\boldsymbol{P}_0$ which is propagated in the recursion, and is valid immediately prior to the next measurement. If $\hat{x}$ contains measured values (as discussed in the previous point), and there is some confidence in the measurement noise covariance matrix $\boldsymbol{R}$, then the corresponding elements of $\boldsymbol{P}_0^*$ could be initialized to those of $\boldsymbol{R}$.

  In linear, stationary systems, the initial values of $\boldsymbol{P}_0$ are not critical, as the feedback action will drive them to convergence. However, in non-stationary or nonlinear systems, more care is required, because the convergence properties of K and P will then vary with the initial conditions.

- Sometimes, the plant model will contain a nonlinear output equation, of the form $y = f(x)$, where f represents a set of known nonlinear functions. In this case, the *extended KF* can be used, in which the full nonlinear output equation is used to calculate the measurement prediction $\hat{z}$. It is then found that a Jacobian matrix of the form $\partial f(x_{k+1})/\partial x_{k+1}$ can be used, in place of C, to calculate K_{k+1} (see Section 14.4 for a discussion of Jacobian matrices).

Example 9.7 *Investigation of a Kalman filter for a simple system*

Consider the simple open-loop system shown in Figure 9.22(a). Assume that this system has an electrical input signal, and that we are going to drive it with a unit step (1 V). The input signal is corrupted by noise of 0.1 V r.m.s., and the output measurement by noise of 1.414 V r.m.s. Both these noise signals are of zero mean value. We wish to estimate x_1 and x_2, as they are not otherwise available.

It is no use trying to use our previous observer designs. A full-order observer might do a little filtering for us, depending upon the positions of its eigenvalues, but the output noise is of rather large amplitude with respect to the signal level, and some of the noise will inevitably be passed onto the estimates. A reduced-order observer will pass all the output noise onto the estimate of x_1, unattenuated (because it will treat x_1 as a measurable signal, and not actually estimate it at all).

In designing a Kalman filter for this system, we have the advantage that we have specified the noise signals, and can therefore tell the KF the variances with some accuracy. This is not the usual state of affairs, and we shall point out the likely consequences. We also have a perfect plant model, so the KF knows this too. Again, in practice, the model used to design the KF will only be approximately that of the plant, so mismatch will occur here, too. The reader will find it instructive to run a simulation of a system such as this, including the KF, and experiment with different levels of error in the plant model, and the noise descriptions in Q and R (actually, q and r will both be scalars in this example, as there is only one noise source affecting the plant, and only one affecting the output measurements). Space does not permit such an investigation to be presented here, but it is easy to do, given the initial simulation program which is in the m-file *fig9_22.m* on the accompanying disk (MATLAB (Appendix 3) and its Control Systems Toolbox will be needed to run the file).

The KF is to be implemented by Equations (9.36) to (9.39), so we need to specify a discrete-time state-space model for the plant. A suitable noise-free continuous-time model is:

$$A = \begin{bmatrix} 0 & 1 \\ 0 & -1 \end{bmatrix}, \quad b = \begin{bmatrix} 0 \\ 1 \end{bmatrix}, \quad c = [1 \quad 0], \quad d = 0$$

and the MATLAB control toolbox command *c2d*, with a sampling period of 50 ms, gives:

$$\Phi = \begin{bmatrix} 1 & 0.0488 \\ 0 & 0.9512 \end{bmatrix}, \quad \Delta = \begin{bmatrix} 0.0012 \\ 0.0488 \end{bmatrix}$$

We also need to specify Γ, r, Q and an initial value for P, before we can begin. Consider the noise-coupling matrix Γ. This couples one noise signal into two states, so it will be 2×1. There is no noise to be added to x_1 (we are considering the 1.414 V r.m.s. signal to be purely output measurement noise), and the 0.1 V r.m.s. noise signal feeds directly onto $\dot{x}_2$ (the reader should draw the simulation diagram of the continuous-time state-space model if it is difficult to visualize this fact). This implies that

$$\Gamma = \begin{bmatrix} 0 \\ 1 \end{bmatrix}$$

Remember, however, that this Γ is for the *continuous-time* version of Equation (9.30), because it feeds onto $\dot{x}_2(t)$. We need a version of Γ to use with the *discrete-time* model, as

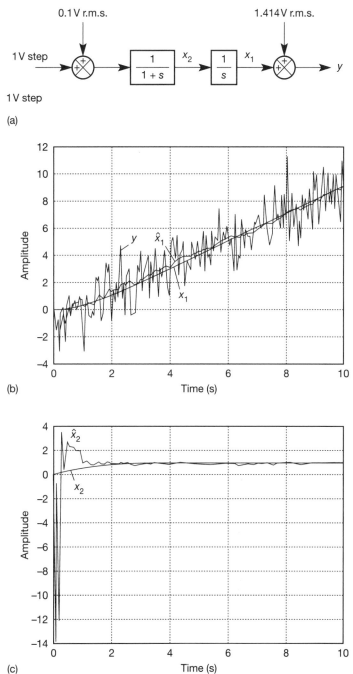

Figure 9.22 (a) A simple second-order system. (b) Kalman filter performance in estimating x_1. (c) Kalman filter performance in estimating x_2.

given in Equation (9.30). Using the MATLAB *c2d* command with A and Γ, rather than A and b, results in Φ as given above (which will always be the case, because Φ depends only upon the A matrix and the sampling period), and

$$\Gamma = \begin{bmatrix} 0.0012 \\ 0.0488 \end{bmatrix}$$

which is the discrete-time version for use in our Kalman filter equations (and happens to be the same as Δ, because the continuous-time version of Γ happened to be the same as b – this would not usually be the case).

The measurement noise is a scalar quantity in this case (only one output), so the covariance matrix R becomes the scalar variance r. Since the r.m.s. value of the noise is 1.414 V, and the variance is the m.s. value, we can say that $r = 2$.

The model noise only feeds state x_2. Q is usually a diagonal matrix with the variance of each noise signal in the appropriate diagonal position. In our case, the variance of a 0.1 V r.m.s. signal is 0.01 V^2, so we have only $q_{22} = 0.01$. (We call it q_{22} to show where it would be, if we had a full 2×2 Q matrix. In fact, the mathematics will give an identical result if we use

$$\Gamma = \begin{bmatrix} 0 & 0.0012 \\ 0 & 0.0488 \end{bmatrix} \quad \text{and} \quad Q = \begin{bmatrix} 0 & 0 \\ 0 & 0.01 \end{bmatrix}$$

but there is little point in such added complication.)

The initial value of P should be chosen to show that we have no confidence in the initial state estimates (that is, the KF must have to do some converging onto the correct values). We therefore specify large initial error covariances:

$$P_0 = \begin{bmatrix} 10\,000 & 0 \\ 0 & 10\,000 \end{bmatrix}$$

(2×2 for two states).

We now know everything we need to perform the iteration of Equations (9.36) to (9.39). These will provide the Kalman gain matrix K_{k+1} at each time step, together with the best state estimate $\hat{x}_{k+1|k+1}$ and the estimate error covariance matrix P, whose elements should get smaller as time progresses.

The system was simulated in MATLAB, by executing the discrete-time plant model (from zero initial conditions) and the KF of Equations (9.36) to (9.39) (using the plant model output as z at each time step) using the m-file *fig9_22.m*. At each time step, the appropriate values of suitable noise signals were added to the input and output as shown in Figure 9.22(a). All the signals were stored for plotting, and the results are as follows.

Figure 9.22(b) shows the noisy plant output y, the actual behaviour of state x_1 (according to the plant simulation) and the Kalman filter estimate denoted simply as $\hat{x}_1$. It can be seen that $\hat{x}_1$ has been extracted from the noisy output, so that it follows the actual value of x_1 reasonably closely. The Kalman gain begins at zero (no correction to the estimate) and then homes in on its final value as time progresses (more on this below). This is why the estimate of x_1 faithfully follows the measured output for the first few samples, before beginning to settle nearer to the correct value. The precise forms of y and $\hat{x}$ will vary according to the actual noise signal used in the simulation – if it is a random signal generated by the computer (using the MATLAB *rand* command, for example), they will also vary from one run to the next on the same machine, so the reader should not expect to duplicate Figure 9.22 precisely!

Figure 9.22(c) similarly shows the noisy state x_2, with the estimate $\hat{x}_2$ from the KF. Note that x_2 is not particularly noisy, because the level of the input noise is relatively low, and it is also filtered somewhat by the first block in the system. Again the agreement is good after a transient settling period. It takes longer for this estimate to home in on the correct value, because the KF has no measurement (not even a noisy one) of this signal.

After 200 iterations of the simulation, the Kalman gain matrix had not finished converging, but the values were

$$K = \begin{bmatrix} 0.007 \\ 0.0001 \end{bmatrix}$$

These small values show that the filter considers the magnitude of the measurement noise to be so bad, relative to the system noise, that it should make only very small corrections to the states estimated by the 'predictor'. The final error covariance matrix is

$$P = \begin{bmatrix} 0.0135 & 0.0002 \\ 0.0002 & 0.0003 \end{bmatrix}$$

the small values suggesting that the KF has high confidence in the estimates.

As we suggested earlier, the reason this KF works so well is that we were able to feed it with an ideal plant model and accurate details of the noise signals. If you perform the simulation and try changing a few things, you should discover the following.

- If r is increased, or q_{22} is reduced, then the Kalman gains decrease, as the filter thinks that either the output noise is even worse, or there is less noise on x_2, as the case may be. Of course, this is not true in this example, because we have not changed the noise at all; but the incorrect value of r or q_{22} makes it appear so, and the KF reacts accordingly as it knows no better. In fact, this gives a better result in this case, because the KF is using a perfect plant model – therefore, the predictions will be excellent, and the less correction it makes to them, the better! The values in P therefore reduce in this case, although in practical examples they might be expected to increase, reflecting the likelihood of poorer estimates from a noisier environment.

- If r is reduced (or q_{22} increased), the filter expects less output noise (or more state noise). It therefore places more trust in the output measurement relative to the states, and the Kalman gains increase, so that more correction is applied to the estimates from the measurements. Again, this is not the right thing to do in this example, because the model is already perfect and these 'corrections' actually make the estimates worse. If we move too far in this direction, the estimates will actually follow the noise signals. Once again, the point is made that the noise is still the same, and it is our poor representation of it that has caused poor filter performance. The conclusion is that Q (q_{22} in this case) and r need to be as accurate as possible. Many other trials can be made, such as using incorrect values in the plant model, but space does not permit us to report the results here.

As a final point, note that the MATLAB control toolbox command *[k,p,e] = dlqe(phi,gamma,c,q,r)*; will return the *stationary* discrete-time Kalman gain matrix in k. This can then be used in a non-recursive scheme, saving a lot of computation. The resulting state error covariance matrix is returned in p and, in this case, both are in reasonable agreement with the values quoted from the recursive simulation above. However, it takes about 50 seconds for the time-varying gains to achieve very good agreement with the stationary gains.

In general, the stationary Kalman gain computed by MATLAB should always agree with the value obtained from the time-varying filter (Equations (9.36) to (9.39)) after it has converged (that is, the steady-state value), but the estimation error covariance matrices will *not* generally agree. This is because of the large *initial* differences which might exist in Kalman gain between the stationary and time-varying approaches, leading to different convergence behaviour. The agreement of the error covariance matrices in the present example arises because the Kalman gains were always small.

Example 9.8 **Design of a Kalman filter for the antenna-positioning system with feedback loops and a tracking integrator**

This is a much more complex system, for which it is not so easy to interpret the results as it was for the simple example above. It is shown in Figure 9.23, and takes the basic form of Figure 9.17, but with the discrete-time plant model and Kalman filter of Figure 9.21 replacing the plant and reduced-order observer. The plant states and output are corrupted by noise. The filter outputs $\hat{x}_{k+1|k+1}$ are then fed back to the plant input via the previously derived SVF gains [160 68.8 7.2] (see Example 5.5 in Section 5.4.4), and the tracking integrator with a gain of 300 is used, as before. This time, it is represented by a discrete-time model (see below).

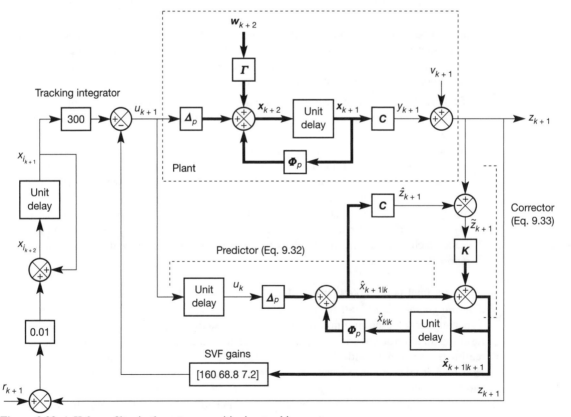

Figure 9.23 A Kalman filter in the antenna-positioning tracking system.

As in Example 9.7, this simulation has the advantage that we shall specify our own noise, so the variances will be accurately known to the KF, and the plant model used to derive the filter gains will be perfect. All the comments made in Example 9.7 on these matters therefore also apply here.

As to the noise, we shall assume that the output y is corrupted by noise of 0.05 rad r.m.s., the drive system output x_3 is corrupted by noise of 0.1 Nm r.m.s. and the load velocity x_2 by noise of 0.1 rad s^{-1} r.m.s. We assume no noise on x_1 other than the measurement noise.

These specifications suggest

$$\boldsymbol{\Gamma} = \begin{bmatrix} 0 & 0 & 0 \\ 0 & 1 & 0 \\ 0 & 0 & 1 \end{bmatrix}, \quad \boldsymbol{Q} = \begin{bmatrix} 0 & 0 & 0 \\ 0 & 0.01 & 0 \\ 0 & 0 & 0.01 \end{bmatrix} \quad \text{and} \quad r = 0.0025$$

In this example, we continue with these as given, in order to avoid potential confusion over the elements of $\boldsymbol{\Gamma}$ and $\boldsymbol{Q}$. However, it would be equally permissible to restrict them to the non-zero parts of $\boldsymbol{Q}$ (as in the previous example) by omitting the first column of $\boldsymbol{\Gamma}$, and the first row and column of $\boldsymbol{Q}$ (note that care must be taken when deciding on such simplifications, to ensure that the term $\boldsymbol{\Gamma Q \Gamma}^{\mathrm{T}}$ remains conformable for multiplication and also gives a result matching the dimensions of the rest of Equation (9.36)).

The discrete-time models of the plant and tracking integrator were obtained for a sampling interval of 10 ms. For reference, they are:

$$\boldsymbol{\Phi}_P = \begin{bmatrix} 1 & 0.01 & 0 \\ 0 & 0.99 & 0.0097 \\ 0 & 0 & 0.9512 \end{bmatrix}, \quad \boldsymbol{\Delta}_P = \begin{bmatrix} 0 \\ 0.0002 \\ 0.0488 \end{bmatrix}, \quad \boldsymbol{\Phi}_I = 1, \quad \boldsymbol{\Delta}_I = 0.01$$

Recall that, in the previous example, the continuous-time version of $\boldsymbol{\Gamma}$ also had to be converted to a discrete-time model. In the present example, the noise signals have been specified as feeding directly onto the states, *not* onto the state derivatives. This means that $\boldsymbol{\Gamma}$ is correct for the discrete-time model as it stands.

The estimate error covariance matrix was initialized to

$$\boldsymbol{P}_0 = \begin{bmatrix} 10\,000 & 0 & 0 \\ 0 & 10\,000 & 0 \\ 0 & 0 & 10\,000 \end{bmatrix}$$

Running the simulation in the MATLAB m-file *fig9_24.m* on the accompanying disk gave the results in Figure 9.24. The Kalman gain matrix again converged onto its final values after about 100 iterations (that is, after 1 second of the simulation). The final gain vector and estimation error covariance matrix were:

$$\boldsymbol{K} = \begin{bmatrix} 0.174 \\ 1.658 \\ 0.1106 \end{bmatrix}, \quad \boldsymbol{P} = \begin{bmatrix} 0.0004 & 0.0041 & 0.0003 \\ 0.0041 & 0.0924 & 0.0071 \\ 0.0003 & 0.0071 & 0.1047 \end{bmatrix}$$

$\boldsymbol{K}$ is found to be in good agreement with the MATLAB control systems toolbox command *[k, p, e] = dlqe(phi, gamma, c, q, r)*, but $\boldsymbol{P}$ differs a little, as we might expect, due to the fact that our simulation uses a time-varying filter, while the MATLAB result is for

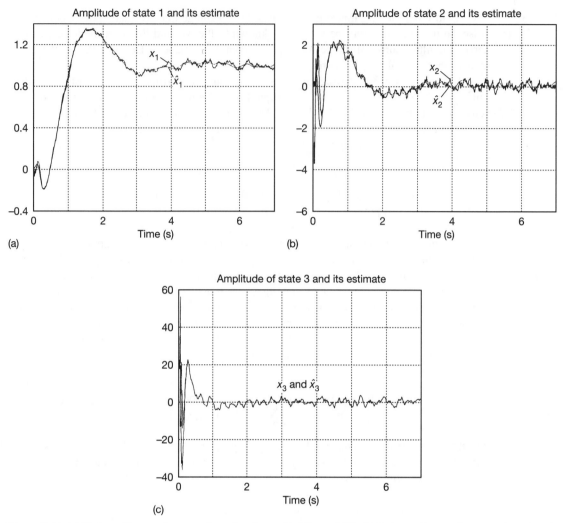

Figure 9.24 (a) Kalman filter performance in estimating x_1 for Figure 9.23. (b) Kalman filter performance in estimating x_2 for Figure 9.23. (c) Kalman filter performance in estimating x_3 for Figure 9.23.

a stationary one; the initial differences in Kalman gain lead to different convergence behaviour.

It can be seen from the Kalman gain vector that the gains for states 1 and 3 are much lower than that for state 2. As we discussed in Example 9.7, for this ideal system it would actually be better if all the gains were zero, as the perfect plant model used for the KF design ought to give perfect estimates from the 'predictor' part of the KF, therefore requiring no correction. However, the KF cannot know this, and works only on the information we feed it, namely the noise covariances. Nevertheless, the performance can be seen to be good. Figures 9.24(a) and (c) show that the estimates of x_1 and x_3 follow the correct values rather closely. Indeed, it looks as though x_3 is estimated precisely, but this is a function of the vertical scaling of the plots. In fact, the errors between x_3 and its estimate are of the same order of magnitude as those between x_1 and its estimate.

Looking at Figure 9.24(b), particularly with reference to the vertical scaling, shows that x_2 is not estimated as well as the other two states, the errors being an order of magnitude larger (although it is still subjectively 'good'). This is because the filter is working harder to try to correct this estimate, due to the higher gain in k_{21}. It therefore picks up more of the output noise (readers with access to MATLAB or a similar package can easily prove this, by looking at the cross-correlation between y and $\hat{x}_2$). Of course, the picture is made much more complicated by the presence of the various feedback loops.

The reader might also have noted that the behaviour of the states does not follow the expected trajectories in the early parts of the responses, some non-minimum-phase type behaviour being apparent (compared with Figure 9.19, for example). This behaviour is very variable from one simulation run to the next, as it depends heavily upon the first few values of the (random) noise sequences. In fact, the responses of Figure 9.24 represent a rather bad case. In any event, the initial transient performance is not of great concern, as we would give the estimates time to settle in practice, before trying to use them.

Variation of the filter parameters has the same qualitative effects as we found in Example 9.7, and the same comments apply. The effects are quite general for Kalman filters:

- If r is increased, or the magnitudes of the elements of Q are reduced, then the Kalman gains decrease, as the filter thinks that either the output noise is even worse, or there is less noise on x_2 and/or x_3 as the case may be. The effect is that less correction from the output measurement is applied to the state estimates from the 'predictor'.

- If r is reduced (or the elements of Q are increased), then the filter expects less output noise (or more state noise). It therefore places more weight on the output measurement relative to the states, and the Kalman gains increase, so that more correction from the measurements is applied to the estimates.

If the KF is replaced by a full-order observer in this system, with the specified noise signals, we discover just how good the KF is in comparison, especially in the estimates of x_2 and x_3. The only advantage of the full-order observer is that it is less susceptible to large initial transient errors than the KF (but, as mentioned previously, we would not put any trust in the estimates of any kind of observer until it had had time to settle, so this advantage is not of any practical significance). If the system model parameters used to generate the KF (or observer) are in error compared with the actual plant, then the relative performance of the KF becomes even better.

9.9 Conclusions

In this chapter, we investigated various ways of estimating the values of system state variables which are not directly measurable. We also investigated, in theory and by simulation examples, the inclusion of the various kinds of estimator into feedback control schemes, for both regulating and tracking systems. We discovered that the estimator and the feedback loops can be designed in isolation from each other, making for an easier design procedure.

In the case of the full-order observer, the design procedure is easy, but all the system states are estimated, leading, in general, to inefficient and potentially inaccurate implementation. The only occasions on which such a scheme would normally be used are when the measurable signals are relatively noisy, and we might prefer slightly filtered estimates to the real thing. Even in such systems, it is just as easy to design a Kalman filter using tools such as MATLAB.

The reduced-order observer has a more complex design procedure, but leads to implementations that use as many measurable states as possible, and estimate only the unmeasurable ones. The resulting schemes are therefore of lower order than full-order schemes, and are likely to give more accurate results. Such a scheme will almost always be preferred over a full-order design.

The Kalman filter is used in place of either of the previous observers in situations where the measurable signals and the plant states are corrupted by noise to the extent that the other types of observer will not work satisfactorily. The full time-varying KF is algorithmic, operating as a recursive function of the system model and the statistical properties of the noise signals. Normally, for linear, stationary systems, a stationary KF can be computed offline and then applied just like an observer, resulting in slightly poorer dynamic behaviour, but much reduced computational effort compared with the full time-varying KF. The MATLAB control toolbox contains tools for designing such Kalman filters in continuous or discrete time, and for various cases we have not considered, such as the case in which the D matrix is present. The KF can work very well, but it can be difficult to specify correctly the required information about the noise signals affecting the system, especially for complex high-order plants.

9.10 Problems

9.1 Show that Figure 9.6(b) is an alternative representation of Figure 9.6(a) (this occurs towards the end of Example 9.1).

9.2 (a) Test the observability of the system:

$$A = \begin{bmatrix} 0 & 1 & 0 \\ 2 & 3 & 1 \\ 0 & 0 & 2 \end{bmatrix}, \quad B = \text{anything},$$

$$c = \begin{bmatrix} 1 & 0 & 0 \end{bmatrix}$$

(b) Consider the output of the system, calculated from $y = cx$. How do you reconcile the result of part (a), together with the definition of observability (Section 5.3.2), with this result?

9.3 Consider the system:

$$A = \begin{bmatrix} 0 & 1 & 0 \\ 0 & 0 & 1 \\ 0 & 0 & 0 \end{bmatrix}, \quad b = \begin{bmatrix} 0 \\ 0 \\ 1 \end{bmatrix}, \quad c = \begin{bmatrix} 1 & 0 & 0 \end{bmatrix}$$

(a) Design (if possible) a full-order observer for the system, placing the observer poles at $s = -4$ and $s = -4 \pm 4j$.

(b) Draw a block diagram of the resulting system, without using vectors or matrices.

(c) If you have access to MATLAB and the control systems toolbox (or a similar package), repeat part (a) with computer assistance. Verify that the resulting observer F matrix gives the desired pole locations. *Hint*: Use the dual of the *place* command to calculate g – see towards the end of Example 9.3.

9.4 (a) For the system of Problem 9.3, design a reduced-order observer having eigenvalues with real parts at $\lambda = -5$, and a damping ratio of 0.707.

(b) Now add a feedback system to place the closed-loop poles of the overall scheme (excluding the observer poles) at $\lambda = -1 \pm 2j$ and $\lambda = -2$.

(c) Draw a diagram of the overall scheme, with no vectors or matrices.

(d) What would you expect the steady-state performance of the final system to be to a step input? Convert the scheme to a tracking system. *Note:* The results of Problems 9.5 to 9.9 may also be converted to tracking systems if you need the practice, but this is not strictly anything to do with observers as such.

9.5 In the system of Figure P9.5, only the lines shown carry measurable signals.
(a) Obtain a state-space model for the plant.
(b) Design a feedback regulator for the plant of part (a), such that the closed-loop poles are positioned at $s = -1$ and $s = -1 \pm j$. Use a full-order observer, but note that x_1 and x_2 are both measurable and can, if you choose, be regarded as outputs of completely separate subsystems.
(c) If you have MATLAB and the control systems toolbox, repeat the exercise and see how much easier it is!

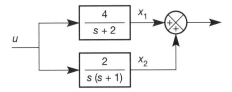

Figure P9.5 System for Problems 9.5 and 9.6.

9.6 For the plant of Figure P9.5, and the state-space model obtained in part (a) of Problem 9.5, repeat part (b) of Problem 9.5 using a reduced-order observer, and compare the resulting systems. If you have used x_1 and x_2 as defined in Figure P9.5, you may find it useful to define x_1 and x_2 as outputs of separate subsystems, since both are measurable.

9.7 A system has the LTF model

$$\frac{Y}{U}(s) = \frac{2s + 14}{s^3 + 10s^2 + 29s + 20}$$

Making connections to u and y only, design a state variable feedback regulator to place the closed-loop poles in a Butterworth configuration (see Equation (5.14) with an associated frequency of 6 rad s^{-1}. Note that one of the open-loop poles is at $s = -5$.

9.8 (a) Consider the system shown in Figure P9.8. Making connections only to the signal paths shown in the figure, design a regulator to move the unstable open-loop poles to $s = -1 \pm 2j$, without affecting the third pole location.
(b) Draw a diagram of your solution which contains no vectors or matrices.
(c) By block diagram reduction, or by using MATLAB or a similar package, show that the diagram of part (b) has the required closed-loop poles (if you use block diagram reduction methods, be careful not to cancel any poles and zeros, otherwise you will not see the observer pole – why is this?).

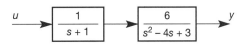

Figure P9.8 System for Problem 9.8.

9.9 Figure P9.9 shows a position controller which contains a variable rate feedback constant K. Design a closed-loop regulator for the plant, such that the closed-loop performance is characterized by poles placed at $s = -2$ and $s = -1 \pm j$. You may assign any value you choose to the constant K, but you may not make any connections to signals inside the indicated subsystem.

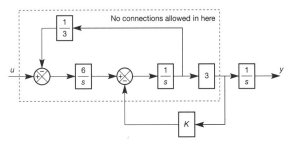

Figure P9.9 Position control system for Problem 9.9.

9.10 In any application of a Kalman filtering scheme, computer assistance and implementation would always be used. However, a useful insight into the operation of such filters can be gained by carrying out by hand the first few iterations of a relatively simple problem, such as that solved in Example 9.7. Even so, it will be necessary (unless great tedium is to be endured) to use the assistance of a computer program, or a programmable calculator, to generate the random noise signals and to carry out the recursive calculations. In addition, to really get a feel for what is happening, it will be necessary to rerun the calculations with different choices for Q and R, so it is definitely a hard job by hand. If MATLAB and its control systems toolbox are available, things are greatly eased, and the file *fig9_22.m* on the accompanying disk can be used as a template and suitably modified in several areas (do not alter the original file – use a copy).

A system has a transfer function

$$G(s) = \frac{3}{s^2 + 2.2s + 3}$$

This can be converted by direct programming to the state-space model

$$A = \begin{bmatrix} 0 & 1 \\ -3 & -2.2 \end{bmatrix}, \quad b = \begin{bmatrix} 0 \\ 1 \end{bmatrix}, \quad c = [3 \quad 0],$$
$$d = 0$$

With this state-space model, the arrangement of the plant is such that states x_1 and x_2 are corrupted with uncorrelated additive white noise of amplitudes 0.2 V r.m.s. and 0.1 V r.m.s. respectively. State x_2, in addition, is subject to the same noise signal as x_1, but attenuated by a factor of 10.

The output of the system is provided by a measuring sensor, which also causes additive white noise, of amplitude 0.1 V r.m.s., at the output.

Investigate the behaviour of a suitable Kalman filter in estimating the states of the system.

Points to note:

- You will need to convert the state-space model to discrete time, using a suitable sampling interval (see Section 3.6.2, or use the MATLAB Control system toolbox *c2d* command).

- From the problem statement, you can work out the matrix Γ which couples the noise into the states as in Equation (9.30). The information given applies to the *continuous-time* system but, since the noise feeds the states directly (*not* the state derivatives), you will not need to convert the system (A, Γ) to discrete time in this case.

- It will be necessary to evaluate sufficient samples to cover at least 2–3 s, in order to see the behaviour of the system.

10 *Multivariable systems in the frequency domain*

In this chapter the reader will learn:

- that multivariable systems (that is, systems with more than one input and/or more than one output) can be handled using Laplace transfer functions combined with matrix methods, and not only by using the state-space approach
- that there is more than one frequency domain approach to MIMO analysis and design
- how to apply two or three methods in practice, using computer-aided control system design (CACSD) techniques.

10.1   Preview

Readers who have studied Chapters 1 to 4 should understand this chapter.

The early chapters introduced various means of modelling, analysing and designing controllers for engineering systems. One of the major approaches involved the use of Laplace transfer functions (LTFs), and a big disadvantage was that it is not clear how to handle multi-input-multi-output (or 'MIMO', or 'multivariable') systems by these methods. The state-space methods could be used, but they also suffered from certain drawbacks. This chapter addresses the use of frequency-domain techniques, based on LTFs, for analysing multivariable systems.

No other part of the book relies on this material, so it can be omitted if desired. For readers who wish to study only one technique from this chapter, the characteristic locus method may be easier for newcomers to apply, since it requires less experience on the part of the designer than does the inverse Nyquist array, in order to obtain reasonable results.

NEW MATHEMATICS FOR THIS CHAPTER

In Equation (2.96) (Section 2.7), the concept of a transfer function matrix (TFM) was briefly introduced. In this chapter, we reinforce this concept, and then use TFMs widely. This is because the multivariable frequency-domain methods combine the benefits of Laplace transfer functions with the natural ability of matrix methods to handle MIMO systems. For readers who have not studied Chapters 5 and 9 and require a refresher on matrix methods, Appendix 1 contains all the necessary material, and now would be a good time to read it. Some more advanced aspects of matrix algebra are also introduced for the first time, as they are needed. These include the notion of the McMillan form of a matrix, singular values, condition number and the Perron–Frobenius eigenstructure of a system.

Much of the mathematics of multivariable systems in the frequency domain is extremely difficult to apply without computer assistance, usually due to the huge volume of calculation required, rather than to any intrinsic difficulty of the mathematics itself. For real-world examples, these methods rely heavily upon computer-assisted control system design (CACSD) environments. Here, we use MATLAB, its Control Systems Toolbox and its Multivariable Frequency Domain Toolbox (MVFDTB – see Ford *et al.* (1990)).

10.2 Introduction

Multivariable systems (systems with multiple inputs, multiple outputs, or both) have always posed an interesting problem for the control engineer. In Chapter 2, we pointed out that one of the major advantages of the time-domain control techniques (using state-space methods) over the frequency-domain techniques (using Laplace transfer functions) was their ability to handle such systems.

As the state-space methods were being developed in the 1960s, it initially appeared that they would be able to do everything that the frequency-domain methods could do, and much more besides. However, as time passed, it became clear that the state-space methods gave rather fragile results in some cases. The most simplistic reasoning for this is that the frequency-domain methods are graphical in nature, and rely only on input–output models of the system. The design process is iterative, the control engineer typically inspecting a suitable frequency response or root locus plot, designing a controller graphically, investigating the effect on the response plot and iterating around such a procedure until satisfied. The results are clearly approximate, are inherently treated as such and tend to be reasonably robust in the face of modelling errors and certain unmodelled plant disturbance signals.

State-space methods, on the other hand, use internal models of systems and synthesize controllers using mathematical algorithms based upon the models. If the models turn out to be poor, then so is the control likely to be.

For multivariable systems, however, the state-space methods seemed the ideal way out, since the number of inputs and/or outputs makes no difference to the design methods; it just alters the size of the matrices. In applying the state-space methods to aircraft and space vehicle control, some notable successes were achieved. This is still one of the major areas of application of state-space methods, because system models in aerospace tend to be relatively accurate, so the synthesized controllers can work quite well.

Unfortunately, unlike the aerospace applications, many industrial processes defy cost-limited attempts to obtain very accurate models (especially accurate *linear* models). So, when state-space methods are applied to industrial problems, they meet with mixed fortunes – sometimes working well, and sometimes being outperformed by the standard PID controller.

Furthermore, when time-domain (state-space) controllers do work well, they can prove rather fragile. For example, disturbance rejection can often be poor; and the failure of any of the feedback loops might be disastrous, leading to total instability. Since the mid-1980s, or thereabouts, a lot of work has been done on improving the robustness of such controllers, but a detailed description is beyond the scope of this text (see Chapter 13 for an introduction to the area of robust control, and further references).

In an effort to apply the elegant matrix methods of time-domain control to multivariable systems, without incurring the apparent disadvantages, workers in the UK investigated the generalization of the standard frequency-domain techniques (Bode and Nyquist) to multivariable systems. The aim was to combine the matrix algebraic approach to multivariable system analysis (thus coping with MIMO systems), with the graphical *design* (as opposed to *synthesis*) approaches of frequency-domain controller selection (thus removing some of the reliance on very accurate plant modelling).

The results included the Inverse Nyquist Array (INA) (Rosenbrock, 1974) and the Characteristic Locus (CL) (MacFarlane and Kouvaritakis, 1977) techniques. These are each heavily dependent upon a CACSD environment for their use, as they are true *design* methods, in which the engineer tries a controller, assesses the results, improves the design and tries again until satisfied. The amount of computation involved, and the amount of data a designer would have to handle, are significantly greater than for single-input-single-output (SISO) systems. Thus, it was not until interactive computer-graphics terminals (and, later, powerful personal computers) became readily available, that the methods began to find wide acceptance.

Other methods of analysing and designing multivariable systems also exist, but space limitations do not allow us to consider more than one additional technique here. The interested reader might investigate the sequential return difference method (Mayne, 1979), dyadic expansion methods (Owens, 1978) and principal gain and phase methods (Postlethwaite *et al.*, 1981). In addition the specialized texts on multivariable systems by Maciejowski (1989), O'Reilly (1987), Patel and Munro (1982) and Owens (1978) will all prove excellent references.

10.3 Frequency-domain description of multivariable systems

Imagine an industrial process in which two streams of liquid feedstock, one hot and the other cold, are poured into a vessel, where they are mixed. The mixture is continuously drawn off from the vessel (but at a variable flowrate) into the next part of the process. It is necessary to control both the temperature of the mixture in the vessel, and its depth, so that the following part of the process is fed under constant head and constant temperature conditions. The available means of control is by flow control valves on the two feedstock pipes. This system has two control inputs (the two incoming flowrate setpoints) and two outputs (the temperature and level in the vessel).

It does not require much imagination to see that this is actually a rather difficult control problem, due to the *interaction* present in the system. For example, say the vessel temperature is too high. The rate of flow of the hot feedstock could be reduced, or that of the cold one increased, in order to correct the temperature error. However, either of these actions would cause an error in the level of liquid in the vessel.

Similarly, if the liquid level falls, then in order to increase it, one of the incoming flowrates might be increased. Whichever one we choose will then cause a temperature error.

In both these cases, what is actually required is a suitably chosen change made to *both* the feedstock flowrates simultaneously. This chapter uses transfer function matrix models, containing the LTFs from each process input to each process output, in order to design such controllers.

Another two-input-two-output system, which we shall use as a case study for this chapter, is shown in Figure 10.1. It is a pneumatic laboratory rig, built to simulate a particular industrial process, not dissimilar from the mixing process described above. The rig comprises four pressure vessels of differing dynamic response, which are fed in pairs from voltage inputs via force-balance E-to-P (voltage to pressure) converters. The rig has two outputs which are each functions

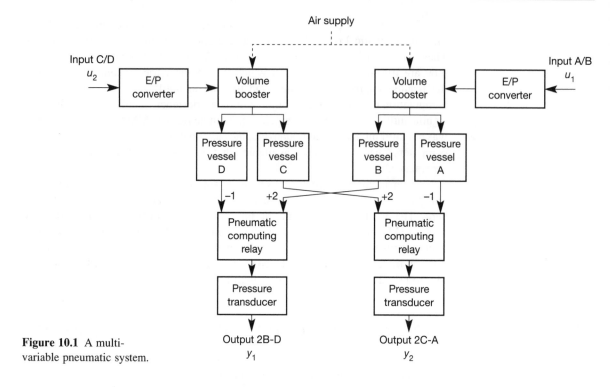

Figure 10.1 A multi-variable pneumatic system.

of the pressures in two of the pressure vessels, and are converted to voltage signals by pressure transducers. Since the outputs are the result of combining pressures from pairs of vessels which are *not* fed from the same input, there is bound to be significant interaction in the system. Driving the system from either input will have a significant effect on both outputs simultaneously. The other system components are the volume boosters, which are simply power amplifiers to boost the very small flows from the E-to-P converters, so that they become capable of supplying the pressure vessels; and the computing relays, which are pneumatic components giving an output pressure which is twice the (gauge) pressure on one input minus the (gauge) pressure on the other.

One way of obtaining a model of a system like this is to apply a voltage step to each system input in turn, record the output behaviour, and fit transfer function models to each of the resulting responses as described in Section 3.8. For more complex systems, or systems for which a step input is not allowable, other methods are available (see Chapter 8).

Example 10.1 *Obtaining a model of the pneumatic system*

Applying steps of 1 volt to each input in turn of the system of Figure 10.1 yielded the responses of Figure 10.2. In this figure we have four response curves. These represent the behaviour at each of the two outputs in response to unit step signals at each of the two inputs. The 'fuzziness' of the traces is due to noise.

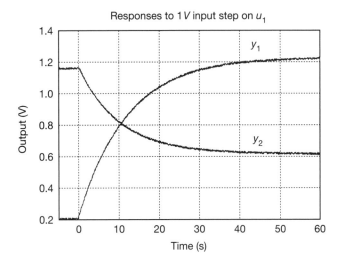

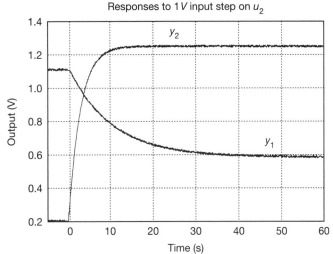

Figure 10.2 Step responses for the system of Figure 10.1.

The curves appear to represent first-order responses but, as noted in Section 3.8, this need not necessarily be the case. A rapid analysis of the system suggests the following:

- The four pressure vessels are simple closed cylinders, fitted with flow restrictors in their inlets. They should therefore be adequately representable by simple (first-order) lags, based on a resistance-plus-capacitance type of lumped-parameter model.

- E/P force-balance transducers tend to be adequately represented by very 'fast', underdamped, second- or third-order models. In this case, these dynamics should be completely 'swamped' by the relatively slow dynamics of the pressure vessels.

- The volume boosters, computing relays and pressure transducers will all have their own dynamics but, in every case, these should be very fast compared with the components discussed above.

- There is nothing to suggest non-minimum-phase behaviour, or significant time delays.

These considerations suggest that first-order models for the four responses may, indeed, be adequate. We shall try fitting such models and, if the fit is good, we shall proceed. Using the methods of Section 3.8 to fit four simple first-order transfer function models to these curves gives the following results:

$$\frac{Y_1}{U_1}(s) = \frac{1.02}{11.76s + 1} = g_{11}(s) \ \text{(say)} \qquad \frac{Y_1}{U_2}(s) = \frac{-0.52}{10.1s + 1} = g_{12}(s) \ \text{(say)}$$

$$\frac{Y_2}{U_1}(s) = \frac{-0.54}{10.4s + 1} = g_{21}(s) \ \text{(say)} \qquad \frac{Y_2}{U_2}(s) = \frac{1.04}{2.6s + 1} = g_{22}(s) \ \text{(say)}$$

$$(10.1)$$

Thus, in Equation (10.1), the transfer function $g_{21}(s)$ represents the relationship between output 2 (y_2) and input 1 (u_1), and refers to the 'falling' trace in the upper half of Figure 10.2. The other three transfer functions can similarly be related to the response curves, the ordering of the subscripts of $g_{jk}(s)$ always specifying the transfer function between output j and input k (see also Section A1.1.1). When these four transfer functions are arranged in a matrix, we obtain $G(s)$, the *transfer function matrix* (TFM) of the system:

$$\begin{bmatrix} g_{11}(s) & g_{12}(s) \\ g_{21}(s) & g_{22}(s) \end{bmatrix} = G(s) = \begin{bmatrix} \dfrac{1.02}{11.76s + 1} & \dfrac{-0.52}{10.1s + 1} \\ \dfrac{-0.54}{10.4s + 1} & \dfrac{1.04}{2.6s + 1} \end{bmatrix} \qquad (10.2)$$

With additional definitions $u(s) = [U_1(s) \ \ U_2(s)]^{\mathrm{T}}$ and $y(s) = [Y_1(s) \ \ Y_2(s)]^{\mathrm{T}}$, we also have:

$$y(s) = G(s)u(s) \qquad (10.3)$$

Equation (10.3) is a general TFM description of any linear multivariable system. $y(s)$ will have as many rows as the system has outputs, $u(s)$ will have as many rows as the system has inputs and the size of $G(s)$ will be (number of outputs) × (number of inputs), in common with any other matrix in a block diagram (Appendix 1). Equation (10.3) suggests a general block diagram representation as shown in Figure 10.3(a), and for a two-input-two-output system such as ours, the expansion of Figure 10.3(a) yields Figure 10.3(b). The heavy lines in Figure 10.3(a) denote vector (multivariable) signal paths, as in our previous state-space work. Note again that the ordering of an element's subscripts specifies the output then the input of $G(s)$ which the element connects internally. All contributions to each output are simply summed. This can be seen to be true from Figure 10.3(b), and from a mathematical viewpoint too: Equation (10.3) for our system can be written:

$$\begin{bmatrix} Y_1(s) \\ Y_2(s) \end{bmatrix} = \begin{bmatrix} g_{11}(s) & g_{12}(s) \\ g_{21}(s) & g_{22}(s) \end{bmatrix} \begin{bmatrix} U_1(s) \\ U_2(s) \end{bmatrix}$$

that is,

$$Y_1(s) = g_{11}(s)U_1(s) + g_{12}(s)U_2(s) \quad \text{and}$$

$$Y_2(s) = g_{21}(s)U_1(s) + g_{22}(s)U_2(s) \qquad (10.4)$$

(a)

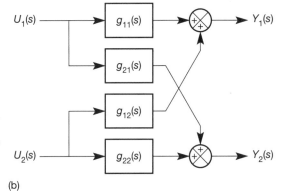

(b)

Figure 10.3 (a) The general transfer function matrix description. (b) The transfer function matrix description expanded for a 2×2 system.

which can be seen to be in agreement with Figure 10.3(b). Substituting the individual transfer function elements from Equation (10.2) into Figure 10.3(b) would give a block diagram model of the system of Figure 10.1.

Later, we shall see that the TFM model of Equation (10.2) is a fairly accurate description of this system. However, in addition to the approximations made above, it is also worth mentioning the issue of linearity. In order to minimize the effects of the noise, we made relatively large changes to the system inputs with respect to normal operating changes (the total input range is 10 V, so our change is 10 per cent). At the same time, we have been careful to keep the signals small enough to avoid any nonlinear effects such as saturation. We have then modelled the system from the resulting responses. This has given us a linear model of the plant. However, this linear model is only valid for small changes around whatever operating point we used in our tests. From Figure 10.2 it can be seen that we used two separate operating points: $y_1 = 0.2$ V and $y_2 = 1.15$ V for the elements in column 1 of the TFM (from the upper half of Figure 10.2); and $y_1 = 1.1$ V and $y_2 = 0.2$ V for the elements in column 2 of the TFM (from the lower half of Figure 10.2). In practice, if we apply control methods which call for inputs much larger than the ones we used in the tests, the system will become nonlinear (largely due to saturation of the E-to-P converters, which can only produce a certain maximum output pressure, no matter how hard we drive them). For methods of handling nonlinear systems, see Chapter 14.

Note that there are other possible model structures for our system. Figure 10.3(b) has arisen naturally because of the way we tested the system and derived our model. It is sometimes called a *p-canonical form* of model. There is also, for example, the so-called *v-canonical form*, shown in Figure 10.4. Since the input and output signals in Figures 10.3(b) and 10.4 are identical (if they model the same plant), it is easy to write down the equations of Figure 10.4, compare them with Equation (10.4), and solve for the unknown transfer function blocks in Figure 10.4 in terms of the known ones of Figure 10.3(b). In this way, the transformation from

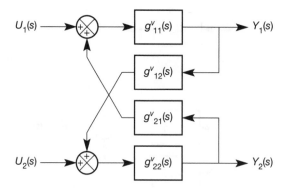

Figure 10.4 The v-canonical form of Figure 10.3(b).

one form to the other will be found (see Problem 10.1). To avoid possible confusion, we shall never use this second form – we shall standardize on that of Figure 10.3(b).

We shall also implicitly assume from now on that any system we consider is 'square' in the sense that the number of outputs and the number of inputs are equal, thus giving a square TFM. This will often be the case in practice. If it is not, then (for cases having an excess of outputs) either some outputs must be left uncontrolled at this stage or extra inputs must be found which are independent of the existing ones (usually difficult); or (for systems with an excess of inputs) some inputs may not be used or new independent outputs may be defined. In order to help decide which inputs or outputs to ignore, and also which input may best be paired with which output for control purposes, techniques such as the relative gain array may be helpful (due to Bristol – described in Maciejowski (1989)).

10.3.1 Relationship with state-space models

Any linear system which can be modelled as a TFM can also have an infinite number of state-space descriptions (see Section 2.5.1). Conversely, any multivariable state-space model can be converted to a TFM model. The conversion is easily accomplished via the Laplace transform.

Consider the standard state-space model

$$\dot{x}(t) = Ax(t) + Bu(t)$$
$$y(t) = Cx(t) + Du(t)$$

Taking Laplace transforms in the state equation and setting initial conditions to zero (because TFMs, like all Laplace transfer functions, are only defined for zero initial conditions) gives:

$$sx(s) = Ax(s) + Bu(s) \quad \text{or} \quad [sI - A]x(s) = Bu(s)$$

$$\text{or} \quad x(s) = [sI - A]^{-1}Bu(s)$$

Substituting this into the Laplace transformed version of the output equation, we obtain:

$$y(s) = C[sI - A]^{-1}Bu(s) + Du(s)$$

Also, we know that $y(s) = G(s)u(s)$ where $G(s)$ is the TFM, and is therefore related to the state-space description by

$$G(s) = C[sI - A]^{-1}B + D$$

10.4 Feedback control of multivariable systems – an intuitive approach

The approach to be developed in this section works for our pneumatic system, but is not of much practical use in general – for reasons we shall discover. However, it does give a motivation for study of the better methods and introduces some of the basic ideas behind these. It is based upon our previous knowledge of frequency-domain control, which suggests that a suitable approach might be to place some kind of compensator in front of the plant, and close unity negative feedback loops (UNF) around the forward path thus formed. The arrangement we are suggesting is shown in Figure 10.5.

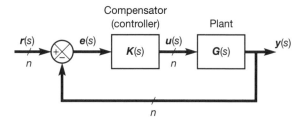

Figure 10.5 A multi-variable compensator in a unity negative feedback arrangement.

In Figure 10.5, we assume that every signal path in the system carries n signals. The plant has n inputs and outputs, so $G(s)$ is an $n \times n$ TFM, and the controller must therefore also be an $n \times n$ TFM. How do we design the controller TFM $K(s)$?

The single problem that makes the design of controllers for multivariable systems different from that for SISO controllers is that of interaction. In general, applying a control signal to a single input in an effort to obtain certain output behaviour will cause responses at more than one output, such as those we saw in Figure 10.2. This makes the controller design a very difficult proposition, as it is hard to predict what control action we would need to apply to several inputs simultaneously, in order to get just one output to behave as we wish. To get several outputs to behave as we wish is, of course, even more of a problem.

The general approach of all multivariable design methods is therefore to remove (or at least greatly to reduce) the effects of such interaction. Ideally, this will leave us with a system in which any given output will respond to only one input. It will then be possible to apply any desired SISO design technique to each of the independent input–output pairs thus formed. For example, in Figure 10.3(b), if we could somehow remove the effects of $g_{21}(s)$ and $g_{12}(s)$, we would be left with only the two SISO paths via $g_{11}(s)$ and $g_{22}(s)$. We could then use Bode, Nyquist, Nichols or root locus analysis to design compensators for these two independent loops in the usual way, one at a time; or even just close UNF loops around the plant and apply PID control (Section 4.5.2) empirically, tuning for acceptable performance.

Another thing to note about this scenario is that with the effects of $g_{21}(s)$ and $g_{12}(s)$ gone, the remaining system has a *diagonal* TFM ($g_{21}(s)$ and $g_{12}(s)$ set to zero in Equation (10.2)). The converse is true – any system with a diagonal TFM will suffer no interactions and is, effectively, a set of independent SISO paths. We therefore initially seek to make the overall TFM of the system (including whatever compensation we add to it) diagonal. After that, we can, in principle, tune the performance with additional single-loop controllers as usual.

If the inverse of the TFM $G(s)$ were to exist, then setting the compensator $K(s)$ in Figure 10.5 to be $G(s)^{-1}$ would give a forward path:

$$y(s) = G(s)K(s)e(s) = G(s)G^{-1}(s)e(s) = Ie(s) = e(s) \qquad (10.5)$$

(see Section A1.1.2 to confirm that the order of multiplication of the terms in a series TFM is always *against the signal flow*).

Equation (10.5) implies a system not only with no interaction, but also with no dynamics! This is strictly unachievable, but we shall proceed with the idea. Having a system which is now conceptually just a set of direct connections from each input to the corresponding output, we could add another compensator and UNF loops to provide the required closed-loop dynamics. We have therefore effectively split $K(s)$ of Figure 10.5 into two parts as shown in Figure 10.6. The first part, $K_p(s)$, is the precompensator containing $G^{-1}(s)$; the second part, $K_d(s)$, will be chosen to provide the required dynamic performance. An example will clarify this.

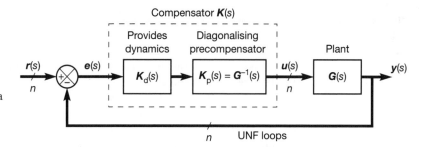

Figure 10.6 Structure of a controller including a diagonalizing precompensator.

Example 10.2 *The pneumatic system with a diagonalizing precompensator*

We consider the pneumatic system of Figure 10.1, whose model is given by Equation (10.2) and Figure 10.3(b), and we shall design for it a control scheme of the form of Figure 10.6. In line with our idea of diagonalizing the compensated plant TFM, we firstly choose the precompensator $K_p(s) = G^{-1}(s)$. This is not a pleasant quantity to evaluate, even for a system as simple as ours. It also has other drawbacks, which we discuss after the example. Using the method of adjoint matrix divided by determinant to calculate the inverse (see Section A1.1.3) generates a sixth-order common denominator polynomial, and seventh-order terms in the remaining numerator polynomial matrix. However, these are all found to have four common roots which, when cancelled out, leave the second-order common denominator and third-order numerator terms as follows:

$$K_p(s) = G^{-1}(s) = \begin{bmatrix} \dfrac{1.02}{11.76s+1} & \dfrac{-0.52}{10.1s+1} \\[2ex] \dfrac{-0.54}{10.4s+1} & \dfrac{1.04}{2.6s+1} \end{bmatrix}^{-1}$$

$$\approx \frac{1}{102.84s^2 + 17.71s + 0.78} \begin{bmatrix} k_{11}(s) & k_{12}(s) \\ k_{21}(s) & k_{22}(s) \end{bmatrix} \tag{10.6}$$

with:

$$k_{11}(s) = 1284.7s^3 + 360.0s^2 + 33.6s + 1.04$$

$$k_{12}(s) = 165.4s^3 + 93.6s^2 + 12.9s + 0.52$$

$$k_{21}(s) = 166.8s^3 + 94.8s^2 + 13.2s + 0.54$$

$$k_{22}(s) = 278.6s^3 + 161.5s^2 + 23.6s + 1.02$$

Since this precompensator should completely diagonalize the plant, such that $G(s)K_p(s) = I$, the system with $K_p(s)$ in place now appears to have vanished, as shown in Figure 10.7. Furthermore, since we now have two completely separate SISO loops, the remaining matrix $K_d(s)$ must be diagonal. Elements $k_{d_{12}}(s)$ and $k_{d_{21}}(s)$ are therefore zero and do not appear in Figure 10.7.

To calculate the contents of the remaining elements of $K_d(s)$, we need to specify the *open-loop* transfer functions which will lead to the required closed-loop performance in each loop. The procedure for each element is identical. Considering $k_{d_{11}}(s)$, we assume that the required transfer function will be a rational polynomial in s, that is,

$$k_{d_{11}}(s) = \frac{kn(s)}{kd(s)}$$

In closed-loop, we shall then obtain:

$$\frac{Y_1}{R_1}(s) = \frac{k_{d_{11}}(s)}{1 + k_{d_{11}}(s)} = \frac{kn(s)}{kd(s) + kn(s)}$$

This can be compared with the required closed-loop transfer function for loop 1, to fix the elements $kn(s)$ and $kd(s)$.

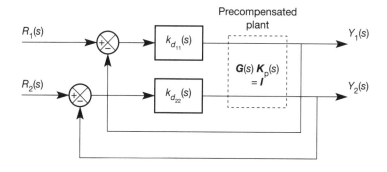

Figure 10.7 A 2×2 system with diagonalizing precompensator in place.

Let us arbitrarily choose to specify for each loop a closed-loop step response having second-order characteristics with $\omega_n = 1$ rad s^{-1} and $\zeta = 0.5$ (corresponding to about 1.6 s 10–90 per cent rise time, and 15 per cent overshoot, from the standard curves in Figure 3.19). The closed-loop transfer function for each loop must therefore be

$$\frac{Y}{R}(s) = \frac{\omega_n^2}{s^2 + 2\zeta\omega_n s + \omega_n^2} = \frac{1}{s^2 + s + 1} = \frac{kn(s)}{kd(s) + kn(s)}$$

from which we see that $kn = 1$ and $kd(s) = s^2 + s$. The required $K_d(s)$ is therefore

$$K_d(s) = \begin{bmatrix} \dfrac{1}{s(s+1)} & 0 \\ 0 & \dfrac{1}{s(s+1)} \end{bmatrix}$$

and the overall compensator is:

$$K_p(s)K_d(s) = \frac{1}{102.84s^2 + 17.71s + 0.78} \begin{bmatrix} k_{11}(s) & k_{12}(s) \\ k_{21}(s) & k_{22}(s) \end{bmatrix} \begin{bmatrix} \dfrac{1}{s(s+1)} & 0 \\ 0 & \dfrac{1}{s(s+1)} \end{bmatrix}$$

$$= \frac{1}{102.84s^4 + 120.55s^3 + 18.49s^2 + 0.78s} \begin{bmatrix} k_{11}(s) & k_{12}(s) \\ k_{21}(s) & k_{22}(s) \end{bmatrix} \quad (10.7)$$

where $k_{11}(s)$ and so on remain as given in Equation (10.6).

Figure 10.8(a) shows the step responses of the open-loop plant, according to the model of Equation (10.2). The agreement with the plant traces of Figure 10.2 is seen to be good. Figure 10.8(b) shows the responses with the compensator of Equation (10.7) placed in front of the plant model, and UNF loops closed around it. The plant input signals u_1 and u_2 are also shown. It can be seen that the desired output responses are achieved, apparently with no interaction. The input signal u_1 moves rather excessively in response to the input on r_1, but is within the plant constraints. However, since the acceptable signal ranges are only 0–10 V, the plant input would saturate if a much greater step were applied, or if the 1 V step was applied to r_1 with an existing input signal level greater than about 2 V on u_1. We are therefore quite close to the plant limits for 1 V input steps (for smaller steps, or smoothly changing signals, the situation would be better).

Note that Figure 10.8 was created by the MATLAB m-file *fig10_8.m* on the accompanying disk. The actual plots are produced by a command *mv2step*. This is not a standard MATLAB command, but is the custom-written m-file *mv2step.m* on the accompanying disk. This takes a state-space model of a two-input-two-output system in the variables a, b, c and d, together with a vector of time values in t, and directly produces plots such as Figure 10.8(a).

The application of our intuitive method in Example 10.2 was fairly successful. It also works reasonably well when applied to the real plant. However, it is not generally of much use for the following reasons:

- It relies upon the controller containing the inverse of the plant TFM, $G^{-1}(s)$. Firstly, this inverse may not exist. Secondly, if it does exist, it will almost always

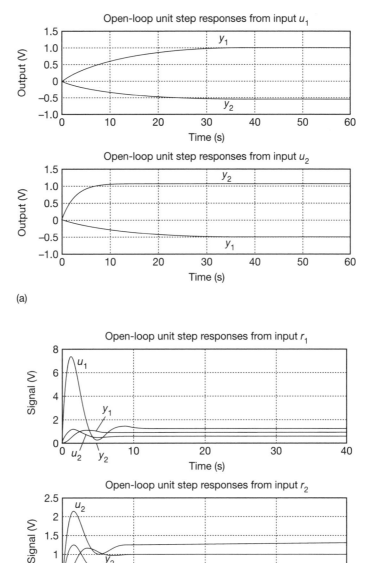

Figure 10.8 (a) Open-loop model step responses. (b) Step responses with diagonalizing controller.

lead to a compensator which cannot be built. For example, Equation (10.6) represents four compensator elements each having numerator order 3 and denominator order 2, and hence being unrealizable (all real systems must have more poles than finite zeros, or at least an equal number to a first approximation). We unwittingly overcame this in Example 10.2 by designing the other part of the compensator so that it had a sufficient excess of poles over zeros to correct this deficiency by the time we reached Equation (10.7). For high-order plants, this would be undesirable.

- The order of compensators designed by this method will always be high. This is because the compensator must firstly contain the inverse of all the open-loop plant dynamics, and secondly add all the required closed-loop dynamics. High-order compensators are harder to implement, and less likely to be robust than lower-order ones.

- Even if cancellation of the plant dynamics is apparently perfect (as in Example 10.2), plant performance will not be as expected since $G(s)$ will only be an approximate model of the plant, and 'cancellation' of the real plant's dynamics will therefore only be approximate. The uncancelled parts of the plant's dynamics will cause some interaction to return.

- We have made no assessment of the stability of such a control scheme. As we shall now discover, this is an area where we must tread rather carefully, and intuition is of little help.

10.5   Stability of multivariable feedback systems

The purpose of this section is not to give a rigorous treatment of this subject, which is actually rather difficult. Here we seek to illustrate that stability problems which we would be unlikely to predict from our knowledge of SISO system behaviour can easily arise in multivariable systems. The message is always to expect trouble!

10.5.1   Gain space analysis and the McMillan form of G(s)

In this section, we make use of two examples from Patel and Munro (1982), which also appear in O'Reilly (1987). However, in the present section the descriptive text has been rewritten and expanded.

Consider an open-loop TFM given by

$$
G(s) = \begin{bmatrix} \dfrac{1}{s+1} & \dfrac{2}{s+3} \\[2mm] \dfrac{1}{s+1} & \dfrac{1}{s+1} \end{bmatrix}
\tag{10.8}
$$

and let us arrange this in a very simple system, comprising a set of closed loops with an adjustable negative feedback gain in each loop. This is shown in Figure 10.9, where F is a diagonal matrix of constant feedback gains; that is, $F = \text{diag}\{f_{11}, f_{22}\}$ in this case, or $F = \text{diag}\{f_{11}, f_{22}, \ldots, f_{nn}\}$ in the general case.

By choice of F, we can obtain any desired loop gains and any combinations of open or closed loops (simply setting $f_{ii} = 0$ makes the ith loop open).

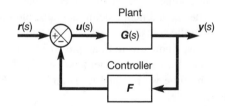

Figure 10.9 Simple constant feedback multivariable controller.

If we had a SISO system comprising a constant negative-feedback gain around a first-order lag, we could guarantee that the closed-loop system would always be asymptotically stable because the closed-loop transfer function would just be a different first-order lag. Our multivariable system is just two such loops, with first-order cross-coupling. We would probably not intuitively expect any problems.

We can define the closed-loop TFM for the system of Figure 10.9 as $H(s)$, such that

$$y(s) = H(s)r(s)$$

By analogy with the equivalent SISO system (but taking care over the order of the matrix multiplications), Problem 10.2 shows that

$$H(s) = [I + G(s)F]^{-1}G(s) \tag{10.9}$$

Returning to our example, if we make $f_{11} = 50$ and $f_{22} = 0$, then $H(s)$ has all its poles in the left-hand half of the s-plane and the system is asymptotically stable. However, if we set $f_{11} = f_{22} = 10$, we find that the system is unstable in closed-loop. It turns out that it is impossible to set up two high-gain loops around this seemingly simple system, without causing instability. This can be shown clearly in a graphical manner, by plotting the stable region in the *gain space*. This is an n-dimensional space containing points defined by the elements on the diagonal of F. Thus, in our 2×2 case, the gain space is two-dimensional (that is, a plane), and is effectively just a plot of f_{22} vs. f_{11}. The first quadrant of the gain space, where all the (diagonal) elements of F are greater than zero, corresponds to negative feedback in all loops (from Figure 10.9).

Figure 10.10 shows the gain space for the system of Equation (10.8) and Figure 10.9, with the approximate regions of stability as the feedback gains vary. The figure suggests that, for stability, we can have a high gain in either loop, but not

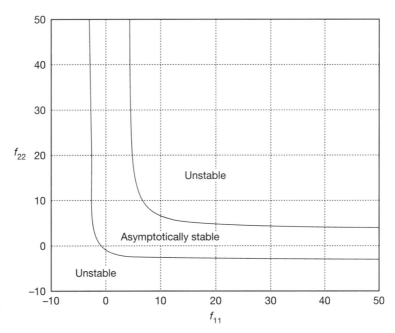

Figure 10.10 Gain space for the system of Equation (10.8) and Figure 10.9.

in both loops simultaneously. However, the system has high integrity in the sense that if one feedback loop is broken (f_{11} or f_{22} falls to zero), the system remains stable.

What has happened to cause this unexpected behaviour is that, due to the interaction, the poles and zeros of the closed-loop system are not what they might be expected to be from an inspection of the open-loop TFM. In SISO systems, we expect the closed-loop zeros to be the same as the open-loop ones, and the closed-loop poles to move. In MIMO systems, cancellations and appearances of poles and zeros often occur due to the interactions.

Unfortunately, even for a system as simple as this one, it is not easy to visualize, in physical terms, what is happening. If the equations of the feedback system are written down (neglecting the s-dependencies for clarity) they are

$$y = Hr, \quad \text{or} \quad y_1 = h_{11}r_1 + h_{12}r_2 \quad \text{and} \quad y_2 = h_{21}r_1 + h_{22}r_2$$

where $H = [I + GF]^{-1}G$ (Equation (10.9)).

The presence of the feedback loop has caused the appearance of the $[I + GF]^{-1}$ term multiplying G in Equation (10.9). Performing the algebra will show that several zeros appear in the closed-loop TFM, H, whereas there were none in the open-loop TFM, G. Also, the poles of H will differ, at least in number and multiplicity, from those of G. In this particular example, every element of H has a third-order denominator, h_{11} has a second-order numerator, h_{12} has a first-order numerator while h_{21} and h_{22} both have third-order numerators.

In general, these extra zeros frequently arise in multivariable feedback systems, but although their behaviour is important, it is beyond the scope of this text. Patel and Munro (1982) devote an entire chapter to the poles and zeros of multivariable systems.

By now, readers should be convinced at least that multivariable systems require a more thoughtful approach than might at first seem to be the case. In particular, the obvious poles and zeros of the open-loop TFM are usually not those which really determine the system's behaviour in closed-loop.

In order to be able to predict something about closed-loop performance from the open-loop TFM (as is our wont in SISO analysis), we need to be able to see the *effective* 'poles' and 'zeros' of the open-loop TFM, with a view to multivariable feedback control, rather than the obvious poles and zeros of its individual elements.

Although it is not the recommended method from a numerical computational viewpoint, one method is to reduce $G(s)$ to its so-called *McMillan form*. The 'poles' and 'zeros' of $G(s)$ from a multivariable viewpoint are then all the obvious poles and zeros of the McMillan form. Obtaining the McMillan form is not difficult and a derivation of the following result can be found in Section A1.5. The McMillan form of $G(s)$ as given in Equation (10.8) is (Section A1.5):

$$M(s) = \begin{bmatrix} \dfrac{1}{(s+1)(s+3)} & 0 \\ 0 & \dfrac{s-1}{s+1} \end{bmatrix}$$

Thus, from a multivariable feedback viewpoint, $G(s)$ effectively has two poles at $s = -1$, one pole at $s = -3$ and one zero at $s = +1$. Note that the numerical values of the poles in a McMillan form are the same as those of the corresponding

$G(s)$, but that they differ in number and grouping. The zeros in a McMillan form are the *transmission zeros* of $G(s)$. It is this zero at $s = +1$ that causes the trouble. In effect, since it is a right half-plane zero, cross-multiplications in the evaluation of $H(s)$ can produce a right half-plane (that is, unstable) pole under the appropriate conditions.

For systems in general, if the McMillan form of $G(s)$ contains any right half-plane zeros, then it will not be possible to use a full set of high-gain loops around the system. This would have allowed us to predict, at least qualitatively, the kind of result found in Figure 10.10.

As another example of the strange things that can happen, consider the same arrangement (Figure 10.9), but this time with the open-loop TFM:

$$G(s) = \begin{bmatrix} \dfrac{s-1}{(s+1)^2} & \dfrac{5s+1}{(s+1)^2} \\ \dfrac{-1}{(s+1)^2} & \dfrac{s-1}{(s+1)^2} \end{bmatrix} \tag{10.10}$$

The gain space for this system, together with an approximation to the regions of stability, is shown in Figure 10.11.

This time, we note the following features:

- For most values in F, increasing gain in one loop allows increasing gain in the other.

- The system is of low integrity, since failure of either feedback loop will make the system unstable (assuming a reasonable level of gain in the other loop).

- The system can be stable with small amounts of *positive* feedback.

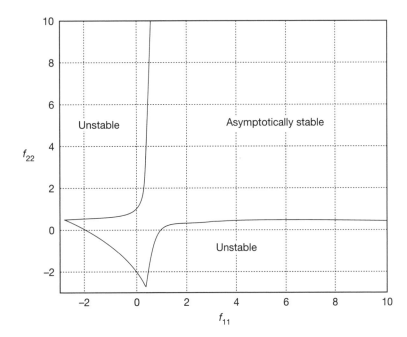

Figure 10.11 Gain space for the system of Equation (10.10) and Figure 10.9.

The McMillan form of this $G(s)$ is

$$M(s) = \begin{bmatrix} \dfrac{1}{(s+1)^2} & 0 \\ 0 & \dfrac{s+2}{s+1} \end{bmatrix}$$

from which we see that, despite the non-minimum-phase terms in the open-loop TFM (Equation (10.10)), there will be no non-minimum-phase behaviour with the feedback loops closed. Note also the counter-intuitive possibility of stability with either positive or negative feedback.

10.5.2 The basis of a Nyquist-type stability criterion for multivariable systems

The familiar Nyquist stability criterion for SISO systems can be extended to the general (multivariable) case. The mathematics of this extension become somewhat involved, so we shall not delve too deeply. This section simply introduces the background which allows this extension. Further details will be found in the following section on the characteristic locus method, and in the multivariable system references such as Maciejowski (1989).

We have already seen, in Figure 10.5, one possible arrangement of a multivariable control scheme. Let us generalize this further by regarding the compensator matrix $K(s)$ as representing the cascaded effects of several separate compensator matrices (there were two in Example 10.2, and there will be several in later examples). The transfer function of the entire forward path (the original system plus all its compensators) is then given by $G(s)K(s)$. Let this be represented by

$$Q(s) = G(s)K(s) \tag{10.11}$$

giving the notation of Figure 10.12(a) if unity negative feedback (UNF) is assumed. In the inverse Nyquist array technique (Section 10.7), the input–output behaviour of the system represented by $Q(s)$ is considered as it stands. In the characteristic locus method (Section 10.6), a more 'algebraic' approach is taken, which works with what are effectively the eigenvalues and eigenvectors of $Q(s)$ (which will, of course, be frequency-dependent quantities). To introduce the structures used in the CL method, and to provide the basis for the stability tests used in both methods, some further analysis of the system of Figure 10.12(a) is required. It will be necessary to use the similarity transform and spectral decomposition in this analysis, so Sections A1.6 and A1.7 should be read if this is unfamiliar.

The open-loop TFM $Q(s)$ in Equation (10.11) (in common with any other square matrix) can be regarded as the result of applying a similarity transform (see Section A1.6) to a diagonal matrix containing its eigenvalues (an example is given in Section A1.7). The eigenvalues of $Q(s)$ will be frequency-dependent, and are known as the *characteristic values* of $Q(s)$. If we let $q_1(s)$, $q_2(s)$, $\ldots$, $q_n(s)$ represent these characteristic values of $Q(s)$, then the appropriate diagonal eigenvalue matrix is

$$\mathrm{diag}\{q_1(s), q_2(s), \ldots, q_n(s)\}$$

Combined open-loop TMF
of plant and all compensators

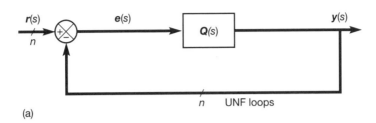

(a)

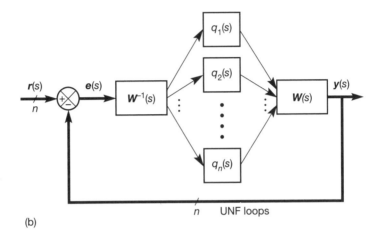

Figure 10.12 (a) Multi-variable controller in a unity negative feedback system. (b) Expanded forward path for Figure 10.12(a).

(b)

and the similarity transform is given by

$$Q(s) = W(s)\text{diag}\{q_1(s), q_2(s), \dots, q_n(s)\}W^{-1}(s) \tag{10.12}$$

where $W(s)$ is the appropriate eigenvector matrix (or modal matrix, that is, the matrix whose columns are the eigenvectors of $Q(s)$ corresponding to each characteristic value). Equation (10.12) now suggests a forward path for Figure 10.12(a) as shown in Figure 10.12(b).

This looks more complicated than it did before, but the similarity transform has some useful properties. One is that the eigenvalues of the forward paths of the systems of Figures 10.12(a) and (b) are identical. The modes of response are therefore unchanged – it is a *similar* system in the mathematical sense.

More importantly for our present interests, there is no interaction between the blocks $q_1(s)$, $q_2(s)$, and so on in the forward path of Figure 10.12(b); and it transpires that *the dynamic behaviour of these blocks alone is sufficient to assess the stability of the entire system*. So, by considering a set of SISO Nyquist-type criteria for these independent blocks, we can, in principle, investigate the closed-loop stability of the multivariable system.

10.6 The characteristic locus (CL) method

10.6.1 A multivariable Nyquist criterion and non-interacting control

As we saw in the previous section, the forward path eigenvalues of the system of Figure 10.12(a) $(q_1(s), q_2(s), \ldots, q_n(s))$ are called the system's *characteristic values*. The SISO Nyquist criterion is generalized to apply to these values as follows (but, in practice, CACSD packages such as MATLAB do all the hard work).

As the usual Nyquist D-contour (Section 4.3.1) in the complex plane is traversed once in a clockwise direction, each characteristic value (being a function of s) will trace out a Nyquist plot in the s-plane (which is called a *characteristic locus*). Furthermore, the forward path characteristic polynomial, given by $\det[s.I - Q(s)]$, will also trace out a locus in the s-plane.

Subject to constraints in certain unusual cases, the net number of clockwise encirclements of the critical $(-1, 0)$ point made by the locus of the *open-loop* characteristic polynomial, $\det[s.I - Q(s)]$, is equal to the number of unstable open-loop poles of the system. Let this number of encirclements be called n_0.

Also, if the characteristic locus due to $q_i(s)$ makes a net number n_i of clockwise encirclements of the $(-1, 0)$ point, then the total number of such encirclements (n_t) made by all the characteristic loci will be given by $n_t = n_1 + n_2 + \ldots + n_n$.

The Nyquist criterion from the SISO case tells us that the net number of clockwise encirclements of the $(-1, 0)$ point due to the *closed-loop* characteristic polynomial is equal to the number of poles of the closed-loop system in the right-hand half of the s-plane. This must be zero for closed-loop stability, and is given by:

$$\begin{bmatrix} \text{net number of closed-} \\ \text{loop clockwise} \\ \text{encirclements} \end{bmatrix} = \begin{bmatrix} \text{total number} \\ \text{of clockwise} \\ \text{encirclements} \end{bmatrix} + \begin{bmatrix} \text{number of} \\ \text{unstable open-} \\ \text{loop poles} \end{bmatrix}$$

The generalization of this to the multivariable case, using the numbers of encirclements defined previously, is (Owens, 1978): $0 = n_0 + n_t$ or $-n_t = n_0$.

In words, 'for closed-loop stability, the total number of *anti*clockwise encirclements of the $(-1, 0)$ point by all the characteristic loci must balance the number of unstable open-loop poles'. Note that the anticlockwise direction is implied by the negative sign of n_t.

Returning now to our initial concern for non-interacting control, if a UNF loop were to be placed just around the ith characteristic value $q_i(s)$ in the forward path of Figure 10.12(b), the resulting transfer function for the ith path between the $W^{-1}(s)$ and $W(s)$ matrices would be given by

$$h_i(s) = \frac{q_i(s)}{1 + q_i(s)} \tag{10.13}$$

After some mathematics, it can be shown (see, for example, Owens (1978)) that the *closed-loop* TFM for the entire system of Figure 10.12, $H(s)$ say, can be written

$$H(s) = W(s).\text{diag}\{h_1(s), h_2(s), \ldots, h_n(s)\}.W^{-1}(s) \tag{10.14}$$

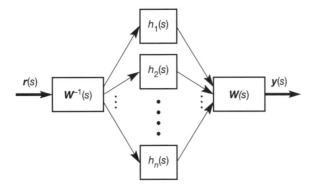

Figure 10.13 Equivalent form (closed loop) for Figure 10.12.

which tells us that Figure 10.12(b) can therefore be simplified to appear as shown in Figure 10.13. Non-interacting control may be achieved in a number of ways with reference to Figure 10.13.

Firstly, the only interaction occurs in the matrix $W(s)$ and its inverse. If we could arrange for $W(s)$ to be diagonal at all frequencies of interest ($W^{-1}(s)$ would then also be diagonal), there would be no interaction. Looking back at Figure 10.5, and remembering that $W(s)$ contains the eigenvectors of $Q(s)$ which, in turn, is equal to $G(s)K(s)$, it seems that $K(s)$ ought to be able to be chosen such that $W(s)$ is given the appropriate properties. In practice, unfortunately, this cannot usually be done for two reasons:

- $W(s)$ will generally vary in a complicated way with frequency.
- It is not understood how the eigenvectors of a matrix product vary with those of either component, so we do not know how to choose $K(s)$ to get the desired result.

We therefore use an approximation to this procedure – described in Section 10.6.2.

Secondly, we might remove interaction by ensuring that all the characteristic loci have the same gain and phase behaviour over the frequency range of interest. The reasoning for this is as follows. If $q_1(s), q_2(s), \ldots, q_n(s)$ all have the same gain and phase behaviour, then it must be the case that $q_1(s) = q_2(s) = q_3(s)$ and so on. This then implies that $h_1(s) = h_2(s) = h_3(s)$ and so on (Equation (10.13)). Thus, $\text{diag}\{h_1(s), h_2(s), \ldots, h_n(s)\} = h_1(s)I$. Substituting into Equation (10.14) gives the closed-loop TFM $H(s) = W(s)h_1(s)IW^{-1}(s) = Ih_1(s)$, which is non-interacting. Again, in practice, it is unrealistic to seek to achieve this accurately for normal systems (for similar reasons to those above). However, approximations are again useful as discussed in Section 10.6.2 below.

Thirdly, the use of high compensator gains in $K(s)$ can help to achieve the same effect. The reason for this is as follows. High compensator gains mean large values in $K(s)$, which imply large values in $Q(s) = G(s)K(s)$ and therefore large characteristic values. If the magnitude of $q_i(s) \gg 1$, then the magnitude of $h_i(s) \approx 1$ (Equation (10.13)). If this is true for all i, it follows that the diagonal matrix $\text{diag}\{h_1(s), h_2(s), \ldots, h_n(s)\} = I$ and so, from Equation (10.14), $H(s) = I$. This apparently has the added attractions that not only would there be no closed-loop interaction, but also an instantaneous response between each input and its output, and zero steady-state errors. In practice, of course, this is unachievable, the approach being of some use, but limited by the onset of instability, as we might expect.

As a corollary of this third point, steady-state errors to step inputs will be small only if the magnitude of $q_i(0) \gg 1$, $1 \leq i \leq n$ (that is, the d.c. gain in each loop is high), so that $H(0) = I$.

10.6.2 Design steps in the CL method

In all the preceding discussion, we have talked of $Q(s)$ and its characteristic values as if we knew what they were – that is, we knew the contents of the compensator $K(s)$ which we have not yet designed. This is not a problem, because the design method is iterative. We try a compensator, see its effect, and then alter it, or add further compensators until we are satisfied (in the accustomed frequency-domain control design manner). The effort involved in attempting to apply this procedure by hand is far too great. It is essential to use a CACSD package to handle the maths and present the graphical results, leaving us to concentrate on the design decisions.

A typical CL design sequence suitable for use with MATLAB (plus its Control Systems and Multivariable Frequency Domain Toolboxes, CSTB and MVFDTB respectively) is given below. The design procedure is first presented in outline for clarity, then some explanation of the steps follows. Finally a case-study example is worked. It should be noted that there is no need to stick to the design methodology presented here; the CACSD tools can be applied in any imaginative way we wish. However, the procedure outlined here has been found to work well.

It transpires that the three suggestions made in Section 10.6.1 as to the removal of interaction map quite neatly onto high, intermediate and low frequency ranges (relative to the system dynamics). The frequency range of interest is therefore typically divided into these three ranges, and the following design steps are applied.

- Step 1: A compensator is designed to make $W(s)$ approximately diagonal at high frequencies. In fact, in the 'standard' procedure, an algorithm is used which selects W as a *constant* matrix. Let us call this matrix K_h. The compensated forward path is then given by $G(s)K_h$.

- Step 2: A second compensator is designed at an intermediate frequency. The aim of this step is to design a compensator which itself has the same *structure* as Figure 10.13 (but different matrix contents). In this compensator, the matrices $W(s)$ and $W^{-1}(s)$ are chosen to be *constant* approximations to the eigenvector (modal) matrix (and its inverse) of the previously compensated plant $G(s)K_h$. They are usually called A and B respectively (but must not be confused with the similarly named matrices used in state-space descriptions).

  Figure 10.14 shows the resulting structure of the plant with high- and mid-frequency compensators in place. The diagonal dynamic elements (let us call them $k_1(s), k_2(s), \ldots, k_n(s)$) are chosen by the usual frequency-domain methods (by performing n single-loop designs) to give the n resulting characteristic loci similar gain and phase behaviour over the intermediate frequency range, but without spoiling the previous high-frequency compensation.

- Step 3: Low-frequency compensation is then added as required. This may just be a diagonal matrix of constant gains to balance the loci at steady state. More often, it will be a diagonal set of proportional-plus-integral-type elements, designed to give unity steady-state gain to step inputs for example, but chosen so as not to

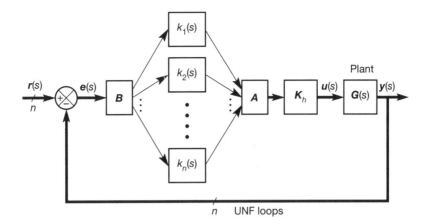

Figure 10.14 High and intermediate frequency compensators in place.

spoil the previous compensation. In some cases, a second compensator (or even more) of the form of step 2 (above) may also be employed.

10.6.3 Notes on the 'steps' of the CL method

In Step 1 above, we design a *constant* compensator. The reason for this is that the dynamic compensator required to diagonalize $W(s)$ properly is usually much too complicated, and may not be able to be found at all. We therefore choose one frequency in the 'high-frequency' range and design a compensator there which approximately (in some sense) diagonalizes $W(s)$, but contains only constant elements. We hope that the beneficial effects obtained from this compensator will spread over a useful range of frequencies to either side of the design value, but this cannot be guaranteed. Use of CACSD allows rapid checks of this, so that we can try a different design frequency if we are not satisfied.

In the MATLAB MVFDTB (and some other packages) the method used is encapsulated in an algorithm called *align* (Edmunds and Kouvaritakis, 1979). If all the $q_i(s)$ are of the same order (that is, the high-frequency phase shifts, neglecting transport lags, are identical), the *align* command in MATLAB will appear to align the CL with each other at high frequencies. However, this is not the reason for its name. Rather, it seeks to align the eigenvectors of the compensated plant with the basis vectors of the space in which they move. A simple example will show why this is a good idea.

Example 10.3 *A brief example of the eigenvectors of general vs. diagonal matrices*

By way of a brief illustration, consider the constant matrix investigated in Section A1.7, that is,

$$Q = \begin{bmatrix} 1 & -\frac{2}{3} \\ 3 & 4 \end{bmatrix}$$

This matrix was found to have eigenvectors given by

$$e_1 = \begin{bmatrix} 1 \\ -3 \end{bmatrix} \quad \text{and} \quad e_2 = \begin{bmatrix} 1 \\ -1.5 \end{bmatrix}$$

Geometrically, we can depict these in a vector space as shown in Figure 10.15(a). The eigenvectors are not aligned with the basis vectors.

On the other hand, the diagonal matrix used in Section A1.7 was

$$Q = \begin{bmatrix} 3 & 0 \\ 0 & 2 \end{bmatrix}$$

This matrix is easily shown to have eigenvectors given by

$$e_1 = \begin{bmatrix} 0 \\ 1 \end{bmatrix} \quad \text{and} \quad e_2 = \begin{bmatrix} 1 \\ 0 \end{bmatrix}$$

These are shown in Figure 10.15(b), and obviously *are* aligned with the basis vectors.

This illustrates the general point, that if a matrix's eigenvectors are aligned with the basis set, it will be a diagonal matrix. Hence using the *align* command to achieve this gives a non- interacting system at the design frequency.

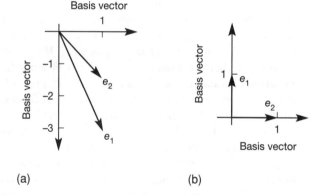

Figure 10.15 (a) Eigenvectors of a non-diagonal matrix. (b) Eigenvectors of a diagonal matrix.

(a)   (b)

One more point of importance is that since the *align* command works with eigenvectors, it works only in terms of directions, and cannot provide scaling information. In MATLAB, the consequence of this is that the compensator matrix designed by the algorithm has columns of arbitrary sign. The resulting compensated loci must therefore be inspected and, if necessary, the signs must be altered to give stable behaviour relative to the $(-1, 0)$ point (that is, the correct number of encirclements).

Further assistance is provided by MATLAB in seeing how effective the compensation of the *align* algorithm has been. The MATLAB MVFDTB command *fmisalg* allows the user to calculate and display (in a 'Bode-type' plot) the *misalignment angles* between the system eigenvectors and the basis set. This display can be used to discover over what range of frequencies compensation has been effective (because the misalignment angles should be small). Unfortunately, though, the misalignment angles are not always a very good indicator of likely performance. This is because small misalignment angles are a sufficient condition for low interaction, but not a necessary condition.

In Step 2 in the previous section, a more complicated controller is designed at an intermediate frequency. Again, in the interests of tractable calculations and simple, realizable compensators, a single frequency is chosen and the constant matrices A and B (Figure 10.14) are derived to approximate (in some sense) $W(s)$ and $W^{-1}(s)$ at this frequency. Again, we hope that any beneficial effects will spread over a usefully wide frequency range. The MATLAB MVFDTB contains an algorithm called *facc* which generates the required A and B matrices at the specified frequency. The name of the algorithm derives from the phrase 'Approximate Commutative Controller'. 'Approximate' is used because A and B are constant, but really should be functions of s. 'Commutative' is used because if a compensator is designed in this way it will commute (in the matrix-multiplication sense) with the TFM of the high-frequency compensated plant. That is, if a compensator of the *form* of Figure 10.13 is added to the high-frequency compensated plant (as in Figure 10.14), the result would be the same mathematically, if the mid-frequency compensator were to be placed *after* the plant, rather than before it. This would not be a good idea in practice, however (see Section 10.7.3). The result is easily proved by taking the eigenvalue decomposition of $Q(s)$ for the high-frequency compensated plant, and either pre-multiplying or post-multiplying by the approximate commutative controller:

$$W(s)\text{diag}\{k_1(s),\ldots,k_n(s)\}W^{-1}(s)\,.W(s)\text{diag}\{q_1(s),\ldots,q_n(s)\}W^{-1}(s)$$
$$= W(s)\text{diag}\{q_1(s),\ldots,q_n(s)\}W^{-1}(s)\,.W(s)\text{diag}\{k_1(s),\ldots,k_n(s)\}W^{-1}(s)$$

We also note that in this result, the dynamic elements of the approximately commutative controller $(k_1(s), k_2(s)$, and so on in Figure 10.14) combine with the characteristic values so that $q_1(s)k_1(s)$, $q_2(s)k_2(s)$, and so on are non-interacting. It is up to us to design these dynamic elements by the normal frequency-domain techniques, on the basis that the matrices A and B have reduced the problem to a set of independent single-loop designs. One point that must be made is that the compensators chosen must not interfere with the high-frequency compensation already carried out in Step 1. Therefore, they must be such that their high-frequency gains all tend to unity (for example, lead-lag type elements). The process is aided by CACSD functions (such as the MATLAB MVFDTB *phlag* command) for the design of such compensators.

Step 3 in the previous section is probably self-explanatory, but again care must be taken not to upset the compensator designed for intermediate frequencies. For example, if integral action is to be used, the compensator cannot take the usual simple integral form of $1/s$. Rather, a proportional + integral form such as $(s+1)/s$ might be used instead since, at relatively high frequencies, this tends to 1. Of course, the numerator coefficients will probably not all be unity. If it transpires that different P + I elements are needed in each loop then, strictly, another approximate commutative controller should be used, as in Step 2. This is to prevent any non-symmetrical cross-coupling effects reintroducing interaction.

10.6.4 Robustness of the controlled system

Being based upon frequency-domain techniques, we might expect the CL method to give more robust results than state-space controllers. Nevertheless, the compensators do involve a number of approximations, and also rely upon inverses and eigenvector calculations based on (possibly uncertain) plant models. We therefore need some means of assessing the likely robustness of the resulting controller in the face of modelling errors and so on.

Since our calculations are mainly based upon matrices, and the 'worst' things we do are to take inverses and find their eigenstructures, it might seem that so long as the matrices involved can be shown to be non-singular, then all will be well; and a few simulation studies (which any CACSD package can do) should confirm this.

Unfortunately, neither the determinant of a matrix nor its eigenvalues (both of which we use) are particularly good indicators of *how near* to singularity it may be. In other words, perhaps just a very small error in the measured or calculated plant parameters could cause the matrix calculations to succeed when, in fact, they should fail. We would then be applying a very delicate controller to our plant.

A widely quoted example considers the following simple constant matrix, in which the quantity ε is arbitrarily small:

$$\begin{bmatrix} -1 & 0 \\ \dfrac{1}{\varepsilon} & -1 \end{bmatrix}$$

This matrix has a determinant of 1 and two eigenvalues, each being -1. It would therefore appear to be non-singular, and we might have no qualms about using it (and its inverse) in a control scheme.

In fact, the matrix is *ill-conditioned*. If a disturbance (or a computer rounding error) caused element (1,2) to assume the arbitrarily small value ε, the matrix would immediately become singular.

A better guide for use in multivariable system design is to study the *singular values* (or strictly, in the s-domain, *principal gains*) of the system's return-difference matrix $T(s) = (I + \text{open-loop TFM})$.

Dropping the s-dependency, the singular values of T are given by the square roots of the eigenvalues of T^*T, where T^* is the transpose of the complex-conjugate of T. (If T is real, then $T^* = T^{\mathrm{T}}$.)

For the example above, let

$$T = \begin{bmatrix} -1 & 0 \\ \dfrac{1}{\varepsilon} & -1 \end{bmatrix}$$

Therefore

$$T^*T = \begin{bmatrix} -1 & \dfrac{1}{\varepsilon} \\ 0 & -1 \end{bmatrix} \begin{bmatrix} -1 & 0 \\ \dfrac{1}{\varepsilon} & -1 \end{bmatrix} = \begin{bmatrix} 1 + \left(\dfrac{1}{\varepsilon}\right)^2 & -\dfrac{1}{\varepsilon} \\ -\dfrac{1}{\varepsilon} & 1 \end{bmatrix}$$

The characteristic equation (CE) of this matrix is

$$\lambda^2 + \lambda\left[-2 - \left(\frac{1}{\varepsilon}\right)^2\right] + 1 = 0$$

from which the eigenvalues λ_1 and λ_2 are found to be:

$$\lambda_1, \lambda_2 = \frac{2\varepsilon^2 + 1 \pm (4\varepsilon^2 + 1)^{1/2}}{2\varepsilon^2}$$

The singular values are then $\sigma_1 = \sqrt{\lambda_1}$ and $\sigma_2 = \sqrt{\lambda_2}$.

If we take the binomial expansion of the term in parentheses and allow ε to tend to zero, we obtain

$$\lambda_1 \approx \frac{2\varepsilon^2 + 1 + 1 + 2\varepsilon^2}{2\varepsilon^2} \quad \text{and} \quad \lambda_2 \approx \frac{2\varepsilon^2 + 1 - 1 - 2\varepsilon^2}{2\varepsilon^2}$$

The two singular values therefore tend to infinity and zero respectively. The number of zero singular values tells us the rank deficiency of the matrix (1 in this case – so we conclude that the matrix is singular).

This also means that as ε tends to zero, σ_1 and σ_2 diverge. However, if ε tends to infinity, the CE becomes $\lambda^2 - 2\lambda + 1 = 0$, so that both singular values tend to 1. The divergence is an indication that the matrix is ill-conditioned (*condition number* = $\max(\sigma)/\min(\sigma)$ and should be small for a well-conditioned matrix – with a minimum value of unity).

The MATLAB CSTB command *sigma* allows us to plot the singular values of a TFM (or, to be precise, its state-space equivalent) against frequency. We can investigate these for zero values and divergence, and thus get some idea of how robust the design might be from a numerical viewpoint. See Section 10.8.2 for how the Perron–Frobenius approach can also assist in robustness analysis, and Chapter 13 for more on robust control.

10.6.5 *Characteristic locus case study using MATLAB*

In Example 10.1 (Section 10.3), we developed a TFM model of a pneumatic laboratory rig. In Example 10.2 (Section 10.4), we designed a diagonalizing controller for it. We now attack the same problem using the CL approach. Later, we shall also try the inverse Nyquist array approach (Section 10.7.5) and the Perron–Frobenius approach (Section 10.8.3) and compare all the results. The MATLAB commands for Example 10.4 are to be found in the m-files *fig10_16.m* to *fig10_20.m* on the accompanying disk, suitably commented. Here, we concentrate on the design procedure, with only the barest outline of how MATLAB achieves the results, as an indication of a typical CACSD methodology.

Example 10.4 *The pneumatic system with a controller designed by the CL method*

We already have a TFM model of the plant from Example 10.1. For numerical reasons, MATLAB works with the equivalent state-space model, which it easily obtains by effectively combining the state-space models of the individual TFM elements, obtained by direct programming. The MATLAB multivariable frequency domain toolbox (MVFDTB) command *mvtf2ss* does this.

Once we have the state-space model, MATLAB can obtain for us what is known as the *multivariable frequency response* (MVFR) matrix via the *mv2fr* command. The MVFR matrix is a data structure containing the frequency response data for every element of the original TFM, and is used as the basis for many of the MVFDTB commands. MATLAB requires a set of frequency values to obtain this, so we use the command *logspace* to set up a vector of 100 points, logarithmically spaced between 0.01 and 100 rad s^{-1}.

From the MVFR matrix, we use MATLAB to calculate the data for the CL plots using the *feig* command. These are typically plotted as either Nyquist or Bode plots. We shall use the Bode presentation for this design, and the CL plots of the uncompensated plant appear in this form in Figure 10.16.

From Figure 10.16 we see that the system is very stable (infinite phase margins), but that there is likely to be interaction in closed-loop. This is indicated by the non-coincidence of the two CL plots, indicating that $q_1(s) \neq q_2(s)$ at any frequency, so the system is not likely to be diagonal at any frequency (see Section 10.6.1). In fact, we know that there is interaction from our earlier investigations (see, for example, Figure 10.2). We apply the CL method to try to remove the interaction. We shall not present plots of the misalignment angles, because they do not appreciably improve during the design of this particular system.

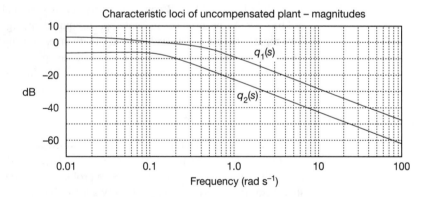

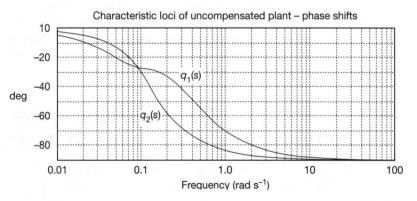

Figure 10.16
Characteristic loci of the uncompensated pneumatic plant.

High-frequency compensation

First, we use the MATLAB MVFDTB *align* command to try to improve things at high frequencies. Looking at the plots of Figure 10.16, it is clear that the order of the two CL plots is the same, so that the magnitude plots are parallel at high frequencies. This implies that if we can line them up with one another at any frequency in this region, they should be lined up at all higher frequencies too, hopefully removing high frequency interaction. The expedient of applying a ruler to the plots suggests that a frequency of 4 rad s^{-1} is far enough into this region to have the desired effect.

The *align* command requires us to provide the components of the frequency response data (that is, the elements of the MVFR matrix) which exist at the design frequency. These can be found by using the MATLAB command *find* to give the index into the frequency vector (and hence into the corresponding MVFR matrix) of the first element greater than 4 rad s^{-1}. This turns out to be the 66th element, and corresponds to $w \approx 0.423$ rad s^{-1}. The MVFDTB command *fgetf* is then used to extract the 66th set of data from the MVFR matrix. The procedure is clear in the m-file *fig10_17.m* on the accompanying disk.

Using the *align* command with the 66th element of the MVFR matrix returns a high-frequency compensator matrix

$$K_{\mathrm{h}} = \begin{bmatrix} 52.74 & 6.828 \\ 6.855 & 11.50 \end{bmatrix}$$

This now needs fitting into the structure of Figure 10.14. However, since we have not yet designed the approximately commutative controller (A, B and the various $k_i(s)$), it looks more like Figure 10.5 for the present.

In order to combine the constant compensator matrix K_{h} with the plant TFM $G(s)$, the MVFDTB command *fmul* is used to form the combined MVFR matrix. This then effectively becomes the MVFR matrix of $Q(s) = G(s)K_{\mathrm{h}}$ as in Figure 10.12(a). We can then inspect the compensated CL (that is, of $Q(s)$) to see what we have achieved. The CL plots appear in Figure 10.17, where we see that the CL plots do indeed now line up with one another at high frequencies.

In order to see what the effect on the step responses has been, MATLAB needs the state-space equivalent of Figure 10.12(a) to perform a simulation. This is easily done using the MVFDTB commands *mvser* (to form the state-space model of the series connection of K_{h} and the original state-space version of the plant TFM), and *mvfb* (to close the unity negative-feedback loops). The custom-written *mv2step* command on the accompanying disk displays the resulting responses, shown in Figure 10.18.

These results look quite promising (especially the behaviour of y_1), but note that the rise times of the step responses are quite fast compared with Figure 10.8. This implies that we may be asking for an unrealistic performance improvement (requiring larger input signals than the plant could handle, for example). However, we shall not investigate this yet, because it would probably need some kind of dynamic compensation to correct for such a fault without spoiling our high frequency result. As an initial attempt, it will be easier to try applying such compensation after the design for non-interaction is complete. We shall therefore carry on with the design procedure, and then try to slow things down at the end if necessary.

Mid-frequency compensation

We now need to design the approximate commutative controller (ACC) which appears in front of the plant and the high-frequency compensator in the forward path of Figure 10.14.

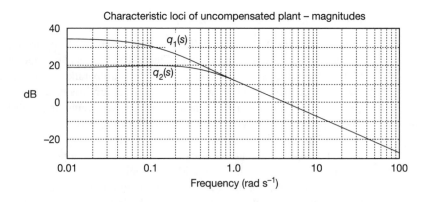

Figure 10.17
Characteristic loci of the
high-frequency
compensated plant.

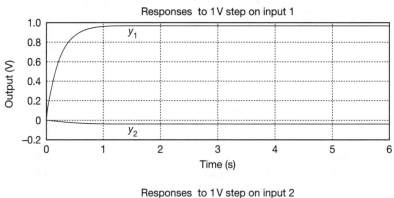

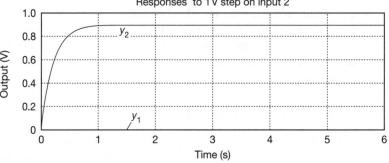

Figure 10.18 Step
responses of the high-
frequency compensated
plant.

The MATLAB *facc* command will provide for us the state-space model of the entire ACC, but we need to provide this command with quite a lot of detail.

Firstly, the *facc* command needs to know the MVFR matrix for the high-frequency compensated plant, and the vector of frequency points at which it was evaluated. We already have all these data from the previous steps.

Secondly, the *facc* command needs to know the particular value in the frequency vector at which we would like the ACC to be designed (remember, it is an approximation to be made at a single frequency and we hope its effects will spread over a useful frequency range).

Thirdly, it needs to know the dynamic compensators $k_1(s)$ and $k_2(s)$ which we want to include in the two loops between the constant matrices $\boldsymbol{B}$ and $\boldsymbol{A}$ (see Figure 10.14). These compensators need to be specified by us, but MATLAB can help.

Inspection of the magnitude plots of Figure 10.17 shows that we need to try to line up the CL plots at frequencies below about 2 rad s^{-1}. In the interests of high d.c. gains (which we shall need for good steady-state behaviour), a first attempt might be to try to leave the upper plot (the CL of $q_1(s)$) unchanged, and bring the lower plot up to meet it. Bearing in mind that the *facc* command will design the $\boldsymbol{B}$ and $\boldsymbol{A}$ matrices so as to remove (in theory, at least) interaction from the plant, we only need concern ourselves with the single-loop designs for $k_1(s)$ and $k_2(s)$. We shall therefore specify a constant compensator $k_1 = 1$ to keep the CL for loop 1 unchanged, and a suitable dynamic compensator $k_2(s)$ for insertion into loop 2 so that the CL for that loop rises to meet that for loop 1 over as wide a frequency range as we can achieve. At the same time, it is vital that $k_2(s)$ has unity gain above about 2 rad s^{-1}, otherwise we shall spoil the high-frequency compensation we have already achieved.

From Figure 10.17, we see that the compensator $k_2(s)$ therefore needs to provide a gain increase of about 15 dB at low frequencies, falling to zero dB by about 2 rad s^{-1}. We could pick a number of frequency values in between these limits, measure the required gain increase at each, and identify the required compensator transfer function from the resulting Bode plot. However, since the behaviour of the plots in Figure 10.17 is smooth and fairly gentle, it looks as though a single lag-lead type compensator might be sufficient, and MATLAB can easily design this for us.

The MATLAB MVFDTB contains a command *phlag* which will design the lag-lead compensator. We need to specify the required d.c. gain increase (15 dB), and the upper (that is, the lead) corner frequency of the compensator. From Figure 10.17, it appears that this needs to be at about 0.35 rad s^{-1}, since this is the frequency at which the gain of loop 2 is 3 dB below that of loop 1 (but note that if the lower break frequency of the compensator is close to this frequency, the inspection of Figure 10.17 for the 3 dB frequency would then have to take into account the 'interference' remaining from the lower break point at any particular frequency – see Table 3.3).

As to the frequency at which we wish the ACC to act, if we are to use this simple lag-lead compensator, the most suitable frequency is likely to be that at which the maximum phase shift is required. The phase shifts of Figure 10.17 suggest that this is at a frequency of around 0.15 rad s^{-1}. It is possible to get MATLAB to calculate the d.c. gain required, and the frequency at which the phase difference between the CL plots is a maximum; but we shall content ourselves with our estimates from Figure 10.17. For use of the *facc* command, we use the *find* function, as before, to obtain the nearest element in the frequency vector to this value. It turns out to be the 30th value in the frequency vector, at 0.1485 rad s^{-1}.

Using the MATLAB MVFDTB *phlag* command, as suggested above, gave a lag-lead compensator with the transfer function

$$k_2(s) = \frac{s + 0.35}{s + 0.0622}$$

We now have all the data required by the *facc* command, and it can design the ACC for us as an overall state-space model including A, B, k_1 and $k_2(s)$ (the only extra comment for readers trying this in MATLAB is that k_1 must be specified as s/s rather than 1, because the *facc* command insists on dynamic compensators).

The state-space compensator returned by the *facc* command is converted to MVFR form using the *mv2fr* command, and cascaded with the high-frequency compensated plant using the *fmulf* command (note that we used *fmul* previously, as the high-frequency compensator was just a constant matrix). This then gives the MVFR matrix of the overall forward path of Figure 10.14, which leads to the CL plots of Figure 10.19(a). Using the *mvser* and *mvfb* commands to form the closed-loop state-space model, as before, then leads to the step responses of Figure 10.19(b).

The CL plots are now well lined up at all frequencies, and the time responses show an almost complete lack of interaction (there is a very slight amount visible on a computer screen). We can therefore now treat the system fairly confidently as two independent SISO loops. The only remaining problems are the small steady-state errors, and the fact that we may be driving the plant too hard, as mentioned previously.

In obtaining the ACC using MATLAB, the A and B matrices for use in Figure 10.14 are transparent to the user – we immediately obtain the state-space model of the entire ACC as a single entity including A, B and the dynamic compensator designed by *phlag* above. In order to see what the compensator looks like, it is a simple matter to edit the MATLAB *facc* command to make it output these two matrices for inspection (MATLAB toolbox commands are simply ASCII m-files full of other MATLAB commands). The compensator matrices turned out to be:

$$A = \begin{bmatrix} 8.0627 & 0.0919 \\ 0.5666 & 1.0127 \end{bmatrix} \quad \text{and} \quad B = \begin{bmatrix} 0.1247 & -0.0113 \\ -0.0698 & 0.9893 \end{bmatrix}$$

The approximate nature of the compensation is illustrated by the fact that, even at a single frequency, these are not quite inverses of one another. For example,

$$A^{-1} = \begin{bmatrix} 0.1248 & -0.0113 \\ -0.0698 & 0.9938 \end{bmatrix}$$

Steady-state compensation

Here, we adopt the usual solution for the removal of steady-state errors, namely integral action. However, in general we ought not simply to apply integrators to each loop, as that may spoil the compensation already carried out at high- and mid-frequencies. Rather, we use proportional plus integral control, since a compensator of the form

$$1 + \frac{K_I}{s} = \frac{s + K_I}{s}$$

will have unity gain at high frequencies, thus leaving the existing compensation largely unaffected for a suitable choice of K_I.

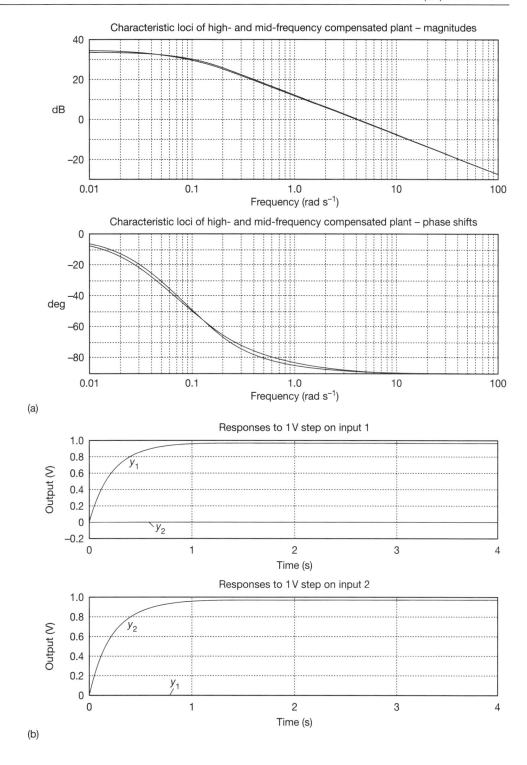

Figure 10.19 (a) Characteristic loci of the high- and mid-frequency compensated plant. (b) Step responses of the high-and mid-frequency compensated plant.

From Figure 10.19(a), the CL plots have a break frequency at about 0.1 rad s^{-1}. If we give our P+I controllers (one in each loop) a zero at about 0.1 rad s^{-1}, we should then get CL plots more or less linearly falling with frequency. To do this, we need a low-frequency dynamic compensator, placed before the ACC in the forward path of Figure 10.14, and having a TFM

$$\begin{bmatrix} \dfrac{s+0.1}{s} & 0 \\ 0 & \dfrac{s+0.1}{s} \end{bmatrix}$$

Of course, since our compensation thus far is not perfect in terms of the removal of interaction, the addition of these single-loop compensators may lead to the return of some interaction.

The use of MATLAB to add this compensator follows exactly the same patterns as for the previous work, and the results appear in Figure 10.20. The CL magnitude plots show the predicted form, now having infinite d.c. gain. The phase plots are not shown, as they are almost constant at −90 degrees, except around 0.1 rad s^{-1} where they vary by a few degrees. The step responses (Figure 10.20(b)) show an almost total lack of interaction and no steady-state error. In addition, they are much faster than our original specification of a 1.6 s rise time (see Example 10.2).

It is now necessary to check the demands we are placing on the plant inputs. The procedure for displaying the inputs in MATLAB is explained in the comments in the m-file *fig10_20.m* on the accompanying disk, and in Section 3.9 of Appendix 3. Basically, it involves adding two rows of zeros to the original *C* matrix of the open-loop plant's state-space model, and a 2 × 2 identity matrix underneath the original *D* matrix. This has the effect of generating two new outputs, which are equal to the inputs, and are then automatically displayed by the commands generating the step responses.

When this is done it is found that plant input 1 is initially driven to about 53 V in response to the 1 V step on reference input 1, and plant input 2 to about 11.5 V in response to the 1 V step on reference input 2. This is fine in simulation, but the real plant input limits are 0–10 V, so they would be saturated, and the behaviour would be nonlinear and not as predicted. In order to overcome this, we need to lower the gains. This is done simply by adding a diagonal gain matrix in front of the low frequency (P+I) compensator.

Strictly, since different gains are needed in the two loops, another ACC should be used to isolate their effects and avoid reintroducing interaction. However, in this simple case, a little experimentation showed that the gain matrix

$$\begin{bmatrix} 0.18 & 0 \\ 0 & 0.8 \end{bmatrix}$$

would limit the plant inputs to the physically acceptable voltage range. However, the dynamic behaviour of loop 1 then became much too slow, having a rise time of around 3 s (loop 2 rise time remained below 1 s).

Since we are allowed up to 15 per cent overshoot according to our specification, we can try changing the P+I compensators to allow some overshoot in the interests of obtaining a faster rise time. Changing the P+I compensator matrix to

$$\begin{bmatrix} \dfrac{s+0.5}{s} & 0 \\ 0 & \dfrac{s+0.5}{s} \end{bmatrix}$$

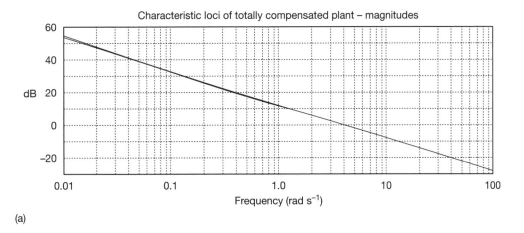

(a)

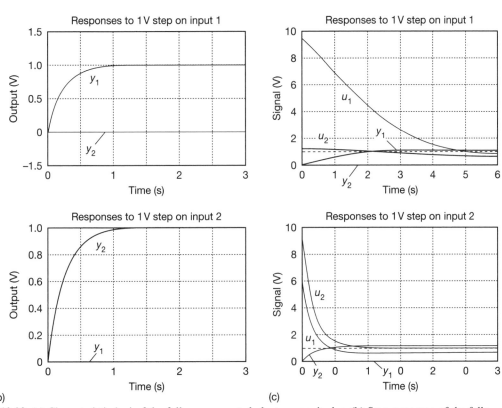

(b)   (c)

Figure 10.20 (a) Characteristic loci of the fully compensated plant – magnitudes. (b) Step responses of the fully compensated plant (CL method). (c) Step responses with modified compensation, showing the plant inputs.

while retaining the diagonal gain matrix above, just about meets the specifications. The responses are shown in Figure 10.20(c). In practice, we would try more sophisticated dynamic compensation elements for this plant, in order to achieve the specification without coming so near to the input limits (using another ACC). At present, note that although y_1 does overshoot by only about 15 per cent, it then takes longer than the six seconds shown in Figure 10.20(c) to return to unity. Also, if the 1 V step were to be applied to input 1 when the plant input 1 was already much greater than zero, the 10 V plant input limit would be reached (however, a 1 V step is not typical of normal operation – recall that a large test step was chosen to avoid confusion with noise).

The overall result

To find the overall compensator which we need to apply to the plant, we must combine all the separate parts. The order of multiplication of these (against the signal flow) will be:

$$K_h AK(s)B[P + I][\text{d.c. gain}], \quad \text{where:}$$

$$K_h = \begin{bmatrix} 52.74 & 6.828 \\ 6.855 & 11.50 \end{bmatrix}, \quad A = \begin{bmatrix} 8.0627 & 0.0919 \\ 0.5666 & 1.0127 \end{bmatrix}, \quad K(s) = \begin{bmatrix} 1 & 0 \\ 0 & \dfrac{s+0.35}{s+0.0622} \end{bmatrix}$$

$$B = \begin{bmatrix} 0.1247 & -0.0113 \\ -0.0698 & 0.9893 \end{bmatrix}, \quad P + I = \begin{bmatrix} \dfrac{s+0.5}{s} & 0 \\ 0 & \dfrac{s+0.5}{s} \end{bmatrix}$$

$$\text{d.c. gain} = \begin{bmatrix} 0.18 & 0 \\ 0 & 0.8 \end{bmatrix}$$

Performing the multiplication leads to the overall compensator:

$$\frac{1}{s^2 + 0.0622s} \begin{bmatrix} 9.4837s^2 + 5.2892s + 0.2737 & 5.4295s^2 + 5.7315s + 1.5084 \\ 1.2326s^2 + 0.6486s + 0.0161 & 9.1572s^2 + 7.9444s + 1.6829 \end{bmatrix}$$

In Section 10.9, this result is compared with those obtained by the other methods.

10.7 The inverse Nyquist array (INA) method

The characteristic locus method of Section 10.6 is based upon an eigenvalue-type analysis of the system. The INA method, on the other hand, is representative of non-eigenvalue methods, and works by consideration of input–output relationships – which is more in line with the philosophy of SISO frequency-domain control.

10.7.1 Diagonal dominance

The underlying philosophy of the INA method is that if $Q(s)$ in Figure 10.12(a) can be kept 'approximately diagonal' (defined below) at all frequencies of interest, then SISO design procedures can be satisfactorily applied at all frequencies. Remembering that $Q(s) = G(s)K(s)$ (Figure 10.5), this will again be achieved by suitable choice of $K(s)$.

We saw earlier that a requirement to make $Q(s)$ strictly diagonal leads, in general, to very complicated compensators, difficult mathematics and other undesirable effects (for example, unrealizable or unstable compensators). The INA method replaces this requirement with the much less stringent requirement that $Q(s)$ must be *diagonally dominant* at all frequencies of interest. So long as this can be achieved, SISO compensators (designed on the basis of one-per-loop) are likely to work satisfactorily.

The definition of diagonal dominance can be stated mathematically as follows (if this looks formidable, the reader can ignore it, and read the definitions given in English afterwards). An $n \times n$ matrix $Q(s)$ is diagonally dominant if

$$|q_{ii}(s)| > r_i(s) \quad \text{for} \quad 1 \leq i \leq n \quad \text{and for all values of } s \quad (10.15)$$

where

$$r_i(s) = \sum_{j=1, j \neq i}^{n} |q_{ij}(s)| \quad \text{for row dominance}$$

or

$$r_i(s) = \sum_{j=1, j \neq i}^{n} |q_{ji}(s)| \quad \text{for column dominance}$$

In words, 'a matrix is row (or column) diagonally dominant if the magnitude of each element on the leading diagonal is greater than the sum of the magnitudes of all the other elements in the same row (or column)'. In the case of a frequency-dependent matrix (letting $s = j\omega$), this must be tested at every frequency. The row dominance test is used predominantly with the *direct* Nyquist array (not studied here), and the column dominance test with the *inverse* Nyquist array. The reason for this will emerge later.

The structure of controller considered in the INA method is similar to that of Figure 10.12(a), the only difference being that we now also consider a diagonal matrix of feedback gains $F = \text{diag}\{f_1, f_2, f_3, \ldots, f_n\}$, rather than simple unity negative feedback loops. As before, $Q(s) = G(s)K(s)$, so we now have the arrangement of Figure 10.21.

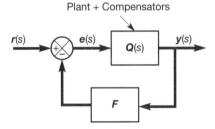

Plant + Compensators

Figure 10.21 Multivariable controller for a system using individual feedback gains.

The closed-loop transfer function matrix (CLTFM) for this system is given by $H(s) = [I + Q(s)F]^{-1}Q(s)$ (cf. Equation (10.9)), so the relationship between the open- and closed-loop TFMs ($Q(s)$ and $H(s)$ respectively) is not particularly straightforward. However, by a suitable choice of $K(s)$, we can ensure that $Q(s)^{-1}$ exists. In this case, we can invert the previous relationship, to give:

$$H^{-1}(s) = Q(s)^{-1}[I + Q(s)F] = Q(s)^{-1} + F \quad (10.16)$$

which is a much more tractable result. We therefore choose to work with the inverse matrices in the remainder of this section.

If $H(s)$ was purely diagonal, then $H^{-1}(s)$ would also be diagonal, with each element being simply the reciprocal of the corresponding element of $H(s)$.

However, since $H(s)$ is to be made only diagonally dominant, non-zero (but relatively small) off-diagonal elements are permitted to remain in $H(s)$. Therefore a diagonal element of $H^{-1}(s)$ will not usually be simply the reciprocal of the corresponding element of $H(s)$. We therefore need a different notation for an inverse matrix, which will allow us to distinguish between $h_{ii}^{-1}(s)$ as element (i, i) of $H^{-1}(s)$, and $h_{ii}^{-1}(s)$ as the reciprocal of element (i, i) of $H(s)$. It is customary to use $\hat{H}(s) = H^{-1}(s)$ for this purpose. The quantity $h_{ii}^{-1}(s)$ now unambiguously refers to the reciprocal of element (i, i) of the matrix $H(s)$, while the quantity $\hat{h}_{ii}(s)$ refers to element (i, i) of the matrix $\hat{H}(s) = H^{-1}(s)$.

Now, as the Nyquist D-contour is traversed once in a clockwise direction, the element (i, j) of the inverse forward path TFM, that is, $\hat{q}_{ij}(s)$, will trace out a contour which is similar to an inverse Nyquist plot. If such a plot is drawn for all elements of $\hat{Q}(s)$, we end up with the *inverse Nyquist array* (INA) for the system.

Note that although these plots will look like inverse Nyquist plots, they do not correspond with anything physically measurable in the real world. Also, although the full array of plots is considered for investigation of *interaction*, only the *diagonal* elements are considered for *stability* analysis, any off-diagonal terms being assumed to be negligibly small due to the requirement for diagonal dominance. Thus, we must apply compensators to achieve diagonal dominance first, and consider stability afterwards.

Once diagonal dominance is achieved, stability information follows from an analysis of encirclements of critical points, as outlined in Section 10.6.1. However, in the INA approach, the critical point for the ith diagonal element of the INA is the point $(-f_i, 0)$, where f_i is the feedback gain in the ith loop (directly analogous to the interpretation of inverse Nyquist plots for SISO systems – see Section 3.5.2).

10.7.2   Gershgorin's theorem

In order to assess the progress being made in trying to remove the interaction from the system in the INA approach, we need a simple technique for investigating the diagonal-dominance properties of the system. Such a technique is provided by *Gershgorin's theorem*.

Clearly, the condition for which we must design is when all the feedback loops are closed. In SISO frequency-domain methods, we analyse and design the behaviour of the *closed-loop* system by interpreting the *open-loop* plot. Extending this idea to the case of the INA means that the closed-loop behaviour of the ith loop $(1 \leq i \leq n)$, should be inferred from the open-loop plot for the ith loop, with all the other loops closed. However, the inverse Nyquist-like plots of the INA are *all* drawn for the open-loop elements of $\hat{Q}(s)$ (the inverse TFM).

If the TFM was purely diagonal, there would be no interaction, and interpreting the INA plots individually would be correct. However, due to the interaction which will remain (at whatever stage of the design we have reached), the precise form of each plot when all the other loops are closed is unknown – it will differ from that shown in the INA with all loops open.

Gershgorin's theorem states that the inverse Nyquist-like plot for the ith loop open and all other loops closed will always lie in the *band* formed by the union of a number of circles superimposed on the inverse Nyquist-type plot already obtained for the ith diagonal element of $\hat{Q}(s)$ (that is, the plot for $\hat{q}_{ii}(s)$). Although we cannot be certain exactly where the plot is, we can therefore narrow down its location to a known band. This remains true, whatever values of feedback gains are used in the other loops.

To obtain this *Gershgorin band*, at a selection of points (frequencies) along the inverse Nyquist-like plot for $\hat{q}_{ii}(s)$, we draw circles centred on the plot, and of radii $r_i(s)$ as given by Equation (10.15), but using elements $\hat{q}_{ij}(s)$ rather than $q_{ij}(s)$. The radii of the circles will generally vary from point to point along the plot for $\hat{q}_{ii}(s)$ as s varies. Figure 10.22 shows an example of a single diagonal element of an INA plot, with Gershgorin circles superimposed. The radii of these circles are calculated from the other elements of the INA (the other plots, not shown in Figure 10.22) in the same row or column, at each frequency point where a circle is centred. The calculated (open-loop) plot is the line indicated by $\hat{q}_{ii}(s)$, but the actual plot with all loops closed except loop i could theoretically lie anywhere inside the Gershgorin band formed by the union of the circles. Packages such as MATLAB can draw any required number of Gershgorin circles on the INA so that the Gershgorin bands can be inspected.

The interpretation of the Gershgorin bands is as follows. We initially seek to achieve *open-loop (diagonal) dominance*, by methods to be suggested below. This has been achieved when the Gershgorin bands for none of the diagonal elements of the INA enclose the origin of the complex plane. The loop shown in Figure 10.22 would pass this test, as none of the circles encloses the origin. However, all other diagonal elements of the INA (similar plots to Figure 10.22, but omitted from the figure) must also pass the test before the system as a whole can be declared diagonally dominant.

After diagonal dominance has been achieved, we can select appropriate values for the feedback gains $(f_i, 1 \leq i \leq n)$ to achieve *closed-loop stability*. For this to be achieved, the Gershgorin bands must not touch the negative real axis between the origin and the point $(-f_i, 0)$. This follows reasonably intuitively from Equation (10.16) because, for an approximately diagonal system, it is approximately true that $\hat{h}_{ii}(s) = \hat{q}_{ii}(s) + f_i$, so a simple shift of f_i relates the open- and closed-loop plots. The critical point therefore moves from the origin to $(-f_i, 0)$. Note, again, the uncertainty due to interaction, which is why we must consider the boundary of the Gershgorin band, rather than the drawn plot of $\hat{q}_{ii}(s)$, in order to cover the worst

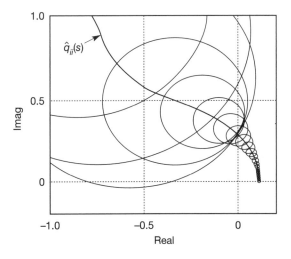

Figure 10.22 An example of an inverse Nyquist-like plot with Gershgorin circles.

possible scenario. Figure 10.22 suggests that loop *i* of the system for which the figure was drawn would be stable for a feedback gain of up to about 0.3, so long as the other loops are already diagonally dominant, too. In practice, since the band only just cuts the negative real axis, the position of the plot would have to be right at one extreme of its possible range for instability to occur. It is therefore very likely that a higher feedback gain could be used (probably any positive gain we wish, for this system), but a ceiling of 0.3 is all we can *guarantee* to cover the worst case.

10.7.3 Achieving diagonal dominance

We are trying to find a compensator $\hat{K}(s)$ so that the inverse forward path TFM $\hat{Q}(s) = [G(s)K(s)]^{-1} = \hat{K}(s)\hat{G}(s)$ is diagonally dominant. Usually a cascaded series of precompensators will result, as in the case of the CL method. Once diagonal dominance has been achieved, single-loop compensators can be added in the normal way. Post-compensators (that is, compensators placed *after* the plant in the forward path) can also be used in theory. This is not useful in practice, because a compensator placed *after* the plant 'scrambles' the system outputs, so that only some linear combinations of the outputs are controlled as required, rather than the plant outputs themselves. Therefore only precompensators are used in a practical INA design.

In an effort to achieve diagonal dominance, the following methods may prove useful. However, it must be said that experience also assists greatly in the use of this method, which relies more on the decisions of the designer than does the CL method. Note that the Perron–Frobenius approach of Section 10.8 reduces to some extent this need for experience.

Initial inspection

When the INA is viewed, it appears as a matrix (that is, array) of inverse Nyquist-like plots, one per TFM element. At the outset, it is sometimes obvious whether or not it is likely to be diagonally dominant. For example, a comparison of the magnitudes of the extremes of the off-diagonal plots with those of the diagonal term in the same row (or column) will often show that the diagonal term cannot possibly be larger than the sum of the off-diagonal ones, so the system cannot then be diagonally dominant, and some compensation will be needed. However, it is only possible to study the endpoints of the plots in this non-rigorous way. Even simple two-input-two-output systems with low-order dynamics are quite capable of showing non-dominant behaviour at mid-frequencies, even though both the high- and low-frequency endpoints of the plots may be satisfactory. It is always worth drawing Gershgorin bands as a check, even if the system superficially appears to be satisfactory.

Elementary row operations

Assuming the system is not dominant at this stage, one can often envisage a set of elementary row operations which will help to make it so; for example, interchanging a pair of rows to bring one element of large magnitude onto the diagonal of the INA while moving a smaller component into the off-diagonal region. Swapping rows in this way is easily accomplished by using a compensator matrix having a single entry of 1 in each row and column (known as a permutation matrix). A simple numerical example follows showing the effects of row interchange (using a pre-multiplying matrix) and column interchange (using a post-multiplying matrix):

$$\begin{bmatrix} 0 & 1 & 0 \\ 1 & 0 & 0 \\ 0 & 0 & 1 \end{bmatrix} \begin{bmatrix} 1 & 9 & 3 \\ 7 & 2 & 6 \\ 5 & 4 & 8 \end{bmatrix} = \begin{bmatrix} 7 & 2 & 6 \\ 1 & 9 & 3 \\ 5 & 4 & 8 \end{bmatrix}$$

interchanges rows 1 and 2

$$\begin{bmatrix} 1 & 9 & 3 \\ 7 & 2 & 6 \\ 5 & 4 & 8 \end{bmatrix} \begin{bmatrix} 0 & 1 & 0 \\ 1 & 0 & 0 \\ 0 & 0 & 1 \end{bmatrix} = \begin{bmatrix} 9 & 1 & 3 \\ 2 & 7 & 6 \\ 4 & 5 & 8 \end{bmatrix}$$

interchanges columns 1 and 2

Another common type of constant compensator is a unit matrix with some non-zero off-diagonal terms added. This has the effect of adding multiples of rows (or columns) to other rows (or columns). For example:

$$\begin{bmatrix} 1 & 2 & 0 \\ 0 & 1 & 0 \\ 0 & 0 & 1 \end{bmatrix} \begin{bmatrix} 1 & 9 & 3 \\ 7 & 2 & 6 \\ 5 & 4 & 8 \end{bmatrix} = \begin{bmatrix} 15 & 13 & 15 \\ 7 & 2 & 6 \\ 5 & 4 & 8 \end{bmatrix}$$

adds twice row 2 to row 1

$$\begin{bmatrix} 1 & 9 & 3 \\ 7 & 2 & 6 \\ 5 & 4 & 8 \end{bmatrix} \begin{bmatrix} 1 & 2 & 0 \\ 0 & 1 & 0 \\ 0 & 0 & 1 \end{bmatrix} = \begin{bmatrix} 1 & 11 & 3 \\ 7 & 16 & 6 \\ 5 & 14 & 8 \end{bmatrix}$$

adds twice column 1 to column 2

Several such simple compensators and permutation matrices can be cascaded to achieve the desired result. The CACSD package can then be asked to combine them all into one and invert the result for real-world application as discussed next.

A note on precompensators, row operations, column dominance and the INA

A point which is worth stressing is that restricting ourselves to the use of precompensators in the INA approach means that we can only carry out *row operations* on the INA. At first sight, this might seem to follow directly from the example above, but a little more thought perhaps makes it seem *incorrect*, and it is important to try to remove the confusion.

In the real world, if we add a precompensator matrix K to a plant matrix G (dropping any s-dependency for clarity), then we have the forward path arrangement shown in Figure 10.5. The overall forward path transfer function is then given by $Q = GK$ (working against the signal flow, as usual). It therefore seems that placing a precompensator K in front of the plant ought to lead to *post-multiplication* by the compensator matrix, thus performing *column* operations as in the example above.

While this argument is correct when working in the real world, we are to perform our design in the *inverse* Nyquist domain. Since we are to work with the inverse plant and compensator matrices we actually have

$$\hat{Q} = (GK)^{-1} = \hat{K}\hat{G} \tag{10.17}$$

Viewed in this light Equation (10.17) shows that when we add a precompensator K to the real-world plant, it is the same as *pre*-multiplying the *inverse* Nyquist array by $\hat{K}$, thus performing row operations on it. Conversely, if we decide that a certain row operation will improve the INA, and pre-multiply the INA by an appropriate matrix to achieve this aim, then this matrix is $\hat{K}$ and must be inverted before application as the precompensator K in the real world.

As a consequence, when plotting Gershgorin circles on the INA, we normally choose to plot *column* dominance circles. The reason for this is that, although precompensators which perform elementary row operations (including row interchange) can affect either row or column dominance (or both), we shall also want to use *diagonal* precompensators sometimes. These cannot affect row dominance (because they have the same effect on every element in a given row), so we shall have to view column dominance in such cases to see any change.

D.C. diagonalization

If $\hat{G}(s)$ exists, it may be worth trying a precompensator which will diagonalize $G(s)$ at d.c., namely $\hat{K} = G(0)$. Again note the possible confusion due to our working with the *inverse* Nyquist array. We apply the precompensator $G(0)$ to the inverse Nyquist array, but it therefore becomes $\hat{G}(0)$, as required, for application to the real-world plant.

Other approaches

Other methods tend to be more complicated. For example, $\hat{K}$ may be chosen as a constant precompensator which best (in a least-squares sense) diagonalizes $\hat{G}(s)$ at some frequency of interest. See Owens (1978) for some more suggestions. There is also the Perron–Frobenius approach, which we discuss in Section 10.8.

Dynamic compensation

As mentioned above, once dominance is achieved as indicated by the failure of any of the Gershgorin bands to enclose the origins, a set of SISO compensators can be designed (individually) to give the required dynamic performance in each loop. Simulation runs using the CACSD package will help to confirm the suitability of the resulting controller. Note that, if different compensators are used in different loops, then diagonal dominance may be lost because of the imperfect removal of interaction before such compensators were applied.

10.7.4 Stability considerations revisited – Ostrowski bands

Once diagonal dominance has been achieved by the methods of Section 10.7.3, a further inspection of the Gershgorin bands allows appropriate feedback gains to be chosen for adequate stability. As discussed in Section 10.7.2, for the system to be closed-loop stable, the Gershgorin band for the ith diagonal element of the INA ($1 \le i \le n$) must not touch the segment of the negative real axis between the origin and the point $(-f_i, 0)$.

Once such stability has been achieved, more accurate information can be found as to the actual location of the inverse Nyquist-type plot for $\hat{q}_{ii}(s)$. This is achieved by applying a theorem of Ostrowski which states that, so long as diagonal dominance and stability have already been confirmed by the Gershgorin bands, a 'shrinking factor' can be applied to focus the bands, giving the *Ostrowski bands*. The calculation of the shrinking factor can be found in Maciejowski (1989). The Ostrowski bands are often sufficiently narrow to get a good idea of the gain and phase margins of the closed-loop system in the usual (SISO) way (see Section 3.5.2). MATLAB can plot the Ostrowski bands.

10.7.5 Inverse Nyquist array case study using MATLAB

In Example 10.4 (Section 10.6.5), we developed a controller for a pneumatic laboratory rig using the CL approach. We now apply the INA method to the same problem so that we can compare the results. The MATLAB commands for Example 10.5 are to be found in the m-files *fig10_23.m* to *fig10_27.m* on the accompanying disk, suitably commented. Here, we concentrate on the design procedure, with only the barest outline of how MATLAB achieves the results, as an indication of a typical CACSD methodology. For more details of the MATLAB commands, inspect the above files.

Example 10.5 *The pneumatic system with a controller designed by the INA method*

The acquisition of the plant TFM model, its conversion to a state-space model, the setting up of a suitable vector of frequency values and the calculation of the corresponding frequency response data as the multivariable frequency response (MVFR) matrix were all carried out in Example 10.4. It must be said that the INA plots for this particular plant are all of the same rather unexciting shape, thus making application of the method somewhat easier than it normally is. Nevertheless, the principles of application of the INA method are all demonstrated and the result gives a useful point of comparison with the other methods we apply to the same plant.

MATLAB plots the INA from the inverse of the MVFR data, and there is a MATLAB MVFDTB command *finv* to calculate this. The MATLAB MVFDTB has a command *fget* which extracts the information for one element of the INA from this inverse MVFR data (that is, the inverse frequency response information for one TFM element), a command *plotnyq* which produces the INA plot for this element, and a command *fcgersh* which plots the Gershgorin circles for (column) dominance when invoked for the diagonal elements of the INA.

Since the design process is iterative, we need to display the INA a large number of times. It therefore makes sense to write a MATLAB m-file to construct the entire 2×2 INA in one go, and add Gershgorin circles as required. This custom-written command is referred to as *inap4* in this example, and appears as the m-file *inap4.m* on the accompanying disk.

Initial investigation

Using the custom *inap4* command with the inverse MVFR data for the uncompensated pneumatic plant produces the INA of Figure 10.23. Figure 10.23(a) shows the whole of the chosen frequency range, with Gershgorin circles plotted at every 4th frequency value, and Figure 10.23(b) shows an expanded version of the low-frequency end of the plots, with Gershgorin circles every 7th frequency point.

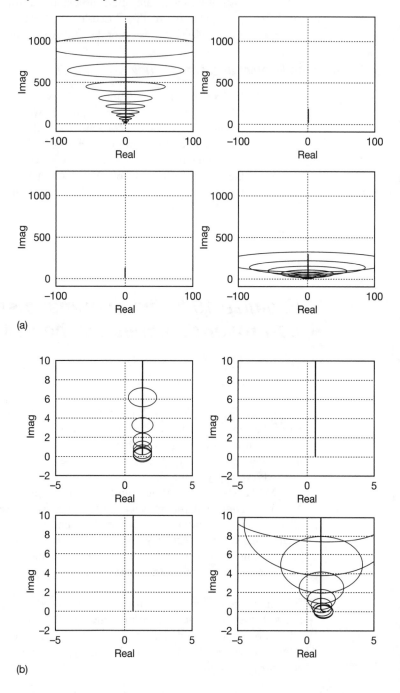

Figure 10.23 (a) Inverse Nyquist array of the uncompensated pneumatic plant – full frequency range. (b) Inverse Nyquist array of the uncompensated pneumatic plant – low-frequency range only, to show origin areas.

Without looking at the Gershgorin bands, the system appears already as though it might be diagonally dominant since, at both the high- and low-frequency extremities of the plots (Figures 10.23(a) and (b) respectively), the diagonal elements are clearly larger than the off-diagonal ones. The Gershgorin bands confirm that this is so at all the frequencies for which they are plotted, as they do not enclose the origins of the plots. Furthermore, the Gershgorin bands do not touch the negative real axes at any frequency, so the system would remain stable for any feedback gains.

The INA theory suggests that we could immediately accept that the plant is diagonally dominant, and begin to design single-loop compensators for it. However, we can tell from the size of the Gershgorin bands that there could be a lot of interaction in the system (especially at higher frequencies, where the circles are larger, but also at d.c.), and we know this to be true from Figure 10.2. We would therefore be wise to try to reduce this interaction first – in other words, the system is already diagonally dominant, but the dominance properties can probably be improved.

Row operations to improve dominance

From the INA of Figure 10.23(a), it is apparent that all four plots are predominantly the same shape (there is more comment on this point shortly). Therefore, it should be possible to make the off-diagonal elements almost vanish at high frequencies, by (for example) subtracting a proportion of row 1 from row 2 which will reduce element (2,1) to zero, and then subtracting a portion of row 2 from row 1 which will make element (1,2) reduce to zero.

Beginning with the intention of reducing element (2,1) to zero at high frequencies, we could get MATLAB to tell us the magnitudes of the plots for elements (1,1) and (2,2) at the high-frequency endpoints, and calculate the appropriate factor (and this is done in the m-file *fig10_24.m* on the accompanying disk). However, the simple expedient of measuring the plots of Figure 10.23(a) suggests that if we subtract $0.12 \times$ (row 1) from row 2, then element (2,1) ought to vanish at high frequency.

We also need to consider the effect of this proposed compensation on element (2,2), which will be affected in precisely the same way as (2,1), and we must also consider the effects at the d.c. ends of the plots in row 2. Subtracting $0.12 \times$ (element (1,2)) from element (2,2) at the high-frequency end (Figure 10.23(a)) will only reduce element (2,2) by about 6 per cent. Since we predicted that element (2,1) will vanish at high frequencies, the row dominance of row 2 should therefore be significantly improved overall at high frequencies. We, however, are plotting column dominance circles, so it is element (1,1) which will show the improvement, while element (2,2) will actually become worse as we have reduced its size relative to element (1,2) (but the second compensator will correct this). From Figure 10.23(b), the effects at the low-frequency end should be to reduce the d.c. value of element (2,1) by about 25 per cent, and that of element (2,2) by about 5 per cent, so dominance should be affected in the same way (qualitatively) as at high frequencies.

On paper, our predicted results seem promising, so we apply the required precompensator to the INA to subtract $0.12 \times$ (row 1) from row 2, namely

$$\hat{\boldsymbol{K}}_1 = \begin{bmatrix} 1 & 0 \\ -0.12 & 1 \end{bmatrix}$$

In MATLAB, it is simply a matter of using the MVFDTB command *fmul* to pre-multiply the existing inverse MVFR matrix by $\hat{K}_1$, and then using the custom *inap4* command to repeat the plotting.

The result appears in Figure 10.24, where element (2,1) has almost disappeared, as predicted. Since the Gershgorin circles shown are for column dominance (for reasons discussed previously), we see larger circles on element (2,2) because it is smaller than before, but the circles on element (1,1) have almost vanished (note the change in real axis scaling compared with Figure 10.23(a)).

We hope that the reduction in the width of the Gershgorin bands indicates a reduction of interaction in the time responses. However, this does not necessarily follow. For example, if the INA plots were actually in the positions shown in Figure 10.23(a), rather than somewhere else in the Gershgorin bands, the improvement would not necessarily be great.

At this point, it is worth stressing two aspects of the procedure we are following. Firstly, it is only easy to predict the effects of row operations in this way when all the elements of the INA are predominantly the same shape (as they are in this example, apart from a slight curve at the extreme low-frequency end of element (2,2)). If the plots are different in shape and/or direction, it becomes necessary to think about both the real and imaginary components at the ends of the plots, and to work with vector measurements from the origin of the *s*-plane (we have only considered the imaginary parts at high frequencies, as the real parts look negligible in comparison, for this particular system).

Secondly, without the use of MATLAB (or another CACSD package) it is only possible to consider the endpoints. There is no guarantee that the distribution of frequency points will be the same along the different elements of the INA, so it is impossible even to estimate the effects on dominance at intermediate points, unless the position of each frequency value on each plot is known. The INA plots for many systems will exhibit extremely strange behaviour at intermediate frequencies when applying row operation precompensators, even if the endpoint frequencies behave as predicted.

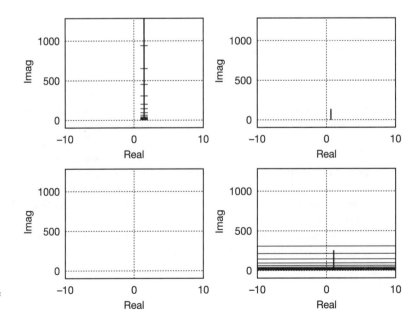

Figure 10.24 Inverse Nyquist array of the pneumatic plant with the first precompensator.

We shall not plot the time responses yet, but continue with the second stage of precompensation; namely, subtracting a suitable proportion of row 2 from row 1, so as to cause element (1,2) to vanish, in the same way as we did for element (2,1). If we achieve this, the system should be much improved, and the circles should vanish on element (2,2) also. By measurements made on Figure 10.24, subtracting $0.64 \times$ (row 2) from row 1 is approximately correct. This should reduce element (1,2) more or less to zero at high frequency, but have a negligible effect on element (1,1) due to the small size of element (2,1). Inspecting the d.c. ends of the plots in Figure 10.24, although rather inaccurate on the scale given, indicates that subtracting $0.64 \times$ (row 2) from row 1 will not spoil the dominance properties at d.c. Applying a second precompensator

$$\hat{K}_2 = \begin{bmatrix} 1 & -0.64 \\ 0 & 1 \end{bmatrix}$$

to the INA results in Figure 10.25 (showing the origin areas only on this occasion).

From Figure 10.25, we see that element (1,2) has been greatly reduced in size (but not quite as successfully as was element (2,1)), and that the circles are now much reduced as a result. On the higher-frequency parts of the plots, which do not appear in Figure 10.25, the circles on plot (1,1) remain at the diameter shown, at all frequencies (because the magnitude of element (2,1) is more or less constant with frequency), but those on element (2,2) increase with frequency, eventually reaching about the same diameter as the circles on element (1,1).

Using the approach we have taken, this is really as far as we can go with row operations. In efforts to improve dominance further, there would be little point in subtracting any proportion of row 2 from row 1, or of row 1 from row 2, because the off-diagonal terms are already so small that no real improvement would result. Indeed, noting that element (2,1) has a magnitude of about 0.45 at all frequencies, we could easily spoil the dominance altogether if we did anything to increase this. Simply multiplying row 2 by some factor greater than unity to make element (2,2) larger with respect to element

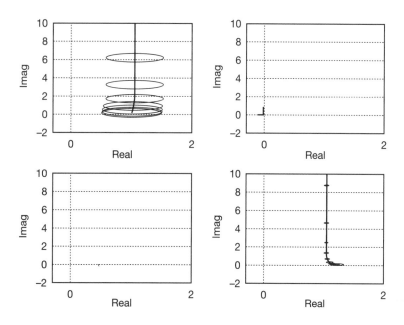

Figure 10.25 Inverse Nyquist array of the pneumatic plant with two precompensators.

(1,2) is therefore not an option. We could multiply row 1 by such a factor, to improve the dominance of element (1,1) further, but the resultant increase in the size of element (1,2) would have some detrimental effect on element (2,2).

Time responses after the row operations

Let us now see how the plant will behave with the two compensators we have designed so far. Recall that we have been adding precompensators to the INA so, in the *inverse* Nyquist domain, we have formed a forward path given by $\hat{K}_2\hat{K}_1\hat{G}$. For use in the real world, this needs inverting, when it becomes $G[\hat{K}_2\hat{K}_1]^{-1} = GK_1K_2$. To see the effect in MATLAB, we therefore form the constant precompensator K_1K_2, and then use the *mvser* and *mvfb* functions, as in Example 10.4, to build the forward path state-space model of the compensated plant, and apply unity negative-feedback loops to it. The custom-written *mv2step* command on the accompanying disk then generates Figure 10.26.

Comparing Figure 10.26 with Figure 10.2, we see that the responses are now significantly faster, and that interaction has been much reduced. However, there is still some interaction, and there are large steady-state errors to be corrected.

Dynamic single loop compensation

Having progressed as far as we are able in improving the diagonal dominance, we now treat the system as two independent single loops and design a dynamic compensator for each loop in an effort to improve the performance. The addition of integral control to each loop should allow removal of the steady-state errors, and the use of proportional plus integral control will additionally allow more control over the amount of overshoot and the other time-domain measures in which we shall be interested.

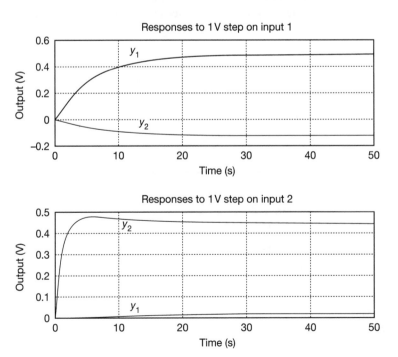

Figure 10.26 Step responses of the pneumatic plant with two precompensators.

Initially, we try a 'default' unity-gain P + I compensator applied in the same way as in the 'Steady-state compensation' section of Example 10.4, and having a TFM given by

$$K_3(s) = \begin{bmatrix} \dfrac{s+1}{s} & 0 \\ 0 & \dfrac{s+1}{s} \end{bmatrix}$$

Figure 10.27(a) shows the responses. The steady-state errors have all been removed, but the response of loop 1 is very sluggish, and the response of neither loop is fast enough to meet the original specification of a 1.6 s 10–90 per cent rise time.

In an effort to speed up the response, the gains need to be raised in each loop. Viewing the behaviour of the plant inputs (as discussed in Example 10.4) shows that a gain of about 8 units can be inserted in each loop if the inputs are to be allowed to move as far as they did in Example 10.4 (Figure 10.20(c) shows the plant inputs generated previously by the CL controller). Inserting gains of 8 units allows the response in each loop to meet the rise time criterion, but still results in too much overshoot in loop 1 (the specification called for 15 per cent). Keeping the proportional gain in loop 1 (and the overall gain in loop 2) at 8 units, but halving the integral gain in loop 1, gives the P + I compensator TFM

$$K_3(s) = \begin{bmatrix} \dfrac{8s+4}{s} & 0 \\ 0 & \dfrac{8s+8}{s} \end{bmatrix}$$

The resulting step responses are shown in Figure 10.27(b). Figure 10.27(c) shows the same responses, but with the plant input demands also displayed.

The performance specification is easily achieved in loop 2, but we have a 10–90 per cent rise time of about 1.7 s in loop 1 (as opposed to 1.6) and an overshoot of about 19 per cent (as opposed to 15). More careful tuning, or a more complicated compensator in loop 1, would achieve the specification, but the existing performance is broadly comparable with that of the CL approach, and we have done sufficient to illustrate the method. Figure 10.27(c) shows that, as for the previous CL result, we require large input signals, and again we rely on the 1 V step test being unrepresentative of normal operation.

Incidentally, reviewing the INA of this compensated system shows that the dominance properties have been spoiled by the addition of the P + I elements. This is not unusual, especially if different compensation is applied in each loop, as we have done. However, the resulting compensator is much simpler than that of the CL controller (see Section 10.9), so there is still scope for rediagonalization and further 'loop-shaping' if desired.

The overall result

In order to find the overall compensator that we need to apply to the plant, we must remember that we have been working in the inverse domain. As far as the INA is concerned, it has precompensators $\hat{K}_1$ and $\hat{K}_2$ (the P + I compensator $K_3(s)$ being applied only in the real world, after diagonal dominance had been achieved). In the inverse Nyquist domain, the overall constant compensator is therefore $\hat{K}_2\hat{K}_1$ (remember, we

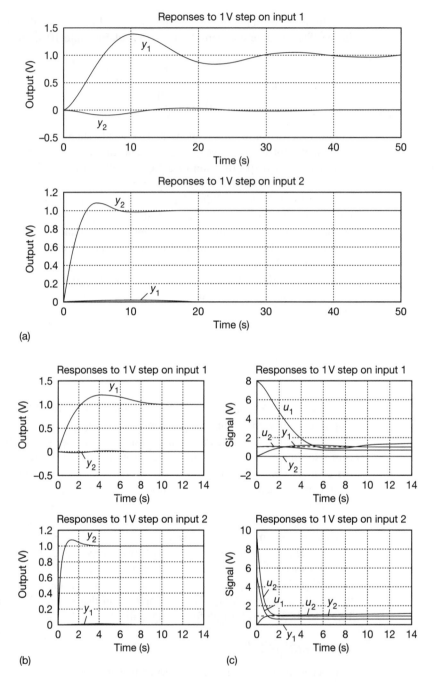

Figure 10.27 (a) Step responses of the fully compensated plant with unity $P+I$ gains. (b) Step responses of the finally-tuned compensated plant (INA method). (c) As Figure 10.27(b), but also showing the plant input demands.

applied first $\hat{K}_1$ and then $\hat{K}_2$ as compensators *pre*-multiplying the INA). In the real world, this needs inverting, so the required overall compensator is:

$$[\hat{K}_2\hat{K}_1]^{-1}K_3(s) = \left\{ \begin{bmatrix} 1 & -0.64 \\ 0 & 1 \end{bmatrix} \begin{bmatrix} 1 & 0 \\ -0.12 & 1 \end{bmatrix} \right\}^{-1} \begin{bmatrix} \dfrac{8s+4}{s} & 0 \\ 0 & \dfrac{8s+8}{s} \end{bmatrix}$$

Performing the algebra gives the overall result:

$$\frac{1}{s} \begin{bmatrix} 8s+4 & 5.12s+5.12 \\ 0.96s+0.48 & 8.6144s+8.6144 \end{bmatrix}$$

This is compared with our other results in Section 10.9.

In Example 10.5, from the outset, we adopted the approach of using row operations on the INA to achieve diagonal dominance. It will sometimes be better to try an initial compensator which diagonalizes the plant TFM at d.c. (that is, $G(0)$ applied to the INA, or $G(0)^{-1}$ applied in the real world – see Problem 10.4), or to try some other method (see Section 10.7.3). The method of performing row operations on the INA and then adding any required dynamic compensation can work well but, as we have seen, needs some experience to be able to apply it really effectively (especially on systems larger than 2×2 and systems whose INA elements are of strange or differing shapes). Another problem is that scaling inputs and changing their order of connection (which is effectively what many row operations do) can have a rather unpredictable effect on the dominance properties of some systems. A third approach to multivariable system design, allied to the INA, is now presented.

10.8 *The Perron–Frobenius (P–F) method*

10.8.1 *Introduction*

This method involves the use of the Perron–Frobenius eigenvalue and eigenvectors of the system. In our context, the P–F eigenvalue is effectively the largest positive eigenvalue of a certain special diagonalization of the inverse TFM of the system. It is therefore frequency-dependent, and can be plotted as (for example) a Bode plot. We use the inverse TFM so that we can apply the resulting precompensator to the *inverse* Nyquist array of Section 10.7. In MATLAB, the inverse of the MVFR matrix (see Example 10.4 in Section 10.6.5) is used.

The approach is useful, in that a simple inspection of the P–F eigenvalue tells us whether or not it is possible to make a system diagonally dominant at any particular frequency, using a purely diagonal compensator. If so, inspection of an associated P–F eigenvector then helps to design the required diagonal compensator. We give some practical details here, but for more details of the theoretical background see Maciejowski (1989) or Munro (in O'Reilly (1987), Chapter 13). For the MATLAB implementation, see the entry for *fperron* in the MVFDTB manual (Ford *et al.*, 1990).

10.8.2 P–F eigenvalue plots

As mentioned above, the P–F eigenvalue can be plotted against frequency as a Bode diagram. In MATLAB, the inverse of the plant MVFR matrix (including any compensators already applied) and the associated vector of frequency points are required in creating the special diagonalized inverse MVFR matrix needed by the *fperron* command. When the P–F eigenvalue is calculated for the original plant of Example 10.5, the result appears as shown in Figure 10.28 (this was plotted by the MATLAB m-file *fig10_28.m* on the accompanying disk, which contains sufficient comments to show how it was done).

The significance of Figure 10.28 is as follows. For any system, over any frequency range where the P–F eigenvalue plot is below 2 (that is, below 6 dB), there exists a purely diagonal precompensator which, when applied to the system, will make it diagonally column dominant (on the INA). In general, the further the P–F eigenvalue is below the value of 2 (6 dB) at any particular frequency, the better should be the resulting dominance properties at that frequency. Conversely, the nearer to 2 the P–F eigenvalue gets, the harder it will be to achieve dominance. If the P–F eigenvalue plot rises above 2 at any frequency, then a more general compensator (that is, not purely diagonal) will be needed, if dominance can be achieved at all. In such cases, it may be worth adding a d.c.-diagonalizing compensator (that is, $G(0)^{-1}$) to the plant, and then trying the P–F method again.

In Figure 10.28, the P–F eigenvalue plot is less than 2 (6 dB) at all frequencies of interest. This means that the system can be made diagonally dominant at all frequencies of interest, using a purely diagonal compensator. This is no surprise, as Example 10.5 has shown it to be diagonally dominant in any case, even before the application of any compensation. We shall use the P–F approach to design the appropriate compensator to improve the dominance.

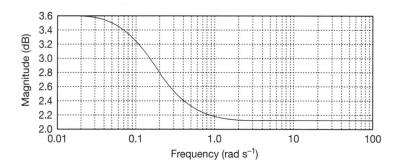

Figure 10.28 Perron–Frobenius eigenvalue plot for the pneumatic plant.

10.8.3 Use of the P–F eigenvectors in compensator design

To pursue the design, it is necessary to obtain not only the P–F eigenvalue, but also the corresponding left- and right-hand eigenvectors (readers unfamiliar with the concept of left-hand eigenvectors should consult Section A1.2.1). These P–F eigenvectors will also be functions of frequency. We shall therefore have eigenvectors each of which has a number of elements equal to the dimension of the system (that is, the number of inputs and outputs), and these elements will each have one value per frequency point. In the case of our 2×2 pneumatic plant example, we shall have left- and right-hand eigenvectors, each having two such frequency-varying elements.

It has been shown (see, for example, Maciejowski (1989)) that the behaviour of the elements of the required compensator to make the INA of the system diagonally dominant is the same as the behaviour of the corresponding elements of the left-hand P–F eigenvector against frequency. Therefore, we can plot the Bode diagrams (for example) of the elements of the left-hand P–F eigenvector, and design dynamic (Laplace transfer function) precompensators which approximate to the same Bode diagrams (the right-hand P–F eigenvector would be used to design a post-compensator for the INA, if we were to allow such a thing in practice; or a precompensator for a *direct* Nyquist array).

Since an eigenvector is correct for any scaling (it only specifies a 'direction' in space), we may as well fix one element of it to be a constant over frequency, thus reducing by one the number of dynamic compensators to be implemented. We might choose to normalize the left-hand P–F eigenvector so that element 1 is always unity. This is done by dividing every element of the left-hand P–F eigenvector at each frequency value, by the value of element 1 at that frequency. If we use MATLAB, the MVFDTB *fperron* command will give us the P–F eigenvector data as well as the P–F eigenvalue data.

Example 10.6 *The pneumatic system with a controller designed by the P–F approach*

Figure 10.28 was discussed in Section 10.8.2, and shows the P–F eigenvalue plot for the system of Example 10.1, whose INA was investigated in Example 10.5. Figure 10.29 shows a plot of the magnitude of element 2 of the left-hand P–F eigenvector against frequency. Note that this has been normalized to make element 1 unity (0 dB) at all frequencies as discussed above.

According to the P–F method, the INA will become diagonally dominant at all the frequencies shown in Figure 10.29, if we apply to loop 2 a precompensator whose frequency response is the same shape as that of Figure 10.29 (that is, of element 2 of the left-hand P–F eigenvector, having already inspected Figure 10.28 to prove that this is possible to achieve). We note from the appearance of Figure 10.29 that a first-order lag-lead compensator might approximate to this curve. As in the case of the CL method (Example 10.4), the MATLAB MVFDTB command *phlag* will design such a lag-lead compensator. We only need to specify the required gain change, the upper break frequency

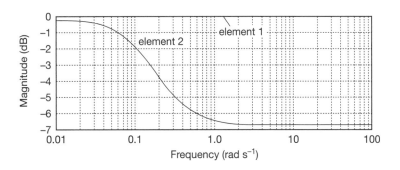

Figure 10.29 Left-hand Perron–Frobenius eigenvector plot for the pneumatic plant.

and the high-frequency gain required of the compensator (this time we do not want the default value of zero dB at high frequencies, which applied in Example 10.4).

We see from Figure 10.29 that the plot for element 2 'begins' at about -0.28 dB and then falls to about -6.68 dB. The upper break frequency appears to be at 0.2 rad s^{-1} (where the gain is 3 dB above the final value), but the lower break frequency is only slightly lower than this value, so the two will interfere with one another. The actual upper break frequency is therefore probably nearer to 0.26 rad s^{-1}. We therefore instruct the *phlag* command to design a lag-lead compensator, having a gain change of -6.4 dB, and settling at -6.68 dB, with an upper corner frequency of 0.26 rad s^{-1}.

The resulting compensator for row 2 of the INA is given by

$$\frac{0.4634s + 0.1205}{s + 0.1243}$$

and its frequency response fits quite well over the plot of Figure 10.29 (in a more difficult case, we may have to try increasing the order of the compensator to match the P–F eigenvector better, for example by using a number of cascaded lead-lag and lag-lead type elements, perhaps coupled with least-squares fitting techniques, such as those used in Section 8.3). We next build up a diagonal precompensator matrix to apply to the INA. In our case, this will have element (1,1) set to unity, and element (2,2) will contain the compensator just designed. In MATLAB this can be done in the same way as for our previous additions of compensators, giving the resulting INA of Figure 10.30(a). In comparison with Figure 10.25 (the comparable point in the previous INA design procedure using row operations) we see that the Gershgorin bands are now wider at low frequency, implying (perhaps) more interaction, but that they are much better balanced between loops 1 and 2. The resulting step responses are rather similar to those of Figure 10.26, except that in response to the step on input 2, y_2 rises to about 0.58 unit and settles back to about 0.48 unit, while the response of y_1 looks like that of y_2 in the upper plot of Figure 10.26, but settles rather more quickly.

As in Example 10.4, we now invert the compensator applied to the INA, and then proceed to add single-loop compensators to speed up the responses and remove the steady-state errors. After a little experimentation, the P $+$ I compensators contained in the TFM

$$\begin{bmatrix} \dfrac{8s + 4}{s} & 0 \\ 0 & \dfrac{4s + 4}{s} \end{bmatrix}$$

gave the step responses of Figure 10.30(b). Figure 10.30(c) shows the same responses, but with the plant input demands also displayed. The responses can be compared with those of Figure 10.27, and will be found to be slightly better in some respects and slightly worse in others, but very similar overall. We could improve the responses by further tuning, but the main point about this design has been made: namely, that we did not need the experience required to perform row operations on the INA.

The overall result

To obtain the overall compensator, we simply need to multiply the two precompensators. However, we must remember that the P–F precompensator was obtained for the *inverse*

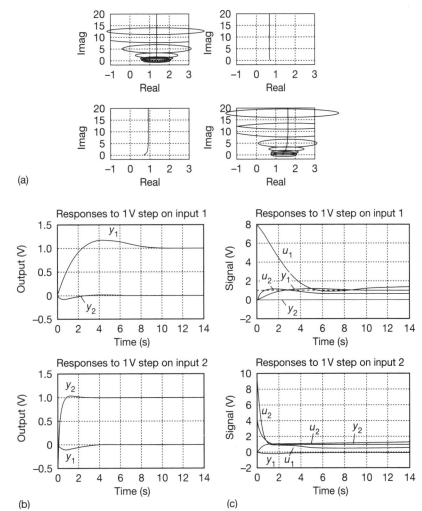

Figure 10.30 (a) INA of the pneumatic plant with P–F precompensator. (b) Step responses of the fully compensated plant (P–F method). (c) As Figure 10.30(b), also showing the plant input demands.

Nyquist array, so we need its inverse in the real world. The overall compensator is therefore given by

$$\begin{bmatrix} 1 & 0 \\ 0 & \dfrac{0.4634s + 0.1205}{s + 0.1243} \end{bmatrix}^{-1} \begin{bmatrix} \dfrac{8s + 4}{s} & 0 \\ 0 & \dfrac{4s + 4}{s} \end{bmatrix}$$

Carrying out the algebra gives the result:

$$\frac{1}{s^2 + 0.26s} \begin{bmatrix} 8s^2 + 6.0803s + 1.0401 & 0 \\ 0 & 8.6319s^2 + 9.7048s + 1.0729 \end{bmatrix}$$

which will now be compared with the previous results.

10.9   Points of comparison

In trying to summarize a few differences between the approaches considered, it must be remembered that neither the CL method nor the INA approach will be 'best' in all situations. Because the CL method concentrates entirely on eigenstructure and characteristic directions, while the INA method concentrates on overall input–output relationships, the approaches are fundamentally different. Therefore, the most appropriate one will be dependent upon the physical configuration of the system to be controlled.

Also, since CACSD packages simply present us with all the tools to apply either method, without prescribing at all *how* they should be applied, we need not keep them separate. For example, if we chose, we could design compensators using the INA display, but use the characteristic loci or singular values, and so on to check on interaction and stability.

Both the CL and the INA techniques have been used successfully in many practical applications. They do possess certain drawbacks, though. Each of the methods is capable of producing controllers which are not particularly robust in the face of disturbances. The fact that each method makes approximations does, however, have the advantage that the resulting compensators consist largely of constant matrices. They therefore give very much simpler results than the full diagonalization approach proposed in Section 10.4. The INA approach using row operations is likely to give the simplest compensator in this sense, but if the P–F approach is used, dynamic elements will be introduced leading to a similar level of complexity (roughly speaking) as in the CL method.

When using the INA approach, it is sometimes found that changes in the ordering or scaling of input and output variables can drastically affect the dominance of the system in unexpected ways. Note that the use of permutation matrices to interchange rows (or columns) of a TFM has the effect of reordering the system outputs (or inputs). The comment is therefore of direct relevance to this approach. In addition, for systems larger than 2×2, the achievement of diagonal dominance by row operations can prove a frustrating exercise unless one is both skilled and patient. Use of the P–F approach removes this uncertainty at the expense of more complicated controllers (that is, dynamic, rather than simply numeric). Finally, the achievement of diagonal dominance, although allowing the design of single-loop compensators, does not necessarily imply low interaction. Interaction can often be further reduced by increasing the feedback gains (where possible), but watch out then for unacceptable steady-state behaviour!

The CL approach also has some disadvantages. Firstly, the algorithm for alignment of the characteristic directions with the basis vectors at high frequency does not always have as great an effect as one might wish (always inspect the time responses afterwards). Secondly, the design of the approximate commutative controller is often very approximate. The CL method also requires the freedom to specify every individual element of the compensator $K(s)$. Thus no constraints can be put on the form of $K(s)$ (in the INA method, some constraints can often be placed on the form of $K(s)$ if desired).

Example 10.7 **Comparison of the numerical results of the approaches in this chapter**

We have already discussed some general points of comparison between the methods we have studied. Finally, we compare the numerical results we have obtained in Examples 10.2, 10.4, 10.5 and 10.6 for the application of the various approaches to the control of the pneumatic plant. As we have already said, the results of such a comparison may be different for a different plant. We make no comment about the time responses obtained, because they were all broadly similar and, in any case, we did not tune them as finely as we could, so any conclusions might be misleading.

In Example 10.2, we attempted complete diagonalization of the plant, and obtained a final compensator (made up of one constant and one dynamic precompensator):

$$\frac{1}{102.84s^4 + 120.55s^3 + 18.49s^2 + 0.78s} \begin{bmatrix} k_{11}(s) & k_{12}(s) \\ k_{21}(s) & k_{22}(s) \end{bmatrix}$$

with: $k_{11}(s) = 1284.7s^3 + 360.0s^2 + 33.6s + 1.04$

$\qquad k_{12}(s) = 165.4s^3 + 93.6s^2 + 12.9s + 0.52$

$\qquad k_{21}(s) = 166.8s^3 + 94.8s^2 + 13.2s + 0.54$

$\qquad k_{22}(s) = 278.6s^3 + 161.5s^2 + 23.6s + 1.02$

In Example 10.4, we used the characteristic locus approach, and the overall compensator we obtained (made up of four constant and two dynamic precompensators) was:

$$\frac{1}{s^2 + 0.0622s} \begin{bmatrix} 9.4837s^2 + 5.2892s + 0.2737 & 5.4295s^2 + 5.7315s + 1.5084 \\ 1.2326s^2 + 0.6486s + 0.0161 & 9.1572s^2 + 7.9444s + 1.6829 \end{bmatrix}$$

The overall compensator resulting from application of the inverse Nyquist array method in Example 10.5 (made up of two constant and one dynamic precompensators) was found to be:

$$\frac{1}{s} \begin{bmatrix} 8s + 4 & 5.12s + 5.12 \\ 0.96s + 0.48 & 8.6144s + 8.6144 \end{bmatrix}$$

Finally, in Example 10.6, we applied the Perron–Frobenius method, and obtained an overall compensator (made up of two dynamic precompensators):

$$\frac{1}{s^2 + 0.26s} \begin{bmatrix} 8s^2 + 6.0803s + 1.0401 & 0 \\ 0 & 8.6319s^2 + 9.7048s + 1.0729 \end{bmatrix}$$

The main differences between these results (setting aside possible performance differences for reasons already discussed) are in their complexity.

The result from the diagonalization method is obviously the most complicated in form, having third-order numerator terms and a fourth-order common denominator. It would therefore be the most difficult and the least robust to apply in practice. The INA result is the simplest, being only first order throughout, although it does involve the construction of some cross-coupling (off-diagonal) terms. It is also the simplest in structure, being largely made up of constant precompensators. The P–F result is second order throughout, but is easy to apply because it is a diagonal compensator, requiring no cross-coupling terms. The CL result is also second order, but does contain cross-coupling terms. It is also by far the most complex in terms of structure, but this is of little consequence if it is to be generated by a CACSD package.

These points of comparison will be found to apply in general to problems to which all four methods can be applied. In summary, the direct diagonalization method is not recommended, and will often not work at all. The INA method is likely to give a relatively simple compensator, but needs experience to apply. The P–F approach removes some of this requirement for design experience and gives compensators which, although probably of higher order than the INA method using row operations, are simpler (that is, diagonal) in structure; however, it can only be applied if purely diagonal compensation is a possibility for the plant in question (a d.c.-diagonalizing compensator may make this possible, if it was not originally so). The CL method gives results of greater complexity than the INA approach, but the compensator can be easier for a non-specialist to design by following a well-defined set of guidelines.

10.10 Conclusions

In this chapter we have studied a number of methods which allow us to handle multivariable systems from a frequency domain, rather than a state-space standpoint.

The main methods we studied were the characteristic locus approach, which attempts to diagonalize the system by considering its eigenstructure, and the inverse Nyquist array approach, which aims at a more approximate form of diagonalization by achieving diagonal dominance. In each case, the aim was to remove the interaction from the plant so that, finally, a set of independent compensators could be designed – one per loop – by normal SISO methods.

We also looked at an intuitive approach based on the notion of exact diagonalization by cancellation of the plant dynamics, but noted that this would be impractical except in very simple cases. In addition, we investigated the Perron–Frobenius methods, which allow the study of frequency-dependent robustness of multivariable systems designed by any method. They also predict whether a system can be made diagonally dominant by purely diagonal compensation and, if so, the form of compensator required to achieve this via the INA approach. This removes one of the difficulties of the latter approach, namely the design experience needed to perform row operations successfully.

Throughout the chapter we used an example of a real pneumatic plant, applying each method to it in turn. Although this was too simple a system really to bring out the best or worst of any of the methods, it did give a useful vehicle for demonstrating the techniques and illustrating the pitfalls. We also found that the application of any of these techniques cannot sensibly be achieved by hand, so we concentrated on a CACSD approach using MATLAB.

10.11   Problems

10.1   Figure 10.1 shows a two-input, two-output pneumatic system. One block diagram representation of it appears in Figure 10.3(b), and the corresponding transfer function matrix (TFM) is given in Equation (10.2). Obtain for this system the alternative *v*-canonical TF elements corresponding to the block diagram of Figure 10.4 (as suggested towards the end of Section 10.3).

10.2   For the system shown in Figure 10.9, show that the closed-loop TFM $H(s)$, as in $y(s) = H(s)r(s)$, is given by $H(s) = [I + G(s)F]^{-1}G(s)$ (Equation 10.9).

10.3   Show that the McMillan form of the TFM

$$G(s) = \begin{bmatrix} \dfrac{s-1}{(s+1)^2} & \dfrac{5s+1}{(s+1)^2} \\ \dfrac{-1}{(s+1)^2} & \dfrac{s-1}{(s+1)^2} \end{bmatrix}$$

(Equation (10.10)) is given by

$$M(s) = \begin{bmatrix} \dfrac{1}{(s+1)^2} & 0 \\ 0 & \dfrac{s+2}{s+1} \end{bmatrix}$$

10.4   The inverse Nyquist array for the system whose TFM is given in Equation (10.2) appears in Figure 10.23. Rather than carrying out row operations on this INA, as we did in Example 10.5 (Section 10.7), design a precompensator to diagonalize the system at d.c. (as suggested towards the end of Section 10.7.3). Evaluate the effect of this precompensator on the INA (the use of a CACSD package will make this easier but, if you do not have access to one, you should at least be able to discuss the effects of your compensator on the endpoints of the various plots).

What further compensation do you consider to be needed after your d.c.-diagonalizing compensator has been added?

10.5   It is an instructive exercise to carry out parts (a) and (b) of this problem with no computer assistance, as a successful series of compensators can be found by careful consideration of the endpoints of the plots provided in Figure P10.5.

A 9th-order, three-input-three-output system has a transfer function matrix defined by

$$G(s) = \frac{NUM(s)}{den(s)}$$

where $den(s)$ is the common denominator polynomial:

$$\begin{aligned} den(s) = {} & 2.5(s+1)(s+2)^4(s+3)^4 \\ & - 0.2(s+1)(s+2)(s+3)^7 \\ & - 0.2(s+1)^6(s+2)^3 \\ & - 0.64(s+1)^3(s+2)^3(s+3)^3 \\ & + 0.08(s+1)^6(s+3)^3 \end{aligned}$$

and $NUM(s)$ is the numerator matrix:

$$num_{11}(s) = 50(s+2)^3(s+3)^3 - 4(s+3)^6$$

$$num_{21}(s) = -10(s+1)^3(s+2)^3 + 4(s+1)^3(s+3)^3$$

$$num_{31}(s) = -16(s+1)^3(s+3)^3$$

$$num_{12}(s) = -20(s+1)^3(s+2)^3 + 16(s+2)^3(s+3)^3$$

$$num_{22}(s) = 50(s+1)(s+2)^4(s+3) - 16(s+1)^3(s+2)^3$$

$$num_{32}(s) = -20(s+1)(s+2)(s+3)^4 + 8(s+1)^6$$

$$num_{13}(s) = 2(s+1)^3(s+3)^3 - 20(s+2)^3(s+3)^3$$

$$num_{23}(s) = -5(s+1)(s+2)(s+3)^4 + 4(s+1)^3(s+2)^3$$

$$num_{33}(s) = 25(s+1)(s+2)(s+3)^4 - 2(s+1)^6$$

Figure P10.5(a) shows the inverse Nyquist array for the system, for a range of frequencies from 0.001 rad s^{-1} to 10 rad s^{-1}. Figure P10.5(b) shows an expanded version of the origin area, with Gershgorin column dominance circles superimposed.

(a) Discuss the dominance properties of the system, and their likely effects upon step responses from each input.

(b) Design a suitable constant precompensator (or a series of such compensators) which will improve the dominance properties as much as possible.

To verify the results of (a) and (b), and to attempt parts (c) and (d), a package such as MATLAB and the MVFDTB is required (if using these, try the MATLAB *help* system to look at the files *mv3step.m* and *inap9.m* on the accompanying disk). It is interesting to note how small an improvement the use of such a package brings to part (b), compared with a careful pen-and-paper solution. Such manual solutions are generally not so successful. Why can they succeed for this system?

(c) Compare the closed-loop behaviour of the system with your precompensator (or precompensators) in place, with the original performance.

(d) Design a suitable set of dynamic SISO compensators to give as little transient interaction as possible and zero steady-state

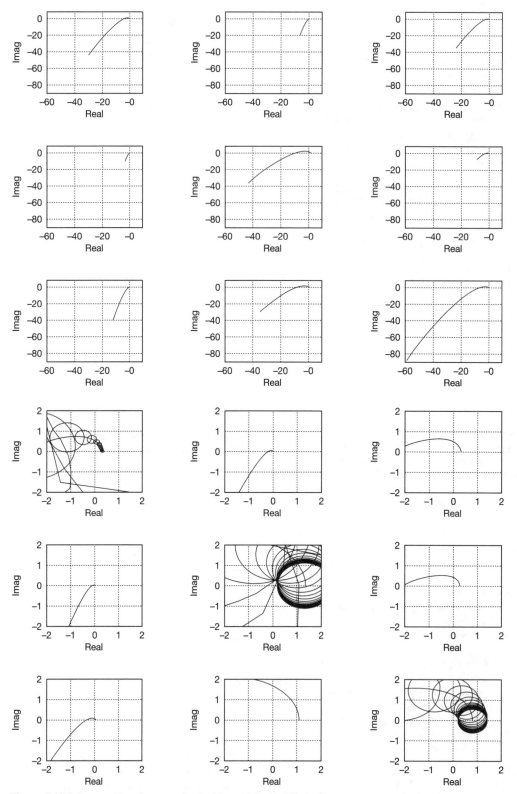

Figure P10.5 Inverse Nyquist array for Problem 10.5. (a) Whole frequency range; (b) low frequency range.

error in each output, following step signals applied to each input.

10.6 This problem requires a CACSD environment equivalent to MATLAB plus the MVFDTB. It exercises the application of the INA method to a more awkward system, in which the elements of the INA have different shapes and orders of magnitude. A two-input-two-output system has a TFM

$$
G(s) = \begin{bmatrix} \dfrac{1}{s+2} & 3 \\ \dfrac{s+1}{s^2+2s+1} & \dfrac{5}{s^2+0.8s+4} \end{bmatrix}
$$

Use the inverse Nyquist array approach to design a non-interacting (as much as possible) controller for this plant, having zero steady-state errors following step inputs. Write down the TFM of the complete compensator which would need to be applied to the real-world (that is, not inverse) plant to implement your design.

10.7 Repeat Problem 10.6 using the characteristic locus approach. Again, the CACSD environment will be found essential. Note that in this case, the *align* algorithm does *not* result in 'parallel' loci at high frequencies. Inspect the misalignment angles and singular values, and keep checking the step responses, in order to measure progress. Write down the TFM of the complete compensator which would need to be applied to the plant to implement your design. Compare your experiences with Problems 10.6 and 10.7, and also compare the results achieved.

10.8 Reconsider the 9th-order, three-input, three-output system of Problem 10.5. Use the characteristic locus approach to design a non-interacting controller for this plant, having zero steady-state errors following step inputs. The CACSD environment mentioned in the preceding two problems is again essential.

You may have found this problem rather challenging. Did you find the INA method easier for this particular problem? Compare your results with those of Problem 10.5.

10.9 Repeat Problem 10.6 using the Perron–Frobenius approach. Again, the CACSD environment will be found essential. Note that in this case, the P–F eigenvalue plot for the original system shows that no simple diagonal compensator exists. It is therefore necessary to take some initial steps (for example, applying a precompensator to achieve d.c. diagonalization), before the method can be applied. Write down the TFM of the complete compensator which would need to be applied to the plant to implement your design. Compare this solution with those of Problems 10.6 and 10.7.

10.10 Again consider the 9th-order, three-input, three-output system of Problem 10.5. Use the Perron–Frobenius approach to design a non-interacting controller for this plant, having zero steady-state errors following step inputs. A suitable CACSD environment is again essential. You may find this easier than either Problem 10.5 or Problem 10.8. How do the results compare?

11 Adaptive and self-tuning control

11.1   Preview

Readers should be able to understand this chapter if they have studied Chapters 1 to 4, Section 5.8 and Chapter 7. In addition, it would be useful to have covered Chapter 8.

We have seen that the usual practice for designing a controller is to obtain a model of the plant, to design a controller on the basis of the plant model, to test it, possibly by simulation prior to implementation, and to tune it after installation. This approach usually works well provided that the parameters of the plant do not change with time or with plant load. Unfortunately, they often do, as the following examples indicate.

- The mass of the object being moved by a robot manipulator will have a considerable effect on the dynamics of the closed-loop system, and will mean that a controller which is well tuned for an intermediate value of the mass will be less well tuned if an extreme value is used, and may even result in an unstable system.

- Changes in the thermal capacity and emissivity of metal components being heated in a modern

furnace with 'fast' linings can cause big changes in closed-loop performance.

- Variations in the width, thickness and hardness of steel strip processed in a rolling mill mean that the mill may have to be set up very differently from one customer order to the next, if good performance is to be obtained. Variations during rolling also mean that the setup has to be changed from one pass to the next in a multi-pass (reversing) mill.

- The action of the control column of an aircraft (one without automatic controls to maintain the 'feel' of the column) varies considerably in its sensitivity as the speed and altitude of the aircraft vary. A skilled pilot becomes accustomed to the variation, and is nevertheless able to fly the plane successfully; but difficulties can arise in the design of autopilots, if the required controller parameters do vary considerably throughout the flight envelope.

In this chapter we shall consider:

- the limitations of control using fixed-structure, fixed-parameter controllers
- the notion of adaptation
- adaptation by gain-scheduling (and by scheduling other parameters)
- online identification and self-tuning control
- other forms of adaptive control
- variable-structure control.

NEW MATHEMATICS FOR THIS CHAPTER

If the reader has not studied Section 9.8 and Appendix 6 (on the Kalman filter), there will be some new statistical ideas concerning random signals. These are reviewed as they are needed.

11.2   The need for adaptation

This has already been reviewed briefly in Section 11.1. There are a number of solutions to the problem of keeping a controller 'in tune', as the parameters of the system it is controlling vary. They range from 'common-sense' approaches to much more mathematical ones. The extra complexity of the more mathematical approaches is often justified by lesser hardware requirements, and more reliable operation. Note, however, that it is very difficult to *prove* the stability properties of controllers whose parameters can vary as time passes, therefore only fairly restricted adaptation may be allowed in some applications. Before considering any of these matters further, a concrete example will illustrate the degradation of performance which can easily occur due to plant parameter variation, and which gives rise to the need for adaptation in the controller.

Example 11.1 *Performance degradation of a positioning system with varying load*

A simple control system which positions a load of mass m has an open-loop transfer function of

$$\frac{Y(s)}{U(s)} = \frac{40}{ms^2 + 10s + 20}$$

It is controlled in closed-loop with unity negative feedback by a PID controller having a gain K_c of 1, a derivative action time T_d of 0.1 s and an integral action time T_i of 0.5 s. The mass m can vary from 0.2 kg to 5 kg, with an average value of 1 kg, and we shall investigate the effect of this load variation using MATLAB (Appendix 3).

   Figure 11.1 shows the arrangement. The investigation is performed using the MATLAB mfile *fig11_2.m* on the accompanying disk. The file inserts the chosen values of m into the plant transfer function, one at a time, and in each case closes a unity negative feedback PID control loop around the plant, and generates the response using the MATLAB *step* command. The PID controller transfer function is

$$\frac{U(s)}{E(s)} = K_c\left[1 + T_d s + \frac{1}{T_i s}\right] = 1 + 0.1s + \frac{1}{0.5s} = \frac{s^2 + 10s + 20}{10s}$$

Note that this idealized PID controller is strictly unrealizable because it is 'improper' (the numerator is of higher order than the denominator). See Section 4.5.2, leading to Equation (4.23), for more comment on the ideal PID representation. However, when it is combined

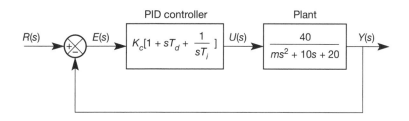

PID controller   Plant

$R(s)$ $E(s)$ $K_c[1 + sT_d + \dfrac{1}{sT_i}]$ $U(s)$ $\dfrac{40}{ms^2 + 10s + 20}$ $Y(s)$

Figure 11.1 Arrangement of a closed-loop system.

with the plant transfer function, the combination becomes 'strictly proper' (higher order denominator than numerator) and the MATLAB *step* command will then accept it.

Figure 11.2 shows the results, from which the following points are noted:

- The response when $m = 1$ kg is quite rapid and overshoot-free using these PID parameters.

- At the lower mass, $m = 0.2$ kg, the performance is degraded slightly, but not sufficiently to be a problem in practice.

- When the mass increases to $m = 5$ kg, considerable performance deterioration occurs. The initial movement becomes noticeably slower and an overshoot of approximately 30 per cent occurs.

In situations such as Example 11.1, we may be able to obtain better performance on average by setting the controller parameters to suit a load in the middle of the performance envelope (perhaps tuning to suit a mass $m = 2$ kg in the example). However, an alternative approach is to cause the controller to set up its own parameters to suit the changing plant parameters. There are three main ways of achieving this:

(1) By 'gain scheduling'. This approach is applicable to situations where the plant parameters are caused to change by changing plant load, as in Example 11.1. Referring back to the illustrations in Section 11.1, it is also widely used in flight control systems and rolling mills.

(2) By the method known as 'model-reference adaptive control', in which the controller adjusts its parameters in order to make the closed-loop plant performance resemble as closely as possible that of a closed-loop model which has the desired performance.

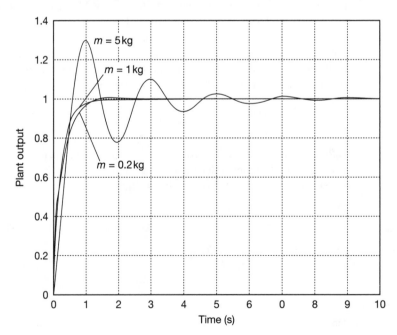

Figure 11.2 The effect of changing the plant load without changing the controller.

(3) By 'self-tuning control'. This type of control mimics, online, the procedure of a human control engineer in identifying the plant, estimating its parameters, and determining those of the controller to suit. It does this by monitoring the plant input and output in order to determine any parameter changes, and recalculating the controller parameters accordingly.

11.3 Gain-scheduling

The term 'gain-scheduling' means that the gain (and/or other parameters) of the controller is adjusted to appropriate values calculated from a knowledge of the plant load, or of other measurable changes in operating conditions, such as changes in incoming material.

In the case of Example 11.1, the mass of the load could perhaps be determined by some form of load-cell arrangement as it entered the plant. The load-cell output would be conditioned so as to give a value of the mass to the control computer, which would then calculate (or, perhaps, load from a look-up table) the appropriate values of the controller parameters. Alternatively, a value of the mass determined at an earlier stage of the process could be used; it could be entered manually by the human operator, or supplied when required by a supervisory computer system (which would be in communication with both the computer executing the gain-scheduling controller, and the computer that 'knows' the mass earlier in the process). Manual entry would be simplest, but the direct weighing approach, although more costly in terms of hardware, would be less susceptible to error.

For the system of Example 11.1, the plant transfer function was

$$\frac{Y(s)}{U(s)} = \frac{40}{ms^2 + 10s + 20}$$

and the PID controller transfer function was

$$\frac{U(s)}{E(s)} = \frac{s^2 + 10s + 20}{10s}$$

Therefore, for a mass of $m = 1$ kg, the controller cancels the plant poles and gives an overall forward-path transfer function of $4/s$. In this simple case, it would be possible to achieve such pole cancellation for all values of a mass, dangerous though it may be, by making the PID transfer function

$$\frac{U(s)}{E(s)} = \frac{ms^2 + 10s + 20}{10s}$$

A little thought will reveal that the value of the derivative action time T_d will therefore need to be equal to $0.1m$ seconds, rather than just 0.1. Practical situations are rarely as simple to work out as this, but the principle is the same.

In this instance, we are aiming to provide the same closed-loop dynamics, whatever the value of the mass m. Such an approach may prove difficult in practice because of actuator saturation; if the actuator is working at the limit of its performance to accelerate a 1 kg mass at a particular rate, it will not be able to accelerate a 5 kg mass at that rate. This point will be more fully discussed in a later section.

11.4 Self-tuning control

Self-tuning controllers have three main elements: a digital controller, a parameter estimator, and a controller synthesizer. A common configuration is shown in Figure 11.3 and that configuration will be assumed in this chapter, but arrangements which perform some or all of the compensation in the feedback loop are also used. The operation is as follows.

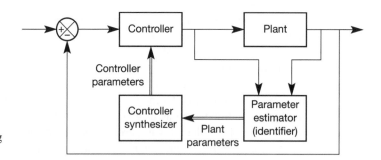

Figure 11.3 A self-tuning controller arrangement.

At the design stage, the plant is modelled (either from first principles or by identification) in order to find a general form for its model – just as would be done if the normal identification process were to be used. It is particularly important, for example, to ascertain the order of the plant and the magnitude of its transport lag, if any – self-tuners for plants having variable transport lags are possible, but are much more difficult to implement, and are outside the scope of this book. An online identifier is then produced which automates the plant parameter estimation and allows it to be performed continuously during plant operation, or on demand. This section will assume that the online identifier will produce a linear, discrete-time model of the form given in Equation (11.1).

$$y_n = \sum_{r=1}^{p} a_r y_{n-r} + \sum_{r=1}^{q} b_r u_{n-r} \tag{11.1}$$

This is known as an ARMA model (Auto-Regressive Moving Average) and is a type of discrete-time model we have considered before (see, for example, Section 2.8.1). a and b are column vectors of lengths p and q, containing the model parameters (coefficients). We shall assume that we are confident of the *order* of this model (that is, that the numbers of parameters, p and q, are correct), but that the parameter values (the elements of a and b) may vary. In principle, the parameters can be determined during system operation, by examining the time history of the plant input u_n and output y_n. The parameter determination is done by the identifier block in Figure 11.3, usually by a method based on least-squares fitting. Such a method will be described shortly. The mathematics performed by the identifier appears somewhat frightening at first sight, but it is much easier than it looks actually to implement in software, especially if a computer language with good matrix-handling ability (such as MATLAB or, for real-time control, appropriate implementations of 'C', Fortran or some other suitable technical language) can be used.

The resulting plant model is used by the controller synthesizer to determine appropriate controller parameters for good closed-loop performance. A pole-placement approach (see Chapters 5 and 7) will be described in this section, but approaches based on optimal control methods have also been used (we describe optimal controllers in Chapter 12). Such approaches to self-tuning control are described in many references, including Åström and Wittenmark (1995) and Wellstead and Zarrop (1991).

The digital controller itself differs from those described in Chapters 5 and 7 only in that its parameters are not constant, but are updated at regular intervals. An important point to note in selecting and using a commercial self-tuning controller is that some tune continuously during the operation of the plant, while others only tune when a 'tuning' button is pressed; the latter normally apply a step input in order to ensure that the plant is responding sufficiently vigorously to obtain a reliable set of parameters. As will emerge later, it proves that the true online type of self-tuner is liable to produce erroneous values for the plant parameters (and hence the controller parameters), if the plant input and output settle to a constant steady state.

The basis of the least-squares fitting normally used by the parameter estimator (often called simply the identifier) is that it adjusts the estimated plant parameter values, such that the mean value of the square of the difference between the actual plant output and the output predicted by the plant model is minimized. If we use the subscript 'm' to indicate the value predicted by the model, the aim is therefore to minimize:

$$\text{mean square error} = \frac{1}{n} \sum_{r=1}^{n} [y(r) - y_m(r)]^2 \tag{11.2}$$

where r is the number of the time step in the discrete-time process.

From a practical viewpoint, working online, it would be better if we had to use only the latest values of y and y_m to modify the existing estimates of the parameters, rather than having to start from the beginning each time (as is implied by Equation (11.2)). Fortunately, a method known as *Plackett's algorithm* exists for performing the parameter estimation interval by interval. A derivation of the algorithm is given in Appendix 7, and the original work is reported in Plackett (1950) and Norton (1986). The resulting equations are:

$$\boldsymbol{P}_n = \boldsymbol{P}_{n-1} - \frac{\boldsymbol{P}_{n-1}\boldsymbol{x}_{n-1}\boldsymbol{x}_{n-1}^{\mathrm{T}}\boldsymbol{P}_{n-1}}{1 + \boldsymbol{x}_{n-1}^{\mathrm{T}}\boldsymbol{P}_{n-1}\boldsymbol{x}_{n-1}} \tag{11.3}$$

$$\boldsymbol{\theta}_n = \boldsymbol{\theta}_{n-1} - \frac{\boldsymbol{P}_{n-1}\boldsymbol{x}_{n-1}(\boldsymbol{x}_{n-1}^{\mathrm{T}}\boldsymbol{\theta}_{n-1} - y_n)}{1 + \boldsymbol{x}_{n-1}^{\mathrm{T}}\boldsymbol{P}_{n-1}\boldsymbol{x}_{n-1}} \tag{11.4}$$

where:

$\boldsymbol{P}_n$ is the *covariance matrix* of the estimation error. The concept of a covariance matrix is developed in Sections A6.1.2, A6.2.3 and A6.2.4. It is used here as a measure of how trustworthy the parameter estimates are. Initially the terms of the covariance matrix are unknown. The usual practice is to start the algorithm with $\boldsymbol{P}_n$ having large numbers on its leading diagonal and zeros elsewhere. This 'tells' the identifier that the estimates are likely to be considerably in error, and that it must make strenuous efforts to correct them.

θ_n is a vector containing the latest parameter estimates. It is made up of the vectors a and b in Equation (11.1), stacked on top of each other (Section A1.3 introduces this type of *partitioned* vector).

x_n is a vector containing present and past values of the plant output y and input u stacked on top of each other, such that the latest value of y is given by

$$y_n = x_{n-1}^T \theta_{n-1}$$

which is equivalent to Equation (11.1), as the vector θ contains a and b, and x contains the corresponding values of y and u.

At each sampling instant, we first update θ_n by Equation (11.4), and then use Equation (11.3) to update P_n.

The following pseudo-code explains how the equations would be used in practice. The quantities *temp*, *temp1* and so on are intermediate values in the computation (scalar or vector as appropriate).

Start:

Initialize P_{n-1} with large numbers on the leading diagonal and zeros elsewhere

Initialize θ_{n-1} with estimated coefficients

Initialize x_{n-1} with actual values or zeros

Loop:

Input the latest values of u and y (u_n and y_n)

Combine u_n and y_n as x_n (see above)

Form $temp = x_{n-1}^T P_{n-1}$

Form $temp1 = 1 + (temp \times x_{n-1})$

Form $temp2 = x_{n-1}^T \theta_{n-1} - y_n$

Form $temp3 = P_{n-1} x_{n-1} \times temp2/temp1$

Form $\theta_n = \theta_{n-1} - temp3$

Form $temp4 = x_{n-1} x_{n-1}^T$ (note that this is the matrix product)

Form $temp5 = P_{n-1} \times temp4$

Form $temp6 = (temp5 \times P_{n-1})/temp1$

Form $P_n = P_{n-1} - temp6$

Set $x_{n-1} = x_n$

Wait for next sampling interval (or synthesize the controller first – see later)

Goto Loop

If storage were at a premium in a small system, reuse of some of the temporary quantities would be arranged – for example, *temp3* could be reused instead of *temp5*.

When Plackett first developed the algorithm in 1950, computers were in their infancy and it was not possible to use the algorithm in an online identification context. Even trying it out in practice with manual calculations must have been very laborious. Such experimentation can now readily be performed with a computer package such as MATLAB (Appendix 3), which will be used in this chapter to demonstrate the capabilities (and limitations) of many of the methods which will be described.

The work on identification and self-tuning control will be demonstrated in action using results from the MATLAB m-file *fig11_4.m* on the accompanying disk. It simulates the action of a second-order plant controlled in the closed-loop by a self-tuning Dahlin controller, with its ringing pole removed (see Section 7.6). The method of calculating the controller coefficients is fully described in Section 11.5. A Dahlin controller was chosen for the demonstration as it is probably the easiest in respect of online coefficient determination. We will see later that producing an apparently more straightforward self-tuning PID controller is actually more difficult!

Example 11.2 *Using Plackett's algorithm to identify a second-order type 0 plant*

The plant transfer function in the m-file *fig11_4.m* is that from Example 11.1, namely:

$$\frac{Y(s)}{U(s)} = \frac{40}{ms^2 + 10s + 20}$$

The value of m (the mass in the modelled process) is set to 1 kg initially, is changed to 5 kg after 20 seconds and is changed to 0.2 kg after 40 seconds, in a total simulation time of 60 seconds. A sampling interval of 0.05 s was chosen and the m-file was run using a low-frequency square wave applied to the setpoint. No controller synthesizer is included in this m-file, so the controller is fixed during these runs – the only variation is in the value of mass. The run was subsequently repeated for a single initial step input and for a sinusoidal input (the m-file allows selection of these inputs from an on-screen menu).

The structure of the z-transform of the plant plus zero-order hold can be deduced to be (see Section 5.8.2):

$$\frac{Y(z)}{U(z)} = \frac{b_1 z^{-1} + b_2 z^{-2}}{1 - a_1 z^{-1} - a_2 z^{-2}} \tag{11.5}$$

The results are shown in Figure 11.4, and the calculated ('correct') values of the parameters are given in Table 11.1.

It is noteworthy that the identified coefficients differ, depending upon the input signal used. The results for the square-wave input (Figure 11.4(a)) are in broad agreement with the expected results from Table 11.1 for the first two values of mass (that is, at the ends of the time periods 0 to 20 seconds, and 20 to 40 seconds), but not for the third (40 to 60 seconds). This is largely because the parameters are slow to converge onto their final values – at 60 seconds, there is still a lot of change to come.

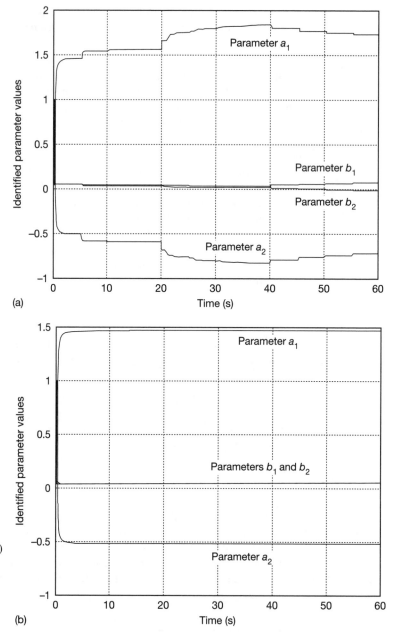

(a)

(b)

Figure 11.4 (a) Identified parameter values using a repeated step (square-wave) input. (b) Identified parameter values using a single-step input. (c) Identified parameter values using a sinewave input.

In the case of the single step input (Figure 11.4(b)), the lack of excitation of the plant as the step settled resulted in the parameter values obtained for the original mass of 1 kg persisting even when the mass changed.

With the sinusoidal input (Figure 11.4(c)), the problem is that only one frequency is present. This violates the assumptions made in deriving Plackett's algorithm and consequently causes biased parameter values. The identifier notes that a change in parameters has occurred, but merely produces a further set of incorrect values.

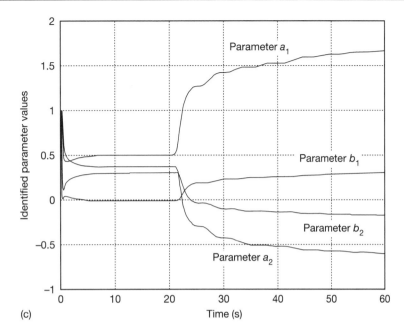

Figure 11.4 (*Continued*) (c)

Table 11.1 Calculated values for the coefficients of the model of Equation (11.5).

Coefficient	a_1	a_2	b_1	b_2
True value for $m = 1.0$ kg	1.567	−0.607	0.042	0.036
True value for $m = 5.0$ kg	1.895	−0.905	0.0097	0.0093
True value for $m = 0.2$ kg	0.992	−0.082	0.124	0.056

For such identifiers to operate correctly, the input must not settle and must contain a good range of frequencies (that is, it must be '*persistently exciting*'). The single step eventually settled, whereas the sinewave, though it did not settle, only contained one frequency. The repeated step (square wave) input had a much higher harmonic content and so was able to produce a more accurate set of estimated coefficients.

For comparison purposes, a system having the identified parameters (from the portion of each input test corresponding to a mass $m = 1$ kg) was simulated, and compared with the response of a model with the 'correct' parameters from Table 11.1. The results appear in Figure 11.5, where we see that the step response of a system having the identified parameters from the square-wave input test is indistinguishable (on the scale of this figure) from that of the actual plant. The parameters from the test using a sinewave input cause a response that is initially slightly in error. Those from the single-step test appear to be closer to the correct result, but Figure 11.4(b) shows that the identified values are nearest to those relating to the original situation with m having a value of 1 kg, suggesting that the identifier made no further progress after the system settled down, so the results would be far worse for the other values of m.

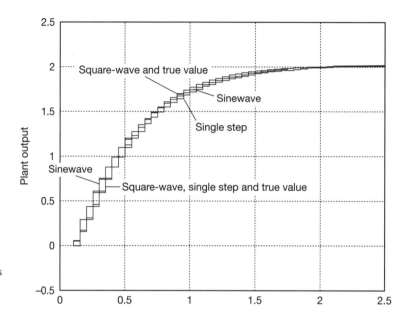

Figure 11.5 Step responses using the identified parameters.

11.4.1   The forgetting factor

It is possible to make the identifier converge faster, achieve better parameter estimates, and react faster to plant changes, by the use of a 'forgetting factor'. This idea arises from the fact that the size of the elements in the covariance matrix P in Plackett's algorithm is a measure of how much in error the estimated coefficients in the vector θ are (that is, a_1, a_2, b_1 and b_2 in Example 11.2). The more they are in error as measured by the covariance matrix, the more the coefficients in the vector θ will be changed at each iteration to correct their values.

It is therefore possible to speed up the convergence by artificially increasing the magnitude of the elements of the covariance matrix. This can be achieved by dividing them all by a number less than 1 at each iteration. The divisor is called the 'forgetting factor' and a value between 0.9 and 0.99 is usual. If it is made much less than 0.9, instability of the identifier is likely to result if the plant input settles. In physical terms, the forgetting factor has the effect of causing the identifier to 'forget' the effects of old data, and concentrate more on present data. The result of using a forgetting factor is demonstrated by the following example.

Example 11.3 *Repeating Example 11.2, using a forgetting factor of 0.9*

The 'experiments' of Example 11.2 were repeated, except that a 'severe' forgetting factor of 0.9 was used. The m-file *fig11_6.m* used for this example is identical to *fig11_4.m*, apart from this. The results were as follows.

For the single-step input, the results improve, but are still only good for the initial value of mass. The sinewave input gives good results for the 5 kg mass, but not for the other two masses. The most worrying aspect of the response to this input is that at 40 seconds, when the mass changes from 5 kg to 0.2 kg, the identifier tends to instability, giving transients of plus and minus several hundred units in the parameter estimates. This behaviour can continue in such identifiers, leading to total instability (a phenomenon known as '*estimator blow-up*').

In the case of the square-wave input, the accuracy is very good. The identified parameter values are shown in Figure 11.6, and the final values in each 20 s time period are all in agreement with those of Table 11.1 to three decimal places. The correct parameter values are also reached quickly (after about five seconds from the start, and after about one second after each mass change).

The conclusion is that very successful parameter estimation can be performed with the aid of a forgetting factor of less than unity, but only if an appropriately changing input is present. A lesson to be learned is that a self-tuning controller must incorporate a check for unreasonable values of the estimated parameters within its software, so as to avoid blow-up.

It is instructive to examine the final contents of the covariance matrices in the square-wave input tests with both values of the forgetting factor. With a forgetting factor of 1 (Example 11.2), the final covariance matrix was:

$$
\boldsymbol{P}_{F/F=1.0} = \begin{bmatrix} 0.7520 & -0.7163 & 0.0131 & -0.0698 \\ -0.7163 & 0.6844 & -0.0113 & 0.0648 \\ 0.0131 & -0.0113 & 0.0119 & -0.0073 \\ -0.0698 & 0.0648 & -0.0073 & 0.0190 \end{bmatrix}
$$

With a forgetting factor of 0.9, the final covariance matrix was:

$$
\boldsymbol{P}_{F/F=0.9} = \begin{bmatrix} 9.3114 & -8.4904 & -0.1803 & -1.4618 \\ -8.4904 & 7.7480 & 0.1584 & 1.3263 \\ -0.1803 & 0.1584 & 0.0469 & -0.0031 \\ -1.4618 & 1.3263 & -0.0031 & 0.2740 \end{bmatrix} \times 10^5
$$

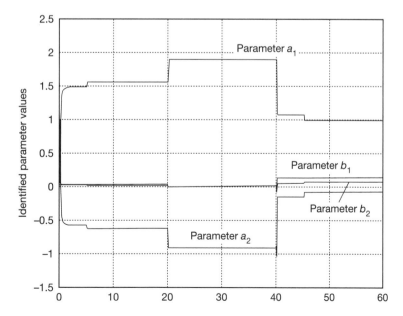

Figure 11.6 Parameter variation with a square-wave input to the closed-loop system (forgetting factor = 0.9).

Since the covariance matrix is a measure of the accuracy of our parameter estimates, and since large values in the matrix mean poor accuracy, it seems illogical that the values for a forgetting factor of 0.9, which gives excellent parameter-estimate accuracy, are so much larger than those for the *less* effective forgetting factor of 1.0 (which had not finished converging, and so would eventually be even smaller than the values above). The reason is simple; the forgetting factor of 0.9 increases the covariance matrix values by over 11 per cent at each iteration, which amounts to an overall factor of $(1/0.9)^{1200}$, or 8.11×10^{54}, over the 1200 iterations made by *fig11_4.m* and *fig11_6.m*. The terms of the covariance matrix become very small indeed if we divide them by the latter figure. That figure also makes it easy to see why forgetting factors introduce the risk of causing the identifier to become unstable.

11.4.2 Some other practical points

Straightforward ordinary least-squares, the basis of Plackett's algorithm, can have other limitations in practice. The following extensions and modifications can improve its operation considerably.

- In practice, the magnitude of some members of the vector θ is considerably smaller than others. A given actual error in those parameters will therefore constitute a greater percentage error than for the larger terms. The percentage errors can be balanced in terms of their effect on the mean-square error by a weighting matrix W (Norton, 1986).

- If noise is present in the measurements of the output y, the algorithm may not work well. If the noise is Gaussian, least-squares should average it out, but true Gaussian noise does not exist! It is possible by a method known as 'extended least-squares' to allow for the noise on the basis that it is linearly filtered white noise. The method relies on the calculation of the output noise as the difference in the predicted and observed values of the output y (*temp2* in the pseudo-code algorithm given earlier in Section 11.4). The vector x is now extended to include previous values of the noise, and the vector θ is likewise extended to contain coefficients giving a prediction of the contribution of those noise values to the predicted latest value of y. More complicated noise models are sometimes used (see Norton (1986), Warwick (1988) and Wellstead and Zarrop (1991)).

- The numerical properties of Plackett's algorithm are such that rounding errors in the computation become a problem if fixed-point arithmetic, or even short-wordlength floating-point, are used. This is less of a problem now that appropriate fast and cheap processors are available – it is often possible and economic simply to use a 40-bit floating-point wordlength, which is usually adequate. An alternative approach is based on matrix factorization and square-roots (typically using a 'U-D factorization'); the equations can be square-rooted, effectively much reducing the percentage errors (see Norton (1986) or Wellstead and Zarrop (1991)).

11.5 The controller synthesizer

The easiest approach to controller synthesis is by the method of pole placement in the z-plane, as the plant is effectively identified in z-transform form. The method works by specifying the closed-loop pole positions required, and calculating the required controller poles. The Dahlin controller is one such controller type and it will be used as an example to illustrate the principle; the synthesis calculation will be modified as described in Section 7.6, to eliminate the 'ringing pole' to which the Dahlin controller is inherently susceptible.

We begin with Equation (11.1), and we assume that the identified parameter vector $\boldsymbol{\theta}$ has been split up to yield the vectors $\boldsymbol{a}$ and $\boldsymbol{b}$ containing the coefficients $a_1 \ldots a_p$ and $b_1 \ldots b_q$ of that equation. Equation (11.1) effectively corresponds to a z-transform representation of the plant as in:

$$G(z) = \frac{Y(z)}{U(z)} = \frac{\displaystyle\sum_{n=1}^{q} b_n z^{-n}}{1 - \displaystyle\sum_{n=1}^{p} a_n z^{-n}} \tag{11.6}$$

Also, in Section 7.6, when a Dahlin controller is causing the closed-loop unit step response to follow an exponential $(1 - e^{-\beta T})$, substituting Equation (7.10) into Equation (7.5) gives the z-transform of the required Dahlin controller as:

$$D(z) = \frac{1 - e^{-\beta T}}{(z - 1)G(z)}$$

Substituting Equation (11.6) into this result gives:

$$D(z) = \frac{(1 - e^{-\beta T})\left[1 - \displaystyle\sum_{n-1}^{p} a_n z^{-n}\right]}{(z - 1)\displaystyle\sum_{n=1}^{q} b_n z^{-n}}$$

Relating the above equation to the simple second-order example, we obtain:

$$D(z) = \frac{(1 - e^{-\beta T})(1 - a_1 z^{-1} - a_2 z^{-2})}{(z - 1)(b_1 z^{-1} + b_2 z^{-2})}$$

and we notice that the denominator term containing b_1 and b_2 has a pole at a negative value of z. In our case, the pole will be at $z = -b_2/b_1$. This is the 'ringing pole' in the controller, and we remove it (as in Section 7.6) by replacing it by an equivalent static gain of $(b_1 + b_2)$. We also note that a factor of z^{-1} must be extracted from this term to leave only the pole. The expression for $D(z)$ therefore becomes:

$$D(z) = \frac{U(z)}{E(z)} = \frac{(1 - e^{-\beta T})(1 - a_1 z^{-1} - a_2 z^{-2})}{(z - 1)z^{-1}(b_1 + b_2)}$$

where $U(z)$ and $E(z)$ are the z-transformed plant input (controller output) and error in the feedback loop (controller input) respectively (see Figure 11.1).

The discrete-time equation for the controller can therefore be found as:

$$U(z)(1 - z^{-1}) = \frac{(1 - e^{-\beta T})}{(b_1 + b_2)} E(z)(1 - a_1 z^{-1} - a_2 z^{-2})$$

or

$$u_n = u_{n-1} + \frac{1 - e^{-\beta T}}{b_1 + b_2} e_n - a_1 \frac{1 - e^{-\beta T}}{b_1 + b_2} e_{n-1} - a_2 \frac{1 - e^{-\beta T}}{b_1 + b_2} e_{n-2}$$

These results can be used by the controller synthesizer and digital controller blocks in the self-tuner.

Example 11.4 *The use of self-tuning for the second-order plant used in the earlier examples*

The MATLAB m-file *fig11_7.m* on the accompanying disk is used with the self-tuning disabled for the first run and enabled for the second. This is selected by an on-screen prompt when the file is run. For the first tests, the square-wave input which produced the best identification is used with a forgetting factor of 0.9 as in Example 11.3. Figure 11.7

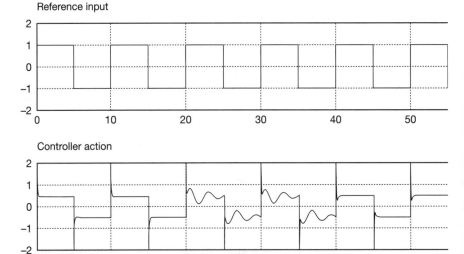

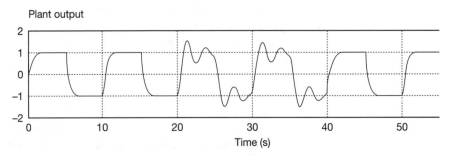

Figure 11.7 Plant output and controller action for the system of Example 11.4 with the controller synthesizer disabled.

shows the response with the self-tuning disabled and Figure 11.8 that with the self-tuning enabled. The following conclusions can be drawn:

- The general form of the response is good when the controller is properly tuned, giving a reasonably rapid movement to the demand value without significant overshoot.

- There is considerable deterioration of the performance of the system without the self-tuning (Figure 11.7) in the middle period when the mass is 5 kg. This deterioration is largely eliminated when the self-tuning is enabled (Figure 11.8).

- The only disadvantage of the self-tuning here is that a considerable overshoot occurs during the period when the mass is 5 kg (it actually extends beyond the plot of Figure 11.8 to ±10 units). This overshoot can be prevented by more rigorous conditions in the 'software jacketing'; this term relates to the checks incorporated in practical self-tuners to detect unreasonable parameter values. The only check used in *fig11_7.m* is for a steady-state plant gain between 0 and 10 (it is actually 2.0 for most of the time, but does approach a magnitude of 10 during the initial transient). It is possible for the steady-state gain to be reasonable despite a set of parameters which lead to a poor controller; one of the end-of-chapter problems asks the reader to modify *fig11_7.m* to incorporate better checks for poor controller parameters.

- Examination of the controller action shows a considerable variation in its maximum value between the extreme values of the load mass; from less than 1 unit for a mass of

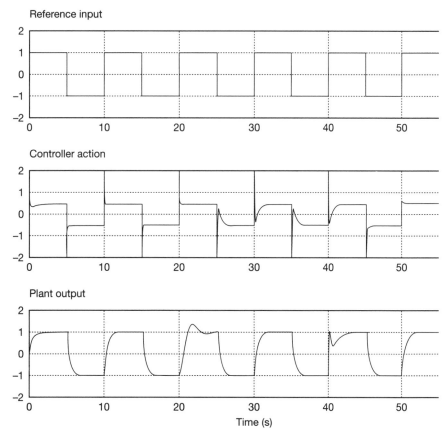

Figure 11.8 Plant output and controller action for the system of Example 11.4 with the controller synthesizer enabled.

0.2 kg to about 10 units (see above) for a mass of 5 kg (Figure 11.8). This variation is to be expected, because we are trying to achieve the same closed-loop dynamics for a wide range of mass – so Newton's second law suggests that a wide range of force will be required. There are considerable operational advantages in having unvarying closed-loop dynamics but that does imply that the speed of response to any input is restricted to that achievable by the plant and actuator with the *maximum* load; so the response at lower loads is certain to be less rapid than it could have been. A possible compromise solution is discussed in respect of self-tuning PID controllers in the next section.

11.6   The controller synthesizer: self-tuning PID

The apparently simpler PID controller is in fact more difficult to implement in self-tuning mode than is the pole-placement type. The two usual Ziegler–Nichols approaches for determining the PID parameters are the process reaction curve method which works from the step-response, or the method of increasing the controller gain until oscillation occurs and calculating the required PID controller parameters (K_c, T_d and T_i) from the resulting gain and frequency. It would be possible to determine the open-loop step-response from the identified plant parameters and use it to determine the Ziegler–Nichols parameters R and L as used in Section 3.3.4 (it is quite easy to program their determination, though that operation is quite computer time-consuming). The method, however, assumes that the open-loop step response settles without overshoot.

A MATLAB m-file *figspid.m* has been included on the accompanying disk to demonstrate the operation of such a self-tuner; its identifier is identical to that of *fig11_7.m*, but its controller synthesizer operates on the above principle, using the MATLAB function *max* to determine the maximum slope of the response and the time at which it occurs. A simple geometrical calculation then determines K_c, T_d and T_i. A problem encountered in running this file is that it works much better at reasonably short sampling intervals than at longer ones, probably because the Ziegler–Nichols parameter L is determined more accurately if the increments of time are shorter. A practical version would benefit from a variable sampling interval determined by the software, perhaps as a fixed fraction of L.

It would likewise be possible to use Jury's test (Leigh, 1992), or an approach based on frequency response, to predict the gain required to cause closed-loop oscillation and the gain required to sustain that oscillation; but that approach would not work with a second-order plant as oscillation would only then occur as a result of the digitization, and at an unrealistically high gain and frequency.

An alternative approach is to proceed as a control engineer might in tuning the controller by hand; by observing the *closed-loop* step overshoot with no derivative or integral action in use (or a known, but not necessarily ideal, amount of both), and the corresponding damped frequency, and using their values in conjunction with simple heuristic rules to determine suitable values for K_c, T_d and T_i. It is semantically debatable whether the resulting system is self-tuning, rather than another form of adaptive control, as no plant transfer function or correspondingly precise representation is produced, but the closed-loop response could itself be argued to be a system model of sorts.

There is at least one commercial self-tuning controller which operates on a related 'pattern recognition' principle and is successful in many practical situations; it has the further advantage that it waits for a naturally-occurring transient rather than needing to have a step specially applied (Kraus and Myron, 1984). Again a MATLAB m-file *fig11_9.m* has been provided on the accompanying disk to demonstrate the operation of this kind of approach in respect of a 'press the button to tune' self-tuner. The file prompts for the value of mass, and a controller test gain, and determines the closed-loop step response for the test gain value k_t. It then determines the times to, and the value of the response at, the first peak and the first trough. It then proceeds as follows.

(1) The peak times are used to determine the damped natural frequency f_d.

(2) The height of the first peak (measured from zero) and the height of the first trough above zero are used to calculate the ratio of the former divided by the latter. This decay ratio is then subtracted from the empirically-selected value of 8 to give the quantity *mmn* in the m-file. The frequency at which oscillation would occur with increased gain, and the gain required, are then calculated from the following equations.

$$f_n = \frac{f_d}{\left[1 - \dfrac{mmn^{3.5}}{1110}\right]^{1/3.5}}$$

and

$$kosc = \frac{k_t}{\sqrt{\left[1 - \dfrac{mmn^2}{55}\right]}}$$

(3) The numerical constants in these equations have been obtained by trial and error and appear to work well for a selection of practical plant transfer functions.

(4) The PID controller constants are based on the Ziegler–Nichols 'oscillation' relations given in Section 3.3.4. However, the values have been adjusted for better responses here, and could be further adjusted in practice if the considerable step overshoot given by the Ziegler–Nichols recommendations is too large. We have used $K_c = 0.5 \times kosc$, $T_d = 0.25/f_n$ and $T_i = 1/f_n$.

(5) A simulation using the calculated PID parameters is performed; in each case, the step input is rate-limited to a rate of rise of $1/[5(T_d - T)]$ units per second (T is the sampling period). Rate-limiting is a common technique in digital PID controllers, whether self-tuning or not, in order to restrict overshoot and keep controller action to a reasonable level while allowing a large enough value of T_d to be used to provide reasonable damping, and enough gain to be used to counteract steady-state disturbances satisfactorily. The actual rate limit is again chosen for experiential reasons.

(6) It is unlikely that precisely this method is used by any commercial self-tuning PID controllers, but it is included as an example of what can be achieved by a mathematically unsophisticated approach, if it is allowable that the self-tuning

facility should only operate initially on a 'press the button to tune' basis. Its effectiveness in conjunction with the example system is shown by Figure 11.9. A sampling interval of 0.05 s was used for each value of m, and it is clear that good responses are obtained, except that the response for the mass of 5 kg is somewhat slower than the others. It would be possible to modify *fig11_9.m* to overcome some of its remaining deficiencies – an opportunity to do so is given in the end-of chapter problems, where appropriate guidance is given.

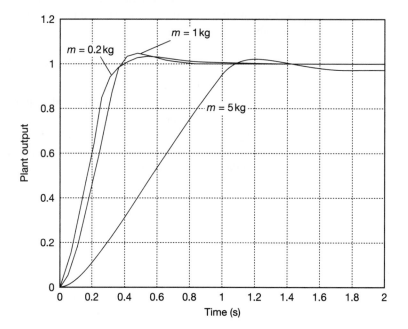

Figure 11.9 Responses of the closed-loop system using the controller parameters from the MATLAB m-file *fig11_9.m*.

11.7 Model-reference adaptive control

Model-reference adaptive control (MRAC) is an alternative to self-tuning control, in which the controller is updated without the intermediate identification operation. The principle is shown by Figure 11.10 which shows two possible configurations. The aim is to cause the closed-loop output to be the same as that of a model G' fed with the same input as the actual closed-loop system, so that the desired closed-loop transfer function $G'(z)$ (or $G'(s)$) is specified.

Figure 11.10 (a) One arrangement of a model-reference adaptive control scheme. (b) Another arrangement of a model-reference adaptive control scheme.

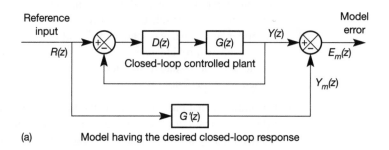

(a)   Model having the desired closed-loop response

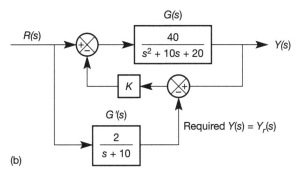

Figure 11.10 (*Continued*)   (b)

In Figure 11.10(a), the parameters of the controller $D(z)$ are adjusted to minimize the 'model error' $E_m(z)$. The arrangement is a simplified version of that reported in Banks (1986) in which a feedback-path controller was also included. A matrix algebra-based approach was used to perform the controller parameter adjustment. Simple gradient optimization methods could also be used.

Figure 11.10(b) shows a simple (in principle) alternative possibility, in which the feedback signal, rather than the controller parameters, is adjusted by the gain K (the *adaptation gain*) to produce the required closed-loop performance (Leigh, 1992). The closed-loop transfer function in s can be shown (by block diagram algebra) to be

$$\frac{Y(s)}{R(s)} = \frac{G(s)[1 + KG'(s)]}{1 + KG(s)}$$

which reduces, if K is large, to $G'(s)$. The closed-loop transfer function can therefore apparently be made equal to $G'(s)$ simply by making K large. The following example shows how the method might work out in practice.

Example 11.5 *Application of a MRAC system*

Investigate the notion of MRAC as described above, with reference to the system shown in Figure 11.10(b).

In this instance, the arrangement was simulated using SIMULINK (Appendix 4), using values of K of 1 (zero would correspond to open-loop conditions with no adaptation), 5 and 100. The SIMULINK block diagram input is shown in Figure 11.11, and the plant outputs in response to a unit step input are shown in Figure 11.12.

The response at a K value of 1 is biased towards that of $G(s)$ (it would settle at 2 for $K = 0$), while that at a K value of 100 closely resembles the response of $G'(s)$ to the same input; the response at a K value of 5 is intermediate in nature. Both the latter responses show a decaying oscillation at quite a high frequency, which is likely to be a problem in practice for two reasons: such frequencies may excite unmodelled plant dynamics and will in any case demand a very high sampling frequency to cope with the effects.

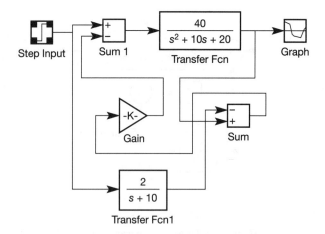

Figure 11.11 SIMULINK block diagram input for Example 11.5.

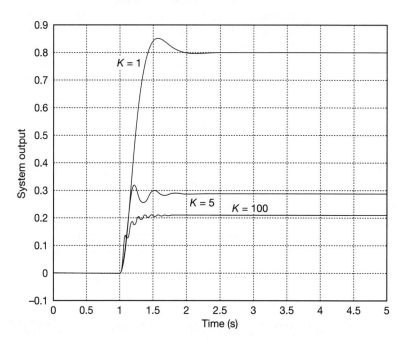

Figure 11.12 Step responses for the system of Example 11.5 with the values of the gain K indicated.

MRAC has encountered more stability problems in practice than has self-tuning control and, as a result, it is less widely used in real situations.

11.8   Variable-structure control

In our treatment of self-tuning control, 'online identification' applied only to the parameter-estimation operation; the general structure of the plant model was predetermined and remained constant throughout. In practice, it is often desirable to be able to change that structure, and the general strategy of the control, to meet changing circumstances. For example, a hydraulic position-control system faced with the task of moving a mass from A to B will probably operate best by controlling for constant acceleration up to a maximum speed determined by pump limitations (and safety); then maintaining that maximum speed until the mass is near B; and only then going into closed-loop 'position mode' to accomplish the final approach. In this instance, the first and last periods will require a plant model of the usual variety (probably the same model in each case), while the middle one will need no model at all if it does simply require the hydraulic valve to remain fully open. So both the controller and the identification arrangement will change in nature with the circumstances. In this example, the changes could be arranged on a common-sense basis, with the identifier only operating in the initial and final periods of the move and an appropriate controller synthesizer operating in each of those periods.

Another manifestation of variable-structure control is 'sliding-mode' control, in which a controlled variable in a nonlinear system can be caused to 'slide' along a desired phase-plane trajectory (see Section 14.5). A simple example is a thermostat-equipped domestic immersion heater in which power is on until the temperature rises to a temperature T_1, and then switches off until the temperature falls to a slightly lower temperature T_2. The closer together these two temperatures, the faster is the switching frequency of the thermostat. If the temperatures were equal, the tank would remain at precisely that temperature – at the cost of considerable thermostat wear if the thermostat were electromechanical. In some practical situations (but not the example just quoted), it is realistically possible to use the principle. Section 14.5.4 introduces the topic, and Slotine and Li (1991) discuss it in more detail.

11.9   Conclusions

In this chapter we have reviewed the broad area of adaptive controllers – control systems which adapt themselves to their changing surroundings. The specific areas we discussed included adaptation by gain scheduling (changing the controller parameters in a pre-programmed manner, dependent upon operating conditions), model reference adaptive control (in which the closed-loop system is made to follow the response of a prespecified model responding to the same input) and variable-structure control (in which not only the controller parameters, but also the structure or strategy of the controller are allowed to vary).

The bulk of the chapter concentrated on self-tuning control, in which a controller structure is chosen (we used a Dahlin controller and a PID controller as examples), but its parameters are tuned during use, to reflect changing performance requirements as the plant parameters change. The tuning may be accomplished by using a parameter identifier to estimate the values of the changing plant parameters (we used Plackett's algorithm, based on a least-squares approach), and a controller synthesizer to tune the controller based upon the identified plant model.

We presented a number of examples with simulation results, from which we drew conclusions about the performance of the various controllers.

11.10 Problems

With this subject, it is difficult to include numerical examples for manual calculation. For that reason, some suggested experiments with the MATLAB m-files mentioned and used in the chapter are given. The reader is reminded to copy the files first, as the examples suggest modifying them in various ways.

11.1   Using the m-file *fig11_2.m*, choose suitable parameters for the PID controller which will minimize the spread of the closed-loop system performance as the mass *m* varies *without* re-tuning the controller. What is the resulting maximum overshoot?

11.2   Using the m-file *fig11_7.m*, carry out the following experiments:

(a)   With the forgetting factor set to 1.0 and the self-tuning disabled (that is, unchanged controller parameters), investigate the effect of using sampling intervals of 0.01 s and 0.2 s rather than the default 0.05 s. Repeat the test with a forgetting factor of 0.9. Was there any difference in estimation accuracy or stability? Remember that the plant parameters are dependent on sampling interval – the MATLAB function *c2dm* will do the *z*-conversion for you.

(b)   Try (a) again with the self-tuning enabled.

(c)   Now try (a) and (b) with a small quantity of noise (Gaussian, generated by the MATLAB random-number generator – try *help rand*)

injected. How much noise is tolerable before significant performance degradation occurs?

11.3   Norton (1986) and other authors state that, for full identification of a system with a sinusoidal input, at least two frequencies must be used.

Modify *fig11_4.m* such that, when a sinewave input is selected, it contains two frequencies – and see whether it works better with such an input.

11.4   Examine the 'controller synthesizer' part of *fig11_7.m* and find the 'software jacketing' instructions. Replace or supplement them with instructions which allow new controller parameters to be calculated only when no plant parameter has changed by more than 1 per cent in the previous interval. Try the resulting file (for example, for a square-wave input and a forgetting factor of 0.9). Are stability and control improved significantly?

11.5   Run *figspid.m* trying a sampling interval of 0.05 s initially. Is stability improved by de-tuning the controller parameters from the Ziegler–Nichols values (for example, making the controller gain K_c equal to $0.8/(RL)$ instead of $1.2/(RL)$)? Again, see if improving the 'software jacketing' is beneficial.

11.6   Try out *fig11_9.m* investigating in particular the effect of the sampling interval and of using a higher-order plant. Can it readily be modified to select its sampling interval automatically?

12 Optimal control

12.1   Preview

This chapter can be understood following a study of Chapters 1 to 4, and Sections 1 to 7 of Chapter 5. No other chapter in the book depends upon a knowledge of the contents of this chapter, so it can be omitted if desired.

The only topic in this chapter is an introduction to the design of optimal controllers. In fact, the optimal design process presented here will be found to be one more technique for designing state variable feedback controllers – but from a different viewpoint from that adopted previously. In principle, the design seeks to minimize some specified performance index, so as to obtain minimum rise time, minimum settling time, minimum control energy expenditure, maximum product yield, or some similar criterion. It will be discovered that this is often difficult to achieve in precisely that form in practice, but that the design methodology is, nevertheless, extremely useful, and *can* be used to obtain such desired results if used pragmatically.

In this chapter, the reader will learn:

- what 'optimal' means in this context, with practical examples
- that an optimal controller is another form of state variable feedback
- the use of performance indices
- one method of deriving the optimal controller
- that optimal control in its ideal theoretical form is rarely of practical use, but that the underlying ideas can be routinely applied in a useful practical design approach
- that optimal control can be successfully applied using computer packages, without necessarily understanding every line of the derivation of the methodology.

NEW MATHEMATICS FOR THIS CHAPTER

Mostly, the mathematics in this chapter involves matrix algebra, of the kind already encountered in Chapter 5. For readers in need of some revision of those techniques, the first few sections of Appendix 1 contain all that is required. Vector–matrix quadratic forms are new in this chapter, and are introduced as they are needed.

12.2 Introduction

In theory, optimal control differs from all the other techniques in this book, in that it seeks to control the plant so as to get the best possible (that is, *optimal*) performance from the plant. In order to achieve this remarkable aim, those aspects of the plant's behaviour which it is desired to control (optimize) are incorporated into a mathematical expression (a *performance index*), and the controller design process synthesizes a controller which will minimize that expression. For example, for a very rapid response, the rise time and settling time of the system could be incorporated in a performance index, and the controller would drive the plant in such a way as to minimize these.

In practice, things do not work out quite so neatly, as will be revealed later (for example, the minimization of rise time and settling time would probably *maximize* control energy usage and hence operating costs; also, unless the plant model is very accurate, the performance will not be as 'optimal' as the design predicts). However, the ideas underlying optimal control are good, and the resulting design methods can be used for the rapid development of controllers with useful properties. Many such controllers are operating successfully around the world.

Optimal control, as studied in this chapter, is based on state variable models of systems. In Section 5.4, the design of state variable feedback (SVF) systems by *pole-placement* was presented, and some ways of choosing where the closed-loop poles should be placed were mentioned. Optimal control methods provide another way of placing the closed-loop poles of a system, in order to achieve the desired behaviour. In this case, the designer does not know the closed-loop pole locations that will result in the optimal control of the plant. Instead, the poles are placed by the optimal control design procedure, in locations which seek to make the resulting closed-loop behaviour the best possible (hence *optimal*), in whatever sense has been specified in advance.

It will be shown that the optimal controller, in the cases considered here, turns out also to be SVF. Thus, these optimal control methods are one more way of choosing the contents of the feedback matrix K in an SVF scheme.

In the design of single-input-single-output (SISO) SVF schemes, Chapter 5 showed that there was a unique feedback vector (k) which would place the closed-loop poles as required. In the multivariable (MIMO) case, Chapter 5 showed that there could be a large number of degrees of freedom in choosing the feedback gains. For example, in a system having n_s states and m inputs, the dimensions of the feedback matrix K are $m \times n_s$, so that it contains ($m \times n_s$) feedback gains. However, only n_s of these are needed to position the closed-loop poles (which are n_s in number). Optimal control is one way of constructively using up the extra degrees of freedom in satisfying more complicated control objectives.

Also in Chapter 5, it was noted that even for SISO systems, spare degrees of freedom associated with *uncontrollable* modes would also lead to an infinite number of sets of possible feedback gains. If optimal control methods are applied to systems that are not completely controllable, then the uncontrollable poles cannot, by definition, be moved by the controller. However, so long as the system is stabilizable (see Section 5.3.3), an optimal controller may be able to make constructive use of these spare degrees of freedom in seeking to obtain the best possible control using the subset of poles which *are* controllable. Later, MATLAB (Appendix 3) will be used to design optimal controllers, and the commands which perform the optimal design will cope with systems that are not completely controllable.

12.3 What does 'optimal' mean?

What is 'optimal' (that is, the best possible) is defined by specifying a *performance index* (PI), which the controller must seek to minimize. The PI is sometimes called a 'performance criterion', 'objective function' or 'cost function'. In principle, there is nothing particularly special about the contents of any given PI, and many different ones could be chosen for the same system to give different results. It would therefore be expected that PIs would look very different for different control schemes. However, in practice, many PIs tend to look very similar, for reasons which will become apparent shortly. A controller design can be attempted which will minimize any desired PI, so long as:

- the PI can be written as a mathematical expression which is capable of minimization;
- the PI is measurable (otherwise there will be no indication of success);
- the PI is as simple as it can be made (otherwise the mathematics is too difficult).

The quantities that it would be desirable to use in a PI to be minimized might include:

- time to move from one operating condition to another;
- amount of control effort required (the amplitude or energy content of the control signals);
- steady-state errors;
- transient errors;
- cost of operation;
- product yield (in this case, the desire would presumably be to *maximize* yield, so a suitable function of ($-$yield) could be included in the index to be minimized).

Unfortunately, the necessity to use simple and mathematically expressible PIs really rules out anything but the simplest combinations of some of these terms.

What goes into a real PI?

The most common *terms* to be found in real PIs are functions of the squares of the errors in the system and of the squares of the control inputs. The most common *function* used in PIs is the integral over time (which will penalize steady-state errors, as discussed in Section 4.5.2).

Squares are used partly to ensure that the resulting expressions are never negative (so that finding minima over time is not complicated by sign changes); and partly because the square of a signal variable is often an indication of some kind of *energy* transfer. Thus, minimizing the square minimizes energy expenditure – and hence cost. For example, if a system has a voltage input signal $v(t)$ feeding into a constant resistance R, then the energy expended during control is given by (power dissipated) × (time), namely

$$\frac{v^2(t)}{R} \cdot t$$

(having units of $V^2s\ \Omega^{-1}$). If $v^2(t)$ is integrated with respect to time, a result having units of V^2s is obtained (that is, the area under the $v^2(t)$ vs. time graph). Minimizing such an integral, so long as R is constant, will therefore minimize energy expenditure.

Similarly, in a mechanical system, if $x_3(t)$ represents a velocity, then its inclusion as $x_3^2(t)$ in an integral PI will seek to minimize instantaneous kinetic energy, so long as the mass of the moving object is constant.

For a general control signal $u(t)$ as shown in Figure 12.1(a), a suitable term for use as a PI (which is usually called J) might therefore be

$$J = \int_{t_0}^{t_f} u^2(t) \cdot dt$$

where the integral is taken over some time interval of interest (often t_0 is some given starting time and $t_f = \infty$ – see Section 12.4.7 for details of how the infinity is handled), and the value of J will therefore be the area under the graph in Figure 12.1(b). As before, if $u(t)$ represents (for example) a voltage or current signal, then minimizing J will minimize energy expenditure.

Three examples will now be used, which will indicate practical items to include in PIs, and also some of the problems in formulating them.

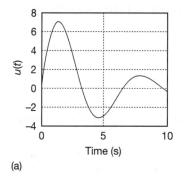
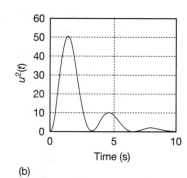

Figure 12.1 (a) A control signal $u(t)$ vs. time. (b) $u^2(t)$ vs. time.

(a)

(b)

Example 12.1 *A journey by road*

Say that it is desired to drive 200 km from town A to town Y, arriving in the minimum possible time. The very simplest approximation to such a problem follows from the assumptions listed below (a rather more realistic treatment follows later, but the principle to be described here is directly applicable to some engineering systems, such as overhead crane control):

● The road is straight and level.

- There are no junctions.
- The vehicle has a given maximum acceleration rate.
- The vehicle has a maximum deceleration rate, equal and opposite to its acceleration rate.

Intuitively, the minimum-time (that is, *optimal*) strategy is to accelerate as rapidly as possible to the halfway point, and then brake as heavily as possible for the remaining distance.

Assuming that the route has town V at its midpoint, the result would be as shown in Figure 12.2. There are some rather obvious problems with the control policy of Figure 12.2.

- Although it minimizes time, it will maximize cost and exhaust emissions.
- It assumes unlimited velocity. For a fairly high-performance saloon, a realistic acceleration capability would be 0 to 30 m s^{-1} (108 km hour^{-1}) in 10 s. If this is assumed to be a constant, sustained acceleration of 3 m s^{-2} and applied to the journey of Figure 12.2, the velocity at town V would be about 774 m s^{-1} (2786 km hour^{-1}). The entire journey would take about 516 s (less than nine minutes)!
- Obviously some constraint on the maximum velocity is required, as shown in Figure 12.3.

The journey time is related to the reciprocal of the area under the velocity–distance graphs, so it now takes much longer (a more realistic 2 hours and 13 minutes, in fact). Fuel consumption and emissions would also be far more acceptable in this velocity-limited journey.

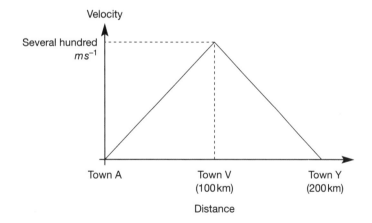

Figure 12.2 The simplest minimum-time solution to the travel problem.

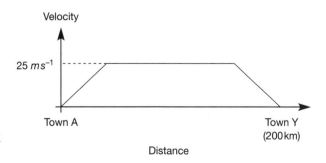

Figure 12.3 A velocity-limited solution to the travel problem.

The conflict between time and fuel consumption in Example 12.1 is representative of a common state of affairs in control system design. It almost always happens that one aspect of system performance can be improved only at the expense of another. For example, to minimize fuel consumption the journey would not be made at all, so the arrival time would be infinite. In practice, limits would be placed on the acceleration and braking rates to conserve fuel, in addition to the limit already imposed on maximum velocity.

To design a better controller, both travel time and fuel consumption should be included in a PI, with suitable weightings attached to each. This problem in revisited in Section 12.4.1.

Example 12.2 *A contrived state-space example to introduce the standard PI*

To see how a very common type of PI could arise mathematically, suppose that a system is expressed in the usual state-space form

$$\dot{x} = Ax + Bu$$

$$y = Cx$$

and that this system has five states and two inputs.

Say that it is desirable in this system for x_5 to track x_1 and for x_3 to track x_4. Also, it is three times more sensible (perhaps for reasons of cost) to use input u_2 for control than u_1. A PI might therefore be set up as follows (to be minimized by the controller):

$$J = \int_{t_0}^{t_f} [(\text{tracking errors}^2) + (\text{weighted inputs}^2)] \, dt$$

$$= \int_{t_0}^{t_f} [(x_1 - x_5)^2 + (x_4 - x_3)^2 + u_2^2 + 3u_1^2] \, dt \qquad (12.1)$$

Note that the factor of three appears in front of u_1 in order to penalize u_1 three times as heavily as u_2 (because the 'area' contributed by u_1 to the PI will then be relatively three times as great as that due to u_2). In addition, the effort the controller will put into minimizing each of the state errors is identical, and is the same as that it will apply to constraining u_2. If that were the wrong thing to do, the weightings on the state errors $(x_1 - x_5)$ and $(x_4 - x_3)$ would be changed from unity to some more appropriate values.

The PI of Equation (12.1) can be made much neater using vector–matrix notation. The term to be integrated, when multiplied out, is:

$$x_1^2 - 2x_1x_5 + x_5^2 + x_4^2 - 2x_3x_4 + x_3^2 + u_2^2 + 3u_1^2$$

Although it is only obvious to a matrix algebra expert, this can be rewritten in a very neat form by taking the vector $x^T = [x_1 \quad x_2 \quad x_3 \quad x_4 \quad x_5]$ as a factor out of the left-hand side of all the terms containing the state variables, and the vector $u^T = [u_1 \quad u_2]$ as a factor out of the left-hand side of all the terms containing the inputs, as follows (multiplying it all out will confirm that it is correct):

$$= \begin{bmatrix} x_1 & x_2 & x_3 & x_4 & x_5 \end{bmatrix} \begin{bmatrix} 1 & 0 & 0 & 0 & -1 \\ 0 & 0 & 0 & 0 & 0 \\ 0 & 0 & 1 & -1 & 0 \\ 0 & 0 & -1 & 1 & 0 \\ -1 & 0 & 0 & 0 & 1 \end{bmatrix} \begin{bmatrix} x_1 \\ x_2 \\ x_3 \\ x_4 \\ x_5 \end{bmatrix} + \begin{bmatrix} u_1 & u_2 \end{bmatrix} \begin{bmatrix} 3 & 0 \\ 0 & 1 \end{bmatrix} \begin{bmatrix} u_1 \\ u_2 \end{bmatrix}$$

Calling the 5×5 matrix $\boldsymbol{Q}$ and the 2×2 matrix $\boldsymbol{R}$ then gives:

$$\boldsymbol{x}^\mathrm{T} \boldsymbol{Q} \boldsymbol{x} + \boldsymbol{u}^\mathrm{T} \boldsymbol{R} \boldsymbol{u}$$

which, when substituted into Equation (12.1), gives:

$$J = \int_{t_0}^{t_f} [\boldsymbol{x}^\mathrm{T} \boldsymbol{Q} \boldsymbol{x} + \boldsymbol{u}^\mathrm{T} \boldsymbol{R} \boldsymbol{u}] \, dt \tag{12.2}$$

The terms in the brackets in Equation (12.2) are called *quadratic forms* and are quite common in matrix algebra. Note that $\boldsymbol{Q}$ and $\boldsymbol{R}$ are matrices chosen to apply the desired weights to the various states and inputs. $\boldsymbol{Q}$ and $\boldsymbol{R}$ will always be square, symmetric matrices (see Problem 12.1). Note also that J will always be a *scalar* quantity, whatever the sizes of $\boldsymbol{Q}$ and $\boldsymbol{R}$.

In the analysis leading to Equation (12.2), the $\boldsymbol{Q}$ matrix had some off-diagonal terms because the differences between states, that is $(x_1 - x_5)$ and $(x_4 - x_3)$, were to be minimized. This is a fairly unusual situation, and normally each state or input is simply weighted relative to the others. In that case, the $\boldsymbol{Q}$ and $\boldsymbol{R}$ matrices will be diagonal (as was the $\boldsymbol{R}$ matrix in Example 12.2). More will be said in Section 12.4.3 about specifying $\boldsymbol{Q}$ and $\boldsymbol{R}$.

Example 12.3 *A machine tool drive*

Imagine a numerically controlled machine tool which is required to track a varying reference signal as some shape is machined out of a block of metal. The parameters to be minimized in the control of the tool's position response would include (in order of importance):

- overshoot (there must be none)
- steady-state errors
- time to settle at a new position in response to a setpoint change.

Assuming a simple SISO setup, a PID controller could be applied and tuned for a good response (Section 4.5.2). Alternatively, the system could be modelled in state-space form.

Assuming a second-order model for simplicity, tool position and velocity (for example) could be used as the state variables. These could be fed back in an SVF system to position the closed-loop eigenvalues (poles) at locations which should give a suitable response (Section 5.4).

A third approach would be to include the two states and the system input in a PI, such as the one in Equation (12.2). The Q and R matrices could then be selected for suitable closed-loop behaviour, and an optimal control design procedure applied. It will emerge later that this would also result in an SVF scheme, but the feedback gains would be chosen (by the design algorithm, not by the human designer) to place the closed-loop eigenvalues at locations giving a closed-loop performance which would minimize the given PI. The designs for this example (both PID and optimal) are carried out in Section 12.6.1, and compared. Links with pole-placement SVF controllers are also mentioned in Section 12.4.3.

For now, note that if Q is diagonal (as mentioned after Example 12.2), and if it is specified as (say)

$$\begin{bmatrix} 4 & 0 \\ 0 & 1 \end{bmatrix}$$

then the controller is basically being asked to exert four times as much effort to minimize errors in x_1 (tool position, say) as in x_2 (tool velocity).

12.3.1 The linear quadratic regulator

When SVF systems were designed previously (in Section 5.4), the resulting designs were initially *regulators*; that is, controllers which try to return their state vector (and hence outputs) to zero in the face of disturbances. Since the optimal controllers to be designed in this chapter are also SVF systems, they will be optimal *regulators*.

This is precisely what is needed, once a system is at some steady operating condition. If the *deviations* from this operating condition are taken as the outputs and states, then they should indeed be maintained at zero in the face of disturbances. However, in systems such as that of Example 12.3, the output position must follow a varying reference signal – so a *tracking* system is required; and the regulator must be modified to achieve this (there is a discussion of regulators and trackers in Sections 5.4.2 and 5.4.3).

Another general point is that the earlier list of ideal items which might be included in a PI, even if applied only to linear systems, usually leads to optimal control problems with no analytical solution. Such problems can only be solved numerically and are computationally intensive and thus unsuitable for online control. However, if attention is restricted to *linear* systems (in state-space form, for example) and if only simple *quadratic* PIs (such as the one in Equation (12.2)) are used, then the resulting *linear-quadratic regulator* (LQR) problem does have known solutions. One such solution will be developed later.

An LQR design philosophy

The decision to choose simple quadratic PIs means that the PIs which can be used in practice are seldom a good representation of what we would *really* like to minimize. This is fundamentally important in the interpretation of what an 'optimal' controller actually does. At best, there must be a compromise between the ideal requirements and a soluble problem. Therefore, in practical application, interest is directed *not* primarily towards the minimization of the given PI as such, but rather to obtaining the required closed-loop performance by 'tuning' Q and R until the minimization of the resulting PI leads to a suitable result on the plant (see Section 12.4.3). With modern computer-assisted control system design (CACSD) environments, it is quite easy to go through an iterative cycle of design, simulation, and re-design until a satisfactory result is obtained. The MATLAB (Appendix 3) Control Systems Toolbox for example, has a single command *lqr* for designing an optimal regulator, given only the matrices Q and R, and the state-space model of the plant.

The LQR methods can also be extended to the so-called LQG case, in which the presence of noise (which is assumed to be Gaussian-distributed) is taken to be corrupting the process. Such systems normally involve the use of a Kalman filter (Section 9.8) as a state estimator, combined with an optimal controller.

12.4 Dynamic programming

How is the minimization of the chosen PI to be achieved by the controller? It has already been hinted that CACSD systems will be the normal route in practice. However, to understand better some of the issues involved, the derivation of an optimal control design method will be presented, before simply applying the computer results. It is possible to apply the method successfully without a full understanding of the underlying mathematics, but it is not possible without a good understanding of the part played by the Q and R matrices in the PI, by the feedback matrix in the resulting SVF scheme and of why the particular PI of Equation (12.2) is normally used. These points, and the basic principles of optimization itself, become clearer as one of the methods is derived.

One possible approach to the optimal control problem is *dynamic programming*. This is not the most elegant approach, but it is probably the easiest for the newcomer to follow with full understanding (some alternative methods are mentioned in Section 12.5).

12.4.1 Optimization – the basic ideas

To introduce some of the required ideas, think again about the problem of driving from Town A to Town Y (Example 12.1). A glance at a road map may show a very large choice of possible routes. Which is best (that is, 'optimal')?

The first requirement is for a definition of 'best' – that is, the specification of the performance index. Let us initially choose to minimize *distance* travelled. Comments about minimizing *time*, as Example 12.1 sought to do, follow later.

Picking a few of the possible routes from the map, with intermediate towns B to X, and measuring the various distances in kilometres, may give the simplified map of Figure 12.4.

The obvious way to find the shortest distance between Town A and Town Y is to add up the distances on each of the possible routes, and select the smallest. Unfortunately, there are 70 possible routes having seven additions per route, so it would be necessary to do 490 additions and make 69 comparisons to find the optimum. If 15 possible sets of intermediate destinations had been chosen, rather than seven, the requirements increase to 193 000 additions and almost 13 000 comparisons. There must be a better way than this!

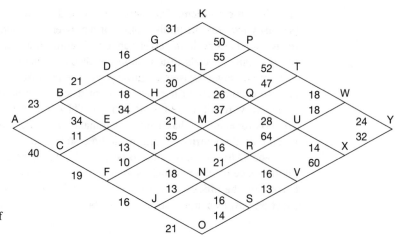

Figure 12.4 Distances (km) between the centres of towns A to Y.

Dynamic programming is a good way of solving this kind of problem. It is based upon ideas introduced by R. Bellman in the 1950s. Fundamental to the method is the *principle of optimality*. One statement of this principle is that if a path from A to Y is the optimal path which minimizes some particular PI, and if G is an intermediate point along the path, then the path from G to Y also minimizes the same PI.

For example, in Figure 12.5, if the optimal way to get from state x_0 to state x_f is as shown, and this minimizes some PI

$$J = \int_{t_0}^{t_f} f(x) \, . \, dt$$

then the path from x_1 to x_f minimizes

$$J = \int_{t_1}^{t_f} f(x) \, . \, dt$$

Thus, the *last part* of an optimal path is itself optimal. Note that this does not necessarily apply to the *first* part. For example, the optimal path from x_0 to x_1 may be different from the path in Figure 12.5 if the state trajectory is no longer constrained to continue to x_f after reaching x_1. Similarly, the optimal path in Figure 12.4 may no longer pass through Town G if it is no longer constrained to proceed to Town Y afterwards, but has some other final destination.

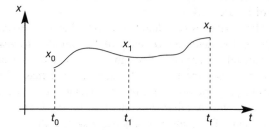

Figure 12.5 An optimal state trajectory.

This principle allows us to solve the travel problem (Figure 12.4) rather more elegantly by *working backwards* from the end. It is worth noting here that travel is only allowed from Town A to Town Y in a left-to-right direction in Figure 12.4. In addition, this map has been arranged so that there is a maximum of only two choices at each node (town), namely the town can be left either 'upwards' or 'downwards'. If there were many more choices at each node, it would not alter the procedure at all – it would just make the job more complicated. Eventually, it will be shown how to apply this method to systems in state-space form.

Returning to Figure 12.4, and remembering that it has been decided to work *backwards* from the end, if the route arrived at Town W there would then be no choice; travel must be *downwards* with an associated cost of 24 km. This operation is denoted D24 in Figure 12.6. Similarly, from Town X there is no choice; travel can only be upwards at a cost of 32 km (written U32).

On the other hand, arriving at Town U, there is a choice to be made. The journey can either proceed upwards to W, adding the new cost of 18 km to the optimal cost already known for the second part of the route after W (giving a new total of 42 km); or we can go downwards via X , adding 14 km to the cost already known for the route after X (giving 46 km in total). The best (that is, shortest distance) of these two possibilities is via W, so at Town U the optimal decision is to turn upwards on the map with a cost of 42 km (U42).

This process is continued until Town A is reached. The optimal route is then given by the U and D pointers in the *forward* direction, as shown in Figure 12.6. The optimal path is apparently via Towns B, D, G, L, and so on and has an associated cost of 187 km.

There are some general points to be drawn from this example.

The quantity minimized in this problem was the *distance* travelled. It is unlikely that the route in Figure 12.6 will also minimize the journey *time*. Perhaps it uses a lot of minor roads, whereas one of the other possible routes might be via motorways. To cope with that problem, the 'costs' (distances in Figure 12.4) would be altered to include weightings penalizing the slower roads and areas of known major roadworks, and then the dynamic programming would be repeated. Effectively, this changes the PI to include both distance and time.

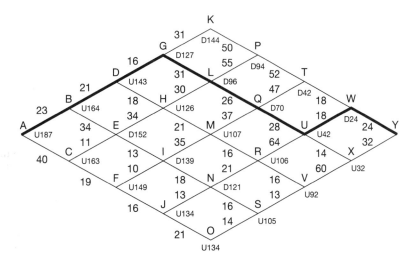

Figure 12.6 The map of Figure 12.4 with optimal route marked.

Sensitivity to errors

Optimal controllers can be very sensitive to small changes in the PI, or to modelling errors. Suppose that the optimal route leaves Town A on its north side, while the other possible route (via Town C) leaves from the south side. Suppose that we actually live on the south side of Town A, 9 km from the centre. This means that, for us, the distance taken from the map between the centres of Towns A and B ought to be increased to 32 km, while that between Towns A and C should be decreased to 31 km. If the last step of the dynamic programming is redone, with this small modification, it will be found that the optimal path changes drastically (the result should be a route via C, E, I, N, S and so on).

However, even this route may not truly be the best. Perhaps Town I is very large, and it would be foolish to drive through the centre of it. Perhaps a diversion in the direction of Towns F and J would be better. Again, the costs on the map could be altered to reflect this kind of knowledge. The point being made is that the optimal controller can only operate on the information it is given. Inaccurate or incomplete models lead to sub-optimal results, and practical models are always inaccurate and incomplete to some degree. This is another reason why the theoretical possibility of setting out to minimize a single unique PI is usually unrealistic.

Because of the way the travel problem has been formulated, several things which would be possible in real life are not possible in the solution obtained. For example, in the model it is not possible to drive from Town E to Town D. The model also fails to recognize the existence of any route except the ones chosen, so there is the possibility that there may be a better solution than the optimal solution of the model of Figure 12.6. Again, the point is that optimal control only gives truly optimal results where the system modelling is very good.

Extra endpoints and constraints

There was one endpoint in the model of Figure 12.4 (Town Y). This is an unnecessary simplification and, in general, there can be as many as necessary. For example, say that it is only desired to visit the area in which Town Y lies, and the precise destination is not of importance (perhaps Town Y is in a large national park). Extra endpoints could be added to the map, each connected to suitable intermediate towns by appropriate distances, and then exactly the same methods as before could be applied. As an example of what would happen, there could be a choice of two final destinations from Town W (Town Y, plus one of the new towns). If the optimal route led us to Town W, we should then end up in whichever of the two was the nearer (not necessarily Town Y, as before). In more general terms, this might correspond to a system having a target state in some region of the state-space, rather than one particular target state vector.

Finally, *constraints* can be added to the problem such as, 'we must travel via Town M to pick up a friend'. The optimal route *after* Town M is already known from Figure 12.6 (as it is for any town on the map), so the dynamic programming method is therefore applied working backwards from Town M to Town A, after deleting the (now unallowable) routes via G, L, J or N. Figure 12.7 shows the result. The penalty for adding the constraint is apparently an extra 3 km. Adding constraints leads to poorer performance than the ideal (the ideal is the optimum, so a constraint cannot *improve* it). However, as in this case, the addition of constraints can often simplify the problem by removing a number of decisions (or allowable

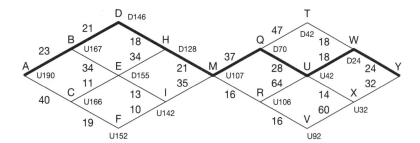

Figure 12.7 Constrained optimal solution of Figure 12.4.

system states, or allowable input or output signal ranges, for example) from the procedure.

12.4.2 Application to dynamical systems (LQ control)

The travel example is an interesting illustration, but how can it be applied to linear dynamical systems (state-space models, for example) with quadratic PIs?

First, note that the problem can be approached in either continuous time or discrete time. In packages such as MATLAB, either can be used with equal ease (specify the required Q and R matrices, with either the continuous- or discrete-time state-space model as desired, and use either the *lqr* command or the *dlqr* command respectively). The discrete time derivation is perhaps the easier to follow on paper, so that is the one adopted for analysis here. However, the final result will be equally applicable either to continuous or discrete systems.

Assume therefore, that a discrete-time state-space model of the process to be controlled has been obtained, of the form:

$$x_{n+1} = \Phi x_n + \Delta u_n$$

$$y_{n+1} = C x_{n+1}$$

where n represents the nth time step through the discrete process. To convert to time values, simply multiply the current value of n by the sampling interval (see Section 2.8.2 for an introduction to such a model).

The PI to be minimized will be assumed to be of the same form as the continuous-time version (Equation (12.2)). In discrete time, a summation is substituted for the integral operator, with the sum taken over sufficient time steps to cover the time interval of interest.

Note that the endpoint has to be handled carefully. Say that the control action is to take place over N time steps. This is realistic for several processes – for example, any system that must achieve its final value within a certain time; or a digital missile guidance controller, which is expected to operate for a certain number of time steps and then cease to exist! (For the present, the case when $t_f = \infty$, implying that $N = \infty$, will be ignored. It is discussed in Section 12.4.7.)

At the endpoint (that is, at the Nth step) the end of the experiment has *arrived*. There is no more opportunity to apply control inputs. If the target state has not been reached, it is too late. For this reason, there should be no value of u_N in the PI, because it does not exist. However, it *will* usually be important to include x_N so as to penalize incorrect final states (that is, steady-state errors). Because it has this specific purpose, the value of Q at step N may be chosen to be different from that at the other time steps.

These considerations lead to the discrete-time version of the quadratic PI, as given in Equation (12.3) below. Note that a factor of $\frac{1}{2}$ has also appeared. This is to make the mathematics easier later. It obviously scales J; but a simple scaling operation on J has no effect upon the value of $\boldsymbol{u}$ which will produce the minimum value of J.

$$J_N = \tfrac{1}{2}\sum_{n=0}^{N-1}(\boldsymbol{x}_n^{\mathrm{T}}\boldsymbol{Q}\boldsymbol{x}_n + \boldsymbol{u}_n^{\mathrm{T}}\boldsymbol{R}\boldsymbol{u}_n) + \tfrac{1}{2}\boldsymbol{x}_N^{\mathrm{T}}\boldsymbol{Q}_N\boldsymbol{x}_N \tag{12.3}$$

The optimal control problem can now be stated as, 'find the sequence of control input vectors $\boldsymbol{u}_0, \boldsymbol{u}_1, \boldsymbol{u}_2, \dots, \boldsymbol{u}_{N-1}$ which will drive the system so as to minimize the performance index J subject to the following constraints:

$$\boldsymbol{x}_{n+1} = \boldsymbol{\Phi}\boldsymbol{x}_n + \boldsymbol{\Delta}\boldsymbol{u}_n$$

$\boldsymbol{x}_0$ must be given

($\boldsymbol{u}_n$ may also be constrained)'. Note also that:

- the analysis can cope with time-varying (that is, non-stationary) systems by replacing $\boldsymbol{\Phi}$ and $\boldsymbol{\Delta}$ with $\boldsymbol{\Phi}_n$ and $\boldsymbol{\Delta}_n$;
- it is not very easy to include constraints on $\boldsymbol{u}_n$. Instead, it is more usual to perform a design with no constraints on $\boldsymbol{u}$, and then to check (by simulation) whether the behaviour of $\boldsymbol{u}$ is satisfactory. If not, changes can be made to the contents of the $\boldsymbol{R}$ and/or $\boldsymbol{Q}$ matrices in the PI, and the design repeated (following the philosophy outlined in Section 12.3.1). Later examples illustrate this.

Now the dynamic programming is carried out, working *backwards* from the endpoint because of the principle of optimality. At each stage, the superscript ^{o} is used to indicate *optimal* values (as in $\boldsymbol{u}_n^o$ and J_n^o). If the reader is prepared to take on trust the resulting equations which define the controller (Equations (12.12) to (12.15) below), this derivation may be omitted. However, it does give useful insights into why the method works.

When $n = N$, the endpoint is being considered, and there is no longer the opportunity to exercise any control over what is happening (as discussed before). Therefore, the optimal (minimum) value of the PI of Equation (12.3) is given by:

$$J_N^o = \tfrac{1}{2}\boldsymbol{x}_N^{\mathrm{T}}\boldsymbol{Q}_N\boldsymbol{x}_N \tag{12.4}$$

Stepping backwards in time by one step, $n = N - 1$, implying that the instant being considered is now the beginning of the last time step. Now there is a decision to be made, because the control input, $\boldsymbol{u}_{N-1}$ can be varied. It is therefore necessary to minimize the PI with respect to $\boldsymbol{u}_{N-1}$ so as to find the *optimal* control, $\boldsymbol{u}_{N-1}^o$, necessary to take the process to the end of the last time step in an optimal manner. Writing the PI for step $N - 1$ from Equation (12.3), and substituting from Equation (12.4) for the last term on the RHS, gives:

$$J_{N-1} = \tfrac{1}{2}(\boldsymbol{x}_{N-1}^{\mathrm{T}}\boldsymbol{Q}\boldsymbol{x}_{N-1} + \boldsymbol{u}_{N-1}^{\mathrm{T}}\boldsymbol{R}\boldsymbol{u}_{N-1}) + J_N^o \tag{12.5}$$

To find the minimum value J^o_{N-1} and the corresponding optimal control u^o_{N-1}, it is necessary to differentiate Equation (12.5) with respect to u_{N-1} and equate it to zero. To tie this in somewhat with the previous example, in the travel problem, arrival at Town W marked the beginning of the last step. The quantity J^o_{N-1} corresponds with the 24 km between Towns W and Y, and the optimal control u^o_{N-1} corresponds with the direction to take ('downwards'). There is no direct analogy for J^o_N in the travel problem, as the possibility of terminal errors did not arise.

It is not easy to follow the vector calculus in the minimization of Equation (12.5), so the problem will be solved by analogy with a similar scalar problem. In scalar terms, Equation (12.5) becomes

$$J_{N-1} = \tfrac{1}{2}\left(qx^2_{N-1} + ru^2_{N-1}\right) + J^o_N$$
$$= \tfrac{1}{2}\left(qx^2_{N-1} + ru^2_{N-1}\right) + \tfrac{1}{2}q_N x^2_N$$

However, there is also the constraint equation $x_N = \phi x_{N-1} + \delta u_{N-1}$ and substitution of this gives

$$J_{N-1} = \tfrac{1}{2}\left(qx^2_{N-1} + ru^2_{N-1}\right) + \tfrac{1}{2}q_N(\phi x_{N-1} + \delta u_{N-1})^2 \tag{12.6}$$

So

$$\frac{\partial J_{N-1}}{\partial u_{N-1}} = ru_{N-1} + q_N(\phi x_{N-1} + \delta u_{N-1})\delta$$

Equating this to zero gives the minimum, so the optimal control is given by:

$$u^o_{N-1} = -\frac{q_N \phi \delta}{r + q_N \delta^2}\, x_{N-1} = -k_{N-1} x_{N-1} \tag{12.7}$$

Realizing that the term multiplying x_{N-1} is a constant, and writing it as k_{N-1}, proves that the optimal control for this type of PI is a state variable feedback regulator (because Equation (12.7) represents a numerical gain, k, feeding back the state, x, with no additional reference input). The differences between this and the previous SVF designs are:

- The feedback gains are calculated according to a completely different set of conditions.
- The feedback is time-varying (there will be a different k_n at each time step).

Such an optimal regulator, which will aim to return the system state (and hence output) to zero following a disturbance, is called a *Linear Quadratic Regulator* (LQR); the LQR scheme is as shown in Figure 12.8.

To proceed with the analysis, the result of Equation (12.7) is now substituted back into Equation (12.6) to find the optimized PI at step $N-1$ as:

$$J^o_{N-1} = \tfrac{1}{2}\left(qx^2_{N-1} + rk^2_{N-1} x^2_{N-1}\right) + \tfrac{1}{2}q_N(\phi x_{N-1} - \delta k_{N-1} x_{N-1})^2$$
$$= \tfrac{1}{2}[q + rk^2_{N-1} + q_N(\phi - \delta k_{N-1})^2]x^2_{N-1}$$

thus

$$J^o_{N-1} = \tfrac{1}{2} p_{N-1} x^2_{N-1} \tag{12.8}$$

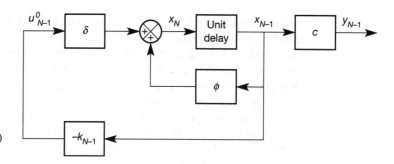

Figure 12.8 The linear
quadratic regulator (LQR)
– scalar case.

where

$$p_{N-1} = q + rk_{N-1}^2 + q_N(\phi - \delta k_{N-1})^2 \qquad (12.9)$$

The thought of carrying on this procedure for however many steps there may
be (N) is not appealing. However, taking just one more backward step, a pattern will
emerge, which can be generalized to every step in the procedure.

As an indication of the result to be obtained, note that p_N will need to be
initialized. If it is chosen to be equal to q_N then Equation (12.4) can be written
$J_N^o = \frac{1}{2} p_N x_N^2$ in the scalar case, the similarity to Equation (12.8) being obvious.

Taking the next backward step, $n = N - 2$. At this instant, the optimal value of
J (that is, J_{N-2}^o) will be the minimized component due to x_{N-2} and u_{N-2} (to be found
by differentiation, as above) plus the optimized value for the remaining time steps
($n = N - 1$ and $n = N$) which has already been calculated as J_{N-1}^o in Equation
(12.8). Note that J_{N-1}^o already contains within it the value of J_N^o – see Equation
(12.5). This is analogous to the way in which the optimal cost at each node of the
travel problem automatically included the costs of all subsequent sections of the
optimal path.

Continuing with the analysis then, at step $n = N - 2$ the optimal cost can be
derived in the same way as Equation (12.6) was derived, and using Equation (12.8)
to eliminate J_{N-1}^o:

$$J_{N-2} = \frac{1}{2}(qx_{N-2}^2 + ru_{N-2}^2) + J_{N-1}^o$$

$$= \frac{1}{2}(qx_{N-2}^2 + ru_{N-2}^2) + \frac{1}{2}p_{N-1}x_{N-1}^2$$

but

$$x_{N-1} = \phi x_{N-2} + \delta u_{N-2}$$

so

$$J_{N-2} = \frac{1}{2}(qx_{N-2}^2 + ru_{N-2}^2) + \frac{1}{2}p_{N-1}(\phi x_{N-2} + \delta u_{N-2})^2 \qquad (12.10)$$

Again, it is necessary to minimize this by differentiating and equating to zero.
However, since the form of Equation (12.10) is identical to that of Equation (12.6),
so will the solution be. Equation (12.10) is merely written one time step earlier,
and with p_{N-1} instead of q_N. By direct analogy with Equations (12.7), (12.8) and
(12.9), the result is therefore:

$$u_{N-2}^o = -\frac{p_{N-1}\phi\delta}{r + p_{N-1}\delta^2} x_{N-2} = -k_{N-2}x_{N-2}$$

and

$$J_{N-2}^o = \tfrac{1}{2} p_{N-2} x_{N-2}^2$$

where

$$p_{N-2} = q + r k_{N-2}^2 + p_{N-1}(\phi - \delta k_{N-2})^2$$

Comparing these equations with the earlier ones, the emerging pattern can now be seen, and can be generalized for the nth step as follows:

Initially, p_N is set equal to q_N, and no value exists for u_N^o.

The procedure is then stepped *backwards in time*, from $n = N - 1$ to $n = 0$.

For step n, the result is:

$$u_n^o = -k_n x_n$$

where

$$k_n = \frac{1}{r + p_{n+1}\delta^2} \, p_{n+1}\phi\delta \tag{12.11}$$

Also,

$$p_n = q + r k_n^2 + p_{n+1}(\phi - \delta k_n)^2$$

and

$$J_n^o = \tfrac{1}{2} p_n x_n^2$$

Equations (12.11), when applied from $n = N - 1$ to $n = 0$, yield the sequence of optimal controls $u_{N-1}^o, u_{N-2}^o, \ldots, u_1^o, u_0^o$ and the corresponding (minimum) values for the performance criterion $J_{N-1}^o, J_{N-2}^o, \ldots, J_1^o, J_0^o$. The control would then be applied by using the feedback gains $k_0, k_1, k_2, \ldots, k_{N-1}$ in the *forward* time direction, that is, at steps $n = 0, n = 1, \ldots, n = N - 1$ respectively.

It is finally necessary to remember that the solution to the optimal control problem given in Equations (12.11) is a *scalar* solution to an equivalent *vector–matrix* problem. Equations (12.11) must now be generalized to the vector–matrix case. Strictly, it needs to be verified that the analogy back to the vector–matrix case from the scalar solution *can* be made. However, this has been proved by others to be allowable in this case.

As a reminder, for a general multivariable system having m inputs, p outputs and n_s states, the sizes of the vectors and matrices involved are as follows (these follow from the fact that a matrix in a block diagram is of size (outputs) × (inputs) as discussed in Section A1.1.1; and from the requirements for the various multiplications in the equations to be conformable):

$$\boldsymbol{u} \text{ is } m \times 1, \quad \boldsymbol{x} \text{ is } n_s \times 1, \quad \boldsymbol{y} \text{ is } p \times 1, \quad \boldsymbol{P}, \boldsymbol{Q}, \boldsymbol{\Phi} \text{ are } n_s \times n_s,$$

$$\boldsymbol{\Delta} \text{ is } n_s \times m, \quad \boldsymbol{R} \text{ is } m \times m, \quad \boldsymbol{K} \text{ is } m \times n_s \text{ and } J \text{ is scalar}$$

Rewriting Equations (12.11) so that the dimensions of the vectors and matrices are conformable for multiplication gives the general results for the nth step as follows:

Initially, $\boldsymbol{P}_N$ is set equal to $\boldsymbol{Q}_N$, and no value exists for $\boldsymbol{u}_N^o$.

The procedure is then stepped *backwards in time*, from $n = N - 1$ to $n = 0$.

For step n, the result is:

$$u_n^o = -K_n x_n \tag{12.12}$$

where

$$K_n = [R + \varDelta^T P_{n+1} \varDelta]^{-1} \varDelta^T P_{n+1} \varPhi \tag{12.13}$$

Also,

$$P_n = Q + K_n^T R K_n + [\varPhi - \varDelta K_n]^T P_{n+1} [\varPhi - \varDelta K_n] \tag{12.14}$$

and

$$J_n^o = \tfrac{1}{2} x_n^T P_n x_n \tag{12.15}$$

The control would then be applied by using the matrices of feedback gains $K_0, K_1, K_2, \ldots, K_{N-1}$ in the *forward* time direction, that is, at steps $n = 0$, $n = 1, \ldots, n = N - 1$ respectively, as for the scalar case.

Equations (12.12) to (12.15) can easily be verified to be at least dimensionally correct by substituting the dimensions of the various vectors and matrices into the products, and ascertaining that the results are consistent.

Equation (12.14) is called a *Discrete Matrix Riccati Equation* (DMRE).

Equations (12.12) to (12.15) are in a reasonable form to be applied recursively to generate the control sequence in a digital computer control scheme. However, many sources use a version of the DMRE which does not include the value of K_n. As a bit of gentle matrix-algebra practice, the conversion to such a form is considered in Problem 12.2.

Continuous-time result

All the analysis of this section has been carried out in discrete time. To give a more complete picture, the result for the *continuous-time* case is given here without proof. See Anderson and Moore (1989) for the derivation (but note that potentially confusing differences in notation will be found).

$$u^o = -Kx$$

where

$$K = R^{-1} B^T P$$

and P is the positive-definite solution of the algebraic MRE:

$$PA + A^T P + Q - PBR^{-1} B^T P = 0$$

Both the DMRE, and the AMRE in this continuous-time result, are normally insoluble by analysis and must be solved numerically using a computer. There is the possibility of hand solution only for very simple problems, examples of which appear at the end of the chapter.

12.4.3 Choice of Q and R matrices

The preceding derivation assumed the use of a quadratic PI of the form of Equation (12.2). The same method could be applied to other forms of index, but all the mathematics would have to be repeated for the new form (which may not necessarily be possible). In addition, care must be taken that the chosen PI is reliably minimizable (for example, it must not have local minima which could lead to a false 'optimum'). Also, the stability properties of an optimal controller using a different form of PI would need to be examined.

Stability

The form of PI used above will always yield a stable controller, even if it is sub-optimal (due to modelling errors, for example). It has been shown (see, for example, Brogan (1991)) that, so long as R is diagonal, the closed-loop system will remain stable for any values of feedback gains from about half of the designed values up to arbitrarily large values. Another way of expressing this good robustness property is given by Anderson and Moore (1989), who show that the Nyquist plot of an optimal controller of this kind avoids a circle of radius 1 centred on the $(-1,0)$ point. These stability margins are very conservative in frequency-domain control terms, and there is a price to be paid for this. It can be shown that the high-frequency response of LQR controllers rolls off at only 20 dB per decade, so the noise rejection properties are not particularly good.

Structure of Q and R

Returning to the quadratic index of Equation (12.2), whatever the choice of elements in Q and R, they require the following mathematical properties so that the PI will have a guaranteed and well-defined minimum, and so that the DMRE will have a solution. Both must be symmetric, Q must be positive semidefinite and R must be positive definite (see Section A1.4 for clarification of this terminology).

Q and R are usually chosen to be purely *diagonal* matrices for two reasons. Firstly, it becomes easy to ensure the correct 'definiteness' properties (the numbers on the leading diagonal of Q must then simply be non-negative, and those of R must be positive). Secondly, the diagonal elements penalize individual states or inputs. This makes the choice of elements easier because the physical significance of the elements is clear for the inputs, and also for the states if the states have been chosen as physically meaningful variables.

It should also be noted that the choice of Q is not unique. Several different Q matrices will yield the same controller, and there is an equivalent diagonal version of Q in every case, so nothing is sacrificed in general by making Q diagonal. In addition, one should always be on the lookout for other helpful features in the structure of the system model. For example, if certain states in the system are fairly closely related dynamically (perhaps just separated by a relatively fast time constant), there may be little to be gained by applying weightings to more than one of these states in a PI. We could perhaps just set one of the corresponding diagonal elements of Q to a positive quantity, and the other to zero. The significance of the states would need to be assessed to ensure that this approach was acceptable in such a case, and to choose which state to weight.

The relative weightings chosen for Q and R determine the relative emphasis placed upon reducing errors and saving control energy. If it were important to minimize control effort at all costs, then the numbers in R would be made much greater than those in Q, for example. This would usually give slow performance, but the exact effect is hard to predict, and an iterative design process would be used as outlined in the next paragraph.

In many cases, as was mentioned in Section 12.3.1, the PI which is minimized is really artificial. The minimization of it is just the means to the end of achieving acceptable control of the plant. This being the case, it is normal practice to select initial values for Q and R from considerations such as those above, carry out the design of the feedback matrix, and then *simulate* the resulting system and view the responses of the states, inputs and outputs. Values in Q and/or R are then changed as required, and the procedure repeated. Using modern CACSD environments (such as MATLAB) it is very easy to investigate quite a range of values for Q and R, and to select those that give the best response. The design thus becomes an iterative process. Section 12.6 contains two worked examples.

Links with pole-placement: the optimal root locus

If the designer has some idea of the kind of closed-loop pole locations (eigenvalues) that will give a suitable response, then use can be made of this knowledge. CACSD packages such as MATLAB will report the closed-loop eigenvalues which will result from applying the optimal controller for any given choice of Q and R. This can assist in the choices of these matrices as follows. The contents of R might be fixed (for example) and an initial set of values chosen for Q. A scalar multiplier can then be applied to Q which is swept through a range of values. For each value of this multiplier the resulting closed-loop eigenvalues can be stored and then the entire set can be plotted as a root locus diagram (see Section 4.4). This diagram then shows the locus of the closed-loop poles of the system as Q is varied. From it, a suitable Q matrix may perhaps be chosen. This procedure is illustrated in the example of Section 12.6.2. Of course, the same procedure can be applied to R, or to any individual element of either Q or R.

Other general points

Two more general points which might assist in the choice of Q and R are:

- Assuming that Q and R are diagonal, then doubling (for example) every element in R will have precisely the same effect on the closed-loop step response as halving every element in Q.
- If the system has only a single input, R will reduce to a scalar (r). Therefore, following from the last point, all possible effects can be obtained by ignoring r (setting it to unity, for example) and tuning only the elements of Q in the PI

$$J = \int_{t_0}^{t_f} [x^T Q x + u^2] \, dt$$

There is further comment in the following sections relevant to the choice of Q and R.

12.4.4 *Systems in which the output response is to be weighted*

In some systems, it may be more appropriate to include functions of the *output* in the PI, rather than the system state. In principle, this is very straightforward and the appropriate PI is:

$$J_N = \tfrac{1}{2} \sum_{n=0}^{N-1} (y_n^{\mathrm{T}} Q y_n + u_n^{\mathrm{T}} R u_n) + \tfrac{1}{2} y_N^{\mathrm{T}} Q_N y_N \tag{12.16}$$

Since $y = Cx$, the quadratic forms such as $y_n^{\mathrm{T}} Q y_n$ may be written as $x_n^{\mathrm{T}} C^{\mathrm{T}} Q C x_n$. Thus, the PI becomes

$$J_N = \tfrac{1}{2} \sum_{n=0}^{N-1} (x_n^{\mathrm{T}} V x_n + u_n^{\mathrm{T}} R u_n) + \tfrac{1}{2} x_N^{\mathrm{T}} V_N x_N$$

where $V = C^{\mathrm{T}} Q C$ and $V_N = C^{\mathrm{T}} Q_N C$, and so all the previous methodology applies after choosing Q to give the required weightings to y, and then substituting V for Q during the analysis.

One problem with this approach is that the number of outputs is likely to be less than the number of states, but the resulting controller will still feed back the states. This means that the controller will have fewer degrees of freedom available to it than in the regulator case because Q in Equation (12.16), whose elements are to be tuned, is of lower order than the V matrix which generates the controller. The results may therefore be inferior.

12.4.5 *Regulator systems and tracking systems*

It has been pointed out that the design discussed thus far has been that of a *regulator* system. The consequence of this is that in Figure 12.8 there is no reference input, and the desired steady state is $x = 0$.

Proof of regulator action

To see why this should be so, consider the direct multivariable equivalent of Figure 12.8. The discrete-time state equation would then be (from the figure) $x_N = \Phi x_{N-1} + \Delta u_{N-1}^o$. Now, for any useful regulator, a steady state will eventually be reached, in which $x_{N-1} = x_N$. Substituting this into the state equation gives $[I - \Phi] x_N = \Delta u_N^o$. But, from Equation (12.12), $u_N^o = -K_N x_N$, therefore:

$$[I - (\Phi - \Delta K_N)] x_N = 0 \tag{12.17}$$

The significance of Equation (12.17) is that the term in square brackets can be shown to be non-zero if R is positive definite, which must always be the case (see Problem 12.3). Given the presence of the identity matrix, it is also most unlikely that the term will be singular, other than in exceptional cases. The conclusion from Equation (12.17) is therefore that x_N, the state vector at steady state, must be zero. From Figure 12.8, if $x_{N-1} = x_N$ is zero, then so must y_N and u_N^o be. This confirms the action of a regulator.

Tracking system design by alteration of the performance index

To obtain an optimal tracking system, a PI similar in form to Equation (12.16) might be used, but employing terms which weight the deviation of the output from the reference signal (that is, the tracking error). This implies the use of $[\mathbf{y}_n - \mathbf{r}_n]$ in Equation (12.16) instead of $\mathbf{y}_n$, where $\mathbf{r}_n$ is the applied reference input (setpoint) to the closed loop scheme as shown in Figure 12.9 (the matrix $\mathbf{G}$ is required primarily to match the number of inputs and reference signals).

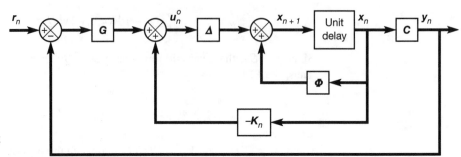

Figure 12.9 A tracking arrangement.

The analysis of this system from an optimal control viewpoint is possible, but is not easy (for more detail, see Anderson and Moore (1989)). This analysis will not be repeated here, but it can be seen that the type of performance criterion would be

$$J_N = \tfrac{1}{2} \sum_{n=0}^{N-1} ([\mathbf{y}_n - \mathbf{r}_n]^{\mathrm{T}} \mathbf{Q}[\mathbf{y}_n - \mathbf{r}_n] + \mathbf{u}_n^{\mathrm{T}} \mathbf{R} \mathbf{u}_n)$$

$$+ \tfrac{1}{2} [\mathbf{y}_N - \mathbf{r}_N]^{\mathrm{T}} \mathbf{Q}_N [\mathbf{y}_N - \mathbf{r}_N]$$

and that $\mathbf{y} = \mathbf{C}\mathbf{x}$ can be substituted, if desired, to obtain

$$J_N = \tfrac{1}{2} \sum_{n=0}^{N-1} ([\mathbf{x}_n^{\mathrm{T}} \mathbf{C}^{\mathrm{T}} - \mathbf{r}_n^{\mathrm{T}}] \mathbf{Q}[\mathbf{C}\mathbf{x}_n - \mathbf{r}_n] + \mathbf{u}_n^{\mathrm{T}} \mathbf{R} \mathbf{u}_n)$$

$$+ \tfrac{1}{2} [\mathbf{x}_N^{\mathrm{T}} \mathbf{C}^{\mathrm{T}} - \mathbf{r}_N^{\mathrm{T}}] \mathbf{Q}_N [\mathbf{C}\mathbf{x}_N - \mathbf{r}_N] \tag{12.18}$$

When the brackets are multiplied out, it is found that there are four extra terms in Equation (12.18), compared with the index for the LQR case, hence the difficulty. In addition, the number of diagonal elements in $\mathbf{Q}$ is now equal to the number of outputs, which is normally lower than the number of states. This would therefore not give complete control over all the system dynamics. If this were particularly important, any 'uncontrolled' states could be defined as outputs by adding the associated state variables (or combinations of them) to the output vector. They would then be controlled together with the original set of outputs. However, since the original outputs are also linear combinations of the state variables, great care would be necessary to make sure that any new outputs defined in this way were independent of the original ones, otherwise conflicting control objectives would result.

Simple tracking system design for servo-like systems

For systems whose models contain pure integrators, depending upon where these integrators are, there may be a simple way in which a tracking system can be obtained without having to alter the previous analysis in any way. This was described in Section 5.4.6, and effectively involves moving the point of application of the reference vector, so that it occurs *before* the SVF gain matrix. The setpoints are then references for the *states* rather than the outputs, as shown in Figure 12.10. It is important to remember that this will only work for systems containing pure integrators in the forward path; and then only if non-zero references are applied only to states which are the outputs of pure integrators whose inputs can become zero at steady state.

If such an approach is taken, a reference signal for the *states* must be provided, which is chosen to cause the *outputs* to behave as required. If the model of the system is chosen so that the outputs correspond with state variables (typically, but not necessarily, with a C matrix of the form $C = [I \ \ 0]$), then this is straightforward. The states which coincide with outputs can be given the required reference signals (so long as they are fed via pure integrators), and the other elements of the reference vector can be set to zero. For C matrices of other forms, especially forms in which the outputs are combinations of more than one system state, this will be harder to achieve. Special care must be taken to ensure that the C matrix is accurate if this method is used, since the output is effectively open-loop via C, with no possibility of correction for modelling errors.

As a final point, note that the controller will now be sub-optimal with respect to the initial PI, because the configuration is no longer the regulator for which the optimal design was carried out. However, given that the advocated philosophy is that of using the LQR design machinery as a means to obtaining reasonable control, and not as a means of minimizing a specific PI, this is of no consequence. The tracking system design would be simulated and Q and R tuned until satisfactory results are obtained.

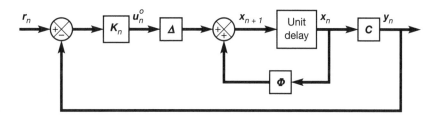

Figure 12.10 Optimal reference tracker for a restricted class of systems.

12.4.6 *Further consideration of steady-state errors*

If unacceptable steady-state errors cannot be removed by the methods of Section 12.4.5, then some kind of integral action must be added, as was discussed for simple SVF schemes (see Section 5.4.4).

There is more than one way of achieving such integral control. One method is to add extra states to the system model which are the time integrals of the real states of the system. If this extra state vector is called z (not to be confused with the z-transform operator), then

$$z = \int x \, dt \quad \text{or} \quad \dot{z} = x$$

In discrete time, this translates to the usual type of difference equation

$$z_{n+1} = z_n + Tx_n$$

where T is the sampling interval.

The system now has the new states z, as well as the old states x. For reasons arising from the mathematics (to do with singular matrices), it is often the case that z will not be able to contain all the possible integral states. In other words, a subset of the integrals of x will often have to be chosen to make up z. If the integrals of all the states are included in z, and if the new and old state equations are combined, an overall state-space model is obtained as:

$$\begin{bmatrix} x \\ z \end{bmatrix}_{n+1} = \begin{bmatrix} \Phi & 0 \\ IT & I \end{bmatrix} \begin{bmatrix} x \\ z \end{bmatrix}_n + \begin{bmatrix} \Delta \\ 0 \end{bmatrix} u_n$$

which, in turn, can be written as follows, where the tildes represent the augmented vectors and matrices:

$$\tilde{x}_{n+1} = \tilde{\Phi}\tilde{x}_n + \tilde{\Delta}u_n$$

Precisely the same analysis as before can then be applied, and Equations (12.12) to (12.15) will give the results in terms of the augmented matrices and vectors. Before interpreting the results, it should be noted that some of the matrices have changed size. Also, the elements of Q which weight the new (integral) states are typically chosen to be an order of magnitude less than the weights on the original states, to ensure reasonable stability. The sizes of the various quantities in the augmented model are (assuming a complete set of integral states):

$$u \text{ is } m \times 1, \quad y \text{ is } p \times 1, \quad R \text{ is } m \times m \quad \text{and J scalar as before}$$

but now

$$\tilde{x} \text{ is } 2n_s \times 1, \quad \tilde{P}, \tilde{Q}, \tilde{\Phi} \text{ are } 2n_s \times 2n_s, \quad \tilde{\Delta} \text{ is } 2n_s \times m \quad \text{and}$$

$$\tilde{K} \text{ is } m \times 2n_s$$

Therefore, K in Equations (12.12) and (12.13), which are repeated here for convenience, feeds back $2n_s$ states to m inputs.

$$\tilde{K}_n = [R + \tilde{\Delta}^T\tilde{P}_{n+1}\tilde{\Delta}]^{-1}\tilde{\Delta}^T\tilde{P}_{n+1}\tilde{\Phi} \tag{12.13}$$

and

$$u_n^o = -\tilde{K}_n\tilde{x}_n \tag{12.12}$$

If $\tilde{K}$ is now partitioned (see Section A1.1.3) $[K_n^P \quad K_n^I]$ where K_n^P, K_n^I are of size $m \times n_s$, then Equation (12.12) becomes:

$$u_n^o = -[K_n^P \quad K_n^I]\begin{bmatrix} x \\ z \end{bmatrix}_n$$

so that K_n^P contains the gains feeding back the states, while K_n^I contains the gains feeding back the integrals of the states. Figure 12.11 shows the resulting scheme. The position of the reference input is not shown, since it depends upon the design,

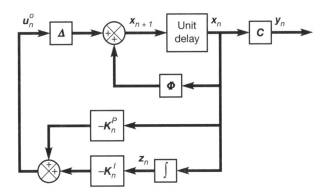

Figure 12.11 An LQR
system with integral action.

as discussed in Section 12.4.5. Note also that strictly an integrator should not be
shown on a discrete-time diagram. It has been done to show the *function* of the
feedback. It could be replaced by a discrete-time version using $z_n = z_{n-1} + Tx_{n-1}$,
where x_{n-1} would be generated from x_n by a unit delay.

12.4.7 The infinite horizon case

Many discrete-time control systems do not only operate over the finite number of
time steps N considered previously. Therefore, for most systems, the upper limit on
the integral of the PI should be infinity (since the operating time of the plant is very
long compared with the sampling interval). This is allowable in the continuous-time
case where the PI contains integrals, but infinite summations are not very helpful in
the discrete-time case. If there is an infinite 'time-to-go' (infinite control horizon),
how can we progress backwards through the dynamic programming procedure from
the end, so as to find the optimal SVF gains?

Fortunately, there is a way out of this difficulty for stationary (non-time-
varying) plants, and the result is simpler to implement than before. Assuming that
the controlled system is stable (which should always be the case), then eventually
all the states will be driven to steady values. Then, all rates of change will be zero,
implying that eventually $x_{n+1} = x_n$, $u_{n+1} = u_n$ and so on, working forwards in time.
In addition, since nothing is changing any more, the solution to the matrix Riccati
equation (P) and the feedback gains in K will also converge to steady values.

In the finite horizon regulator case with $n = 0, 1, \ldots, N - 1$, the final values in
K are zero. It is also usually found that the values in K are more or less constant
from the time of application of control (t_0) until a few plant time constants before
the end of the operating period (t_f), when the decay to zero begins. The controller
therefore spends most of its time applying the same feedback gains to the plant.

In the infinite horizon case $t_f = \infty$, and so the decay of the feedback gains to
zero never occurs (that is, a time corresponding to 'a few plant time constants
before the end of the operating period' is never reached). This means that the
controller gains are fixed in the infinite horizon case, as K rather than K_n. This
greatly simplifies the implementation, and is the type of controller designed by
MATLAB, for example.

There are various ways of finding K. One way is to set N to a large number and
begin to apply Equations (12.13) and (12.14) backwards in time as before, until K_n

converges to the constant matrix K. This would then be used in exactly the same way as a normal SVF gain matrix.

Analytical solutions leading to K also exist. These usually involve looking at the DMRE given in Problem 12.2, and arguing that in the steady state $P_n = P_{n+1} = P$ (say). Thus, the DMRE becomes

$$P = Q + \Phi^T P \Phi - \Phi^T P \Delta [R + \Delta^T P \Delta]^{-1} \Delta^T P \Phi$$

This is called an *Algebraic Matrix Riccati Equation* (AMRE), and can (with some difficulty) be solved analytically for P (the procedure is closely related to that used in deriving the Kalman filter gains from Equation (A6.4) in Appendix 6). Given the solution for P, Equation (12.13), with the n and $n + 1$ subscripts omitted, then gives K.

The solution for P is not given here, nor is the result quoted, because it needs more mathematical background to be meaningful. For the interested reader, Phillips and Nagle (1990: Section 10.6) and Ogata (1987: Section 7.3) both begin with versions of the DMRE identical to that above (except for notation changes, and an error in the positioning of a bracket in Equation (10.46) of Phillips), so the derivation can be followed in either text without undue difficulty (although it does involve some extra mathematics not covered in this text).

12.5 Other approaches

Although many of the comments made in this chapter (especially in Sections 12.4.3 to 12.4.7) are applicable to optimal control schemes derived by many methods, the only method actually used in the derivation has been dynamic programming. Many other methods lead to the same result. Since this text is really interested in using the result, rather than deriving it for its own sake, these will not be discussed here. The reader interested in other approaches should look through some of the texts in the bibliography for topics such as the following (some of which are interrelated). The texts by Anderson and Moore (1989), Brogan (1991) and Ogata (1987) contain relatively wide-ranging material.

- Pontryagin's minimum principle (or just, 'the Minimum Principle')
- The calculus of variations
- Solution via the Hamiltonian matrix
- Solution via Lyapunov (or Liapunov) methods
- Minimization via Lagrange multipliers

12.6 Two design examples using MATLAB

MATLAB (Appendix 3) is used to illustrate the design procedure here, but MATLAB is just the chosen tool to apply the design methodology. The same steps can be applied using any other suitably featured CACSD package. The *lqr* command which designs LQR controllers in MATLAB uses a Hamiltonian method, with some numerical refinements. However, it is not necessary to know that in order to be able to use it, as the examples below will show. The knowledge that *is* necessary is the rôles of the Q and R matrices, and the philosophy of using the LQR design method, which have been outlined in the earlier sections. Both these examples are SISO systems for simplicity, but the application to multivariable systems follows the same pattern.

Note that continuous-time models are used in these examples, even though all the previous analysis was done in discrete time. This is of no consequence, because there is a parallel continuous-time derivation of the same SVF-type result to that derived in discrete time above. In these examples, CACSD system results will be used, rather than results derived on paper, so it is simply a matter of selecting the CACSD tool that works in continuous time and gives a controller matrix for the continuous-time case (the *lqr* command in the case of MATLAB).

The practical points discussed previously apply equally to either the continuous- or discrete-time case. An infinite control horizon will also be assumed in these examples, as the controller can operate for very long periods compared with the system dynamics. Therefore the controller matrices will be stationary, rather than time-varying. The MATLAB *lqr* command automatically designs this type of control.

12.6.1 Case study – an electro-hydraulic machine tool drive

Figure 12.12 shows a highly simplified model of an electro-hydraulic machine tool-positioning system. The workpiece is assumed to be being rotated on the bed of a vertical milling machine. The system shown in the figure controls one axis of the movement of the tool itself.

It is likely that a straightforward two- or three-term controller would be considered for such a system. This will be designed first; then the problem will be solved using an optimal controller for comparison purposes, and to provide almost the simplest possible design exercise.

The requirements for the control of this system are simply stated. The tool position must have:

- zero overshoot,

- zero steady-state error,

- the fastest possible rise time which does not violate the overshoot condition.

In addition, there are some practical constraints on signal levels which must not be violated because saturation effects would come into action on the real plant, leading to nonlinear behaviour. The d.c. gain of the servovalve block can be seen to be 0.05 m s^{-1}/V. In this system, the range of servovalve control voltage is ± 12 V, which therefore leads to maximum velocities of ± 0.6 m s^{-1}. In addition, the normally expected range of movement at the tool during operation is ± 15 mm, and the plant has been

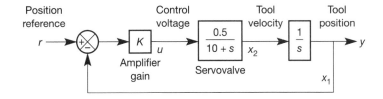

Figure 12.12 An electro-hydraulic tool-positioning system.

designed with the expectation that this will correspond to the ±12 V control range (a separate offset, not shown in the model, is used to move the tool tip to its starting point).

Using MATLAB for the design, it is possible to arrange things so that the built-in *step* command can be used, without modification, to generate the step responses (this approach is unnecessary, but does have certain attributes of convenience). The problem is that this command applies a *unit* step which, for the present system, will be assumed to represent 1 m of tool movement. Of course, this is totally impractical, but since the model of the system is linear, the size of the input step is immaterial. However, to check that the real-world signals do not exceed the stated limits, the model responses will need to be interpreted as follows:

- The input to the solenoid valve must be in the range ±800 V (12 V × 1 m step/15 mm).
- The velocity must remain in the range ±40 m s^{-1} (800 V × d.c. gain of 0.05 m s^{-1}/V).
- The output must reach 1 m as soon as possible, and with zero overshoot.

The simplest investigation is to examine the feasibility of control by varying the amplifier gain, K. The closed-loop LTF is given by:

$$\frac{y}{r}(s) = \frac{0.5K}{s^2 + 10s + 0.5K}$$

This can be compared with the standard second-order form (see Equation (3.19) in Section 3.2.2, for example). It can be seen from the denominator term in s that $2\zeta\omega_n = 10$. For the fastest rise time with zero overshoot, $\zeta = 1$ so ω_n must be 5 rad s^{-1}. The numerator term then gives the gain $K = 50$.

The resulting step response is the standard critically damped second-order one (see Figure 3.19 in Section 3.2.2). It has no overshoot, no steady-state error and a 10–90 per cent rise time of about 665 ms.

Since the forward path gain of 50 will apply 50 V to the solenoid valve when the unit step is applied (that is, before the feedback loop begins to react to reduce the initial error), and since the unit-step limit is 800 V, the performance should be fairly poor compared with the best possible (because only 6.25 per cent of the control range is being used – but if any more is used by increasing K, then overshoot will occur). In addition, given that the model will be inaccurate, perhaps even this gain will give some overshoot. As we expect to reject this design, we shall not pursue it further.

Next, a proportional plus derivative (PD) controller will be tried (using the inherent integral action in the plant for the 'I' term of the PID controller). The gain K in Figure 12.12 would ideally be subsumed into a controller of the form $(K_P + sK_D)$. This is, of course, physically unrealizable without the addition of some poles. In ideal analyses, these poles are usually assumed to be so 'fast' as to be negligible. Packages like MATLAB are rather more realistic, and will not allow the modelling of a zero without at least one accompanying pole to counter the otherwise infinite high-frequency gain. For this investigation, a single additional controller pole will be assumed (the minimum allowable) and it will be placed 100 times as 'fast' as the controller zero. The 'real' PD controller transfer function will therefore be

$$\frac{K_P\left(1 + s\,\dfrac{K_D}{K_P}\right)}{1 + s\,\dfrac{0.01K_D}{K_P}}$$

There is another practical problem. If this is used in the forward path of Figure 12.12 as it stands, then the derivative term (even with the pole which has been added) will tend to call for very large control actions in response to the step input test. This would immediately saturate the plant actuator and give a nonlinear response which will not be seen in the linear model. To avoid this problem, the commonly used solution will be adopted of putting the derivative term in a feedback path. This works because the control signal is (ideally):

$$(K_P + sK_D)(r - y) = K_P(r - y) + K_D sr - K_D sy$$

and, so long as the input (r) is not in the act of altering, the term $K_D sr$ can be ignored (since $sr \equiv dr/dt$ and is therefore zero if r is not changing). The actual arrangement is therefore as shown in Figure 12.13.

For this system, K_P was chosen as 800, since this would cause the maximum permissible signal to appear transiently at the plant input for a unit step test signal. K_D was then increased to a level which removed the overshoot due to the high value of K_P. The resulting value was $K_D = 60$. With this controller, the 10–90 per cent rise time improved to about 170 ms. No separate plot of the resulting performance is given, because it was almost indistinguishable from that of Figure 12.14 which appears later.

A likely drawback to the PD controller is its noise performance. A derivative gain of 60 will tend to amplify any noise appearing at the output. Further simulation studies would be necessary in practice.

Although all the 'headroom' in the plant input has been used up, the velocity signal is still a long way below the allowable maximum of 40 ms^{-1}. An optimal controller will now be designed for comparison, and may be able to use some of this freedom (but since the velocity is 'tied' to the voltage input by a d.c. gain of 0.05 and a time constant of only 100 ms, and all the voltage 'headroom' has been used, it is not likely that any great improvement will be forthcoming from a linear controller).

In principle, all the design constraints should be combined in a performance index, and then an optimal controller should be designed to minimize it. In practice, the standard performance index of the form

$$J = \int_{t_0}^{\infty} [x^\mathsf{T} Q x + u^\mathsf{T} R u] \, dt$$

is used. Since for a single-input system, R will reduce to a scalar, it can be fixed at unity as discussed in Section 12.4.3. Q will be a 2×2 weighting matrix for the (two) states.

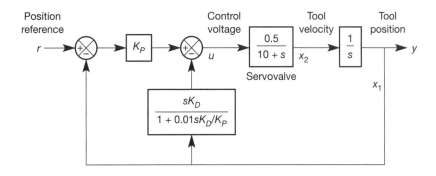

Figure 12.13 Tool-positioning system with a practical PD controller.

A state-space model for the open-loop system between u and y (Figure 12.12) is therefore generated, having as its states the measurable quantities x_1 (position) and x_2 (velocity). Such a model is given by:

$$A = \begin{bmatrix} 0 & 1 \\ 0 & -10 \end{bmatrix}, \quad b = \begin{bmatrix} 0 \\ 0.5 \end{bmatrix}, \quad c = [1 \quad 0], \quad d = 0$$

The MATLAB Control Systems Toolbox *lqr* command can then be used as:

$\gg[k,p,e] = lqr(a,b,q,1)$

to design the controller. Note that this command performs a *continuous-time* design. The *dlqr* command would be used with discrete-time models.

a,b are the state-space model of the plant (using the convention of all-lower-case characters in MATLAB)

q is the 2×2 matrix of state weightings in the performance index (yet to be chosen)

1 is the scalar input weighting (r, fixed at unity, as above)

k will return the designed optimal (stationary) state feedback matrix

p will return the steady-state solution of the Riccati equation

e will return the eigenvalues of the resulting closed-loop-plant with the designed optimal control in place

The MATLAB statements used in the design are to be found in the m-file *fig12_14.m* on the accompanying disk, the general approach being as follows (there are alternative possibilities):

- Set up the state-space model.
- Set $r = 1$, and initialize Q to a 2×2 unit matrix, thus giving equal weightings to everything initially.
- Rearrange the plant model so that the states are available as outputs for feeding back (the MATLAB *feedback* command only works with outputs, not states).
- Use the *lqr* command to design the SVF feedback matrix (a vector in this SISO case).
- Connect up the system as a simple tracking system, as discussed in Section 12.4.5.
- Impose a state reference of a unit step on x_1 (position) and zero on x_2 (velocity).
- Display the step responses.
- Tune Q and repeat the previous four steps until a satisfactory result is obtained.

Note that this is just a way of choosing the feedback gains in a state variable feedback controller. Nothing meaningful is optimized in a mathematical sense, but our understanding of the optimal control methodology is used to design a suitable controller for the plant, given the initial constraints.

Table 12.1 contains the tuning sequence used in this design. Note that the steady-state error is zero in every case.

Fixing $r = 1$ and trying $Q = \text{diag}(1, 1)$ as suggested, gave row 1 in Table 12.1. The performance is very poor, and the input signal is clearly being penalized far too much compared with the states. The states should therefore be penalized significantly more. This will allow the input to rise to much higher levels in an effort to control the states more rapidly.

Table 12.1 Design iterations for the optimal tool-positioning controller.

| | PI weightings | | | Rise time | Max. input (V) | Max. vel. (ms^{-1}) |
Row	q_{11}	q_{22}	r	(ms)	u	x_2
1	1	1	1	45 seconds	1	negligible
2	1000	1000	1	2500	33	0.72
3	20000	20000	1	2200	150	0.95
4	60000	20000	1	1300	250	1.6
5	60	20	0.001	1300	250	1.6
6	600	20	0.001	405	780	4.4
7	600	10	0.001	295	780	5.5
8	600	1	0.001	135	780	8.0
9	600	2	0.001	155	780	7.5

Trying $Q = \text{diag}(1000, 1000)$ gave row 2 in the table. The input now rises to 33 V (remember that 800 V is allowed *in the model* for a unit step test). The rise time is still too long. The same procedure is therefore repeated.

Trying $Q = \text{diag}(20000, 20000)$ gave row 3. This did not give much improvement in rise time, but the input now reached 150 V.

Since the step is applied in order to control x_1, it now seemed sensible to carry on increasing q_{11} only, until the maximum allowable control input was reached, and then see if any further improvement might be possible by tuning q_{22}. Accordingly, trying $Q = \text{diag}(60000, 20000)$ gave row 4 in the table. Also at this stage, to avoid the numbers being quite so large, it was decided to make use of the fact that the location of the minimum point of the PI is not altered by identical scaling operations upon Q and r. Therefore, all the numbers were divided by 1000, giving row 5 in the table. Progress is still encouraging, so this step is repeated.

Row 6 was achieved with $Q = \text{diag}(600, 20)$ (and $r = 0.001$ now). The maximum permissible input voltage has now been reached. The velocity signal still peaks at quite a low value (remember that 40 ms^{-1} is allowable in these tests). The velocity signal (x_2) will reach higher values if it is penalizing *less* with respect to everything else in the performance index. Thus, q_{22} should now be reduced.

Trying $Q = \text{diag}(600, 10)$ gave row 7 in the table. The maximum velocity has been allowed to rise, and so the rise time is improved as a result.

Trying to move further in this way and using $Q = \text{diag}(600, 1)$ gave row 8. There is still a lot of headroom left in the velocity signal, and this is the best rise time yet seen. Unfortunately, though, the output response exhibited a very slight overshoot under these conditions. This means that the relaxing of the constraints upon velocity has gone too far.

$Q = \text{diag}(600, 2)$ with $r = 0.001$ gave row 9 in the table. Clearly this is the best that can be achieved.

The resulting performance is shown in Figure 12.14, and is almost identical to that achieved with the PD controller. Thinking about the structure of the optimal controller, this is not surprising. The optimal gain matrix giving rise to row 9 of Table 12.1 was given by $k = [775 \quad 54]$. In terms of the block diagram of a *tracking* system, this implies the system of Figure 12.15.

Figure 12.15 can be compared directly with Figure 12.13, in which the forward path gain (K_P) was set at 800, very close to the value in Figure 12.15. Also in Figure 12.13, a realistic differentiator was used for the 'D' term of the controller. If an ideal differentiator had been used, that feedback block would simply have contained sK_D. Applying sK_D to x_1

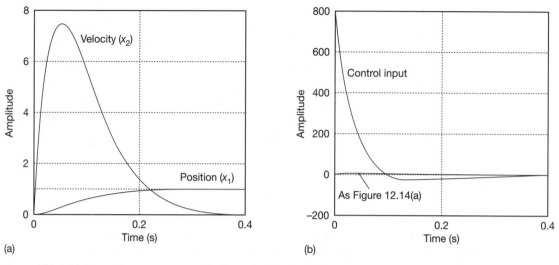

Figure 12.14 (a) State and output responses for the optimal tool-positioner. (b) Control signal for the optimal tool-positioner.

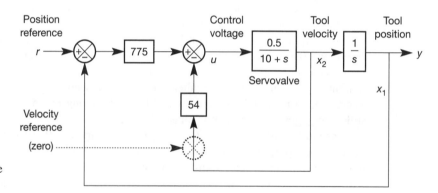

Figure 12.15 Optimal tracking controller for the tool-positioning system.

(as in Figure 12.13) is identical to applying K_D (without the s) to x_2 (because x_2 is the derivative of x_1). This effectively gives a pure velocity feedback block with a gain (K_D) which was chosen as 60 for the PD controller. This would then be identical in structure to Figure 12.15 and, again, the numerical value is very similar. Hence the similar performance.

In the final analysis, for this system there is therefore nothing to choose between these designs so long as the pure velocity feedback is used in the PD controller – otherwise, with the implementation of Figure 12.13, poor noise performance would be expected, and the optimal tracker would prove superior (and simpler to implement). Note that this similarity has only resulted because such a simple system was used. There were only as many degrees of freedom available to the optimal controller as to the PD controller. In higher-order designs and (especially) in multivariable designs, the greater number of degrees of freedom available to the optimal design procedure are likely to allow better performance to be obtained in many cases. In addition, the optimal control is always easy to implement, having no dynamic elements (unless an observer is needed).

Manual solution

This particular example is possible to solve by hand, being so simple. Any real system of higher than second order, and most second-order systems too, are too complicated to solve without computer assistance. However, in this case, to lend credibility to the continuous-time solution of the LQR problem (given at the very end of Section 12.4.2) and also to confirm the MATLAB result just obtained, see Problem 12.4.

12.6.2 Case study – the antenna-positioning system

Finally, the antenna-positioning system used as an example in several chapters will be reconsidered. Figure 12.16 repeats the open-loop system block diagram, for convenience.

In section 5.4.4, the required performance was obtained by choosing the SVF gains to give specified closed-loop eigenvalues. Now the optimal controller will select the feedback gains, based on the PI to be minimized. Of course, there must be some PI (that is some contents of Q and R) which, when minimized, would lead to the same set of feedback gains which the previous pole-placement design method generated. However, trying to find this PI will be of no physical significance.

Instead, the objective will be to obtain the fastest possible rise time in response to a step input, with no steady-state error, and no more than 5 per cent overshoot.

A state-space model for the system of Figure 12.16 has previously been obtained as:

$$A = \begin{bmatrix} 0 & 1 & 0 \\ 0 & -1 & 1 \\ 0 & 0 & -5 \end{bmatrix}, \quad b = \begin{bmatrix} 0 \\ 0 \\ 5 \end{bmatrix}, \quad c = [1 \quad 0 \quad 0], \quad d = 0$$

For this system, the steady-state gains of the drive system and the load dynamics are both unity. Therefore an input of 1 V at u will produce (at steady-state) 1 N m of torque at x_3 and 1 rad s^{-1} of velocity at x_2. The real-world control range at u is ± 15 V. So long as this is not exceeded, the torque and velocity will therefore never exceed ± 15 N m and ± 15 rad s^{-1} respectively. As modelled, these figures will not even be exceeded transiently, because the dynamic blocks are only first order; so the relationships between control voltage, torque and velocity cannot be oscillatory. However, the relationship between the *closed-loop* reference signal and these quantities may be oscillatory, due to the feedback action.

The limiting figure for torque is acceptable, but the velocity value is too great for this plant. The maximum allowable is ± 55 deg s^{-1} (that is ± 0.96 rad s^{-1}), and the intention of the plant designer was that the relatively slow dynamics of the load would easily allow the velocity to be restricted to this value with feedback in place (clearly, the open-loop system could theoretically reach 15 rad s^{-1} after a few seconds of acceleration at full input voltage, but this is physically prevented by an arrangement incorporating a viscous coupling, in order to avoid damage).

Figure 12.16 Open-loop block diagram of the antenna-positioning system.

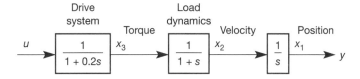

The range of output (position) motion allowed in this plant is approximately $\pm\pi/2$ rad (that is ±90 degrees). Therefore, if a unit step response from the simulated model is to correspond to this limit, then the previously discussed signal limits all need to be divided by $\pi/2$ to obtain the corresponding simulation-world limits.

To sum up, the requirements in simulation for a unit (full output) step test are:

- zero steady-state error
- overshoot not to exceed 5 per cent
- voltage at u to be in the range ±9.5 V (from $15 \text{ V} \times 2/\pi$)
- velocity at x_2 to be in the range ±0.6 rad s^{-1} (from 0.96 rad s$^{-1} \times 2/\pi$)
- minimum rise time (10–90 per cent of the final value) commensurate with the above.

The previous SVF design would fail to meet almost all these criteria, but no comparison can be made, as no such constraints were set at the time of that design. Bearing in mind that the dominant time constant of the open-loop plant is 1 second (due to its inertia, and some slippage in the viscous coupling), very fast performance is unlikely to be obtained within these constraints.

Using exactly the same philosophy as in the previous example, the design process begins with r set to 1 and Q set to diag(1, 1, 0) in the performance criterion (r is still scalar, because there is a single input, Q is now 3×3 because the plant model is third order). The reason that q_{33} is set to zero is that there is no point in constraining the torque (x_3) in any way, since it can never exceed its limit of 15 Nm if u is restricted to 9.5 V, due to the d.c. gain of 1 Nm V^{-1} in the intervening first-order element.

The same sequence of MATLAB commands is used to design and test the controller as was used in the previous section. Only the sizes of the matrices alter. The same general procedure is also followed for varying Q. The sequence of design steps can be repeated by editing and running the m-file *fig12_18.m* on the accompanying disk.

Row 1 of Table 12.2 might suggest that u is being penalized too much compared with the states (it reaches only 1.9 V, and the rise time is long), so the states are progressively penalized more until, in row 3, the velocity (x_2) has reached its allowed maximum.

In this design there is an extra degree of freedom compared with the previous example, so it is possible to comply with the maximum velocity constraint simultaneously with the maximum control input constraint. To achieve this, the input must be allowed to

Table 12.2 Design iterations for the optimal antenna positioner.

Row	PI weightings q_{11}	q_{22}	q_{33}	r	Rise time (seconds)	Max. input (V) u	Max. torque (Nm) x_3	Max. V (rad s^{-1}) x_2
1	1	1	0	1	3.3	1.9	0.7	0.36
2	10	10	0	1	2.4	3.2	1.5	0.57
3	20	20	0	1	2.3	4.5	1.9	0.60
4	90	170	0	1	3.0	9.5	2.6	0.60
5	10	7	0	1	2.1	3.2	1.6	0.60
6	8	4.5	0	1	2.0	2.8	1.5	0.60
7	5	0.6	0	1	1.8	2.2	1.3	0.60
8	4.5	0	0	1	1.7	2.1	1.3	0.60

increase further (by penalizing the states more), but x_2 must also be penalized more heavily relative to x_1, so as to prevent the maximum velocity from exceeding its limit. Row 4 of Table 12.2 shows a set of weightings which achieve this, and the resulting system (according to a simulation) exhibits zero overshoot. However, note that the rise time is *worse* than it was in row 3. This means that, for this system, controlling to the limiting values given for the signals does *not* produce the best performance in terms of minimum rise time.

In order to improve the rise time compared with row 3 of Table 12.2, the weighting q_{11} can be reduced, so as to allow x_1 to move more freely. Rows 5 to 8 of the table show this procedure. At each stage, the weighting q_{22} was adjusted so that the maximum velocity was attained, and the overshoot was checked (by the simulation in the m-file) to be within 5 per cent. The final result (row 8) was achieved when it was only just possible to reduce q_{22} far enough to allow x_2 to rise to its limit. Any further reduction of q_{11} then leads to a system in which the maximum velocity is never reached, so the rise time is lengthened.

The feedback gains for row 8 of the table are $k = \begin{bmatrix} 2.12 & 1.42 & 0.25 \end{bmatrix}$ (as reported by the m-file), resulting in the system of Figure 12.17. The references for torque and velocity are zero, and they are omitted from the figure. According to the MATLAB *lqr* command, the resulting closed-loop eigenvalues are positioned at approximately $-1.1 \pm j0.9$ and -5, and Figure 12.18 shows the responses. In Figure 12.18(b) the position response is identical to that of Figure 12.18(a), only the axis scaling having altered. All the other responses eventually settle at zero.

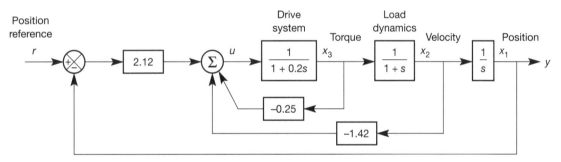

Figure 12.17 Optimal tracking controller for the antenna-positioning system.

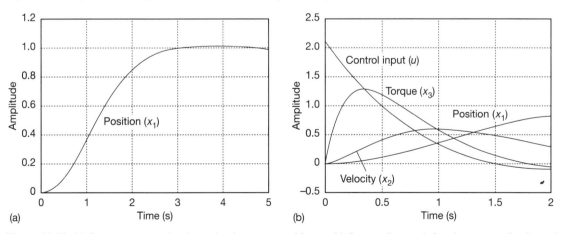

Figure 12.18 (a) Output response for the optimal antenna positioner. (b) State and control signal responses for the optimal antenna positioner.

The optimal root locus

Although the design of this section has been carried out by physical reasoning plus trial-and-error, the use of a root locus showing the variation of the closed-loop eigenvalues with Q or R (or individual elements of these) can sometimes be useful, especially if the designer has some idea of the closed-loop pole locations that would give the kind of performance required. To show how such a root locus can be obtained, the m-file *fig12_19.m* on the accompanying disk uses the MATLAB code for the example of this section, and adds the appropriate commands. The result appears in Figure 12.19.

Figure 12.19 shows the loci of the closed-loop eigenvalues of the optimally controlled scheme, with r fixed at 1, and Q beginning at diag(0.1, 0, 0) and being multiplied by 1, 2, 3, ..., 90, to end at diag(9, 0, 0). Two of the three loci are clearly visible, but the locus on the negative real axis does not move far from the value -5 for this range of values in Q. The three loci pass through the solution previously obtained, which is shown in the figure.

If the root locus is drawn for a much greater range of values in Q, it will always be found that the branches become asymptotic to lines of constant damping ratio in the s-plane (the radial lines in Figure 12.19). For the present system, and for very

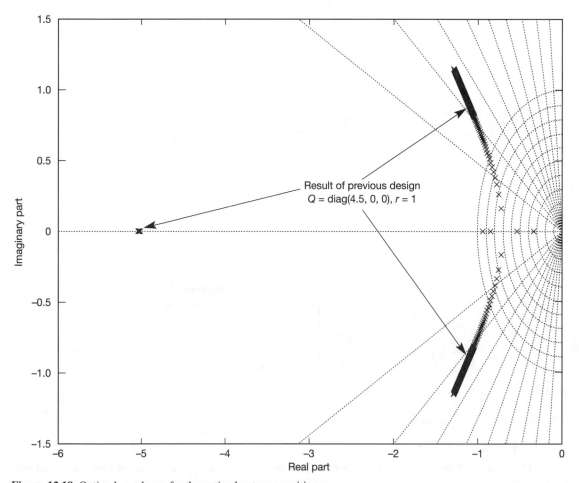

Figure 12.19 Optimal root locus for the optimal antenna positioner.

high values of q_{11}, the complex conjugate loci are found to be asymptotic to a damping ratio of about 0.52.

12.7 Summary

In this chapter, the notion that there may be a 'best' way to control a plant has been investigated. What is 'best' has been represented by a *performance index* (PI), the minimization of which leads to the required controller. Unfortunately, it was also discovered that the mathematics necessary to solve the resulting optimization problem may not be possible. For this reason, it is usual to employ a standard PI containing *quadratic forms* which apply weightings to the relative importance of restricting the excursions of the individual state variables and control inputs.

Since the standard PI will not usually be the thing which ideally we would like to minimize, its minimization was not regarded as an end in itself. Instead, the use of the machinery of the *linear quadratic regulator* (LQR) design procedure was presented as one step in an iterative *design* process using CACSD tools. In this process, the elements in the PI were systematically tuned until a satisfactory performance was achieved.

In addition, a number of ways were discussed by which the resulting *regulator* design could be made to behave as a *tracking* system.

The method was applied to two simple case-study examples, using values of actual voltages, torques, and so on to guide progress.

12.8 Problems

12.1 In the text it was stated that the Q and R matrices in a performance index can always be symmetric matrices with no loss of generality. Prove that this is the case for a general matrix Q in a quadratic form $x^T Q x$.

Hint: Any general matrix Q can be rewritten as $Q = Q_{\text{sym}} + Q_{\text{skew}}$ where:

Q_{sym} is a symmetric matrix given by

$$Q_{\text{sym}} = \tfrac{1}{2}(Q + Q^T)$$

and Q_{skew} is a skew-symmetric matrix given by

$$Q_{\text{skew}} = \tfrac{1}{2}(Q - Q^T)$$

12.2 By substituting K_n from Equation (12.13) into Equation (12.14), show that Equation (12.14) may be rewritten as:

$$P_n = Q + \Phi^T P_{n+1} \Phi$$
$$-\Phi^T P_{n+1}^T \Delta [R + \Delta^T P_{n+1} \Delta]^{-1} \Delta^T P_{n+1} \Phi$$

(J_n^o can still be obtained by Equation (12.15) as before). This equation for P_n is another DMRE and is used in some texts and papers. Some authors omit the transpose operator from P_{n+1}. This is quite in order, since P is symmetric.

12.3 Equation (12.17) gave the steady-state equation of the closed-loop LQR system with feedback matrix K in place, as $[I - (\Phi - \Delta K_N)] x_N = 0$. It was subsequently argued that so long as the square bracket could be proved to be non-zero and full-rank, then the correct regulatory action would occur (that is, $x_N = 0$). Prove that the term in the square bracket must be non-zero.

Hint: The only way the bracketed term could be zero is if $(\Phi - \Delta K_N) = I$. Try substituting this into Equation (12.14) in the steady-state case. By considering the definiteness properties of Q and R you should be able to show that the result is impossible (assume Q to be diagonal for simplicity).

12.4 For the hydraulic machine tool example in Section 12.6.1, take the continuous-time state-space model and the final values of

$$Q = \begin{bmatrix} 600 & 0 \\ 0 & 2 \end{bmatrix}$$

and $r = 0.001$, and use the continuous-time solution to the LQR problem (given at the very end of

Section 12.4.2) to derive the optimal feedback gains.

Hint: Assume P to be symmetric (it always should be) and write out the full AMRE with the elements of P as unknowns. Equate elements on each side of the AMRE and solve for the elements of P. The values of K then follow directly and should agree with the MATLAB result obtained in Section 12.6.1 (namely, $k = [775 \quad 54]$).

12.5 Repeat Problem 12.4, if the system now has a second input. Assume that the A and Q matrices remain unchanged and the B matrix becomes

$$\begin{bmatrix} 0.5 & 0 \\ 0 & 0.5 \end{bmatrix}$$

Assume that a suitable R matrix is

$$\begin{bmatrix} 0.001 & 0 \\ 0 & 0.001 \end{bmatrix}$$

Go as far as possible by hand, and then use computer assistance to overcome any difficulties.

12.6 Towns A to L are connected by a number of permitted routes having weightings as given in Table P12.6. These weightings include distance, the class of road and the number of junctions and other hazards. They are calculated so that the route with the lowest score will give the best compromise between cost and travelling time. It is only permissible, for the purposes of this problem, to travel in the directions implied by the information in the table.

(a) Find the best route from A to L.
(b) Find the worst route from A to L.
(c) Find the best route from A to L if the traveller is constrained to call at town H.

Table P12.6 Weighted distances for Problem 12.6.

Route between towns	Weighted distance	Route between towns	Weighted distance
A to B	3	E to I	3
B to C	5	F to G	9
B to D	5	G to J	3
C to E	8	H to J	2
C to F	1	H to K	5
D to E	9	H to L	10
D to I	10	I to K	6
E to G	2	J to L	9
E to H	5	K to L	4

12.7 A system has a discrete-time model

$$x_{n+1} = \Phi x_n + \Delta u_n \quad \text{with} \quad \Phi = \begin{bmatrix} 1 & 0.01 \\ 0 & 0.9 \end{bmatrix} \quad \text{and}$$

$$\Delta = \begin{bmatrix} 0.005 & 0.001 \\ 0 & 0.004 \end{bmatrix}$$

(a) Find the infinite-horizon solution to the linear quadratic regulator problem for this system, given that it is twice as costly to use u_1 for control as u_2, it is five times more important to control x_1 correctly than x_2 and it is ten times more important to control x_1 correctly than it is to minimize energy expenditure at u_1. You will find it necessary to use a programmable calculator or a computer, as the equations need to be iterated at least 250 times to reach a reasonably steady solution.

(b) If you have access to MATLAB and the Control Systems Toolbox, or a similar package, use it to verify your design (for example, use the MATLAB *dlqr* command).

13 An introduction to robust control

13.1   Preview

This chapter can be understood if Chapters 1 to 5 have been studied. Some mathematical ideas from Chapters 10 and 12 are used, but it is not necessary to read them first.

Throughout this text, controllers have been designed based upon mathematical models of the plant to be controlled. It has often been stressed that these models are approximate. This chapter begins to address the question, 'how can the uncertainty in the model be included in the design procedure, and how does it affect the performance of the controlled system?'.

There is more than one way of answering this question, depending upon the type of model being considered and the type of control scheme being designed. Therefore, this chapter does not seek to propose any particular robust design *method*. Rather, it presents a collection of tools and *concepts*, which can be applied to several of the techniques elsewhere in the book to give robust control systems. This chapter might therefore seem rather more theoretical than the others. This feeling is heightened by the lack of worked examples and end-of-chapter problems. This latter is partly to save space, and partly because the working of the problems would require the use of *two* more MATLAB toolboxes (which are referenced later, for the interested reader). This was not felt justified in this chapter.

In this chapter, the reader will learn:

- that plant–model mismatch can often be handled in design procedures
- that the effects of plant–model mismatch on performance can sometimes be evaluated
- that there are various ways of evaluating robustness, for both frequency-domain and time-domain systems
- that many of the methods of this chapter can be applied to multivariable systems.

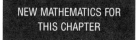

NEW MATHEMATICS FOR THIS CHAPTER

Some parts of this chapter may appear rather mathematical, because they use notation that is not used elsewhere in the text. All this notation is explained as it is met, but it can still seem strange on a first reading. It is concerned with the norms of quantities, and their maximum, minimum and supremum values. Other than that, singular values are encountered (they have been used in Chapter 10), as are Lyapunov equations and Riccati equations, which are also briefly encountered elsewhere in the book. Parseval's theorem is necessary only to follow one of the derivations, and is stated in a section of its own.

13.2   Introduction

In a general engineering context, the addition of controls must be justified in terms of profitability, or in terms of personnel or environmental safety. A control system should be effective and efficient, and remain so throughout the life of the plant. Needless to say all equipment deteriorates with time and the idealized model (or *nominal model*) used to design the control system will, at best, only indicate approximate plant behaviour. For the control system, this model–plant mismatch can result in significant changes in dynamics from those originally predicted. Of course model–plant mismatch can occur for many reasons other than plant deterioration; using a linear model to represent a nonlinear plant, making simplifying assumptions which effectively means neglecting fast dynamic phenomena, inaccuracies in estimating model parameters, and so on. A control system which can cater for such changes would be classified as being *robust* and a control system design procedure which allowed for such uncertainties as a *robust design method*.

Robust design methods cover analysis and synthesis techniques. A typical robust analysis design method would use some standard control technique to find a controller that will produce an acceptable closed-loop design based on the nominal plant model. Robust analysis would then establish whether the performance remained acceptable if the plant (or, as discussed later, an input) differs from nominal. The alternative synthesis approach uses robust methods to find a plant controller that meets some predetermined specification for the nominal model and its anticipated uncertainty.

Uncertainty in the nominal model is accounted for by considering a range (or family) of models which will encompass all likely variations. If it can be shown that a particular control system will stabilize all plants within the family, then that controller has *robust stability*. However, for a control system to be of practical use, stability guarantees alone are insufficient, since some stable realizations may give totally inadequate performance. For a controller to have *robust performance* the realization with the 'worst' stability characteristics must meet the desired performance specifications.

Apart from the uncertainty regarding the model's ability to represent the plant, it is necessary to consider the various inputs disturbing the plant. Indeed, implicit in many performance specifications are assumptions relating to the type of disturbances. For example, a sudden, or step, change is often assumed. In practice the hardware that goes to make up engineering plant seldom, if ever, experiences such a disturbance, although it might experience something approximating a step. It is easily shown that a system designed to give a good response to a step change will not necessarily have a good response to a gradually increasing (or decreasing) change, or ramp. If the assumed disturbance is inappropriate, then it is likely that there will be a deterioration in control system performance. Again, a robustness analysis can be used to indicate that a control system is tolerant to disturbance uncertainty by considering a family of disturbances to which the nominal disturbance belongs.

13.3 The control system design model

In this chapter the discussion, although often directly applicable to multivariable systems, will concentrate on continuous single-input, single-output plant which may be analysed in the frequency domain. For this reason, the main emphasis is on transfer function models. Any differences will be highlighted in the text.

Figure 13.1 shows the standard feedback configuration. The plant is a physical piece of hardware with fixed dynamics and a finite input signal range, which will impose limits on the performance of the closed-loop system. Its output $y(s)$ is a function of the manipulable input $u(s)$ and any external disturbance $d(s)$. A perfect transducer (one having no dynamics) measures the output variable at $y(s)$ and the resulting measurement, possibly contaminated with noise $n(s)$, is fed back to the controller. Ignoring transducer dynamics is permissible if the speed of response of the transducer is significantly faster than that of the plant; that is, the transducer dynamics are assumed to be included with those of the plant. At this point it will suffice to say that this is often the case. The controller consists of a pre-filter $P(s)$ which modifies the setpoint $r(s)$ and a forward path compensator $K(s)$ which generates the signal $u(s)$.

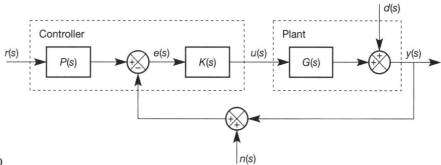

Figure 13.1 Standard SISO feedback configuration.

What is not shown in Figure 13.1 is the nominal plant model $G_m(s)$, used to design the controller elements $P(s)$ and $K(s)$. To demonstrate how the nominal plant model influences the controller design, assume that the compensator $K(s)$ may be represented as shown in Figure 13.2, such that

$$K(s) = \frac{C(s)}{1 - G_m(s)C(s)} \tag{13.1}$$

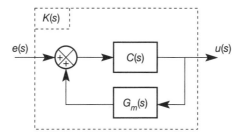

Figure 13.2 A representation of $K(s)$ for Figure 13.1.

Using this assumed compensator and, after some algebraic manipulation, it is possible to transform the standard feedback configuration into the two-degree-of-freedom IMC (Internal Model Control) configuration of Figure 13.3. $C(s)$ is called the IMC compensator and, for comparison purposes, this element has been moved from the forward to the feedback and pre-filter paths as shown. The IMC pre-filter $P'(s)$ is identical to the quantity $P(s)C(s)$.

At this stage it is important to realize that for analysis purposes Figures 13.1 and 13.3 are identical provided $K(s)$ and $C(s)$ are related through Equation (13.1). However, the IMC structure clearly demonstrates the importance of a 'good' design model and additionally, the function of the compensator and pre-filter.

The transfer function that shows the influence of the disturbance, noise and any setpoint on the output for both the standard feedback and two-degree-of-freedom IMC configuration, is given by

$$y = \frac{1 - G_m C}{1 + C(G - G_m)} d + \frac{GP'}{1 + C(G - G_m)} r - \frac{G}{1 + C(G - G_m)} n$$

(13.2)

For clarity the Laplace operator 's' is omitted in the above equation.

The following observations relating to Figure 13.3 (and by inference to Figure 13.1) may now be made:

If the plant and nominal plant model are identical ($G(s) = G_m(s)$ in Equation (13.2)) and (to simplify the discussion at this stage) it is assumed that $n(s) = 0$, then

(1) The feedback loop compensates solely for plant disturbances. From Equation (13.2), perfect disturbance rejection would be obtained if $G_m(s)C(s) = G(s)C(s) = 1$.

(2) Setpoint tracking (the closeness with which y follows r) depends solely on the pre-filter $P'(s)$. Again from Equation (13.2) perfect setpoint tracking would be obtained if $G(s)P'(s) = 1$. Looking at Figure 13.3, it is evident that when the model and plant dynamics are identical the system is open-loop; that is, feedback only occurs when $d(s)$ is not equal to zero. Consequently, the only element affecting tracking is the pre-filter.

There are various practical reasons that make perfect tracking and perfect disturbance rejection impossible. However, the above observations

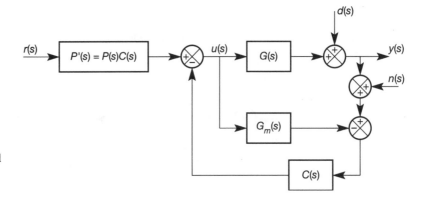

Figure 13.3 Two-degree-of-freedom internal model control (IMC) configuration.

demonstrate the closed-loop system's disturbance rejection properties and indicate the need for a pre-filter. In practice, a pre-filter would only be required if the frequency components of $r(s)$ and $d(s)$ were well separated (see point 4 below).

(3) The plant dynamics dictate the compensator and pre-filter dynamics. That is, the ideal compensator and pre-filter are both explicit functions of the plant dynamics: see the previous two points.

(4) When the system does not require a pre-filter, the compensator $C(s)$ is moved back into its forward path position to give the standard single-degree-of-freedom IMC control structure shown in Figure 13.4. Simple block diagram manipulation shows that the disturbance $d(s)$ may be combined with $r(s)$ and, assuming their dynamics are similar, both can be treated in an identical manner.

If the plant and nominal plant model are not identical $(G(s) \neq G_m(s))$, then:

(5) Both plant disturbance and modelling errors are fed back and the compensator $C(s)$ must be detuned to accommodate the additional modelling errors.

Before leaving this section note that if $G(s)$ is stable and $G_m(s) = G(s)$, then the closed-loop system is stable if and only if $C(s)$ is stable. Further, $G(s)$ need not be linear since all the results are equally applicable to the nonlinear case.

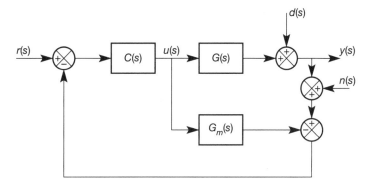

Figure 13.4 Single-degree-of-freedom internal model control (IMC) configuration.

13.3.1 Sensitivity and Bode's sensitivity index

Advanced section

This section provides some background theory on control system sensitivity analysis. It is not essential to the main theme of the chapter and therefore, for an initial reading, the reader should turn directly to Section 13.4.

In a general context, sensitivity is the incremental change in the value of some function due to an incremental change in some parameter on which the function is dependent. So, for example, a straight line could be considered to be some function y of x, or

$$y(x) = mx + c$$

and the sensitivity of $y(x)$ to x is m, the slope of the line. To generalize this concept consider a function $q(a)$. At the specific value $\hat{a}$ let

$$\hat{q} = q(\hat{a})$$

Assume a small change in the parameter $\hat{a}$ to $\hat{a} + \delta a$, such that

$$\hat{q} + \delta q = q(\hat{a} + \delta a)$$

Using a Taylor series expansion around the nominal value $\hat{a}$ (as in Section 2.4.5), gives

$$q(\hat{a} + \delta a) = q(\hat{a}) + \left.\frac{dq}{da}\right|_{a=\hat{a}} \delta a + \cdots$$

Hence for small values

$$\delta q \approx \left.\frac{dq}{da}\right|_{a=\hat{a}} \delta a$$

and the function

$$f(a) = \frac{dq}{da}$$

is called the sensitivity function (and is analogous to m, the slope of the line in the graph $y(x) = mx + c$). If $f(a) = 0$, then the property q is insensitive at the nominal point. If $f(a) \neq 0$, then q is sensitive.

Now consider the sensitivity of a transfer function $G(s, a)$. If a logarithmic scale is used to describe $G(s, a)$ (similar to that used to produce Bode plots) then the result is Bode's sensitivity index, which is defined as

$$f(a) = \frac{d \ln G(s, a)}{d \ln a} = \frac{\dfrac{dG(s, a)}{G(s, a)}}{\dfrac{da}{a}}$$

and gives a measure of the relative change of $G(s, a)$ and a. To apply this index to the standard feedback configuration, Figure 13.1, is to compare the relative sensitivity of the closed-loop system's output relative to that of the plant. If $d(s)$ and $n(s)$ are both zero, then

$$\frac{y(s)}{r(s)} = G_{yr}(s, a) = \frac{G(s, a)K(s)P(s)}{1 + G(s, a)K(s)}$$

and Bode's sensitivity index for the system is

$$\frac{\dfrac{dG_{yr}(s,a)}{G_{yr}(s,a)}}{\dfrac{dG(s,a)}{G(s,a)}} = \frac{dG_{yr}(s,a)}{dG(s,a)} \cdot \frac{G(s,a)}{G_{yr}(s,a)}$$

$$= \frac{(1+G(s,a)K(s))K(s)P(s) - G(s,a)K(s)P(s)K(s)}{(1+G(s,a)K(s))^2}$$

$$\times \frac{(1+G(s,a)K(s))}{K(s)P(s)} = \frac{1}{1+G(s,a)K(s)} = S(s,a)$$

This index is similar to the sensitivity function used throughout this chapter. However, sensitivity design objectives and robust design objectives are different. Sensitivity looks at the dependence of closed-loop performance on parameter variations relative to the nominal values. Robustness, however, requires that the closed-loop system performance is acceptable for all possible parameter values.

13.4 The design problem

Although all the relationships to be derived in this section could be obtained for the IMC structure (Morari and Zafiriou, 1989), the more familiar standard feedback configuration, Figure 13.1, will be analysed. In most cases the IMC relationships may be obtained by comparing terms in the appropriate IMC and feedback equations.

Rewriting Equation (13.2) specifically for the structure of Figure 13.1 gives

$$y = \frac{1}{1+GK}d + \frac{GKP}{1+GK}r - \frac{GK}{1+GK}n \tag{13.3}$$

If the *sensitivity function* $S(s)$ (see Section 13.3.1) is defined as:

$$S(s) = \frac{1}{1+G(s)K(s)} \tag{13.4}$$

and the *complementary sensitivity function* $T(s)$ as:

$$T(s) = \frac{G(s)K(s)}{1+G(s)K(s)} \tag{13.5}$$

then Equation (13.3) becomes

$$y(s) = S(s)\,d(s) + T(s)P(s)r(s) - T(s)n(s) \tag{13.6}$$

The following design objectives are evident:

(1) For disturbance rejection (irrespective of the quality of the plant model $G_m(s)$) the sensitivity function $S(s)$ must be as small as possible. If $S(s) = 0$, there is perfect disturbance rejection; that is, $d(s)$ does not affect $y(s)$ in Equation (13.6).

(2) For good setpoint tracking the complementary sensitivity function $T(s)$ must be finite.

(3) For noise rejection $T(s)$ must be as small as possible. If $T(s) = 0$ in Equation (13.6), then there would be perfect noise rejection.

Items 2 and 3 are in conflict, and it would appear that some compromise is required. It may be argued that for some systems the speed of response is so slow that noise (electrical noise) could be almost completely eliminated by filtering, thus eliminating the need to make $T(s)$ very small. Unfortunately, though, the problem is more complicated than that indicated by Equation (13.6).

First, note that

$$S(s) + T(s) = 1 \tag{13.7}$$

and then find the equation describing how the control signal, $u(s)$, is affected by the setpoint, noise and disturbance inputs. In practice $u(s)$ is associated with power amplification (needed to drive the plant) and hence should be kept small if the running costs are to be minimized. From Figure 13.1, and the complementary sensitivity equation, Equation (13.5), it is found that

$$u(s) = K(s)e(s) = K(s)\, \frac{1}{1 + G(s)K(s)}\, [P(s)r(s) - n(s) - d(s)]$$

$$= \frac{T(s)}{G(s)}\, [P(s)r(s) - n(s) - d(s)] \tag{13.8}$$

Therefore if $u(s)$ is to be small then $T(s)$ must be small. In view of Equation (13.7), it is impossible for both $T(s)$ and $S(s)$ to be small and so there will be difficulties in meeting the stated design objectives.

The control system design problem may now be defined as that of finding the controller $K(s)$ that gives the best compromise between the conflicting sensitivity and complementary sensitivity requirements.

13.4.1 The design problem for multivariable systems

For multivariable systems the design problem is identical to that described earlier in Section 13.4, except that the sensitivity function, Equation (13.4), becomes:

$$S_y(s) = [I + G(s)K(s)]^{-1} \tag{13.9}$$

and is called the output sensitivity function. From Figure 13.1 and Equation (13.3), it can be seen that Equation (13.9) gives the sensitivity function seen at the output $y(s)$.

Similarly, the multivariable equivalent of Equation (13.8) is

$$u(s) = [I + K(s)G(s)]^{-1}K(s)[P(s)r(s) - n(s) - d(s)] \tag{13.10}$$

The first term on the right-hand side of Equation (13.10) is called the input sensitivity function, that is

$$S_u(s) = [I + K(s)G(s)]^{-1}$$

and gives the sensitivity function seen at the input $u(s)$.

For single-input-single-output systems $S_y(s) = S_u(s) + S(s)$ since the positions of $K(s)$ and $G(s)$ are interchangeable in the sensitivity functions. However, with multivariable plant the relative positions of $\mathbf{K}(s)$ and $\mathbf{G}(s)$ become important (see Section A1.1.2). In particular, if the number of plant inputs differs from the number of outputs then the dimensions of $\mathbf{S}_y(s)$ will not be the same as the dimensions of $\mathbf{S}_u(s)$.

13.5 The generally accepted frequency-domain solution

With single-input-single-output systems, for example, it is normal practice to define a set of frequency-domain performance specifications which, if satisfied, would ensure satisfactory closed-loop performance (see Chapter 3). In particular, bandwidth is specified, since it can be measured from a closed-loop Bode plot (note that the closed-loop Bode plot and a Bode plot of $T(j\omega)$, Equation (13.5), are identical) and would indicate the noise suppression characteristics of the closed-loop system. In essence, a bandwidth specification requires that for noise suppression $T(j\omega)$ must be small at high frequencies. Intuitively, a general solution to the control system design problem can be obtained in the frequency domain, by making $S(j\omega)$ small at some frequencies and $T(j\omega)$ small at others. Exactly which frequencies, as always, depends on the plant dynamics.

All physical engineering plants will be strictly proper, that is, having more poles than zeros, so that:

$$\lim_{s \to \infty} |G(s)| = 0 \tag{13.11}$$

and at high frequencies (where noise predominates) the magnitude of the plant's frequency response will therefore be small. At low frequencies (those frequencies at which the reference and disturbance signals are concentrated) the plant's magnitude will be finite, except possibly at zero frequency ($\omega = 0$) where the magnitude would be infinite if the system's type number was one or more.

In terms of the design objectives of Section 13.4, $S(j\omega)$ should be made small at low frequencies. This is achieved by ensuring that the controller gain $|K(j\omega)|$ is large at low frequencies or, to be pedantic, that $|G(j\omega)K(j\omega)| \gg 1$, see Equation (13.4).

To minimize high-frequency noise and controller effort, $T(j\omega)$ should be small which, from Equation (13.5), indicates that $|G(j\omega)(K(j\omega)| \ll 1$. Since it is assumed that the system is strictly proper, then at very high frequencies $|G(j\omega)|$ will be small and provided $|K(j\omega)|$ does not increase with frequency $T(j\omega)$ will also be small. In practice, some additional controller gain reduction may be required.

For multivariable plant the generalizations of the Bode magnitude plot are the singular value plots (singular values are defined in Section 10.6.4). In terms of a system's sensitivity and complementary sensitivity functions the nominal performance specifications are:

- For disturbance rejection and setpoint tracking, the maximum singular value of $S(j\omega)$, $\bar{\sigma}(S(j\omega)) \ll 1$ at low frequencies
- For noise suppression, $\bar{\sigma}(T(j\omega)) \ll 1$ at high frequencies.

13.6 Modelling model uncertainty

The simplest models of uncertainty that can be defined within a plant model are normally classified as being additive or multiplicative, see Figure 13.5.

Additive uncertainty is defined as

$$G(j\omega) = G_m(j\omega) + l_a(j\omega) \tag{13.12}$$

where

$$|l_a(j\omega)| \le l_a(\omega)$$

That is, the true plant model $G(j\omega)$ can be equated to the nominal plant $G_m(j\omega)$, plus some uncertainty $l_a(j\omega)$. Further, the gain of the uncertainty at any frequency will be less than or equal to the positive frequency dependent constant $l_a(\omega)$.

Multiplicative uncertainty is defined as

$$G(j\omega) = G_m(j\omega)(1 + l_m(j\omega)) \tag{13.13}$$

where

$$|l_m(j\omega)| \le l_m(\omega)$$

and, for both descriptions to be equivalent, it is necessary that

$$l_m(j\omega) = \frac{l_a(j\omega)}{G_m(j\omega)}$$

or

$$l_m(\omega) = \frac{l_a(\omega)}{|G_m(j\omega)|}$$

Additive uncertainty is particularly useful in Nyquist, or polar plots, where the uncertainty can be represented by a series of discs of radius $l_a(\omega)$ centred on the $G_m(j\omega)$ loci. Multiplicative uncertainty is useful in Bode diagrams, where the dB gains of the nominal model and the uncertainty can be combined to give the worst case.

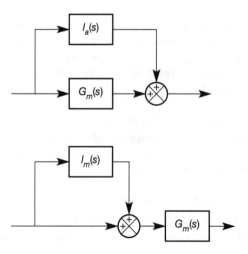

Figure 13.5 Additive and multiplicative uncertainties.

13.6.1 Modelling input uncertainty

Figure 13.6 shows the general input uncertainty model in which $v(s)$ is used to represent all external inputs (setpoint changes or disturbances) entering the feedback loop. The signal $v(s)$ is generated by first passing $v*(s)$, a mathematically bounded and normalized function, through the input weighting block, $W(s)$. This model can represent both specific inputs (for example, impulses or steps), as well as families (or sets) of bounded inputs.

For an impulse (a specific input), $v*(s)$ is set equal to one and the input weighting block is a pure unity gain. A step is generated by again setting $v*(s)$ equal to one (thus ensuring the required mathematical bound), and $W(s)$ to $1/s$. In both cases,

$$v(s) = W(s) \tag{13.14}$$

Sets of bounded inputs may also be represented by Figure 13.6. For example, if

$$v*(s) = \frac{\sqrt{2\alpha}}{s + \alpha} \quad \alpha > 0$$

and

$$W(s) = \frac{s + \beta}{s\sqrt{2\beta}} \quad \beta > 0$$

then all the mathematical constraints will be satisfied and the resulting input $v(s)$ can be used to represent a step ($\alpha = \beta$), or a step modified by a lead ($\alpha > \beta$), or a step modified by a lag ($\alpha < \beta$). In this case,

$$v(s) = W(s)v*(s) \tag{13.15}$$

Figure 13.6 A general input uncertainty model.

13.7 Robustness – a worst-case analysis

The simplest robustness technique available is that of looking for the worst case.

Example 13.1 *A simple robustness analysis*

Consider the contrived example of a plant which can be fairly accurately modelled by the equation

$$G_m(s) = \frac{e^{-0.5s}}{s + 1}$$

That is, $G_m(s) \approx G(s)$. It is further assumed that the sensor measuring the output $y(s)$ is known to be accurate to 10 per cent up to 1 rad s^{-1}, but then deteriorates linearly up to 20 rad s^{-1}, beyond which point it could have errors of 100 per cent. Since the sensor is part of the feedback loop there is model uncertainty. The given uncertainty description is multiplicative, although this will be converted to additive form (so that a Nyquist analysis may be performed), using the relationship $l_a(\omega) = l_m(\omega)|G_m(j\omega)|$, following Equation (13.13) above.

Figure 13.7 shows the polar plot for the nominal plant model and a compensator gain $K(s) = 2$. The plot was produced by the MATLAB m-file *fig13_7.m* on the accompanying disk, and a few data points are shown in Table 13.1. It can be seen that the closed-loop system based on the nominal plant model will be stable, with a gain margin of 1.95 (at a phase crossover frequency of 3.8 rad s^{-1}) and a phase margin of 70.38° (at a gain crossover frequency of 1.7 rad s^{-1}).

Table 13.1 also gives the additive uncertainty, and this is shown in Figure 13.7 in the form of a series of discs superimposed on the nominal system's polar plot. The true plant model $G(s)$ will produce a polar plot which will be contained in the region defined by these discs. Since this defined region has the $(-1,0)$ point to its left, the system has robust stability. However, the gain margin for the worst case is reduced to 1.53, and a complete analysis, in each case looking at the worst case, would indicate similar deteriorations in all the performance indicators.

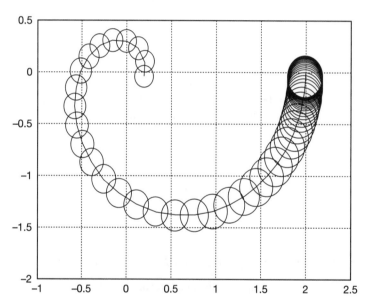

Figure 13.7 Polar plot with additive uncertainty.

Table 13.1 Polar plot data for the robustness example.

ω (rad s^{-1})	Gain	Phase (deg)	Additive uncertainty
0.01	1.99	−0.86	0.2
0.81	1.55	−62.5	0.16
1.22	1.27	−85.4	0.14
1.81	0.97	−114.0	0.13
2.7	0.69	−148.0	0.13
4.0	0.48	−191.0	0.12
6.0	0.33	−253.0	0.11
9.0	0.22	−341.0	0.11

A worst-case analysis is by definition conservative. Improvements are possible if more information regarding the uncertainty can be obtained; that is, if structured model uncertainty rather than the assumed unstructured uncertainty is available. If it was known that the sensor, although inaccurate, could never give values less than the true values, then the uncertainty becomes structured and the uncertainty band, defined by the discs in Figure 13.7, could be reduced. In general, to say that the uncertainty is unstructured indicates that several sources of uncertainty have been described by a single perturbation (for multivariable plant the single perturbation will be a full matrix having the same dimensions as the plant). Further, the analysis does not explicitly cater for particular inputs, or input uncertainty. The techniques described in the following sections are more flexible and can also be used to synthesize robust controllers.

13.8   Defining the H_2 and H_∞ approach

(These are pronounced 'H two' and 'H infinity'.) Solving the block diagram of Figure 13.1 for the error signal gives the expression (easily visible from Equation (13.8)):

$$e(s) = S(s)[P(s)r(s) - d(s) - n(s)] \tag{13.16}$$

which, as seen, is related to the sensitivity function $S(s)$ and the external inputs, any one of which may be represented by $v(s)$. Since $v(s)$ is bounded, then provided the closed-loop system is stable, the error signal will be bounded.

A measure of the area bounded by the error signal in the time domain is the integral of the error squared (such measures were also discussed early in Section 12.3), namely

$$\|e(t)\|_2^2 = \int_0^\infty e^2(t)\, dt \tag{13.17}$$

The left-hand side of this equation is referred to as 'the 2-norm of the error squared'. One solution to the control problem is to find the controller, $K(s)$, that will minimize the 2-norm for one specific input, $v(s)$. Using Parseval's theorem (which is introduced in Section 13.8.1, below) to transform Equation (13.17) into the frequency domain, and then replacing the error signal with the product of the sensitivity function and the specific input function (see Equations (13.16) and (13.14) respectively), gives

$$\min_K \|e(t)\|_2 = \sqrt{\left[\min_K \frac{1}{2\pi} \int_{-\infty}^\infty |S(j\omega)W(j\omega)|^2\, d\omega \right]} \tag{13.18}$$

The LHS of this equation is read as 'the minimum value of the 2-norm of the error, over all possible values of K'.

Equation (13.18) is known as the *Linear Quadratic*, or the H_2-*Optimal Control* problem. The 'H' comes from the term 'Hardy space', and is a mathematical description of the domain within which the solution is sought. Simply, the H_2-optimal controller minimizes the average magnitude of the weighted sensitivity function, and the robust H_2-optimal controller performs this minimization for the 'worst' plant within the family of plants. As presented, the controller which satisfies one specific input is unlikely to satisfy other practical design requirements. Typically, the response will be too oscillatory and have a large overshoot. For these reasons some designers would use $W(j\omega)$ as a 'tuning parameter' that would be adjusted until satisfactory performance was achieved. Alternatively, and perhaps additionally, further terms may be included into the minimization problem. For example, including the control signal $(+u(j\omega)^2)$ penalizes excessive control effort in a similar way to that described in Section 12.3.

What makes robust H_2-optimal control so attractive is the existence of very effective computer-based solutions, and its ability to deal with multi-input-multi-output systems. In terms of control system design, the technique is flawed in that minimization of the average error is not a particularly good design criterion. However, there is an alternative, the so-called H_∞-*Optimal Control* defined as

$$\min_K \|e(t)\|_\infty = \min_K \sup_\omega |S(j\omega)W(j\omega)| \qquad (13.19)$$

The 'sup' in this equation (pronounced 'soup') is the supremum, and it means that the overall result is the least upper bound. Hence, the H_∞-optimal controller minimizes the maximum magnitude of the weighted sensitivity function evaluated over the frequency range ω; comparable, in principle, to making a Bode magnitude plot have the lowest possible peak magnification.

Equation (13.19) can be manipulated to provide both robust stability and performance. Like the H_2-optimal control problem, $W(j\omega)$ is thought of as a tuning parameter and computer-based solutions exist, as do extensions to the multi-input-multi-output case. For control system design the H_∞ measure for single-input-single-output systems is equivalent to the closed-loop system's peak magnification, M_{peak}. Since M_{peak} and the frequency at which M_{peak} occurs are good indicators of a system's response, the method is proving a useful design tool.

13.8.1 Parseval's Theorem

Advanced section

This section can be omitted, if Equation (13.18) is taken 'on trust'.

Consider the problem of finding the value of the integral

$$I = \int_{-\infty}^{\infty} x(t)y(t)\, dt$$

where both $x(t)$ and $y(t)$ are Laplace transformable. Using the inverse Laplace transform, $y(t)$ may be written as

$$y(t) = \frac{1}{2\pi j} \int_{-j\omega}^{j\omega} Y(s)e^{st}\, ds$$

On substitution into the expression for I, this gives

$$I = \frac{1}{2\pi j} \int_{-\infty}^{\infty} \int_{-j\infty}^{j\infty} x(t)Y(s)e^{st} \, ds \, dt$$

$$= \frac{1}{2\pi j} \int_{-j\infty}^{j\infty} Y(s) \int_{-\infty}^{\infty} x(t)e^{st} \, dt \, ds$$

Noting that the inner integral gives the Laplace transform $X(-s)$, the original integral may be written as

$$I = \frac{1}{2\pi j} \int_{-j\infty}^{j\infty} X(-s)Y(s) \, ds$$

Parseval's theorem is obtained if $y(t) = x(t)$, to give the following:

$$\int_{-\infty}^{\infty} x(t)^2 \, dt = \frac{1}{2\pi j} \int_{-j\infty}^{j\infty} X(s)X(-s) \, ds$$

The steps required to obtain Equation (13.18), which were listed in the previous section, can then be followed.

Parseval's theorem enables time-domain integrals of stable systems to be evaluated in the Laplace or frequency domains. For example, when $X(s)$ has all its poles in the finite left half portion of the s-plane, the method of residues indicates that the value of the integral is equal to the sum of the residues of $X(s)X(-s)$ evaluated at all of its poles in the left half of the s-plane. Hence if $x(t) = 3e^{-2t}$, then $X(s) = 3/(s+2)$ and

$$X(s)X(-s) = \frac{3}{s+2} \cdot \frac{3}{-s+2} = \frac{-9}{(s+2)(s-2)}$$

From Parseval's theorem

$$\int_{0}^{\infty} x^2(t) \, dt = \left[\frac{-9}{s-2}\right]_{s=-2} = \frac{9}{4}$$

and may be verified by integration of $9e^{-4t}$.

13.8.2 Calculating the H$_2$ and H$_\infty$ norms

Advanced section

For completeness, this section will consider the H$_2$ and H$_\infty$ norms for constant vectors and matrices, in addition to time functions and systems. Some parts of this may seem rather abstract, as it is mainly a catalogue of results. After the first few, most are not required again, so the section can be omitted once the going gets hard, if desired.

Norms are a mathematical measure which enable the comparison of objects belonging to the same set. For illustrative purposes, consider the set of all points in the normal Euclidean space. In this example the objects are points and the Euclidean norm provides a distance measure. That is, the distance of a point P with coordinates $P(x, y, z)$ from the origin, coordinates $P(0, 0, 0)$, could be defined by the Euclidean norm

$$\|P(x, y, z)\|_2 = \sqrt{x^2 + y^2 + z^2}$$

Alternatively, the distance of the point P from the origin could be defined in terms of the maximum distance of P from the origin along any of its principal directions, that is

$$\|P(x, y, z)\|_\infty = \max(|x|, |y|, |z|) \qquad (13.20)$$

For example, the point $P(-1, 2, -3)$ has a Euclidean norm (2-norm) of $\sqrt{(14)}$ and an ∞-norm, Equation (13.20) of 3.

The 2-norm is a particular generalization of the Euclidean norm for an n-dimensional space and the ∞-norm is the n-dimensional space generalization of Equation (13.20). Other generalizations of the Euclidean norm are possible (in particular the Frobenius norm). However, the 2-norm of the $n \times 1$ vector x is

$$\|x\|_2^2 = \sum_{i=1}^{n} |x_i|^2$$

and the ∞-norm of the vector is

$$\|x\|_\infty = \max_i |x_i|$$

The Euclidean norm (or Frobenius norm) of an $n \times n$ matrix A is defined by

$$\|A\|_F = \left[\sum_i \sum_j |a_{ij}|^2 \right]^{1/2}$$

and the 2-norm (or spectral norm) by

$$\|A\|_2 = \max_{x \neq 0} \frac{\|Ax\|_2}{\|x\|_2}$$

or equivalently by

$$\|A\|_2 = \max_{\|x\|_2 = 1} \|Ax\|$$

Geometrically, $\|x\|_2 = 1$ is the set of all vectors of unit length (the unit sphere) and $\|A\|_2$ (the induced 2-norm of the linear map A) is the maximum magnification of the elements of this set by the operator A. It can be shown that the 2-norm of A is in fact the largest singular value of A, that is

$$\|A\|_2 = \bar{\sigma}(A)$$

(see Section 10.6.4 for a definition of the singular values).

In contrast the ∞-norm of the matrix A is defined as the largest row sum of the matrix, that is

$$\|A\|_\infty = \max_i \sum_{j=1}^{n} |a_{ij}|$$

For an $n \times 1$ vector time function $\boldsymbol{x}(t)$ the square of the 2-norm is defined as

$$\|\boldsymbol{x}(t)\|_2^2 = \int_0^\infty \boldsymbol{x}(t)^{\mathrm{T}} \boldsymbol{x}(t) \, dt$$

and gives the energy of the time function (the integral of the instantaneous power over time), as described early in Section 12.3. The concept of $\boldsymbol{x}^{\mathrm{T}}\boldsymbol{x}$ being a vector equivalent of the scalar x^2 was also introduced in Chapter 12.

The ∞-norm of $\boldsymbol{x}(t)$ (assuming all the signals $x_i(t)$ are bounded) is

$$\|\boldsymbol{x}(t)\|_\infty = \sup_{t \geq 0} \max_i |x_i(t)|$$

and gives the peak value of the signal with the largest peak amplitude.

When defining system norms it will be assumed that G represents the system transfer function (or TFM in the multivariable case) $G(j\omega)$ or $G(s)$ (which version is appropriate should be obvious from the context). It will be further assumed that G has the state-space realization $(\boldsymbol{A}, \boldsymbol{B}, \boldsymbol{C}, \boldsymbol{0})$. The H$_2$ norm is defined as

$$\|G\|_2^2 = \frac{1}{2\pi} \int_{-\infty}^\infty |G(j\omega)|^2 \, d\omega$$

and could be thought of as giving the system's RMS output value when the input is white noise. In general, its value can be calculated from

$$\|G\|_2^2 = trace(\boldsymbol{B}^{\mathrm{T}} \boldsymbol{W}_o \boldsymbol{B}) = trace(\boldsymbol{C} \boldsymbol{W}_c \boldsymbol{C}^{\mathrm{T}})$$

where the trace of a matrix is defined in Section A1.2, and $\boldsymbol{W}_o$ and $\boldsymbol{W}_c$ are known as the observability and controllability Grammians respectively, and are found by solving the following equations, known as Lyapunov equations:

$$\boldsymbol{A}^{\mathrm{T}} \boldsymbol{W}_o + \boldsymbol{W}_o \boldsymbol{A} + \boldsymbol{C}^{\mathrm{T}} \boldsymbol{C} = \boldsymbol{0}$$

$$\boldsymbol{A} \boldsymbol{W}_c + \boldsymbol{W}_c \boldsymbol{A}^{\mathrm{T}} + \boldsymbol{B} \boldsymbol{B}^{\mathrm{T}} = \boldsymbol{0}$$

(these can be seen to be duals of each other, as were the matrices used in controllability and observability rank tests in Section 5.3).

The H_∞ norm is defined as

$$\|G\|_\infty = \sup_{x \neq 0} \frac{\|G\boldsymbol{x}\|_2}{\|\boldsymbol{x}\|_2}$$

or equivalently, when the system is stable, by

$$\|G\|_\infty = \sup_\omega |G(j\omega)|$$

and is simply the maximum magnification (the M_{peak} value) of the system's frequency response. As yet, there appears to be no method of calculating the ∞-

norm directly for multivariable systems. Rather, it is numerically approximated from the following matrix (known as a Hamiltonian matrix):

$$
H_\gamma = \begin{bmatrix} A & \dfrac{BB^{\mathrm{T}}}{\gamma} \\[2ex] -\dfrac{C^{\mathrm{T}}C}{\gamma} & -A^{\mathrm{T}} \end{bmatrix}
$$

Again, it is assumed that the system is stable and G has the state-space realization $(A, B, C, 0)$. If and only if H_γ has no imaginary eigenvalues (complex eigenvalues with real parts $= 0$), then

$$
\|G\|_\infty < \gamma
$$

The above is solved by repeatedly selecting and adjusting γ until a minimum value is found for which the Hamiltonian has no imaginary eigenvalues. Using MATLAB, the method can be easily tested for any asymptotically stable second-order system. It will be found that the minimum value of γ is the system's M_{peak} value.

13.9 Robust stability

Assume that the combination of the controller and nominal plant model is closed-loop stable, and consider a band of polar plots consisting of $G_m(s)K(s)$ surrounded by the family of all possible plant realizations $G(s)K(s)$. Nyquist's stability criterion indicates that provided all the plants within the family have the same number of right half plane poles and that the polar band does not include the point $(-1, 0)$, the system has robust stability. That is, stability is guaranteed provided the distance from the $(-1, 0)$ point in the $G(s)K(s)$-plane to any point on the $G_m K(j\omega)$ locus is larger than the uncertainty. Hence, from Figure 13.8, a system with multiplicative uncertainty $l_m(\omega)$ has robust stability if

$$
|1 + G_m K(j\omega)| > |G_m K(j\omega)| l_m(\omega) \qquad \text{for all } \omega \tag{13.21}
$$

or

$$
|T_m(j\omega)| l_m(\omega) < 1 \qquad \text{for all } \omega \tag{13.22}
$$

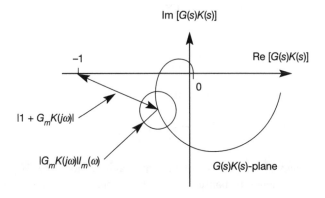

Figure 13.8 Polar plot showing the parameters which determine robust stability.

where $T_m(s)$ is the complementary sensitivity function for the nominal plant $G_m(s)$, that is

$$T_m(s) = G_m(s)K(s)(1 + G_m(s)K(s))^{-1}$$

In terms of the H_∞ norm (see Equation (13.19) and the associated discussion), a system has robust stability if

$$\|T_m(j\omega)l_m(\omega)\|_\infty < 1$$

that is, if

$$\sup_\omega |T_m(j\omega)l_m(\omega)| < 1 \qquad (13.23)$$

for unstructured multiplicative uncertainty (where the uncertainty can be represented by a series of discs on the plant's nominal open-loop polar plot). Equation (13.23) is both necessary and sufficient for robust stability, provided the nominal model produces a stable closed-loop system.

13.9.1 General robust stability for unstructured uncertainty

Figure 13.9 shows a standard one-degree-of-freedom closed-loop control system containing the nominal plant model $G_m(s)$. Assuming the closed-loop system is stable, the ability to determine the smallest uncertainty that would destabilize the system would be useful. In Figure 13.9 this uncertainty, $\Delta(s)$ with input b' and output a', can be connected to the system at points a and b. As indicated, $\Delta(s)$ represents unstructured multiplicative uncertainty on the plant's input (comparable with $l_m(s)$ in Figure 13.5). Now by letting $M(s)$ represent the closed-loop transfer function between input a and output b, that is

$$M(s) = -K(s)G(s)[1 + K(s)G(s)]^{-1}$$

Figure 13.9 can be drawn in the M–Δ structure shown in Figure 13.10. Note that the M–Δ structure can be used to represent plant with various forms of additive and multiplicative unstructured uncertainty. Also, in this instance, $M(s) = -T(s)$ (for the multivariable case $\boldsymbol{M}(s)$ would equal $\boldsymbol{T}_I(s)$, the input complementary sensitivity function).

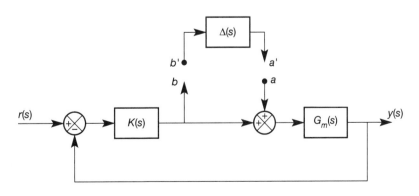

Figure 13.9 Standard closed-loop control arrangement, with removable uncertainty.

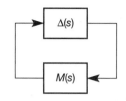

Figure 13.10 The M–Δ structure.

Perhaps the simplest way to determine the stability of Figure 13.10 is to use the small gain theorem. This states that if $M(s)\Delta(s)$ is stable, the closed-loop system will remain stable provided

$$|M(s)\Delta(s)| < 1$$

Essentially, the small gain theorem requires that the $M(j\omega)\Delta(j\omega)$ locus remains within the unit circle in the $M(j\omega)\Delta(j\omega)$-plane. This is considerably more conservative than the Nyquist stability criterion which would require that the $M(j\omega)\Delta(j\omega)$ locus does not enclose the -1 point. It is useful, however, in that it provides bounds on $\Delta(s)$ which, if satisfied, would guarantee closed-loop stability.

Since $|M(s)\Delta(s)| = |M(s)|\,|\Delta(s)|$ closed-loop stability is again guaranteed if

$$|M(s)|\,|\Delta(s)| < 1$$

or if the uncertainty

$$|\Delta(s)| < 1/|M(s)| = 1/|T(s)|$$

For multivariable plant it can be shown that the system of Figure 13.10 is stable for all perturbations $\Delta(\bar{\sigma}(\Delta) \leq 1)$ if and only if

$$\|M\|_\infty < 1$$

Typically, for multivariable systems, Δ will be a block diagonal matrix containing all the modelling uncertainties along its leading diagonal (as either elements or sub-matrices). Under these conditions, it is always possible to ensure that $\bar{\sigma}(\Delta) \leq 1$ by scaling each of the uncertainties and then including the factors in $M(s)$.

13.9.2 Standard system representations for robustness studies

Figure 13.11 shows the standard method of system representation for robustness studies. In this figure, it is assumed that the plant, $P_m(s)$, and controller, $K(s)$, are known accurately and that all plant uncertainties are contained in $\Delta(s)$. If the plant and controller are grouped together to form a transfer function (or transfer function matrix), then the M–Δ structure, shown in Figure 13.10, is obtained. Alternatively, the plant and its uncertainties could be grouped together to form the two-port representation shown in Figure 13.12, which is used, for example, in the MATLAB μ-analysis and synthesis toolbox for H_∞ studies (Balas *et al.*, 1991).

The two-port representation, Figure 13.12, shows the plant $P_{m\Delta}(s)$ and its controller $K(s)$. Note that the plant has two inputs u and w, and two outputs z and y. Of the inputs, u represents the control inputs emanating from the controller and w represents the various exogenous inputs – inputs like noise and disturbance which are not manipulated by the controller. Similarly, on the output side z represents

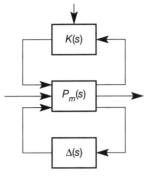

Figure 13.11 Standard system representation for robustness studies.

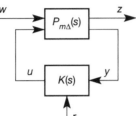

Figure 13.12 Two-port system representation.

those outputs which are to be controlled and y those outputs which are used for feedback purposes. Also shown in Figure 13.12 is a reference input r. For regulation problems r is set equal to zero, but for tracking problems r would be combined with w and become another exogenous input. Essentially, the two-port representation is multivariable even when the original system has only a single input and a single output. For example, the two-port representation of Figure 13.1 is shown in Figure 13.13. Solving the closed-loop relationship between the exogenous plant inputs $w(s)$ and outputs $z(s)$ gives the relationship

$$
\begin{bmatrix} y_p(s) \\ u_p(s) \end{bmatrix} = \begin{bmatrix} S(s) & -T(s) & T(s)P(s) \\ -\dfrac{T(s)}{G(s)} & -\dfrac{T(s)}{G(s)} & \dfrac{T(s)P(s)}{G(s)} \end{bmatrix} \begin{bmatrix} d(s) \\ n(s) \\ r(s) \end{bmatrix}
$$

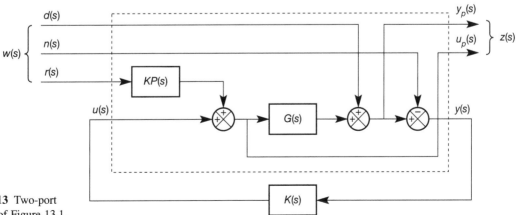

Figure 13.13 Two-port equivalent of Figure 13.1.

and provides a compact means of expressing Equations (13.6) and (13.8) which respectively give the output and control signal for Figure 13.1.

Now the two-port representation can be manipulated directly to obtain closed-loop relationships and used with either transfer function or state-space models. By setting r equal to zero (or combining r with w), the transfer function equations for Figure 13.12 are

$$z(s) = P_{zw}w(s) + P_{zu}u(s)$$
$$y(s) = P_{yw}w(s) + P_{yu}u(s)$$

and

$$u(s) = K(s)y(s)$$

where the elements P_{ij} are the transfer function between output i and input j. Using the normal rules of algebra to eliminate u and y gives the relationship

$$z(s) = T_{zw}w(s)$$

where the closed-loop transfer function

$$T_{zw} = P_{zw} + P_{zu}K(s)[I - P_{yu}K(s)]^{-1}P_{yw}$$

and is called the *linear fractional transformation* (LFT).

If the plant is represented in state space form then the governing equations are

$$\dot{x} = Ax + B_1w + B_2u$$
$$z = C_1x + D_{11}w + D_{12}u$$
$$y = C_2 + D_{21}w + D_{22}u$$

and the plant model would then be represented using the packed-matrix notation, to give

$$P(s) = \begin{bmatrix} A & B_1 & B_2 \\ C_1 & D_{11} & D_{12} \\ C_2 & D_{21} & D_{22} \end{bmatrix}$$

13.10 Robust performance

A system's performance, the way in which it rejects introduced errors, may be measured in many ways but is invariably related to its external inputs and the sensitivity function. Following on from Section 13.8, robust performance measures will be developed for the H$_\infty$ control problem.

The ∞-norm of the weighted sensitivity function is defined as

$$\|SW\|_\infty \equiv \sup_\omega |SW| \tag{13.24}$$

and the H$_\infty$-optimal control problem is given by Equation (13.19).

If, for a particular W, the optimal value of Equation (13.24) is σ, then Equation (13.19) implies that $|S| \le \sigma |W|^{-1}$ for all frequencies. For design purposes, this suggests a bound on the sensitivity function

$$|S(j\omega)| < |W(j\omega)|^{-1} \tag{13.25}$$

which the H_∞-optimal controller aims to satisfy. If a controller is found such that

$$\|SW\|_\infty < 1 \tag{13.26}$$

then Equation (13.25) is satisfied. The above condition gives nominal performance.

For robust performance, Equation (13.26) must be satisfied for the 'worst' plant in the family, hence

$$\|SW\|_\infty = \sup_\omega |SW(j\omega)| < 1 \quad \text{for all possible plants} \tag{13.27}$$

Again using geometric arguments based on the polar plot of Figure 13.8, note that at any given frequency the system's actual frequency response, $GK(j\omega)$, would be contained within the disc representing the region of uncertainty, hence in general

$$|1 + GK(s)| \ge |1 + G_m K(s)| - |G_m K(s)|l_m \quad \text{for all possible } G(s) \tag{13.28}$$

or

$$|S| = \left| \frac{1}{1 + GK(s)} \right| \le \frac{|S_m|}{1 - |T_m|l_m} \quad \text{for all possible } G(s) \tag{13.29}$$

where $S_m = (1 + G_m K(S))^{-1}$ is the sensitivity function for the nominal plant. Equation (13.27) now becomes

$$\frac{|S_m W|}{1 - |T_m|l_m} < 1 \quad \text{for all } \omega \tag{13.30}$$

or

$$|T_m l_m| + |S_m W| < 1 \quad \text{for all } \omega \tag{13.31}$$

which is the required measure for robust performance.

Note that robust performance implies robust stability, Equation (13.23), and nominal performance, Equation (13.26). Also, due to the relationship between the sensitivity and complementary sensitivity functions, improving stability causes a deterioration in performance and vice versa.

13.11 Concluding comments

It will be seen from this chapter that robust control is not a particular design technique. Rather, it is an approach to the design of control systems which recognizes the limitations of using a mathematical model, which will inevitably contain uncertainties, to represent real plant. Therefore, this chapter has concentrated on providing design concepts which can be used with the various design techniques presented in this text. By choice, there has been an emphasis on frequency-domain rather than state-space methods. The reason for this is that the

MATLAB toolboxes dealing with robust control use a predominantly frequency-domain approach involving structured singular values and loop shaping. These are the 'Robust Control Toolbox' (Chiang and Safonov, 1988) and the 'μ-Analysis and Synthesis Toolbox' (Balas *et al.*, 1991).

For further reading, these MATLAB toolbox manuals contain much useful material, as do the texts by Garcia and Morari (1982), Lunze (1989), Maciejowski (1989) and Morari and Zafiriou (1989).

14 Nonlinear systems

14.1   Preview

This chapter can be understood following a study of Chapters 1 to 5. Its purpose is to provide an introductory, but wide-ranging, study of nonlinear systems. In fact, some people have regarded this chapter as more of a 'nonlinear systems' course than a 'nonlinear control' course. The justification for this is as follows.

The whole of the rest of this book (and of most other control engineering textbooks) concentrates on controller design using linear system models. The reason for this is very simple: all the techniques using Bode plots, Nyquist diagrams, Nichols charts, Laplace transfer function models, state-space models, optimal controllers, root locus plots, block diagram reduction and so on, *only work with linear system models.* Furthermore, if such techniques work for one linear system model, they will usually be applicable to similar models of any system which can be suitably approximated by a linear model, making them very widely useful.

Unfortunately, all real plants are nonlinear to some extent, and sometimes a given plant's dynamic behaviour cannot be represented adequately by a

linear model. In such cases, nonlinearities have to be introduced into the plant model to obtain suitably realistic results. All the previously mentioned techniques then cease to work. Even worse, there is no general analysis and design technique that can be applied to *all* nonlinear system models.

It is therefore necessary to study a selection of nonlinear analysis methods, so that the most appropriate one can be chosen to suit any given nonlinear model. The first four sections of the chapter are introductory and should all be studied. The remaining sections describe various different techniques.

In this chapter, the reader will discover:

- the kinds of physical system components that give rise to nonlinear behaviour
- how the behaviour of nonlinear systems differs from linear behaviour
- that the methods of other chapters cannot be directly applied to nonlinear systems
- that there is no generally applicable technique for handling nonlinear systems
- how to obtain and use approximate linear models of nonlinear processes
- an introduction to some nonlinear analysis techniques.

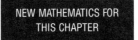 NEW MATHEMATICS FOR THIS CHAPTER

The Fourier transform is used in the production of describing functions. Jacobian matrices and Lyapunov functions are introduced as they become necessary.

14.2 Introduction

All dynamical systems are nonlinear to some extent. This means that it cannot simply be assumed that they obey the principle of superposition (see Section 2.3). For example, if the separate application of input signals u_1 and u_2 to the input of a single-input-single-output nonlinear system produces output signals y_1 and y_2 respectively, then the application of a combined input $(u_1 + u_2)$ may not necessarily produce the output $(y_1 + y_2)$.

As an initial example, a universally encountered form of nonlinearity is the *saturation* effect, in which some system variable is prevented from exceeding some limiting value, however large an input is applied. Every real system is capable of exhibiting this behaviour, since all physical signals have an upper limit on their magnitude. Figure 14.1 illustrates the input–output characteristic of such a system. Common occurrences of this behaviour are in mechanical systems in which end stops limit movement (such as hydraulic cylinders, or rotating components which are limited to a certain maximum angle of rotation), and in electrical and electronic systems whose output voltage levels cannot exceed their supply voltage (such as amplifiers).

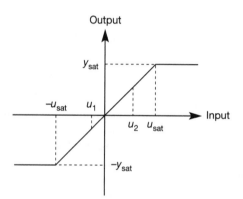

Figure 14.1 Input–output characteristic of a system element with saturation.

14.2.1 Superposition and linearization

In Figure 14.1, two input levels u_1 and u_2 are shown, such that u_1, u_2 and $(u_1 + u_2)$ are all within the range $[-u_{\text{sat}}, u_{\text{sat}}]$. For these inputs, the system will behave linearly (superposition will apply). However, if the magnitude of any input signal to this system exceeds u_{sat} the output magnitude will not be able to exceed y_{sat}, however large an input is applied. Superposition clearly does not apply in that case.

This gives one indication of how nonlinear systems might be handled. So long as the signal levels are kept within a certain range, the system of Figure 14.1 will behave linearly. Very many systems have approximately linear behaviour for suitably restricted signal ranges. This is basically the approach taken in all linear control work – restrict the signal levels, and ignore the nonlinear effects. Fortunately this works very well in many cases. It is fortunate, because most of the previous techniques studied for the systematic analysis and design of control systems only work with linear system models.

If a system contains significant nonlinear elements, a linear model will not give a good representation of its behaviour. The model must then be modified by including nonlinear effects until the model gives a close enough approximation to the real world. Once the model has become nonlinear, all the linear control techniques begin to collapse. Previous approaches which can no longer be applied include the use of Laplace transfer functions; block diagram reduction techniques; Bode, Nyquist and Nichols frequency response plots; s-plane analysis such as root locus; z-transform models; the Routh and Nyquist criteria for stability analysis; state-space models of the form $\dot{x} = Ax + Bu$; state variable feedback, state observers and optimal control based upon such models.

Even more problematic is the fact that there is no analysis approach with which to replace all these things, which will work well with all nonlinear systems. Therefore every nonlinear system has to be treated on its own merits.

There are, however, two fundamentally different approaches which allow the use of at least some of the previous techniques.

The first approach is to 'ignore' the nonlinearity (as has already been suggested), but to ensure that this is done deliberately and with full knowledge of the limitations of the modelling. This is the approach of 'linearization' and it works for many systems. It is effectively what has been done in the rest of the book. In this chapter we discover how to obtain linearized models of systems which contain known nonlinearities. Once a linearized model exists, *all* the previous linear control techniques can be applied to it. However, it must always be borne in mind that there are differences between the *linear model* on which the analysis and design techniques are being carried out, and the *nonlinear plant* which the model is claimed to represent.

The second approach is to try to include the effects of the nonlinearity in the analysis as accurately as is feasible. Since every nonlinearity is different, this becomes an impossible task if many types of nonlinear system are to be considered. However, several analysis techniques have been developed which have their own advantages and disadvantages in different general classes of situation. Most of these build upon previous linear control knowledge in some way, by modifying linear forms of analysis to deal with nonlinear situations. Before discussing any of these methods, a little more appreciation of nonlinear behaviour is necessary.

14.2.2 Nonlinear system behaviour

In addition to the fact that superposition does not apply, nonlinear systems may also exhibit a range of other effects which are not evident in linear systems. These effects include the following.

Limit cycles and multiple equilibrium points

In a limit cycle, the system exhibits a sustained, repetitive oscillation (not necessarily sinusoidal). All electronic oscillators are *nonlinear* systems, because oscillation cannot be sustained in a *linear* system unless it is simply following an oscillating input – except in the theoretical case of a system model having a pair of conjugate poles on the imaginary axis of the s-plane (it is 'theoretical' because, in the real world, such a pole pair will always drift off the imaginary axis, giving an asymptotically stable system with a very long settling time if they drift to the left, or an unstable system if they drift to the right).

Another example of a limit-cycling system is a space heating scheme in which on-off thermostats sense the temperature (and/or on-off relays control the heat source). At steady state, such a system will be 'hunting' around the setpoint temperature – repeatedly switching on and off. In this system the nonlinear behaviour has been *deliberately* introduced by the designer as a cheap and rugged means of control (thermostats and relays being much less costly than analog temperature sensors and continuously variable heat sources).

Limit cycles can be stable or unstable. If a *stable* limit cycle is disturbed, the system will return to the same pattern of oscillation after the disturbance is removed. If an *unstable* limit cycle is disturbed, the system will leave the limit cycle, and will adopt some other kind of behaviour after the disturbance has passed. This also indicates that a nonlinear system can have more than one kind of steady-state behaviour. A stable *linear* system with no input will always settle to a steady state in which the output and state variables are zero (a pure integrator is considered unstable for this definition); but a *nonlinear* system may have several possible stable steady-state conditions for zero input (known as *equilibrium points*), the one adopted being dependent upon the initial conditions from which the system was released.

For linear systems, various ways of determining system stability have been studied, which applied to all linear systems. Unfortunately, there is no general technique for determining the stability characteristics of nonlinear systems which, as has been indicated, can be quite complex. The most appropriate technique must be chosen for the particular system under investigation.

Chaotic behaviour

In some nonlinear systems (even some quite simple systems in the discrete-time case), another form of sustained oscillations can occur in which the oscillations do not seem to be repetitive, but are pseudo-random (even though the system model is deterministic). This behaviour is known as *chaos*. Another characteristic of such systems is that very small changes in initial conditions (for example) which one might dismiss as negligible, can lead to drastically different system behaviour in the medium and long term. This is one of the reasons for the difficulty (maybe impossibility) of accurate long-term weather and economic forecasting.

Input-dependent stability

In a stable linear system, if the input is limited to some maximum magnitude, the output will also remain within some magnitude limit (this is called bounded-input-bounded-output behaviour). Nonlinear systems may not behave like this. For example, a nonlinear system may settle to a steady output value following the application of a step input of 2 units, but may be unstable for input steps of 4 units and 0.5 unit. The same kind of thing can happen in response to parameter variations within the system.

Asynchronous quenching

A good example of the kind of stability just discussed occurs in public address systems. The system may behave perfectly until the presenter suddenly speaks more loudly than usual, and may then break into uncontrollable sustained oscillation (caused by acoustic feedback from the loudspeakers to the microphone). Once established, the oscillation is normally removed by breaking the feedback loop; for example, by switching off the microphone for a few seconds, or cutting the amplifier gain for a period, when the system will return to normal operation. This is another example of a limit cycle (the sustained oscillation) and also an example of a system with more than one equilibrium point (with no voice input the system can be either quiescent, or oscillating).

Even more strangely, under certain circumstances the oscillation can be removed by the application of a different input (such as a loud hand-clap close to the microphone). This is known as *asynchronous quenching*, in which a limit cycle at some frequency can be removed by forcing the system with a different frequency.

Harmonic generation

The output of a linear system will always be of the same frequency as the input, even though it will differ in magnitude and phase. Nonlinear system outputs may contain harmonics and sub-harmonics of the input frequency (for example, an input signal at a frequency of 5 Hz may give rise to components of 25 Hz and/or 1 Hz in the output signal). This is why the frequency response plots cannot be used with nonlinear systems – for such plots to be meaningful, the gain and phase shift must be plotted with both input and output at the same single frequency.

Jump resonance

In a system that exhibits this effect, if a sinusoidal input signal is swept through a range of increasing frequencies, there will be a discontinuity in the frequency response of the system at some frequency. As the input frequency is decreased, there may be a discontinuity at some different frequency as indicated in Figure 14.2.

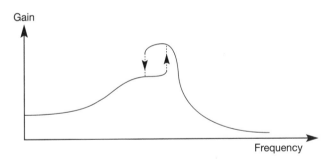

Figure 14.2 The jump resonance phenomenon.

14.2.3 True nonlinear analysis and computer simulation

If the approach of linearization (Section 14.2.1) is adopted, then clearly any subsequent analysis using the linear model will not be able to predict any of the effects noted in Section 14.2.2. Any control system designed on the basis of a linear model may therefore not function at all well if effects such as these can occur during the operation of the real system. To some extent, this problem can be avoided by *simulation studies*, in which the linearly designed controller is applied to a full nonlinear computer simulation of the plant to be controlled, and tested over a wide range of initial conditions, input signals and disturbances. The major drawbacks are that it is difficult to decide when the nonlinear simulation is sufficiently accurate and to know when the simulation studies have covered every conceivable mode of plant behaviour (indeed there may be some inconceivable effects).

If the nonlinear effects of Section 14.2.2 are to be included in the analysis (because they are too significant to be ignored) then either computer simulation must be used, or techniques aimed specifically at the analysis of nonlinear systems must be applied (with computer simulation normally being used to test the results in any case). A selection of such nonlinear analysis techniques is introduced in this chapter, covering both time-domain and frequency-domain approaches. Knowledge of the analytical techniques also aids in choosing suitable ranges of initial conditions for computer simulation runs.

In the mainframe era of the 1960s and early 1970s, analog computers were cheaper than digital ones, and all dynamical simulation was originally carried out using such machines. There are still some advantages to analog simulation, including the facts that it works without signal sampling (that is, it is continuous, like the real world) and it works in real time. These attributes are especially valuable for 'stiff' systems (systems having a mixture of relatively fast and relatively slow dynamics, neither of which can be ignored for some reason) and also for 'hardware in the loop' simulations in which a controller, perhaps programmed into a microcomputer, is tried out on a continuous-time analog simulation of the plant; which is usually a more realistic test than using the equivalent discrete-time digital simulation. There is also something valuable in the engineering 'feel' for a system which is obtained by adjusting the coefficient potentiometers and seeing the immediate effects upon the system response, although modern interactive digital systems are approaching the same level of convenience.

Nevertheless, since the end of the 1970s digital computing power has become very cheap (although the software is not necessarily so), while reasonably-sized analog computers are now relatively expensive. The majority of dynamical simulation is therefore now carried out using digital simulation. The passage of time has made life easier here, too. Originally, one would have to convert the system model to difference equations and program numerical integration routines to solve these from scratch using languages such as Fortran. After a while, a number of pre-written function libraries became available which could be linked together to build simulations. Next came specialized simulation languages such as ACSL (Mitchell and Gauthier, 1987) and TSIM (Cambridge Control, 1988) which are in widespread use. Now we have computer-assisted control system design (CACSD) environments such as MATLAB (Appendix 3) and SIMULINK (Appendix 4) which make the job easier still.

For these reasons, methods of analog computing will not be presented, nor will the setting up of digital simulations other than to run under CACSD environments. For more detail on simulation see Charlesworth and Fletcher (1974) or Rieder and Busby (1986).

14.3 Nonlinear system elements

There are many causes of nonlinear behaviour in systems. In electrical systems we seldom hesitate to use the linear model $I = V/R$ to represent a resistor (I is the current in amperes, V the potential difference in volts and R the resistance in ohms –

R is assumed constant for the model to be linear). However, as with all models, this one is approximate as Figure 14.3(a) shows. If a high potential difference is applied across the resistor it will heat up sufficiently to cause the resistance to change somewhat (the precise form of the change depending upon the material of the resistor and its mode of construction). Because the resistance alters, the original $I = V/R$ model will no longer predict the correct current – the system has become nonlinear. If the potential difference increases too much, the resistor will burn and become open-circuit. When working at very high frequencies, it also becomes necessary to take account of the inductance and capacitance of the resistor.

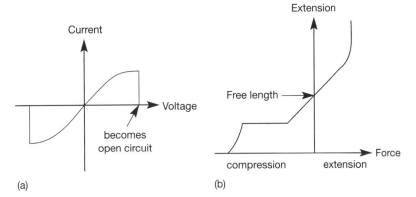

Figure 14.3 Nominally linear components subjected to large signals: (a) resistor, (b) spring.

Simple mechanical elements also become nonlinear if wide signal ranges are used. Consider a linear spring modelled by $x = f/K$ (x is the extension or compression in metres, f the applied force in newtons and K the spring constant in newtons per metre – K is assumed constant for a linear model. If the spring is extended too far, the elastic limit of the material will be exceeded, K will effectively alter and the model will no longer be correct (Figure 14.3(b)). If it is extended much too far, the spring will break. If the spring is compressed too far it will become coil-bound, leading to the saturation characteristic in the lower left of Figure 14.3(b). If the compressive force is increased further the spring will eventually be crushed to a useless mass from which it will not recover.

These two simple examples serve to illustrate the point that all systems are nonlinear. However, they are both essentially linear over their designed working ranges (the central areas in Figure 14.3, with the origin of Figure 14.3(b) shifted to coincide with the free length of the spring). This chapter is more concerned with systems which are nonlinear in their working ranges, as these are the ones which will cause the most problems.

As mentioned before, some nonlinearities are deliberately introduced by the designer. On-off control was mentioned in Section 14.2.2, and adaptive controllers such as those discussed in Chapter 11 also result in nonlinear systems. In what follows, the interest switches to the unintentional nonlinearities inherent in the plant itself, since it is these for which compensators must be designed.

14.3.1 Continuous nonlinearities

These can be defined as nonlinear elements whose input–output characteristics can be described by analytic functions and are continuously differentiable (output with respect to input). Such elements are quite common and include the flow vs. opening characteristics of hydraulic valves, the resistance vs. temperature characteristics of thermistors, the characteristics of rubber springs and the characteristics of several types of measuring transducer (all assumed to be operating within their designed working ranges). Figure 14.4 shows a nonlinear transducer characteristic as a representative continuous nonlinearity.

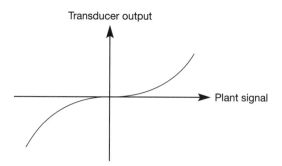

Figure 14.4 A continuous nonlinearity.

Continuous nonlinear elements may also be deliberately introduced as 'linearizing feedback'. One example is the use of thermistors (resistors designed to have a large and predictable change in resistance as temperature alters) as feedback elements to stabilize the output level of electronic oscillator circuits based upon operational amplifiers (if the output level increases, so does the self-heating of the thermistor, thus lowering its resistance and lowering the amplifier gain so as to restore the output level). Problem 14.1 contains an example.

14.3.2 Discontinuous nonlinearities

The input–output characteristics of these cannot be modelled by analytic functions and the derivatives of output with respect to input contain singularities. Linearization in the sense in which we shall apply it later is therefore not straightforward, and it is more usual to seek to simulate or analyse systems containing such nonlinearities (or to linearize using a computer package). In this section, several of the common types of discontinuous nonlinear element are defined, with indications as to where they might arise.

First, there is the class of *single valued nonlinearities*. This means that for each value of input signal to the nonlinear element, there is only one possible output value (except at switching points). As one might imagine, this can simplify analysis of the effects.

Saturation

This has already been mentioned in Section 14.2 (Figure 14.1). It is worth pointing out, however, that Figure 14.1 depicts *hard* saturation. There is also the possibility of *soft* saturation, as shown in Figure 14.5(a). Whereas hard saturation is characteristic of mechanical end stops and simple transistor amplifiers, soft

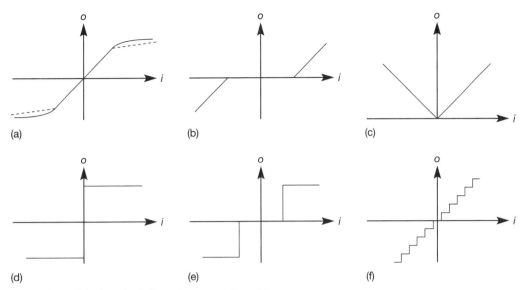

Figure 14.5 A selection of single-valued discontinuous nonlinearities.

saturation is associated with valve or magnetic amplifiers, and variable-rate springs. In the case of the valve amplifier, this soft saturation leads to the generation of fewer harsh harmonics as the amplifier saturates on fast transients, and may be one reason why some audio enthusiasts and many stage musicians prefer the 'valve' sound. Note also that the effect is often approximated by the dashed line in Figure 14.5(a), which is why soft saturation has been included as a discontinuous nonlinearity.

Deadzone (or deadspace)

Figure 14.5(b) shows the I/O characteristic of an element which cannot respond to small signals, but is otherwise considered to be linear.

A classic example of this occurs in hydraulic servovalves and pneumatic pilot valves. In order to be able to shut off the flow of fluid, the valve spool has to be slightly larger than the port over which it slides, so as to leave no gap. Therefore, in order to open the valve and start the flow, the input signal must cause the spool to move a small distance *before* the edge of the port begins to be uncovered and the flow can begin. If the input signal to the valve is less than this small amount, the port will remain closed and no flow will result.

Electronic diodes also approximate this behaviour when forward biased, where a potential difference equal to the small forward voltage drop (typically 0.6 V) must be applied before any appreciable current begins to pass.

This kind of behaviour can sometimes be overcome, but always at some expense (for example, by the use of dither signals applied to hydraulic valves, or by the use of ideal diode circuits using operational amplifiers).

Absolute value detector

Figure 14.5(c) shows an element which fails to detect the polarity of the input signal. Examples include electrical full-wave rectifiers (where this behaviour is a fundamental part of the operation of the system), anemometers (which measure wind speed, but cannot determine the direction of the wind) and some kinds of a.c. transducer.

Ideal relay

Figure 14.5(d) shows the characteristic of an ideal relay. The output changes between two distinct states (on or off, open or closed, ± some voltage level, for example) at a particular value of input. An electronic comparator can be made to approximate this behaviour. A real electromechanical relay may also approximate this behaviour under some conditions, but usually its behaviour is represented by more complex models.

Relay with deadzone

The first enhancement of the electromechanical relay model is to add some deadzone, as shown in Figure 14.5(e). Any real electrical relay will need a certain amount of current to be flowing in its coil before sufficient magnetizing force is developed to pull the armature in. Therefore, input signals (voltages or currents) of less than a certain critical value will have no effect on the output.

Quantization

With digital controllers becoming ever more important, it is common for system variables to be processed by analog-to-digital converters (ADCs). The outputs of ADCs have a minimum resolution of one bit of the digital wordlength, so that a continuously increasing analog input gives rise to the staircase form of output shown in Figure 14.5(f). In other words, the input can have any value on the continuous scale, but the output is restricted to the set of discrete values shown.

A rather more traditional element exhibiting this kind of behaviour is the wire wound potentiometer, in which continuous rotation of the input shaft causes one turn at a time to be contacted by the wiper, and thus one discrete step at a time of resistance change.

Hysteresis

This is a *double-valued nonlinearity* in which, for some particular value of input, there could be one of two possible output signals.

Figure 14.6(a) shows a switching characteristic with hysteresis. This means that the switching point occurs at a different value of input signal depending upon whether the input is increasing or decreasing.

The usual origin of this type of behaviour (although there are others) is in magnetic materials, where the classic magnetization curve appears as shown in Figure 14.6(b), the input signal being the strength of the magnetizing field, and the

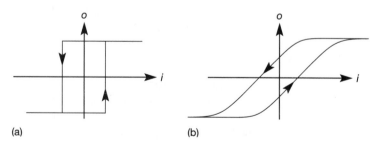

Figure 14.6 Hysteresis characteristics.

(a) (b)

output the intensity of the resulting magnetization. The tendency of the material to remain magnetized when the magnetizing force is removed causes the shape of the curve. The sharpness of the 'corners' and the width of the central loop depends upon the material.

Since electromechanical relays work on the principle of magnetizing and demagnetizing metallic cores, they also exhibit this behaviour. A more realistic relay model includes both deadzone and hysteresis as shown in Figure 14.7. Hysteresis is often deliberately introduced into relay systems (comparators, thermostats, and so on) to avoid rapid hunting around the operating point.

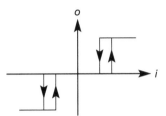

Figure 14.7 Input–output characteristic of an electromechanical relay.

Backlash

This is an important *multi-valued nonlinearity*, in which there could be many different values of output corresponding to any given input value. It is usually a very unpleasant element to have to analyse, and a good candidate for simulation studies.

Figure 14.8 shows this effect. It is essentially a *moving* deadzone such that, whenever the input signal changes direction, there is a deadzone during which the output signal remains constant, before it reverses and begins to follow the input again. The position of the horizontal lines on the characteristic (and hence the output value) therefore depends entirely upon the input signal level at which reversal occurs.

The most common cause of this kind of behaviour is clearance in mechanical linkages (due to wear, for example) and, particularly, in geartrains. Figure 14.9 shows part of a pair of meshing gearwheels. The upper wheel is connected to the input shaft, and is driving the lower wheel in a clockwise direction through the contact of the gear teeth at the face 'A'. What happens when the input shaft reverses depends upon whether the behaviour is governed by the *friction* of the load or its *inertia*. Let us first imagine that *friction* is the dominant factor.

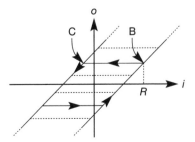

Figure 14.8 A backlash characteristic.

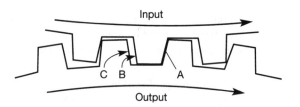

Figure 14.9 Meshing gear teeth with backlash.

When the input shaft stops, the output shaft will also stop (due to the high assumed friction). Say this occurs at point 'R' on Figure 14.8. However, when the input shaft begins to rotate in the opposite direction, the output shaft will initially remain stationary, because the input shaft has to take up the 'play' in the gears as the teeth move from position 'B' to position 'C' (shown on both Figures 14.8 and 14.9), before the teeth of the input gear again come into contact with the output gear at point 'C' and movement of the output shaft in the reverse direction begins.

If the *inertia* of the output governs the behaviour, when the input shaft stops the output shaft will continue to rotate until point 'C' in Figure 14.9 contacts point 'B', and will then stop (it may even bounce). As soon as the input shaft begins to turn in the opposite direction, the output shaft will now follow it. The effect upon Figure 14.8 is to make the horizontal lines become vertical instead.

Friction

This is a universally encountered effect in any system with moving mechanical components, but is actually rather complex. Figure 14.10(a) shows idealized versions of three components of friction as follows, and Figure 14.10(b) is a more realistic representation of the combination of the effects.

Static friction (commonly abbreviated to 'stiction') is represented by the force required to get a component moving initially. Once the component is in motion, its effect is gone. One example is the force required to break the seal between a hydraulic valve spool and the cylinder in which it moves (often overcome by using low amplitude dither signals to keep the spool moving, which also helps overcome the deadzone inherent in such valves). Another example is the force required to *start* a heavy object sliding across your desk.

Viscous friction is often the dominant part of the characteristic, and is the force which opposes motion proportionally to the velocity involved. Simple linear lumped-parameter models of mechanical components only include this term, for example $f = B.v$ (where f is the frictional force in N, B is the coefficient of friction

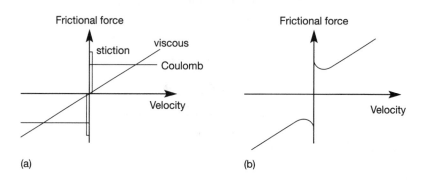

Figure 14.10 Frictional characteristics.

(a)

(b)

in N/(ms^{-1}) and v is the velocity in ms^{-1}). This is approximately the force required to *keep* a moving object sliding across your desk at constant speed.

Coulomb friction is the component of frictional force which is independent of velocity. This is the component which allows a car with a manual gearbox to be held with the clutch, stationary on an uphill slope, more or less *independently* of engine speed.

Some of the methods of handling nonlinear systems will now be presented. For more detail on all of the techniques outlined here, consult specialized texts such as Slotine and Li (1991), Atherton (1982) or Khalil (1992).

14.4 *Linearization*

If a linear model of a real plant can be obtained, which is acceptably accurate over a suitable range of operating conditions (even though the real plant will be nonlinear), then all the previous linear control analysis and design techniques can be applied to the linear model. There is therefore a strong motivation for obtaining linear models. In all the rest of this book, and in all other linear control texts, this is the approach which has been adopted (explicitly or implicitly) in obtaining the transfer functions or state-space models used in analysis and design. These are all linear approximations of systems which must become nonlinear if driven hard enough.

In using such models it must be ascertained that they remain valid over a suitable range of operation of the plant. If they do not, then either more linear models are needed for different operating regions (so that the linear model to be used is selected depending upon the operating conditions), or some form of explicitly nonlinear analysis or simulation must be used. It can also be inferred from these comments that linearization will often work better for regulator systems which, by definition, should have restricted operating ranges, than for tracking (servo) systems in which the operating range may be quite large.

A reminder will now be given of how approximating linear models can be obtained in the scalar case, and then the techniques will be generalized to the multivariable case.

14.4.1 *Single-input-single-output (SISO) systems*

The linearization of such systems was introduced in Section 2.4.5. However, that section was denoted optional. If it was omitted, it should be studied now. In summary, the following is always the basic approach (even for multivariable systems):

- Choose an operating point about which to linearize the system.
- Expand the nonlinear system equations as a Taylor series about the selected operating point.
- Recast the system model in terms of *deviations from the operating point.*
- Assume these deviations to be 'small', so that the Taylor series can be truncated after the first term, leaving a linear model.
- A rather more physical view of the procedure is that an operating point is chosen on the nonlinear characteristic, the origin of the characteristic's graph is moved to that operating point, and the characteristic is then approximated by the tangent

to the characteristic at the operating point (now the origin – so there is no offset term).

In Section 2.4.5, Example 2.10 linearized a simple pendulum, Example 2.11 linearized the flow of fluid through a valve and Example 2.12 linearized a product of two variables.

14.4.2 Multivariable (state-space) linearization

The state-space models considered so far have always been linear. However, there is no reason why a state-space description should not arise in nonlinear form. The problem with nonlinear state-space models is that some (or all) of the A, B, C and D matrices will no longer contain only pure numbers, but may also contain functions of states, time, outputs and/or inputs. This means that all the previous linear techniques for handling state-space models collapse, as they were based on the assumption of purely numerical system matrices.

To see how such models can arise, consider the simple system shown in Figure 14.11. This contains not only the usual integrators, summers and gains, but also some nonlinear damping involving a signal multiplier, and a measuring transducer with a cube law characteristic. This is a SISO system, but it could just as well be multivariable (several inputs and/or outputs) – the arguments and techniques are identical.

From Figure 14.11, the system equations are:

$$\dot{x}_1 = x_2$$
$$\dot{x}_2 = -2x_1^3 - 1.5x_1x_2 + u \tag{14.1}$$

Although these are state equations, they cannot be put into the usual state-space form giving constant matrices. If they were to be forced into the usual form, there would be more than one possible result. As an example, one result would be

$$\dot{x} = \begin{bmatrix} 0 & 1 \\ -2x_1^2 - 1.5x_2 & 0 \end{bmatrix} x + \begin{bmatrix} 0 \\ 1 \end{bmatrix} u$$

which is in the usual form (that is, $\dot{x} = Ax + bu$), but A does not contain only pure numbers (the other possibility would be $a_{21} = -2x_1^2, a_{22} = -1.5x_1$). In fact, this form is not of much use, since none of the previous techniques can be applied to it. Therefore, the model will be generalized and then examined to see how it might be

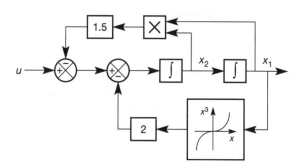

Figure 14.11 A nonlinear system in state-space form.

linearized, returning to the example later. In functional form, Equation (14.1) becomes:

$$\dot{x}_1 = f_1(\boldsymbol{x}, \boldsymbol{u})$$

$$\dot{x}_2 = f_2(\boldsymbol{x}, \boldsymbol{u})$$

where $f_i(\boldsymbol{x}, \boldsymbol{u})|_{i=1,2}$ are functions of all the state and input variables in the system. Thus, for an nth-order system having m inputs, the general state equation would be:

$$\dot{x}_i = f_i(x_1, x_2, \ldots, x_n, u_1, u_2, \ldots, u_m) = f_i(\boldsymbol{x}, \boldsymbol{u})|_{i=1,\ldots,n} \tag{14.2}$$

In addition, $\boldsymbol{x}$ and $\boldsymbol{u}$ will be functions of time, but the notation which would show this is omitted for clarity. Finally, if the notation $\boldsymbol{f}$ is allowed to represent the *vector* of functions $f_1, f_2, \ldots, f_n$, then Equation (14.2) can be written in the general compact form:

$$\dot{\boldsymbol{x}} = \boldsymbol{f}(\boldsymbol{x}, \boldsymbol{u}) \tag{14.3}$$

The usual, linear, state-space model is a special case of Equation (14.3).

Methods exist for handling the nonlinear state equation (Equation (14.3)), but they are complicated and somewhat impractical. Therefore, a linearized version of Equation (14.3) in its general form will be found, which can then be used to generate a model which looks like the usual state-space model. Once this has been achieved, all the previous state-space methods can be applied to such a model but, as in the scalar case, care must be taken over the range of operating conditions for which the linear model is assumed to be valid.

Equation (2.29), in Section 2.4.5, was previously used to linearize a scalar model in a single variable. Precisely the same equation can be applied to a model containing more than one variable, as in Example 2.12. In this case, the partial derivatives of the function (which is the system model) are taken with respect to each of the variables in the model. As a reminder, consider the following restricted case of Equation (14.3), comprising a single function f in two variables (x_1 and u_1):

$$\dot{x} = f(x_1, u_1)$$

This would be expanded about an operating point (x_{1_0}, u_{1_0}), using Equation (2.29), as:

$$\dot{x} = f(x_1, u_1) = f(x_{1_0}, u_{1_0}) + \left. \frac{\partial f(x_1, u_1)}{\partial x_1} \right|_{x_{1_0}, u_{1_0}} (x_1 - x_{1_0})$$

$$+ \left. \frac{\partial^2 f(x_1, u_1)}{\partial x_1^2} \right|_{x_{1_0}, u_{1_0}} \frac{(x_1 - x_{1_0})^2}{2!} + \left. \frac{\partial f(x_1, u_1)}{\partial u_1} \right|_{x_{1_0}, u_{1_0}} + \cdots$$

$$+ \left. \frac{\partial f(x_1, u_1)}{\partial u_1} \right|_{x_{1_0}, u_{1_0}} (u_1 - u_{1_0}) + \left. \frac{\partial^2 f(x_1, u_1)}{\partial u_1^2} \right|_{x_{1_0}, u_{1_0}} \frac{(u_1 - u_{1_0})^2}{2!} + \cdots$$

$$\tag{14.4}$$

Assuming that (for small deviations from the operating point) the second and higher derivatives are negligible, and dropping the dependency on x_1 and u_1 in the partial derivatives for clarity, so that $\partial f(x_1, u_1)$ is simply written as ∂f, Equation (14.4) becomes:

$$\dot{x} = f(x_1, u_1) = f(x_{1_0}, u_{1_0}) + \left.\frac{\partial f}{\partial x_1}\right|_{x_{1_0}, u_{1_0}} (x_1 - x_{1_0})$$

$$+ \left.\frac{\partial f}{\partial u_1}\right|_{x_{1_0}, u_{1_0}} (u_1 - u_{1_0})$$

Writing $f(x_{1_0}, u_{1_0}) = \dot{x}_0$ then leads directly to:

$$\dot{x} - \dot{x}_0 = \left.\frac{\partial f}{\partial x_1}\right|_{x_{1_0}, u_{1_0}} (x_1 - x_{1_0}) + \left.\frac{\partial f}{\partial u_1}\right|_{x_{1_0}, u_{1_0}} (u_1 - u_{1_0}) \tag{14.5}$$

It is now straightforward to apply this result to the multivariable case of Equation (14.3). The operating point becomes an operating point in the state space defined by a particular state vector x_0 and a particular input vector u_0. Note that the operating point may be a single point in the state space – in which case u_0 will often be 0 (in regulating systems), or it can be a reference *trajectory* defined by some time-varying $x_0(t)$ and $u_0(t)$, which the system is expected to follow (in tracking systems).

For any one of the functions f_i (that is, $\dot{x}_i$) in Equation (14.3), the result will look like Equation (14.5) and will contain one first derivative term for each of the n state variables and each of the m input variables. In addition, there will be one such complete equation for each of the functions f_i, $i = 1, \ldots, n$. The result for the entire general model of Equation (14.3) therefore appears as:

$$\dot{x}_1 - \dot{x}_{1_0} = \left.\frac{\partial f_1}{\partial x_1}\right|_{x_0, u_0} (x_1 - x_{1_0}) + \cdots + \left.\frac{\partial f_1}{\partial x_n}\right|_{x_0, u_0} (x_n - x_{n_0})$$

$$+ \left.\frac{\partial f_1}{\partial u_1}\right|_{x_0, u_0} (u_1 - u_{1_0}) + \cdots + \left.\frac{\partial f_1}{\partial u_m}\right|_{x_0, u_0} (u_m - u_{m_0})$$

$$\dot{x}_2 - \dot{x}_{2_0} = \left.\frac{\partial f_2}{\partial x_1}\right|_{x_0, u_0} (x_1 - x_{1_0}) + \cdots + \left.\frac{\partial f_2}{\partial x_n}\right|_{x_0, u_0} (x_n - x_{n_0})$$

$$+ \left.\frac{\partial f_2}{\partial u_1}\right|_{x_0, > u_0} (u_1 - u_{1_0}) + \cdots + \left.\frac{\partial f_2}{\partial u_m}\right|_{x_0, u_0} (u_m - u_{m_0})$$

$$\vdots \qquad\qquad\qquad\qquad\qquad\qquad\qquad \vdots \tag{14.6}$$

$$\dot{x}_n - \dot{x}_{n_0} = \left.\frac{\partial f_n}{\partial x_1}\right|_{x_0, u_0} (x_1 - x_{1_0}) + \cdots + \left.\frac{\partial f_n}{\partial x_n}\right|_{x_0, u_0} (x_n - x_{n_0})$$

$$+ \left.\frac{\partial f_n}{\partial u_1}\right|_{x_0, u_0} (u_1 - u_{1_0}) + \cdots + \left.\frac{\partial f_n}{\partial u_m}\right|_{x_0, u_0} (u_m - u_{m_0})$$

This result can now be made much neater by adopting some notational changes. First, the superscript * is used to denote deviations from the nominal (operating point) values (and is not to be confused with the notation for the complex conjugate transpose of a matrix, used elsewhere). Thus:

$$\dot{x}_i^* = \dot{x}_i - \dot{x}_{i_0}, \quad x_i^* = x_i - x_{i_0}, \quad u_i^* = u_i - u_{i_0} \tag{14.7}$$

A briefer notation for the partial derivatives at the operating point can also be adopted:

$$\frac{\partial f_2}{\partial x_1}\circ = \frac{\partial f_2}{\partial x_1}\bigg|_{x_0, u_0} \quad \text{etc.} \tag{14.8}$$

Substituting the changes indicated by Equations (14.7) and (14.8) into Equation (14.6) then gives:

$$\dot{x}_1^* = \frac{\partial f_1}{\partial x_1}\circ x_1^* + \cdots + \frac{\partial f_1}{\partial x_n}\circ x_n^* + \frac{\partial f_1}{\partial u_1}\circ u_1^* + \cdots + \frac{\partial f_1}{\partial u_m}\circ u_m^*$$

$$\dot{x}_2^* = \frac{\partial f_2}{\partial x_1}\circ x_1^* + \cdots + \frac{\partial f_2}{\partial x_n}\circ x_n^* + \frac{\partial f_2}{\partial u_1}\circ u_1^* + \cdots + \frac{\partial f_2}{\partial u_m}\circ u_m^*$$

$$\vdots \qquad\qquad\qquad \vdots \tag{14.9}$$

$$\dot{x}_n^* = \frac{\partial f_n}{\partial x_1}\circ x_1^* + \cdots + \frac{\partial f_n}{\partial x_n}\circ x_n^* + \frac{\partial f_n}{\partial u_1}\circ u_1^* + \cdots + \frac{\partial f_n}{\partial u_m}\circ u_m^*$$

Finally, the *Jacobian Matrices* $\boldsymbol{J}_x$ and $\boldsymbol{J}_u$ are introduced, whose elements are constants, where:

$$\boldsymbol{J}_x = \begin{bmatrix} \frac{\partial f_1}{\partial x_1}\circ & \frac{\partial f_1}{\partial x_2}\circ & \cdots & \frac{\partial f_1}{\partial x_n}\circ \\ \frac{\partial f_2}{\partial x_1}\circ & \frac{\partial f_2}{\partial x_2}\circ & \cdots & \frac{\partial f_2}{\partial x_n}\circ \\ \cdot & \cdot & \cdots & \cdot \\ \cdot & \cdot & \cdots & \cdot \\ \frac{\partial f_n}{\partial x_1}\circ & \frac{\partial f_n}{\partial x_2}\circ & \cdots & \frac{\partial f_n}{\partial x_n}\circ \end{bmatrix} \quad \boldsymbol{J}_u = \begin{bmatrix} \frac{\partial f_1}{\partial u_1}\circ & \frac{\partial f_1}{\partial u_2}\circ & \cdots & \frac{\partial f_1}{\partial u_m}\circ \\ \frac{\partial f_2}{\partial u_1}\circ & \frac{\partial f_2}{\partial u_2}\circ & \cdots & \frac{\partial f_2}{\partial u_m}\circ \\ \cdot & \cdot & \cdots & \cdot \\ \cdot & \cdot & \cdots & \cdot \\ \frac{\partial f_n}{\partial u_1}\circ & \frac{\partial f_n}{\partial u_2}\circ & \cdots & \frac{\partial f_n}{\partial u_m}\circ \end{bmatrix}$$

$$\tag{14.10}$$

Equations (14.9), therefore, finally become:

$$\dot{\boldsymbol{x}}^* = \boldsymbol{J}_x \boldsymbol{x}^* + \boldsymbol{J}_u \boldsymbol{u}^* \tag{14.11}$$

Equation (14.11) is now a linear state-space model, to which all the previous techniques can be applied. It represents the nonlinear system $\dot{x} = f(x, u)$ and is valid for small deviations (x^* and u^*) from the operating point (x_0, u_0).

The output equation $y = Cx + Du$ can still be applied directly if it is linear (it will automatically be linear if each output is simply a state variable, for example). If it is nonlinear, then it can be treated in the same way as the state equation to obtain a linearized version.

An example will dispel some of the mystique.

Example 14.1 Linearized model for the system of Figure 14.11

Earlier (Equation (14.1)), a nonlinear model of the system shown in Figure 14.11 was obtained as:

$$\dot{x}_1 = f_1(x_1, x_2, u) = x_2$$

$$\dot{x}_2 = f_2(x_1, x_2, u) = -2x_1^3 - 1.5x_1x_2 + u$$

To linearize this, the Jacobian matrices (Equation (14.10)) are evaluated:

$$J_x = \left.\begin{bmatrix} \dfrac{\partial f_1}{\partial x_1} & \dfrac{\partial f_1}{\partial x_2} \\[2mm] \dfrac{\partial f_2}{\partial x_1} & \dfrac{\partial f_2}{\partial x_2} \end{bmatrix}\right|_{x_0, u_0} = \begin{bmatrix} 0 & 1 \\ -6x_{1_0}^2 - 1.5x_{2_0} & -1.5x_{1_0} \end{bmatrix} \quad \text{and}$$

$$J_u = \left.\begin{bmatrix} \dfrac{\partial f_1}{\partial u} \\[2mm] \dfrac{\partial f_2}{\partial u} \end{bmatrix}\right|_{x_0, u_0} = \begin{bmatrix} 0 \\ 1 \end{bmatrix}$$

(14.12)

When the operating point values (x_0, u_0) are substituted into Equation (14.12), the resulting (constant) Jacobian matrices can be used in the linear model, as in Equation (14.11) (valid for small deviations from the operating point).

It is now necessary to specify the operating point. Since the forward path of Figure 14.11 contains pure integrators, any *steady* operating point must have the integrator inputs at zero. Under these circumstances, the nonlinear damping feedback term ($= 1.5x_1x_2$) will be zero (because x_2 must be zero), so the feedback from the transducer must balance the steady-state input signal. Specifying an operating point at $u = 1$, the steady-state output at the operating point must therefore be $x_1 = 0.7937$, in order for the input to the first integrator to be zero.

Such operating points can also be found by simulation. Figure 14.12(a) shows a MATLAB (Appendix 3) simulation of the nonlinear system from zero initial conditions with an input step of $u = 1$. The steady-state values are the same as those predicted by physical reasoning above. The MATLAB m-file *fig14_12.m* on the accompanying disk will duplicate this result.

Note that, since nonlinear systems can have multiple equilibrium points and initial-condition-dependent stability, it will by no means always be the case that a MATLAB simulation from zero initial conditions will just happen to end up at the desired operating point for every system. SIMULINK (Appendix 4) can be used more easily than MATLAB to find the operating point values of nonlinear systems. This is the outline procedure:

- Build up a SIMULINK block diagram of the system of Figure 14.11 (with 'inport' and 'outport' blocks from the 'connections' library at the input and output respectively; and a 'Fcn' block from the 'nonlinear' library, containing u^3 for the cubic function).
- Save it with a filename *nlmod* (for example).
- From the MATLAB prompt, issue the command:

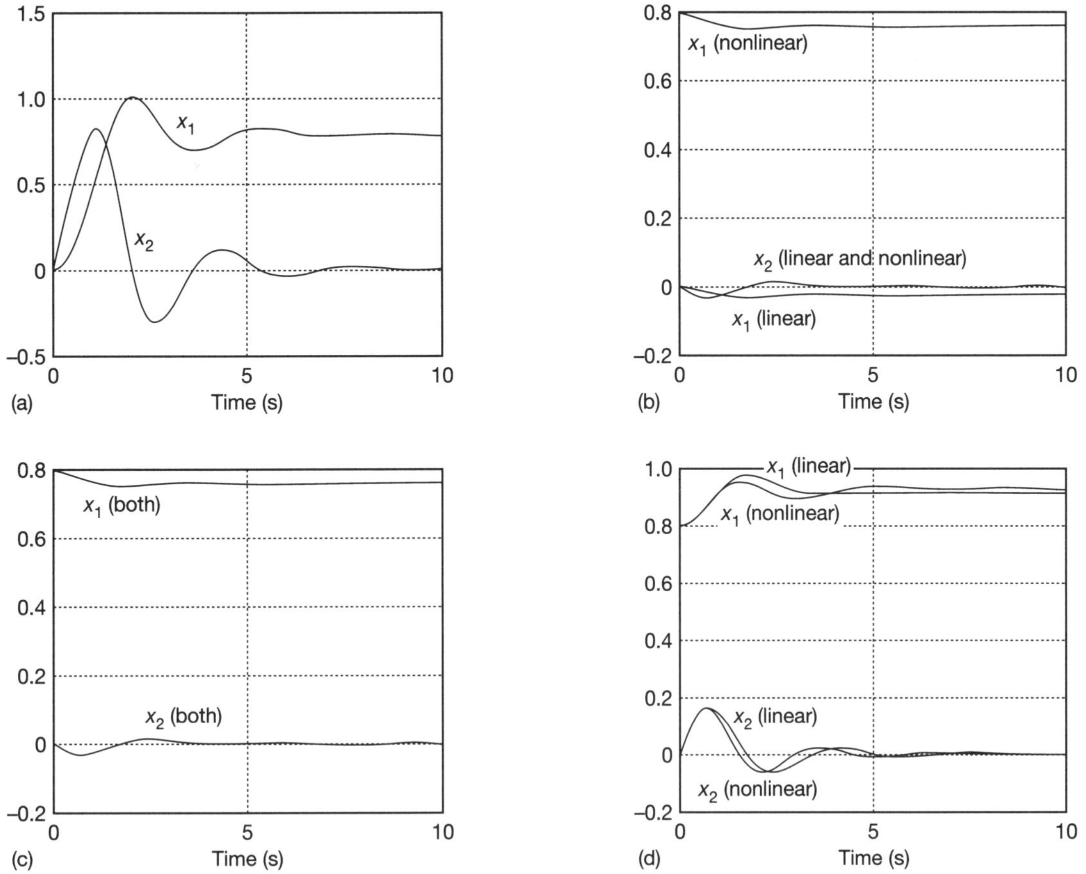

Figure 14.12 Linear and nonlinear model responses of the system of Figure 14.11.

>[xo, uo, yo, dx] = trim('nlmod',[1 1]',1,1,[ ],1,[ ]) % finds the operating point

>[jx, ju, c, d] = linmod('nlmod',xo,uo) % finds the state-space model at the
% operating point given by xo and uo

Assuming that SIMULINK is available, use the MATLAB commands *help trim* and *help linmod* for more detail. In outline, the [1 1]' in the *trim* command is an initial guess at the values in x_0, the first of the two unity elements is the specified value of u_0, while the second is an initial guess at y_0 (the plant output at the steady operating point). The three elements [ ],1,[ ] are two empty matrices which give the algorithm permission to vary the values of x_0 and y_0 in its search for the steady operating point, and a unity element which tells it that the specified value of u_0 is fixed, and must not be allowed to vary.

Returning to the MATLAB simulation in the m-file *fig14_12.m*, using $u_0 = 1$ and $x_0 = [0.7937 \quad 0]^T$ in Equations (14.12) and (14.11) gives the required linearized model

$$\dot{x}^* = \begin{bmatrix} 0 & 1 \\ -3.78 & -1.19 \end{bmatrix} x^* + \begin{bmatrix} 0 \\ 1 \end{bmatrix} u^* \tag{14.13}$$

for deviations from this operating point.

To test the model, the same nonlinear MATLAB simulation that produced Figure 14.12(a) is rerun, but this time with the initial condition $x_0 = [0.7937 \quad 0]^T$ and $u_0 = 1$. With these conditions, the system ought to remain in this condition and, in fact, it does so. An extra input step is therefore superimposed upon u_0 to represent a deviation from the operating point. This same input (deviation) can be applied to the linear model of Equation (14.13), allowing comparison of the resulting behaviour.

If a step of -0.1 unit is applied to the linear model, and an *additional* step of -0.1 unit to the operating point of the full nonlinear simulation, Figure 14.12(b) results. The responses for x_2 are more or less identical on these scales, but the linear model response for x_1 is based on zero, while the full nonlinear response is based on 0.79 (the operating point). This is entirely correct, because the linear model is showing the *deviations* from the operating point, and the operating point for x_2 just happens to be zero, while that for x_1 is not.

To get a direct comparison, the operating point values need to be added to the responses generated by the linear model. Doing this produces Figure 14.12(c), from which it is seen that the linear model is a reasonable approximation to the plant (although it does not look quite so good on a larger-scale plot).

Moving away from the operating point causes the accuracy of the linear model to fall off. For example, a step of $+0.5$ unit with respect to the operating point produced Figure 14.12(d). Use of MATLAB, or a similar package, to experiment with the models proves quite interesting. For example, the performance of the linear model deteriorates much more rapidly for negative steps away from the operating point, than for positive steps. Also, with no nonlinear damping at all, this system is unstable and produces a very nonlinear response.

As a final comment in this section, it is worth noting that the production of the linearized model by SIMULINK, mentioned during the example, is done by numerical perturbation of the full nonlinear simulation. This approach allows SIMULINK (or other packages, such as ACSL – Mitchell and Gauthier (1987)) to obtain the Jacobian matrices equally well for systems which contain discontinuous nonlinearities, and which therefore could not be handled by the analytical Taylor series method used in this section. Of course, the linearized approximation is likely to be of rather limited applicability in such systems, if the effects of the discontinuous element are significant.

14.5   *Phase plane analysis*

This is the first method which aims to analyse nonlinear behaviour, rather than 'ignoring' it by linearization. This method is studied first because:

- it is a time domain method, so the results are easily visualized;
- it is relatively straightforward to derive and apply;
- the results it produces are a useful way of depicting nonlinear system behaviour, which will be used in other sections.

Approximations and limitations involved in this method are:

- The linear part of the plant is limited to second order. There is no reason in principle why higher orders cannot be considered, but it is difficult to interpret the results. Once the second-order approximation is made, no other approximations are involved except that it is a graphical technique, with the normal limitations that implies if drawn by hand.

- The parameters of the plant must be stationary (non-time-varying).

- General input signals cannot be handled. However, initial conditions are automatically included, so it is normally possible to investigate the effects of step or ramp inputs.

- The method is traditionally applied 'on paper' to discontinuous nonlinearities, although it is equally possible to investigate continuous ones. Normally only one nonlinear element can be handled. If more than one nonlinear element is to be considered, their effects may have to be combined, or computer simulation should be used.

- One advantage of the phase plane approach is that it can handle very severely nonlinear systems, so long as the other constraints are met.

- These days, a computer simulation would normally be used to obtain the same kind of information which is available from phase plane plots, so the need actually to produce such plots is diminishing. However, phase plane analysis still gives valuable insights into system behaviour, and a rapid check on the validity of simulation results. It is useful both for these reasons, and because it can give a good idea of nonlinear system behaviour when there is no computer or CACSD package available.

14.5.1   *The phase plane*

Imagine a mass suspended on a long linear spring in the air, and at rest. If the mass is now pulled downwards by one metre and then released, the resulting behaviour can be visualized – it will be a damped oscillation which will eventually end when the mass is back at its starting point and at rest. Defining downward displacement as positive, the *displacement* will trace out the curve which begins at $+1.0$ in Figure 14.13(a).

At the same time, the *velocity* of the oscillating mass will behave as shown in the second trace in Figure 14.13(a). The velocity begins to increase in a negative direction because the mass initially accelerates upwards, which has been defined as the direction of negative displacement.

This behaviour is well approximated by a second-order linear model (so long as the spring is not stretched too far, and so long as the damping effects are more or less linear). If a state-space model of the system were to be written, the position (displacement) and velocity of the mass would probably be chosen as the state variables (because they are physically meaningful, measurable and give rise to a reasonable model structure). The captions on Figure 14.13(a) assume that displacement has been chosen as state x_1 and velocity as x_2. The model is derived in Example 14.2, below.

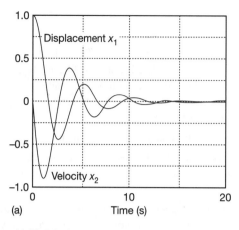

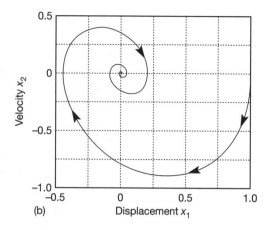

Figure 14.13 Displacement plotted against velocity.

Since there are only two states in this model, the state space is a *plane*, with axes x_1 and x_2. The behaviour of Figure 14.13(a) can therefore be directly transferred to this plane to give a picture of the state trajectory of the system as shown in Figure 14.13(b), showing how the state vector of the system varies as time passes (time increases along the curve). The state vector begins at $x = \begin{bmatrix} 1 & 0 \end{bmatrix}^T$ and ends at $x = \begin{bmatrix} 0 & 0 \end{bmatrix}^T$ with the obvious oscillations in between. The same behaviour is visible in this plot as in Figure 14.13(a). For example, velocity (x_2) is zero whenever displacement (x_1) is a maximum or minimum (that is, the mass is reversing).

The plane in Figure 14.13(b) could be called a state plane. However, a special condition is also operating here, namely that $x_2 = \dot{x}_1$. The plot is therefore of some system variable (in this case displacement) versus its derivative. A state plane with such axes is called a *phase plane*, and the plot of a variable versus its derivative is called a *phase plane trajectory* (PPT).

The form of the PPT depends entirely upon the system and the initial conditions on *x*. For example, if there were zero damping in the mass–spring system, the PPT would be a circle centred on the origin. The radius of the circle would depend upon the point from which the mass was released. Several different release points plotted on the same phase plane would yield a family of concentric circles centred on the origin. Such a family of PPTs is called a *phase portrait*. Similarly, a phase portrait could be obtained in Figure 14.13(b) by plotting more PPTs for several different release points.

PPTs and phase portraits can yield much useful information about nonlinear systems including stability (which, as described in Section 14.2.2, can be difficult to determine analytically), switching points, limit cycles, overshoot and steady-state error. In addition, it is possible to timescale PPTs so as to determine all the usual time-domain measures such as rise time, peak time and settling time. This can be done by hand (Section 14.5.5), but it is normally much better to obtain timescale information from computer simulations if accurate results are required.

We now look at how PPTs can be produced, and at some of their characteristics.

14.5.2 The method of isoclines

Careful study of Figure 14.13(b) reveals some information which will help in the construction of PPTs in general. Since the variable of interest (x_1) is always plotted on the horizontal axis and its derivative (x_2) on the vertical axis, some conclusions can be drawn about the *direction* a PPT might take in the plane. For example, if x_2 is positive then x_1 must be increasing, while if x_2 is negative x_1 must be decreasing. This implies that PPTs must circle in a generally *clockwise* direction. It cannot be said that they will always circle the origin, though. They may well be centred on some non-zero value.

Of even greater interest is the fact that whenever x_2 is zero then x_1 must be constant. This means that whenever any PPT crosses the horizontal axis $(x_2 = 0)$ it must be vertical. In any PPT, the horizontal axis is therefore a line joining points where the PPT has infinite slope. A line joining points of equal slope is called an *isocline*, so the horizontal axis is an isocline with an associated slope of infinity. If several other isoclines could be found for any given system, each with a known associated slope, then the PPT could be sketched following the principle shown in Figure 14.14. This follows, because the initial conditions for the system would be known (giving the starting point in the phase plane). It has also been determined that the PPT must progress from the starting point in a generally clockwise direction. The starting point and the slopes associated with the isoclines therefore fix the route of the PPT. Obviously the more isoclines there are, the more accurate will be the result. Note that isoclines are not necessarily straight lines. For systems with continuous nonlinearities, they may be curves.

How can these isoclines be found? First, a linear system will be considered, in order to establish the method. Remember that only second-order systems with no general input can be considered, and that it will always be the case that $x_2 = \dot{x}_1$. The state equation for the linear part of any system for which a PPT can be drawn will therefore be of the form of Equation (14.14), perhaps with the addition of constant terms to represent step or ramp inputs (although these might also be represented by the initial conditions on x).

$$\dot{x} = \begin{bmatrix} 0 & 1 \\ a_{21} & a_{22} \end{bmatrix} x \tag{14.14}$$

It is not strictly necessary for a_{12} to be unity (in other words, x_2 could be a scaled version of the derivative of x_1), but it will always be assumed equal to unity in this

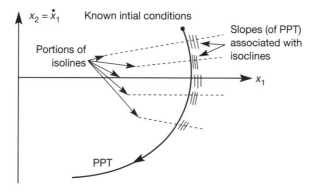

Figure 14.14 The method of isoclines.

text. Note that this is a standard companion form state-space model (see Equation (2.51) in Section 2.5.1).

Now, from Figures 14.13(b) and 14.14, the slope N of a PPT at any point is given by

$$N = \frac{dx_2}{dx_1} \quad \text{which can be rewritten} \quad \frac{dx_2/dt}{dx_1/dt} \quad \text{i.e.} \quad N = \frac{\dot{x}_2}{\dot{x}_1}$$

(14.15)

Equation (14.15) is used to determine the equations of isoclines for the system under consideration. For this simple linear system, the state equations of Equation (14.14) can be substituted directly into Equation (14.15), to give

$$N = \frac{\dot{x}_2}{\dot{x}_1} = \frac{a_{21}x_1 + a_{22}x_2}{x_2}$$

This can then be rearranged to give

$$x_2 = \frac{a_{21}}{N - a_{22}} x_1$$

(14.16)

Equation (14.16) shows that the isoclines for a purely linear (second-order) system are always straight lines passing through the origin of the phase plane. The model coefficients a_{21} and a_{22} will be known, so it is merely necessary to choose a number of values of N (the slope of the PPT when it crosses the isocline) and use Equation (14.16) to work out the equation of the isocline for each value of PPT slope. The isoclines can then be drawn in the phase plane and, given the initial conditions on x, the approach of Figure 14.14 can be used to sketch the PPT. Care must be taken with regard to the accuracy of drawing, since any errors are cumulative along the PPT. There are more accurate geometrical techniques for PPT construction (Atherton, 1982) but the assumption is that hand-drawn PPTs will be used only for approximate investigations nowadays, so they will not be described here. Almost all the PPTs in the figures were computer-generated by the MATLAB m-files on the accompanying disk.

Example 14.2 *PPT for a linear mass–spring–damper system*

Consider the example used in Section 14.5.1 and Figure 14.13. Let the mass be M kg, the spring stiffness be K Nm^{-1} and the viscous damping coefficient of the air be B N/(ms^{-1}). If positive displacement x_1 (m) is measured downwards from zero at the equilibrium position, then the restraining force due to the spring will be Kx_1 N (opposing downward motion), the damping force will be $B\dot{x}_1$ N (opposing motion) and the force accelerating the mass will be $M\ddot{x}_1$ N. For equilibrium in the face of no input other than initial conditions, these forces must sum to zero. The equation of motion is therefore

$$-Kx_1 - B\dot{x}_1 = M\ddot{x}_1$$

Given that $x_2 = \dot{x}_1$, this equation can be rearranged to give the state-space model

$$\dot{x}_1 = x_2$$

$$\dot{x}_2 = -\frac{K}{M}x_1 - \frac{B}{M}x_2 \qquad (14.17)$$

Now, let the mass be 1.9 kg, the spring stiffness be 3.0 Nm^{-1} and the damping coefficient be 1.2 N/(ms^{-1}). This gives $\dot{x}_2 = -1.58x_1 - 0.63x_2$ in Equation (14.17), so that the coefficients a_{21} and a_{22} in Equation (14.14) are -1.58 and -0.63 respectively, for this system.

Substituting into Equation (14.16) gives the isocline equation

$$x_2 = \frac{-1.58x_1}{N + 0.63}$$

into which values of N can be inserted to find several isoclines. For example, if $N = \infty$, $x_2 = 0$, giving the horizontal axis, as expected. However, if $N = 1$ then $x_2 = -0.969x_1$, so the straight line passing through the origin and the point $(1, -0.969)$ is an isocline with which is associated a slope of 1, which the PPT must adopt every time it crosses this isocline. This isocline is shown in Figure 14.15 along with several others obtained using the given values of N (that is, the slope of the PPT associated with the isocline) in the equation above.

Given that the mass is released from a displacement of $x_1 = 1$ m, and from rest ($x_2 = 0$), the PPT can be sketched in as shown in Figure 14.15. In fact, the result should be identical to that of Figure 14.13(b), as this is the system which produced that figure, too.

From the PPT in Figure 14.15, information such as the following can be obtained:

- The system is stable (it settles to a steady state).

- The system settles after about three complete cycles of oscillation (which is also in agreement with Figure 14.13(a)).

- The system comes to rest at zero displacement and zero velocity.

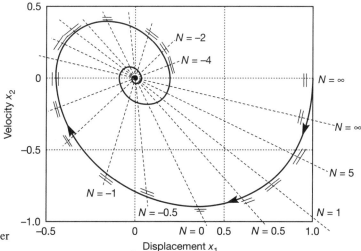

Figure 14.15 PPT for a linear mass–spring–damper system.

- There is a peak overshoot to about -0.44 m in the displacement, that is, about 44 per cent since the system is effectively responding to a unit step input.

- Since the system model is second order, the damping ratio can be ascertained from standard curves (see Figure 3.19 in Section 3.2.2), where it is found that 44 per cent overshoot corresponds with a damping ratio of about 0.25. This can also be checked from the original model, because a standard second-order system in companion form gives a plant matrix

$$\begin{bmatrix} 0 & 1 \\ -\omega_n^2 & -2\zeta\omega_n \end{bmatrix}$$

  Comparing the model stated after Equation (14.17) with this, it must be the case that $\omega_n^2 = 1.58$ and $2\zeta\omega_n = 0.63$. Solving these gives $\zeta = 0.25$, which agrees with the result from the PPT and the standard curves.

- The peak velocity is about -0.9 ms^{-1} (that is, 0.9 ms^{-1} in an upwards direction).

- The PPT can also be time-scaled, but that is left until Example 14.6.

It can be seen from this simple linear example that the PPT is a useful representation of second-order system behaviour.

14.5.3 Application to nonlinear systems

This topic is best studied by example. However, one or two points are worth noting at the outset. If the behaviour at the *output* of a plant is of interest, it makes sense to plot a PPT of the output versus its derivative.

If the response to a step input is being considered, this will normally cause no problem. However, if the response to a ramp input is sought, it may well be the case that the output will never settle (it will probably follow the ramp), so a PPT of the output might then be of limited use.

In such cases, it is often preferable to plot the PPT of some other variable which *does* settle (for example, the error signal in the feedback loop of a linear second-order system should settle to a steady value in response to a ramp input, even though the output will be ramping). The output behaviour can then be inferred from the behaviour of this other variable.

Also note that the location of the nonlinear element makes no difference to the method. It can be an actuator at the input, a nonlinear transducer, or something (such as backlash) at the plant output. However, it may sometimes be difficult to use the method for a particular combination of type of nonlinearity, type of input signal and nonlinear element location. Example 14.4 will eventually illustrate this.

Example 14.3 *A system with input saturation*

Figure 14.16 shows a plant controlled by an amplifier and actuator with a gain of 2, which saturates at ± 2 units. What is the overshoot at the output in response to an input step of 4 units? How has the saturation affected the response? What would be the maximum input step size which could be applied without the nonlinear element saturating?

First, it can be seen that the nonlinear actuator will cause the system behaviour to divide into three distinct classifications. Each will be considered separately.

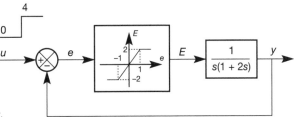

Figure 14.16 A system with a saturating actuator.

(1) When the actuator is not saturated, the system will be purely linear, with a gain of 2 from the slope of the actuator characteristic.

(2) When the actuator is saturated at the upper limit, the linear part of the plant will have a constant input of 2 units at all times.

(3) When the actuator is saturated at the lower limit, the linear part of the plant will have a constant input of -2 units at all times.

First, in the unsaturated condition, the forward path is linear with a gain of 2 from the actuator, so the closed loop transfer function of the plant becomes:

$$\frac{Y}{U}(s) = \frac{2}{2s^2 + s + 2}$$

Cross-multiplying, taking inverse Laplace transforms with zero initial conditions and substituting $u = 4$ for the step input gives

$$2\ddot{y} + \dot{y} + 2y = 2u = 8$$

Now, letting $x_1 = y$ (because the output is of interest) and $x_2 = \dot{x}_1$ (as always), gives:

$$2\dot{x}_2 + x_2 + 2x_1 = 8 \quad \text{or, in state space form:}$$

$$\dot{x}_1 = x_2$$

$$\dot{x}_2 = -x_1 - 0.5x_2 + 4$$

From Equation (14.15), the isocline equation is therefore:

$$N = \frac{\dot{x}_2}{\dot{x}_1} = \frac{-x_1 - 0.5x_2 + 4}{x_2} \tag{14.18}$$

leading to:

$$x_2 = \frac{-x_1}{N + 0.5} + \frac{4}{N + 0.5} \tag{14.19}$$

Equation (14.19) is the equation of a straight line with a non-zero intercept on the x_2 axis. A convenient way to draw these is to substitute zero values for x_1 and x_2 in turn. When x_1 is zero, Equation (14.19) gives

$$x_2 = \frac{4}{N + 0.5}$$

When x_2 is zero, Equation (14.19) yields $x_1 = 4$.

Therefore, every isocline for this system passes through the point (4,0), and has an intercept on the x_2 axis given by

$$x_2 = \frac{4}{N + 0.5}$$

The first two rows of Table 14.1 give a few values, and the resulting isoclines can be seen centred around $x_1 = 4$, $x_2 = 0$ in Figure 14.17(a). Note that the region of the phase plane over which these isoclines are valid is not yet known. It can be seen from Figure 14.17(a) that they have only been drawn in the range $3 \leq x_1 \leq 5$. The proof that this is the correct range will emerge as the other regions are considered.

In the second region (actuator saturated at the upper limit), the input to the plant is a constant +2 units, so the output is given by

$$Y(s) = \frac{2}{2s^2 + s}$$

Table 14.1 Isocline parameters for Figure 14.17.

N		-2	-1	-0.5	-0.25	0	0.25	1	∞
x_2 (linear intercept)		$-8/3$	-8	∞	16	8	$16/3$	$8/3$	0
x_2 (upper sat. intercept)		$-2/3$	-2	∞	4	2	$4/3$	$2/3$	0
x_2 (lower sat. intercept)		$2/3$	2	∞	-4	-2	$-4/3$	$-2/3$	0

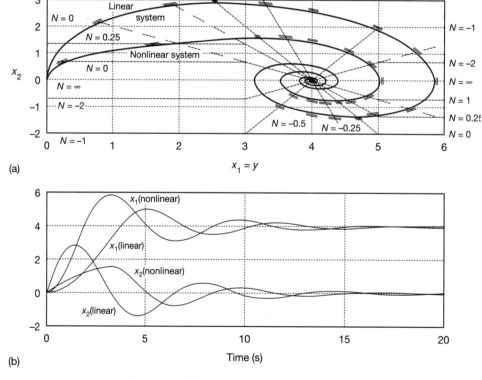

(a)

(b)

Figure 14.17 PPT and time responses for y in the system of Figure 14.16.

Cross-multiplying and taking inverse transforms as before gives $2\ddot{y} + \dot{y} = 2$. Substituting x_1 and x_2 exactly as before results in:

$$\dot{x}_1 = x_2$$
$$\dot{x}_2 = -0.5x_2 + 1$$

The isocline equation is therefore

$$N = \frac{\dot{x}_2}{\dot{x}_1} = \frac{-0.5x_2 + 1}{x_2}$$

leading to

$$x_2 = \frac{1}{N + 0.5}$$

The isoclines in this region are therefore horizontal lines at values of x_2 determined by N.

It is also necessary to work out where this region is in the phase plane. From Figure 14.16, in order for the upper saturation limit to be reached, $e > +1$. Since $u = +4$, this implies that $x_1 = y < +3$. The x_2 values for these horizontal isoclines are in the third row of Table 14.1, and they appear in the left-hand half of Figure 14.17(a).

For the third case, where the actuator is in the lower saturation region, the reader should be able to show that application of an identical approach yields the isocline equation

$$x_2 = \frac{-1}{N + 0.5}$$

valid for the region of the phase plane where $x_1 > 5$. These values are given in the last row of Table 14.1, and appear in the right-hand portion of Figure 14.17(a).

The region over which the linear system isoclines (calculated earlier) are valid is now known. Note also that the isoclines with the same associated slopes neatly meet up at the boundaries of the three regions. This does *not* happen for all systems, but when it does it is a good indication that the solution is proceeding correctly. In addition, once we realize that this is going to happen for any particular system, it can save a lot of calculation. For example, in the present system, once the central isoclines had been calculated, and the boundaries of the regions had been found, the rest could have been drawn in immediately with no further calculation, after working out that the first horizontal isocline behaved in this way.

The PPT itself is plotted in precisely the same way as before. Zero initial conditions are assumed (that is, the response begins at the origin of the phase plane). Note also that if the plotting of the PPT had been begun as soon as the isoclines for the left-hand and central regions had been obtained, it would have been discovered that those for the right-hand region need not be calculated at all, as the PPT barely enters that zone.

To get a comparison with the performance of the system in the absence of saturation, the set of isoclines corresponding with linear (unsaturated) operation is extended over the entire plane, and used for a second PPT as shown. More isoclines would be needed to obtain the accuracy shown, especially in the early parts of the trajectories, but they have been omitted for clarity.

Now the initial questions can be answered. The peak value of $y (= x_1)$ from the PPT of the nonlinear system in Figure 14.17(a) is seen to be about 5 units. The overshoot is therefore about 25 per cent. In the absence of the nonlinear (saturation) effect, the peak value from the PPT of the linear system is about 5.86 units (46.5 per cent overshoot). In this system, the presence of the saturation has therefore tended to stabilize the system, reducing overshoot (and settling time to within any given percentage of the final value). Note that this is not a general effect of all nonlinearities (for example, backlash can often have a very destabilizing effect, and so might saturation in other circumstances).

The behaviour is confirmed by the time responses for the system which are given in Figure 14.17(b). The nonlinear behaviour is particularly evident in the response of x_2.

The question regarding the maximum input step which would not cause saturation cannot be answered from the PPT. This kind of question is generally best answered by simulation. However, in this case, since the system is linear for small input signals, the answer could be found by linear analysis of the behaviour of e, by analysing the response of

$$\frac{E}{U}(s) = \frac{2s^2 + s}{2s^2 + s + 2}$$

This will not be done rigorously here, but intuitively, since y cannot change instantaneously, any input step is initially passed unaltered (because the feedback signal has not yet moved from zero) to the nonlinear element. Any step of magnitude greater than 1 unit will therefore cause saturation. The error transfer function represents an underdamped system, with a steady-state gain of zero. An input of 1 unit will therefore cause an immediate error of 1 unit, followed by an underdamped transient response in the error signal in the opposite direction, aiming at zero. Since there is some damping in the system, this cannot overshoot by 100 per cent, so it will not exceed -1 unit. The error response will therefore remain in the linear region for input steps of less than 1 unit in magnitude.

Example 14.4 *A system with output deadzone*

This example is deliberately chosen because it is awkward, and illustrates some of the pitfalls of the approach. Figure 14.18 shows a second-order system with a deadzone at the output. Say that it is desired to examine the output behaviour in response to a ramp input of 1 unit per second.

Consideration of the system structure indicates that at steady state, y will be continuously increasing in response to the ramp input. The PPT for y would therefore be difficult to interpret. However, since the linear part of the system is second order, the error e *will* settle to some constant value. The PPT for e rather than y might therefore be considered, and the behaviour of y inferred from it.

However, a worse problem then occurs. The nonlinear element will divide the phase plane into three regions, as in Example 14.3, but it is not possible to tell where they are.

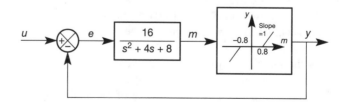

Figure 14.18 A system with an output deadzone.

For example, in the positive 'linear' region, it is evident that $m > 0.8$, but it is not possible to convert this information into a condition on e and $\dot{e}$ so as to determine where this is valid on the phase plane for e. There are other problems too, but the conclusion is that this system cannot be handled by the method of isoclines. Of course, a computer simulation could easily be performed, and e plotted against $\dot{e}$ to produce the PPT, but that is not the purpose here.

In order to be able to produce a PPT for this system by hand using the method of isoclines, the problem will have to be modified! Let us examine the response to a *unit step* input instead.

With a step input, the output should eventually settle if the system is stable, so an attempt to draw the PPT for y would be made. Unfortunately, there is now yet another problem. In the deadzone, the value of y is fixed at zero, and so $\dot{y}$ must be zero, too. However, when y leaves the deadzone, $\dot{y}$ must immediately adopt the value of $\dot{m}$, thus producing discontinuities in the PPT. These, together with the fact that the PPT for y vs. $\dot{y}$ will always return to the origin when the deadzone is entered, will make it difficult to interpret.

Again, recourse is made to plotting a PPT for e rather than y. However, a moment's thought shows that since e is related to y by a simple subtraction from the (constant) input, the same qualitative behaviour will be evident in e and $\dot{e}$ as in y and $\dot{y}$.

Finally then, it is decided to draw the PPT for the linear plant output signal m, and to infer the behaviour of y from it. As usual, the three portions of the nonlinear characteristic are considered separately:

In the deadzone, $y = 0$, so $e = u - 0 = 1$. Therefore,

$$m = \frac{16}{s^2 + 4s + 8}$$

Cross-multiplying, taking inverse Laplace transforms, letting $x_1 = m$ and $x_2 = \dot{m}$ leads to:

$$\dot{x}_1 = x_2$$
$$\dot{x}_2 = -4x_2 - 8x_1 + 16$$

Using Equation (14.15) to calculate the isocline equation as before gives:

$$x_2 = \frac{-8x_1}{N+4} + \frac{16}{N+4}$$

that is, the isoclines in the deadzone are straight lines passing through the point $(2,0)$ and having x_2 axis intercepts given by $16/(N+4)$. These isoclines are valid over the range $-0.8 \le m \le 0.8$.

For the positive 'linear' region, $m > 0.8$. In this region, the system has a unity negative feedback loop, but the signal fed back is *not* equal to m. Rather, it is $y = m - 0.8$. This implies that

$$m = \frac{16}{s^2 + 4s + 8}(u - y) = \frac{16}{s^2 + 4s + 8}(u - m + 0.8)$$

Proceeding in the usual manner leads to $\ddot{m} + 4\dot{m} + 8m = 16(u - m + 0.8) = 28.8 - 16m$.
Making the usual substitutions for x_1 and x_2 and using Equation (14.15) to find the isocline

equation then gives

$$x_2 = \frac{-24x_1}{N+4} + \frac{28.8}{N+4}$$

These are straight lines passing through the point (1.2,0) and having x_2 axis intercepts given by $28.8/(N+4)$.

It is unlikely that m will ever become < -0.8 in response to a (positive) unit step, so no isoclines will be calculated for the negative 'linear' region.

Table 14.2 gives a few values from the equations above and Figure 14.19(a) shows the resulting PPT (only relatively few isoclines are shown). It can be seen that the peak overshoot in m (x_1) is about 18 per cent. However, the percentage overshoot in the output (y) will differ from this, because m has to reach a value of 0.8 units before y begins to move. The value of y will be zero until m reaches 0.8, and $(m - 0.8)$ thereafter. The final value of y will therefore be $1.2 - 0.8 = 0.4$, and the peak value of y will therefore be about $1.42 - 0.8 = 0.62$, giving an overshoot in y of about 55 per cent.

Figure 14.19(b) shows the time responses, confirming these findings. Also shown is the response of the linear-only system (that is with a deadzone of zero) for comparison.

Table 14.2 Isocline parameters for Figure 14.19.

N	-6	-4	-2	-1	0	1	2	4	12	∞
x_2 (deadzone intercept)	–	–	–	–	4	3.2	2.7	2	1	0
x_2 (pos. linear intercept)	-14.4	∞	14.4	9.7	7.2	5.8	4.8	3.6	1.8	0

Example 14.5 *A system controlled by an ideal relay*

Consider the system of Figure 14.20(a). Here a relay controls the plant, and it will switch whenever the sign of e changes. The plant input is therefore always either 1 or -1 unit, and this is represented as $\text{sgn}(e) = 1$ when e is positive and $\text{sgn}(e) = -1$ when e is negative. Analysing the error signal for step inputs to the system,

$$e = u - y = u - \frac{3\,\text{sgn}(e)}{2s^2 + s}$$

For step inputs, cross-multiplying and taking inverse Laplace transforms with zero initial conditions (as in the previous examples) yields $2\ddot{e} + \dot{e} = -3\,\text{sgn}(e)$. Letting $x_1 = e$ and $x_2 = \dot{e}$ gives:

$$\dot{x}_1 = x_2$$

$$\dot{x}_2 = -0.5x_2 - 1.5\,\text{sgn}(e)$$

Using Equation (14.15) to find the isocline equation gives:

$$N = \frac{\dot{x}_2}{\dot{x}_1} = \frac{-0.5x_2 - 1.5\,\text{sgn}(e)}{x_2}$$

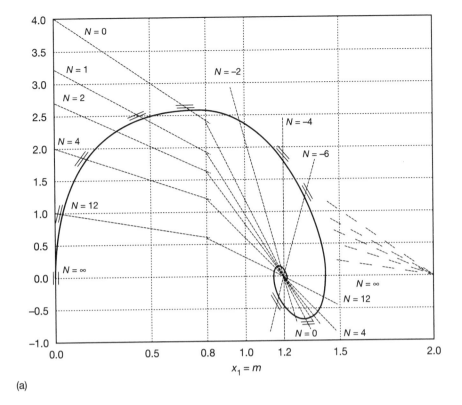

(a)

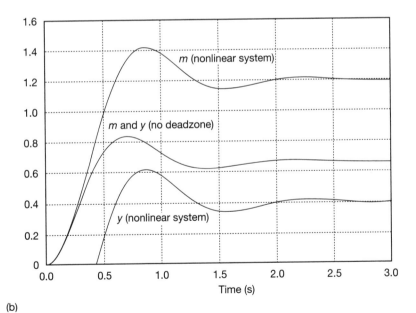

Figure 14.19 PPT and time response for m in the system of Figure 14.18.

(b)

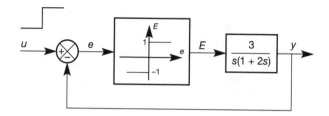

(a)

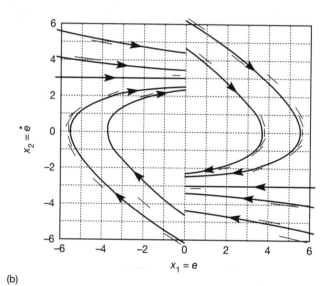

Figure 14.20 An ideal
relay system and forms of
the PPT. (b)

and so

$$x_2 = \frac{-1.5\,\text{sgn}(e)}{N + 0.5} \tag{14.20}$$

This shows that every isocline is horizontal, and that the value of x_2 depends upon N and the sign of the error. Note that the value of u (the input step size) does not figure in this equation at all. However, it will reappear when the initial conditions are calculated (note also that if the denominator of the plant transfer function contained a non-zero constant term, then u would contribute directly to Equation (14.20), and so would x_1, so the isoclines would not be horizontal).

Table 14.3 gives some values for the cases when the error is positive ($\text{sgn}(e) = 1$) and negative ($\text{sgn}(e) = -1$). These lead to Figure 14.20(b), which shows some possible part-trajectories. The actual behaviour depends upon the initial conditions, but the trajectories will always follow the patterns shown in the various regions of Figure 14.20(b).

Table 14.3 Isocline parameters for Figures 14.20(b) and 14.21.

N	-3	-2	-1	$-7/8$	$-3/4$	$-1/2$	$-1/4$	$-1/5$	$-1/8$	0	$1/4$	1	2	∞
x_2 (error positive)	3/5	1	3	4	6	$-\infty$	-6	-5	-4	-3	-2	-1	$-3/5$	0
x_2 (error negative)	$-3/5$	-1	-3	-4	-6	∞	6	5	4	3	2	1	3/5	0

In operation, the relay will switch every time e crosses zero. This means that the PPT will switch repeatedly from one set of trajectories to the other. Figure 14.21(a) shows about 15 seconds' worth of an actual response to an input step of 4 units. The starting point is found from $e = u - y$, so for a step input and a constant zero initial output $e = u$ and $\dot{e} = 0$ initially. The resulting behaviour at the output y (over 20 seconds) is shown in Figure 14.21(b).

A system such as this will end up in a limit cycle, because the signal at E must be either $+1$ or -1, so the system can never settle. However, as Figure 14.21(b) shows, the amplitude of the limit cycle can be very small. Also, from Figure 14.20(b), it can be seen that any initial condition will eventually lead to the same limit cycle, since starting points in the ranges (e positive and $\dot{e} > -3$) and (e negative and $\dot{e} < +3$) will give the same qualitative behaviour as in Figure 14.21(a); while for any other starting point, the magnitude of the slope associated with the isoclines is always between 0.5 and zero, and always leads the trajectory towards the $x_2 = +3$ or $x_2 = -3$ isoclines as appropriate. Once either of these isoclines is reached, the trajectory will slide along it until e changes sign, when the behaviour reverts to the former set of isoclines.

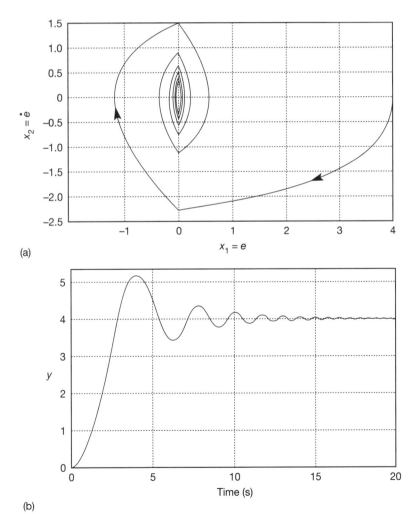

(a)

(b)

Figure 14.21 Responses of the system of Figure 14.20.

14.5.4 Optimal switching

The digital computer simulation leading to Figure 14.21 in Example 14.5 is not easy to perform correctly. The reason is that the switching of the relay (being assumed to be ideal) is exceedingly fast compared with the dynamics of the rest of the system. Therefore, a very small discrete time step is required if a digital simulation is to be successful in predicting the result (there were 20 000 data points per plot in the data file which produced the figure, although only every fifth point was plotted). The instant at which the relay switches in the simulation can be up to one time step late compared with the real (continuous-time) system. For the system of Example 14.5, if the time step is more than a millisecond or so, this effect becomes visible as a clockwise skewing of the switching line (the vertical axis in Figures 14.20(b) and 14.21(a)). This simulates the effect of a time delay, and affects the whole response in a potentially misleading manner.

Of course, in the real system there *will* be some time delay, since no relay (even a solid state one) is instantaneous. If a delay of 100 ms is introduced into the relay characteristic, the PPT is modified as shown in Figure 14.22(a), and the time response as in Figure 14.22(b). This time, there is no doubt about the limit cycle. Its amplitude is indicated in Figure 14.22(a), and can be confirmed from Figure 14.22(b) (it becomes only slightly smaller if further points are plotted).

Figure 14.22 shows that the performance has been changed significantly by moving the switching line. Although the result is qualitatively worse than that of

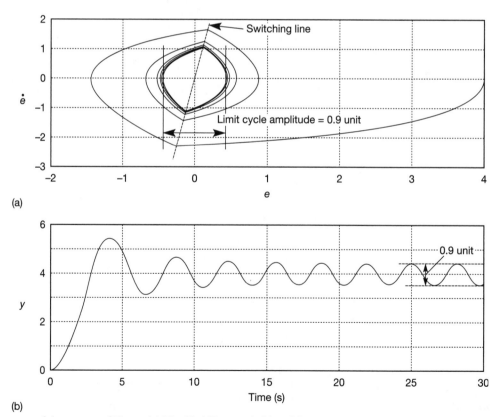

(a)

(b)

Figure 14.22 Responses of the system of Figure 14.20 with 100 ms switching delay.

Figure 14.21 (the time delay having its usual destabilizing influence), this leads to consideration of the possibility that the switching line might be moved deliberately in an effort to obtain better performance.

So long as the definition $e = u - y$ is maintained, and so long as the input to the linear part of the plant is either $+1$ or -1, the general phase portrait of Figure 14.20(b) holds true. This means that some function of the designer's choice (it might be called a compensator) can be inserted between the signal e and the relay as shown in Figure 14.23, but the trajectories of Figure 14.20(b) could still be used. Moving the switching line simply 'covers up' some parts of the portrait while 'uncovering' others. For example, in Figure 14.22(a), the PPT in the vicinity of the points $(-0.1, -2.25)$ and $(0.1, 1.75)$ is following parts of the phase portrait which were not visible in Figure 14.20(b), because of the different positions of the switching lines in the two figures.

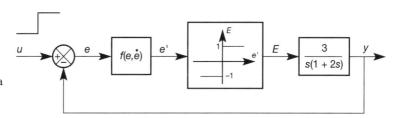

Figure 14.23 Addition of a nonlinear controller to the plant.

The time delay leading to Figure 14.21 could have been modelled in this way as $e' = f(e, \dot{e}) = e + 0.1\dot{e}$ for a 0.1 s delay (that is the value of e, plus its rate of change multiplied by the time for which it persists). This fits the form of Figure 14.23, and the switching line in Figure 14.22(a) can be seen to follow the relationship for $e' = 0$ (that is, $\dot{e} = -10e$).

If the switching line could be rotated anticlockwise rather than clockwise, better performance might be expected, as the isoclines would then always give a PPT ending at the origin. Consider Figure 14.24, in which the switching line is a

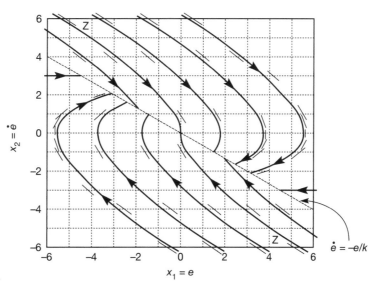

Figure 14.24 The effect on the PPT of rotating the switching line anticlockwise.

straight line of slope $-1/k$ (where k is some positive constant), with the phase portrait of Figure 14.20(b) superimposed and suitably extended. In order to make this happen, the relay needs to give an output as follows:

$$+1 \text{ when } \dot{e} > -\frac{1}{k}e, \text{ and } -1 \text{ when } \dot{e} < -\frac{1}{k}e$$

The relay in Figure 14.23 now switches when $e' = f(e, \dot{e}) = 0$, so a compensator function $e' = f(e, \dot{e}) = k\dot{e} + e$ is required.

This is actually a familiar compensator. Since $e = u - y$, it follows that $\dot{e} = -\dot{y}$ for step inputs. The relay input is therefore given by $e' = e + k\dot{e} = e - k\dot{y}$. This is rate feedback, as shown in Figure 14.25, and much used in linear controllers (see, for example, Section 4.5.3 or Section 12.6.1). It is equivalent to adding some phase lead control or derivative action. The signal $\dot{y}$ may be available directly, as shown.

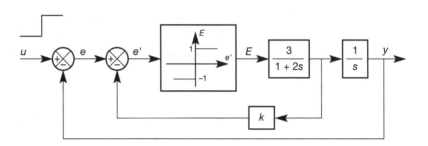

Figure 14.25 The system with rate feedback.

Alternatively, if y is a position signal in a servo, then $\dot{y}$ might be generated using a tacho. If none of these is possible, a linear observer might be used (Chapter 9).

Figure 14.26 shows the responses for a k value of 1/3. The initial part of the response is, as expected, the same as that in Figure 14.21. However, when the switching line is reached, the PPT switches to the 'other' set of curves as shown in Figure 14.24. The performance is obviously improved, and the PPT eventually 'slides' to the origin along the switching line. In fact, the PPT oscillates from one side of the switching line to the other during this sliding (the relay 'chatters' between its two states) and ends up in a small-amplitude limit cycle near the origin (due to the ideal relay characteristic), but these phenomena are invisible on the scale of the plot. The magnitude of the chattering depends upon the slope of the switching line and the linear system dynamics.

To make the behaviour better still (minimum settling time), the compensator in Figure 14.23 could be chosen such that $f(e, \dot{e})$ actually implemented the curves of the PPTs marked 'Z' in Figure 14.24. This is called *optimal switching*, and the system will then settle from any initial condition after a single switching of the relay. As soon as the PPT encounters the switching line (Z), the PPT actually becomes the switching line, so it should slide to the origin without any chattering. In practice, non-ideal characteristics of the relay may cause chattering, but the response will be much improved. Also, the small deadzone inherent in relays (or the hysteresis deliberately built into electronic comparators) will tend to quench the chattering and the limit cycle at the origin giving a stable steady state, but a small

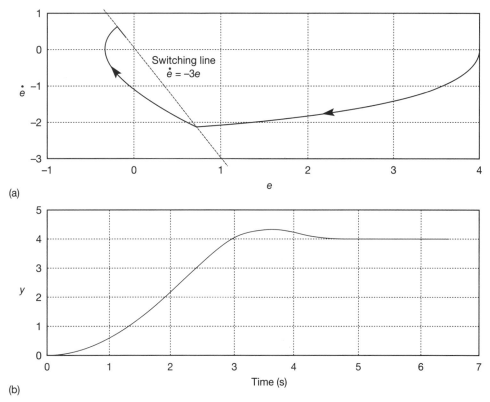

(a)

(b)

Figure 14.26 Responses of the system of Figure 14.25.

steady-state error. On–off (or 'bang–bang') time-optimal controllers of this type can be designed using a Hamiltonian approach to optimal control, based on Pontryagin's maximum principle (Banks, 1986).

14.5.5 Timescaling PPTs

This is a rather inaccurate procedure, and if reliable timescaling is needed a computer simulation will always be preferable. However, the procedure outlined here will give a quick guide when required.

Since, for PPTs, it is always the case that

$$\frac{dx_1}{dt} = x_2, \quad \text{then} \quad \frac{dt}{dx_1} = \frac{1}{x_2}$$

Integrating this with respect to x_1 gives

$$t = \int \frac{1}{x_2} \, dx_1$$

This means that the time to any point on the PPT is the area up to that point under the curve of $1/x_2$ versus x_1. Therefore, the x_1 axis can be divided into a number of segments, small enough that the corresponding behaviour of x_2 is adequately represented, the value of x_2 corresponding to each segment can be measured, converted to $1/x_2$ and the resulting areas can then be evaluated by hand, totalling them to give t.

Example 14.6 *Timescaling a PPT*

The timescaling procedure is applied (roughly) to the PPT of Figure 14.13(b), in order to find the time to the first peak. The first two columns of Table 14.4 show the information taken from the figure. The fourth column converts to $1/x_2$. Here, there is a problem due to the fact that x_2 is sometimes zero. This can be avoided by taking the mean of two adjacent x_2 values in each case, before calculating $1/x_2$. The fifth column shows the area of the resulting segment of the plot of $1/x_2$ versus x_1 (note that the width of each segment is a constant $x_1 = 0.1$). The time is shown in the sixth column as the cumulative sum of the values in the fifth (that is, the integral). It can be seen to be in approximate agreement with Figure 14.13(a). For example, the time when x_1 first crosses zero can be found from the table as $t = 1.48$ s, and the time to the first peak is approximately the final value in the table, $t = 2.21$ s. The agreement with Figure 14.13(a) would be better for a smaller x_1 increment.

Table 14.4 Timescaling for the PPT of Figure 14.13(b).

x_1	x_2	Mean of each pair of x_2 values	Magnitude of $1/x_2$	Area $= 1/x_2$ $\times 0.1$	Time
1.0	0	–	–	–	–
0.9	−0.54	−0.27	3.7	0.37	0.37
0.8	−0.68	−0.61	1.64	0.16	0.53
0.7	−0.79	−0.74	1.35	0.14	0.67
0.6	−0.84	−0.82	1.22	0.12	0.79
0.5	−0.86	−0.85	1.18	0.12	0.91
0.4	−0.90	−0.88	1.14	0.11	1.02
0.3	−0.90	−0.90	1.11	0.11	1.13
0.2	−0.86	−0.88	1.14	0.11	1.24
0.1	−0.84	−0.85	1.18	0.12	1.36
0	−0.79	−0.82	1.22	0.12	1.48
−0.1	−0.72	−0.75	1.33	0.13	1.61
−0.2	−0.62	−0.67	1.49	0.15	1.76
−0.3	−0.48	−0.55	1.82	0.18	1.94
−0.4	−0.25	−0.37	2.70	0.27	2.21

14.5.6   Classification of PPTs – critical points and Lyapunov's first method

This section ends with some general guidelines on typical forms of PPTs for various configurations of the linear part of the system. The presentation of such general guidelines is only feasible because of the limitation to second-order systems. These standard forms give an immediate guide to the general form of the PPT for the linear part of any second-order system, and sometimes allow an estimate to be made of the form of the PPT for systems with quite complicated nonlinear behaviour.

In general, a *critical point* (or equilibrium point) of a system is defined as a point in the state space at which all derivatives are zero. It is therefore a point at which the system might theoretically *come to rest*. However, although this will be possible in a carefully constructed simulation, it may not always be possible in the real world. If attention is restricted to two dimensions for simplicity, it is found that the PPT of any system either ends at, begins at, or circles around a critical point. For systems with no forcing input, it is the case that $\dot{x} = Ax = 0$ at the critical points (this is true for higher order systems too, of course). In PPTs, the relationship $\dot{x}_1 = x_2$ has always been used, with x_2 plotted vertically, so it can be concluded that the critical points in PPTs will be on the horizontal axis. Their actual location depends upon the system dynamics and any initial conditions which may be applied.

In the neighbourhood of a critical point, and with the origin of the phase plane moved onto the critical point (so that the state vector contains deviations from that point), the representation $\dot{x} = Ax$ is a linearized model of the nonlinear system. It should therefore be a good approximation to the true behaviour of the system, so long as operation does not move far from the critical point. The dynamic behaviour is determined by the eigenvalues of the A matrix, as shown in Table 14.5 (the curves are qualitative indications only). There are various special cases involving repeated and zero eigenvalues (see, for example, Leigh (1983)).

In the case of the stable and unstable nodes, the two straight lines through the origin of each plot are the *eigenvectors* of the linear part of the system. The one which is an asymptote for the other PPTs is called the *slow eigenvector* and corresponds to the eigenvalue having the smaller magnitude. The other is the *fast eigenvector* because it corresponds to the larger eigenvalue, and therefore the solutions move faster in this direction (because $e^{-\lambda_1 t}$ decays faster than $e^{-\lambda_2 t}$ if λ_1 is greater than λ_2).

A linear second-order system has only one critical point (for instance, Example 14.2 and Figure 14.15 are for a system whose only critical point is a stable focus). On the other hand, a nonlinear system can have multiple critical points. A simple example is the pendulum considered in Example 2.10 and Figure 2.23 (see Section 2.4.5). This has two kinds of critical point at which the system can come to rest. One is the obvious one in which the mass hangs vertically below the pivot and intuitively this also is a stable focus. These will occur at angles of $\theta = 0, \pm 2\pi, \pm 4\pi, \ldots$ radians.

The other kind of critical point is the unstable equilibrium condition when the pendulum is balanced with the mass vertically *above* the pivot. Consideration of Table 14.5, with a little imagination of the various possibilities of the behaviour of position and velocity (x_1 and x_2 respectively) as the pendulum oscillations approach this second critical point, will reveal that it is a saddle point. These will occur at $\pm \pi, \pm 3\pi, \ldots$ radians.

Information like this can be used to sketch a rough PPT for the system. This is actually an application of *Lyapunov's first method*. The method is applicable to systems of any order. It derives the local stability of a nonlinear system in the region of each of its critical points, by determining the eigenvalues of the linearized model about each critical point.

In general, if the eigenvalues of the linearized model about a critical point all have negative real parts, the critical point will be stable and, for initial conditions sufficiently close to the critical point, the state vector of the nonlinear system will converge to the point as time passes. If any eigenvalue has a positive real part, the critical point will be unstable, and the nonlinear system state vector will diverge from it. However, if the linearized model around a critical point has a zero eigenvalue, even if all the other eigenvalues have negative real parts, the linearized model allows no conclusion to be drawn about the stability of the nonlinear system.

Table 14.5 Critical point classifications.

Name	Phase portrait	Eigenvalues	Comments
Stable node		Both real Both negative	PPTs approach monotonically
Unstable node		Both real Both positive	PPTs leave monotonically
Saddle point		Both real One negative, one positive	PPTs approach and leave monotonically
Stable focus		Complex Real parts negative	PPTs approach in an oscillatory manner
Unstable focus		Complex Real parts negative	PPTs leave in an oscillatory manner
Centre (vortex)		Imaginary	PPTs are ellipses

For second-order systems, the results are as given in Table 14.5. Although the predicted behaviour will change as operation moves away from a critical point (that is, as the linearized model becomes inaccurate), it should follow smooth curves. Therefore if critical points are reasonably close together, the behaviour of a nonlinear system can be envisaged by connecting up the known phase portraits for the linearized system at each critical point.

For the pendulum, the stable foci and saddle points occur at the locations determined above. Application of the method described above gives a resulting phase portrait of the type indicated in Figure 14.27 (the figure is not intended to be accurate). The figure repeats indefinitely along the real axis, and such behaviour is also correct from physical considerations.

For example, if the pendulum begins in the stable condition at $x_1 = \theta = 0$ radian, and is then given a large initial positive velocity ($x_2 = \dot{\theta} =$ some large positive number of rad s^{-1}), then it will follow a trajectory such as that uppermost in Figure 14.27. When the saddle point at 180° is reached (that is, when $x_1 = \pi$ radians – the saddle point on the right of Figure 14.27), the velocity is still sufficiently high to carry the pendulum right past the vertical and into the next section of the PPT (where $x_1 > 180°$, beyond the right-hand edge of Figure 14.27).

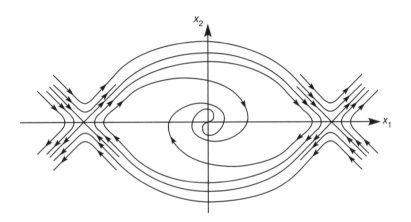

Figure 14.27 Indicative PPTs for a system with many critical points (the pendulum).

14.6 Lyapunov's second (direct) method

This is also a time-domain approach to nonlinear system analysis. Its purpose is to examine the stability of any system whose model can be expressed in a general state-space form. The investigation of stability is one of the most important requirements of any control system design, but particularly so for nonlinear systems where stability is a rather complex topic. In this method, no approximations are made (other than the initial modelling). It is applicable to:

- systems of any order
- linear systems
- nonlinear systems with any number of nonlinearities
- time-varying systems
- multivariable systems
- systems whose models are not entirely numerical – that is, it can be used in principle with unknowns in the state space model, and therefore assist in

controller design by allowing the unknowns to be calculated to give whatever stability characteristics are required.

This is the nearest yet to a general approach to nonlinear system analysis, so it is an extremely powerful technique. However, there are some limitations:

- It is not directly applicable to systems with discontinuous nonlinearities (because a state-space model is required).

- Application of the method involves a search for a mathematical function of the system variables which has certain properties (to be described below). This search can be extremely difficult. Although the method is based on state-space models, there is no 'synthesis' approach for finding these functions; intuition and experience are needed.

14.6.1 Lyapunov stability

Section 14.5.6 outlined Lyapunov's first method, in which the stability of a nonlinear system in the vicinity of a critical point is determined by calculating the eigenvalues of a linear model of the system (linearized about that critical point). The second method allows the determination of stability without having to solve the system equations (that is, without having to find the eigenvalues – hence it is called the *direct* method). This is useful for nonlinear and/or time-varying systems, where solution of the state equations is usually impractical.

The general state-space model $\dot{x} = f(x)$ is used, where the functions in f can be linear or nonlinear. Subject to certain constraints, f can also be time-varying (Slotine and Li, 1991).

As with Lyapunov's first method, the critical point under investigation will be considered to be located at the origin of the state space (that is, where $x_1 = x_2 = \ldots = x_n = 0$). Analysis of any critical point to which this does not apply can be carried out after firstly shifting the axes of the state space so that the origin corresponds with the critical point. This is done by substituting $x - x_c$ for x in the state-space model, where x_c contains the values of the state variables at the critical point.

Firstly, the main kinds of stability possible for nonlinear systems are defined. The definitions are illustrated in the phase plane in Figure 14.28, which obviously

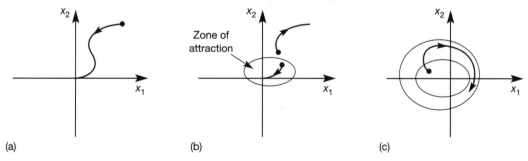

Figure 14.28 Types of nonlinear stability. (a) Global asymptotic stability; (b) local asymptotic stability; (c) Lyapunov stability.

applies to second-order systems. For third-order systems, the plane areas in Figure 14.28(b) and (c) become three-dimensional surfaces, while for higher orders they become abstract hypersurfaces. However, the ideas and the subsequent mathematics apply to systems of any order.

Globally asymptotically stable (or stable 'in the large' – Figure 14.28(a)) describes a system whose state vector decays to the origin and remains there as time passes, in response to any initial condition (this is the kind of stability which applies to linear systems, for example Figure 14.13). It is worth noting that this definition can only be fulfilled for systems having a single critical point at which the system can come to rest.

Locally asymptotically stable (or stable 'in the small' – Figure 14.28(b)) describes a system whose state vector decays to the origin and remains there as time passes, so long as the initial condition vector was within a certain 'distance' of the origin (this is not intended to imply a circle, merely to give a physical 'feel' for the behaviour). In other words, there is some region around the origin of the state space in which initial condition vectors will lead to asymptotically stable responses. This is called the *zone of attraction* to the origin. If the system is released from initial conditions outside this region, the response may be unstable or, if it is stable, the state vector may never again approach the origin. The simple pendulum is an example of this kind of system. In practical systems, asymptotic stability is often the requirement. If the system is not globally asymptotically stable, then the zone of attraction may have to be made sufficiently large that no disturbance can fall outside it.

Lyapunov stable (Figure 14.28(c)) describes a system in which the state vector will remain in some region around the origin as time passes (but will not necessarily ever approach the origin), so long as the initial condition vector was within a certain 'distance' of the origin. In other words, there is some region around the origin of the state space in which initial condition vectors will lead to responses which cause the state vector to remain within some other region around the origin. Limit cycles are a good example of this kind of behaviour – a limit-cycling system could not be called 'stable' in the same sense as linear system stability is defined, but nor is it unstable in the sense that the magnitude of the state vector can increase without limit as time passes. Rather, the oscillation is contained within some region of the state space, and the system is said to be 'stable in the sense of Lyapunov'. Systems with this kind of stability are not usually designed deliberately (except for electronic oscillators), but it is useful for describing system behaviour, and is also used in formal mathematical definitions of asymptotic stability.

14.6.2 Lyapunov functions and their application

If it is desired to investigate the stability of the state-space model of any system, however complex, it would be very helpful if some function of all the system states could be found which could be shown to be always decreasing in magnitude (that is, towards zero) as time passes. The existence of such a function, if it was constructed in a suitable manner, would then imply that eventually the state vector would reach zero, and the system would be asymptotically stable.

As an example, for mechanical or electrical systems, an equation for the total energy stored by the system might be written in terms of the state variables. The energy stored can only be a positive (or zero) quantity. If it can be shown, for some particular system, that the rate of change of energy stored is always negative, then that system must eventually lose all its stored energy and come to rest at a critical point. The fact that the rate of change of stored energy is always negative therefore guarantees asymptotic stability (but not necessarily global asymptotic stability, as will become apparent later).

Example 14.7 *Stability analysis of the simple pendulum*

Consider again the pendulum of Figure 2.23 (Section 2.4.5). The equation of motion for this arrangement has already been derived in Example 2.10 as Equation (2.30), repeated here for convenience:

$$m \frac{d^2\theta}{dt^2} + \frac{mg}{l} \sin \theta = 0$$

This simple model ignores frictional effects, so it is conservative – if the pendulum is given some initial displacement and released, it will oscillate for ever. To make the model more realistic, the friction due to the pivot and air must be included, since this is the term which dissipates the energy and causes the oscillations to die away. The simplest way to do this is to include the resulting viscous damping torque, which might be represented as $T = B \, d\theta/dt$, Nm (rad/s)$^{-1}$, where B is the damping coefficient. Multiplying the original equation of motion (above) by l^2, so that it is expressed in terms of torques, and including this damping torque term, then gives the improved equation of motion:

$$ml^2 \frac{d^2\theta}{dt^2} + B \frac{d\theta}{dt} + mgl \sin \theta = ml^2\ddot{\theta} + B\dot{\theta} + mgl \sin \theta = 0 \qquad (14.21)$$

Now consider the energy in the system. At any instant, this comprises the potential energy and the kinetic energy. The potential energy is given by mgh, where h is the vertical height of the bob above the rest position, namely $h = l - l \cos \theta$. The kinetic energy is $\frac{1}{2}mv^2$, where v is the tangential velocity, namely

$$v = l \frac{d\theta}{dt} = l\dot{\theta}$$

for θ in radians. Calling the total energy $V(\theta)$ therefore gives:

$$V(\theta) = mgl - mgl \cos \theta + \tfrac{1}{2}ml^2\dot{\theta}^2 \qquad (14.22)$$

Differentiating Equation (14.22) with respect to time gives the rate of change of energy:

$$\dot{V}(\theta) = mgl\dot{\theta} \sin \theta + ml^2\dot{\theta}\ddot{\theta} = \dot{\theta}(mgl \sin \theta + ml^2\ddot{\theta})$$

Substituting from Equation (14.21) for the term in parentheses:

$$\dot{V}(\theta) = -B\dot{\theta}^2 \qquad (14.23)$$

Since Equation (14.23) is always negative, this shows that the energy is always decreasing and therefore the system will be asymptotically stable (but it has not yet been shown that the stable point will be at the origin). From a physical viewpoint, it also shows that it is the damping term (B) which is responsible for dissipating the energy. If there were no damping then Equation (14.23) would be zero. This means that the total energy would be conserved and repeatedly converted, without loss, from potential to kinetic energy and back again, according to Equation (14.22). The system would then *not* be asymptotically stable as it would oscillate for ever in a limit cycle. However, it would be Lyapunov stable.

Another point worthy of note is that Equation (14.22) (the V function) cannot be negative, and can only be zero if $\cos \theta = 1$ and $\dot{\theta} = 0$. This could happen at angles of $\theta = \pm n2\pi$, where $n = 0, 1, 2, 3 \ldots$, that is, with the bob stationary directly below the pivot. This corresponds to the only *stable* critical points of the system as discussed in Section 14.5.6. If the possible values of θ are limited to the range $-\pi < \theta < \pi$, then the single critical point at $\theta = 0$ is found. However, the system will not necessarily settle at this particular critical point unless the initial value of *velocity* ($\dot{\theta}$) is also constrained as discussed at the end of Section 14.5.6. This shows that in this case, analysis of stability in this way applies in the vicinity of critical points; it does not necessarily tell us about stability in the large. The approach will now be generalized, and this system will be considered further in Example 14.9.

Lyapunov's second method is a generalization of the approach of Example 14.7 to any system expressible in state-space form, even if energy is not a meaningful concept for the system under investigation.

Let R represent some region of the state space including the origin (a critical point), and let $V(x)$ be some scalar function of the system states (in Example 14.7, choosing $x_1 = \theta$ and $x_2 = \dot{\theta}$ would make $V(\theta)$ such a function). The function $V(x)$ is said to be *positive definite* in the region R, so long as it has continuous partial derivatives and is always greater than zero for any state vector in R, except that $V(0) = 0$. If the function $V(x)$ is never less than zero, but can be zero for some *non-zero* value of x in the region R, then $V(x)$ is said to be *positive semi-definite*. *Negative definite* and *negative semi-definite* functions are defined in the same way, but exchanging 'greater than' and 'less than' as one would expect.

If $V(x)$ is positive definite in the region R, and if $\dot{V}(x)$ is negative semi-definite in the region R, then $V(x)$ is called a *Lyapunov function*.

If a Lyapunov function exists for a system, it is Lyapunov stable in the region R. Furthermore, if such a function can be found for which $\dot{V}(x)$ is negative definite, rather than negative semi-definite, then the system is asymptotically stable in the region R. If $\dot{V}(x)$ is positive definite in the region R, the origin is an unstable critical point. This latter statement allows attempts at proving instability if stability cannot be proved.

Some of these ideas are illustrated in Figure 14.29 for a second-order system. Contours of a Lyapunov function V are drawn for three different values of V. If $\dot{V}$ is

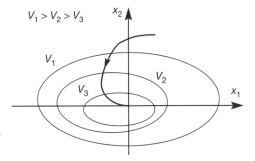

Figure 14.29 Illustration of a Lyapunov function.

always negative except at the origin (negative definite), any state trajectory must always move from an outer curve to an inner one, ending up at the origin. If $\dot{V}$ is negative except that it can become zero somewhere other than at the origin (negative semi-definite), the state trajectory would not reach the origin, but would remain some distance from it, possibly in a limit cycle, as the dissipation of 'energy' in the system would then have ceased (analogous to the behaviour of the pendulum in Example 14.7 with zero damping).

Example 14.8 **Stability investigation and controller design for a different system**

Consider the stability of the system shown in Figure 14.30(a), ignoring the external inputs u_1 and u_2 for the present. This system is described by the state-space model:

$$\dot{x}_1 = -4x_1 - x_1 x_2^2 + x_3$$
$$\dot{x}_2 = x_1 - x_2^2 \tag{14.24}$$
$$\dot{x}_3 = x_1$$

If a Lyapunov function for this system can be found, its stability is guaranteed as described above. In the search for such a function, first choose a $V(x)$ which is positive definite, and then test it to see whether $\dot{V}(x)$ is negative semi-definite (or preferably negative definite).

Let us try $V(x) = ax_1^2 + bx_2^2 + cx_3^2$ since this is guaranteed to be positive definite so long as a, b and c are all positive constants. Note that every state variable must appear in $V()$ if it is to have a chance of being positive definite. For example, if x_2 did not appear, then $V()$ would be zero everywhere along the x_2 axis (where $x_1 = x_3 = 0$), making it positive semi-definite only. This $V(x)$ can be differentiated with respect to time using the rule

$$\frac{df(x)}{dt} = \frac{df(x)}{dx} \frac{dx}{dt}$$

which gives:

$$\dot{V}(x) = 2ax_1\dot{x}_1 + 2bx_2\dot{x}_2 + 2cx_3\dot{x}_3 \tag{14.25}$$

To discover whether or not $V(x)$ is a Lyapunov function for the system of Equation (14.24), it is necessary to substitute Equations (14.24) into Equation (14.25) and test for negative definiteness. The substitution yields:

$$\dot{V}(x) = -8ax_1^2 - 2ax_1^2x_2^2 + 2(a+c)x_1x_3 + 2bx_1x_2 - 2bx_2^3 \tag{14.26}$$

If Equation (14.26) can be shown to be negative definite, the system has been proved to be asymptotically stable (or Lyapunov stable, if Equation (14.26) is only negative semi-definite). Unfortunately, neither is the case here. The first two terms in Equation (14.26) are both negative definite. However, the last term can be seen to be negative definite only so long as x_2 is constrained to be positive. Following from this, x_1 would have to be negative in the penultimate term, and x_3 positive in the third term. This set of conditions cannot be guaranteed and $V(x)$ is therefore not a Lyapunov function for this system.

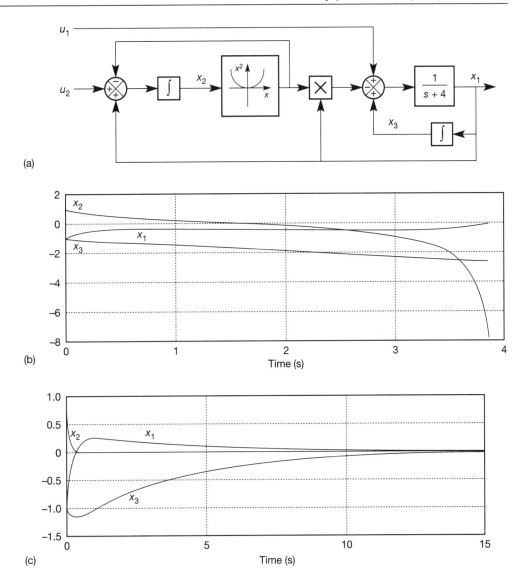

Figure 14.30 A nonlinear system, with open-loop response and response with SVF. (a) The open-loop system; (b) the open-loop response; (c) the response with state variable feedback.

In fact, these restrictions (that is, $x_1 < 0$, $x_2 > 0$ and $x_3 < 0$) show that the origin of the state space is precisely at one corner of the region in which Equation (14.26) would be negative definite (that is, the cubic region of the state space defined by $x_1 < 0$, $x_2 > 0$ and $x_3 < 0$) and therefore the real system could not be stable, since an arbitrarily small disturbance on x at the origin could take it out of the stability region.

It is most important to note that *the system has not been proved to be unstable*; we have only failed to prove that it is stable.

In order to avoid this kind of conclusion and arrive at a V function which *is* a Lyapunov function for the system, it is necessary to be able to remove all such products as $x_1 x_2$ and $x_1 x_3$ from Equation (14.26), since these are always indefinite. It can be seen that no possible choice of a, b and c (all of which must be positive, remember) can achieve this here. A plot of the system response for an arbitrary initial condition vector $x = [-1 \quad 1 \quad -1]^T$ is given in Figure 14.30(b), and the predicted instability does occur, especially once x_1 becomes positive (slightly beyond the RHS of the plot), following which the response of x_2 rapidly becomes vertical. The simulation is to be found in the MATLAB m-file *fig14_30.m* on the accompanying disk.

A different choice of V function might yield a better result and, in general, if one such as that tried above does not work, it might next be wise to try including functions such as $2ex_1x_2$ in the V function. This means that a and b (for example) are no longer constrained to be positive, but e must be chosen such that the overall V function remains positive definite (a test for which is given in Section 14.6.3 below). This modification would add a term $2ex_1\dot{x}_2 + 2e\dot{x}_1x_2$ to the RHS of Equation (14.25). Substitution of the state equations then often gives the possibility of choosing e so that Equation (14.26) is negative definite. In the present case, this is not very successful either, so the system will be *assumed* to be unstable for practical purposes (note again that this has not actually been *proved*), and an investigation will be carried out to see how the Lyapunov approach can assist in the design of a stabilizing controller.

The state-space model is cast in the usual way, so that external inputs can be applied as shown in Figure 14.30(a). If state variable feedback is applied from the states x_1 and x_2 via a feedback matrix $-K$ (see Section 5.4), this implies that these system inputs become:

$$\begin{bmatrix} u_1 \\ u_2 \end{bmatrix} = -Kx = -\begin{bmatrix} k_{11} & k_{12} & k_{13} \\ k_{21} & k_{22} & k_{23} \end{bmatrix} x = \begin{bmatrix} -k_{11}x_1 - k_{12}x_2 - k_{13}x_3 \\ -k_{21}x_1 - k_{22}x_2 - k_{23}x_3 \end{bmatrix}$$

making the system equations:

$$\dot{x}_1 = -(4 + k_{11})x_1 - x_1x_2^2 - k_{12}x_2 + (1 - k_{13})x_3$$

$$\dot{x}_2 = (1 - k_{21})x_1 - x_2^2 - k_{22}x_2 - k_{23}x_3$$

$$\dot{x}_3 = x_1$$

Substituting these into Equation (14.25) yields the new result:

$$\dot{V}(x) = -2a(4 + k_{11})x_1^2 - 2ax_1^2x_2^2 - (2ak_{12} - 2b(1 - k_{21}))x_1x_2$$
$$+ (2a(1 - k_{13}) + 2c)x_1x_3 - 2bx_2^3 - 2bk_{22}x_2^2 - 2bk_{23}x_2 \quad (14.27)$$

Although this looks complex at first sight, it is better than the previous result, as the feedback gains in K can now be chosen to remove the unwanted product terms in Equation (14.27). In some cases an infinite number of choices exists, each of which yields a different performance. Just one choice will be used here to illustrate the method.

Any negative definiteness of Equation (14.27) will not be spoiled by choosing $k_{11} = k_{12} = k_{23} = 0$. It may appear that k_{22} could also be set to zero, but notice that the term in x_2^3 will be positive if x_2 is negative. Therefore k_{22} must be retained as a 'tuning' parameter so that the overall set of terms involving x_2 can be made negative definite (if possible). With these choices, Equation (14.27) becomes:

$$\dot{V}(x) = -8ax_1^2 - 2ax_1^2x_2^2 + (2b(1 - k_{21}))x_1x_2$$
$$+ (2a(1 - k_{13}) + 2c)x_1x_3 - 2bx_2^3 - 2bk_{22}x_2^2 \quad (14.28)$$

The first two terms in Equation (14.28) are negative definite. The awkward products can now be removed from the third and fourth terms by choosing $k_{21} = 1$ and $k_{13} = 1 + c/a$. This just leaves the last two terms, which can be written $-2bx_2^2(x_2 + k_{22})$, and will therefore be negative definite so long as $x_2 > -k_{22}$. The larger we make k_{22}, the greater will be the stability margins of the plant, but the harder we shall drive the input. Choosing $a = b = c = 1$, and $k_{22} = 10$ gives:

$$K = \begin{bmatrix} 0 & 0 & 2 \\ 1 & 10 & 0 \end{bmatrix} \text{ and}$$

$$\dot{V}(x) = -8x_1^2 - 2x_1^2x_2^2 - 2x_2^3 - 20x_2^2 \tag{14.29}$$

Overall, Equation (14.29) is negative semi-definite (because it does not contain x_3 and will therefore be zero anywhere on the x_3 axis) for all x_1 and for $x_2 > -10$. A plot of the controlled system's response appears in Figure 14.30(c) for the same initial conditions as Figure 14.30(b). Different choices for K would yield better performance, but the stability guaranteed by the Lyapunov approach is clearly visible and is actually asymptotic, so the Lyapunov analysis was pessimistic.

Example 14.8 showed that there can be many Lyapunov functions for a system. It is entirely possible that there is a different Lyapunov function for the system of Example 14.8 which would prove a much greater stability region, or even asymptotic stability; so the analysis may be rather conservative. This is the major problem with the approach – how is the search for Lyapunov functions to be made and, if one is found, is it the best? Unfortunately, there is no easy answer, since every nonlinear system can have unique properties. If the system is an electrical/mechanical one, in which the total energy can be calculated, then choosing $V(x)$ as the energy in the system (as in Example 14.7) will often work. For more general systems, or for a 'second opinion' on an energy-type system, the form of function used in Example 14.8 is also a good candidate (as is the modification to it which was suggested there). Leigh (1983) gives some other suggestions, but often it depends upon intuition. Note again that the failure to find a Lyapunov function does not imply instability; it may just be that the search has not been sufficiently thorough.

To find the region of stability, it is necessary to discover the region of the state space over which the chosen Lyapunov function is valid. In Example 14.8 this was found as part of the exercise. However, in Example 14.7 it was not done at all, merely noting that stability depended upon the initial velocity. With the present level of knowledge, that investigation can be repeated.

Example 14.9 *The simple pendulum of Example 14.7 revisited*

In Example 14.7, it was shown that one of the requirements for $V(\theta)$ in Equation (14.22) to be positive definite was that θ be restricted to the range $-\pi \leq \theta \leq \pi$ (except that this terminology was not used at that time). It was also noted that the stability behaviour depended upon the initial value of the velocity $\dot{\theta}$.

This section has shown that, in order to prove asymptotic stability, $V(\theta)$ must be positive definite, *and* $\dot{V}(\theta)$ negative definite. This allows limits to be placed on the range of values of $\dot{\theta}$ which can appear in Equation (14.22) if the system is to be asymptotically stable. It is necessary to find the 'contour' in the state space corresponding to the maximum allowable value of $V(\theta)$ (similar to the contours shown in Figure 14.29). Any trajectory remaining within the contour will then yield an asymptotically stable solution.

Using the definitions $x_1 = \theta$ and $x_2 = \dot{x}_1 = \dot{\theta}$, Equation (14.22) becomes:

$$V(\boldsymbol{x}) = mgl - mgl \cos x_1 + \tfrac{1}{2}ml^2 x_2^2 \tag{14.30}$$

It is known from Example 14.7 that the limits on x_1 are $\pm\pi$. It is also known that $x_2 = 0$ at the critical points represented by these limits (the saddle points shown in Figure 14.27). Substitution of either limit into Equation (14.30) yields $V(\boldsymbol{x}) = 2mgl$. For a contour of constant $V(\boldsymbol{x})$, substitution of this value into Equation (14.30) gives:

$$x_2 = \sqrt{\frac{2g(1 + \cos x_1)}{l}} = 2 \cdot \sqrt{\frac{g}{l}} \cdot \cos\frac{\theta}{2}$$

For any specific value of l, this curve can be plotted in the state space (the phase plane in this case). The resulting contour is shown in Figure 14.31.

If any initial conditions or disturbances cause responses which remain within the closed curves, such responses will settle asymptotically to the origin of the particular critical point concerned (that is, to $x_1 = 0$, $\pm2\pi$, $\pm4\pi$, and so on as appropriate), as was predicted from physical arguments earlier. What the analysis does not show is that the system is also stable for *any* initial conditions – but will not necessarily settle at the 'nearest' critical point if released outside the closed curves of Figure 14.31 (the critical point which will finally capture the trajectory depends upon the initial velocity and the point of release).

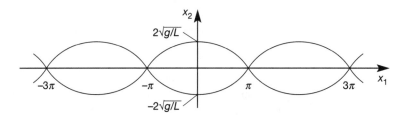

Figure 14.31 Regions of asymptotic stability for the simple pendulum.

14.6.3 The Lyapunov approach for linear systems and the Lyapunov equation

Lyapunov's direct method is also applicable to linear systems. For linear systems, a V function can always be chosen which has the following quadratic form. It can then be tested to see whether it is a Lyapunov function (although written for a different context, Example 12.2 gives a brief introduction to matrix quadratic forms):

$$V(\boldsymbol{x}) = \boldsymbol{x}^{\mathrm{T}} \boldsymbol{Q} \boldsymbol{x} \tag{14.31}$$

where $\boldsymbol{Q}$ is some chosen symmetric and *positive definite* matrix ('definiteness' with regard to matrices is covered in Section A1.4).

Quadratic Lyapunov functions can be expanded to test for definiteness. For example, for a second order system (and remembering that $\boldsymbol{Q}$ is symmetric), the Lyapunov function could be written:

$$V(\boldsymbol{x}) = \begin{bmatrix} x_1 & x_2 \end{bmatrix} \begin{bmatrix} q_{11} & q_{12} \\ q_{12} & q_{22} \end{bmatrix} \begin{bmatrix} x_1 \\ x_2 \end{bmatrix} = q_{11}x_1^2 + 2q_{12}x_1x_2 + q_{22}x_2^2$$

Here, q_{11}, q_{22} and q_{12} replace a, b and e respectively, as used in Example 14.8. The definiteness of $\boldsymbol{Q}$ can be tested by the methods in Appendix 1 to give the definiteness properties of $\boldsymbol{V}$.

Having chosen a symmetric and positive definite $\boldsymbol{Q}$, the derivative of Equation 14.31 is tested for negative definiteness, in the same manner as was done for scalar V functions. Differentiating Equation (14.31) with respect to time, and then substituting the state equation

$$\dot{\boldsymbol{x}} = \boldsymbol{A}\boldsymbol{x}$$

gives

$$\dot{V}(\boldsymbol{x}) = \boldsymbol{x}^{\mathrm{T}}\boldsymbol{Q}\dot{\boldsymbol{x}} + \dot{\boldsymbol{x}}^{\mathrm{T}}\boldsymbol{Q}\boldsymbol{x} = \boldsymbol{x}^{\mathrm{T}}\boldsymbol{Q}\boldsymbol{A}\boldsymbol{x} + \boldsymbol{x}^{\mathrm{T}}\boldsymbol{A}^{\mathrm{T}}\boldsymbol{Q}\boldsymbol{x} = \boldsymbol{x}^{\mathrm{T}}\boldsymbol{P}\boldsymbol{x}$$

where

$$\boldsymbol{P} = \boldsymbol{Q}\boldsymbol{A} + \boldsymbol{A}^{\mathrm{T}}\boldsymbol{Q} \tag{14.32}$$

Equation (14.32) is called a Lyapunov equation. If $\boldsymbol{P}$ turns out to be negative definite with our choice of $\boldsymbol{Q}$, the system will be asymptotically stable. Of course, for a linear system, this also means that it will be *globally* asymptotically stable, as there will be only one critical point.

Unfortunately, approaching the problem this way is likely to provide little useful information, because any randomly chosen $\boldsymbol{Q}$ matrix can fail to give the correct properties in $\boldsymbol{P}$, even if the system is stable.

The properties of the Lyapunov equation (Equation (14.32)), and theorems associated with it (Atherton, 1982), imply that it is better to begin with a negative definite and symmetric $\boldsymbol{P}$ matrix, and effectively solve Equation (14.32) for the corresponding $\boldsymbol{Q}$. If this turns out to be positive definite, the system is asymptotically stable. Moreover, this always works for a stable system, so the test is conclusive if performed this way round, and $\boldsymbol{P}$ can be chosen simply as $\boldsymbol{P} = -\boldsymbol{I}$. However, the solution of Equation (14.32) is often not straightforward. Several nonlinear control texts, for example Atherton (1982) and Slotine and Li (1991), give more detail; but beware some notational changes.

14.7 The describing function method

This is the first of three methods to be studied which work in the *frequency domain*. The describing function (DF) approach allows the use of polar plots in the analysis of nonlinear systems. Again, the technique aims to investigate the stability conditions for the system under investigation, particularly the prediction of the existence and frequency of limit cycles.

As with all the methods for nonlinear systems, this approach has its advantages and disadvantages, which determine the type of system to which it is best applied.

To a limited extent, this approach and the phase plane approach are complementary, some of the weaknesses of one being covered by the strengths of the other. The DF approach:

- is applicable to systems whose linear part is of any order (as opposed to the phase plane method, which is limited to second order). In fact, the DF approach works better for higher order systems than lower in general
- works for continuous and discontinuous nonlinearities
- allows for the use of simple polar plots, and the familiar interpretation of frequency response information.

Approximations and limitations are as follows. There are methods which allow most of these to be relaxed (Atherton, 1982), but these lead to much more complicated procedures than the one to be described here. In addition, the last-mentioned approximation is fundamental, and is often the major limitation of the DF approach:

- Normally, only a single nonlinearity is considered.
- The nonlinear element cannot be time-varying.
- The input–output characteristic of the nonlinearity should be symmetrical about the origin (as are all those shown in Figures 14.5 to 14.8 and 14.10). It is also preferable that they exhibit odd symmetry (which would exclude Figure 14.5(c)), and only such cases will be considered here.
- The DF approach works best for systems where the nonlinearity is not too severe, because it is an extension of linear techniques (as opposed to the phase plane approach which is usable for grossly nonlinear systems).
- The approach assumes that the linear part of the plant acts as a very good low-pass filter (which is why the method tends to work better for systems with higher-order linear parts).

14.7.1 The describing function

Strictly, this approach is the *sinusoidal input describing function*, which distinguishes it from the more complicated procedures hinted at in the previous section. However, since this is the only approach to be covered, the simple abbreviation 'DF' will suffice.

Figure 14.32 shows the type of system to be considered. $G(j\omega)$ represents the

Figure 14.32 System configuration for basic describing function analysis.

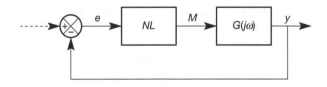

transfer function of the linear part of the plant, and *NL* the nonlinearity. Note that the application of an external forcing input is not allowed (in any case, the DF will not reveal much about the transient response of the system, only its stability properties).

With the restrictions shortly to be defined, it makes no difference whether the NL element is in the position shown, or at the output side of $G(j\omega)$, or in the feedback path. In addition, multiple NL elements may be able to be included under certain conditions (given below after Example 14.10).

The problem in analysing the behaviour of Figure 14.32 from a frequency domain standpoint is that the approaches of Bode and Nyquist do not work for the nonlinear element. However, under certain conditions, a frequency-response type of model can be generated to represent the NL element, and then a modified version of the Nyquist criterion used in linear control can be applied. The fundamental approximation which allows this analysis is now given. For any system not meeting these conditions, the method is inappropriate.

It is assumed that if the system can exhibit a limit cycle, then the signal at e during the limit cycle will be *sinusoidal*. Since the output from the NL element (M) will almost certainly *not* be sinusoidal, this seems unlikely at first sight. However, the signal at M will be periodic (in a limit cycle) and can therefore be represented as a Fourier series of sinusoidal components, beginning with a non-oscillatory (bias) term, and adding components of ever increasing frequency until sufficient accuracy is reached. For example,

$$M(t) = \frac{M_0}{2} + M_1 \sin(\omega t + \phi_1) + M_2 \sin(2\omega t + \phi_2) + \cdots \quad (14.33)$$

where M_0 is the amplitude of the bias component, M_1 of the fundamental, M_2 of the second harmonic and so on, and ϕ_1, ϕ_2, and so on are the phase shifts at the various frequencies.

The assumption that e is sinusoidal is now justified as follows. For signals in real systems, it is the case that the amplitude of the Fourier components (M_1, M_2 and so on) generally decreases as frequency increases. Furthermore, the linear part of any real plant $G(j\omega)$ will have low-pass filter characteristics to some degree (these both follow from the fact that no real system can respond to infinitely high frequencies). The basic assumption is that these two conditions combine so that, however irregular the waveform at M, only the (sinusoidal) fundamental component of it will appear back at e having traversed the control loop. To ensure that there is no bias component, attention is restricted to nonlinear elements whose characteristics exhibit odd symmetry, as mentioned earlier.

The describing function (DF) is then defined as the ratio of the fundamental component of the NL element output (M) to the sinusoidal NL element input signal (e); in other words, it is the transfer function (strictly, the frequency response function) of the NL element, considering only a sinusoidal input and the fundamental component of the output. This will become clearer below when some DFs are derived.

This transfer function analogy may now be used as follows, but the analogy must not be pushed too far. Representing the NL element in Figure 14.32 by a describing function $N(j\omega)$, the closed loop frequency response function can be seen to be

$$\frac{G(j\omega)N(j\omega)}{1 + G(j\omega)N(j\omega)}$$

A stability rule derived from the characteristic equation can now be used, in the same way as is done in linear analysis:

$$1 + G(j\omega)N(j\omega) = 0 \quad \text{so} \quad \frac{1}{G(j\omega)} = -N(j\omega)$$

For stability, the frequency response of

$$\frac{1}{G(j\omega)}$$

(that is, the inverse Nyquist plot of the linear part of the system) must encircle the points given by $-N(j\omega)$. The $(-1, 0)$ point of linear stability analysis is therefore replaced by the (negated) frequency response plot of the NL element, namely $-N(j\omega)$, which normally also turns out to be a function of signal *amplitude*.

14.7.2 *Evaluation of describing functions*

This is rather tedious, and can usually be avoided by looking up DFs in tables such as Table 14.6 (to be discussed later) or the far more comprehensive tables in Atherton (1982). Here, the method of derivation and a couple of examples are given.

If the NL element input signal is given by $e(t) = E_m \sin \omega t$, whose period (T) is defined as $T = 2\pi/\omega$, then the Fourier series expansion of the NL element's output waveform is given by (see, for example, Balmer (1991)):

$$M(t) = \frac{A_0}{2} + \sum_{n=1}^{\infty} (A_n \cos n\omega t + B_n \sin n\omega t)$$

where

$$A_n = \frac{2}{T} \int_{t_0}^{t_0+T} M(t) \cos n\omega t.dt \qquad (14.34)$$

and

$$B_n = \frac{2}{T} \int_{t_0}^{t_0+T} M(t) \sin n\omega t.dt$$

Note that Equations (14.33) and (14.34) are identical by letting $M_n = \sqrt{(A_n^2 + B_n^2)}$ and

$$\phi_n = \tan^{-1} \frac{A_n}{B_n}$$

Restricting attention to NL elements whose input–output characteristics have odd symmetry about the origin, there will be no bias term in Equation (14.34) $(A_0 = 0)$. This is because any positive bias component generated by the NL element in response to the positive half-cycle of the input will be precisely cancelled by that generated by the negative half-cycle.

Taking into account the additional assumption that all harmonics higher than the fundamental are negligible, the expression then becomes:

$$M(t) \approx A_1 \cos \omega t + B_1 \sin \omega t = M_1 \sin(\omega t + \phi_1)$$

where

(14.35)

$$M_1 = \sqrt{A_1^2 + B_1^2} \quad \text{and} \quad \phi_1 = \tan^{-1} \frac{A_1}{B_1}$$

Equation (14.35) will now be used to derive the describing functions of two common NL elements.

Example 14.10 **DF for a saturation element**

Figure 14.33 shows the NL element characteristic, and its input and output waveforms. Notice that the output of the NL element saturates at a time t_s, when the input has reached an amplitude of E_s. The gain of the central portion of the NL characteristic is K, so this generates a saturated output level of KE_s.

Since the output is in phase with the input, there can only be sine terms in Equation (14.35), so it is immediately apparent that $A_1 = \phi_1 = 0$. The only quantity to calculate is therefore B_1, and the appropriate expression from Equation (14.34) is used to evaluate this, namely:

$$B_1 = \frac{2}{T} \int_0^T M(t) \sin \omega t . dt$$

The term to be integrated will always be positive and the symmetry of the resulting waveform means that the integral need only be performed over 1/4 of the period, giving:

$$B_1 = \frac{8}{T} \int_0^{T/4} M(t) \sin \omega t . dt$$

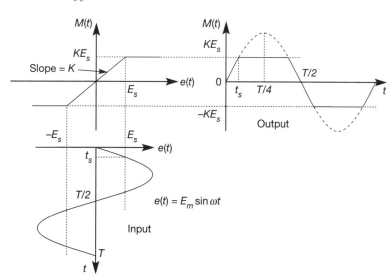

Figure 14.33 Input–output behaviour of a saturation element.

From Figure 14.33, $M(t) = KE_m \sin \omega t \ (0 \le t \le t_s)$, and $M(t) = KE_s \ (t_s \le t \le T/4)$.
Therefore,

$$B_1 = \frac{8}{T} \left[\int_0^{t_s} KE_m \sin^2 \omega t.dt + \int_{t_s}^{T/4} KE_s \sin \omega t.dt \right]$$

$$= \frac{8}{T} \left[\int_0^{t_s} KE_m(0.5 - 0.5 \cos 2\omega t).dt + \int_{t_s}^{T/4} KE_s \sin \omega t.dt \right]$$

$$= \frac{8}{T} \left\{ KE_m \left[0.5t - \frac{0.25}{\omega} \sin 2\omega t \right]_0^{t_s} - KE_s \left[\frac{\cos \omega t}{\omega} \right]_{t_s}^{T/4} \right\}$$

After substituting the limits into the results of the integrations, and then making the
substitutions $\omega T/4 = \pi/2$, $\sin 2\omega t = 2 \sin \omega t \cos \omega t$, and $E_m \sin \omega t_s = E_s$, the final result
becomes:

$$B_1 = \frac{2K}{\pi} (E_m \omega t_s + E_s \cos \omega t_s)$$

which, from Equation (14.35) with $A_1 = 0$, is the magnitude M_1 of the fundamental
component of the output signal from the NL element.

The definition of the DF is the ratio between this magnitude and that of the signal
input to the NL element so, for the saturation element, the DF is:

$$N(j\omega) = \frac{B_1}{E_m} = \frac{2K}{\pi} \left(\omega t_s + \frac{E_s}{E_m} \cos \omega t_s \right) \tag{14.36}$$

The DF result of Equation (14.36) has no phase shift component. This tends to be
the case for single-valued nonlinearities. Double-valued and multi-valued NL
elements such as hysteresis and backlash generate phase shifts. Note also that the
term E_s/E_m is the same thing as $\sin(\omega t_s)$. This can lead to different representations
of the DF, such as the frequency-independent ones which will be used later.

If the system contains more than one NL element, one might expect that their
DFs could be combined by multiplication, as in the case of other frequency
response functions. However, this is not generally allowable, because the DFs
describe nonlinear elements and so superposition does not apply (for example, the
output of the first NL element would be non-sinusoidal, thus breaking the rules at
the input of the second NL element, and making its DF invalid). The rules for
multiple NL elements are therefore that the DFs *can* be multiplied together, but
only if the NL elements are separated by low-pass elements, which filter out
everything except the fundamental. If this is not the case, the NL characteristics
must be combined, and the DF evaluated for the combined characteristic (see
Problems 14.8 and 14.9(a)).

For some NL elements, the DF can be plotted against signal amplitude. This
gives an insight into the correlation between the DF and the physical behaviour of
the NL element. Again, various approaches are possible. One approach begins by
normalizing Equation (14.36), so that the DF is evaluated in terms of the quantity
E_m/E_s (that is, the ratio between the peak amplitude of the NL input signal
and the amplitude at which saturation occurs). Making the substitution

$\cos \omega t_s = \sqrt{1 - \sin^2 \omega t_s}$ in Equation (14.36), and then substituting $\sin(\omega t_s) = E_s/E_m$ leads to:

$$N(j\omega) = \frac{2K}{\pi}\left(\omega t_s + \frac{E_s}{E_m}\sqrt{1 - \sin^2 \omega t_s}\right)$$

$$\Rightarrow N\left(\frac{E_m}{E_s}\right) = \frac{2K}{\pi}\left(\sin^{-1}\frac{1}{E_m/E_s} + \frac{\sqrt{1 - \frac{1}{(E_m/E_s)^2}}}{E_m/E_s}\right) \quad (14.37)$$

If $E_m/E_s < 1$, then obviously the peak value of the input signal (E_m) is less than E_s, and will not saturate the NL element (by inspection of Figure 14.33), and its DF (which is just the effective gain, since there is no phase shift) will then simply be K. For $E_m/E_s > 1$, saturation will occur, and Equation (14.37) then applies. Plotting $N(E_m/E_s)$ for a selection of values of E_m/E_s gives the plot of Figure 14.34.

The plot shows that for $E_m/E_s = 1$, the DF (the NL element gain) is K. As the value of E_m increases with respect to E_s, the saturation effect comes into play, and the DF (and hence the output amplitude) is therefore reduced (as shown in Figures 14.33 and 14.34). The higher the ratio E_m/E_s becomes, the greater is the effect of the saturation, and the further the DF (gain) falls. Finally, for very large values of E_m, the saturation will effectively clip off almost all the signal, so that the NL output (and hence its gain, and hence the DF) is approximately zero in comparison.

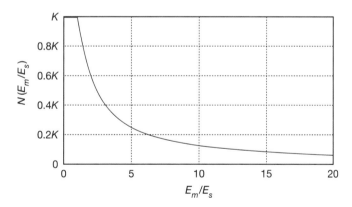

Figure 14.34 DF for a saturation element.

Example 14.11 **DF for an ideal relay element**

Figure 14.35 shows the NL element characteristic with the input and output waveforms. The approach is identical to that of Example 14.10. For this NL element, it can again be seen that the output is in phase with the input, so that there can only be sine terms in Equation (14.35), and therefore $A_1 = \phi_1 = 0$, leaving just B_1 to be calculated using the appropriate expression from Equation (14.34), as in Example 14.10.

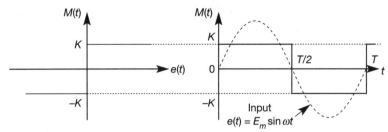

Figure 14.35 Input–output behaviour of an ideal relay element.

The symmetry of the output waveform means that again the integral need only be performed over 1/4 of the period, giving

$$B_1 = \frac{8}{T} \int_0^{T/4} M(t)\sin \omega t.dt$$

From Figure 14.35, it can be seen that $M(t) = K$ ($0 \le t \le T/4$). Therefore, the expression is:

$$B_1 = \frac{8}{T} \int_0^{T/4} K \sin \omega t.dt = \frac{8K}{T} \left[-\frac{\cos \omega t}{\omega} \right]_0^{T/4}$$

After substituting the limits into the result of the integration, and then making the substitutions $\cos (\omega T/4) = \cos(\pi/2) = 0$ and $\omega T = 2\pi$, the final result is that

$$B_1 = \frac{4K}{\pi}$$

As before, this represents the magnitude M_1 of the fundamental component of the output signal from the NL element. To obtain the DF, the ratio between this magnitude and that of the signal input to the NL element is again needed so, for the ideal relay element, a DF is obtained of:

$$N = \frac{4K}{\pi E_\mathrm{m}}$$

In the DF of the ideal relay, the dependency on E_m indicates that the effective gain of the relay falls as E_m (the peak magnitude at the relay input) increases. This makes sense because the relay output is always $\pm K$, so if E_m is very large, the output becomes negligible in comparison, giving a very low gain.

The DFs evaluated in Examples 14.12 and 14.13, together with those of a few other common NL elements, appear in Table 14.6. In that table, an input signal of $e(t) = E_\mathrm{m} \sin \omega t$ is always assumed. Note that all the single-valued NL elements give a DF which has no phase shift component, because the output signal will always be in phase with the input. However, in the case of the hysteresis and backlash elements this is not so. Phase shifts will occur between the input and output of the NL elements (see Example 14.13, below), and so both the A_1 and B_1 terms exist in the Fourier expansion (Equation (14.35)), and are given in the table accordingly. Note especially the division by E_m when combining A_1 and B_1 to calculate N for the hysteresis and backlash elements (this is easily overlooked). See Atherton (1982) for more comprehensive tables.

Table 14.6 Some common describing functions for an input signal $e(t) = E_m \sin \omega t$.

Input–output characteristics	Describing function
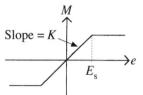	$N = K, \quad E_m \leq E_s$ $N = \dfrac{2K}{\pi}\left(\sin^{-1}\dfrac{E_s}{E_m} + \dfrac{E_s}{E_m}\sqrt{1-\left(\dfrac{E_s}{E_m}\right)^2}\right), \; E_m > E_s$
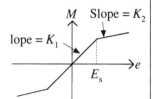	$N = K_1, \quad E_m \leq E_s$ $N = \dfrac{2K_2}{\pi}\left\{\dfrac{\pi}{2} + \left(\dfrac{K_1}{K_2}-1\right)\left[\sin^{-1}\dfrac{E_s}{E_m} + \dfrac{E_s}{E_m}\sqrt{1-\left(\dfrac{E_s}{E_m}\right)^2}\right]\right\}, \; E_m > E_s$
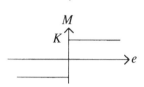	$N = \dfrac{4K}{\pi E_m}$
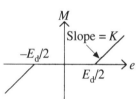	$N = 0, \quad E_m \leq E_d/2$ $N = \dfrac{2K}{\pi}\left[\dfrac{\pi}{2} - \sin^{-1}\dfrac{E_d}{2E_m} - \dfrac{E_d}{2E_m}\sqrt{1-\left(\dfrac{E_d}{2E_m}\right)^2}\right], \; E_m > E_d/2$
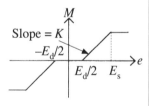	$N = 0, \; E_m \leq E_d/2; \; N =$ as for deadzone, $E_d/2 < E_m \leq E_s; \;$ otherwise $N = \dfrac{2K}{\pi}\left[\sin^{-1}\dfrac{E_s}{E_m} - \sin^{-1}\dfrac{E_d}{2E_m} + \dfrac{E_s}{E_m}\sqrt{1-\left(\dfrac{E_s}{E_m}\right)^2} - \dfrac{E_d}{2E_m}\sqrt{1-\left(\dfrac{E_d}{2E_m}\right)^2}\right]$
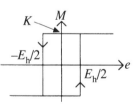	$N = 0, \; E_m \leq E_h/2; \quad$ otherwise $N = \dfrac{\sqrt{A_1^2 + B_1^2}}{E_m}, \quad \phi = \tan^{-1}\dfrac{A_1}{B_1}$ with: $A_1 = -\dfrac{2E_h K}{\pi E_m}, \quad B_1 = \dfrac{4K}{\pi}\sqrt{1-\left(\dfrac{E_h}{2E_m}\right)^2}$
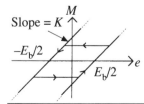	$N = 0, \; E_m \leq E_b/2; \quad$ otherwise N and ϕ as for hysteresis, with: $A_1 = -\dfrac{E_b K}{\pi}\left(2-\dfrac{E_b}{E_m}\right), B_1 = \dfrac{KE_m}{\pi}\left[\dfrac{\pi}{2} + \sin^{-1}\left(1-\dfrac{E_b}{E_m}\right) + \dfrac{E_b}{E_m}\left(1-\dfrac{E_b}{E_m}\right)\sqrt{\dfrac{2E_m}{E_b}-1}\right]$

14.7.3 Stability analysis using the DF

At the end of Section 14.7.1, it was stated that the fundamental approach is to plot the inverse Nyquist plot of the linear part of the plant, and investigate its intersections with the frequency response function of (minus) the DF. This is illustrated by two examples based on the antenna-positioning system, before some more general comments conclude the topic.

Example 14.12 *The antenna positioner with a softly saturating control amplifier*

Figure 14.36(a) shows the system under consideration. The linear part of the plant is the transfer function used before (see, for example, Figure 2.48 in Section 2.7), with a d.c. gain of 10. The DF of the NL element is given in the second row of Table 14.6 and, with $K_1 = 1$, $K_2 = 0.5$ and $E_s = 2$, becomes:

$$N = \frac{1}{2} + \frac{1}{\pi} \left(\sin^{-1} \frac{2}{E_m} + \frac{2}{E_m} \sqrt{1 - \frac{4}{E_m^2}} \right)$$

For values of $E_m < 2$, this equation is invalid (because $E_s = 2$ in Table 14.6), and $-N = -1$. For values of E_m between 2 and ∞, the locus of $-N$ (that is, the DF) is plotted in Figure 14.36(b), together with the inverse Nyquist plot of the linear part of the plant.

If the inverse Nyquist plot passed entirely to the left of the DF (that is, $-N$), the system would be stable. However, this is not the case unless $E_m > 14$. For $E_m < 14$, the system will be unstable. Therefore, even with no input except for an arbitrarily small initial condition, the system will tend to oscillate and the oscillations will increase until E_m

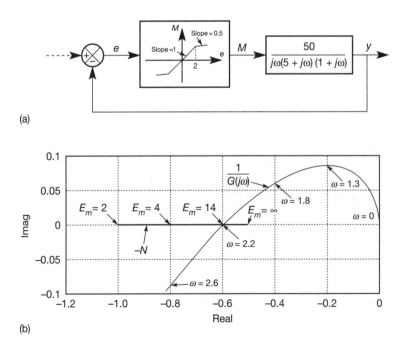

(a)

(b)

Figure 14.36 The DF approach applied to the antenna positioner with saturating amplifier.

exceeds 14 units. At this point, the system becomes stable. In a stable system with no input, any oscillations will tend to die away. The oscillations therefore decrease until E_m again becomes less than 14 units. Instability then returns.

It can thus be seen that the system will settle into a limit cycle with $E_m = \pm 14$ units. It is also evident that this is a stable limit cycle, as the system will return to it if disturbed in either direction. Since the system must operate somewhere on the inverse Nyquist plot, and also on the DF, the limit cycle occurs at the intersection of the two, and its frequency is therefore predicted to be 2.2 rad s^{-1}.

The predicted amplitude and frequency can be confirmed by simulation, using the MATLAB m-file *fig14_36.m* on the accompanying disk, but it is necessary to use a very short sampling period (1 or 2 ms) to represent the discontinuity of the NL element sufficiently accurately. Also the oscillation is quite slow to build up, taking about 90 seconds to settle into the limit cycle from an initial condition of 0.1 unit on the output. The simulation therefore takes a long time to run, and generates a huge quantity of data. An analog computer would do better here, or a variable step numerical integration method in SIMULINK (Appendix 4).

Finally, note that the DF is useful in design too. It can be seen from Figure 14.36 that the system would be unconditionally stable if the inverse Nyquist plot crossed the negative real axis anywhere to the left of the point $(-1, 0)$, since the DF would then be enclosed for any value of E_m. This can be achieved by increasing the apparent gain of the plot by a factor of $1/0.6 = 1.67$. However, since this is an *inverse* Nyquist plot, the gain of the system must be *reduced* by this factor to make it unconditionally stable. Any numerator term lower than $50/1.67 = 30$ would achieve this.

Example 14.13 *The antenna positioner with backlash in the output drive*

Figure 14.37(a) shows the same linear system, but with a small amount of backlash in the output drive gear. The NL element also has an inherent gain factor of 0.3 as shown on its characteristic (representing an additional gearbox reduction ratio).

The relevant part of the inverse Nyquist plot of the linear part of the system appears as

$$\frac{1}{G(j\omega)}$$

in Figure 14.37(b). It is, of course, identical to that in Figure 14.36(b), since the linear part of the system is the same. The general describing function for the backlash element is given in the last row of Table 14.6. With the values $E_b = 0.1$ and $K = 0.3$ from Figure 14.37(a), the Fourier components are

$$A_1 = \frac{0.003}{\pi E_m} - \frac{0.06}{\pi}$$

and

$$B_1 = \frac{0.3 E_m}{\pi} \left[\frac{\pi}{2} + \sin^{-1}\left(1 - \frac{0.1}{E_m}\right) + \frac{0.1}{E_m}\left(1 - \frac{0.1}{E_m}\right)\sqrt{20 E_m - 1} \right]$$

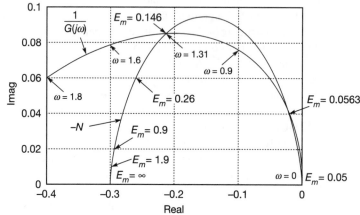

(a)

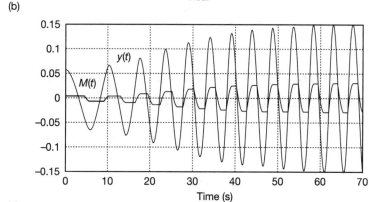

(b)

Figure 14.37 The DF
approach applied to the
antenna positioner with
output backlash.

(c)

E_m still refers to the peak value of the input sinusoid to the NL element, so in the
arrangement of Figure 14.37(a) E_m refers to the peak value of y, the linear system output.
The Fourier components given above must be calculated for several values of E_m and then
used to calculate the magnitude and phase shift of the DF at each value of E_m using

$$N = \frac{\sqrt{A_1^2 + B_1^2}}{E_\mathrm{m}}, \quad \text{and} \quad \phi = \pi + \tan^{-1}\frac{A_1}{B_1} \text{ from Table 14.6}$$

Note the addition of π radians (180°) to the values of ϕ. This is done so as to yield $-N$
directly for plotting. Plotting the resulting values of $-N$ on top of the inverse Nyquist plot
gives the curve shown in Figure 14.37(b). The curve begins at $E_\mathrm{m} = 0.05$, since the DF
is only valid for $E_\mathrm{m} > E_\mathrm{b}/2$ (for smaller values of E_m, the output would never overcome
the backlash, assuming the initial condition to be at the origin of its characteristic).

As E_m increases, the curve is traversed in an anticlockwise direction, ending at the point $(-0.3, 0)$ for $E_m \to \infty$. This endpoint also makes physical sense since, for very large input signals, the effect of the backlash becomes negligible, leaving only the inherent gain of 0.3.

It can be seen that the inverse Nyquist plot intersects the DF twice. For values of $E_m < 0.0563$, the inverse Nyquist plot (only just) encloses the DF, and so the system is (only just) stable. However, it would be physically impossible to operate the system in a manner which would keep E_m below this value, bearing in mind that no output will be obtained at all until E_m exceeds 0.05. In any case, the models are unlikely to be sufficiently accurate to allow reliance on the difference between 0.05 and 0.0563.

As soon as E_m does exceed 0.0563, the inverse Nyquist plot no longer encloses the DF, so instability sets in. This will make E_m tend to increase until it reaches the value 0.146 shown in Figure 14.37(b). At this point, the inverse Nyquist plot again encloses the DF, so the system becomes stable. Following the same arguments as in Example 14.12, a stable limit cycle would be expected to develop with an amplitude of ± 0.146 units, at a frequency of 1.31 rad s^{-1} (this corresponds to a period of oscillation of about 4.8 s).

Figure 14.37(c) (generated by the MATLAB m-file *fig14_37.m* on the accompanying disk) shows a time response of y and M from an initial condition of $y = 0.06$ (just enough to get out of the deadband initially) and $M = 0$. Again, the simulation requires a rather small time step size, but not as small as Example 14.12. The figure is plotted for a time step of 10 ms, but even this involved 7000 points per plot. A smaller time step does result in a final amplitude nearer to the predicted value of ± 0.146, rather than the ± 0.15 shown, but again, it is unrealistic to seek this kind of accuracy. The frequency of the limit cycle can be seen to have been correctly predicted. Note that the frequency gradually alters until the limit cycle is established – this is another kind of behaviour typical only of nonlinear systems.

As in the previous example, Figure 14.37(b) is useful for design, in that it allows the determination of how far the linear system's gain should be reduced so that the inverse Nyquist plot would enclose the DF under all circumstances.

This section ends with a few general points which these examples could not bring out. First, nonlinearities do not always lead to instability and limit cycles (as was seen in Example 14.3). If Figure 14.36 had contained a deadzone element (the fourth line in Table 14.6) rather than the soft saturation, the DF would again occupy a linear segment of the negative real axis, but would start at the origin for $E_m = E_d/2$, and end at $(-1, 0)$ for $E_m \to \infty$ (assuming $K = 1$). The direction of increasing E_m would therefore be opposite to that in Figure 14.36. This means that the system would be stable so long as E_m remained below the level at the intersection of the DF and the inverse Nyquist plot (assuming the system parameters to be such that an intersection existed). However, for higher values of E_m the system would be totally unstable, so it would not be very robust.

A NL element having both a deadzone and saturation (the fifth line in Table 14.6) has a DF plot which begins at the origin for $E_m = E_d/2$ and then, as E_m increases, moves out along the negative real axis until $E_m = E_s$, when it reverses and returns to the origin as $E_m \to \infty$. Such a system, in which the inverse Nyquist plot of the linear part intersects the DF, will remain stable for relatively low values of E_m, but go into a stable limit cycle if some value of E_m is exceeded.

A system with a hysteresis nonlinearity (the penultimate line in Table 14.6) has a DF plot which is a curve in the second quadrant, similar to the backlash one in Figure 14.37(b), except that its base is on the imaginary axis rather than the real axis. It starts on the imaginary axis above the origin, when $E_m = E_h/2$, at a value of $8K/(\pi E_m)$. As E_m increases, it traces out a semicircular type of curve, arriving at the origin as $E_m \to \infty$. If the inverse Nyquist plot of the linear part of the system intersects this curve, limit cycle operation is again likely to result.

Other types of NL element give rise to similar phenomena. For example, a system with both hysteresis and a deadzone will have a DF plot which begins and ends at the origin, making a loop in the second quadrant as E_m varies.

14.8   Popov's method

Popov's method is the second frequency-domain technique for the study of nonlinear systems. Like Lyapunov's time-domain approach (Section 14.6), it allows the system stability to be determined without having to investigate the nonlinear equations of the system. In fact, Popov's method can be derived from a Lyapunov viewpoint (Khalil, 1992). It can also be derived using functional analysis (Banks, 1986). Popov's method is an excellent example of the fact that a method which is easy to apply will only be applicable to a rather restrictive class of situations!

Popov's method is based upon the assumption that the characteristic of the nonlinear (NL) element in the system is constrained to lie in a sector of the first and third quadrants, as indicated in Figure 14.38. Since this is the only information that it is necessary to know about the NL element, some potentially significant behaviour may clearly be overlooked. However, Popov's method gives a result which does guarantee the asymptotic stability of the system (if it is indeed stable). The conclusion is that this stability guarantee must be rather conservative – in other words, if it were possible to take into account the finer detail of the NL element, then stability could doubtless be proved over a wider range of conditions than Popov's result would suggest. Note that the same limitations apply to Lyapunov's method, the circle criterion (Section 14.9) and any other method which does not involve investigation of the actual nonlinear equations.

Since the full nonlinear equations can rarely be solved anyway, Popov's method often turns out to be the most easily applicable and useful result, so long as its restrictions are met. In practice it is often used in preference to the Lyapunov

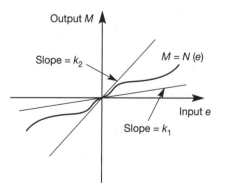

Figure 14.38 A sector-bounded nonlinear characteristic.

approach, because of the difficulty of finding a Lyapunov function in the latter method. Indeed, for one particular class of commonly used Lyapunov functions (called Lur'e functions and involving a quadratic and integral term), it transpires that Popov's result is, in any case, the best that can be achieved (Atherton, 1982).

Here, Popov's method will be regarded as applying to systems of the form of Figure 14.32, and the major points to note about the method include:

- Its use is limited to sector-bounded NL elements such as that in Figure 14.38. This covers a very wide range of practical NL elements, but does imply some specific limitations:

  – applies to a single NL element

  – applies only to NL elements whose characteristics have odd symmetry

  – applies only to NL elements with no hysteresis, backlash or similar behaviour which would take the characteristic into the second or fourth quadrant.

Additional points are:

- It applies only to continuous or piecewise continuous NL elements. This means that the slope of the NL characteristic is not permitted to be infinite at any point, so NL elements such as relays must be specified with a finite (but very steep) slope to the switching characteristic.
- The NL element must be stationary (non time-varying).
- The linear part of the plant must not have any poles in the right-hand half of the *s*-plane.
- Any reference or disturbance inputs to the system must be of a generally impulsive type (that is, they are not considered at all, or they must be such that they tend to zero as time passes).
- The results may be rather conservative.

The method is applied as follows.

In Figure 14.38, the lines bounding the sector in which the NL characteristic lies have equations: $M = k_1 e$ and $M = k_2 e$, where $k_1 < k_2$ are both positive constants. For the NL characteristic to lie between these lines, it must be the case that

$$k_1 e \leq N(e) \leq k_2 e \quad \text{or, alternatively} \quad k_1 \leq \frac{N(e)}{e} \leq k_2 \qquad (14.38)$$

If Popov's method is to be applied, the linear part of the system must have no poles in the right-hand half of the *s*-plane, and the NL element must meet the criterion in Equation 14.38.

The constant k_1 cannot be zero if the system has imaginary-axis poles. The reason for this is that $k_1 = 0$ implies that the output from the NL element could be zero at some non-zero value of e. The forward path would then be effectively open-loop, so the system poles would be those of $G(j\omega)$ (see Figure 14.32) and, if some of these were on the imaginary axis, the system could then not be asymptotically stable.

Forcing k_1 to be positive ensures that the output from the NL element can be zero only when e is zero. With the earlier restriction on the input function tending to zero as time passes, this implies that asymptotic stability will then be achieved if the overall system is proved to be stable.

If the linear part of the system is asymptotically stable (that is, $G(j\omega)$ has no poles on the imaginary axis, but all its poles are strictly in the left half of the s-plane) then Equation (14.38) can be relaxed somewhat, in that k_1 can then be zero.

So long as the above conditions are met, Popov's criterion states that the overall system will be globally asymptotically stable so long as there is a positive number q and a second arbitrarily small positive number δ, such that the following inequality holds:

$$\text{Re}[(1 + j\omega q)G(j\omega)] + \frac{1}{k_2} \geq \delta \qquad \omega \geq 0 \qquad (14.39)$$

A graphical test, similar in principle to the Nyquist criterion for linear systems, is available to show whether or not Equation (14.39) is fulfilled. This is how it works.

A modified Nyquist diagram for the linear part of the system is constructed as follows. The real part of $G(j\omega)$ is used as it stands, but the imaginary part is first multiplied by ω at each frequency point. Thus, the Nyquist diagram for $G^*(j\omega)$ is plotted, where

$$G^*(j\omega) = \text{Re}[G(j\omega)] + j\omega \, \text{Im}[G(j\omega)]$$

The diagram is called a *Popov plot* or *Popov locus*. Using this equation for $G^*(j\omega)$ in Equation (14.39) gives (see Problem 14.5):

$$\text{Im}[G^*(j\omega)] \leq \frac{1}{q} \text{Re}[G^*(j\omega)] + \frac{1}{qk_2} - \frac{\delta}{q} \qquad \omega \geq 0 \qquad (14.40)$$

For q positive, and δ positive and arbitrarily small, Equation (14.40) implies that the Nyquist plot of $G^*(j\omega)$ must lie entirely to the right of a straight line of slope $1/q$ and real axis intercept $-1/k_2$.

The truth of this condition can easily be checked graphically and, if true, the system will be globally asymptotically stable. Note that, because of the conservative nature of the result, if a system fails this test *it will not necessarily be unstable*. Equation (14.40) is a sufficient condition for stability, but not a necessary condition.

Example 14.14 *The antenna positioner with saturating control amplifier analysed by Popov's method*

This example revisits the system of Example 14.12 and Figure 14.36(a).

The NL element can be seen to fit the sector-bounded limitation of Figure 14.38 with k_1 any arbitrarily small positive constant, and $k_2 \geq 1$. The linear part of the system is stable (not asymptotically stable due to the pole at the origin, but this is acceptable since

$k_1 > 0$) and the NL element is not time-varying. The NL element is also piecewise continuous. Popov's method is therefore applicable to this system.

Figure 14.39(a) shows the Popov locus for the linear part of the system (that is, the Nyquist plot of $G^*(j\omega) = \text{Re}[G(j\omega)] + j\omega\,\text{Im}[G(j\omega)]$), together with a Popov line passing through the point $(-1/k_2, 0) = (-1, 0)$, and of arbitrary slope $1/q = 1$.

It can be seen that there is no positive constant q (that is, no positive slope between horizontal and vertical) for which the Popov line could be drawn in a manner which would avoid the Popov locus while passing through the point $(-1/k_2, 0) = (-1, 0)$. Therefore this test fails, and we cannot conclude that the system is asymptotically stable. Unfortunately, because of the conservative nature of the Popov test, we can draw *no conclusion* about the stability of the system – we cannot conclude that it is unstable, only that it has not been proved to be stable. In Example 14.12, the describing function (DF) approach showed that a limit cycle will develop, but no such information is revealed by the Popov test.

It is possible to proceed further, though. Let the constant k_2 (and hence the real intercept $-1/k_2$) be regarded as a variable parameter. Examination of the Popov locus then allows a prediction of the maximum value of k_2 for which asymptotic stability *will* be guaranteed. In Figure 14.39(b) a Popov line is drawn for $k_2 = 0.6$ and $q = 1.2$. It can be seen that under these conditions the system will be asymptotically stable. The system is therefore stable for any $k_2 < 0.6$.

This ties in very well with the information gained from the DF method in Example 14.12. At the end of that example, it was discovered that the system should be unconditionally stable (no limit cycle) if the d.c. gain in the linear part of the system was reduced by a factor of 0.6. The simpler Popov approach has now suggested that the slope of the central part of the NL characteristic (that is, k_2) should be reduced by a factor of 0.6. (Note that this is not absolutely identical to reducing the linear d.c. gain by a factor of 0.6 – to do that, the slopes of the soft-saturated parts of the NL characteristic would also need reducing by 0.6, but the Popov result shows that to be unnecessary.)

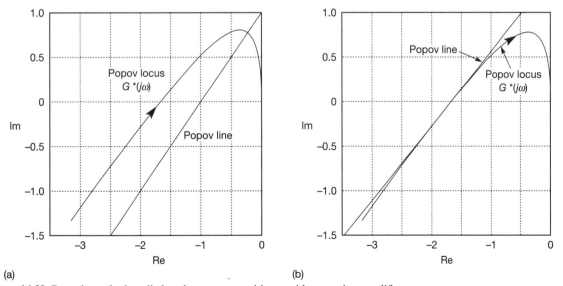

(a) (b)

Figure 14.39 Popov's method applied to the antenna positioner with saturating amplifier.

In addition, the Popov result shows that the system would be asymptotically stable for *any other* NL characteristic which remained in a sector bounded by the slopes $k_2 = 0.6$ and $k_1 =$ an arbitrarily small positive number. This information was *not* available from the DF method. In the DF method, it would be necessary to look up, or derive, the DF of any different NL element whose effect was of interest, and apply the method for each one separately.

14.9 Zames' circle criterion

This is the last method to be studied for nonlinear systems. It is another frequency-domain method and is a development of Popov's method described in the previous section. The salient features of the method are similar to those of Popov's method, with the extra points:

- The circle criterion allows the use of a normal (unmodified) Nyquist plot of the linear part of the system.
- The NL element may be time-varying.
- The NL element characteristic must be odd, but is not restricted to the first and third quadrants only.
- The linear part of the plant may be unstable in open-loop.
- The results can be *extremely* conservative.

The NL element is again considered to have a characteristic bounded by the sector shown in Figure 14.38. If the NL element is time-varying, it must remain within the sector at all times. The linear part of the plant may have right half-plane poles, but its transfer function must be strictly proper (that is, the denominator order must be higher than the numerator order, or the state-space model must have no D matrix) and it must be both controllable and observable (that is, there must be no mode cancellations).

The normal linear system Nyquist criterion is then applied, but with the $(-1, 0)$ point replaced by a circular disc. This disc is centred on the negative real axis, and cuts that axis at the points $-1/k_1$ and $-1/k_2$ as shown in Figure 14.40(a). The interpretation of the plot depends upon the relationship between k_1 and k_2 in Figure 14.38 as follows:

- For $k_2 \geq k_1 > 0$ (as shown in Figure 14.38), the overall system will be globally asymptotically stable if the Nyquist plot of $G(j\omega)$ does not touch the disc, and encircles it anticlockwise a number of times equal to the number of right half-plane poles of $G(j\omega)$ (Figure 14.40(a)). Note that this is precisely the Nyquist criterion for linear systems, but with the disc replacing the $(-1, 0)$ point.
- For k_2 positive and $k_1 = 0$, the overall system will be globally asymptotically stable if the Nyquist plot of $G(j\omega)$ remains to the right of a vertical line through the point $(-1/k_2, 0)$ (Figure 14.40(b)). This is similar to the Popov result for the same case.

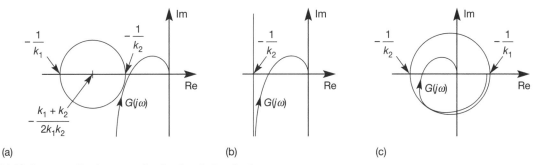

Figure 14.40 Some application scenarios for the circle criterion.

- For k_2 positive and k_1 negative, the overall system will be globally asymptotically stable if the Nyquist plot of $G(j\omega)$ remains inside the disc (Figure 14.40(c)).
- For k_2 and k_1 both negative, the overall system will be globally asymptotically stable if the Nyquist plot of $G(j\omega)$ does not touch the disc, and encircles it clockwise a number of times equal to the number of right half-plane poles of $G(j\omega)$.

Example 14.15 *The antenna positioner with saturating control amplifier analysed by the circle criterion*

Consider again the system of Examples 14.14 and 14.16, shown in Figure 14.36(a).

For the circle method, k_1 could be regarded as being zero, and the test of Figure 14.40(b) could be applied. However, one look at the Nyquist plot of the linear part of the plant (Figure 14.41(a)) shows that a vertical line at $-1/k_2 = -1$ would give very poor conclusions. In order to avoid the Nyquist plot, Figure 14.41(a) indicates that the vertical

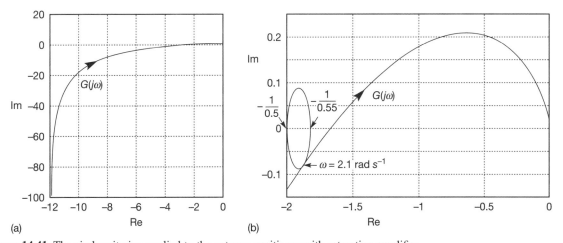

Figure 14.41 The circle criterion applied to the antenna positioner with saturating amplifier.

line would have to be placed somewhere around $-1/k_2 = -12$, namely $k_2 = 0.08$. Since both the describing function method and Popov's method have shown that k_2 can be as high as 0.6, the value of k_2 suggested now is far too conservative and would lead to extremely poor performance.

Tightening up the sector onto the NL characteristic will give a better result. It is not necessary to set $k_1 = 0$. Rather, as large a value for k_1 as possible should be used, so as to generate a disc of as small a radius as possible, which can then be fitted up against the Nyquist plot to find the maximum permissible value of k_2 for stability. In the present case, the Nyquist plot must not encircle the disc, as there are no unstable open-loop poles.

From the NL element characteristic in Figure 14.36(a), it can be seen that the lower sector bound on the nonlinear characteristic could be taken as $k_1 = 0.5$, since this would never intersect the saturation characteristic, no matter how large e were to become. The test of Figure 14.40(a) can therefore be applied with $-1/k_1 = -2$. Looking at the portion of the Nyquist plot near the origin, shown in Figure 14.41(b), it is now clear that a disc can be fitted as shown (it is mathematically circular – only the plot scaling makes it physically not so!). The maximum value of k_2 for guaranteed stability is seen to be 0.55, which is close to the previous results (0.6), but more conservative.

Finally, although an example will not be given, note that there is a generalized version of the circle criterion. In this version, one circle is drawn for each frequency value used in calculating the Nyquist plot. The centre of each circle is at a location in the s-plane given by:

$$(\text{real, imag}) = \left(-\frac{k_2 + k_1}{2k_1 k_2}, \quad \omega q \frac{k_2 - k_1}{2k_1 k_2} \right) \quad \text{for } k_2 > k_1 \quad (14.41)$$

and the points of intersection with the negative real axis are as shown in Figure 14.40(a). Choosing $q = 0$ gives the criterion as previously used.

The general version of the criterion states that the system is globally asymptotically stable if a real value of q can be found such that the previously stated stability criterion is met at each individual frequency value. This is obviously much harder to test, but gives a more accurate (that is, less conservative) result. For instance, in Example 14.14, the circle drawn in Figure 14.41(b) touches the Nyquist plot at a point corresponding to a frequency of about 2.1 rad s^{-1}. Using this value in Equation (14.41), a value of q could be chosen which would allow a larger value of k_2 to be specified by pulling the centre of the circle vertically upwards, while maintaining the real axis intercept point $-1/k_1$ and the real axis intercept point for the modified value of $-1/k_2$. It would then be necessary to check that the chosen values of q and k_2 predicted stability at every other frequency value.

14.10 Summary

In this chapter it was stressed that all real systems are nonlinear to some degree, and that the linear analysis tools developed in the rest of the book are applicable only to linear models which are an approximation to the real world. Approximate linear models of known nonlinear systems were obtained by linearization, for both SISO and MIMO systems. Several types of behaviour which might occur in nonlinear systems, but which are not predictable by linear analysis methods, were discussed.

Nonlinear systems cannot be solved in any general way, so it is fortuitous indeed that linear approximations are often sufficiently accurate for practical designs. When this is not the case, recourse must be made to the study of a system model with sufficient nonlinearities added to make it an acceptable representation of the real world. There is no general approach to the study of such models, so several of the more important techniques have been presented, from which the most appropriate can be selected.

In the time domain, the method of producing and analysing phase plane trajectories was studied, as were the two Lyapunov approaches. In the frequency domain the describing function (DF) approach, Popov's method and the circle criterion were all introduced. It was also mentioned several times that computer simulation is the major technique for handling nonlinear systems, but that one still needs recourse to analytical methods to check the feasibility of results and to suggest likely ranges for initial conditions. The problem with simulation is knowing when all possible operating regimes have been covered.

For simulation using digital computers, the choice of integration algorithm and time step size is crucial for the accurate simulation of systems with discontinuous nonlinearities – it is very easy to be misled. Evidence of this has been presented in the examples, and MATLAB m-files on the accompanying disk show how most of the results were generated. Analog computer simulation does not share these problems, but such techniques have not been covered, due to the (unfortunate) relative scarcity of analog computers these days.

The phase plane (PP) and DF approaches each take into account the shape of the nonlinear characteristic, and therefore tend to give relatively accurate results for the classes of systems to which they apply. In addition, these two methods are complementary to some extent, one working better where the other falls short. However, both make considerable approximations in order to make the analysis possible, thus restricting the number of systems to which they truly apply. The major approximation of the PP method is that the linear part of the system can be adequately represented by a second-order model, while the major approximation of the DF technique is that the linear part of the plant is a sufficiently good low-pass filter that only the fundamental frequency of any oscillating signal leaving the nonlinear element will survive the trip around the feedback loop. The PP method is the only one giving transient response information, the DF method being used to investigate limit cycles.

The Lyapunov direct method and the Popov and circle methods are also used primarily to investigate stability. The Popov and circle methods do not even consider the shape of the NL characteristic, and therefore tend to give conservative stability estimates. Systems will usually be stable over wider ranges of conditions than the results of these methods suggest, but this can never be *guaranteed*. Furthermore, because of this conservatism, if these methods fail to predict stability, it cannot be assumed that the system is unstable. In the case of the Lyapunov approach, it can be very difficult to find a Lyapunov function, but in that case it still cannot be said that the system is not stable – only that stability has not been proven. Due to this difficulty, the Popov or circle methods are often preferred, despite their inaccuracy. The Popov method is usually the preferred technique for classes of system to which it applies, due to the ease of application (especially when computing power is available to produce the Popov locus).

The specialist references (Atherton, 1982; Khalil, 1992; Slotine and Li, 1991) give extensions to several of the techniques, which remove some of the restrictions which were imposed to allow the introductory treatment in this text. Extensions and specific methods for multivariable nonlinear systems will also be found there.

14.11 Problems

14.1 (a) Consider a first-order system whose state equation is nonlinear, and is given by $\dot{x} = x^2 + u$. Show that the addition of state variable feedback using a nonlinear measuring transducer, such that the signal fed back (and subtracted from a reference input) is x^2, will linearize the state equation.

(b) What would be the practical problems with the implementation of this solution?

14.2 The system of Figure 14.20(a), using an ideal relay as the control element, was analysed by the phase-plane trajectory approach in Example 14.5. A real electromechanical relay will have a deadzone (because it will not switch in response to very small signals). Its characteristic will then be approximately that of Figure 14.5(e) – note that no switching delay is being included.

Introduce a deadzone of ± 0.25 unit into the ideal relay in Figure 14.20(a). This will give a third region in the phase plane compared with Example 14.5. Produce the PPT for this new arrangement in response to an input step of 4 units from zero initial conditions. Comment upon the effect of the deadzone upon the system (compared with Example 14.5), and produce a plot of the system output (y) vs. time.

14.3 Derive the describing function of the hysteresis element given in Table 14.6.

14.4 Derive the describing function of the backlash element given in Table 14.6. Note that this is a deceptively difficult problem. It requires considerable thought about signal magnitudes and polarities if the correct answer is to be obtained.

14.5 Equation (14.39) gave a relationship used in the development of Popov's method, namely:

$$\text{Re}[(1 + j\omega q)G(j\omega)] + \frac{1}{k_2} \geq \delta \qquad \omega \geq 0$$

where $G(j\omega)$ is the frequency response function of the linear part of the system under investigation, ω is the angular frequency (rad s^{-1}) and k_2, q and δ are constants.

Popov's method is actually applied to a modified form of this equation, however, as given in Equation (14.40). To obtain this, the frequency response function $G(j\omega)$ is replaced by the modified form $G^*(j\omega) = \text{Re}[G(j\omega)] + j\omega\,\text{Im}[G(j\omega)]$.

Make this substitution in Equation (14.39), and show that Equation (14.40) results, namely:

$$\text{Im}[G^*(j\omega)] \leq \frac{1}{q}\,\text{Re}[G^*(j\omega)] + \frac{1}{qk_2} - \frac{\delta}{q} \qquad \omega \geq 0$$

14.6 A linear system defined by

$$G(s) = \frac{15(s + 0.4)(s + 1)}{s^2(s + 2)(s + 0.3)(s + 4.5)}$$

has an actuator which suffers from a large deadzone. The system is included in a feedback loop as shown in Figure P14.6.

(a) Use the describing function approach to predict the behaviour of the system. A frequency range from about 0.1 rad s^{-1} to about 1.4 rad s^{-1} will be found to be adequate for the polar plot of the linear part of the system.

(b) If you have a suitable CACSD package, use simulation to verify the predictions of part (a). You will find that the results do not agree as closely as one might wish. Why is this?

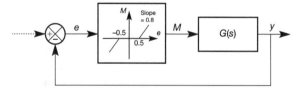

Figure P14.6 System for Problem 14.6.

14.7 A linear system with the transfer function

$$G(s) = \frac{25}{(1.5s^2 + 0.4s + 15.5)(s^2 + 0.3s + 3)}$$

is to be put into a closed-loop arrangement with a backlash nonlinearity in its output as shown in Figure P14.7.

(a) What would the closed-loop behaviour be like if the nonlinearity was absent, and the feedback was therefore directly from $q(t)$?

(b) Using the describing function approach, discuss fully the behaviour of the system as shown in Figure 14.7. Mention appropriate amplitudes and frequencies of the signals at both $q(t)$ and $y(t)$.

(c) If the system is not unconditionally stable in the form shown in Figure 14.7, design a simple gain to be put in front of the linear part of the plant, which would make it unconditionally stable.

(d) If you have access to SIMULINK (Appendix 4) or a similar package, confirm your findings by simulation (in this case, the agreement should be very close). Something like SIMULINK is the only method that can *easily* be used here. It is possible to program the simulation in any technical language, but in lower-level languages (even MATLAB), the correct programming of the backlash element is not a simple task. Also, simulation with fixed time step algorithms is time-consuming and generates huge quantities of data, because it is very important to tie down accurately the instants at which the backlash characteristic changes direction (perhaps to sub-millisecond resolution). Even slight mis-timing of these events can have significant effects on the simulated behaviour.

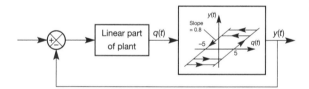

Figure P14.7 System for Problem 14.7.

14.8 Rather than deriving describing functions from scratch, the DF for a nonlinear characteristic which can be represented as the sum of two (or more) other characteristics can be obtained from the sum of the DFs of the other characteristics. Using this rule:

(a) Show that the DF of a deadzone (fourth row of Table 14.6) can be obtained from the DF of a saturation element (first row of Table 14.6) subtracted from a linear gain K.

(b) Derive the DF of a deadzone with saturation (fifth row of Table 14.6) from the sum of the DFs of two suitably specified deadzone characteristics (fourth row).

14.9 Figure P14.9 shows a linear system element with a mechanical actuator. The actuator contains an output element which is constrained by preloaded springs, giving the nonlinear characteristic shown.

(a) By adding a suitable gain and a relay DF, show that the DF for the nonlinear element is given by

$$N = \frac{2}{\pi}\left(\pi + \frac{1}{E_\text{m}}\right)$$

(b) If

$$G(s) = \frac{0.25}{s(2.5s^2 + s + 2)}$$

investigate the stability behaviour of the system with no driving input other than initial conditions inside the system. If there is a limit cycle, estimate its amplitude and frequency.

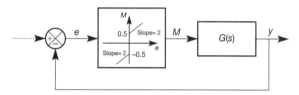

Figure P14.9 System for Problem 14.9.

14.10 Consider the system of Figure P14.10, with

$$G(s) = \frac{7}{s^2 + 3s + 2}$$

and a step input of 4 units applied at r.

(a) Using Lyapunov's first method (see Section 14.5.6), predict the form of the phase plane trajectory of (the linear part of) the system, by evaluating the closed-loop eigenvalues of the system with no deadzone present.

(b) What would be the expected percentage overshoot and steady-state output for the zero-deadzone configuration of the system used for part (a)?

(c) Plot the phase plane trajectory for the system including the deadzone, and compare the form of the PPT and its overshoot and steady-state characteristics with those predicted for the zero-deadzone system in part (b). Account for any differences.

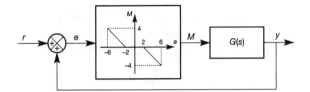

Figure P14.10 System for Problem 14.10.

14.11 For the system shown in Figure P14.6, the deadzone characteristic now has unity slope (rather than the value of 0.8 given in Figure P14.6), and

$$G(s) = \frac{10}{s^3 + s^2 + 5s + 1}$$

(a) Use Popov's method to predict the stability properties of the system.

(b) If you have access to MATLAB and its Control Systems Toolbox (or a similar package), test the predictions.

(c) If the system *is not* guaranteed to be stable, what modification to the slope of the deadzone characteristic would be required in order for stability to be guaranteed by Popov's method? If it *is* guaranteed stable, what change in the slope of the deadzone characteristic could be tolerated before the guarantee was lost?

(d) Repeat part (c) with reference to modifications to the d.c. gain of $G(s)$, rather than the deadzone slope.

(e) What difference would it make to your previous analysis of this system if the deadzone width was changed to ±0.1 unit, or to ±5 units. What does this tell you about the method?

(f) If $G(s)$ contained a pure integrator ($1/s$), how would the specification of the problem need to be altered, before we could tackle it using Popov's method?

14.12 Repeat all parts of Problem 14.11 for the same $G(s)$, but with the nonlinear element replaced with the one shown in Figure P14.12. In part (c), keep the saturation level unaltered, just change the slope.

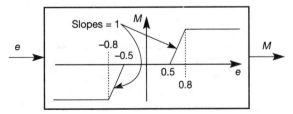

Figure P14.12 Nonlinearity for use in Problem P14.12.

14.13 Obtain a linear model of the pendulum of Example 2.10 in Section 2.4.5, but linearized about the *unstable* critical point $\theta = \pi$ radians. Compare the result with that of Example 2.10.

14.14 Figure P14.14 shows a nonlinear plant in which two signals are multiplied together, and a third signal is generated by a measuring transducer having a square-law characteristic.

(a) Obtain the steady-state operating point of this plant when u_1 is a step input of size 3 units, and u_2 is a step input of size 9 units. Use the state variables shown in the figure. If there is more than one possible operating point, use one having positive values for the state variables.

(b) Obtain a linearized model of the plant, about the steady operating point determined in part (a).

(c) If you have access to MATLAB and its Control Systems Toolbox, or to SIMULINK or some similar package, compare the linearized model with the nonlinear model for a range of input signals. How far from the operating point do you think the linear model remains valid?

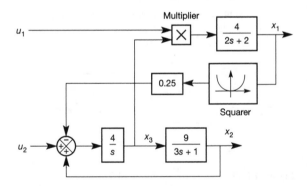

Figure P14.14 System for Problem 14.14.

14.15 Figure P14.15 shows a nonlinear plant containing three signal multipliers (one of which is connected as a squarer).

(a) Obtain the steady-state operating point of this plant when u_1 is a step input of size +14 units, u_2 is a step input of size −2 units and u_3 is a step input of size −4/3 units. Use the state variables shown in the figure, and any others you find to be necessary.

(b) Obtain a linearized model of the plant, about the steady operating point determined in part (a).

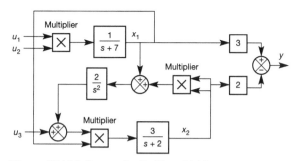

Figure P14.15 System for Problem 14.15.

(c) Inspect the eigenvalues of the linearized model (that is, of J_x) and hence comment on its likely usefulness. Also, consider the implications for the full nonlinear model.

(d) You will need computer assistance for this part. Using the methods of Chapter 5 (via the MATLAB *place* command) a state-variable feedback regulator can be designed for the linearized model, which keeps the stable eigenvalues in their original locations, but moves the unstable ones to (say) $-6 \pm 6j$. A gain matrix which achieves this (there are infinitely many others) is

$$K = \begin{bmatrix} 0 & 0 & 0 & 0 \\ -0.0442 & -0.1812 & -0.8356 & -0.603 \\ 0.1156 & 1.2277 & 2.7717 & 2.6627 \end{bmatrix}.$$

Apply this SVF gain matrix both to the linear model (for which it was designed) and also to the full nonlinear model. Compare the linearized model with the nonlinear model for a range of input signals. How far from the operating point do you think the linear model remains valid? Note that the steady operating point moves, as a result of applying the SVF, to $x_0 = [-3.9354 \quad 1.9838 \quad -0.2607 \quad 0]^{\mathrm{T}}$.

14.16 For the system of Figure P14.14, but ignoring the inputs, use Lyapunov's direct method to investigate its stability properties. If you fail to find a suitable Lyapunov function, try reinstating the inputs and adding state variable feedback from x to u via a 2×3 matrix of gains K, such that $u = -Kx$. Use Lyapunov's direct method to investigate the stability properties for this new configuration (which should prove easier).

14.17 For the system shown in Figure P14.6, the deadzone characteristic now has unity slope

(rather than the value of 0.8 given in Figure P14.6), and

$$G(s) = \frac{10}{s^3 + s^2 + 5s + 1}$$

(a) Use the circle criterion to predict the stability properties of the system.

(b) If the system *is not* guaranteed to be stable, what modification to the slope of the deadzone characteristic would be required in order for stability to be guaranteed by the circle method? If it *is* guaranteed stable, what change in the slope of the deadzone characteristic could be tolerated before the guarantee was lost? Compare your result with that of Problem 14.11(c).

(c) Repeat part (b) with reference to modifications to the d.c. gain of $G(s)$, rather than the deadzone slope. Compare your result with that of Problem 14.11(d).

14.18 Again consider a system as in Figure P14.6, and with the same $G(s)$ as in Problem 14.17. However, this time replace the nonlinear element with that shown in Figure P14.18. The range of operation of the system is such that the horizontal part of the saturation characteristic is guaranteed never to be reached.

(a) Use the circle criterion to investigate the stability of the system.

(b) If the system *is not* guaranteed to be stable, design a modification to the nonlinear characteristic which will make it so. Note that the gain must always change at an input value of ± 6 units, and the output must still saturate at ± 9 units if the input reaches ± 30 units (though you may still assume that it will not).

(c) Achieve the result of part (b) by leaving the nonlinear characteristic in its original form, and designing a gain reduction for $G(s)$.

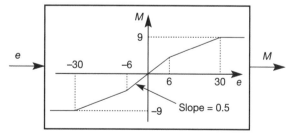

Figure P14.18 Nonlinearity for use in Problem P14.18.

Appendix 1
Matrix algebra relevant to control systems

A1.1 Elementary matrix algebra review

There are often situations in which the control engineer needs to handle simultaneously many equations containing many variables. For example, a system to be analysed or controlled may have several inputs and outputs. In order to describe the system mathematically, it will probably be necessary to have at least one equation relating each input to each output, so the number of equations can rapidly become quite large. In such situations (and others) it is best to use the techniques of matrix algebra to handle the equations. This reduces the complexity of the notation, in that many equations can be encapsulated in a single expression. It also eases the solution of the equations, because computers are very good at solving matrix equations.

Many readers will be very familiar with the normal algebra required for manipulating single equations, yet be unfamiliar with the matrix algebra which is introduced here – or they may have once known the matrix algebra, but forgotten it due to lack of use.

This appendix is an introduction to, and handy reference for, the basic operations of matrix algebra. It also introduces some more advanced topics in linear algebra which are required for the study of some branches of control engineering later in the text. It may be best to skip these later topics until their use is required.

The early sections of this appendix are written from an unusual angle which links the matrix operations with the structure of physical systems from a control engineering viewpoint. For greater detail on matrix algebra in general, refer to such texts as Kreysig (1993) or Ayres (1987).

A1.1.1 *Matrices and vectors*

A matrix is simply a rectangular array of numbers, these being referred to as the *elements* of the matrix (matrices whose elements are algebraic expressions will also be used).

From the control engineering viewpoint, there are two fundamental ways of visualizing what a matrix does. An example of each is given below. The first is the usual mathematical interpretation which views a matrix as representing the coefficients in a set of simultaneous equations. The second approach to matrices may be less familiar, but can be extremely useful. It is to regard a matrix as something which operates on a set of input signals to produce a set of output signals. This fits in well with other standard control engineering ideas such as block diagrams and transfer functions. It is worth mastering this second approach, as it is often the most helpful in control systems analysis.

As an example (which is presented, and then explained afterwards), consider a set of three signals, x_1, x_2 and x_3, which are combined in various proportions to form two other signals y_1 and y_2. The equations for this operation might be:

$$\begin{aligned} y_1 &= 2x_1 + 3x_2 - x_3 \\ y_2 &= x_1 - 6x_2 \end{aligned} \tag{A1.1}$$

Now, it is much neater to represent this pair of simultaneous equations in matrix form:

$$\begin{bmatrix} y_1 \\ y_2 \end{bmatrix} = \begin{bmatrix} 2 & 3 & -1 \\ 1 & -6 & 0 \end{bmatrix} \begin{bmatrix} x_1 \\ x_2 \\ x_3 \end{bmatrix} \tag{A1.2}$$

or, even more neatly, as

$$y = Cx, \text{ where } y = \begin{bmatrix} y_1 \\ y_2 \end{bmatrix}, \quad C = \begin{bmatrix} 2 & 3 & -1 \\ 1 & -6 & 0 \end{bmatrix} \text{ and } x = \begin{bmatrix} x_1 \\ x_2 \\ x_3 \end{bmatrix} \tag{A1.3}$$

C is known as a 'two by three' matrix (written as 2×3), because it has two rows and three columns. Sometimes (but not in this text) a mathematical shorthand $C \in \mathcal{R}^{2 \times 3}$ is used, which means, 'the matrix C belongs to the set of real matrices of size 2×3'. The size (or 'dimensions', or 'order') of a matrix is *always* specified in the order (rows $\times$ columns). Upper-case, bold, italic characters are used to represent matrices.

The quantities represented by lower-case bold, italic characters (x and y) are *vectors*. A vector is just a special case of a matrix – it has only one row or column. Thus, in the example above, x is a three-element column vector, and y is a two-element column vector. A geometrical view of vectors is mentioned in Section A1.2.

Equations (A1.2) and (A1.3) show the usual mathematical approach of representing a set of simultaneous equations in matrix form, and the matrix C has arisen naturally as a result. If the reader is totally unfamiliar (or very out of practice) with the method of multiplying out Equation (A1.2) to get back to

Equation (A1.1), that problem will be discussed below. Before that, the second approach to visualizing matrices will be introduced.

Equation (A1.3) states that $y = Cx$. To the control engineer, it seems fairly natural to represent this equation (as with all other equations, whenever possible!) in the form of a block diagram, for example as shown in Figure A1.1 (block diagrams are studied in Chapter 2).

There are one or two things worth noting about Figure A1.1. First, the input path carries three signals and the output path two signals (in figures in the text, the '/' with the number of signals adjacent is often omitted, but wide lines are always used for paths carrying more than one signal). It is *always* the case that, for a matrix in a block diagram, its size is given by (number of outputs) × (number of inputs). Another way of stating this is as follows:

number of rows = number of outputs

number of columns = number of inputs

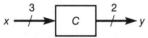

Figure A1.1 Matrices and vectors in a block diagram.

Furthermore, this ordering rule (output first, input second) can *always* be applied to the *elements* of a matrix to generate the corresponding interconnections. This is how it works. Each element of any matrix is referenced by two subscripts. The first gives the row in which the element lies, and the second the column. Thus, for the C matrix, the elements are

$$\begin{bmatrix} c_{11} & c_{12} & c_{13} \\ c_{21} & c_{22} & c_{23} \end{bmatrix}$$

So $c_{13} = -1$, $c_{22} = -6$, etc. Now, applying the ordering rule, element c_{13} (for example) feeds *output 1* of matrix C from *input 3* of matrix C; while element c_{21} feeds *output 2* of matrix C from *input 1* of matrix C; and so on. Wherever more than one contribution is made to the same output, the separate contributions are simply summed. Treating the whole C matrix in this way leads to the alternative representation of Figure A1.2.

If the equations of this diagram are written down, it will be found that they are the same as Equation (A1.1). Thus, the two views of matrix C are shown to be equivalent.

In summary, if the reader is of mathematical inclination, a matrix can be thought of as representing a set of equations, and Figure A1.2 can be drawn from the

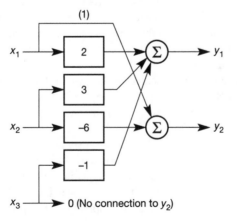

Figure A1.2 A simple vector–matrix equation expanded as a block diagram.

equations. If the reader prefers to think in terms of block diagrams, Figure A1.2 can be drawn directly from the matrix elements, using the ordering rule (output first, input second) on the element subscripts, without ever needing to write out the equations.

The only thing to watch out for is that the ordering rule applies to the outputs and inputs *of the matrix under consideration*, and these will not necessarily be the outputs and inputs of any particular plant. Indeed, if matrix C was in a feedback path around a plant, then the outputs of matrix C (which were called '*outputs*' when generating Figure A1.2) would actually be the *inputs* of the plant, and the *outputs* of the plant would be the *inputs* of matrix C. It is therefore important, when applying this rule, always to do what was done above, that is, mentally refer to 'the outputs *of matrix C*' and 'the inputs *of matrix C*', rather than just 'the outputs' and 'the inputs'. This should avoid confusion.

In this section, it was also seen that the size of a matrix in a block diagram is always given by (number of outputs) × (number of inputs), that is, one row per output from the matrix and one column per input to the matrix.

A1.1.2 Basic matrix operations and properties

Addition and subtraction

In order for two matrices (or vectors) to be added or subtracted, they must be of the same size. It is then just a case of adding or subtracting the corresponding elements. For example:

$$\begin{bmatrix} 1 & 2 & -3 \\ 6 & 14 & 2 \end{bmatrix} + \begin{bmatrix} -4 & 1 & 8 \\ 3 & 2 & 0 \end{bmatrix} = \begin{bmatrix} -3 & 3 & 5 \\ 9 & 16 & 2 \end{bmatrix}$$

and

$$\begin{bmatrix} 1 & 2 & -3 \\ 6 & 14 & 2 \end{bmatrix} - \begin{bmatrix} -4 & 1 & 8 \\ 3 & 2 & 0 \end{bmatrix} = \begin{bmatrix} 5 & 1 & -11 \\ 3 & 12 & 2 \end{bmatrix}$$

In physical terms, the operation of subtraction is shown in Figure A1.3 as an example.

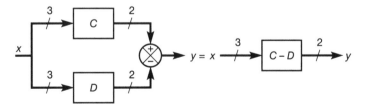

Figure A1.3 Matrix subtraction.

Multiplication by a scalar (that is, by a number)

Each element is multiplied by the scalar. For example:

$$\begin{bmatrix} 1 & 2 & -3 \\ 6 & 14 & 2 \end{bmatrix} 4 = 4 \begin{bmatrix} 1 & 2 & -3 \\ 6 & 14 & 2 \end{bmatrix} = \begin{bmatrix} 4 & 8 & -12 \\ 24 & 56 & 8 \end{bmatrix}$$

In physical terms, this represents a scalar gain applied simultaneously to every input (or every output) of a system.

Multiplication of matrices

Again, the block diagram approach can shed a new light on some of the mathematical rules which may previously have been learned about multiplying matrices. Consider a system as shown in Figure A1.4, in which D has 2 outputs (z) and 3 inputs (x), while C has 2 outputs (y) and 2 inputs (z).

Figure A1.4 Matrix multiplication.

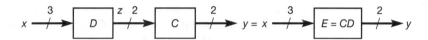

From Section A1.1.1, C must be of size 2×2, and D must be of size 2×3 (remember: outputs $\times$ inputs). From Figure A1.4, $y = Cz$ and $z = Dx$, therefore $y = CDx$, as shown on the right-hand side of Figure A1.4. It can also be seen that the overall matrix relating y and x physically has to be of size 2×3 to relate the two signals of y with the three of x.

Now, the product CD (which is the overall matrix relating y and x) involves multiplying a 2×2 matrix by a 2×3 matrix. Let the result be E, so that $y = Ex$, where

$$\begin{matrix} E & = & C & \cdot & D \\ 2 \times 3 & & 2 \times 2 & & 2 \times 3 \end{matrix}$$

From this, note that the number of columns in C ($=$ second dimension $=$ number of inputs of C) is the same as the number of rows in D ($=$ first dimension $=$ number of outputs of D). It is obvious from Figure A1.4 that the number of outputs of D must be the same as the number of inputs of C, since they both represent the vector of signals z. It is also an unbreakable mathematical rule that, *in order to be able to multiply two matrices together, the number of columns of the first must equal the number of rows of the second*. Or, if the sizes are written down, as was done above, *the inner dimensions must be the same*.

From Figure A1.4, it was noted that the result E will have to be of dimension 2×3. Again, in mathematical terms with which the reader may be familiar, if the sizes of the matrices are written down as above, the size of the product is always given by *the outer two dimensions*. All these things (fortunately) agree, tying together the mathematical rules with what can be seen to be necessary from Figure A1.4.

From these rules, it is apparent that the order of writing down the product $E = CD$ cannot be reversed. In this case, it is said that the matrices are *not conformable* for multiplication if written in the order DC (implying $(2 \times 3)(2 \times 2)$ which will not work according to the rules above). This care which is necessary in the order in which the matrices are written down is true of matrix multiplication in general. Even if C and D were both square (both 2×2 or both 3×3 say; in which case they are conformable if written either way round), the product CD would, in general, differ from DC (the matrices are then said not to *commute*). For products involving non-square matrices (except in the special case where one matrix is of size $(a \times b)$ and the other is of size $(b \times a)$), if the product $\boldsymbol{CD}$ is conformable, then the product $\boldsymbol{DC}$ is simply not defined – it does not exist. The definition of how to calculate a matrix product will now be given, and then the above points will be illustrated by an example.

A matrix product is evaluated one element at a time. To calculate element (i, j) of the product, all the elements in the ith row of the first matrix are multiplied with the corresponding elements in the jth column of the second matrix, and the results are summed. This confirms that the number of elements in each row of the first matrix must be the same as the number in each column of the second matrix, that is, the number of columns of the first must equal the number of rows of the second, as has already been deduced physically.

As an example, consider the product of two matrices

$$\begin{bmatrix} 2 & -1 \\ 1 & 5 \end{bmatrix} \text{ and } \begin{bmatrix} 3 & 2 \\ 4 & -2 \end{bmatrix}$$

According to the rules above, element $(2, 1)$ of the product, for example, is calculated by summing the products of the elements in the second row of the first matrix and the first column of the second matrix, that is, element $(2, 1) = 1 \times 3 + 5 \times 4 = 23$. The whole product is:

$$\begin{bmatrix} 2 & -1 \\ 1 & 5 \end{bmatrix} \begin{bmatrix} 3 & 2 \\ 4 & -2 \end{bmatrix} = \begin{bmatrix} 2 & 6 \\ 23 & -8 \end{bmatrix}$$

If the order of multiplication is now reversed:

$$\begin{bmatrix} 3 & 2 \\ 4 & -2 \end{bmatrix} \begin{bmatrix} 2 & -1 \\ 1 & 5 \end{bmatrix} = \begin{bmatrix} 8 & 7 \\ 6 & -14 \end{bmatrix}$$

then, as suggested above, the result is different. If the matrices had been of dimensions, say, 2×2 and 2×4, then only one product exists, the other being impossible to calculate according to the rules.

In block diagram terms, this means that *matrices must always be multiplied together in the order which works against the signal flow* ('working *backwards*'). Only in this way can the previous mini-derivation leading to the equation $\boldsymbol{y} = \boldsymbol{CDx}$ be satisfied for Figure A1.4.

Multiplying matrices and vectors

The same rules for multiplication apply as for general matrices, and vector–matrix products were used in the previous section. However, if the reader is unfamiliar with such operations, it is possible to become confused by the fact that there is only one element in certain rows or columns. A few examples should clarify matters.

If a 1×3 row vector is multiplied by a 3×2 matrix, the rules suggest that the result should be of dimensions 1×2. Here is an example:

$$\begin{bmatrix} 1 & 3 & 2 \end{bmatrix} \begin{bmatrix} 4 & 2 \\ 2 & -1 \\ 3 & -6 \end{bmatrix} = \begin{bmatrix} 16 & -13 \end{bmatrix}$$

If a 3×2 matrix is multiplied by a 2×1 vector, the expected result is 3×1:

$$\begin{bmatrix} 4 & 2 \\ 2 & -1 \\ 3 & -6 \end{bmatrix} \begin{bmatrix} 1 \\ 2 \end{bmatrix} = \begin{bmatrix} 8 \\ 0 \\ -9 \end{bmatrix}$$

Perhaps more surprisingly, if a 2×1 vector is multiplied by a 1×3 vector, then a 2×3 result should be expected:

$$\begin{bmatrix} 1 \\ 2 \end{bmatrix} \begin{bmatrix} 1 & 3 & 2 \end{bmatrix} = \begin{bmatrix} 1 & 3 & 2 \\ 2 & 6 & 4 \end{bmatrix}$$

and if a 1×2 vector is multiplied by a 2×1 vector a 1×1 (scalar) result should appear:

$$\begin{bmatrix} 5 & 2 \end{bmatrix} \begin{bmatrix} 2 \\ 1 \end{bmatrix} = 12$$

Even these cases obey all the previous rules.

For a geometrical view of matrix–vector multiplication see Section A1.2.

The transpose of a matrix

The transpose of a matrix A is denoted by A^T (or sometimes by A', but not in this text other than in MATLAB examples), and is obtained by interchanging the rows and columns. Thus:

$$\begin{bmatrix} 1 & 4 & 2 \\ 3 & 8 & 9 \\ 5 & 7 & 6 \end{bmatrix}^T = \begin{bmatrix} 1 & 3 & 5 \\ 4 & 8 & 7 \\ 2 & 9 & 6 \end{bmatrix} \quad \text{and} \quad \begin{bmatrix} 1 & 3 & 2 \\ 2 & 6 & 4 \end{bmatrix}^T = \begin{bmatrix} 1 & 2 \\ 3 & 6 \\ 2 & 4 \end{bmatrix}$$

The transpose of a row vector is a column vector, and vice versa. The transpose of a product $(AB)^T$ is given by $B^T A^T$ (the reader can easily try a simple example to become convinced).

In physical terms, there is no analogy to the operation of transposition. It is therefore to be regarded as a mathematical tool, rather than a physical operator.

The determinant of a matrix

This falls into the same category as the transpose – it is a mathematical device which will be needed in analysis on a number of occasions.

The determinant of a matrix is a scalar value which has certain characteristics inherited from the original matrix. It is defined for square matrices only. For a matrix A, the determinant is written as $|A|$, and is calculated for a 2×2 matrix example as:

$$\begin{vmatrix} 1 & 3 \\ 2 & 5 \end{vmatrix} = 1 \times 5 - 3 \times 2 = -1$$

In words, the determinant of a 2×2 matrix is the product of the elements on the *leading diagonal* (top left to bottom right), minus the product of the elements on the other diagonal. For larger matrices, the determinant can be found by breaking the matrix down into 2×2 sub-matrices (known as *minors*) in a certain well-

defined manner, and calculating the determinants of these. For example, for a
3×3 matrix the result is written down as follows (and then explained):

$$\begin{vmatrix} 1 & 4 & 2 \\ 3 & 8 & 9 \\ 5 & 7 & 6 \end{vmatrix} = +1.\begin{vmatrix} 8 & 9 \\ 7 & 6 \end{vmatrix} - 3.\begin{vmatrix} 4 & 2 \\ 7 & 6 \end{vmatrix} + 5.\begin{vmatrix} 4 & 2 \\ 8 & 9 \end{vmatrix}$$

$$= +1.(-15) - 3.(10) + 5.(20) = 55$$

Once the series of three 2×2 determinants is obtained, the previous rule for
evaluating them is followed. Where did they come from though? It will be noticed
that, except for the signs, the numbers multiplying the three 2×2 determinants are
the elements of the first column of the original matrix. The signs were obtained
from an array like this:

$$\begin{bmatrix} + & - & + \\ - & + & - \\ + & - & + \end{bmatrix}$$

which begins with a $+$ for element $(1, 1)$, and then proceeds in the obvious manner
for any size of matrix. Thus the values 1 and 5 have $+$ signs associated with them,
but the value 3 has a $-$ sign. To obtain the 2×2 determinant associated with each
of these elements, the row and column of the original matrix which contain the
element in question are simply deleted. Thus, for example, in the case of the value
5 (element $(3, 1)$), row 3 and column 1 are struck out, leaving the determinant

$$\begin{vmatrix} 4 & 2 \\ 8 & 9 \end{vmatrix}$$

This determinant is called the *minor* of element $(3, 1)$.

In the above example, the determinant of the original 3×3 matrix was
expanded by the first column. However, expanding it by *any* row or column will
give the same result (one of the interesting properties of matrices). To prove the
point, the reader might like to check by picking a different row or column, and
applying the same method (remember the array of $+$ and $-$ signs).

For even larger matrices, the procedure is simply extended. For example, for a
4×4 matrix, expansion would be done by some chosen row or column (hopefully
one containing lots of zeros), to obtain a series of four 3×3 determinants, and then
each of these would be treated as above.

Note that the minor associated with each element of a matrix, when the sign
from the array of $+$ and $-$ signs is also included, is called the *cofactor* of the
element. Thus, in the case of element $(2, 3)$ in the above example, the cofactor of
element

$$(2, 3) = -\begin{vmatrix} 1 & 4 \\ 5 & 7 \end{vmatrix} = 13$$

This definition is needed below.

A1.1.3 *Matrix division – the inverse of a matrix*

Matrix division, as such, is not defined. Rather, if it is desired to solve for x in an equation such as $Ax = b$, then each side is pre-multiplied by the *inverse* of A. Although it is possible to define inverses for rectangular matrices (known as pseudo-inverses), attention will be restricted to the case of square matrices in this text.

The inverse of A is that matrix, written as A^{-1}, which makes the matrix product $AA^{-1} = A^{-1}A = I$. The matrix I is described below, but is analogous to 1 in the equivalent scalar case $a.a^{-1} = a^{-1}.a = 1$.

I is a special matrix called the *unit matrix*, or the *identity matrix*. It is a *diagonal* matrix. That is, it is square, and every element is 0, except for elements on the leading diagonal.

In the case of the unit matrix, all these diagonal elements are 1. Thus, some unit matrices are:

$$I_2 = \begin{bmatrix} 1 & 0 \\ 0 & 1 \end{bmatrix} \quad \text{and} \quad I_4 = \begin{bmatrix} 1 & 0 & 0 & 0 \\ 0 & 1 & 0 & 0 \\ 0 & 0 & 1 & 0 \\ 0 & 0 & 0 & 1 \end{bmatrix}$$

In use, the subscript is usually omitted, as the size is clear from the context.

Readers can soon convince themselves that any matrix or vector, when pre- or post-multiplied by a suitably dimensioned unit matrix (square, remember), remains unchanged (as in the analogous scalar case of multiplication by 1). Therefore, returning to the equation $Ax = b$, premultiplying by A^{-1} gives $A^{-1}Ax = Ix = x = A^{-1}b$ (the RHS), and the equation is solved.

Unfortunately, for some square matrices, an inverse cannot be found. This can lead to severe problems when trying to solve systems of equations. It is therefore important to be able to tell whether or not any given matrix is invertible. Various methods exist, but as a representative example consider a method involving determinants (see previous section). One derivation of the inverse of a matrix A is given by

$$A^{-1} = \frac{\text{adj}(A)}{|A|} \tag{A1.4}$$

The quantity adj(A) is the *adjoint matrix* of A. It is found as the transpose of the matrix of cofactors (refer back to the section on determinants for the definition of a cofactor). For the example above, namely

$$\begin{bmatrix} 1 & 4 & 2 \\ 3 & 8 & 9 \\ 5 & 7 & 6 \end{bmatrix}$$

the matrix of cofactors is given by (some of the elements were calculated in the section on determinants):

$$\begin{bmatrix} -15 & 27 & -19 \\ -10 & -4 & 13 \\ 20 & -3 & -4 \end{bmatrix}$$

and the adjoint is the transpose of this, namely:

$$\begin{bmatrix} -15 & -10 & 20 \\ 27 & -4 & -3 \\ -19 & 13 & -4 \end{bmatrix}$$

The determinant of this matrix has already been calculated as 55 in the section on determinants. Therefore, the inverse, from Equation (A1.4), is this matrix divided by 55, that is,

$$\begin{bmatrix} -0.272 & -0.182 & 0.364 \\ 0.491 & -0.073 & -0.055 \\ -0.345 & 0.236 & -0.073 \end{bmatrix}$$

The result can be checked by multiplying the matrix and its inverse:

$$\begin{bmatrix} 1 & 4 & 2 \\ 3 & 8 & 9 \\ 5 & 7 & 6 \end{bmatrix} \begin{bmatrix} -0.272 & -0.182 & 0.364 \\ 0.491 & -0.073 & -0.055 \\ -0.345 & 0.236 & -0.073 \end{bmatrix}$$

$$= \begin{bmatrix} 1.002 & -0.002 & -0.002 \\ 0.007 & 0.994 & -0.005 \\ 0.001 & -0.005 & 0.997 \end{bmatrix}$$

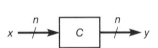

Figure A1.5 A square system, $y = Cx$.

Within the limits of the rounding errors in the inverse, the result is, indeed, I_3.

The work involved in calculating the inverse need not cause concern. It is rarely necessary to do it, as plenty of suitable computer packages exist. It is worth noting that they do not use the approach taken here, but use other methods which are less subject to numerical error, and are computationally more efficient.

Returning to Equation (A1.4), it is evident that the inverse of a matrix cannot be calculated if its determinant is zero. A square matrix with a zero determinant is called a *singular* matrix, and does not have an inverse.

In physical terms, a singular matrix will either be found to have some complete rows or columns of zeros, or some rows (or columns) will be found to be linear combinations of others. If the matrix is viewed in block diagram terms, as shown in Figure A1.5 (which can be expanded to look like a suitably sized equivalent of Figure A1.2), then the first case (rows or columns of zeros) implies that there is no connection either from one input to any of the outputs (for a zero column), or from any input to one of the outputs (for a zero row). The second case (linear dependence of rows or columns) implies that one output is a linear combination of the other outputs.

In either case, it is not possible to derive the values in x by measuring the values of y, because there are fewer than n degrees of freedom available to solve the n resulting equations. The only condition under which the values of x can be calculated from measurements of y is the case where the inverse of C exists (then $y = Cx$ so $x = C^{-1}y$).

A1.1.4   The rank of a matrix

As mentioned before, if a square matrix has a non-zero determinant, it also has an inverse. Such a matrix is said to be of *full rank*.

Technically, the rank of a matrix is the dimension of the largest non-zero determinant which can be extracted from it. This applies as much to rectangular matrices as to square ones but, in the case of a rectangular matrix, the maximum possible rank is the minimum dimension of the matrix. It will be necessary to find the rank of rectangular matrices in connection with controllability and observability studies (introduced in Chapter 5). As an example, the matrix

$$\begin{bmatrix} 1 & 3 & 7 & 2 & 1 \\ 2 & 2 & 3 & 4 & 1 \\ 0 & 4 & 11 & 0 & 1 \end{bmatrix}$$

could have a maximum rank of 3 (because no larger determinant than 3×3 exists in this matrix). This would be full rank for this matrix. It will be found, however, that there is no 3×3 determinant extractable from this matrix, whose value is non-zero. As an example, try columns 2, 4 and 5:

$$\begin{vmatrix} 3 & 2 & 1 \\ 2 & 4 & 1 \\ 4 & 0 & 1 \end{vmatrix} = 0$$

Any other possible 3×3 determinant extracted from the rectangular matrix above is also zero. The reason for this is a lack of linear independence. Row three can be seen to be twice row one minus row two.

It is possible to extract a large number of non-zero 2×2 determinants from the matrix, therefore its rank is 2 (note that just one non-zero 2×2 determinant is sufficient to confirm this). Since its rank could have been 3, this matrix is said to have a *rank deficiency* of 1 (full rank minus actual rank). This number (the rank deficiency) will also be of use in the text.

A1.2   Eigenvalues, eigenvectors and the characteristic equation

The physical significance of these properties of a matrix is discussed in the main text. Here, the method of calculating them is reviewed.

The eigenvalues and eigenvectors of a matrix are particular values which give the result

$$\lambda v = A v \tag{A1.5}$$

where:

A is a square matrix of size $(n \times n)$

λ is an *eigenvalue* of A, and is a scalar quantity

v is an *eigenvector* of A corresponding to λ. It is of size $(n \times 1)$

Equation (A1.5) implies that there are some particular values of v associated with A, for which multiplication of the vector v by the matrix A simply scales each element of v by the scalar factor λ .

A vector can be regarded as a line having magnitude (length) and direction, in a vector space with as many dimensions as the order of the vector (more than three dimensions are difficult to visualize, but the principle still holds good). For example, the vector $v = [1 \ 2 \ 3]^{\mathrm{T}}$ can be represented as in Figure A1.6. In the product Av, the square matrix A operates on v to produce a result having the same dimensions as v, and therefore being plottable in the same vector space (see Figure A1.6).

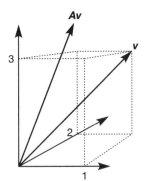

Figure A1.6 Vectors in a vector space.

Equation (A1.5) therefore represents the rather special case in which Av and v point in the same direction, differing only in length, by the factor λ.

To find the eigenvalues and eigenvectors, proceed as follows. First, rearrange Equation (A1.5):

$$\lambda v - A v = 0 \quad \text{or} \quad [\lambda I - A] v = 0 \tag{A1.6}$$

where 0 is a *null vector* (of size $n \times 1$, and with all elements zero).

Note also the necessity to keep the matrix dimensions conformable by inserting the unit matrix I_n to multiply the scalar λ, when extracting v as a factor. This is a common requirement in matrix algebra, and must not be overlooked. It is also important that v was extracted as a factor from the *right-hand side* of the bracket in order to keep the order of multiplication correct.

Equation (A1.6) can be rearranged as $v = [\lambda I - A]^{-1} 0$. This has a trivial solution $v = 0$, but that is not of interest, and therefore it is required that the inverse of $[\lambda I - A]$ should not exist, as the trivial solution will then be avoided. This will be the case when:

$$|\lambda I - A| = 0 \tag{A1.7}$$

Equation (A1.7) is an extremely important equation, which is used a lot in the text, and is called the *characteristic equation* of A. If the use of characteristic equations is recalled from transfer functions in frequency-domain control, it will eventually be seen that this is the same thing in a different guise.

Solving Equation (A1.7) for λ gives the eigenvalues of the matrix. For example, for the matrix

$$A = \begin{bmatrix} 6 & 2 \\ 4 & 1 \end{bmatrix}$$

the characteristic equation (Equation (A1.7)) is:

$$\left| \begin{bmatrix} \lambda & 0 \\ 0 & \lambda \end{bmatrix} - \begin{bmatrix} 6 & 2 \\ 4 & 1 \end{bmatrix} \right| = \begin{vmatrix} \lambda - 6 & -2 \\ -4 & \lambda - 1 \end{vmatrix}$$

$$= (\lambda - 6)(\lambda - 1) - 8 = \lambda^2 - 7\lambda - 2 = 0$$

the solutions for which are found to be $\lambda_1 = 7.275$ and $\lambda_2 = -0.275$.

Returning to Equation (A1.5), and substituting each eigenvalue in turn, gives the eigenvectors. Since these specify the *direction* in which Equation (A1.5) holds, their magnitude is unimportant, and therefore they can be scaled in any desired way, so long as the same is done to each element.

Using λ_1 and A in Equation (A1.5) gives

$$7.275 \begin{bmatrix} v_{11} \\ v_{21} \end{bmatrix} = \begin{bmatrix} 6 & 2 \\ 4 & 1 \end{bmatrix} \begin{bmatrix} v_{11} \\ v_{21} \end{bmatrix}$$

leading to: $7.275v_{11} = 6v_{11} + 2v_{21}$ and $7.275v_{21} = 4v_{11} + v_{21}$. Or, gathering terms: $1.275v_{11} = 2v_{21}$ and $4v_{11} = 6.275v_{21}$. Upon closer inspection, these two equations are found to be identical. They therefore have, as suggested above, an infinite number of possible solutions, obtained from each other by simple scaling factors. Arbitrarily choosing $v_{11} = 1$ yields $v_{21} = 0.638$. Thus, corresponding to the eigenvalue $\lambda_1 = 7.275$ is an eigenvector

$$v_1 = \begin{bmatrix} 1 \\ 0.638 \end{bmatrix}$$

The reader is invited to check that the eigenvector corresponding to $\lambda_2 = -0.275$ is similarly given by

$$v_2 = \begin{bmatrix} 1 \\ -3.138 \end{bmatrix}$$

If the MATLAB *eig* command (see Appendix 3) is used to calculate the eigenvectors of a matrix, it will be found that they are scaled in such a way that they are normalized to unit length. For example, the two eigenvectors above would be given by

$$\begin{bmatrix} 0.843 \\ 0.538 \end{bmatrix} \quad \text{and} \quad \begin{bmatrix} -0.304 \\ 0.954 \end{bmatrix}$$

respectively. The directions are the same, but the lengths have been scaled to unity.

Some matrices have eigenvalues which are very easy to find and lead to system models having particularly useful structures for some purposes. For example, a diagonal matrix (Section A1.1.3) has its eigenvalues on the leading diagonal. If it is drawn as a block diagram element (Section A1.1.1), it represents a series of independent gains (equal to the eigenvalues) with no interaction between them. A *triangular* matrix has all elements below, and to the left of, the leading diagonal equal to zero, for example

$$\begin{bmatrix} 1 & 6 & 5 \\ 0 & -4 & 2 \\ 0 & 0 & 7 \end{bmatrix}$$

Such a matrix also has its eigenvalues on the leading diagonal (that is, the eigenvalues of this matrix are 1, −4 and 7). In block diagram terms, this matrix represents a system in which input 1 only feeds output 1 (but so do inputs 2 and 3), output 3 is fed only from input 3 (but input 3 also feeds outputs 1 and 2), and the interactions between the other signals follow a definite pattern.

Eigenvalues also have other interesting properties. For example, the determinant of any square matrix is the *product* of its eigenvalues. The *trace* of any square matrix (the sum of the elements on the leading diagonal) is the *sum* of its eigenvalues. These two things are easily seen to be true for diagonal or triangular matrices, but also apply to all other square matrices.

A1.2.1 Left-hand eigenvectors

The eigenvectors calculated in Section A1.2 are strictly 'right-hand eigenvectors', because v is a factor on the right-hand side of each term in Equation (A1.5). There is no reason why a similar type of equation $v\lambda = vA$ could not be used instead. The eigenvectors here are called 'left-hand eigenvectors' of A and will have to be row vectors, rather than columns (why is this?).

Normally, if eigenvectors are mentioned (in any context), it is assumed that right-hand eigenvectors are used by default, to the extent that many people are unaware that left-hand ones even exist. However, a use for these left-hand ones is occasionally found.

For the second eigenvalue of the matrix A of Section A1.2, it was discovered that $\lambda_2 = -0.275$. Using this in the left-hand eigenvector equation, above, gives

$$[v_{11} \quad v_{12}](-0.275) = [v_{11} \quad v_{12}]\begin{bmatrix} 6 & 2 \\ 4 & 1 \end{bmatrix}$$

from which the following result is obtained: $-0.275v_{11} = 6v_{11} + 4v_{12}$ and $-0.275v_{12} = 2v_{11} + v_{12}$. Gathering terms yields: $6.275v_{11} = -4v_{12}$ and $2v_{11} = -1.275v_{12}$.

Again, these two equations are found to be identical. This time, it is decided (arbitrarily) to set $v_{12} = 1$, so that $v_{11} = -0.638$, giving the left-hand eigenvector $[-0.638 \ 1]$. This obviously bears a relationship to the right-hand eigenvector obtained for the other eigenvalue (λ_1) in Section A1.2. This need not be pursued here, but the reader can similarly obtain the result for the left-hand eigenvector corresponding to λ_1, and compare it with the right-hand eigenvector calculated for λ_2 in Section A1.2.

A1.3   *Partitioning of matrices*

In general, matrix methods are used as a means of writing system equations (especially state-space models) in a compact way, and carrying out analysis much more elegantly than could otherwise be done. Partitioning is a device which can be used to simplify matrix algebra even further.

It turns out that if a vector–matrix equation is divided (*partitioned*) into sub-matrices (*partitions*), by drawing straight lines through it in a systematic manner, then the basic operations of matrix algebra can be carried out using the separate partitions as if they were single matrix elements. Consider the following equation pre-multiplying a 3 × 2 matrix by a 3 × 3 matrix. Ignoring the partitioning lines, it will be found to be correct:

$$\left[\begin{array}{cc|c} 2 & -4 & 1 \\ -3 & 3 & 0 \\ \hline 1 & 7 & 2 \end{array}\right] \left[\begin{array}{cc} 2 & -1 \\ 3 & 0 \\ \hline 1 & 2 \end{array}\right] = \left[\begin{array}{cc} -7 & 0 \\ 3 & 3 \\ 25 & 3 \end{array}\right]$$

Now allow the lines to divide the matrices into the partitions shown. These can then be named (for example) as follows:

$$\left[\begin{array}{c|c} U & v \\ \hline w & x \end{array}\right] \left[\begin{array}{c} Y \\ \hline z \end{array}\right] = \left[\begin{array}{c} P \\ \hline q \end{array}\right]$$

where

$$U = \begin{bmatrix} 2 & -4 \\ -3 & 3 \end{bmatrix}, \quad v = \begin{bmatrix} 1 \\ 0 \end{bmatrix}, \quad w = \begin{bmatrix} 1 & 7 \end{bmatrix}, \quad x = 2$$

$$Y = \begin{bmatrix} 2 & -1 \\ 3 & 0 \end{bmatrix}, \quad z = \begin{bmatrix} 1 & 2 \end{bmatrix}$$

Performing the matrix multiplication in the partitioned equation gives $P = Uy + vz$ and $q = wY + xz$. Substituting the numerical values of U, v, w, x, Y and z, it is found that

$$P = \begin{bmatrix} -8 & -2 \\ 3 & 3 \end{bmatrix} + \begin{bmatrix} 1 & 2 \\ 0 & 0 \end{bmatrix} = \begin{bmatrix} -7 & 0 \\ 3 & 3 \end{bmatrix}$$

and

$$q = \begin{bmatrix} 23 & -1 \end{bmatrix} + \begin{bmatrix} 2 & 4 \end{bmatrix} = \begin{bmatrix} 25 & 3 \end{bmatrix}$$

which is correct. This partitioning procedure can generally be applied at will. It is only necessary to ensure that the selected partitions are of consistent dimensions for the operations to be carried out.

Addition and subtraction of partitioned matrices follow the obvious pattern. Other operations on partitioned matrices require more care. For example, a partitioned matrix can be transposed, but the individual partitions must also be transposed at the same time. By inserting the numerical values, it can be verified that, for the matrix above,

$$\left[\begin{array}{c|c} U & v \\ \hline w & x \end{array}\right]^{\mathrm{T}} = \left[\begin{array}{c|c} U^{\mathrm{T}} & w^{\mathrm{T}} \\ \hline v^{\mathrm{T}} & x^{\mathrm{T}} \end{array}\right]$$

(the transpose operator on x is not really needed in this example, because it is a scalar). There are more complicated rules for inverses of partitioned matrices.

Eigenvalues of partitioned matrices will sometimes be of use in the text. Specifically, if a partitioned matrix is triangular or diagonal (sometimes called a *block triangular* or *block diagonal* matrix), and all the partitions on the leading diagonal are square, then its eigenvalues will be those of the partitions on the leading diagonal. This is directly analogous to the unpartitioned cases discussed at the end of Section A1.2.

Some of the problems at the end of Chapter 2 use partitioning to obtain combined state-space models of systems made up of subsystems in state-space form.

A1.4 Definiteness of matrices

This topic is only of relevance to parts of Chapters 12 and 14.

A non-zero *scalar* value is clearly positive or negative. For non-zero *matrices*, this concept is replaced by the notion of 'definiteness' in suitable circumstances. For example, a quadratic form $x^{\mathrm{T}}Qx$ (see Equation (12.2), for example) in which the matrix Q is either *positive definite* or *positive semidefinite*, will always have a non-negative scalar result for any non-zero values in x. The difference between semidefinite and definite is the difference between whether the result can or cannot adopt the value zero itself, respectively. Negative definite and negative semidefinite matrices are similarly defined, with the obvious sign changes.

These definiteness properties can therefore be used in an optimal control performance index (for example) to ensure that a non-negative minimum will always exist.

If a matrix is diagonal, and all the diagonal entries are positive, it will be positive definite. If all the diagonal entries are non-negative, it will be positive semidefinite (with similar definitions for negative definite and negative semidefinite). For more general matrices, a test is necessary.

One test is to find the eigenvalues. If they are all real and positive, the matrix is positive definite. If they are all real and non-negative, the matrix is positive semidefinite.

Another test is to examine the *principal minors* of the matrix. These are the determinants of increasing size, each starting with element $(1, 1)$. Thus, for the $n \times n$ matrix Q (with $n > 3$), the principal minors are:

$$q_{11}, \quad \begin{vmatrix} q_{11} & q_{12} \\ q_{21} & q_{22} \end{vmatrix}, \quad \begin{vmatrix} q_{11} & q_{12} & q_{13} \\ q_{21} & q_{22} & q_{23} \\ q_{31} & q_{32} & q_{33} \end{vmatrix}, \ldots, \quad |Q|$$

If all these are positive, Q is positive definite. If they are non-negative, Q is positive semidefinite. The definitions of negative definite and negative semidefinite follow the lines which would be expected. If Q meets none of these definiteness criteria, it is said to be *indefinite*.

A1.5 Obtaining the McMillan form of a matrix

This is not a rigorous mathematical derivation, but gives sufficient information to bridge the gap in Chapter 10 of the text. In some references, this form, or a very similar one, is called the Smith–McMillan form, or the Smith normal form of a matrix.

Starting with the matrix of Equation (10.8):

$$G(s) = \begin{bmatrix} \dfrac{1}{s+1} & \dfrac{2}{s+3} \\ \dfrac{1}{s+1} & \dfrac{1}{s+1} \end{bmatrix}$$

firstly the monic least common multiplier is extracted (monic means that this term must have a unity coefficient for the highest power of s), so as to leave a polynomial matrix:

$$G(s) = \frac{1}{(s+1)(s+3)} \begin{bmatrix} s+3 & 2s+2 \\ s+3 & s+3 \end{bmatrix} \tag{A1.8}$$

The polynomial matrix from Equation (A1.8) is then worked on as follows:

Find the greatest monic common divisor $g_1(s)$ (say) of all the first-order minors. Recall from Section A1.1.2 (the part on determinants) that a *minor* is effectively the determinant of a sub-matrix. The first-order minors are determinants of size 1 (that is, scalars). In the polynomial matrix from Equation (A1.8), there are therefore four such first-order minors (that is, the individual elements). They only have a common divisor of unity, so in this case, $g_1(s) = 1$.

Next, $g_1(s)$ is made to appear in element $(1, 1)$ by row and column operations. In the present case, subtracting $0.5 \times$ (column 2) from column 1, and then multiplying row 1 by 0.5, gives:

$$\begin{bmatrix} 2 & s+2 \\ 0.5(s+3) & s+3 \end{bmatrix} \quad \text{then} \quad \begin{bmatrix} 1 & s+1 \\ 0.5(s+3) & s+3 \end{bmatrix}$$

The next step is to use further elementary operations to obtain the general form ($\times$ is a general entry)

$$\begin{bmatrix} g_1(s) & 0 & 0 & \cdots \\ 0 & \times & \times & \cdots \\ 0 & \times & \times & \cdots \\ \vdots & \vdots & \vdots & \ddots \end{bmatrix}$$

or, in this simpler case,

$$\begin{bmatrix} g_1(s) & 0 \\ 0 & \times \end{bmatrix} = \begin{bmatrix} 1 & 0 \\ 0 & \times \end{bmatrix}$$

This might be done by subtracting $(s + 1) \times$ (column 1) from column 2, then subtracting $0.5(s + 3) \times$ (row 1) from row 2, and finally multiplying row 2 by -1, giving:

$$\begin{bmatrix} 1 & 0 \\ 0.5(s + 3) & 0.5(s + 3)(1 - s) \end{bmatrix} \quad \text{then} \quad \begin{bmatrix} 1 & 0 \\ 0 & 0.5(s + 3)(s - 1) \end{bmatrix}$$

Next, the greatest monic common divisor of all the second-order (that is, 2×2) minors is found, $g_2(s)$. In the present case this is trivial, because there is only one such minor in a 2×2 matrix (the entire matrix is one), but it illustrates how the method would be continued for a larger matrix. For the present case, $g_2(s) = (s + 3)(s - 1)$.

$g_2(s)/g_1(s)$ is then made to appear in element $(2, 2)$ by row and column operations. In this case, this simply means multiplying row 2 by 2, giving

$$\begin{bmatrix} 1 & 0 \\ 0 & (s + 3)(s - 1) \end{bmatrix} \tag{A1.9}$$

For a larger matrix, in order to continue the process the following form would be found:

$$\begin{bmatrix} g_1(s) & 0 & 0 & 0 & \cdots \\ 0 & \dfrac{g_2}{g_1}(s) & 0 & 0 & \cdots \\ 0 & 0 & \times & \times & \cdots \\ 0 & 0 & \times & \times & \cdots \\ \vdots & \vdots & \vdots & \vdots & \ddots \end{bmatrix}$$

Then $g_3(s)$ would be found as the greatest monic common divisor of the third-order minors, $g_3(s)/g_2(s)$ would be made the next diagonal element, and so on.

In the present example, the process is complete by Equation (A1.9), and the end result is obtained by re-including the factor extracted in Equation (A1.8), to give:

$$\boldsymbol{M}(s) = \frac{1}{(s + 1)(s + 3)} \begin{bmatrix} 1 & 0 \\ 0 & (s + 3)(s - 1) \end{bmatrix}$$

or, as in the text,

$$\boldsymbol{M}(s) = \begin{bmatrix} \dfrac{1}{(s + 1)(s + 3)} & 0 \\ 0 & \dfrac{(s - 1)}{(s + 1)} \end{bmatrix}$$

A1.6   The similarity transform

This is an interesting matrix operation which is required at a number of points in the text, and also in the next section. It can be applied to partitioned matrices (Section A1.3). Here, it is introduced from a state-space viewpoint.

Consider the usual state equation $\dot{x} = Ax + Bu$, having a state vector x. It may be desirable to use a different set of states ($\tilde{x}$, say), either for its own sake, or perhaps in order to achieve a different form of the A matrix for some reason.

Now relate this new set of states to the existing set by a transformation matrix T, which must be invertible, such that $x = T\tilde{x}$. Substitution into the state equation then gives $T\dot{\tilde{x}} = AT\tilde{x} + Bu$. Premultiplying each side by T^{-1} then gives the new state equation: $\dot{\tilde{x}} = \tilde{A}\tilde{x} + \tilde{B}u$ where $\tilde{A} = T^{-1}AT$ and $\tilde{B} = T^{-1}B$. Treating the output equation $y = Cx + Du$ in the same way leads to $y = \tilde{C}\tilde{x} + Du$, where $\tilde{C} = CT$.

Careful choice of the transformation matrix T can lead to some interesting effects. For example, if T is chosen to be the eigenvector matrix (*modal matrix*) of A, then $\tilde{A}$ will be found to be a diagonal matrix having the eigenvalues of A on its leading diagonal.

As an example, consider the matrix

$$A = \begin{bmatrix} -3 & -2 \\ 1 & 0 \end{bmatrix}$$

Using the methods of Section A1.2, this is found to have eigenvalues of -2 and -1, and eigenvectors

$$v_1 = \begin{bmatrix} 1 \\ -0.5 \end{bmatrix} \quad \text{and} \quad v_2 = \begin{bmatrix} 1 \\ -1 \end{bmatrix}$$

If a transformation matrix is chosen as

$$T = [v_1 \quad v_2] = \begin{bmatrix} 1 & 1 \\ -0.5 & -1 \end{bmatrix}$$

then

$$T^{-1} = \begin{bmatrix} 2 & 2 \\ -1 & -2 \end{bmatrix}$$

and

$$\tilde{A} = T^{-1}AT = \begin{bmatrix} 2 & 2 \\ -1 & -2 \end{bmatrix} \begin{bmatrix} -3 & -2 \\ 1 & 0 \end{bmatrix} \begin{bmatrix} 1 & 1 \\ -0.5 & -1 \end{bmatrix}$$

$$= \begin{bmatrix} -2 & 0 \\ 0 & -1 \end{bmatrix} = \Lambda$$

where Λ is the diagonal matrix having the eigenvalues of A on the leading diagonal as predicted. The fact that $\tilde{A}$ and A have the same eigenvalues makes them *similar matrices* in the mathematical sense, and this happens for any allowable choice of T, not just the choice leading to a diagonal $\tilde{A}$. This is the reason for the name of the transform. Since the eigenvalues of similar matrices are the same, so are their determinants (which are the products of the eigenvalues).

The operation of the similarity transform is viewed 'in reverse' in the next section. By this, it is meant that any square matrix can equivalently be regarded as being made up of a similarity transform acting on a diagonal matrix of its eigenvalues (Λ). This is easily justified by pre-multiplying both sides of the equation $T^{-1}AT = \Lambda$ (which was derived above) by T, and post-multiplying by T^{-1}.

This leads directly to $A = TAT^{-1}$, which is used in Section A1.7 (with Q and W in place of A and T).

Note that some texts use a notation of the type $\tilde{A} = TAT^{-1}$, where this text uses $\tilde{A} = T^{-1}AT$. This is not an error; it simply means that whereas T is the transformation matrix here, the other text used T^{-1}, beginning from the premise that $\tilde{x} = Tx$, rather than $x = T\tilde{x}$.

There are other interesting choices for T. For a controllable (see Section 5.4), single-input system, T can be chosen to give the matrix $\tilde{A} = T^{-1}AT$ in a controllable companion form which is useful for some purposes. Another of the properties of similar systems is that their controllability is the same, so the original system must be controllable in this case. It is used in Ackermann's method for designing state variable feedback controllers (Section 5.4.5).

A1.7 A non-dynamic example of matrix expansion (spectral decomposition)

This section follows directly from Section A1.6, but is written in language to suit Chapter 10. Consider the matrix

$$Q = \begin{bmatrix} 1 & -2/3 \\ 3 & 4 \end{bmatrix}$$

By the usual methods (Section A1.2), this is found to have eigenvalues given by $\lambda_1 = 3$ and $\lambda_2 = 2$. The corresponding eigenvectors (Section A1.2) are

$$v_1 = \begin{bmatrix} 1 \\ -3 \end{bmatrix} \quad \text{and} \quad v_2 = \begin{bmatrix} 1 \\ -1.5 \end{bmatrix}$$

The eigenvector matrix (*modal matrix*) is thus

$$W = [v_1 \quad v_2] = \begin{bmatrix} 1 & 1 \\ -3 & -1.5 \end{bmatrix} \quad \text{and so} \quad W^{-1} = \begin{bmatrix} -1 & -2/3 \\ 2 & 2/3 \end{bmatrix}$$

If the quantity $W.\text{diag}\{\lambda_1, \lambda_2\}.W^{-1}$ is evaluated (see Chapter 10 and also Section A1.6), the following is obtained:

$$\begin{bmatrix} 1 & 1 \\ -3 & -1.5 \end{bmatrix} \begin{bmatrix} 3 & 0 \\ 0 & 2 \end{bmatrix} \begin{bmatrix} -1 & -2/3 \\ 2 & 2/3 \end{bmatrix} = \begin{bmatrix} 1 & -2/3 \\ 3 & 4 \end{bmatrix} = Q$$

Although not a proof in the mathematical sense, this indicates that a square matrix can be viewed as a decomposition involving a similarity transform (Section A1.6), in which the modal matrix is used as the transforming matrix, operating on a diagonal matrix of the eigenvalues of the original matrix.

Appendix 2
Partial-fraction expansions in inverse Laplace transforms

A2.1   Introduction

In Section 2.5.3, Equation (2.71) defined a Laplace function as a rational polynomial in s:

$$F(s) = \frac{N}{D}(s) = \frac{b_m s^m + b_{m-1} s^{m-1} + \cdots + b_0}{s^n + a_{n-1} s^{n-1} + \cdots + a_0}$$

where $N(s)$ and $D(s)$ are polynomials as shown.

This appendix addresses the problem of how to put $F(s)$ into partial fractions which appear in the Laplace transform tables and thus allow time-domain solutions to be found. The type of partial fraction expansion depends upon the roots of $D(s)$.

A2.2   Cases relevant to hand calculation

For hand calculation only two cases need be considered.

A2.2.1   An unrepeated real, or complex, root of D(s)

If $-p_i$ is an unrepeated pole of $F(s)$, there is a corresponding partial fraction

$$\frac{C_i}{s + p_i}$$

for which it is evident that

$$\mathscr{L}^{-1}\left[\frac{C_i}{s + p_i}\right] = C_i e^{-p_i t} \tag{A2.1}$$

Note that a root of $D(s)$ is a pole of $F(s)$ since, at $s = -p_i$, $F(s)$ becomes infinite.

A2.2.2 A repeated real, or complex, root of D(s)

If $-p_i$ is a pole of order n (that is, if there are n poles all at $s = -p_i$), there is a corresponding partial-fraction expansion of the form

$$\frac{C_{i,n}}{(s+p_i)^n} + \frac{C_{i,n-1}}{(s+p_i)^{n-1}} + \cdots + \frac{C_{i,1}}{s+p_i}$$

and the inverse transform is, from Tables 2.8 and 2.9,

$$\mathscr{L}^{-1}\left[\frac{C_i(s)}{(s+p_i)^n}\right] = e^{-p_i t} \sum_{k=1}^{n} C_{i,k} \frac{t^{k-1}}{(k-1)!} \tag{A2.2}$$

Equations (A2.1) and (A2.2) may be combined so that, if $F(s)$ has a total of N distinct poles, the complete inverse transform is given by

$$f(t) = \sum_{i=1}^{N}\left(e^{-p_i t} \sum_{k=1}^{n} C_{i,k} \frac{t^{k-1}}{(k-1)!}\right) \tag{A2.3}$$

where i denotes the pole, and k the order of the pole in the partial fraction.

A2.3 Cases relevant to machine calculation

For machine calculation, Equation (A2.3) would normally be used only to find the inverse transform of real poles. If a pole is complex, it is more convenient to combine it with its complex conjugate, which will always exist if the coefficients of $D(s)$ are real. Therefore, for a pair of complex conjugate poles, $s = -a + jb$ and $s = -a - jb$, it is usual to write the corresponding partial function in the form

$$\frac{As + B}{(s+a)^2 + b^2}$$

where A and B are real constants.

In this form two further expansions need to be considered.

A2.3.1 An unrepeated pair of complex conjugate roots of D(s)

A quadratic, unrepeated factor of $D(s)$ with poles at $s = -a \pm jb$ contributes a term in the time response $f(t)$ defined by

$$f_1(t) = \mathscr{L}^{-1}\left[\frac{As + B}{(s+a)^2 + b^2}\right]$$

In partial-fraction form this is

$$f_1(t) = \mathscr{L}^{-1}\left[\frac{C_R + jC_I}{s+a+jb} + \frac{C_R - jC_I}{s+a-jb}\right]$$

and the inverse Laplace transform is

$$f_1(t) = (C_R + jC_I)e^{-(a+jb)t} + (C_R - jC_I)e^{-(a-jb)t} \tag{A2.4}$$

Equation (A2.4) may be rearranged, using Equation (2.65) (Section 2.5.2 in the main text) to give

$$f_1(t) = e^{-at}(2C_R \cos bt + 2C_I \sin bt)$$

which is the required inverse transform.

A2.3.2   A repeated pair of complex conjugate roots of D(s)

If $D(s)$ has n repeated quadratic factors of the form $[(s+a)^2 + b^2]^n$, there corresponds a partial-fraction expansion having an inverse Laplace transform given by

$$e^{-at} \sum_{k=1}^{n} \left(\frac{t^{k-1}}{(k-1)!} \left(2C_{R_k} \cos bt + 2C_{I_k} \sin bt\right) \right)$$

This result follows directly from the previous three cases considered.

A2.4   Heaviside's partial-fraction technique

In each of the above four cases the partial-fraction expansion leads directly to the inverse Laplace transform, but it remains to find the coefficients C of the various partial-fraction expansions. This is done by using the procedure which has been credited to the electrical engineer Oliver Heaviside (1850–1925) and called the Heaviside technique. The method consists of four formulae for calculating the inverse Laplace transforms of components in the decomposed form of $F(s)$. Each formula corresponds to one of the four types of factor, already considered, which can arise in the denominator of $F(s)$. They are as follows:

A2.4.1   A distinct linear factor of the form (s + pᵢ)

If the denominator of $F(s)$ contains a linear factor $s + p_i$ (that is, if there is a first-order real pole at $s = -p_i$), there will be a term in $f(t)$ corresponding to this factor which can be computed from the equation

$$f(t)_{s+p_i} = [(s+p_i)F(s)e^{st}]|_{s=-p_i} \tag{A2.5}$$

Thus, the coefficient C_i in Equation (A2.1) may be written as

$$C_i = \lim_{s \to -p_i} [(s+p_i)F(s)]$$

or

$$C_i = \lim_{s \to -p_i} \left[(s+p_i) \frac{N(s)}{D(s)}\right] \tag{A2.6}$$

or

$$C_i = \frac{N(s)}{D'(s)} \Bigg|_{s=-p_i} \tag{A2.7}$$

where $D'(s)$ denotes the differential of $D(s)$ with respect to s.

Coefficients of simple, non-repeated linear factors are called residues; Equations (A2.6) and (A2.7) may be used to find the residue C_i.

A2.4.2 Repeated linear factors of the form $(s + p_i)^n$

If there is a factor $(s + p_i)^n$ in the denominator of $F(s)$, there will be corresponding terms in $f(t)$ which can be computed from the equation

$$f(t)_{(s+p_i)^n} = \frac{1}{(n-1)!} \left[\frac{d^{n-1}}{ds^{n-1}} \left[(s+p_i)^n F(s) e^{st} \right] \right]\Bigg|_{s=-p_i} \tag{A2.8}$$

Alternatively, Equation (A2.8) may be written in the form

$$f(t)_{(s+p_i)^n} = e^{-p_i t} \sum_{k=1}^{n} \left(\frac{t^{k-1}}{(n-k)!(k-1)!} \frac{d^{n-k}}{ds^{n-k}} \left[(s+p_i)^n F(s) \right] \right)\Bigg|_{s=-p_i}$$

Thus, $C_{i,k}$ in Equation (A2.2) is given by

$$C_{i,k} = \frac{1}{(n-k)!} \left(\frac{d^{n-k}}{ds^{n-k}} (s+p_i)^n F(s) \right)\Bigg|_{s=-p_i} \tag{A2.9}$$

The above two methods of finding coefficients are valid for both real and complex poles. However, as has already been indicated, for machine computation it is preferable to use other formulae when the poles are complex.

A2.4.3 A distinct quadratic factor of the form $(s + a)^2 + b^2$

A quadratic unrepeated factor representing a pair of complex conjugate poles at $s = -a \pm jb$ contributes a term in $f(t)$ defined by

$$f(t)|_{(s+a)^2 + b^2} = e^{-at}(2C_R \cos bt + 2C_I \sin bt) \tag{A2.10}$$

where C_R and C_I are the real and imaginary parts of

$$[(s+a-jb)F(s)]|_{s=-(a+jb)} \tag{A2.11}$$

A2.4.4 Repeated quadratic factors of the form $[(s + a)^2 + b^2]^n$

Finally, for each repeated quadratic factor there will be a term of the form

$$f(t)|_{[(s+a)^2+b^2]^n} = e^{-at} \sum_{k=1}^{n} \frac{t^{k-1}}{(k-1)!} (2C_{R_k} \cos bt + 2C_{I_k} \sin bt) \tag{A2.12}$$

where

$$C_{R_k} + jC_{I_k} = \frac{1}{(n-k)!} \left(\frac{d^{n-k}}{ds^{n-k}} [(s+a-jb)^n F(s)] \right)\Bigg|_{s=-(a+jb)} \tag{A2.13}$$

Equations (A2.5), (A2.8) and (A2.10)–(A2.13) are the required Heaviside formulae.

A2.5 Examples

The remainder of this section illustrates with examples some techniques of partial-fraction expansion. An examination of these examples, together with the Laplace transform tables, Tables 2.8 and 2.9, should enable each of the Heaviside formulae to be derived.

Example A2.1 *Partial fraction expansion for unrepeated real poles*

Prove that Equation (A2.6) gives the partial fraction coefficient associated with a linear unrepeated factor of the denominator of $F(s)$.

Let

$$F(s) = \frac{N(s)}{D(s)} = \frac{C_i}{s + p_i} + \text{(all other terms arising from the factors of } D(s))$$

To find the partial fraction coefficient C_i multiply throughout by $s + p_i$ to give

$$(s + p_i) \frac{N(s)}{D(s)} = C_i + (s + p_i) \times \text{(all other terms)}$$

If s is set equal to $-p_i$, then

$$C_i = \left((s + p_i) \frac{N(s)}{D(s)} \right) \Bigg|_{s=-p_i}$$

which is the required solution. In practice, the above equation amounts to what is sometimes called 'the cover-up rule'. The factor $s + p_i$ in the denominator of the rational polynomial $F(s)$ is covered up, and to find the coefficient C_i all the remaining values of s are set equal to $-p_i$.

Example A2.2 *Partial fraction expansion for repeated real poles*

Prove that Equation (A2.9) gives the partial-fraction coefficients associated with a linear repeated factor in the denominator polynomial $D(s)$.

Let

$$F(s) = \frac{C_{i,n}}{(s + p_i)^n} + \frac{C_{i,n-1}}{(s + p_i)^{n-1}} + \cdots + \frac{C_{i,1}}{s + p_i} + \text{(all remaining terms)}$$

Multiplying throughout by $(s + p_i)^n$ and setting $s = -p_i$ will yield the sum $C_{i,n}$. This is exactly the same procedure as used in the previous example, Example A2.1.

To find $C_{i,n-1}$ first multiply throughout by $(s + p_i)^n$ to give

$$(s + p_i)^n F(s) = C_{i,n} + C_{i,n-1}(s + p_i) + (s + p_i)^n \times \text{(all other terms)}$$

Differentiating both sides of the above equation with respect to s gives

$$\frac{d}{ds}\left[(s + p_i)^n F(s)\right] = C_{i,n-1} + n(s + p_i)^{n-1} \times \text{(all other terms)}$$

Evaluating at $s = -p_i$ gives $C_{i,n-1}$ and, by induction,

$$C_{i,k} = \frac{1}{(n-k)!}\left(\frac{d^{n-k}}{ds^{n-k}}(s + p_i)^n F(s)\right)\Bigg|_{s=-p_i}$$

which is the required proof (see Equation (A2.9)).

Example A2.3 *An inverse Laplace transform of mixed type*

Find the inverse Laplace transform of

$$f(s) = \frac{s + 3}{(s + 1)^2(s - 2)}$$

The partial-fraction expansion of $F(s)$ is

$$\frac{s + 3}{(s + 1)^2(s - 2)} = \frac{C_1}{s - 2} + \frac{C_{2,2}}{(s + 1)^2} + \frac{C_{2,1}}{s + 1}$$

The coefficients C_1 and $C_{2,2}$ are easily calculated using the cover-up rule to be 5/9 and $-2/3$ respectively. $C_{2,1}$ is obtained by differentiation:

$$C_{2,1} = \left(\frac{d}{ds}\frac{s + 3}{s - 2}\right)\Bigg|_{s=-1} = -\frac{5}{9}$$

The inverse Laplace transform then follows as

$$f(t) = -\frac{2}{3}te^{-t} - \frac{5}{9}e^{-t} + \frac{5}{9}e^{2t}$$

An alternative way to find the partial-fraction expansion would be to let

$$F(s) = \frac{1}{s + 1}\frac{s + 3}{(s + 1)(s - 2)}$$

Taking a partial-fraction expansion of the right-hand side gives

$$F(s) = \frac{1}{s + 1}\left(\frac{-2/3}{s + 1} + \frac{5/3}{s - 2}\right)$$

$$= \frac{-2/3}{(s + 1)^2} + \frac{5/3}{(s + 1)(s - 2)}$$

A further partial-fraction expansion yields the required result:

$$F(s) = \frac{-2/3}{(s+1)^2} - \frac{5/9}{s+1} + \frac{5/9}{s-2}$$

Example A2.4 *An inverse Laplace transform involving repeated differentiation*

Find the inverse Laplace transform of

$$F(s) = \frac{s+2}{(s-1)^2 s^3}$$

This example is included since it requires repeated differentiation. The expansion of $F(s)$ takes the form

$$F(s) = \frac{C_{1,2}}{(s-1)^2} + \frac{C_{1,1}}{s-1} + \frac{C_{2,3}}{s^3} + \frac{C_{2,2}}{s^2} + \frac{C_{2,1}}{s}$$

The coefficients $C_{1,2}$ and $C_{2,3}$ are readily evaluated using the cover-up rule as 3 and 2 respectively. $C_{1,1}$ and $C_{2,2}$ require a single differentiation:

$$C_{1,1} = \left(\frac{d}{ds} \frac{s+2}{s^3} \right)\Bigg|_{s=1} = -8$$

and

$$C_{2,2} = \left(\frac{d}{ds} \frac{s+2}{(s-1)^2} \right)\Bigg|_{s=0} = 5$$

The final coefficient is obtained by double differentiation. From Equation (A2.9),

$$C_{2,1} = \frac{1}{2!} \left(\frac{d^2}{ds^2} \frac{s+2}{(s-1)^2} \right)\Bigg|_{s=0} = 8$$

from which it follows that the time function is

$$f(t) = 3te^t - 8e^t + 2t^2 + 5t + 8$$

In ending this section, it is worth noting that if a factor p_i is complex, the coefficient C_i is complex, too. This poses no real problem for hand calculation, but it does make machine computation awkward.

Appendix 3
A brief introduction to
MATLAB® and its toolboxes

MATLAB (The Mathworks Inc., 1993a, 1993b – first referenced in Section 1.3.7) is basically a 'number-crunching' mathematical software package, with add-on 'toolboxes' which can be bought to make it more easily usable for specific areas of work (such as control systems work). It is described more fully below, together with the 'bits' of it which are necessary in order to be able to carry out the examples in the text.

It is assumed that it is known how to start MATLAB on whatever machine is being used. With MATLAB versions of 4.0, or later (referred to as 'v4.x' from now on – version 4.2c.1 is current at the time of writing), this will normally be done by selecting an icon from a windows-type operating environment (for example, on a PC, Mac', or workstation). In older versions of MATLAB (for example, version 3.5, referred to as 'v3.x' from now on), MATLAB might still have been set up to run from an icon in a windows-type operating environment, but it is equally likely that it may be necessary to type *matlab* at an operating system prompt, for example. If it is not known how to start MATLAB at this stage, another source of help must be consulted – for example, the MATLAB manuals for the specific installation, or the computer system manager. These notes are to allow a start to be made with using MATLAB, particularly for the examples on the disk accompanying this book, and to answer frequently asked questions. They are not intended to replace other sources of reference.

A3.1   Introduction – what is MATLAB, and what other reference sources are there?

MATLAB is a command-driven, interactive language, aimed at the solution of problems involving vectors and matrices (some of the commands are listed shortly). That may sound rather restrictive, but in fact the opposite is true. Sets of data from real plant can be written in rows or columns, when they immediately become vectors. Several sets of such data form a matrix. The coefficients in a transfer function (or any other equation) can be written as a row of numbers – a vector again. If a plant or a control system is modelled in state-space form, then the model is already written in terms of matrices and vectors. This means that MATLAB is good for providing computer assistance in most branches of science and engineering.

The only data structure which MATLAB uses is a non-dimensioned matrix (or array), the dimensions being adjusted automatically by MATLAB as required. Thus, given a matrix

$$A = \begin{bmatrix} 1 & 2 & 3 \\ 2 & 8 & 4 \\ 1 & 7 & 9 \end{bmatrix}$$

for example, it is very easy to enter the matrix into MATLAB and, by way of an example, to find its inverse and its eigenvalues. All that is necessary is to type the following commands (at the MATLAB prompt $\gg$), terminating each line with a 'Return' (or 'Enter') keystroke (some comments follow the example):

$\gg$ a = [1 2 3   (or, alternatively, $\gg$ a = [1 2 3; 2 8 4; 1 7 9])

  2 8 4

  1 7 9]

$\gg$ inv(a)

$\gg$ eig(a)

Comments: MATLAB is case-sensitive. This can be overridden, but it is recommended that lower-case is used throughout so, in this appendix, *a* is used where **A** would normally appear, for example (vectors and matrices are distinguished purely by their context and size). Lower-case input should certainly always be used for MATLAB commands, otherwise they may not be executed in some versions. For example, to exit MATLAB, the command is:

$\gg$ quit

If $\gg$ *QUIT* is entered instead, it is possible that an error message may result, or the command may simply be ignored, depending upon the setup (version number and operating system).

Two different methods of entering the matrix *a* are shown above. Whichever is preferable to the user will work. It is always possible to enter more than one line's worth of MATLAB commands on a single line, by using semicolons to separate information that ought to be on separate lines.

If the commands above are entered, it will be found that MATLAB displays the result of each one. To prevent this, simply terminate the command with a semicolon. For example, the command

$$\gg a = [1 \quad 2 \quad 3; \quad 2 \quad 8 \quad 4; \quad 1 \quad 7 \quad 9];$$

will enter the matrix *a* into the workspace, but will *not* echo the result.

Many MATLAB commands can be used either with or without left-hand arguments. For example, the command *inv(a)*, above, displayed the inverse of the matrix *a*. If seeing the result was unnecessary, the command could be terminated with a semicolon. What then would be the point of using it? Not much! However, the inverse of *a* might well be required in some later calculations, so it could be stored as a new matrix (say *ai*), by the command:

$$\gg ai = inv(a);$$

It is a reasonably general rule that if a command is used with left-hand arguments it will store data for use later, whereas if it is used without them it will give immediate results (a number, a plot, and so on). More examples will be found later.

Another fairly general rule is that square brackets [ ] are used to delimit vectors and matrices, while parentheses ( ) are used to enclose command argument lists.

Apart from these notes, other reference sources are MATLAB's own help system (see later), textbooks such as Biran and Breiner (1995), Ogata (1994, two titles) and Strum and Kirk (1994), together with the manuals published by the Mathworks (1993a, 1993b) (the suppliers of MATLAB). If Internet access is available, the Mathworks maintain a Web site which includes a list of textbooks based on MATLAB, together with other useful information. It can be found at http://www.mathworks.com.

In many versions of MATLAB, further information is forthcoming after using:

$$\gg intro$$

and $$\gg demo$$

Several MATLAB commands have already been used in this section. Very many more are used in the examples on the disk which accompanies this text. Here is a list of about one hundred which are mentioned by name in the text, with a selection of sections in which they appear. Note that some of these commands are not in MATLAB itself, but are in add-on toolboxes (which are described in the next section). The vast majority of the commands are either in MATLAB, or in the Control Systems Toolbox (see Section A3.2) (the rest are in the Multivariable Frequency Domain Toolbox or SIMULINK). The MATLAB *help* system (see Section A3.3) can be used to provide information on any of these commands. If it fails to do so, the command in question is probably in a toolbox which is unavailable on the reader's system. The indicated text sections, together with the associated files on the accompanying disk, give examples of their use.

acker 5.4.7; *align* 10.6.3, 10.6.5; *augstate* Chapter 5 problems, A3.9;

bode 3.5.1, 3.5.2, A3.6;

A3.2 Toolboxes and m-files

Since MATLAB was written to handle matrices and vectors, and since real-world data can be represented as vectors (columns of values against time, for example), it follows that MATLAB finds wide use in all fields of science and engineering. Various *toolboxes* are available, which extend the functions of MATLAB into specific areas of work. Each has its own manual.

There are many functions in MATLAB of direct use in control engineering. The *Control System Toolbox* (The Mathworks Inc., 1994) (CSTB) was therefore written to make use of these in commands which extend MATLAB's basic facilities. This provides single commands for such things as Bode plots, time responses, state variable feedback design and so on. These are for single-input-single-output (SISO) systems (some CSTB commands will also work with multi-input-single-output systems), represented in any of a number of continuous and discrete forms (Laplace or z transfer function, state space, and so on). MATLAB plus the CSTB will suffice for most of the work in this text.

There are many other toolboxes available for MATLAB. The only other one used in this text is the *Multivariable Frequency Domain Toolbox* (Cambridge Control, 1990) (MVFDTB) used mainly in Chapter 10.

The way these toolboxes work is of some relevance. They are actually written in MATLAB (that is, they use the statements and commands of the MATLAB language). They consist of collections of files called *m-files* (because they have the filename extension .m, short for 'meta'). Anyone can write an m-file. It is just an ASCII file created using any ASCII text editor, and containing a sequence of MATLAB commands, typed exactly as they would be from the keyboard when using MATLAB interactively, with nothing else added.

In MATLAB versions 4.x, the process is made easy by (on a PC system, for example) clicking on *file* in the menu bar, then on *new*, then on *m-file* – which will open a system editor. (To alter an existing m-file, use *file* in the menu bar, then *open m-file* – but remember that the editor's *save* command must be issued before any changes take effect.) Thus, an m-file called garbage.m might be created, containing nothing but the following lines:

a = [1 2 3; 2 8 4; 1 7 9]

inv(a)

eig(a)

and simply entering the filename (without the .m extension) in response to the MATLAB prompt, would execute the commands in the file. Thus

》 garbage

would have exactly the same result as entering the original commands in Section A3.1. The file garbage.m has effectively become a new MATLAB command. All the m-files on the accompanying disk work in this way.

It is exactly the same with the toolboxes. Any toolbox installed on the system simply adds its own collection of m-files to the MATLAB directory structure (see Section A3.4), and they become new MATLAB commands. It is not necessary for the user to know that they are in a toolbox – if they are on the system, they appear just as any other MATLAB commands, and are run by entering their command name at the MATLAB prompt.

Finally, if it is desired to 'crash' running m-files which have gone into a loop, or which are not doing what was intended, the usual things can be tried, such as <CTRL>C and <CTRL>Z (but do not blame anyone except yourself for the results!).

A3.3   The MATLAB help system – and other tips

One disadvantage of MATLAB to the new (or occasional) user is that it is necessary to know the commands before they can be used! However, given some idea of what a required command might be called, or what it needs to do, the online help is reasonable. If the following command is entered:

》 help

in any version of MATLAB, some assistance will be forthcoming. In earlier versions (v3.x), the procedure after that is normally just to type help followed by a command name of interest. For example:

》 help eig

will tell you all about the *eig* command which was used earlier – it will be found that it can give eigen*vector* information as well as eigenvalues.

In later versions (v4.x), that is also the basic way to use help. However, there are alternatives. For example, typing *help* on its own now gives a list of broad help topics, and typing *help topic* (where 'topic' is some interesting item from the list) then gives more detail. For example, in the topic list is a topic *matlab\general*. The command:

》 help general

will then expand this category, giving about six further categories, each containing (say) eight or nine individual commands. One of these is the command *who*. Typing

》 help who

then informs you that the *who* command lists variables currently in the workspace. Executing the command, thus:

》 who

lists them. If the examples above have been followed, the variables a and ai will be listed as current. What they contain can be discovered simply by typing their name at the MATLAB prompt, as in:

》 a

The size of a variable can be found by typing the following command, and so on. 》 *help size* explains this, and the command *whos* is also worth a try:

》 size(a)

If the control toolbox is installed, there will have been a topic toolbox\control in the help list.

》 help control

will list its contents, from which some commands of interest will be noticed (*step*, *bode*, *impulse*, *nyquist* and so on). Again, *help nyquist* (for example) will give more information.

If this method of searching for likely commands is thought too unwieldy, try the *lookfor* command. ≫ *help lookfor* will give information about it but, as an example, to find out how to integrate a function, try:

≫ lookfor integ

This command might result in a list of several suggested commands which could be considered (depending on the toolboxes available). The normal *help* command can then be used to check those of greatest interest.

Finally, *help* can be selected from the menu bar at the top of the MATLAB window, in installations where this is visible.

Help can also be obtained on some syntactical aspects of MATLAB. For example, the colon is a potentially powerful operator in several circumstances and, although it might be rather advanced for new users at this stage, the following command will give the details:

≫ help colon

Note that the command *help* : will list all the possible syntactic and logical topics, although not all have any further help available beyond that which is then given.

To find out even more detail about a command than the help system will give, there are two options. Firstly, the relevant manual may contain more information. Secondly, for many commands (most toolbox commands), the actual MATLAB code is available in the m-file which corresponds to the command name. This can be extremely useful, as it allows the MATLAB commands to be edited so as to make them behave in a different way. For example, the *step* command from the CSTB could be altered, so that it draws grid lines on the step response by default. When doing this, be sure not to overwrite the original MATLAB command – use a copy with a slightly different name.

To look at commands in this way, use the *type* command as below, or simply use any ASCII text editor (which will have to be done in any case, if the command is to be altered).

≫ type step

Note again that, if commands are altered in this way, the modified version must be saved with a new m-file name – do *not* try to overwrite the original command. Note also that MATLAB will only allow m-file-type commands to be viewed and altered in this way – not built-in MATLAB functions. If the m-files on the accompanying disk are to be used as templates for your own work, these too should be copied and renamed – do not overwrite the originals!

Another point worthy of note is that if a standard word processor is used in ASCII mode (or 'text only', or 'nondocument', perhaps) then some WP packages are better than others at really being 'text only'. At least one very well-known WP package is notorious for adding extra characters at the ends of lines in a file which is supposedly pure ASCII text. If a WP package, which has not been used successfully for m-files before, is used to write an m-file, and the m-file does not run as expected, then this aspect should be investigated, as well as checking the MATLAB code.

In Section A3.5, it is shown how to add help text for your own m-files to the built-in help system.

One final note in this section – if a program variable is given the same name as a MATLAB command, the variable name will take precedence. For example, if a variable called *eig* is created, and later the *eig* command is issued, an error message will be generated. If error messages have no obvious cause, it is worth checking to see if a variable with the same name as the command causing the error exists in the workspace (use *who*). To get rid of such a variable, copy it to a variable having a different name, and then clear it from the workspace using (for example):

>> eigv = eig;

>> clear eig

The *eig* command will then work properly again, and the variable *eigv* contains the data that used to be in the variable *eig*.

A3.4   *The MATLAB directory structure – including the working directory and the matlabpath*

This may vary somewhat with version and installation, and is the authors' interpretation of events. The following comments apply largely to v4.x, running under Windows on PCs. If MATLAB is simply to be used as supplied (assuming a correct installation), this section can be omitted. However, if m-files are to be written, or the ones from the accompanying disk are to be installed on a hard disk drive, then some thought must be given as to how MATLAB will find them.

When MATLAB is invoked from the operating system, it will run in a certain working directory, which depends upon the setup. If new m-files are always saved (or copied) to this default directory, all will be OK, but the (lack of) structure will get 'messy' and eventually it may become difficult to distinguish additional files from the original MATLAB installation files. There are various options for keeping things tidier.

MATLAB has a built-in search path for files called, imaginatively, the *matlabpath*. This is initialized each time MATLAB is started. Whenever a command name is issued, MATLAB searches its workspace (in memory) and then every directory in the path until it resolves the command reference (or fails to find it). Toolboxes are usually installed as subdirectories under the MATLAB root directory (for example, below the *matlab* directory), and are automatically added to the built-in *matlabpath* by their installation software. The search path is initialized by the *matlabpath* command. Note that there is no *help* available for this, as it is an operating system environment variable. However, *help path* can be typed instead. In v4.x, the *matlabpath* command will be found in the file *matlabrc.m*, in the root directory of the MATLAB installation. This file can be viewed (and altered) using any ASCII text editor. The command will be found towards the end of the file, and might begin something like this:

matlabpath([...

'C:\MATLAB\toolbox\local',...

';C:\MATLAB\toolbox\matlab\datafun',...

';C:\MATLAB\toolbox\matlab\elfun',...

and continue in a manner dependent upon the selection of toolboxes installed. Maybe the neatest way to organize new work is therefore to put any new m-files into a subdirectory (created by you) which fits this structure (for example, c:\matlab\toolbox\mymfiles), and to add this new subdirectory name to this *matlabpath* command in the obvious manner. MATLAB will then always include a search of the new directory when looking for a command name, so your commands are just as accessible as the original MATLAB ones. It is also possible to alter the search path in any desired way from the MATLAB prompt, using the *path* command, but any such alteration will be forgotten when MATLAB is closed down.

There are other ways of accessing new m-files. For example, maybe selections of m-files are kept in project subdirectories, along with all the other work on the same project. In this case, either the appropriate subdirectories can be added to the *matlabpath* command, as above, or MATLAB can be started, and then the working directory can be changed to the one containing the m-files it is desired to execute. This is done using the *cd* command in the usual way (for MS-DOS machines), for example:

> cd a: % or:

> cd c:\project1\filters

to change to a floppy-disk drive, or a different hard-drive directory, respectively.

Note that in later versions of MATLAB (v4.x), the *cd* command is a MATLAB command (as is the *dir* command to check what is in the new directory). For earlier versions, such commands will need to be issued as operating system commands. This is done by using an exclamation mark to precede the operating system command. For example, on a PC running MATLAB v3.x under MS-DOS:

> !cd c:\project1\filters

> !dir

These also work in v4.x and higher, but may leave an inactive operating system window open every time they are used. To list just the MATLAB m-files and data files in the current working directory, the *what* command can be used.

If running MATLAB on a network in an academic institution, there are likely to be restrictions on directory manipulation commands. For example, teaching staff might place m-files in a file-server directory q:\matlab\toolbox\stafffile which can be read, but not altered. However, what can then be done is to open the files with an ASCII text editor, and save them to a disk of your own in the a: drive (for example) – then they can be altered as necessary and, presumably, the a: drive will already be on the MATLAB search path.

A3.5   *Operations with matrices and vectors – with comments*

Most mathematical operators are entered into MATLAB either just as they would be written on paper, or in a similar way to most other technical computing languages. The good thing about MATLAB is that this also applies to vector and matrix manipulation. For example, to check that the matrices *a* and *ai* entered in Section A3.1 really are inverses of one another, they can be multiplied together, and the result can be inspected to ensure that it is a unit matrix. It is as easy as (the spaces used for layout are optional):

$\gg$ a * ai

The answer will be displayed. If it was not required to display the answer, but rather to store it in the variable *prod* for use later, then the following would be used:

$\gg$ prod = a * ai;

MATLAB always assumes that vector–matrix maths is required by default. For example, to set up a 3 kHz sinewave having 200 sampling intervals in the time range 0 s to 2 ms, these commands can be used (as always, the spacing is optional):

$\gg$ t = [0 : 0.002/200 : 0.002];   % see notes (a) and (c) below

$\gg$ x = sin(2 * pi * 3000 * t);   % see Section A3.6 for ways of
                                      % plotting this

- Note (a): the first command sets up a vector of data points (time values) between 0 and 0.002 s, in steps of 0.002/200 seconds. An alternative would have been to use the *linspace* command to generate the 201 points as follows:

  $\gg$ t = linspace(0, 0.002, 201);   % could use 2e − 3 instead of 0.002

- Note (b): the equation $x = \sin(2\pi f t)$ is evaluated at every point in the time vector *t* by the single command shown, so *x* will also be a 201-element vector (note that *pi* is a built-in MATLAB function, so it can be used like this without prior definition). There is no need for the *do* or *for* loops which would be necessary in other languages.

- Note (c): everything following a % sign is a comment, and is ignored by the interpreter. Such comments can be placed at the end of a command as shown above, or on a line of their own (with % as the first character).

The MATLAB *help* system works by looking for comment lines at the start of the m-file containing the command, and listing them until it comes to the first non-comment line (which may be a blank line). New m-files can therefore be added to the help system automatically. For example, the file *garbage.m* created in Section A3.2 could have been written:

% The command *garbage* sets up a specific 3×3 matrix and lists
% its inverse and eigenvalues.

% The two lines above would appear in response to *help garbage*.

% These lines would not, because of the blank line in between.

% These lines can be used to add more details for the user who is
% prepared to open the file and read its contents.

a = [1 2 3; 2 8 4; 1 7 9]

inv(a)

eig(a)

Because of the vector–matrix mode of operation, there are a couple of operations not found in other languages. The first is to allow element-by-element operations. For example, say that it is required to modulate the 3 kHz sinewave created above (in the variable *x*), with a 600 Hz sinewave. To do this, the data points for the 600 Hz wave would be created at the same time values as the 3 kHz wave:

$\gg$ y = sin(2 * pi * 600 * t);

(Note that this command to form *y* can be issued by using the upward arrow key to recall previous commands, until the one is reached which formed *x*. This can then be edited to look like the command above, and 'Return', or 'Enter' pressed to execute it.)

Now the two waves need to be multiplied, but the problem is that the expressions *x*y* and *y*x* are both invalid, since both *x* and *y* are 1×201 vectors, so the product makes no sense in either case (see Section A1.1.2). What is actually required is to multiply together the individual values of *x* and *y* at each value of time in *t*. MATLAB can be instructed to do this by using a full stop (period), or dot, in conjunction with the multiplication sign:

$\gg$ z = x .* y;

A dot can also be used with the exponentiation operator (as in .^), so as to raise every individual element in a vector or matrix to the same power.

The other MATLAB-specific vector–matrix operation is backward division for solving equations. Say there is an equation $p = Qr$ where

$$p = \begin{bmatrix} 4 \\ 6 \end{bmatrix} \quad \text{and} \quad Q = \begin{bmatrix} 1 & 2 \\ 3 & 4 \end{bmatrix}$$

The following commands will solve for *r*:

$\gg$ p = [4; 6]; q = [1 2; 3 4];

$\gg$ r = q\p

A3.6   Plotting – and hardcopy (also introducing the transpose operator)

This is probably the major area of difference between earlier MATLAB versions (v3.x), and more recent ones (v4.x). If basic plots are all that is required, there is little obvious difference. However, as soon as anything more complex is attempted, the differences begin to become apparent. They are caused by the fact that v4.x uses a graphical environment completely rewritten in an object-oriented form, and based on an extremely flexible graphical user interface. This means that there is great flexibility over the production and presentation of plots, but it does take some getting to grips with! Here, only the very basic operations are discussed. The help system will give more information about every command used, but if it is necessary to get to grips with the later graphics in a serious way, then it is necessary to read the appropriate MATLAB *User's Guide* and *Building a Graphical User Interface* manuals (The Mathworks, Inc., 1993a, 1993c).

To get a plot of the sinewave data created in Section A3.5, it is only necessary to issue the command:

≫ plot(t,x)

Using v4.x, the plot will appear in a separate window and, if it is not visible, it will be hiding behind the command window. It can be brought to the front by any method appropriate to the operating system in use.

The basic plot can be altered in a number of ways. The most common thing to do is to add grid lines. Issuing the command *grid* at the next MATLAB prompt will do this, but it could have all been done at once, thus:

≫ plot(t,x); grid

Altering the plot colour, number and style of gridlines, adding text on the plot grid, and so on is all possible, but not described here. Simple titles and labels can be added to the plot as follows:

≫ title('3kHz Sine Wave')

≫ xlabel('Time (s)')

≫ ylabel('sin(2 * pi * f * t)')

To see all three waveforms from Section A3.5 on the same axes, the following command could be used:

≫ plot(t, [x; y; z]); grid

which forms a matrix having the elements of *x* on row 1, *y* on row 2 and *z* on row 3, then plots them against *t* in whichever way causes the number of points to match up. Thus, the following command, which introduces the use of the single apostrophe as the MATLAB *transpose* operator, would also work (this time stacking *x*, *y* and *z* as three columns prior to plotting):

≫ plot(t, [x' y' z']); grid

If an error message about incompatible rows and columns is encountered during such multiple plots, the most likely cause is that one of the dependent variables is stored as a row, and one as a column. For example, the sinewaves above were created using the (row) time vector, but vector results of most MATLAB commands would appear as columns by default. Using the command *size(x)*, for example, will determine which is which, and the transpose operator can then be used on appropriate variables to correct things.

The last common basic requirement is to examine a small section of the plot. This can be done by using the *axis* command. Note that in MATLAB v3.x, this must be issued *before* the *plot* command, but in the windows-type versions v4.x, it must be issued *afterwards*. To see the section of the existing plot in the x-axis range (0.0008, 0.001) and the y-axis range (0, 0.4), use:

≫ axis([8e–4, 1e–3, 0, 0.4])

Note that, on this scale, the plots are rather crude – this would be cured by using more time values in the *t* vector.

Many MATLAB functions (especially toolbox functions) produce their own plots if issued with *no* left-hand arguments, or produce data for subsequent manipulation if used *with* left-hand arguments. For example, to use the *bode* command from the control systems toolbox to get a frequency response plot of the system model having the Laplace transfer function

$$G(s) = \frac{3s + 4}{s(s^2 + 2s + 8)}$$

proceed as follows:

> num = [3 4];   % numerator coefficients in a vector

> den = [1 2 8 0];   % denominator coefficients in a vector (all
                        % must be present – note 0 coefficient of s^0)

> bode(num,den)

This will produce the standard magnitude and phase plots on the screen. However, it might be preferred to see the magnitude plot separately (for example), in which case it must be produced explicitly. Using the *bode* command with left-hand arguments allows this:

> [mag, pha, w] = bode(num,den);   % semicolon prevents listing of
                                   % all data

> semilogx(w,20*log10(mag)); grid % produce dB plot on
                                   % semilogarithmic axes

Finally, a specific frequency range can be used, if the default one chosen by MATLAB is unsuitable. These are the values in **w**, and it will probably be the case that MATLAB chooses 0.1 to 10.0 rad s⁻¹ for this system. If it was required to use 200 values in the range 0.01 to 100 rad s⁻¹, the *logspace* command would be used to generate a logarithmically spaced set of such points, and these would be used in the *bode* command, rather than allowing MATLAB to set its own default values:

> w = logspace(−2, 2, 200);   % required values for w

> [mag, pha] = bode(num,den,w);   % w now on right-hand side

> semilogx(w,20 * log10(mag)); grid % produce dB plot on
                                     % semilogarithmic axes

At this point, note also that MATLAB has a *ginput* command, allowing data points to be picked off plots. This can be used in various ways (use the *help* system to investigate), but the simple command *ginput* at the next MATLAB prompt will put up a crosshair on the existing plot. This can be moved over the plot (typically using a mouse) to the point of interest, the left-hand mouse button clicked, and then the *return* key pressed, when the data values at the point will be reported. If several points are clicked before pressing *return*, the coordinates of every selected point will be reported.

To get hardcopy of plots, there are basically three options. The simplest is to do a screen dump, by whatever means your computer system uses. This may be the most convenient for earlier versions of MATLAB, but is likely to give the poorest quality. The next method (if using v4.x) is to print from the windows-type environment, by using the *file* menu-bar item in the figure window, and then selecting the *print* option. The final option, usable with any MATLAB version, is to divert hardcopy to a word processor.

To obtain word processor compatible hardcopy of a plot, it is clearly necessary to produce the plot as a data file which the word processor can read. The examples below assume that HPGL format (Hewlett-Packard® graphics language) will be used, either for sending to a plotter, or for importing into a WP with an input filter for such files (for example, Microsoft® Word for Windows™).

In *earlier versions* of MATLAB (v3.x), proceed as follows:

⟩ meta filnam   % produces a device-independent metafile called
                              % filnam.met in the working directory, where
                              % filnam is a name of your choice.

⟩ !gpp filnam /dhpgl /ol % issues an operating system command for the GPP
                              % (graphics post-processor) routine to convert
                              % filnam to the chosen WP format (it will create
                              % filnam.hpgl in this case). Typing gpp at the
                              % operating system prompt gives help.

In *later versions* of MATLAB (v4.x), proceed as follows:

⟩ orient landscape   % necessary to *prevent* the plot appearing sideways.

⟩ print -dhpgl filnam   % produces file filnam.hgl in the working directory
                              % which can be plotted, or imported to
                              % a suitable WP. *help print* gives more info.

Note that different file forms used for such transfers give different results. In the above example, MS® Word for Windows™ will import the HPGL file as a MS® Draw™ file (using *insert* then *picture*) – so it can subsequently be edited at will, just like any other drawing. If some printer file format had been selected instead of HPGL, and imported into the WP, it may well be imported as a (non-editable) bit map instead. The appearance in the document would be more or less the same, but nothing could be done with the image, other than change its size.

A3.7   *Numerical problems*

In common with all languages capable of matrix manipulation, numerical problems due to both rounding errors and ill-conditioning can occur. As an example of ill-conditioning (from the CSTB manual) consider the matrix equation $Ax = b$ where

$$A = \begin{bmatrix} 0.78 & 0.563 \\ 0.913 & 0.659 \end{bmatrix}, \quad b = \begin{bmatrix} 0.217 \\ 0.254 \end{bmatrix}$$

and it is desired to solve for x.

The MATLAB command x=a\b will do this, as mentioned in Section A3.5. The answer is $x = [-1 \ \ 1]^T$. However, if the values in A are slightly perturbed by adding 0.001 to each element in the top row, subtracting 0.002 from element (2, 1) and 0.001 from element (2, 2), MATLAB then gives the result $x = [-5 \ \ 7.3085]^T$.

This does *not* indicate that there is a problem with MATLAB (checking the answer shows it to be correct). The problem is that the matrix A is almost singular, so that it is ill-conditioned with respect to inversion. What it *does* indicate is that (particularly in the case of high-order or multivariable systems) much care is required in the formulation of this kind of problem.

Further discussions in this area will be found in any numerical analysis text. For present purposes, it is sufficient to note that real single-input-single-output problems, together with the choice of algorithms used in MATLAB, rarely cause any difficulties for systems of order less than five (according to the CSTB manual). As with any other computer-aided analysis software, it is obviously beneficial if the user has a ball-park idea of the correct answer, so that such problems should not mislead too much.

A3.8 'Out-of-memory' messages

Trying to run MATLAB (especially the later versions, v4.x) without much system memory can give rise to such messages. The first thing to be sure of is that the system is set up correctly – especially with regard to such matters as environment space in PC systems, for example (consult the release notes and installation notes for your version of MATLAB). Assuming this to be OK, the following matters are worthy of consideration.

Is any other program being run which uses a lot of memory simultaneously (for example, some TSR programs, or word processors for importing MATLAB plots)? If so, these may have to be closed, letting MATLAB have all the memory. Also note that some programs are not very good at releasing memory when they terminate, so it may even be necessary to reboot, and run MATLAB only.

Are there large matrices or vectors in memory which are no longer required? This is often worth checking if, for example, a plot command generates the out-of-memory message. It may be that several long vectors have been generated in the process of creating the plot data, but are not needed actually to produce the plot itself. In this case, the *whos* command can be used to look for the largest variables which are no longer required, and the *clear* command can be used to delete them and free up the memory they were occupying. For example:

> ≫ clear var1 var2 var3

The following command will clear from memory any compiled m-files:

> ≫ clear functions

After issuing such commands, the *pack* command should be used to reorganize the unused memory into contiguous blocks. The MATLAB manual suggests that using *pack* initially may avoid the need actually to clear any variables, but the authors' experience is that normally some variables must be cleared first. Finally, the command below gives some system-specific tips for various platforms:

> ≫ help memory

A3.9 Block diagram manipulation in MATLAB

There is one more technique worthy of mention in this appendix, because it is used in many of the m-files on the accompanying disk.

Apart from plotting frequency responses, and carrying out various design procedures, the major use of MATLAB in this text is in producing simulations – that is, time responses. Often, SIMULINK would be better for this, but MATLAB has deliberately been used for all but one or two of the simulations, so as to make the results (and the underlying design and simulation methods in the m-files) available to readers who may have MATLAB, but not SIMULINK.

Before time responses can be produced for anything but the simplest of systems, it is necessary to connect together the various subsystems making up the whole simulation. Some of these will be in series, and some in feedback paths, and the MATLAB CSTB *series*, *feedback*, *parallel* and *cloop* commands are used for this purpose (the more specialized *connect* and *blkbuild* commands could be used, but have not been in this text).

The problem is that the commands mentioned above only allow connections to the inputs and outputs of subsystems. However, in state-space work, it is often necessary to feed back the *states* of a system, rather than the *outputs*, so special arrangements have to be made to allow this. In the same way, the *step* command generates output responses. When used with left-hand arguments it will also generate the state responses, which can be plotted separately. However, it is often desirable to view the *input* signal to a block in a system, so as to check for unrealistic controller demands, for example. Again, if the block has to be built into an overall block diagram, special steps are necessary to achieve this.

The technique in each case is straightforward. It is to modify the *C* and *D* matrices of the state-space model, so that extra outputs are defined, corresponding with the signals to be plotted or fed back. At first sight, this arbitrary alteration of the system model can appear to be a questionable approach – what right have we to do it? Are we not changing the model of the system being analysed? However, note that the only quantities in the state-space model which will usually be changed in this way are the *C* and *D* matrices. These do not affect the poles (eigenvalues) of the model in any way (those are determined by the *A* matrix, which will never be altered arbitrarily), nor do they affect the manner in which the system inputs drive the model (that is determined by the *B* matrix, which will only very occasionally need to be modified, and then in a manner which does not affect the existing inputs in any way). The only thing the *C* and *D* matrices affect is the generation of the model *outputs* according to the usual output equation $y = Cx + Du$. This means that the zeros of the model could be affected, but they will not be, so long as the original model outputs are preserved.

Any modifications to the *C* and *D* matrices will therefore be made in such a way that they simply generate *extra* outputs which did not exist in the original model. This is equivalent to going to the plant being modelled, and measuring some extra variables which is, of course, acceptable. Figure A3.1 shows a state-space model of a system with three state variables, which is to be arranged in a state-variable feedback arrangement, as shown. The obvious thing is to feed back the three states, x_1, x_2 and x_3, but the MATLAB *feedback* command can only feed back outputs. It is therefore necessary to define all the signals to be fed back as extra (new) outputs, as shown. Also shown is a new output connected directly to the input, which will allow the controller action to be plotted by the *step* command.

In general, to obtain a new output which is a 'straight-through' connection from the input, a row of zeros is added to the bottom of the *C* matrix, and a new row is added to the bottom of the *D* matrix containing a unity element only in the column corresponding to the input which is to become the new output signal (with zeros elsewhere).

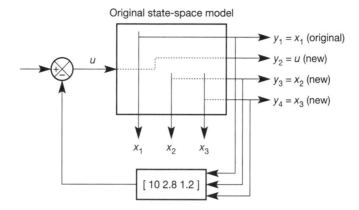

Figure A3.1 Modifying a state-space model by adding new outputs.

For example, consider the model of the antenna-positioning system used in Example 5.4 (coincidentally, in Section 5.4):

$$A = \begin{bmatrix} 0 & 1 & 0 \\ 0 & -1 & 1 \\ 0 & 0 & -5 \end{bmatrix}, \quad b = \begin{bmatrix} 0 \\ 0 \\ 5 \end{bmatrix}, \quad c = \begin{bmatrix} 1 & 0 & 0 \end{bmatrix} \quad \text{and} \quad d = 0$$

This represents a system with three state variables, one input and one output. It is the forward path block in Figure A3.1. Using the variables a, b, c and d to represent A, b, c and d in MATLAB, the commands:

$\gg$ a = [0 1 0; 0 −1 1; 0 0 −5]; b = [0 0 5];

$\gg$ c = [1 0 0]; d = 0;

$\gg$ step(a, b, c, d); grid

will plot the time response (to a unit step input) of the system output. This is the angular position of the antenna (in radians). Since this system has an eigenvalue (a pole) at the origin, the open-loop step response is predominantly a ramp, due to the action of this integrator (a constant input causes the antenna to rotate continuously). If it was desired to see the *state* responses too (the antenna's angular velocity and acceleration), this could be achieved simply by using left-hand arguments in the *step* command:

$\gg$ [y, x, t] = step(a, b, c, d);

$\gg$ plot(t, x); grid

This is fine, but still does not show the input signal. Nor could these state responses be used in a feedback scheme by the MATLAB *feedback* command, for example, since it can only feed back outputs.

To see the plant input on the same trace, the procedure outlined above can be followed. A second plant output is defined, which is made equal to the input. The existing output is not altered in any way. The *step* command will then show this second output (which is the input) too. The existing c and d quantities represent the output equation:

$$y = cx + du = \begin{bmatrix} 1 & 0 & 0 \end{bmatrix} x = x_1$$

This output will now be called y_1, and a second (new) output will be added such that $y_2 = u$ (see Figure A3.1). The new output equations are therefore:

$$y_1 = 1x_1 + 0x_2 + 0x_3 + 0u$$

and $$y_2 = 0x_1 + 0x_2 + 0x_3 + 1u$$

In vector–matrix terms, $y = Cx + du$, where

$$y = \begin{bmatrix} y_1 \\ y_2 \end{bmatrix}, \quad C = \begin{bmatrix} 1 & 0 & 0 \\ 0 & 0 & 0 \end{bmatrix} \quad \text{and} \quad d = \begin{bmatrix} 0 \\ 1 \end{bmatrix}$$

This can be seen to represent the new output equations, and also fits the procedure outlined above. In MATLAB, the modification to the model is easy:

```
≫ c = [c; 0 0 0];        % zero row added to c

≫ d = [d; 1];            % new row added to d with 1 coupling the
                         % input to output 2

≫ step(a, b, c, d); grid
```

At this stage, the system of Figure A3.1 now has two outputs – the original y_1 and the new y_2. Now both outputs are plotted, the second of which is, of course, equal to the step input. MATLAB v4.x plots each output on a separate graph. To see them on the same axes, use:

```
≫ [y, x, t] = step(a, b, c, d);

≫ plot(t, y); grid
```

Similarly, to add state variables as new outputs, in general a row of zeros is added to the bottom of the D matrix for each new output; and a new row is added to the bottom of the C matrix for each new output, which contains a unity element only in the column corresponding to the state variable which is to become the new output signal (with zeros elsewhere).

There is a MATLAB CSTB command *augstate* which will automatically append *all* a system's states as new outputs – this can make things easier, but it may not be necessary to define *all* the states as new outputs in any given case (this command simply adds an identity matrix to the bottom of the C matrix, and a correspondingly sized zero matrix to the bottom of D). In the example above, output y_1 is equal to the state x_1. It would therefore be possible to use this output as x_1 in a state variable feedback system, and define two more extra outputs to be the states x_2 and x_3. Using exactly the same reasoning as in the case of the new output y_2 above, x_2 and x_3 can be made to appear as outputs y_3 and y_4 (see Figure A3.1) by modifying the output equation matrices and vectors to:

$$C = \begin{bmatrix} 1 & 0 & 0 \\ 0 & 0 & 0 \\ 0 & 1 & 0 \\ 0 & 0 & 1 \end{bmatrix} \quad \text{and} \quad d = \begin{bmatrix} 0 \\ 1 \\ 0 \\ 0 \end{bmatrix}$$

Starting from the modified versions of c and d for the two-output system, the MATLAB commands could be written:

$\gg$ c = [c; 0 1 0; 0 0 1]; d = [d; 0; 0];

$\gg$ [y, x, t] = step(a, b, c, d)

$\gg$ plot(t, y); grid

This will now plot the four outputs of the open-loop forward path of Figure A3.1 (no feedback loop is yet in place) in response to a unit step at u, on the same graph as follows:

y_1 is the original output (and is equal to x_1)

y_2 is equal to the system input (a unit step from the *step* command)

y_3 is equal to x_2

y_4 is equal to x_3

Now that all the states are available as outputs, it is easy to build the state variable feedback scheme designed in Example 5.4, and also shown in Figure A3.1. The required feedback vector for feeding back the *state variables* was found to be $k = [10\ 2.8\ 1.2]$. The state variables are available as *outputs* 1, 3 and 4. The MATLAB *feedback* command can feed these back to the input, but also needs a state-space model of the feedback gain vector k. The easiest way to provide this is to specify a model with *empty* a, b and c quantities, and the d quantity equal to k. This is because the d quantity in a state-space model specifies the direct coupling from the input to the output, with no dynamics (which is what the gain vector k is). In MATLAB, an empty matrix is represented by a pair of opening and closing brackets, with nothing between them. Designating the *closed-loop* system state-space model by ac, bc, cc and dc, it can now be generated by MATLAB using:

$\gg$ [ac, bc, cc, dc] = feedback(a, b, c, d, [ ], [ ], [ ], [10 2.8 1.2], ...
    % ... indicates a contuation line −1, [1 3 4]);

$\gg$ [y, x, t] = step(ac, bc, cc, dc)

$\gg$ plot(t, y); grid

The quantities a, b, c, d in the feedback command are the forward path state-space model, *[ ]*, *[ ]*, *[ ]*, *[10 2.8 1.2]* are the feedback path state-space model (the constant feedback gain vector, in this case), the quantity '−1' tells MATLAB to connect the feedback to input 1 of the forward path system (that is, the only input in this example) using negative feedback. The vector *[1 3 4]* specifies the forward path outputs which are to be to be fed back. The resulting plot shows all four outputs of the *closed-loop* system of Figure A3.1 – namely, the three states, and the plant input. The fact that the plant input is displayed makes it possible to assess the control effort being imposed by the controller. With the exception of this new input trace, the other three responses should be identical to those in Figure 5.4 (in Section 5.4).

The approach used in the m-file *fig5_4.m* on the accompanying disk (which produced Figure 5.4) is the same, in principle, to that used above. However, the version in the m-file is slightly simpler, as no input is displayed, and the only forward path outputs are the three state variables, so that it is unnecessary to specify (to the *feedback* command) the vector of outputs to be fed back (the default of feeding back all outputs is sufficient).

Many other m-files on the disk use these kinds of techniques for building up systems models for simulation. Some are more complex than the case above. For example, it may have been necessary to cascade the forward path state-space model in Figure A3.1 with a following forward path block before closing the feedback loop. This would be done using the *series* command, but the **B** matrix of the following system in the forward path would then also require alteration so that the new outputs generated above could be fed into matching (new) inputs on the next forward path block, and then, in turn, to new outputs of that block. The ideas are, however, identical to those worked through above, and the m-files are reasonably heavily commented; so the interested reader should be able to follow what is going on.

Appendix 4
A brief introduction to
SIMULINK®

A4.1   SIMULINK

SIMULINK (The Mathworks, Inc., 1993d) works in conjunction with MATLAB (Appendix 3), and is installed just like a MATLAB toolbox (Section A3.2). The version described here will only run with MATLAB versions 4.0 or later, running under 'Windows'-type operating systems.

SIMULINK is a dynamic system simulation environment. This means that the user enters details of a system into SIMULINK, specifies the inputs which will drive the system, and SIMULINK then predicts the system's behaviour by means of on-screen plots, or data written to files. Of course, MATLAB (without SIMULINK) can also do these things. The main reasons why SIMULINK would often be chosen instead include:

- Systems can be entered simply by 'wiring up' their block diagrams on the computer screen – there is no programming to be done (although a basic familiarity with entering vectors and matrices into MATLAB is helpful for use in the dialogue windows, which are used to specify the parameters of the various blocks).

- System models can be entered equally easily in Laplace transfer function, state-space or discrete-time forms, or any mixture of these.

- SIMULINK includes blocks for nonlinear system elements (such as saturation or backlash – see Chapter 14). It is very tedious to program some of these in a MATLAB (or any other high-level language) simulation. In SIMULINK, the nonlinear block is simply dragged into the diagram and 'wired up' in exactly the same way as any other block.

- If a system model exists as a set of nonlinear differential equations (which can often be the case for complex real-world systems), then the SIMULINK model can be built directly from these equations by writing them into a specially constructed file. Unless the reader has had to cope with such models, it may not be clear how much of an advantage this can be!

- Once a nonlinear system model is entered (by any method), SIMULINK is capable of providing the linearized model of the system about a chosen operating point (if you have SIMULINK, type *help linmod* at a MATLAB prompt for more detail).
- SIMULINK has a good selection of built-in numerical integration algorithms for solving the system's equations to produce the time response. The most appropriate algorithm for the system can therefore be chosen. In MATLAB, the algorithm of the *step* command (for example) might not suit some systems (leading to very long simulation runs, perhaps).

Having said all that, since most of the examples used in this text are of relatively simple systems, it has been decided to stick with MATLAB and the Control systems toolbox (see Appendix 3) for most of the simulations, even though this makes some of those in later chapters more complicated to program than would otherwise be the case. This has been done so that readers with access to MATLAB, but not to SIMULINK, can run nearly all the simulations on the accompanying disk and, perhaps more importantly, have access to them for modification to suit problems of their own.

A4.1.1 Starting SIMULINK

SIMULINK is invoked by first starting MATLAB (see Appendix 3), and then issuing its name as a MATLAB command:

» simulink

This will result in the opening of the SIMULINK control window. Click on *file* then *new* from the menu bar of this control window to obtain a blank worksheet to hold a new system diagram.

Once a system has been entered (as described below), its diagram can be saved to disk (using *file* then *save*). To retrieve the diagram of an existing system which has been saved previously, use *file* then *open*. Note that the name of a previously saved system can also be issued directly as a command at the MATLAB prompt. This will start SIMULINK and open the block diagram of the system automatically, together with whatever SIMULINK settings were in force at the time the system was saved. This might be more convenient if it is desired simply to run the model (however, if further editing of the diagram is to be done, it is better to invoke SIMULINK first, then open the existing model file from the SIMULINK control window as suggested earlier).

The block libraries in the SIMULINK control window can be opened by double-clicking them in the normal way (or whatever is appropriate to the computer system being used), and will be found to contain numerous system elements of the appropriate type (note that a pure time delay can be found as a time function in the nonlinear library, although its Laplace transfer function would be linear). These system elements can be dragged into the worksheet and released (for example, by holding down the left-hand mouse button to do the 'dragging' on PC systems) to form the required system diagram.

A4.2 Building the system diagram

(This section describes work on a PC system, but using other systems should be sufficiently similar for it to be useful.)

As an example, investigate the performance of the system containing a backlash element (see Chapter 14), shown in Figure A4.1. This example is chosen because it would not be easy to study it without a program like SIMULINK; correct programming of the backlash element in MATLAB – or other technical computing languages – is a surprisingly non-trivial task!

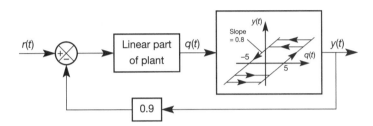

Figure A4.1 A block diagram of a system with backlash at its output.

The linear part of the system has the Laplace transfer function model

$$G(s) = \frac{25}{(1.5s^2 + 0.4s + 15.5)(s^2 + 0.3s + 3)}$$

The input $r(t)$ is to be a step of $+10$ units applied after 0.2 s.

Start with a blank worksheet, as above, and build up the diagram of Figure A4.2, aided by the comments below:

• Use the LH mouse button to drag the required icons into the worksheet from the menus as follows. The Backlash element is from the *nonlinear* collection; the Gain blocks, Transfer Fcn and Sum blocks are from the *linear* collection; the Step Input block is from the *sources* collection; the Scope block is from the *sinks* collection and the Mux block is from the *connections* collection. The purpose of the Mux (multiplexer) block is to allow the display of more than one signal on the (nominally single-channel) Scope.

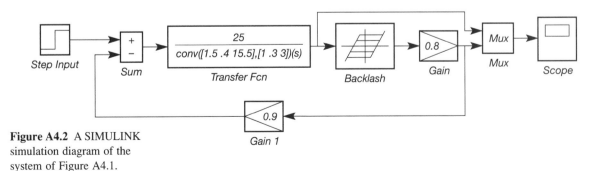

Figure A4.2 A SIMULINK simulation diagram of the system of Figure A4.1.

- The feedback gain block can be made to face the right way by selecting it (one click) and then clicking on *options* and *flip horizontal*. It does not actually matter if it is left facing the wrong way, the effect is purely one of presentation.

- Double-click the Sum block and set the inputs to $+$ and $-$ by typing $+-$ into the *list of signs* dialogue box (the number of inputs can be changed in the same way, if required).

- Double-click the Mux block and set the number of inputs to 2, by overwriting the default 3 in the *number of inputs* dialogue box.

- Double-click the Step Input block. Set the *Step Time* to 0.2 and the *Final Value* to 10.

- Double-click the Gain blocks and set the gains to 0.8 and 0.9 as required (note that the backlash element is assumed to have a gain of unity in the 'linear' part of its characteristic, so the required gain of 0.8 has to be set externally using the extra Gain block).

- Double-click the Backlash element, set the *deadband width* to 10 and the other parameters to zero.

- Double-click the Transfer Fcn block and set the *numerator* to 25, and the *denominator* to the MATLAB command conv([1.5 0.4 15.5], [1 0.3 3]) which saves having to multiply out the coefficient vectors – use *help conv* for details.

- Make the interconnections by dragging lines between the icons with the LH mouse button. Where there are branches in the signal paths, create one path, and then add the next by positioning the cursor on the existing path, and holding down the <CTRL> key while dragging the new connection.

To delete blocks or connections, select them (one click), then press the <DEL> key.

A4.3   Setting the simulation parameters and the scope controls

SIMULINK works by automatically forming the system equations from the block diagram, and then solving them by numerical integration to obtain the time responses. There is a selection of integration techniques available, but the default (fifth-order Runge–Kutta) will suffice for most examples in this text. This is a variable-step algorithm (meaning that it can change the length of its integration time step, depending on the perceived behaviour of the system). Short time steps give better accuracy when the system response is changing rapidly, but long execution times for the simulation. Longer time steps give faster execution times, but less accurate results during transient behaviour. The idea behind variable-step algorithms is that the best of both worlds is obtained by changing the time step according to the system's behaviour – when responses are changing rapidly, the time step will be shortened, and when things are relatively steady, it will be lengthened.

Systems with discontinuous nonlinearities, such as the present one, can easily 'catch out' variable-step algorithms, in that the nonlinear element can suddenly pass through a discontinuity and change its behaviour suddenly, when the system as a whole is exhibiting only very slowly varying behaviour. The algorithm may then have selected a long time step, and the switching point of the nonlinearity may be drastically misplaced in time, giving totally misleading results for the subsequent behaviour of the system. For this reason, the maximum time step which the algorithm is allowed to take must be restricted (in fact, the default value in SIMULINK is too long for many of the purely linear systems in the text too, so the procedure below is to be recommended in every case).

It is always wise to try running a simulation which contains continuous system elements (Laplace transfer functions or state-space models) with both a shorter and a longer Max Step Size setting than the initial selection, to see how sensitive it is to the choice of time step. In general, if the apparent behaviour changes significantly with a shorter maximum time step, then it is not yet short enough to give accurate results. On the other hand, if the behaviour does not change noticeably with either a shorter or a longer maximum time step, then the time step can be lengthened in the interests of faster execution. The author has been involved with simulations of relay control schemes (see Chapter 14) where a *maximum* step size of a few microseconds was necessary for reasonably realistic performance (such simulation is still better done on an analog computer!).

To limit the time step selection to suitable values for the present example, click on *simulation* and *parameters*, and set the Max Step Size to 0.01 (seconds) and the Min Step Size to 0.001 (to save a bit of execution time when possible). Note that the Stop Time is set to 999999 (seconds) in this example – that is, the simulation will run continuously, once started. This is set to a shorter time if it is desired to stop the simulation after some given time.

The Scope block gives an oscilloscope-style display, and it is therefore necessary to have some idea of the axis calibrations. These can be found by trial and error. In the present case, the system will settle into a large-amplitude limit cycle (see Chapter 14), and an adequate display can be obtained as follows.

Double-click the Scope block. A simulated 'scope screen should appear, with sliders to adjust the horizontal and vertical ranges above it. If the sliders do not appear, increase the size of the 'scope window by dragging one corner of it, until they do. To get the correct operating ranges for this system, overwrite the numerical values in the Horizontal and Vertical Range boxes with 20 (seconds) and 30 (units) respectively, and click on OK. This will make the Scope display vanish in some installations, but double-clicking the Scope block should get it back again, with the correct settings intact.

There are other output devices available in the *Sinks* library. The *Scope* has the advantages of simplicity, and speed of execution, but the disadvantage of a fairly crude display. For a better display, but slower execution, the *Graph* block can be substituted – again it is necessary to scale the axes manually. For automatic axis scaling, the *Auto-Scale Graph* block can be used. This removes any need to calibrate the axes, but can significantly increase the run time of the simulation, and can also have the annoying habit of losing the trace at the end of the simulation run in some circumstances. It is also possible to send the simulation results to the MATLAB workspace for further analysis or plotting using MATLAB, as noted in the next section.

A4.4   Running the simulation

Click on *Simulation* and *start*, and (hopefully) watch the fun! The two traces show the input of the backlash block (the smooth trace) and the overall system output (the discontinuous trace). All being well, after several oscillations, the output should gradually build up to a sustained continuous limit cycle. It takes quite a while to get there. Click on *Simulation* and *stop*, when enough has been seen.

If the simulation diagram is saved (using *file* and *save* as usual), it will be saved as an m-file (see Section A3.2) which sets up SIMULINK appropriately when it is rerun as noted previously.

As an alternative (or addition) to plotting during the simulation run, numerical values can be sent to MATLAB, and plotted there in the normal way after the simulation is finished. To do this, drag two or three *To Workspace* blocks from the *sinks* library and connect one to each signal of interest. Double-click on each of them in turn, and name them as you feel fit. After running the simulation, all these variables will be available in MATLAB, and can be plotted in the normal way. The only extra requirement is that MATLAB needs also to know the time values at which the data points were generated. To make SIMULINK supply these, click on *simulation* and *parameters*, and then type t into the Return Variables dialogue box. SIMULINK interprets this as an instruction to send a variable called t to the MATLAB workspace, containing all the time values corresponding to the data points, when the simulation is run. (As an alternative, a copy of the *Clock* icon can be dragged into the workspace from the *sources* library, connected to a *To Workspace* block and the time values will sent to MATLAB in whatever variable name is specified for the *To Workspace* block.)

Note also that the number in the *Maximum number of Steps* dialogue box for the *To Workspace* blocks may have to be increased to get sufficient stored points.

Appendix 5
The 'true' z-transform

A5.1   Introduction

In Chapter 5, we introduced the z-transform, but we did it in a rather non-rigorous manner. We first introduced z^{-1} as a one-time-step-delay operator, and then performed the z-transform by making various substitutions for s into Laplace transfer functions. In this appendix, we give the somewhat more formal background to 'z', required for Chapter 7.

A5.2   The z-transform

This transform is a special case of the Laplace transform, which deals with sampled quantities. Such quantities have non-zero values only at sampling instants. Figure 7.1 shows a digital control loop in which the function $u(t)$ (for example) is shown as being sampled at intervals T.

We assume that the samples actually last for a time τ, which is taken to be short but not zero, as shown by the shaded areas in Figure 7.2.

The next step is to determine the Laplace transform of the series of samples. This is done by treating them as a series of impulses of strength equal to their areas on the graph. The area of the one at time $t = 0$ is $\tau u(0)$ so, since the Laplace transform of an impulse at $t = 0$ is simply the strength of the impulse, it transforms to $\tau u(0)$.

For a general function $f(t)$, this initial impulse would therefore have a Laplace transform $\tau f(0)$. The Laplace transforms of the subsequent 'impulses' are only slightly more difficult. We remember from the Laplace transform real shift theorem (Section 2.5.2 and Table 2.9) that, if the Laplace transform of an undelayed signal is $F(s)$, that of the same signal delayed by a time T is $F(s)e^{-sT}$. So the subsequent 'impulses' transform to $\tau f(T)e^{-sT}$, then $\tau f(2T)e^{-2sT}$, and so on. The transform of the entire series of 'impulses' is the sum of the individual transforms, so the sum will be

$$\tau(f(0) + e^{-sT}f(T) + e^{-2sT}f(2T) + \cdots) \text{ or } \tau \sum_{m=0}^{\infty} e^{-msT} f(mT)$$

In practice, we omit the τ, but make a mental note that it is there. So the z-transform of $f(t)$ is defined as $\sum_{m=0}^{\infty} e^{-msT} f(mT)$.

So where does 'z' come in? We introduce z as a more convenient way of expressing e^{sT}, so e^{sT} becomes z. Our definition therefore becomes

$$Z(f(t)) = \sum_{m=0}^{\infty} z^{-m} f(mT)$$

This means that z-transforms of common functions can be determined by summing infinite series. A simple example follows, but take heart – like Laplace transforms, it is usual to look them up in tables in practice!

Example A5.1 *Find the z-transform of* e^{-at}

Using the above definition,

$$Z(e^{-at}) = \sum_{m=0}^{\infty} z^{-m} e^{-amT}$$

If we think about this expression, it becomes clear that it is a geometric series, that is, each term of it is equal to the previous one times $e^{-aT} z^{-1}$. The first term is 1, so we can use the formula '$a/(1-r)$' for the sum to infinity of the series $a, ar, ar^2 \ldots$. This gives the sum as $1/(1 - e^{-aT} z^{-1})$. If we multiply top and bottom of the fraction by z, the result is $z/(z - e^{-aT})$, which agrees with the transform given in Table A5.1.

The reader may like to try the calculation for a unit step function at time $t = 0$, which should produce $z/(z - 1)$.

A5.3   Useful facts about the z-transform

Since $z = e^{sT}$, frequency response in z can be calculated by putting $s = j\omega$, which gives $z = e^{j\omega T}$, which can be expressed by De Moivre's theorem as $z = \cos(\omega T) + j \sin(\omega T)$. The resulting sums are tedious by hand but easy by computer program or spreadsheet.

There is a result comparable to the Laplace final value theorem to obtain the steady-state value of $f(t)$ from $F(z)$ without having to invert it. It is:

$$\text{steady-state } f(t) = f(\infty) = \lim_{z \to 1} [(z - 1)F(z)]$$

We will verify that the result is sensible by testing it both for a unit step (which should give a final value of 1) and for a decaying exponential e^{-at}, which should give a final value of zero. The expressions for $F(z)$ come from Table A5.1.

Unit step: $F(z) = z/(z - 1)$, so $(z - 1)F(z) = z$, which becomes 1 as z tends to 1.

Exponential: $F(z) = \dfrac{z}{z - e^{-aT}}$, so $(z - 1)F(z) = \dfrac{(z - 1)z}{z - e^{-aT}}$, which becomes zero as z tends to 1.

Table A5.1 Table of z-transform pairs.

Time function $f(t)$	Laplace transform $F(s)$	z-transform $F(z)$ (Sampling interval $= T$)
Unit impulse $\delta(t)$	1	1
Step function $u(t)$	$\dfrac{1}{s}$	$\dfrac{z}{z-1}$
t	$\dfrac{1}{s^2}$	$\dfrac{Tz}{(z-1)^2}$
$\dfrac{t^2}{2}$	$\dfrac{1}{s^3}$	$\dfrac{T^2 z(z+1)}{2(z-1)^3}$
e^{-at}	$\dfrac{1}{s+a}$	$\dfrac{z}{z-e^{-aT}}$
te^{-at}	$\dfrac{1}{(s+a)^2}$	$\dfrac{Tze^{-aT}}{(z-e^{-aT})^2}$
$1-e^{-at}$	$\dfrac{a}{s(s+a)}$	$\dfrac{(1-e^{-aT})z}{(z-1)(z-e^{-aT})}$
$\sin(\omega t)$	$\dfrac{\omega}{s^2+\omega^2}$	$\dfrac{z\sin(\omega T)}{z^2-2z\cos(\omega T)+1}$
$\cos(\omega t)$	$\dfrac{s}{s^2+\omega^2}$	$\dfrac{z[z-\cos(\omega T)]}{z^2-2z\cos(\omega T)+1}$
$e^{-at}\sin(\omega t)$	$\dfrac{\omega}{(s+a)^2+\omega^2}$	$\dfrac{ze^{-aT}\sin(\omega T)}{z^2-2ze^{-aT}\cos(\omega T)+e^{-2aT}}$
$e^{-at}\cos(\omega t)$	$\dfrac{s+a}{(s+a)^2+\omega^2}$	$\dfrac{z[z-e^{-aT}\cos(\omega T)]}{z^2-2ze^{-aT}\cos(\omega T)+e^{-2aT}}$

A5.4 Inverting the z-transform

Like the Laplace transform, the z-transform can be inverted by using tables and, where necessary, partial fractions to arrange the overall transform in a form in which it can be directly inverted from the tables. It must be borne in mind, however, that the inverse is only valid at sampling instants.

An alternative strategy is to produce the inverse as a time series by using the z-transform to obtain a difference equation. Both procedures will be demonstrated by an example.

Example A5.2 *Finding an inverse z-transform (using two different methods)*

Find the inverse z-transform (using the two different methods) of the following expression in z, assuming a sampling interval T of 0.1 s:

$$F(z) = \frac{50}{(z-1)(z^2-1.5z+0.6)}$$

(i) Solution by partial fractions and tables

Table A5.1 gives the following conversions which look as if they might fit the two denominator terms:

$$\frac{z}{z-1} \quad\rightarrow\quad \text{unit step} \tag{A5.1}$$

$$\frac{ze^{-aT}\sin\omega t}{z^2 - 2ze^{-aT}\cos\omega t + e^{-2aT}} \quad\rightarrow\quad e^{-aT}\sin\omega t \tag{A5.2}$$

$$\frac{z^2 - ze^{-aT}\cos\omega t}{z^2 - 2ze^{-aT}\cos\omega t + e^{-2aT}} \quad\rightarrow\quad e^{-aT}\cos\omega t \tag{A5.3}$$

First, $F(z)$ must be divided into partial fractions (see Appendix 2 for a discussion of partial fractions in the context of the Laplace transform).

$$\frac{50}{(z-1)(z^2 - 1.5z + 0.6)} = \frac{Az}{z-1} + \frac{z(Bz + C)}{z^2 - 1.5z + 0.6} \tag{A5.4}$$

Multiplying out the RHS, we are suggesting that:

$$\frac{50}{(z-1)(z^2 - 1.5z + 0.6)} = \frac{z(Bz + C)(z-1) + Az(z^2 - 1.5z + 0.6)}{(z-1)(z^2 - 1.5z + 0.6)}$$

So the denominators agree, and the numerator of the expression must equal 50.

The numerator can be rearranged as $z(-C + 0.6A) + z^2(C - B - 1.5A) + z^3(B + A)$. At first sight, since there is no z^0 term in the numerator, it cannot compare with the constant value 50. The conclusion is that a delay of one sampling interval is in force and that the rearranged numerator is therefore to be multiplied by z^{-1}, giving:

$$(-C + 0.6A) + z(C - B - 1.5A) + z^2(B + A) = 50 \tag{A5.5}$$

and the final version of Equation (A5.4) will then be:

$$\frac{50}{(z-1)(z^2 - 1.5z + 0.6)} = \left[\frac{Az}{z-1} + \frac{z(Bz + C)}{z^2 - 1.5z + 0.6}\right] z^{-1} \tag{A5.6}$$

Now, comparing coefficients of powers of z between the LHS and RHS of Equation (A5.5) gives:

$$z^2: \quad (B + A) = 0 \quad \text{so} \quad B = -A$$

$$z^1: \quad C - B - 1.5A = C - 0.5A = 0 \quad \text{so} \quad A = 2C$$

$$z^0: \quad -C + 0.6A = -C + 1.2C = 50 \quad \text{so} \quad C = 250$$

By substituting the last result back into the previous two, it follows that $A = 500$ and $B = -500$ which, when substituted back into Equation (A5.6), means that we are therefore inverting:

$$\left[\frac{500z}{z-1} + \frac{-500z^2 + 250z}{z^2 - 1.5z + 0.6}\right] z^{-1} \tag{A5.7}$$

The first term is easy (giving a step of height 500 from Equation (A5.1), delayed by one sampling interval as a result of the z^{-1} term on the right-hand side). The second term will take a little more thought.

Comparing the denominators of either of Equations (A5.2) or (A5.3) with that of the second term in Equation (A5.7), we have: $e^{-2aT} = 0.6$, so $-aT = 0.5 \ln(0.6) = -0.2554$.

As T is given as 0.1 s, $a = 2.554$ s^{-1}.

We similarly have $2e^{-aT} \cos(\omega T) = 1.5$, so $\cos(\omega T) = \dfrac{1.5}{2 \times \sqrt{0.6}} = 0.9682$. Thus $\omega T = 0.2527$ rad and $\omega = 2.527$ rad s^{-1}.

The $-500z^2$ in Equation (A5.7), means that we shall have to use the transform pair with numerator $z^2 - ze^{-aT} \cos(\omega t)$ (Equation (A5.3)), giving $-500z^2 + 375z$ on multiplication by -500. We now have $375z$ instead of $250z$, so we must subtract $125z$ by means of the transform pair having a numerator $ze^{-aT} \sin(\omega t)$ (Equation (A5.2)). That numerator evaluates to 0.1937, so we must multiply by $250/0.1937 = 1291$ to obtain the required 250.

The overall inverse is therefore zero up to $t = 0.1$ s (because of the transport lag represented by the z^{-1} multiplying everything) and as follows thereafter:

$$500 - 500e^{-2.554(t-0.1)} \cos[2.527(t - 0.1)] - 1291e^{-2.554(t-0.1)} \sin[2.527(t - 0.1)]$$

(ii) Solution by means of a difference equation

This approach is based on the fact that the inverse of a z-transform expressed as a time series is the same as the impulse response of a discrete-time system whose transfer function is that z-transform. The impulse transforms to 1, that is, $1 + 0z^{-1} + 0z^{-2} + \cdots$

Let the input (impulse) be u and the output (z-transform inverse) be y. So:

$$\frac{y}{u}(z) = \frac{50}{(z - 1)(z^2 - 1.5z + 0.6)}$$

This expression gives $(z - 1)(z^2 - 1.5z + 0.6)y(z) = 50u(z)$.

Multiplying out, we obtain $(z^3 - 2.5z^2 + 2.1z - 0.6)y(z) = 50u(z)$ and multiplying both sides by z^{-3} gives $(1 - 2.5z^{-1} + 2.1z^{-2} - 0.6z^{-3})y(z) = 50z^{-3}u(z)$, or:

$$y(z) = 50z^{-3}u(z) + (2.5z^{-1} - 2.1z^{-2} + 0.6z^{-3})y(z)$$

This equation converts to a discrete-time equation:

$$y_n = 50u_{n-3} + 2.5y_{n-1} - 2.1y_{n-2} + 0.6y_{n-3}$$

which we can use as follows to determine the progress of y in response to our unit impulse input. The following algorithm will produce and output the values of y_n (as will the MATLAB commands given later):

Set all y_n to zero

Set all u_n to zero

Set u_0 to 1

Set $n = 0$

Loop: Calculate $y_n = 50u_{n-3} + 2.5y_{n-1} - 2.1y_{n-2} + 0.6y_{n-3}$

Output n and y_n

Set $y_{n-3} = y_{n-2}$

Set $y_{n-2} = y_{n-1}$

Set $y_{n-1} = y_n$

Set $u_{n-3} = u_{n-2}$

Set $u_{n-2} = u_{n-1}$

Set $u_{n-1} = u_n$

Set $n = n + 1$

Goto *Loop*

The inverse transform was calculated by computer using both methods and the results for the first 20 steps appear in Table A5.2. Good agreement resulted, the slight differences being explicable by numerical rounding errors.

MATLAB can perform this discrete-time simulation in three commands:

$\gg$ num = 50;   % numerator of transfer function

$\gg$ den = [1 -2.5 2.1 -0.6];   % denominator coefficients (descending powers
                                  % of z)

$\gg$ dimpulse(num, den, 40)   % this usage of *dimpulse* plots the response (40
                                  % points)

Table A5.2 Numerical values of the inverse z-transform from Example A5.2.

Time (second)	Results using partial fractions and tables	Results by simulation of difference equation
0	0	0
0.1	0	0
0.2	−0.01457	0
0.3	49.97998	50
0.4	124.9813	125
0.5	207.4865	207.5
0.6	286.2433	286.25
0.7	354.8751	354.875
0.8	410.5683	410.5626
0.9	452.9286	452.9188
1.0	483.0527	483.0408
1.1	502.8225	502.8101
1.2	514.4024	514.391
1.3	519.9102	519.9008
1.4	521.2239	521.217
1.5	519.8896	519.8857
1.6	517.1001	517.0989
1.7	513.7162	513.7177
1.8	510.3141	510.318
1.9	507.2414	507.2473
2.0	504.6736	504.6811

MATLAB could also have multiplied out the denominator, using the convolution command

>> den = conv([1 −1],[1 −1.5 0.6]);

We can compare the results with the tabulated ones using:

>> y = dimpulse(num,den,20)   % this usage of *dimpulse* tabulates the resulting
                              % output

A5.5   *Relationship between the s-plane and the z-plane*

On several occasions, we have plotted the poles and zeros of a system's Laplace transfer function model in the *s*-plane. We have found that knowledge of the *s*-plane locations of the poles and zeros can tell us a lot about how a system will behave in the time domain.

Clearly, a *z*-domain transfer function also has poles and zeros (which would be plotted in the *z-plane*), and it is natural to ask whether the same kind of knowledge is available from these. Indeed it is, and the various contours we discussed in the *s*-plane (of constant natural frequency, constant damping ratio, and so on – see Figure 3.21 in Section 3.2.2) all have their counterparts in the *z*-plane. Since the mapping from the *s*-plane to the *z*-plane is nonlinear, they are not the same shapes as in the *s*-plane. We shall consider these *z*-plane contours and their interpretation in this section.

When the general complex expression for *s* (that is, $s = \sigma + j\omega$) is mapped into the *z*-plane using $z = e^{sT}$, the result is

$$z = e^{\sigma T} e^{j\omega T} \tag{A5.8}$$

Equation (A5.8) shows that *z* is a complex quantity, having magnitude $e^{\sigma T}$ and phase angle ωT radians. This gives us the basis for examining the mapping of the various *s*-plane contours into the *z*-plane.

A5.5.1   *Lines of constant damping factor ($\zeta\omega_n$) – and the stability boundary*

For a system modelled by a dominant pair of second-order poles, lines of constant damping factor ($\zeta\omega_n$) are verticals in the *s*-plane, as shown in Figure 3.21(c). Whatever its derivation, *any* vertical line in the *s*-plane clearly has a constant real part (equal to σ) as ω varies. From Equation (A5.8), such verticals in the *s*-plane therefore map to *z*-plane circles of radius $e^{\sigma T}$ centred on the origin.

For verticals in the left half of the *s*-plane, $\sigma < 0$, so the radius of the corresponding *z*-plane circle is less than unity. For verticals in the right half of the *s*-plane, the *z*-plane circle will be of radius greater than unity. For the *s*-plane imaginary axis, $\sigma = 0$, so the corresponding *z*-plane circle has unity radius.

This means that for stability, whereas all a system's *s*-plane poles must lie in the left-half plane, all its *z*-plane poles must lie within the unit circle. The entire left half of the *s*-plane maps to the unit circle in the *z*-plane.

Progress around any of these circles as frequency varies can be determined in terms of the quantity ωT in Equation (A5.8), which has units of radians. Since *T* is the sampling period, it is related to the (radian) sampling frequency by $T = 2\pi/\omega_s$. A general frequency ω therefore corresponds to a phase angle in the *z*-plane of

$\omega T = 2\pi\omega/\omega_s$ radians. Therefore, when $\omega = 0$, the phase angle is zero, and the corresponding point (on the circle of radius $e^{\sigma T}$) will lie on the positive real axis in the z-plane. When $\omega = \omega_s/4$, the phase angle is $\pi/2 = 90°$, and the point will lie on the positive imaginary axis of the z-plane. Similarly, when $\omega = \omega_s/2$, the point will lie on the negative real axis of the z-plane and when $\omega = \omega_s$, it is back at the positive real axis again.

Clearly, ω can carry on increasing, and more revolutions of the circle occur. Strictly, the behaviour is now more complicated due to aliasing, and each revolution of the circle really ought to be drawn on a new copy of the z-plane (such a multi-sheet 'plane' is called a Riemann surface), but since we assume anti-aliasing filtering to be used, and that signals will not appear at frequencies greater than $\omega = \omega_s/2$, we shall not further describe such behaviour here, and we shall assume that a single copy of the z-plane is sufficient for all our needs.

Stable s-plane pole pairs lying close to the imaginary axis (and therefore having long decay or rise times) correspond to poles on a circle close to the unit circle. We can therefore conclude that the closer a stable z-plane pole is to the unit circle, the longer will be its decay or rise times. Non-real poles in the z-plane have to occur in complex conjugate pairs, for the same reasons as they do in the s-plane.

A5.5.2 Lines of constant frequency – and real s-plane poles

Figure 3.21(a) shows lines of constant frequency as horizontals in the s-plane. According to Equation (A5.8), constant frequency will give lines of constant phase angle (that is, radial lines) in the z-plane. For any given frequency, ω, the angle of the line in the z-plane will be given by ωT radians, where T is the sampling period. Again, we can use the relationship $\omega T = 2\pi\omega/\omega_s$, where ω_s is the (radian) sampling frequency.

As in Section A5.5.1, if $\omega = 0$, the corresponding points lie on the positive real axis in the z-plane. This means that all non-oscillatory (that is, real) s-plane poles map to the positive real z-plane axis. Stable real s-plane poles map to the segment of the positive real z-plane axis in the range zero to unity. The corollary of this is that any z-plane pole which is within the unit circle, but not on the positive real axis, must be associated with oscillatory responses, and must therefore be one of a pair of poles. The difference from the s-plane is that such a pair of poles can appear on the negative *real* axis in the z-plane – so they are not necessarily complex conjugate pairs in the z-domain (but they will either occur as a double negative real pole, or a negative and positive real pole having the same magnitude). We now illustrate this point.

The locations of poles in the z-plane depend on the chosen sampling frequency. Since the negative real axis of the z-plane corresponds to a radial line at an angle of π radians (180°), it can be seen from the relationship $\omega T = 2\pi\omega/\omega_s$ (above) that s-plane poles with an associated frequency value ω will always map onto the negative real z-plane axis, if ω is half the sampling frequency. The sampling *period* for this to occur is therefore $T = 2\pi/\omega_s = \pi/\omega$ seconds.

For example, the s-domain LTF $G(s) = 13/(s^2 + 6s + 13)$ has (s-plane) poles at $s = -3 \pm j2$, therefore lying on the $\omega = \pm 2$ rad s^{-1} horizontal lines in the s-plane. If we were to sample this system at $\omega_s = 4$ rad s^{-1} (that is, using a sampling period of $T = \pi/2$ s), then the z-plane poles would both be negative and real (also, due to the discussion in Section A5.5.1, since the s-plane poles lie on the vertical line $\sigma = -3$, they should both appear on the circle of radius $e^{\sigma T} = e^{-3\pi/2} = 0.009$). Therefore, for this system sampled at intervals of $T = \pi/2$ s, we expect to find a double z-plane pole at $z = -0.009$. This is confirmed by the first MATLAB example in Section A5.5.5.

The case where two (stable) real poles in the z-plane have the same magnitude, but opposite sign, is the case which occurs in the Dahlin controller in Example 7.5. What happens in this case is that the positive pole corresponds to a stable decaying envelope as expected. The negative one clearly has the same envelope, but due to the sampling, its response changes sign at each sampling instant.

A5.5.3 Lines of constant damping ratio

Complex pairs of s-plane poles are associated with lines of constant damping ratio (ζ) which are radial lines in the s-plane (Figure 3.21(b)). Having variable real and imaginary parts, these are a little harder to map into the z-plane, but it can be shown that the result is a logarithmic spiral. We shall not prove this here, but the MATLAB *zgrid* command can superimpose such lines on the z-plane, and the results appear in Figure A5.1.

A5.5.4 Lines of constant undamped natural frequency – and the s-plane origin

The s-plane circles of constant undamped natural frequency (ω_n) (see Figure 3.21(d)) similarly map to strange shapes in the z-plane, and these are also shown in Figure A5.1.

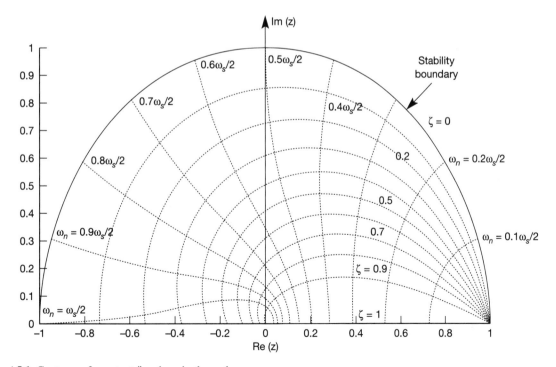

Figure A5.1 Contours of constant ζ and ω_n in the z-plane.

Combining the information of Sections A5.5.1 and A5.5.2 shows that the origin of the s-plane corresponds to the point $(+1, 0)$ in the z-plane, so this is where we would expect to find the z-plane pole of a pure integrator. Inspecting the table of z-transforms (Table A5.1) shows this to be the case.

It is also interesting to note that all the s-plane to z-plane conversions discussed above, together with a careful inspection of Figure A5.1, show that poles in the vicinity of the point $(+1, 0)$ in the z-plane have the same contours and rules associated with them as poles in the vicinity of the s-plane origin. Of course, as we move away from these points, this ceases to be true.

A5.5.5 Tests using MATLAB

It is both easy and instructive to use a package such as MATLAB (Appendix 3) to investigate the migration of poles and zeros between the s-plane and z-plane. This can be done as follows (obviously, you can enter any transfer function you wish, but it may be better to start with single poles to isolate the effects). Note that the *c2dm* command is used with LTF models (*c2d* only works with state-space models), and the conversion is assumed to be done with a zero-order hold by default (other methods can be specified – use *help c2dm* to discover how). Note also that extra zeros will appear in the z-transfer function due to the hold element which is combined with $G(s)$ during the conversion process.

For the system discussed in Section A5.5.2

```
≫ snum = 13;                          % LTF numerator
≫ sden = conv([1  3+2j],[1  3−2j]);   % evaluate LTF denominator
≫ [znum,zden] = c2dm(snum,sden,pi/2); % convert to z at T = π/2 second
≫ roots(zden)                         % list z-plane poles
≫ roots(znum)                         % note new z-plane zero
```

For a more complicated system
For the system

$$G(s) = \frac{(s+1)(s-2)}{s(s-1)(s^2+1.6s+1)}$$

```
≫ snum = conv([1  1],[1  −2]);        % evaluate LTF numerator
≫ sden = conv([1  −1  0],[1  1.6  1]);% and denominator
≫ pzmap(snum,sden),sgrid      % s-plane map, with zeta and wn contours
≫ [znum,zden] = c2dm(snum,sden,0.6);  % convert to z at T = 0.6 second
≫ pzmap(znum,zden),zgrid      % z-plane map, with zeta and wn contours
```

Appendix 6
Random signals and the Kalman filter derivation

A6.1   DEFINITION OF RANDOM SIGNAL QUANTITIES

A6.2   SOME COMBINATIONS OF RANDOM SIGNALS

A6.3   DERIVATION OF THE KALMAN FILTER

There are very many published papers and books containing details of the Kalman filter derivation. However, students often find them too difficult to follow. This is partly due to the fact that the papers often contain notation that the general reader does not understand; and partly due to the fact that they are usually written by people of advanced mathematical capabilities, and tend to leave portions of the derivation to the imagination of the reader – who may not be as mathematically adept. The authors are unaware of any derivation which both follows the notation used in this text, and has no steps omitted; so one is provided here.

It is quite possible, given the knowledge in Section 9.8 of the text, to make use of the Kalman filter without understanding every step of its derivation. However, for the interested reader, this appendix contains the necessary background of random (stochastic) signals to allow the Kalman filter derivation to be developed. The main mathematical part of the derivation itself is also given here.

A6.1   Definition of random signal quantities

The purpose of the Kalman filter in this text is to extract estimates of the state variables of systems from noisy measurements. The noise on the measurements is assumed to be caused by randomly varying disturbance signals. The starting point must therefore be a definition of the various quantities required to allow the handling of such randomly varying signals.

A6.1.1   Mean (or average value, or expected value, or expectation) (scalar case)

For N samples of a single (that is, scalar) time-varying signal $x(t)$, taken at times $t_1, t_2, \ldots t_N$ and having corresponding values $x_1, x_2, \ldots x_N$, the mean (or average) value is simply given by the sum of all the values, divided by the number of samples taken:

$$\bar{x} = \frac{1}{N} \sum_{k=1}^{N} x_k$$

If a guess was required as to what the value of x might be at any given time then, in the absence of any further information about the signal $x(t)$, this mean value is perhaps the best guess that could be made. It is therefore often called the 'expected value' or 'expectation' of x (written as $E[x]$). Remember, it is just the average (mean) value of all the samples:

$$E[x] = \bar{x}$$

A6.1.2 Variance (scalar case)

This is another quantity which is required in the analysis later. It is a measure of the uncertainty involved when guessing the value of a signal from the signal's mean value. It carries information about how far the samples of the signal $x(t)$ (namely $x_1, x_2, \ldots x_N$) are spread around their mean value. A low value of variance implies that most of the samples are generally close to the mean value, so it is relatively certain that the mean value of the signal over all times is a good guide as to the likely value of the signal at any particular time. On the other hand, a high value of variance implies that the individual samples of the signal are scattered widely to either side of the mean, so that the mean value is not a good guide to the likely signal value at any particular time, and the value is therefore relatively uncertain. More will be said about this in Section A6.1.3.

The variance is defined as the square of the 'population standard deviation' σ_x which, in turn, is the root-mean-square (r.m.s.) of the deviations of the samples from the mean, given by:

$$\sigma_x = \sqrt{\frac{1}{N} \sum_{k=1}^{N} (x_k - \bar{x})^2}$$

So,

$$\text{variance} = \sigma_x^2 = \frac{1}{N} \sum_{k=1}^{N} (x_k - \bar{x})^2 = \left[\frac{1}{N} \sum_{k=1}^{N} x_k^2 \right] - \bar{x}^2$$

or, in terms of expectations,

$$\sigma_x^2 = E[(x - \bar{x})^2] = E[x^2] - \bar{x}^2 = E[x^2] - \{E[x]\}^2$$

In words, the variance of a set of measurements (a set of samples of a signal) is the mean of the squares of the deviations of all the individual samples from the mean value of the set. If variance is zero, then the signal must be equal to its mean at every sample.

As discussed in Section 9.8, the Kalman filter is a minimum-variance estimator, the variance in question here being that of the error between the estimated value and the true value of the state vector of a system. The Kalman filter therefore minimizes the uncertainty in the state estimates compared with other estimators. In this sense, it is sometimes referred to as an *optimal state estimator*.

A6.1.3   *Gaussian variables (normal distribution)*

In order to say something about the 'randomness' of a time-varying signal, either its variations of amplitude with time can be considered, or the spectrum of frequencies which make up the signal can be considered, or both. For example, the amplitudes of many samples of the signal might be measured, and from all these measurements the probability of the signal being at any particular level at any given time can be calculated. If these probabilities are plotted against signal level, a graph such as Figure A6.1 might result (infinitely many other shapes are possible). Considering only the overall shape of Figure A6.1, the highest probability is that an individual sample of the signal will be near its mean value $\bar{x}$. The 'bell-shape' of this curve means that the probability that an individual sample of the signal will have an amplitude very much greater (or very much smaller) than the mean value is very small.

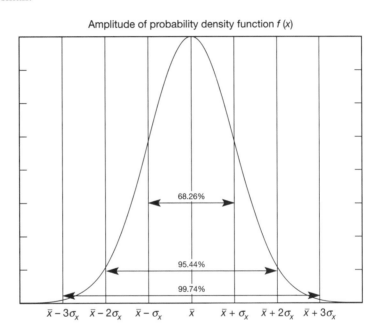

Amplitude of probability density function $f(x)$

68.26%

95.44%

99.74%

$$\bar{x} - 3\sigma_x \quad \bar{x} - 2\sigma_x \quad \bar{x} - \sigma_x \quad \bar{x} \quad \bar{x} + \sigma_x \quad \bar{x} + 2\sigma_x \quad \bar{x} + 3\sigma_x$$

Figure A6.1 A normally distributed (Gaussian) signal.

A smooth curve such as that in Figure A6.1 can be represented by an equation (known as a probability density function). A variable which is described as being 'Gaussian' in probability distribution can be represented by the particular equation

$$f(x) = \frac{1}{\sqrt{2\pi\sigma_x^2}}\, e^{-(x-\bar{x})^2/(2\sigma_x^2)}$$

This particular bell-shaped curve is called a 'normal distribution', and sufficiently large sets of samples of very many real-world quantities (from random noise to examination marks) tend to follow such a distribution.

For this Gaussian distribution, samples of the signal can be expected to be within ± 1, 2 or 3 standard deviations of the mean value, with the probabilities shown in Figure A6.1. If the variance (and hence the standard deviation) is small, the peak in Figure A6.1 will be sharp compared with the data range of the x-axis. If the variance is large, the peak will be broad compared with the data range of the x-axis.

As an example, someone might say that a set of examination results has a mean of 55 per cent, and a standard deviation of 15 per cent (since the variance is just the square of the per unit standard deviation, it could be used instead if preferred). From Figure A6.1, this implies that the mean value ($\bar{x}$) would be at 55 per cent on the horizontal axis. If a large number of candidates had been considered, and a normally distributed set of marks was assumed, then 68.26 per cent of the candidates would be expected to have scores between 40 per cent and 70 per cent. This implies that about 32 per cent of candidates would have marks either lower than 40 per cent, or higher than 70 per cent. If 40 per cent represents a failure, and 70 per cent a first-class result, then on average 16 per cent of candidates might be expected to fail the examination, and 16 per cent to get a first-class result.

A6.1.4 White noise

A 'white noise' signal can be defined as a signal containing all possible *frequency* components at equal levels of probability, that is, it has a *flat* probability distribution as far as frequency is concerned.

If it is Gaussian in terms of its *amplitude* probability density, as discussed above, then its amplitude can be defined in terms of mean and variance (or standard deviation, if preferred). Thus the signal will have a certain mean value, but samples taken at different times may be greater or smaller than this according to the distribution in Figure A6.1. The term 'random' below is taken to refer to such 'white noise' signals.

A6.2 Some combinations of random signals

For reasons of mathematical convenience, it is normal to work with the variance of a signal, rather than its standard deviation. For analysis of systems containing random signals of the kind discussed previously, it is therefore necessary to know what happens to such signals in terms of their mean values and variances, when they are combined in various ways with other signals.

A6.2.1 Multiplication by a scalar

If a random signal $x(t)$, having mean $\bar{x}$ and variance σ_x^2, is multiplied by a constant c and has a bias term (that is, a constant level shift) d added, then we obtain the new (random) variable $y(t)$, given by

$$y(t) = cx(t) + d$$

The mean of the new signal $y(t)$ is given by:

$$E[y] = \bar{y} = c\bar{x} + d$$

and hence the variance of $y(t)$ is given by:

$$E[(y - \bar{y})^2] = \sigma_y^2 = c^2 \sigma_x^2$$

A6.2.2 Addition

If two random signals are added together, such that $y(t) = w(t) + v(t)$, then

$$E[y] = E[w] + E[v], \text{ that is, } \bar{y} = \bar{w} + \bar{v}$$

and

$$E[(y - \bar{y})^2] = \sigma_y^2 = \sigma_w^2 + \sigma_v^2 + 2\sigma_{vw}$$

where $\sigma_{vw} = E[(v - \bar{v})(w - \bar{w})] = covariance$ (see below).

A6.2.3 Multiplication

The multiplication of two random signals, one of which may be time-shifted relative to the other by τ seconds, can lead to the *cross-correlation, auto-correlation* or *covariance* functions as follows:

For random signals $v(t)$ and $w(t)$,

$$E[v(t)w(t)] = \lim_{T \to \infty} \frac{1}{T} \int_0^T v(t)w(t + \tau)\, dt = R_{vw}(\tau)$$

This is the *cross-correlation* function.

If $w(t) = v(t)$, the result becomes:

$$E[v(t)v(t)] = \lim_{T \to \infty} \frac{1}{T} \int_0^T v(t)v(t + \tau)\, dt = R_{vv}(\tau)$$

This is the *auto-correlation* function.

If $v(t)$ and $w(t)$ are zero-mean signals (or have their mean values subtracted), and τ is zero (that is, no time shift), then:

$$E[\{v(t) - \bar{v}\}\{w(t) - \bar{w}\}] = \lim_{T \to \infty} \frac{1}{T} \int_0^T \{v(t) - \bar{v}\}\{w(t) - \bar{w}\}\, dt$$

$$= \text{cov}(vw) = \sigma_{vw}$$

This is the *covariance* function.

Note that if $v(t)$ and $w(t)$ are uncorrelated (that is, they are independent of one another), then $\text{cov}(vw) = $ zero.

A6.2.4 The vector case – the covariance matrix

Now imagine that rather than just one (scalar) random signal $x(t)$ a set (or 'ensemble') of n random signals $x_1(t), x_2(t), \ldots, x_n(t)$ exists, making up a signal vector $x(t)$ as shown in Figure A6.2. At any particular sampling instant k there is therefore a set of values (see Figure A6.2) $x_k = [x_{1_k}\ x_{2_k}\ \ldots\ x_{n_k}]^T$. At the next instant $k + 1$, the set $x_{k+1} = [x_{1_{k+1}}\ x_{2_{k+1}}\ \ldots\ x_{n_{k+1}}]^T$ will exist.

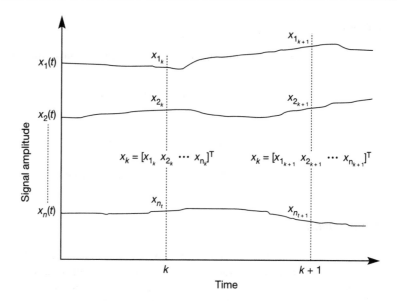

Figure A6.2 A set (ensemble) of random signals.

Associated with such a random signal vector is a *covariance matrix*, containing all the possible covariances between the elements of the vector, hence:

$$
\text{cov}[x(t)] = R = \begin{bmatrix}
\sigma_{x_1}^2 & \sigma_{x_1 x_2} & \sigma_{x_1 x_3} & \cdots & \sigma_{x_1 x_n} \\
\sigma_{x_2 x_1} & \sigma_{x_2}^2 & \sigma_{x_2 x_3} & \cdots & \sigma_{x_2 x_n} \\
\sigma_{x_3 x_1} & \sigma_{x_3 x_2} & \sigma_{x_3}^2 & \cdots & \sigma_{x_3 x_n} \\
\vdots & \vdots & \vdots & \ddots & \vdots \\
\sigma_{x_n x_1} & \sigma_{x_n x_2} & \sigma_{x_n x_3} & \cdots & \sigma_{x_n}^2
\end{bmatrix}
$$

Such a matrix has some special properties which will be useful later, namely:

- From the definition of covariance in Section A6.2.3, $\sigma_{ij} = \sigma_{ji}$, so the matrix is always *symmetric*. Therefore $R^T = R$.
- If the elements $x_1(t), x_2(t), \ldots, x_n(t)$ of the signal vector $x(t)$ are uncorrelated (that is, they are independent of each other), then all such terms as $\sigma_{x_i x_j}|_{i \neq j}$ will be zero. R will then be *diagonal*.
- $R = \mathrm{E}[xx^T]$ (remember $R = \mathrm{E}[x^2]$ for the scalar case of a zero-mean signal).

A6.2.5 Covariance transformation

If $x(t)$ contains zero-mean random variables, and C is a constant matrix which operates on it such that $y(t) = Cx(t)$, what is the covariance matrix of $y(t)$?

$$
\text{cov}[y(t)] = \mathrm{E}[yy^T] = \mathrm{E}[Cxx^T C^T] = C\mathrm{E}[xx^T]C^T
$$

Thus, $\text{cov}[Cx(t)] = CRC^T$, where $R = \text{cov}[x(t)]$.

A6.3 Derivation of the Kalman filter

This is the most heavily mathematical part of the derivation begun in Section 9.8. All the required steps are described here and should be digestible if followed slowly line by line. In Section 9.8, the *estimation error* (Equation (9.35)) was defined as:

$$\tilde{x}_{k+1} = x_{k+1} - \hat{x}_{k+1|k+1} \tag{9.35}$$

Equations (9.30) (for x_{k+1}) and (9.34) (for $\hat{x}_{k+1|k+1}$) from the main text can now be substituted into Equation (9.35) to obtain

$$\tilde{x}_{k+1} = \boldsymbol{\Phi}x_k + \Delta u_k + \boldsymbol{\Gamma}w_k - [I - KC][\boldsymbol{\Phi}\hat{x}_{k|k} + \Delta u_k] - Kz_{k+1}$$

Rearranging, and including Equation (9.31) (the noisy output equation) to eliminate z gives

$$\tilde{x}_{k+1} = \boldsymbol{\Phi}x_k - \boldsymbol{\Phi}\hat{x}_{k|k} + \boldsymbol{\Gamma}w_k + KC[\boldsymbol{\Phi}\hat{x}_{k|k} + \Delta u_k] - K[Cx_{k+1} + v_{k+1}]$$

This can be greatly simplified by again substituting for x_{k+1} from Equation (9.30), then using $x_k - \hat{x}_{k|k} = \tilde{x}_k$ (from Equation (9.35)) to eliminate $\hat{x}_{k|k}$, then multiplying out and re-grouping terms to leave

$$\tilde{x}_{k+1} = [I - KC][\boldsymbol{\Phi}\tilde{x}_k + \boldsymbol{\Gamma}w_k] - Kv_{k+1} \tag{A6.1}$$

We also know that the *covariance matrix* of the estimation error is defined as $P_k = \mathrm{E}[\tilde{x}_k \tilde{x}_k^{\mathrm{T}}]$.

The design of the KF follows from a choice of K such that P_k is minimized, giving the *minimum variance estimator* as mentioned in the text.

To achieve this, some information is required about the noise signals. It is assumed that w_k and v_k have zero mean over all k (if not, the mean values can be subtracted, with suitable modification of the equations).

It is also assumed that w and v are vectors of stationary white noise sequences (that is, their statistical properties are constant with time), having covariances given by:

$$\mathrm{E}[w_k w_k^{\mathrm{T}}] = Q\big|_{\mathrm{all}\,k} \quad \text{and} \quad \mathrm{E}[v_k v_k^{\mathrm{T}}] = R\big|_{\mathrm{all}\,k}$$

Remember that Q and R are symmetric matrices. Also, if the separate noise signals in the vector w are uncorrelated, then Q will be diagonal (similarly for v and R) (from Section A6.2.4). The main text has some comments on practical evaluation of Q and R.

Now, an expression for the covariance of the estimation error at the next step P_{k+1} can be obtained, so that an investigation can be made as to a suitable choice of K to minimize it. First, rewrite Equation (A6.1) so as to separate the noise terms:

$$\tilde{x}_{k+1} = [I - KC]\boldsymbol{\Phi}\tilde{x}_k + [I - KC]\boldsymbol{\Gamma}w_k - Kv_{k+1}$$

which, writing $F = [I - KC]$ for ease of notation is

$$\tilde{x}_{k+1} = F\boldsymbol{\Phi}\tilde{x}_k + F\boldsymbol{\Gamma}w_k - Kv_{k+1}$$

Post-multiplying each side by its transpose (similar to squaring in the scalar case) gives:

$$\tilde{x}_{k+1}\tilde{x}_{k+1}^\mathrm{T} = [F\Phi\tilde{x}_k + F\Gamma w_k - Kv_{k+1}]$$

$$\times [\tilde{x}_k^\mathrm{T}\Phi^\mathrm{T}F^\mathrm{T} + w_k^\mathrm{T}\Gamma^\mathrm{T}F^\mathrm{T} - v_{k+1}^\mathrm{T}K^\mathrm{T}]$$

$$= F\Phi\tilde{x}_k\tilde{x}_k^\mathrm{T}\Phi^\mathrm{T}F^\mathrm{T} + F\Gamma w_k w_k^\mathrm{T}\Gamma^\mathrm{T}F^\mathrm{T} + Kv_{k+1}v_{k+1}^\mathrm{T}K^\mathrm{T}$$

$$+ F\Phi\tilde{x}_k w_k^\mathrm{T}\Gamma^\mathrm{T}F^\mathrm{T} - F\Phi\tilde{x}_k v_{k+1}^\mathrm{T}K^\mathrm{T} + F\Gamma w_k\tilde{x}_k^\mathrm{T}\Phi^\mathrm{T}F^\mathrm{T}$$

$$- F\Gamma w_k v_{k+1}^\mathrm{T}K^\mathrm{T} - Kv_{k+1}\tilde{x}_k^\mathrm{T}\Phi^\mathrm{T}F^\mathrm{T} - Kv_{k+1}w_k^\mathrm{T}\Gamma^\mathrm{T}F^\mathrm{T}$$

Assuming that x, w and v are uncorrelated, the expected value of any product of samples of two different vectors will be zero (see Section A6.2.3). There are such products in the last six terms on the RHS of the above equation, so it can be simplified greatly if it is rewritten in terms of expected values. Using the expectation operator on each side therefore, only the expected values of the first three terms on the RHS remain:

$$\mathrm{E}[\tilde{x}_{k+1}\tilde{x}_{k+1}^\mathrm{T}] = F\Phi\mathrm{E}[\tilde{x}_k\tilde{x}_k^\mathrm{T}]\Phi^\mathrm{T}F^\mathrm{T} + F\Gamma\mathrm{E}[w_k w_k^\mathrm{T}]\Gamma^\mathrm{T}F^\mathrm{T}$$

$$+ K\mathrm{E}[v_{k+1}v_{k+1}^\mathrm{T}]K^\mathrm{T}$$

Each of the four expectation terms is recognizable as previously defined covariance matrices (see Section A6.2.4), so the equation is:

$$P_{k+1} = F\Phi P_k\Phi^\mathrm{T}F^\mathrm{T} + F\Gamma Q\Gamma^\mathrm{T}F^\mathrm{T} + KRK^\mathrm{T}$$

$$= F[\Phi P_k\Phi^\mathrm{T} + \Gamma Q\Gamma^\mathrm{T}]F^\mathrm{T} + KRK^\mathrm{T}$$

Next, defining

$$P_k^* = \Phi P_k\Phi^\mathrm{T} + \Gamma Q\Gamma^\mathrm{T} \tag{A6.2}$$

and reinstating the expansion $F = [I - KC]$, yields

$$P_{k+1} = [I - KC]P_k^*[I - KC]^\mathrm{T} + KRK^\mathrm{T} \tag{A6.3}$$

The values of the matrix K which minimize Equation (A6.3) must now be found, so as to result in the Kalman filter. Equation (A6.3) is called a matrix Riccati equation, and will be encountered again in Chapter 12, on optimal control. It is not obvious how to minimize this equation by choice of K, and at this point most texts and papers (even Healey (1979), who gives an otherwise fairly complete derivation) make one or two rather large leaps!

An initial multiplying out of Equation (A6.3) and a rearrangement to obtain K in slightly more isolated positions gives

$$P_{k+1} = KCP_k^*C^\mathrm{T}K^\mathrm{T} + KRK^\mathrm{T} - P_k^*C^\mathrm{T}K^\mathrm{T} - KCP_k^* + P_k^*$$

or

$$P_{k+1} = K[CP_k^*C^\mathrm{T} + R]K^\mathrm{T} - P_k^*C^\mathrm{T}K^\mathrm{T} - KCP_k^* + P_k^* \tag{A6.4}$$

To understand more easily how to solve Equation (A6.4) for K, an analogy can be drawn with a simpler-looking equation. Visualizing the first term on the right-hand side of Equation (A6.4) as $\boldsymbol{KGK}^\mathrm{T}$, and noting that a matrix term such as $\boldsymbol{KGK}^\mathrm{T}$ is broadly equivalent in its behaviour to a term K^2G in scalar algebra, consider a *scalar* equation:

$$P_{k+1} = K^2G - f(K,G) + P_k^* \tag{A6.5}$$

in which it is desired to solve for K, but the function $f(K,G)$ is not convenient for analysis.

One approach is to investigate the existence of a new variable M, such that

$$P_{k+1} = K^2G - 2KMG + P_k^*$$

This is known as 'completing the square' to get a more useful form for analysis, as K is now completely isolated:

$$P_{k+1} = (K - M)^2G - M^2G + P_k^* \tag{A6.6}$$

Setting Equation (A6.6) = Equation (A6.5) would then eliminate several terms, and hopefully allow a solution for K by a suitable choice of M.

Returning to the vector–matrix case, and bearing in mind that K^2G in the scalar case is equivalent to $\boldsymbol{KGK}^\mathrm{T}$ in the vector–matrix case, Equation (A6.5) is then equivalent in form to Equation (A6.4), with $\boldsymbol{G} = \boldsymbol{CP}_k^*\boldsymbol{C}^\mathrm{T} + \boldsymbol{R}$.

The method leading to Equation (A6.6) suggests that a new equation be written, equivalent to Equation (A6.6), of the form:

$$\boldsymbol{P}_{k+1} = [\boldsymbol{K} - \boldsymbol{M}]\boldsymbol{G}[\boldsymbol{K} - \boldsymbol{M}]^\mathrm{T} - \boldsymbol{MGM}^\mathrm{T} + \boldsymbol{P}_k^* \tag{A6.7}$$

where the matrix $\boldsymbol{M}$ is yet to be specified, and it is desired to solve for $\boldsymbol{K}$ so as to minimize $\boldsymbol{P}_{k+1}$.

Expanding the equation gives:

$$\boldsymbol{P}_{k+1} = \boldsymbol{KGK}^\mathrm{T} - \boldsymbol{MGK}^\mathrm{T} - \boldsymbol{KGM}^\mathrm{T} + \boldsymbol{P}_k^* \tag{A6.8}$$

Now, setting Equation (A6.8) = Equation (A6.4) (with the term $[\boldsymbol{CP}_k^*\boldsymbol{C}^\mathrm{T} + \boldsymbol{R}]$ replaced by $\boldsymbol{G}$) as suggested by the scalar analogy, the first and last terms will cancel, and the two following equalities result, which can be investigated to see if they really are an acceptable solution.

$$\boldsymbol{MGK}^\mathrm{T} = \boldsymbol{P}_k^*\boldsymbol{C}^\mathrm{T}\boldsymbol{K}^\mathrm{T} \tag{A6.9}$$

and

$$\boldsymbol{KGM}^\mathrm{T} = \boldsymbol{KCP}_k^* \tag{A6.10}$$

Recall again that the covariance matrices $\boldsymbol{P}_{k+1}$, $\boldsymbol{Q}$ and $\boldsymbol{R}$ are symmetric. Therefore, from Equation (A6.2), $\boldsymbol{P}_k^*$ must also be symmetric. Since $\boldsymbol{G} = \boldsymbol{CP}_k^*\boldsymbol{C}^\mathrm{T} + \boldsymbol{R}$, $\boldsymbol{G}$ must also be symmetric. From Equation (A6.9), $\boldsymbol{MG} = \boldsymbol{P}_k^*\boldsymbol{C}^\mathrm{T}$, and taking transposes gives $\boldsymbol{G}^\mathrm{T}\boldsymbol{M}^\mathrm{T} = \boldsymbol{CP}_k^{*\mathrm{T}}$.

Since $\boldsymbol{G}$ and $\boldsymbol{P}_k^*$ are symmetric, this implies that $\boldsymbol{GM}^\mathrm{T} = \boldsymbol{CP}_k^*$.

This is in agreement with Equation (A6.10). Therefore, the choice of Equations (A6.9) and (A6.10) is self-consistent, and a matrix $\boldsymbol{M}$ must exist such that Equations (A6.4) and (A6.8) are identical. Hence, from Equation (A6.9):

$$\boldsymbol{M} = \boldsymbol{P}_k^*\boldsymbol{C}^\mathrm{T}\boldsymbol{G}^{-1} = \boldsymbol{P}_k^*\boldsymbol{C}^\mathrm{T}[\boldsymbol{CP}_k^*\boldsymbol{C}^\mathrm{T} + \boldsymbol{R}]^{-1} \tag{A6.11}$$

Now, since it is desired to minimize Equation (A6.7) by choice of K, the quadratic form of Equation (A6.7) shows that the minimum occurs when $K = M$.

Noting that the values of K which minimize P_{k+1} may differ from those which minimize P_{k+2} and so on, this particular matrix will be called K_{k+1}. Therefore from Equation (A6.11):

$$K_{k+1} = P_k^* C^T [C P_k^* C^T + R]^{-1} \tag{A6.12}$$

Writing $K - M = 0$ and $M = K_{k+1}$ in Equation (A6.7), the optimal solution is then obtained:

$$P_{k+1} = -K_{k+1} G K_{k+1}^T + P_k^* = -K_{k+1}[C P_k^* C^T + R] K_{k+1}^T + P_k^*$$

From Equation (A6.12), it can be seen that $K_{k+1}[C P_k^* C^T + R] = P_k^* C^T$. Combining with the previous equation yields:

$$P_{k+1} = -P_k^* C^T K_{k+1}^T + P_k^* = P_k^*[I - C^T K_{k+1}^T]$$

Transposing both sides:

$$P_{k+1}^T = [I - C^T K_{k+1}^T]^T P_k^{*T}$$

which, remembering that P_{k+1} and P_k^* are symmetric, gives:

$$P_{k+1} = [I - K_{k+1} C] P_k^* \tag{A6.13}$$

Equations (A6.2), (A6.12), (9.34) (with varying K) and (A6.13) together form a recursive set which will implement the KF. They are gathered together as Equations (9.36) to (9.39) in the main text, and discussed further there.

$$P_k^* = \Phi P_k \Phi^T + \Gamma Q \Gamma^T \tag{9.36}$$

$$K_{k+1} = P_k^* C^T [C P_k^* C^T + R]^{-1} \tag{9.37}$$

$$\hat{x}_{k+1|k+1} = [I - K_{k+1} C][\Phi \hat{x}_{k|k} + \Delta u_k] + K_{k+1} z_{k+1} \tag{9.38}$$

$$P_{k+1} = [I - K_{k+1} C] P_k^* \tag{9.39}$$

Appendix 7
Derivation of Plackett's algorithm for online least-squares fitting

The derivation starts from the basic concept of ordinary least-squares. In the single-input-single-output case, we start with a series of corresponding values of the actual input and output over time. We then formulate a model of appropriate order and adjust its parameters to minimize the mean value of the square of the difference between the actual output and that predicted by the model. A model of the following form will be used:

$$y_n = \sum_{r=1}^{p} a_r y_{n-r} + \sum_{r=1}^{q} b_r u_{n-r} + v_n \tag{A7.1}$$

where v_n is a random error ('noise' of mean value zero will be assumed).

Apart from the noise term, Equation (A7.1) is identical to Equation (11.1) in Chapter 11. Since the noise term is zero mean, the best guess we could make as to its value at any given instant is zero. The value of y_n *predicted* by the model of Equation (A7.1) will therefore exclude the v_n.

Since we wish to minimize the mean square of the error over the total number of time steps taken (N), we shall be trying to choose the elements of the vectors $\boldsymbol{a}$ and $\boldsymbol{b}$ so as to minimize:

$$\sum_{n=1}^{N} \left(\text{Model } y_n - \text{Actual } y_n\right)^2$$

$$= \sum_{n=1}^{N} \left[\sum_{r=1}^{p} a_r y_{n-r} + \sum_{r=1}^{q} b_r u_{n-r} - \text{Actual } y_n\right]^2 \tag{A7.2}$$

Since we are to minimize the *mean* of the error, the reader may wonder why there is no division by N in Equation (A7.2). A little thought will show that this would make no difference to the *location* of the minimum in terms of the values of $\boldsymbol{a}$ and $\boldsymbol{b}$. It would change the *value* at the minimum, but that is immaterial. The values of the coefficients in the $\boldsymbol{a}$ and $\boldsymbol{b}$ vectors can be found by partial differentiation with respect to each a_r and b_r term, or by numerical minimization by one of the standard methods (for example, maximum gradient or Fibonacci search).

The problem with such an approach in the online context (that is, inside a digital control scheme) is that all the calculations have to be re-performed whenever a new pair of values of y and u occurs. An arrangement for online parameter estimation based on that approach is therefore impractical, since the number of terms over which the summations must be done will soon become extremely large, when the system will effectively come to a halt!

Algorithms have been devised which allow an existing estimate of the a and b values to be updated in the light of the most recent pair of values of y and u only. Many of these are based on Plackett's algorithm, which is explained and derived as follows.

First, we modify Equation (A7.1) in the same way as in Chapter 11, by combining the a and b values into a vector θ (a stacked over b) and sufficient past values of y and u to fulfil Equation (A7.1) as a vector x (y stacked over u). This means that Equation (A7.1) becomes:

$$y_n = x_{n-1}^T \theta + v_n \tag{A7.3}$$

Now consider how the parameter estimator ('identifier') will operate. Using the notation of Plackett (1950 – except that we have substituted x for z so as to avoid confusion with the z-transform), it will update θ each time step by taking a weighted average of a quantity J_n times the previous estimate of θ (namely θ_{n-1}) and a vector k_n times the latest error (defined as the difference between the predicted value $x_{n-1}^T \theta_{n-1}$ and the observed value y_n), giving:

$$\theta_n = J_n \theta_{n-1} + k_n (x_{n-1}^T \theta_{n-1} - y_n) \tag{A7.4}$$

If the noise were zero, and we had estimated θ_{n-1} accurately, the bracket following k_n would be zero. In those circumstances no change in θ would be required. Therefore $J_n = 1$.

Now we see how the least-squares method is incorporated. For the single-input-single-output scenario, we were minimizing the mean-square error between actual and model outputs. In statistical terms, that is the *variance* of one of those outputs assuming the other one to be the true value.

In the online case, the approach is to proceed similarly with the vector θ, assuming the error in our estimation of that vector to be a vector $(\theta_n - \theta)$, where θ is the true value. Again the idea is to minimize the mean-square value of this difference. The fact that the difference is now a vector is overcome by the use the *covariance matrix* of this error P_n (see Section A6.2.4), which is defined as

$$P_n = (\theta_n - \theta)(\theta_n - \theta)^T \tag{A7.5}$$

This is the outer product of the two vectors, thus giving a matrix result. The values on the leading diagonal will be the mean-square errors (variances) of the individual members of θ. Those off the leading diagonal will be products of the error in one term of θ_n and that in another term of θ_n. Two things follow with regard to the off-diagonal terms (Appendix 6):

- If the errors are random, the average values of the off-diagonal terms should be zero.
- The matrix will be symmetric.

Since P_n is a measure of the error in the parameter estimates, the next step is to determine the *expected value* of P_n and minimize it. Substituting Equation (A7.4) (with $J_n = 1$) into Equation (A7.5) gives:

$$P_n = [\theta_{n-1} + k_n(x_{n-1}^\mathrm{T}\theta_{n-1} - y_n) - \theta][\theta_{n-1} + k_n(x_{n-1}^\mathrm{T}\theta_{n-1} - y_n) - \theta]^\mathrm{T}$$

which, by substituting for y_n from Equation (A7.3), becomes

$$P_n = [\theta_{n-1} + k_n(x_{n-1}^\mathrm{T}\theta_{n-1} - x_{n-1}^\mathrm{T}\theta - v_n) - \theta]$$
$$\times [\theta_{n-1} + k_n(x_{n-1}^\mathrm{T}\theta_{n-1} - x_{n-1}^\mathrm{T}\theta - v_n) - \theta]^\mathrm{T}$$

We now perform some matrix algebra to simplify this expression prior to minimizing P_n:

$$P_n = [(I + k_n x_{n-1}^\mathrm{T})(\theta_{n-1} - \theta) - k_n v_n]$$
$$\times [(I + k_n x_{n-1}^\mathrm{T})(\theta_{n-1} - \theta) - k_n v_n]^\mathrm{T}$$
$$= [I + k_n x_{n-1}^\mathrm{T}][\theta_{n-1} - \theta][\theta_{n-1} - \theta]^\mathrm{T}[I + k_n x_{n-1}^\mathrm{T}]^\mathrm{T}$$
$$- k_n v_n[\theta_{n-1} - \theta]^\mathrm{T}[I + k_n x_{n-1}^\mathrm{T}]^\mathrm{T}$$
$$- [I + k_n x_{n-1}^\mathrm{T}][\theta_{n-1} - \theta]k_n^\mathrm{T}v_n + k_n v_n^2 k_n^\mathrm{T}$$

Two things now emerge. Firstly, the product $[\theta_{n-1} - \theta][\theta_{n-1} - \theta]^\mathrm{T}$ is, by definition, the covariance matrix as it existed at time step $n - 1$ (that is, P_{n-1}). Secondly, if we examine the expected value of P_n, we can ignore any term multiplied by the noise v_n as its expected value is zero (note, however, that the square of v_n has a non-zero average value). We therefore have:

$$P_n = [I + k_n x_{n-1}^\mathrm{T}]P_{n-1}[I + k_n x_{n-1}^\mathrm{T}]^\mathrm{T} + k_n v_n^2 k_n^\mathrm{T}$$

and with one further transposition step

$$P_n = [I + k_n x_{n-1}^\mathrm{T}]P_{n-1}[I + x_{n-1} k_n^\mathrm{T}] + k_n v_n^2 k_n^\mathrm{T} \tag{A7.6}$$

The idea is now to choose k_n to minimize P_n given the existing values of everything else. We therefore investigate the variation of P_n as k_n is changed by a small amount Δk_n (note that the latter is a vector of small amounts). Thus Equation (A7.6) becomes:

$$P_n + \Delta P_n = [I + (k_n + \Delta k_n)x_{n-1}^\mathrm{T}]P_{n-1}[I + x_{n-1}(k_n^\mathrm{T} + \Delta k_n^\mathrm{T})]$$
$$+ (k_n + \Delta k_n)v_n^2(k_n^\mathrm{T} + \Delta k_n^\mathrm{T})$$

For a minimum, we need to examine ΔP_n relative to Δk_n. Expanding the previous equation will give rise to several terms which do not involve Δk_n, so we shall omit those for clarity. Also, some terms of the expansion involve both Δk_n and its transpose. Since Δk_n is assumed to be very small, these can also be ignored (roughly equivalent to ignoring the square of a small error in scalar analysis). With these omissions, the terms involving Δk_n in the expansion of the RHS of the previous equation are:

$$\Delta k_n x_{n-1}^\mathrm{T} P_{n-1}(I + x_{n-1}k_n^\mathrm{T}) + (I + k_n x_{n-1}^\mathrm{T})P_{n-1}x_{n-1}\Delta k_n^\mathrm{T}$$
$$+ \Delta k_n v_n^2 k_n^\mathrm{T} + k_n v_n^2 \Delta k_n^\mathrm{T}$$

For a minimum in P_n, ΔP_n must be zero for small changes in k_n, so the following must both be valid from the list of terms above which cause the changes:

$$\Delta k_n [x_{n-1}^\mathrm{T} P_{n-1}(I + x_{n-1}k_n^\mathrm{T}) + v_n^2 k_n^\mathrm{T}] = 0$$

and

$$[(I + k_n x_{n-1}^\mathrm{T})P_{n-1}x_{n-1} + k_n v_n^2]\Delta k_n^\mathrm{T} = 0$$

These two equations give the same solution. Using the second one, and assuming that, since Δk_n contains small (but nonzero) changes, the vector represented by the term in square brackets must be zero, we have:

$$P_{n-1}x_{n-1} + k_n x_{n-1}^\mathrm{T} P_{n-1}x_{n-1} + k_n v_n^2 = 0$$

or

$$k_n[x_{n-1}^\mathrm{T} P_{n-1}x_{n-1} + v_n^2] = -P_{n-1}x_{n-1}$$

Noting that the term in square brackets is scalar, we obtain the solution:

$$k_n = \frac{-P_{n-1}x_{n-1}}{x_{n-1}^\mathrm{T} P_{n-1}x_{n-1} + v_n^2}$$

This equation can also be used to update P_n by substituting it into Equation (A7.6):

$$P_n = \left[I - \frac{P_{n-1}x_{n-1}x_{n-1}^\mathrm{T}}{D}\right] P_{n-1} \left[I - \frac{x_{n-1}x_{n-1}^\mathrm{T} P_{n-1}^\mathrm{T}}{D}\right]$$
$$+ \frac{P_{n-1}x_{n-1}v_n^2 x_{n-1}^\mathrm{T} P_{n-1}^\mathrm{T}}{D^2}$$

where $D = x_{n-1}^\mathrm{T} P_{n-1}x_{n-1} + v_n^2$.

Noting that P is symmetric, so $P^\mathrm{T} = P$, and rearranging and simplifying yields

$$P_n = \frac{\begin{array}{c} D^2 P_{n-1} - 2D P_{n-1}x_{n-1}x_{n-1}^\mathrm{T} P_{n-1} \\ + P_{n-1}x_{n-1}x_{n-1}^\mathrm{T} P_{n-1}x_{n-1}x_{n-1}^\mathrm{T} P_{n-1} + P_{n-1}x_{n-1}v_n^2 x_{n-1}^\mathrm{T} P_{n-1} \end{array}}{D^2}$$

which factorizes to

$$P_n = \frac{\begin{array}{c} D^2 P_{n-1} - 2D P_{n-1}x_{n-1}x_{n-1}^\mathrm{T} P_{n-1} \\ + P_{n-1}x_{n-1}[x_{n-1}^\mathrm{T} P_{n-1}x_{n-1} + v_n^2]x_{n-1}^\mathrm{T} P_{n-1} \end{array}}{D^2}$$

Further, noting that the term in square brackets is D (which is scalar), the equation now simplifies dramatically:

$$P_n = P_{n-1} - \frac{P_{n-1}x_{n-1}x_{n-1}^{\mathrm{T}}P_{n-1}}{D}$$

that is,

$$P_n = P_{n-1} - \frac{P_{n-1}x_{n-1}x_{n-1}^{\mathrm{T}}P_{n-1}}{v_n^2 + x_{n-1}^{\mathrm{T}}P_{n-1}x_{n-1}}$$

We also require an updating equation for θ. We have from Equation (A7.4), with $J_n = 1$:

$$\theta_n = \theta_{n-1} + k_n(x_{n-1}^{\mathrm{T}}\theta_{n-1} - y_n)$$

which, on substituting for k_n from Equation (A7.5), becomes:

$$\theta_n = \theta_{n-1} - \frac{P_{n-1}x_{n-1}(x_{n-1}^{\mathrm{T}}\theta_{n-1} - y_n)}{v_n^2 + x_{n-1}^{\mathrm{T}}P_{n-1}x_{n-1}}$$

In practice, we do not usually know the value of v_n. In consequence, it is usual to set it equal to 1 and work with a 'normalized' covariance, giving the following updating equations used as Equations (11.3) and (11.4) in the main text (Chapter 11):

$$P_n = P_{n-1} - \frac{P_{n-1}x_{n-1}x_{n-1}^{\mathrm{T}}P_{n-1}}{1 + x_{n-1}^{\mathrm{T}}P_{n-1}x_{n-1}}$$

$$\theta_n = \theta_{n-1} - \frac{P_{n-1}x_{n-1}(x_{n-1}^{\mathrm{T}}\theta_{n-1} - y_n)}{1 + x_{n-1}^{\mathrm{T}}P_{n-1}x_{n-1}}$$

Other formulations of these equations exist, and can be found in the works of Plackett (1950) and several subsequent authors.

References

Ackermann J. (1972). Der Entwurf linearer Regelungssysteme im Zustandsraum. *Regelungstech. Prozess-Datenverarb.*, **7**, 297–300

Anderson B. D. O. and Moore J. B. (1979). *Optimal Filtering*. Englewood Cliffs, NJ: Prentice-Hall

Anderson B. D. O. and Moore J. B. (1989). *Optimal Control – Linear Quadratic Methods*. Englewood Cliffs, NJ: Prentice-Hall

Åström K. J. and Wittenmark B. (1990). *Computer-Controlled Systems*, 2nd edn. Englewood Cliffs, NJ: Prentice-Hall

Åström K. J. and Wittenmark B. (1995). *Adaptive Control*, 2nd edn. Reading, MA: Addison-Wesley

Atherton D. P. (1982). *Nonlinear Control Engineering*. Wokingham, UK: Van Nostrand Reinhold

Ayres F. (1987). *Theory and Problems of Matrices*. New York: McGraw-Hill

Balas G. J., Doyle J. C., Glover K., Packard A. and Smith R. (1991) μ-*Tools Manual supplied with* μ-*Analysis and Synthesis Toolbox for MATLAB*. Natick, Massachusetts: The Mathworks, Inc.

Balmer L. (1991). *Signals and Systems*. Hemel Hempstead, UK: Prentice-Hall

Banks S. P. (1986). *Control Systems Engineering*. Hemel Hempstead, UK: Prentice-Hall

Bennett S. (1994). *Real Time Computer Control*, 2nd edn. Hemel Hempstead, UK: Prentice-Hall

Bennett S. and Linkens D. A. (1982). *Computer Control of Industrial Processes*. London: Peter Peregrinus

Biran A. and Breiner M. (1995). *MATLAB for Engineers*. Wokingham, UK: Addison-Wesley

Blackman P. F. (1977). *Introduction to State Variable Analysis*. London: McMillan Press

Bozic S. M. (1979). *Digital and Kalman Filtering*. London: Arnold

Brogan W. L. (1991). *Modern Control Theory*. Englewood Cliffs, NJ: Prentice-Hall

Cambridge Control Ltd (1988). *TSIM User Guide*. Cambridge, UK: Cambridge Control

Cambridge Control Ltd (1990). *Multivariable Frequency Domain Toolbox for Use with MATLAB*. Cambridge, UK: Cambridge Control

Charlesworth A. S. and Fletcher J. R. (1974). *Systematic Analogue Computer Programming*. London: Pitman

Chiang R. Y. and Safonov M. G. (1988). *Robust Control Toolbox for Use with MATLAB – User's Guide*. Natick, Massachusetts: The Mathworks, Inc.

Conte S. D. and de Boor C. (1972). *Elementary Numerical Analysis*. Tokyo: McGraw-Hill

D'Azzo J. J. and Houpis C. H. (1995). *Linear Control System Analysis and Design*, 4th edn. New York: McGraw-Hill

Dutton K. and Barraclough W. A. (1996). A progressive introduction of computer-assistance into the teaching of advanced topics in control engineering. *IEE Engineering Science and Education Journal*, **5** (1), 32–40

Edmunds J. and Kouvaritakis B. (1979). Extensions of the frame alignment technique and their use in the characteristic locus design method. *International Journal of Control*, **29**, 787–96

Elgerd O. I. (1967). *Control Systems Theory*. St. Louis: McGraw-Hill

Ford M. P., Maciejowski J. M. and Boyle J. M. (1990). *Multivariable Frequency Domain Toolbox for Use with MATLAB*. Cambridge, UK: Cambridge Control and the GEC Engineering Research Centre

Franklin G. F., Powell J. D. and Workman M. L. (1990). *Digital Control of Dynamic Systems*, 2nd edn. Menlo Park, CA: Addison Wesley

Franklin G. F., Powell J. D. and Emami-Naeini A. (1994). *Feedback Control of Dynamic Systems*, 3rd edn. Menlo Park, CA: Addison Wesley

Friedland B. (1987). *Control System Design (An Introduction to State-space Methods)*. Singapore: McGraw-Hill

Furuta K., Sano A. and Atherton D. (1988). *State Variable Methods in Automatic Control*. Chichester, UK: Wiley

Garcia C. E. and Morari M. (1982). Internal Model Control – 1. A unifying review and some new results. *Ind. Eng. Chem. Process Des. & Dev.*, **21**, 308–23

Godfrey K. (1993). *Perturbation Signals for System Identification*. Hemel Hempstead, UK: Prentice-Hall

Grimble M. J., Patton R. J. and Wise D. A. (1979). The design of dynamic ship positioning control systems using extended Kalman filtering techniques. *IEEE Conference, Oceans '79*, CA, San Diego

Healey M. (1979). A working man's guide to Kalman filters. *IMC symposium, Kalman filtering in theory and practice*, UK: Teesside Polytechnic

Horowitz P. and Hill W. (1980). *The Art of Electronics*, 2nd edn. Cambridge, UK: Cambridge University Press

Hostetter G. H. (1988). *Digital Control System Design*. New York: Holt, Rinehart and Winston

Kalani G. (1988). *Microprocessor Based Distributed Control Systems*. Hemel Hempstead, UK: Prentice-Hall

Khalil H. K. (1992). *Nonlinear Systems*. New York: Macmillan

Kraus T. W. and Myron T. J. (1984). Self-tuning PID uses pattern recognition approach. *Control Engineering*, June; 106–11

Kreysig E. (1993). *Advanced Engineering Mathematics*, 7th edn. New York: Wiley

Kuo B. C. (1992). *Digital Control Systems*, 2nd edn. Orlando, FL: Saunders/Harcourt Brace Jovanovich

Leigh J. R. (1983). *Essentials of Nonlinear Control Theory*. London: Peter Peregrinus

Leigh J. R. (1992). *Applied Digital Control*, 2nd edn. Hemel Hempstead, UK: Prentice-Hall

Lewis R. W. (1995). *Programming Industrial Control Systems Using IEC* 1131-3. London: Peter Peregrinus

Lunze J. (1989). *Robust Multivariable Feedback Control*. Hemel Hempstead, UK: Prentice-Hall

MacFarlane A. G. J. and Kouvaritakis B. (1977). A design technique for linear multivariable feedback systems. *International Journal of Control*, **25**, 837–74

Maciejowski J. M. (1989). *Multivariable Feedback Design*. Wokingham, UK: Addison-Wesley

Mathworks – see 'The Mathworks'

Mayne D. Q. (1979). Sequential design of linear multivariable systems. In *Proc. IEE*, **126**, 568–72

Mitchell and Gauthier Associates (1987). *ACSL Advanced Continuous Simulation Language Reference Manual*. Concord, MA: Mitchell and Gauthier Associates

Morari M. and Zafiriou E. (1989). *Robust Process Control*. Englewood Cliffs, NJ: Prentice-Hall

Norton J. P. (1986). *An Introduction to Identification*. London: Academic Press

Ogata K. (1987). *Discrete Control Systems*. Englewood Cliffs, NJ: Prentice-Hall

Ogata K. (1994a). *Designing Linear Control Systems with MATLAB*. Englewood Cliffs, NJ: Prentice-Hall

Ogata K. (1994b). *Solving Control Engineering Problems with MATLAB*. Englewood Cliffs, NJ: Prentice-Hall

O'Reilly J., ed. (1987). *Multivariable Control for Industrial Applications*. London: Peter Peregrinus

Owens D. H. (1978). *Feedback and Multivariable Systems*. Stevenage: Peter Peregrinus

Patel R. V. and Munro N. (1982). *Multivariable System Theory and Design*. Oxford: Pergamon

Phillips C. L. and Harbor R. D. (1991). *Feedback Control Systems*, 2nd edn. Englewood Cliffs, NJ: Prentice-Hall

Phillips C. L. and Nagle H. T. (1990). *Digital Control System Analysis and Design*, 2nd edn. Englewood Cliffs, NJ: Prentice-Hall

Plackett R. L. (1950). Some theorems in least squares. *Biometrika*, **37**, 149–57

Postlethwaite I., Edwards J. M. and MacFarlane A. G. J. (1981). Principal gains and principal phases in the analysis of linear multivariable feedback systems. *IEEE Transactions on Automatic Control*, **AC-26**, 32–46

Rieder W. G. and Busby H. R. (1986). *Introductory Engineering Modeling*. New York: Wiley

Rosenbrock H. H. (1974). *Computer Aided Control System Design*. New York: Academic Press

Slotine J.-J. and Li W. (1991). *Applied Nonlinear Control*. Englewood Cliffs, NJ: Prentice-Hall

Smith O. J. M. (1958). *Feedback Control Systems*. New York: McGraw-Hill

Strum R. D. and Kirk D. E. (1994). *Contemporary Linear Systems using MATLAB*. Boston, MA: PWS

The Mathworks Inc. (1993a). *MATLAB for Microsoft Windows – User's Guide*. Natick, MA: The Mathworks Inc.

The Mathworks Inc. (1993b). *MATLAB for Microsoft Windows – Reference Guide*. Natick, MA: The Mathworks Inc.

The Mathworks Inc. (1993c). *MATLAB High-Performance Numeric Computation and Visualization Software – Building a Graphical User Interface*. Natick, MA: The Mathworks Inc.

The Mathworks Inc. (1993d). *SIMULINK Numerical Simulation Software – Reference Guide*. Natick, MA: The Mathworks Inc.

The Mathworks Inc. (1994). *Control System Toolbox for Use with MATLAB*. Natick, MA: The Mathworks Inc.

Warwick K., ed. (1988). *Implementation of Self-tuning Controllers*. London: Peter Peregrinus

Webb J. (1992). *Programmable Logic Controllers*. New York: Merrill (Maxwell Macmillan)

Wellstead P. E. and Zarrop M. B. (1991). *Self Tuning Systems*. Chichester, UK: Wiley

Williams B. W. (1992). *Power Electronics*, 2nd edn. London: Macmillan

Williamson D. (1991). *Digital Control and Implementation – Finite Wordlength Considerations*. Sydney: Prentice-Hall

Ziegler J. G. and Nichols N. B. (1942). Optimal settings for automatic controllers. *Trans. ASME*, **64**(11), 759

Index